The Genetics of Type 2 Diabetes and Related Traits

Jose C. Florez
Editor

The Genetics of Type 2 Diabetes and Related Traits

Biology, Physiology and Translation

Springer

Editor
Jose C. Florez
Diabetes Unit
Center for Human Genetic Research
Massachusetts General Hospital
Programs in Metabolism and Medical & Population Genetics
Broad Institute
Department of Medicine
Harvard Medical School
Boston
Massachusetts
USA

ISBN 978-3-319-01573-6 ISBN 978-3-319-01574-3 (eBook)
DOI 10.1007/978-3-319-01574-3

Library of Congress Control Number: 2015958349

Printed on acid-free paper

This Springer imprint is published by Springer Nature
The registered company is Springer International Publishing AG Switzerland

Preface

When Victor Neel termed type 2 diabetes (T2D) "the geneticist's nightmare," (Neel 1962) it was already evident that T2D harbored a familial component; yet its polygenic genetic architecture, the inherent limitations in the techniques available at the time, and the strong (and growing!) influence of environmental determinants made genetic discovery an arduous task. In the 50 years since then, the public health impact of T2D has only skyrocketed, fueled by the changes in modern lifestyle increasingly adopted by developing societies and the expansion of caloric superabundance.

As a result, T2D and its complications represent one of the most serious challenges to public health in the twenty-first century. We live in the midst of a worldwide pandemic that threatens to undermine the significant gains we have made against cardiovascular disease over the last few decades. Despite its status as one of the oldest known endocrinopathies, the availability of a molecular therapy since 1920, and the existence of over a dozen drug classes approved for the management of the disease, we are largely unable to cure it and are losing the population battle in both the developed and developing worlds.

The population and healthcare costs are enormous and only expected to rise. Stoked by the snowballing obesity epidemic, diabetes affects over 29 million Americans, with more than 80 million at high risk. People with Asian, African, and American indigenous ancestry experience a higher risk, with the worldwide prevalence of diabetes expected to hit 500 million by 2030. In the USA alone, diabetes and its complications account for $245 billion annually. With its concomitant life-threatening complications of cardiovascular disease, renal failure, visual loss, and peripheral vascular disease, T2D can undermine the global economy, with a disproportionate burden on underprivileged groups and low- and middle-income countries.

At the crux of our lack of clinical effectiveness lies our incomplete understanding of its pathogenesis, evolution, and metabolic consequences. Definite causal triggers, the interplay between various risk factors, and specific mechanisms that underlie long-term complications remain largely unknown. Thus, T2D has become the quintessential complex disease, with substantial genetic and environmental components, significant variation in prevalence across ethnic groups, pathophysiological heterogeneity, multiple treatment modalities, and poorly defined interactions with related metabolic phenotypes.

In the midst of many significant advances, key questions remain unanswered: Why do people develop T2D? Why do incidence rates differ across populations drawn from around the world? What causes beta cells to fail? How does obesity influence T2D risk? Can medications be designed that cure, rather than treat, diabetes? If more than one medication is needed, what is the best sequence or combination for specific subgroups?

These pressing questions also represent tantalizing opportunities. Our growing understanding of pathophysiology, the invention and deployment of global technologies that query specific areas of the biological space, and our improved ability to focus on the human as the targeted model system have placed the field of diabetes investigation on the verge of momentous discovery. However, among all valid experimental approaches employed in humans, only two can consistently bypass correlative associations and firmly establish causal inference. Beyond expensive and focused clinical trials, *the genetic method* is unique in ensuring that the exposure of interest (genotype) precedes phenotype, that it is conferred on individuals on a randomized basis at the time of conception, and that it is not in turn affected by the disease process or its treatment. Thus, it can serve as a powerful approach to dissect the nosology of T2D, illuminate its pathogenesis, and identify therapeutic targets through mechanistic insight.

In this volume, we have endeavored to take a contemporary snapshot of a rapidly moving field. In the time passed since Neel's initial cautionary statement, the scientific community has developed methods to measure global genomic variation with great precision, together with the statistical concepts and related analytical techniques that allow us to draw rigorous conclusions. Investigators have coalesced to advance knowledge in a collaborative fashion where needed, introducing appealing notions on the sociology of team science.

All of these ideas are illustrated in this book. The amount of novel information collected here, most of which was simply undreamed of just a decade ago, is staggering. At the same time, and reflecting the dizzying pace of discovery, nascent findings that have emerged most recently may not be fully captured in these chapters, making this reading all the more exciting. Finally, the organization of this work was intended to mirror the collaborative atmosphere that pervades our field, in that every chapter is authored by two or more investigators who hail from different research groups and yet complement each other in style, insight, and perspective.

The initial section, containing seven chapters, centers on fundamental genetic discovery. The initial overview provides a helpful historical viewpoint that will help the nongenetic reader take stock of the chronological evolution of the research enterprise in this area. From the proven effectiveness of genome-wide association studies (GWAS), subsequent contributions touch on the challenges that follow initial associations, the extension of this method to less accessible phenotypes, and the arrival of next-generation sequencing as the harbinger of discovery focused on rare genetic variation.

The second section expands beyond genetics and illustrates how other data sources can inform initial genetic findings. Leveraging population diversity, correlating genetic associations with physiological measurements, learning from genetic variants that have strong phenotypic effects, and incorporating other key influences such as the modulation of gene expression, environmental factors, and our microbial commensals all help place genetic findings in focus and can lead to additional insight.

The third section makes a fundamental point: genetic association, however robust, is only the beginning of a laborious process. Convincing association signals notwithstanding, in most cases the specific DNA sequences that cause the molecular phenotype have not been identified. Indeed, the polymorphisms identified thus far merely signal genomic regions—at times hundreds of kilobases away from known genes—where an association has been found, but do not necessarily represent the causal variants: further fine-mapping and functional studies must be carried out before the true contribution of these loci to T2D can be accurately assessed. Thus, while we can rapidly and systematically uncover new associations, genetic studies do not circumvent the process of refining the associated loci to find the precise "causal" DNA sequences (causal in the sense of having a direct impact on RNA and/or protein quality or quantity that contributes to the diabetic phenotype). Indeed, variants may exert their molecular effects at remote sites even when they are relatively close to other uninvolved genes. Thus, loci identified by GWAS require in-depth sequencing and functional studies of the cellular and molecular effects of genes in that region. Six successful vignettes are described in this section, illustrating the progress we have made in just a few years.

The final section, comprising seven chapters, attempts to bring our current state of knowledge closer to the clinic, acknowledging both its potential and its limitations. It includes chapters on prediction, interaction of genetic variants with drugs or nutrients, and approaches to prevention or to the inference of causality for clinical relevant questions where randomized clinical trials have not produced conclusive answers or cannot be carried out. The epilogue, authored by a trio of long-standing collaborators who have set the pace for our field and whom many of us consider inspiring mentors, paints a realistic but hopeful vision of the future.

This book would not have been possible without the prescience of Andrea Pillmann at Springer in making the initial suggestion that we undertake this initiative, and without Jutta Lindenborn's patience in managing the editorial process. Over the years I have been fortunate to count on the professionalism and support of a superb publisher such as Springer in a variety of editorial projects, and this was no exception. I am most thankful to so many of my colleagues and friends who took time out of their busy professional lives to share their thoughts through eminently readable and informative chapters. Naturally not everyone who should or could have contributed was able to do so, but we have benefited from their wisdom as well, as this book largely reflects the collective body of knowledge garnered by the community over the past decade. Our remembrance goes to those luminaries and pioneers who are no longer with us, such as Alan Permutt, Steve Elbein, and Linda Kao. And finally our mind rests in the smart, competent, and energized trainees we have the pleasure of working with, as they represent the bright future for our field: theirs will be the next edition in this fascinating journey of discovery, as we materialize our heartfelt commitment to ameliorate world suffering by improving human health.

Boston, MA Jose C. Florez

Reference

Neel JV (1962) Diabetes mellitus: a "thrifty" genotype rendered detrimental by "progress"? Am J Hum Genet 14:353–362

Contents

Part I
Genetic Discovery

Chapter 1
Historical Overview of Gene Discovery Methodologies in Type 2 Diabetes

Amélie Bonnefond, Alan R. Shuldiner, and Philippe Froguel

Abstract This initial chapter presents a historical snapshot of the various approaches utilized to discover genes implicated in the pathogenesis of type 2 diabetes.

1.1 Introduction

In 1976, James V. Neel described type 2 diabetes as "a geneticist's nightmare" (Neel 1976). Forty years later, this statement holds true despite a plethora of different approaches and methodologies that have been used to discover genetic etiologies of monogenic and polygenic forms of type 2 diabetes (T2D) (Fig. 1.1); still, almost 30 % of patients presenting with putative monogenic diabetes do not have mutations in known causative genes (Vaxillaire et al. 2012), and all T2D-associated genetic variants identified to date explain less than 15 % of T2D heritability (Morris et al. 2012).

A. Bonnefond
CNRS-UMR8199, Lille Pasteur Institute, Lille, France

Lille University, Lille, France

European Genomic Institute for Diabetes (EGID), Lille, France

A.R. Shuldiner
Program for Personalized and Genomic Medicine, and Division of Endocrinology, Diabetes and Nutrition, School of Medicine, University of Maryland, Baltimore, MD, USA

Regeneron Genetics Center, Regeneron Pharmaceuticals Inc., Tarrytown, NY, USA

P. Froguel (✉)
CNRS-UMR8199, Lille Pasteur Institute, Lille, France

Lille University, Lille, France

European Genomic Institute for Diabetes (EGID), Lille, France

Department of Genomics of Common Disease, School of Public Health, Imperial College London, Hammersmith Hospital, London, UK
e-mail: p.froguel@imperial.ac.uk

J.C. Florez (ed.), *The Genetics of Type 2 Diabetes and Related Traits*,
DOI 10.1007/978-3-319-01574-3_1

Historical Overview of Methodology in Type 2 Diabetes

Amélie Bonnefond[1-3], Alan Shuldiner[4-6], Philippe Froguel[1-3,7]

[1]CNRS-UMR8199, Lille Pasteur Institute, Lille, France;
[2]Lille University, Lille, France;
[3]European Genomic Institute for Diabetes(EGID), Lille, France;
[4]Program in Personalized and Genomic Medicine, and Division of Endocrinology, Diabetes and Nutrition, School of Medicine, University of Maryland, Baltimore, Maryland, USA;
[5]Geriatric Research and Education Clinical Center, Baltimore Veterans Administration Medical Center, Baltimore, Maryland, USA;
[6]Present address: RegeneronGenetics Center, Regeneron Pharmaceuticals Inc., Tarrytown, New York, USA;
[7]Department of Genomics of Common Disease, Schoolof Public Health, Imperial College London, Hammersmith Hospital, London, United Kingdom.

Abbreviations:
GWAS, genome-wide association study
NGS, next-generation sequencing
T2D, type 2 diabetes
WES, whole-exomesequencing
WGS, whole-genome sequencing

Fig. 1.1 (continued)

1.2 Family-Based Linkage Analysis

In 1992, the first successful methodology to identify genes involved in diabetes was targeted or genome-wide linkage analysis (using microsatellites or other highly polymorphic genetic markers) in pedigrees (both consanguineous and non-consanguineous) with maturity-onset diabetes of the young (MODY) and in patients with neonatal diabetes mellitus. In combination with Sanger sequencing of candidate genes under the peak(s) of linkage, and more recently wider sequencing approaches, this strategy has enabled the identification of many genes involved in these monogenic forms of diabetes (Fig. 1.1): *GCK* (Froguel et al. 1992; Froguel et al. 1993), *HNF4A* (Bell et al. 1991; Yamagata et al. 1996a), *HNF1A* (Vaxillaire et al. 1995; Yamagata et al. 1996b), *CEL* (Raeder et al. 2006), *BLK* (Borowiec et al. 2009), *TRMT10A* (Igoillo-Esteve et al. 2013), *WFS1* (Inoue et al. 1998; Bonnycastle et al. 2013), *PCBD1* (Simaite et al. 2014), *INS* (Støy et al. 2007), *SLC19A2* (Labay et al. 1999), *EIF2AK3* (Delépine et al. 2000), *FOXP3* (Bennett et al. 2001), *PTF1A* (Sellick et al. 2004), *GLIS3* (Senée et al. 2006), and *IER3IP1* (Poulton et al. 2011). In polygenic forms of T2D, linkage analyses were far less successful, despite huge efforts of the research community. Only two T2D linkage signals were subsequently found to harbor variants in genes reproducibly associated with T2D (Fig. 1.1): *HNF4A* (Silander et al. 2004) and *TCF7L2* (Grant et al. 2006). However, for reasons that are unclear, it is probable that the T2D-associated common variants in these genes do not explain the original linkage signal.

In patients from consanguineous families, homozygosity mapping (namely, the identification of regions of the genome that are homozygous in affected individuals) through DNA arrays in combination with sequencing of candidate genes within

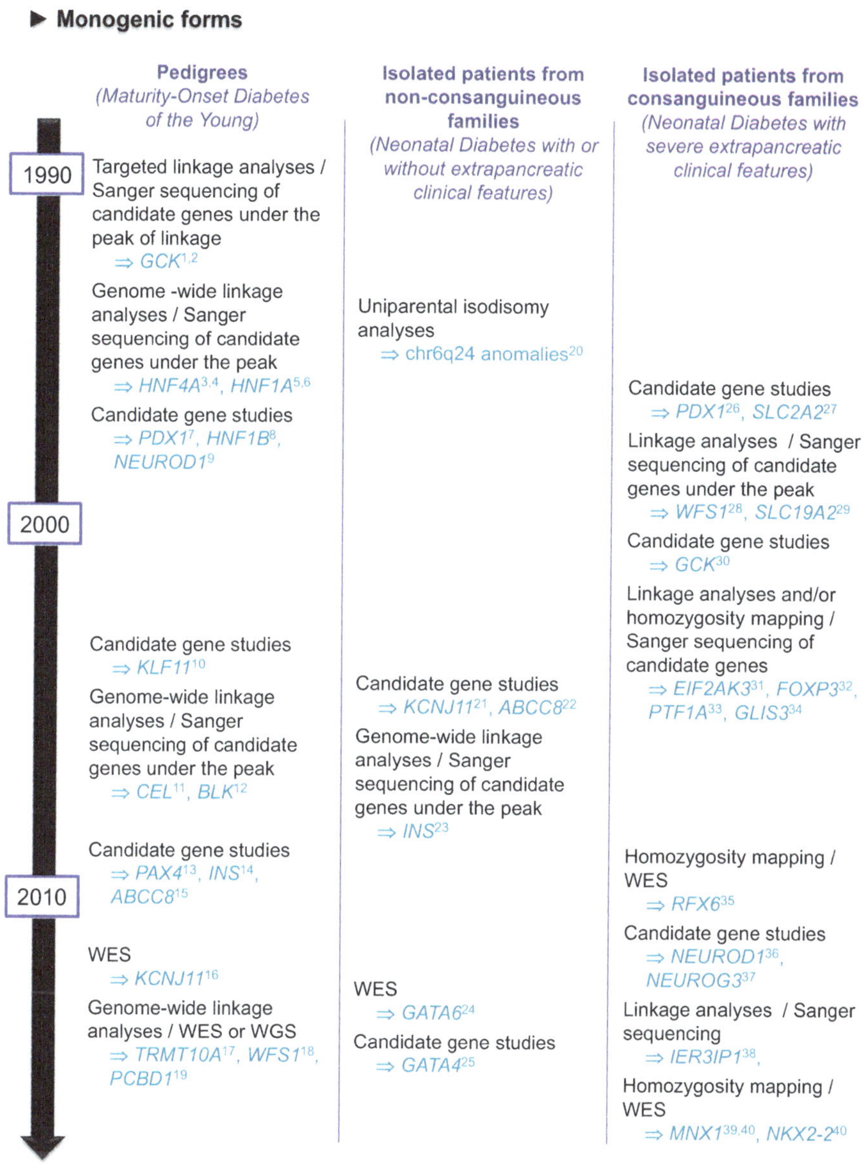

Fig. 1.1 (continued)

homozygous loci has enabled the identification of several genes involved in neonatal diabetes associated with severe extrapancreatic clinical features (Fig. 1.1): *RFX6* (Smith et al. 2010), *MNX1* (Bonnefond et al. 2013; Flanagan et al. 2014) and *NKX2-2* (Flanagan et al. 2014).

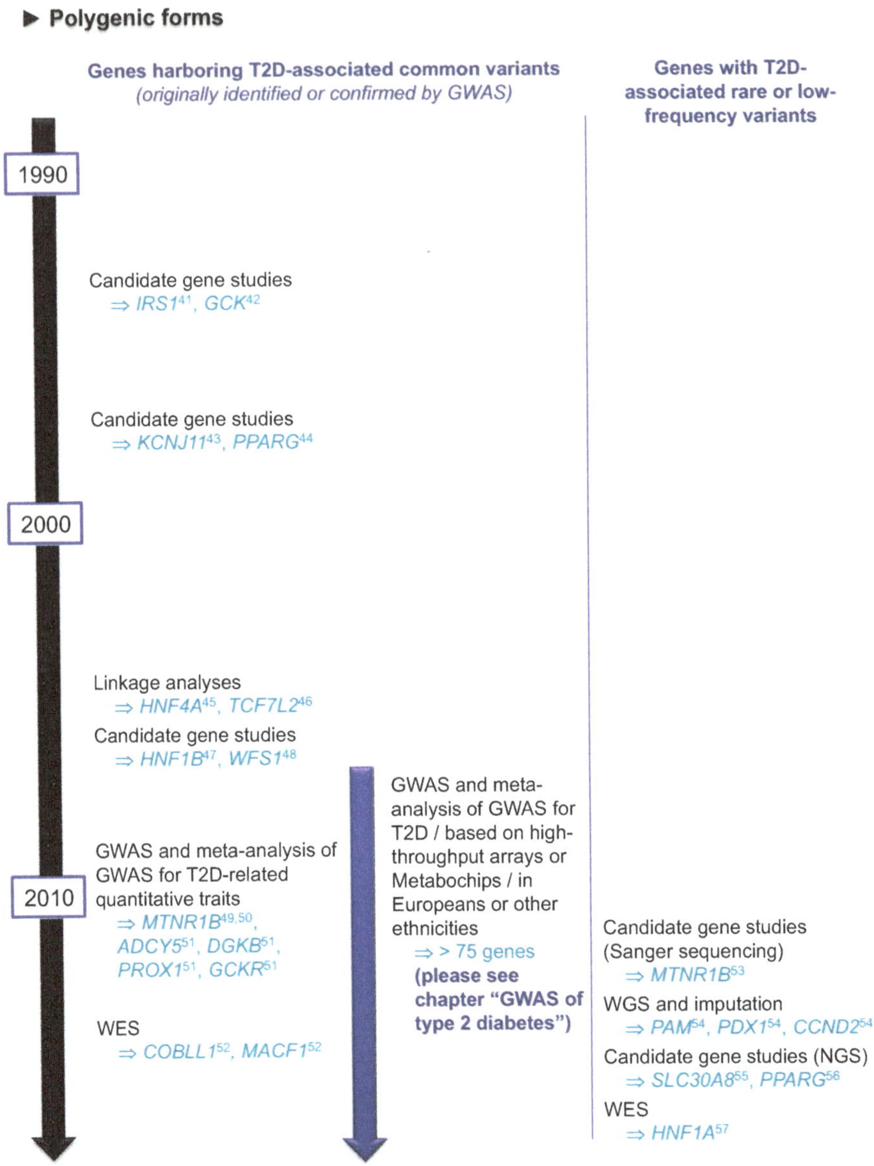

Fig. 1.1 (continued)

References

1. Froguel, P. *et al. Nature* **356,** 162–164 (1992).
2. Froguel, P. *et al.N. Engl. J. Med.* **328,** 697–702 (1993).
3. Bell, G. I. *et al. Proc. Natl. Acad. Sci. U. S. A.* **88,** 1484–1488 (1991).
4. Yamagata, K. *et al. Nature* **384,** 458–460 (1996).
5. Vaxillaire, M. *et al. Nat. Genet.* **9,** 418–423 (1995).
6. Yamagata, K. *et al. Nature* **384,** 455–458 (1996).
7. Stoffers, D. A., Ferrer, J., Clarke, W. L. & Habener, J. F. *Nat. Genet.* **17,** 138–139 (1997).
8. Horikawa, Y. *et al. Nat. Genet.* **17,** 384–385 (1997).
9. Malecki, M. T. *et al. Nat. Genet.* **23,** 323–328 (1999).
10. Neve, B. *et al. Proc. Natl. Acad. Sci. U. S. A.* **102,** 4807–4812 (2005).
11. Raeder, H. *et al. Nat. Genet.* **38,** 54–62 (2006).
12. Borowiec, M. *et al. Proc. Natl. Acad. Sci. U. S. A.* **106,** 14460–14465 (2009).
13. Plengvidhya, N. *et al. J. Clin. Endocrinol. Metab.* **92,** 2821–2826 (2007).
14. Meur, G. *et al. Diabetes* **59,** 653–661 (2010).
15. Bowman, P. *et al. Diabetologia* **55,** 123–127 (2012).
16. Bonnefond, A. *et al. PLoS ONE* **7,** e37423 (2012).
17. Igoillo-Esteve, M. *et al. PLoS Genet.* **9,** e1003888 (2013).
18. Bonnycastle, L. L. *et al. Diabetes* **62,** 3943–3950 (2013).
19. Simaite, D. *et al. Diabetes* **63,** 3557–3564 (2014).
20. Temple, I. K. *et al. Nat. Genet.* **9,** 110–112 (1995).
21. Gloyn, A. L. *et al. N. Engl. J. Med.* **350,** 1838–1849 (2004).
22. Babenko, A. P. *et al. N. Engl. J. Med.* **355,** 456–466 (2006).
23. Støy, J. *et al. Proc. Natl. Acad. Sci. U. S. A.* **104,** 15040–15044 (2007).
24. Lango Allen, H. *et al. Nat. Genet.* **44,** 20–22 (2012).
25. Shaw-Smith, C. *et al. Diabetes* **63,** 2888–2894 (2014).
26. Stoffers, D. A., Zinkin, N. T., Stanojevic, V., Clarke, W. L. & Habener, J. F. *Nat. Genet.* **15,** 106–110 (1997).
27. Santer, R. *et al. Nat. Genet.* **17,** 324–326 (1997).
28. Inoue, H. *et al. Nat. Genet.* **20,** 143–148 (1998).
29. Labay, V. *et al. Nat. Genet.* **22,** 300–304 (1999).
30. Njølstad, P. R. *et al. N. Engl. J. Med.* **344,** 1588–1592 (2001).
31. Delépine, M. *et al. Nat. Genet.* **25,** 406–409 (2000).
32. Bennett, C. L. *et al. Nat. Genet.* **27,** 20–21 (2001).
33. Sellick, G. S. *et al. Nat. Genet.* **36,** 1301–1305 (2004).
34. Senée, V. *et al. Nat. Genet.* **38,** 682–687 (2006).
35. Smith, S. B. *et al. Nature* **463,** 775–780 (2010).
36. Rubio-Cabezas, O. *et al. Diabetes* **59,** 2326–2331 (2010).
37. Rubio-Cabezas, O. *et al. Diabetes* **60,** 1349–1353 (2011).
38. Poulton, C. J. *et al. Am. J. Hum. Genet.* **89,** 265–276 (2011).
39. Bonnefond, A. *et al. Diabetes Metab.* **39,** 276–280 (2013).
40. Flanagan, S. E. *et al. Cell Metab.* **19,** 146–154 (2014).
41. Almind, K. *et al. Lancet* **342,** 828–832 (1993).
42. Stone, L. M., Kahn, S. E., Fujimoto, W. Y., Deeb, S. S. & Porte, D. *Diabetes* **45,** 422–428 (1996).
43. Hani, E. H. *et al. Diabetologia* **41,** 1511–1515 (1998).
44. Deeb, S. S. *et al. Nat. Genet.* **20,** 284–287 (1998).
45. Silander, K. *et al. Diabetes* **53,** 1141–1149 (2004).
46. Grant, S. F. A. *et al. Nat. Genet.* **38,** 320–323 (2006).
47. Winckler, W. *et al. Diabetes* **56,** 685–693 (2007).
48. Sandhu, M. S. *et al. Nat. Genet.***39,** 951–953 (2007).
49. Bouatia-Naji, N. *et al. Nat. Genet.* **41,** 89–94 (2009).
50. Prokopenko, I .*et al. Nat. Genet.* **41,** 77–81 (2009).
51. Dupuis, J. *et al. Nat. Genet.* **42,** 105–116 (2010).
52. Albrechtsen, A. *et al. Diabetologia* **56,** 298–310 (2013).
53. Bonnefond, A. *et al. Nat. Genet.* **44,** 297–301 (2012).
54. Steinthorsdottir, V. *et al. Nat. Genet.* **46,** 294–298 (2014).
55. Flannick, J. *et al. Nat. Genet.* **46,** 357–363 (2014).
56. Majithia, A. R. *et al. Proc. Natl. Acad. Sci. U. S. A.* (2014). doi:10.1073/pnas.1410428111.
57. SIGMA Type 2 Diabetes Consortium *et al. JAMA J. Am. Med. Assoc.* **311,** 2305–2314 (2014).

Fig. 1.1 Historical overview of methodology in type 2 diabetes (including monogenic and polygenic forms)

1.3 Candidate Gene Approaches

Candidate gene studies in which genes thought to be involved in glucose homeostasis are queried for sequence variation have been met with some success. Sequencing in families segregating atypical forms of early-onset nonimmune diabetes or syndromic conditions that include glucose intolerance has resulted in the identification of novel highly penetrant genes causing monogenic diabetes. By contrast, candidate genes for typical T2D involve querying variation in large numbers of unrelated T2D cases and nondiabetic controls to identify sequence variants enriched in the cases. These variants are typically common in the population and exert a small effect on T2D susceptibility. In a few examples, the same gene may harbor variants that have a large effect on function causing monogenic diabetes as well as more common variants that have a smaller effect on function and involved in polygenic T2D. Examples of successful identification of monogenic and polygenic diabetes genes using candidate gene approaches include:

MODY:

1. Key role in pancreatic beta cells: *PDX1* (Stoffers et al. 1997a), *NEUROD1* (Malecki et al. 1999), *KLF11* (Neve et al. 2005), and *PAX4* (Plengvidhya et al. 2007)
2. Genes belonging to a family that includes other genes previously shown to cause MODY, with a putative role in pancreatic beta cells: *HNF1B* (Horikawa et al. 1997)
3. Genes previously found to cause neonatal diabetes mellitus: *INS* (Meur et al. 2010) and *ABCC8* (Bowman et al. 2012)

Neonatal Diabetes Mellitus (Including Syndromic Forms):

1. Key role in pancreatic beta cells: *SLC2A2* (Santer et al. 1997), *KCNJ11* (Gloyn et al. 2004), *ABCC8* (Babenko et al. 2006), and *NEUROG3* (Rubio-Cabezas et al. 2011)
2. Genes previously found to cause MODY: *PDX1* (Stoffers et al. 1997b), *GCK* (Njølstad et al. 2001), and *NEUROD1* (Rubio-Cabezas et al. 2010)
3. Genes belonging to a family that includes other genes previously shown to cause MODY, with a putative role in pancreatic beta cells: *GATA4* (Shaw-Smith et al. 2014)

Polygenic Forms of T2D:

1. Key role in pancreatic beta cells or insulin sensitivity: *IRS1* (Almind et al. 1993), *KCNJ11* (Hani et al. 1998), *PPARG* (Deeb et al. 1998), and *WFS1* (Sandhu et al. 2007)
2. Genes previously found to cause a monogenic form of diabetes: *GCK* (Stone et al. 1996), and *HNF1B* (Winckler et al. 2007)

1.4 Genome-Wide Association Studies (GWAS)

Since 2007, GWAS and meta-analyses of GWAS in Europeans and other ethnicities have been very successful in identifying common single nucleotide polymorphisms (SNPs) significantly associated with T2D (please see Chap. 2). These GWAS have been based on high-throughput DNA microarrays assessing hundreds of thousands to few millions of SNPs across the genome in thousands of T2D cases and nondiabetic controls. GWAS of quantitative traits associated with T2D such as fasting plasma glucose, fasting serum insulin, and 2-h plasma glucose levels during an oral glucose tolerance test have also been quite successful in identifying new T2D-associated loci (Fig. 1.1). More recently, imputation of SNPs known from the 1000 Genomes project but not actually genotyped on the GWAS has boosted the number of analyzed SNPs (>6 million SNPs), enabling the identification of additional T2D-associated loci.

1.5 Next-Generation Sequencing (NGS) Approaches

Most recently, whole exome sequencing (WES) and whole genome sequencing (WGS) have become the most promising methodologies in finding new genes causing monogenic diabetes as well as novel loci associated with T2D risk. WES successfully identified two genes involved in monogenic diabetes (Fig. 1.1): *KCNJ11* in MODY (Bonnefond et al. 2012) (which was known to be mutated in patients with neonatal diabetes) and *GATA6* in neonatal diabetes (Lango Allen et al. 2012). Furthermore, WES performed in 2000 Europeans identified novel T2D-associated common variants in two genes: *COBLL1* and *MACF1* (Albrechtsen et al. 2013). Moreover, WES performed in 3756 Mexicans identified a low-frequency missense variant in *HNF1A* which strongly contributed to T2D risk (SIGMA Type 2 Diabetes Consortium et al. 2014). Finally, WGS of 2630 Icelanders and imputation (either direct imputation based on DNA array genotyping or in silico imputation based on genealogy information) into 11,000 cases and 267,000 controls of Icelandic origin identified T2D-associated low-frequency variants in three genes: *PAM*, *PDX1*, and *CCND2* (Steinthorsdottir et al. 2014).

1.6 Summary and Future Prospects

With increasing knowledge of genetic variation across the human genome, coupled with technological advances to query such variation, dramatic advances have been made in understanding the genetic basis of rare monogenic forms of diabetes as well as more common polygenic T2D. With these insights into the genetic architecture of diabetes, we now understand why family-based linkage analysis approaches were successful in identifying genes causing monogenic forms of diabetes, while large case-control GWAS approaches were more successful in identifying genes and loci for T2D susceptibility. While small-effect variants identified by GWAS do not contribute substantially to individual risk and are not useful clinically to predict who will develop diabetes, identification of these genes has provided new insights into underlying disease mechanisms. NGS applied to both family-based and population-based approaches promises to unveil even greater granularity of the genetic architecture of diabetes, underlying biological mechanisms, and novel approaches for treatment and prevention.

References

Albrechtsen A, Grarup N, Li Y et al (2013) Exome sequencing-driven discovery of coding polymorphisms associated with common metabolic phenotypes. Diabetologia 56:298–310. doi:10.1007/s00125-012-2756-1

Almind K, Bjørbaek C, Vestergaard H et al (1993) Aminoacid polymorphisms of insulin receptor substrate-1 in non-insulin-dependent diabetes mellitus. Lancet 342:828–832

Babenko AP, Polak M, Cavé H et al (2006) Activating mutations in the ABCC8 gene in neonatal diabetes mellitus. N Engl J Med 355:456–466. doi:10.1056/NEJMoa055068

Bell GI, Xiang KS, Newman MV et al (1991) Gene for non-insulin-dependent diabetes mellitus (maturity-onset diabetes of the young subtype) is linked to DNA polymorphism on human chromosome 20q. Proc Natl Acad Sci USA 88:1484–1488

Bennett CL, Christie J, Ramsdell F et al (2001) The immune dysregulation, polyendocrinopathy, enteropathy, X-linked syndrome (IPEX) is caused by mutations of FOXP3. Nat Genet 27: 20–21. doi:10.1038/83713

Bonnefond A, Philippe J, Durand E et al (2012) Whole-exome sequencing and high throughput genotyping identified KCNJ11 as the thirteenth MODY gene. PLoS ONE 7, e37423. doi:10.1371/journal.pone.0037423

Bonnefond A, Vaillant E, Philippe J et al (2013) Transcription factor gene MNX1 is a novel cause of permanent neonatal diabetes in a consanguineous family. Diabetes Metab 39:276–280. doi:10.1016/j.diabet.2013.02.007

Bonnycastle LL, Chines PS, Hara T et al (2013) Autosomal dominant diabetes arising from a Wolfram syndrome 1 mutation. Diabetes 62:3943–3950. doi:10.2337/db13-0571

Borowiec M, Liew CW, Thompson R et al (2009) Mutations at the BLK locus linked to maturity onset diabetes of the young and beta-cell dysfunction. Proc Natl Acad Sci USA 106:14460–14465. doi:10.1073/pnas.0906474106

Bowman P, Flanagan SE, Edghill EL et al (2012) Heterozygous ABCC8 mutations are a cause of MODY. Diabetologia 55:123–127. doi:10.1007/s00125-011-2319-x

Deeb SS, Fajas L, Nemoto M et al (1998) A Pro12Ala substitution in PPARgamma2 associated with decreased receptor activity, lower body mass index and improved insulin sensitivity. Nat Genet 20:284–287. doi:10.1038/3099

Delépine M, Nicolino M, Barrett T et al (2000) EIF2AK3, encoding translation initiation factor 2-alpha kinase 3, is mutated in patients with Wolcott-Rallison syndrome. Nat Genet 25:406–409. doi:10.1038/78085

Flanagan SE, De Franco E, Lango Allen H et al (2014) Analysis of transcription factors key for mouse pancreatic development establishes NKX2-2 and MNX1 mutations as causes of neonatal diabetes in man. Cell Metab 19:146–154. doi:10.1016/j.cmet.2013.11.021

Froguel P, Vaxillaire M, Sun F et al (1992) Close linkage of glucokinase locus on chromosome 7p to early-onset non-insulin-dependent diabetes mellitus. Nature 356:162–164. doi:10.1038/356162a0

Froguel P, Zouali H, Vionnet N et al (1993) Familial hyperglycemia due to mutations in glucokinase. Definition of a subtype of diabetes mellitus. N Engl J Med 328:697–702. doi:10.1056/NEJM199303113281005

Gloyn AL, Pearson ER, Antcliff JF et al (2004) Activating mutations in the gene encoding the ATP-sensitive potassium-channel subunit Kir6.2 and permanent neonatal diabetes. N Engl J Med 350:1838–1849. doi:10.1056/NEJMoa032922

Grant SFA, Thorleifsson G, Reynisdottir I et al (2006) Variant of transcription factor 7-like 2 (TCF7L2) gene confers risk of type 2 diabetes. Nat Genet 38:320–323. doi:10.1038/ng1732

Hani EH, Boutin P, Durand E et al (1998) Missense mutations in the pancreatic islet beta cell inwardly rectifying K+ channel gene (KIR6.2/BIR): a meta-analysis suggests a role in the polygenic basis of Type II diabetes mellitus in Caucasians. Diabetologia 41:1511–1515. doi:10.1007/s001250051098

Horikawa Y, Iwasaki N, Hara M et al (1997) Mutation in hepatocyte nuclear factor-1 beta gene (TCF2) associated with MODY. Nat Genet 17:384–385. doi:10.1038/ng1297-384

Igoillo-Esteve M, Genin A, Lambert N et al (2013) tRNA methyltransferase homolog gene TRMT10A mutation in young onset diabetes and primary microcephaly in humans. PLoS Genet 9, e1003888. doi:10.1371/journal.pgen.1003888

Inoue H, Tanizawa Y, Wasson J et al (1998) A gene encoding a transmembrane protein is mutated in patients with diabetes mellitus and optic atrophy (Wolfram syndrome). Nat Genet 20: 143–148. doi:10.1038/2441

Labay V, Raz T, Baron D et al (1999) Mutations in SLC19A2 cause thiamine-responsive megaloblastic anaemia associated with diabetes mellitus and deafness. Nat Genet 22: 300–304. doi:10.1038/10372

Lango Allen H, Flanagan SE, Shaw-Smith C et al (2012) GATA6 haploinsufficiency causes pancreatic agenesis in humans. Nat Genet 44:20–22. doi:10.1038/ng.1035

Malecki MT, Jhala US, Antonellis A et al (1999) Mutations in NEUROD1 are associated with the development of type 2 diabetes mellitus. Nat Genet 23:323–328. doi:10.1038/15500

Meur G, Simon A, Harun N et al (2010) Insulin gene mutations resulting in early-onset diabetes: marked differences in clinical presentation, metabolic status, and pathogenic effect through endoplasmic reticulum retention. Diabetes 59:653–661. doi:10.2337/db09-1091

Morris AP, Voight BF, Teslovich TM et al (2012) Large-scale association analysis provides insights into the genetic architecture and pathophysiology of type 2 diabetes. Nat Genet 44:981–990. doi:10.1038/ng.2383

Neel JV (1976) Diabetes mellitus — a geneticist's nightmare. In: Creutzfeldt W, Köbberling J, Neel JV (eds) The genetics of diabetes mellitus. Springer, Berlin, pp 1–11

Neve B, Fernandez-Zapico ME, Ashkenazi-Katalan V et al (2005) Role of transcription factor KLF11 and its diabetes-associated gene variants in pancreatic beta cell function. Proc Natl Acad Sci USA 102:4807–4812. doi:10.1073/pnas.0409177102

Njølstad PR, Søvik O, Cuesta-Muñoz A et al (2001) Neonatal diabetes mellitus due to complete glucokinase deficiency. N Engl J Med 344:1588–1592. doi:10.1056/NEJM200105243442104

Plengvidhya N, Kooptiwut S, Songtawee N et al (2007) PAX4 mutations in Thais with maturity onset diabetes of the young. J Clin Endocrinol Metab 92:2821–2826. doi:10.1210/jc.2006-1927

Poulton CJ, Schot R, Kia SK et al (2011) Microcephaly with simplified gyration, epilepsy, and infantile diabetes linked to inappropriate apoptosis of neural progenitors. Am J Hum Genet 89:265–276. doi:10.1016/j.ajhg.2011.07.006

Raeder H, Johansson S, Holm PI et al (2006) Mutations in the CEL VNTR cause a syndrome of diabetes and pancreatic exocrine dysfunction. Nat Genet 38:54–62. doi:10.1038/ng1708

Rubio-Cabezas O, Minton JAL, Kantor I et al (2010) Homozygous mutations in NEUROD1 are responsible for a novel syndrome of permanent neonatal diabetes and neurological abnormalities. Diabetes 59:2326–2331. doi:10.2337/db10-0011

Rubio-Cabezas O, Jensen JN, Hodgson MI et al (2011) Permanent neonatal diabetes and enteric anendocrinosis associated with biallelic mutations in NEUROG3. Diabetes 60: 1349–1353. doi:10.2337/db10-1008
Sandhu MS, Weedon MN, Fawcett KA et al (2007) Common variants in WFS1 confer risk of type 2 diabetes. Nat Genet 39:951–953. doi:10.1038/ng2067
Santer R, Schneppenheim R, Dombrowski A et al (1997) Mutations in GLUT2, the gene for the liver-type glucose transporter, in patients with Fanconi-Bickel syndrome. Nat Genet 17: 324–326. doi:10.1038/ng1197-324
Sellick GS, Barker KT, Stolte-Dijkstra I et al (2004) Mutations in PTF1A cause pancreatic and cerebellar agenesis. Nat Genet 36:1301–1305. doi:10.1038/ng1475
Senée V, Chelala C, Duchatelet S et al (2006) Mutations in GLIS3 are responsible for a rare syndrome with neonatal diabetes mellitus and congenital hypothyroidism. Nat Genet 38: 682–687. doi:10.1038/ng1802
Shaw-Smith C, De Franco E, Lango Allen H et al (2014) GATA4 mutations are a cause of neonatal and childhood-onset diabetes. Diabetes 63:2888–2894. doi:10.2337/db14-0061
SIGMA Type 2 Diabetes Consortium, Estrada K, Aukrust I et al (2014) Association of a low-frequency variant in HNF1A with type 2 diabetes in a Latino population. JAMA 311: 2305–2314. doi:10.1001/jama.2014.6511
Silander K, Mohlke KL, Scott LJ et al (2004) Genetic variation near the hepatocyte nuclear factor-4 alpha gene predicts susceptibility to type 2 diabetes. Diabetes 53:1141–1149
Simaite D, Kofent J, Gong M et al (2014) Recessive mutations in PCBD1 cause a new type of early-onset diabetes. Diabetes 63:3557–3564. doi:10.2337/db13-1784
Smith SB, Qu H-Q, Taleb N et al (2010) Rfx6 directs islet formation and insulin production in mice and humans. Nature 463:775–780. doi:10.1038/nature08748
Steinthorsdottir V, Thorleifsson G, Sulem P et al (2014) Identification of low-frequency and rare sequence variants associated with elevated or reduced risk of type 2 diabetes. Nat Genet 46:294–298. doi:10.1038/ng.2882
Stoffers DA, Ferrer J, Clarke WL, Habener JF (1997a) Early-onset type-II diabetes mellitus (MODY4) linked to IPF1. Nat Genet 17:138–139. doi:10.1038/ng1097-138
Stoffers DA, Zinkin NT, Stanojevic V et al (1997b) Pancreatic agenesis attributable to a single nucleotide deletion in the human IPF1 gene coding sequence. Nat Genet 15:106–110. doi:10.1038/ng0197-106
Stone LM, Kahn SE, Fujimoto WY et al (1996) A variation at position -30 of the beta-cell glucokinase gene promoter is associated with reduced beta-cell function in middle-aged Japanese-American men. Diabetes 45:422–428
Støy J, Edghill EL, Flanagan SE et al (2007) Insulin gene mutations as a cause of permanent neonatal diabetes. Proc Natl Acad Sci USA 104:15040–15044. doi:10.1073/pnas.0707291104
Vaxillaire M, Boccio V, Philippi A et al (1995) A gene for maturity onset diabetes of the young (MODY) maps to chromosome 12q. Nat Genet 9:418–423. doi:10.1038/ng0495-418
Vaxillaire M, Bonnefond A, Froguel P (2012) The lessons of early-onset monogenic diabetes for the understanding of diabetes pathogenesis. Best Pract Res Clin Endocrinol Metab 26: 171–187. doi:10.1016/j.beem.2011.12.001
Winckler W, Weedon MN, Graham RR et al (2007) Evaluation of common variants in the six known maturity-onset diabetes of the young (MODY) genes for association with type 2 diabetes. Diabetes 56:685–693. doi:10.2337/db06-0202
Yamagata K, Furuta H, Oda N et al (1996a) Mutations in the hepatocyte nuclear factor-4alpha gene in maturity-onset diabetes of the young (MODY1). Nature 384:458–460. doi:10.1038/384458a0
Yamagata K, Oda N, Kaisaki PJ et al (1996b) Mutations in the hepatocyte nuclear factor-1alpha gene in maturity-onset diabetes of the young (MODY3). Nature 384:455–458. doi:10.1038/384455a0

Chapter 2
Genome-Wide Association Studies of Type 2 Diabetes

Rob Sladek and Inga Prokopenko

Abstract Genome-wide association (GWA) studies represent the single most effective technique for identifying genetic risk loci causing complex diseases. Since the publication of the first GWA studies for type 2 diabetes (T2D) in 2007, nearly 90 statistically robust risk loci have been identified. The T2D risk loci identified by GWA studies contained several genes that are targets of current diabetic therapies; however, the majority of genes in these loci had not previously been implicated in the pathophysiology of T2D. Mechanistic insights about the physiological role of T2D loci in the disease predisposition have been gained from investigation of their contribution into glycemic trait variability in nondiabetic individuals. Current efforts to identify the causative genetic mutations in these loci and the molecular mechanisms through which they exert their effects have the potential to make far-reaching contributions to our understanding of molecular basis of T2D and the development of novel strategies for patient care.

2.1 Introduction

Type 2 diabetes (T2D) is a common, chronic disorder whose prevalence is increasing rapidly across the globe. Like other complex diseases, T2D represents a challenge for genetic studies aiming to uncover the underlying pathophysiological mechanisms. It is predicted that T2D will affect 592 million individuals by 2035 (Federation 2013) in developed and low- and middle-income countries. While the recent increase in T2D prevalence has been attributed to a sedentary "westernized"

R. Sladek
Department of Human Genetics, McGill University, Genome Quebec and McGill University Innovation Centre, 740, Doctor Penfield Avenue, Montreal, QC, Canada
e-mail: robert.sladek@mcgill.ca

I. Prokopenko (✉)
Department of Genomics of Common Disease, School of Public Health, Imperial College London, Hammersmith Hospital, Burlington Danes Building, Du Cane Road, London W12 0NN, UK
e-mail: i.prokopenko@imperial.ac.uk

J.C. Florez (ed.), *The Genetics of Type 2 Diabetes and Related Traits*,
DOI 10.1007/978-3-319-01574-3_2

lifestyle and changes in diet, a significant proportion of heritable factors also contribute to individual susceptibility (Hu 2011).

The strong family clustering and heritability of T2D and related glycemic traits have motivated a large number of studies to identify genetic factors that cause this disease (Permutt et al. 2005; Stumvoll et al. 2005); despite much effort, by late 2006 only three genetic loci had been reproducibly shown to increase T2D risk [reviewed in Majithia and Florez (2009), McCarthy (2008), Stolerman and Florez (2009)]. The earliest attempts to discover T2D-associated genes used either position- or function-based strategies. In a position-based search, genes are identified within families by studying the co-inheritance of the disease with a set of polymorphic markers whose genomic positions are known. Such "linkage studies" usually identify a genomic region (~10 Mbp) that confers genetic risk; the disease-causative mutations are identified by sequencing transcribed and functional elements of all genes in the target region. In function-based approaches, risk association is tested for common genetic variants in candidate genes involved in T2D pathophysiology. In these studies, variants identified in a small number of patients and control subjects are genotyped in larger case-control samples. Both these approaches are characterized by a number of limitations. Linkage analysis is underpowered to detect low penetrance variants, expected to contribute to T2D susceptibility, given its high population prevalence. Candidate gene studies usually were conducted in samples of insufficient size and their findings had low reproducibility as well as difficulty to select good biological candidates.

Positional strategies have identified putative T2D loci in several large chromosome regions (McCarthy 2003) and in a number of specific genes (Horikawa et al. 2000; Meyre et al. 2005; Silander et al. 2004; Hara et al. 2002); however, none of these associations have been convincingly replicated. The candidate-gene approach generated a large number of positive reports, two of which have been confirmed in independent studies (Table 2.2) (Gloyn and McCarthy 2001). The Pro12Ala variant in the peroxisome proliferator-activated receptor gamma (*PPARG*) gene (Deeb et al. 1998; Altshuler et al. 2000; Lohmueller et al. 2003) and the Glu23Lys variant in the potassium inwardly rectifying channel, subfamily J, member 11 (*KCNJ11*) gene were shown to contribute to T2D risk in multiple studies (Gloyn et al. 2003; Laukkanen et al. 2004). Each of these two common variants contributes only modestly (increasing T2D risk by 15–20 % for each susceptibility allele) to the risk of developing common form of diabetes, while rare variants in both these genes cause monogenic diseases such as familial partial lipodystrophy and neonatal diabetes. Interestingly, these variants occur within pharmacological targets for the thiazolidinedione (PPARG) and sulfonylurea compounds (KCNJ11) used to treat T2D.

2.2 Common Variants: The First Steps Toward Large-Scale Association Mapping

Given the inefficient progress of early T2D gene discovery, the application of genome-wide association (GWA) studies to identify risk loci for T2D and glycemic traits represented a major advance in complex trait genetics. GWA studies are observational epidemiological studies in which genetic risk exposure is measured using hundreds of thousands of genotyping assays. The critical difference between GWAs and other observational epidemiological studies lies in the large number of genetic tests performed to assess exposure in each individual patient. On the one hand, the success of GWA studies relies on the development of technologies capable of screening a large number of polymorphisms (predominantly represented by single-nucleotide polymorphisms, SNPs), as the prior probability that any individual polymorphism will be associated with disease is small. On the other hand, the polymorphisms studied using commercial microarray platforms are not genetically independent and display complex linkage structures that may extend over tens or hundreds of thousands of base pairs. As a result, the success of a GWA study relies on achieving an adequate marker density to model local linkage structures across the genome.

The potential benefits of using GWA studies to discover complex disease risk loci were first demonstrated in a seminal paper by Risch and Merikangas that showed that the analysis of one million variants in the sample of unrelated individuals had greater statistical power than a linkage analysis with a few hundred markers (Risch and Merikangas 1996). In this context, Reich and Lander suggested a theoretical population-genetics model for a relatively simple distribution of susceptibility variants at a disease locus and rephrased the common disease common variant hypothesis (CDCV) to propose that high-frequency variants with low penetrance at disease loci contribute to the largest proportion of disease risk in a population (Reich and Lander 2001). Their theoretical demonstration of the CDCV hypothesis did not provide any expectation about the number of disease loci or their effect sizes in establishing complex disease risk.

The majority of GWA studies performed to identify T2D risk loci have used a case-control study design (Table 2.1), with retrospective longitudinal studies being primarily reserved for validation of previously identified loci. Alternate study designs to detect T2D risk associations are far less common and have included populations with early-onset diabetes (taken as a proxy for more severe illness), longitudinal studies in at-risk populations, and studies in isolated populations. Affected individuals in the genetic discovery cohorts are typically selected carefully using diagnostic criteria established by the American Diabetes Association or World Health Organization that are based solely on blood glucose levels. In contrast, selection of control subjects has been more problematic, with most discovery cohorts including patients based on a single normal blood glucose measurement and absent medical history of glucose intolerance. Many discovery cohorts have excluded patients with monogenic diabetes based on a suggestive

Table 2.1 Major published T2D GWAS and meta-analyses

Study	Ethnicity/origin	*N* cases[a]	*N* controls[a]	Novel loci identified	GWAS or meta-analysis discovery approach	GWAS array	Reference panel for imputation	T2D phenotype definition/other specs
Diabetes Gene Discovery Group (Sladek et al. 2007), Nature	European	694	645	*SLC30A8*, *HHEX/IDE*	GWA	Illumina 300k +	–	Family history of T2D, AAO <45 years, BMI <30 kg/m^2
Finland–US Investigation of NIDDM Genetics (FUSION) (Scott et al. 2007a), Science	European	1161	1174	*CDKN2A/2B*, *IGF2BP2*, *CDKAL1*	GWA	Illumina 300k	–	Partial enrichment for family history
deCODE Genetics (Steinthorsdottir et al. 2007), Nat Genet	European	1399	5275	*CDKAL1*	GWA	Illumina 300k	–	No specific enrichment for family history, young AAO, or BMI
Diabetes Genetics Initiative (Diabetes Genetics Initiative of Broad Institute of H et al. 2007), Science	European	1464	1467	*CDKN2A/2B*, *IGF2BP2*, *CDKAL1*	GWA	Affymetrix 500k	–	Partial enrichment for family history and lean T2D
Wellcome Trust Case–Control Consortium (Zeggini et al. 2007), Science	European	1924	2938	*CDKAL1*, *CDKN2A/2B*, *IGF2BP2*	GWA	Affymetrix 500k	–	Enrichment for family history of T2D, AAO <65 years
DIAGRAM (Zeggini et al. 2008), Nat Genet	European	4549	5579	*JAZF1*, *CDC123-CAMK1D*, *TSPAN8-LGR5*, *THADA*, *ADAMTS9*, *NOTCH2*	GWA M-A	Affymetrix 500k chip/Illumina 317k chip	CEU HapMap Phase 2	–

DIAGRAM (Voight et al. 2010), Nat Genet	European	42,542	98,912	*BCL11A*, *ZBED3*, *KLF14*, *TP53INP1*, *CHCHD9*, *KCNQ1*, *CENTD2*, *HMGA2*, *HNF1A*, *ZFAND6*, *PRC1*, *DUSP9*,	GWA M-A	Mixed	CEU HapMap Phase 2	–
Qi et al. (2010), Hum Mol Gen	European	2591	3052	*RBMS1/ITGB6*	GWA	Affymetrix SNP Array 6.0	–	T2D self-reported cases, confirmed by medical records review
DIAGRAM (Perry et al. 2012), Plos Genet	European	2112 lean; 4123 obese	54,412	*LAMA1* (*lean*), *HMG20A*(*obese*)	GWA M-A	Mixed	CEU HapMap Phase 2	Lean BMI<25 kg/m^2; obese BMI >=30 kg/m^2
DIAGRAM (Morris et al. 2012), Nat Genet	European	34,840	114,981	*ZMIZ1*, *ANK1*, *KLHDC5*, *TLE1*, *ANKRD55*, *CILP2*, *MC4R*, *BCAR1*, *HMG20A*, *GRB14*	GWA M-A	Mixed genome-wide + Illumina CardioMetabochip	CEU HapMap Phase 2	–
Albrechtsen et al. (2013), Diabetologia	Europeans	1000	1000	*COBLL1*, *MACF1*	Whole-exome sequencing association study	8 x exome capture by a NimbleGen 2.1M HD array (target region 34.1 Mb, 21,810 genes) and Illumina GAII sequencing	–	T2D cases with BMI >27.5 kg/m^2 and hypertension
deCODE Genetics (Steinthorsdottir et al. 2014), Nat Genet	Europeans	11,114	267,140	*CCND2*, *PAM*, *PDX1*	GWA	Mixed Illumina	2630 whole-genome sequenced Icelanders	Hospital record, self-reported, HBA1C>6.5%, oral diabetes medication

(continued)

Table 2.1 (continued)

Study	Ethnicity/origin	*N* cases[a]	*N* controls[a]	Novel loci identified	GWAS or meta-analysis discovery approach	GWAS array	Reference panel for imputation	T2D phenotype definition/other specs
Palmer et al. (2012), Plos One	African Americans	965	1029	*RND3/RBM43*	GWA	Affymetrix SNP Array 6.0	CEU and YRI HapMap Phase 2 release 22	T2D and end-stage renal disease (T2D-ESRD), AAO >25, T2D diagnosed ≥5 years before renal replacement therapy
Hanson et al. (2014), Diabetes	American Indians	416	424	*DNER*[b]	GWA	Affymetrix SNP Array 6.0	–	AAO <25 years, enrichment for family history
Unoki et al. (2008), Nat Genet	Japanese	194	1558	*KCNQ1*	GWA	High-density oligo-nucleotide arrays (Perlegen Sciences)/Affymetrix GeneChip	–	T2D and diabetic retinopathy
Yamauchi et al. (2010), Nat Genet	Japanese	4878	3345	*UBE2E2*, *C2CD4A-C2CD4B*	GWA	Illumina HumanHap610-Quad and 550k BeadChip	JPT and CHB HapMap Phase 2	T2D cases
Shu et al. (2010), Plos Genet	Asian	1019	1710	*SPRY2*	GWA	Affymetrix SNP Array 6.0	JPT/CHB and CEU HapMap Phase 2	T2D cases
Kooner et al. (2011), Nat Genet	South Asian	5561	14,458	*GRB14*, *ST6GAL1*, *VPS26A*, *HMG20A*, *AP3S2*, *HNF4A*	GWA	Illumina Infinium BeadChips	Multiethnic HapMap Phase 2	T2D cases

Saxena et al. (2013), Diabetes	Punjabi Sikhs from India	842	774	*SGCG*	GWA	Human 660W-Quad BeadChip panel (Illumina)	Multiethnic HapMap Phase 3	ADA 2004 criteria (plasma glucose 7.0 mmol/l or 2-h postglucose load 11.1 mmol/l), excluded T1D, family history for T1D, MODY, and t2D from hemo-chromatosis or pancreatitis
Tabassum et al. (2013), Diabetes	Indians	1256	1209	*TMEM163*	GWA	Illumina Human610-Quad BeadChips	1000 Genomes Phase 1	T2D by WHO criteria (plasma glucose 7.0 mmol/l or 2-h postglucose load 11.1 mmol/l)
Tsai et al. (2010), Plos Genet	Chinese	995	894	*PTPRD*, *SRR*	GWA	Illumina HumanHap550-Duo BeadChip/ Sequenom iPLEX	–	Age >20 years, excluded T1D, GD, MODY
Cho et al. (2012), Nat Genet	East Asian	6952	11,865	*GLIS3*, *PEPD*, *FITM2-R3HDML-HNF4A*, *KCNK16*, *MAEA*, *GCC1-PAX4*, *PSMD6*, *ZFAND3*	GWA M-A	Mixed	JPT and CHB HapMap Phase 2	T2D cases
Li et al. (2013), Diabetes	Han Chinese	1999	1976	*GRK5*, *RASGRP1*	GWA	Illumina Human660W-Quad BeadChip	JPT and CHB HapMap Phase 2	T2D cases
Ma et al. (2013), Diabetologia	Chinese	684	955	*PAX4*	GWA M-A	Illumina	CHB + JPT 1000 Genomes project	WHO 1998 criteria

(continued)

Table 2.1 (continued)

Study	Ethnicity/origin	*N* cases[a]	*N* controls[a]	Novel loci identified	GWAS or meta-analysis discovery approach	GWAS array	Reference panel for imputation	T2D phenotype definition/other specs
Hara et al. (2014), Hum Mol Gen	East Asian	5976	20,829	*MIR129-LEP*, *GPSM1*, *SLC16A13*	GWA	Illumina 610k	CHB+CHS +JPT 1000 Genomes Phase 1	WHO criteria, excluded GAD positive, monogenic diabetes, other diseases or drugs causing diabetes
Parra et al. (2011), Diabetologia	Hispanic (Mexican-American)	947	343	*C14orf70*	GWA	Affymetrix genome-wide human SNP array 5.0	HapMap Phase 2 combined + HapMap Phase 3 Mexican-American	ADA criteria (FPG $\geq$ 7.0 mmol/l or 2-h OGTT glucose $\geq$ 11.1 mmol/l)
SIGMA Type 2 Diabetes Consortium (2014b), Nature	Mexican	3848	4366	*SLC16A11*/*SLC16A13*	GWA	Illumina OMNI2.5 array	1000 Genomes Phase 1	ADA criteria
DIAGRAM, AGEN-T2D, SAT2D, MAT2D, T2D-GENES, (DIAGRAM Consortium et al. 2014), Nat Genet	Multiethnic	26,488	83,964	*TMEM154*, *SSR1-RREB1*, *FAF1*, *POU5F1-TCF19*, *LPP*, *ARL15*, *MPHOSPH9*	GWA M-A	Mixed	HapMap Phase 2/3	T2D cases

Saxena et al. (2012), Am J Hum Genet	Multiethnic	17,418	70,298	*GATAD2A/CILP2/ PBX4*, *BCL2*	Gene-centric GWA M-A	50k SNP Human CVD BeadChip (ITMAT-Broad-CARe [IBC] array)	–	ADA criteria(FPG ≥7.0 mmol/l or 2-h OGTT or nonfasting glucose ≥11.1 mmol/l), AAO ≥ 25 years

Abbreviations: *AAO* Age at onset, *T1D* Type 1 diabetes, *GD* Gestational diabetes, *MODY* Maturity-onset diabetes of the young, *GWA M-A* Genome-wide association meta-analysis

[a]Sample sizes in discovery analyses

[b]Association reached study-wise significance corrected for multiple testing

family history or specific genetic tests and patients with autoimmune diabetes based on specific serum markers. In many studies, the case and control samples differ significantly in age, in order to avoid selecting individuals who will develop diabetes later in life as control subjects. Comparable environmental exposures are used as basis for selection of both, T2D cases and controls; and in addition to matching affected and control individuals for general ethnic background, ethnic outliers are excluded from analyses. Diabetic individuals usually are of comparable body mass index (BMI) in respect to controls in large genetic studies.

The first successful "hypothesis-free" demonstration of T2D association came from the discovery of an intronic SNP in the transcription factor 7-like-2 (*TCF7L2*) gene, which confers the largest effect on T2D risk reported to date among common variants (Fig. 2.1; Tables 2.1 and 2.2) (Weedon 2007; Grant et al. 2006). These association studies were motivated by the group's earlier demonstration of microsatellite associations in a linkage region on chromosome 10 (Reynisdottir et al. 2003) rather than by functional criteria. In fact, *TCF7L2* encodes a transcription factor within the Wnt signaling pathway whose involvement in T2D pathogenesis remained elusive for many years following the initial genetic studies. Despite this, detailed physiological studies have now demonstrated the importance of the *TCF7L2* locus in β-cell function and insulin secretion in human cohorts (da Silva Xavier et al. 2009, 2012; Dupuis et al. 2010; Dimas et al. 2014) and as a critical regulator of β-cell mass and function (Takamoto et al. 2014) and hepatic carbohydrate metabolism (Boj et al. 2012) in mouse models (see Chap. 15 for details).

2.3 Loci Established Through T2D GWA Studies

The capacity to undertake efficient, large-scale association analyses using hypothesis-free approach through genome-wide studies opened a new wave of discoveries in T2D genetics. Four GWA studies published in 2007 (Diabetes Genetics Initiative of Broad Institute of H et al. 2007; Scott et al. 2007b; Sladek et al. 2007; Zeggini et al. 2007) (Table 2.1) confirmed the strongest association at *TCF7L2*, two previously established signals at *PPARG* and *KCNJ11* and identified six novel loci, at *HHEX/IDE*, *CDKAL1*, *IGF2BP2*, *CDKN2A/2B*, *SLC30A8*, and *FTO* (Frayling et al. 2007; Freathy et al. 2008; Fall et al. 2013). Although it is conventionally used to name the loci by the most credible regional candidate (e.g., *SLC30A8*) rather than the tag SNP showing the strongest association (e.g., rs13266634), these assignments are used as a matter of convenience and do not imply that a mechanistic link has been proven. The association signals found in GWA studies require further investigation through extensive fine mapping and functional characterization to establish causal variants and determine their impact on T2D pathogenesis at a molecular level (Prokopenko et al. 2008).

The first round of published T2D GWA studies has provided both the identification of novel associated loci and the landscape of T2D susceptibility across the

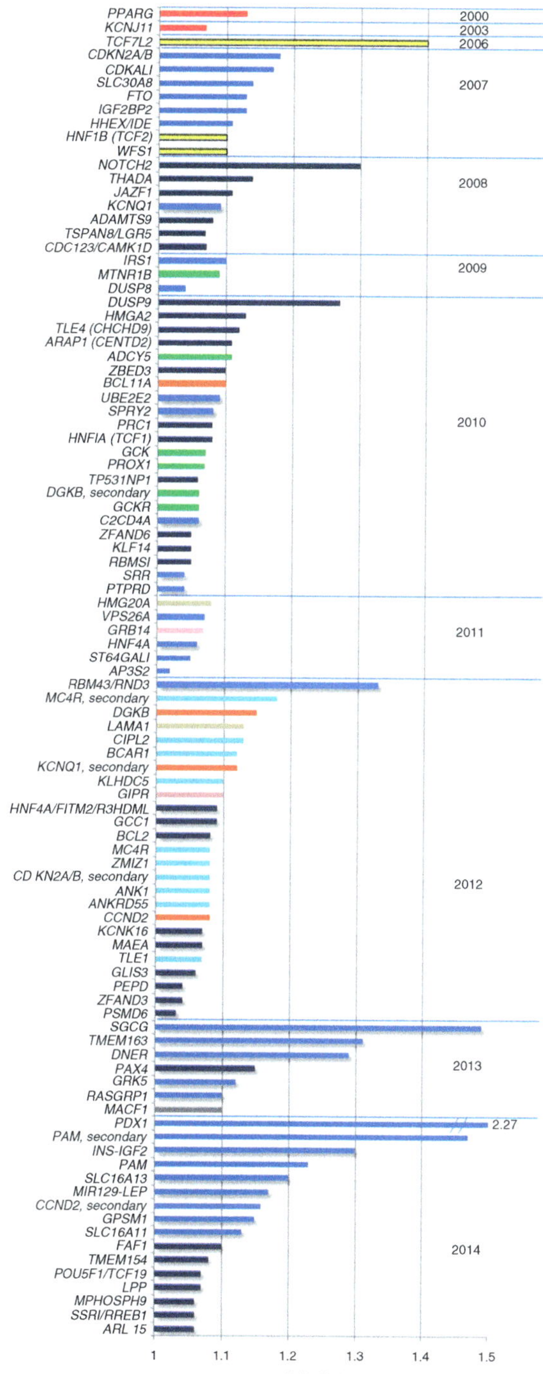

Fig. 2.1 Effect sizes of established T2D-susceptibility loci. Established T2D-susceptibility variants have only modest individual effects. The *x*-axis gives the per-allele odds ratio estimated for European-descent samples or for the ethnic group of discovery, if association was reported after

whole genome, the latter providing a point of reference for the previous equivocal findings accumulated through candidate-gene and linkage studies (Parikh and Groop 2004). Thus, controversial evidence for variants in Calpain-10 (*CAPN10*) and the insulin (*INS*) gene in T2D has not been confirmed by the GWA approach. Similarly, while the 1q chromosome region (30-Mb region near to centromere) contained a number of highly plausible candidates supported by genome-wide linkage analyses in multiple ethnicities, none have been confirmed through association studies and large-scale GWA meta-analyses to date (Prokopenko et al. 2009b; Morris et al. 2012; Replication and Meta-analysis 2014).

Simultaneously with the advent of GWA studies, large-scale replication efforts confirmed two loci highlighted by the candidate-pathway pre-GWA studies. Variants within Wolfram syndrome 1 (*WFS1*) gene and common variants in hepatocyte nuclear factor 1-b (*HNF1B*, also known as *TCF2*) were confirmed as associated with T2D (Franks et al. 2008; Sandhu et al. 2007; Winckler et al. 2007). These variants, along with *KCNJ11* and *PPARG*, provide interesting examples of the convergence between Mendelian and polygenic causes of diabetes, as coding variants in these genes had previously been isolated in families with autosomal dominant inheritance of diabetes (Maturity Onset Diabetes of the Young, MODY) and as part of the multisystem Wolfram syndrome (McCarthy and Hattersley 2008).

As anticipated, GWA studies, by testing hundreds of thousands of genetic variants in parallel, have identified loci with modest effects (Manolio et al. 2009). To contend with the stringent significance thresholds that account for the number of independent tests performed across the genome, identification of additional T2D susceptibility loci required larger population samples, which was achieved by combining existing GWA studies in meta-analyses. The Diabetes Genetics Replication And Meta-analysis (DIAGRAM, http://www.diagram-consortium.org/) consortium carried out the first meta-analysis for T2D (Zeggini et al. 2008) of three GWA studies of European-descent individuals, including ~4500 cases and 5500 controls. Differences in the genotyping platforms used for individual GWA studies were overcome by imputation using a common variant set based on haplotype structure of densely characterized reference samples in HapMap (Consortium IH 2005) and extended the analysis to ~2.2 million SNPs across the genome

---

Fig. 2.1 (continued) mid-2012 (Table 2.1) for each locus listed on the *y*-axis. Loci are sorted by descending order of per-allele effect size within each year. *Colors* highlight the discovery study approach: *red*, candidate gene; *yellow*, large-scale association; *blue*, genome-wide association; *dark blue*, genome-wide association meta-analysis; *sky blue*, genome-wide meta-analysis with Metabochip follow-up; *green*, genome-wide meta-analysis of glycemic traits; *pink*, genome-wide sex-differentiated meta-analysis with larger effects in women; *brown*, genome-wide sex-differentiated meta-analysis with larger effects in men; *hacky*, genome-wide meta-analysis in lean/obese; *gray*, whole-exome sequencing. For loci with sex differentiation, the effect size for the sex with larger effect is presented. *X*-axis lists loci names, labeled by the gene names within region. *Y*-axis shows odds ratio for T2D observed at a given locus. Loci are split by the year of discovery and are ordered from top to bottom by the decreasing OR on T2D risk within each year. *Shadow* is used for loci from studies with discovery including non-European individuals

Table 2.2 T2D-susceptibility loci with genome-wide significant evidence for association

Chromosome	Position (HG19)	Locus (nearest genes)	Effect on glucose homeostasis[a]	Index variant	Effect size (95 %CI)	Risk allele/other alleles	Risk allele frequency	Year first association report	Locus lead SNP, effect size, and allele frequency reference
1	39,835,817	*MACF1*	UC	rs2296172	1.10 (1.06–1.14)	G/A	0.23	2013	Albrechtsen et al. (2013), Diabetologia
1	50,909,985	*FAF1*	UC	rs17106184	1.10 (1.07–1.14)	G/A	NA	2014	DIAGRAM Consortium et al. (2014), Nat Genet
1	120,517,959	*NOTCH2*	UC	rs10923931	1.30 (1.17–1.43)	T/G	0.11	2008	Morris et al. (2012), Nat Genet
1	214,154,719	*PROX1*	BC	rs2075423	1.07 (1.05–1.10)	G/T	0.62	2010	Morris et al. (2012), Nat Genet
2	27,741,237	*GCKR*	IR	rs780094	1.06 (1.04–1.08)	C/T	NA	2010	Morris et al. (2012), Nat Genet
2	43,690,030	*THADA*	BC	rs10203174	1.14 (1.10–1.19)	C/T	0.89	2008	Morris et al. (2012), Nat Genet
2	60,568,745	*BCL11A*	UC	rs243088	1.10 (1.06–1.13)	T/A	0.45	2010	Morris et al. (2012), Nat Genet
2	135,479,730	*TMEM163*	IR	rs6723108	1.31 (1.20–1.44)	T/G	0.87	2013	Tabassum et al. (2013), Diabetes[c]
2	151,637,936	*RBM43/ RND3*	UC	rs7560163	1.33 (1.19–1.49)	G/C	0.14	2012	Palmer et al. (2012), Plos One[c]
2	161,346,447	*RBMS1*	UC	rs7569522	1.05 (1.03–1.07)	A/G	0.44	2010	Morris et al. (2012), Nat Genet
2	165,528,876	*GRB14*	IR	rs13389219	1.07 (1.05–1.10)	A/C	0.64	2011	Morris et al. (2012), Nat Genet
2	227,093,585	*IRS1*	IR	rs2943640	1.10 (1.07–1.12)	C/A	0.63	2009	Morris et al. (2012), Nat Genet

(continued)

Table 2.2 (continued)

Chromosome	Position (HG19)	Locus (nearest genes)	Effect on glucose homeostasis[a]	Index variant	Effect size (95 %CI)	Risk allele/ other alleles	Risk allele frequency	Year first association report	Locus lead SNP, effect size, and allele frequency reference
2	230,522,398	*DNER*	IR[b]	rs1861612	1.29	T/C	0.64	2013	Hanson et al. (2014), Diabetes[c]
3	12,393,125	*PPARG*	IR	rs1801282	1.13 (1.09–1.17)	C/G	0.86	2000	Morris et al. (2012), Nat Genet
3	23,454,790	*UBE2E2*	BC	rs1496653	1.09 (1.06–1.12)	A/G	0.75	2010	Morris et al. (2012), Nat Genet
3	64,090,363	*PSMD6*	UC	rs12497268	1.03 (1.01–1.07)	G/C	0.80	2012	Morris et al. (2012), Nat Genet
3	64,705,365	*ADAMTS9*	UC	rs6795735	1.08 (1.06–1.11)	C/T	0.59	2008	Morris et al. (2012), Nat Genet
3	123,082,398	*ADCY5*	BC	rs11717195	1.11 (1.08–1.14)	T/C	0.77	2010	Morris et al. (2012), Nat Genet
3	185,511,687	*IGF2BP2*	BC	rs4402960	1.13 (1.10–1.16)	T/G	0.33	2007	Morris et al. (2012), Nat Genet
3	186,613,409	*ST64GAL1*	UC	rs17301514	1.05 (1.01–1.09)	A/G	0.13	2011	Morris et al. (2012), Nat Genet
3	187,740,523	*LPP*	UC	rs6808574	1.07 (1.04–1.09)	C/T	NA	2014	DIAGRAM Consortium et al. (2014), Nat Genet
4	1,293,245	*MAEA*	UC	rs6819243	1.07 (1.01–1.14)	T/C	0.96	2012	Morris et al. (2012), Nat Genet
4	6,289,986	*WFS1*	IR	rs4458523	1.10 (1.07–1.12)	G/T	0.57	2007	Morris et al. (2012), Nat Genet
4	153,520,475	*TMEM154*	UC	rs6813195	1.08 (1.06–1.10)	C/T	NA	2014	DIAGRAM Consortium et al. (2014), Nat Genet

5	53,271,420	*ARL15*	UC	rs702634	1.06 (1.04–1.09)	A/G	NA	2014	DIAGRAM Consortium et al. (2014), Nat Genet
5	55,806,751	*ANKRD55*	IR[b]	rs459193	1.08 (1.05–1.11)	G/A	0.70	2012	Morris et al. (2012), Nat Genet
5	76,427,311	*ZBED3*	UC	rs6878122	1.10 (1.07–1.13)	G/A	0.28	2010	Morris et al. (2012), Nat Genet
5	102,338,739	***PAM, secondary***	Insulin secretion	rs78408340	1.47	G/C	0.07	2014	Steinthorsdottir et al. (2014), Nat Genet
5	102,338,811	***PAM***	Insulin secretion	rs35658696	1.23	G/A	0.05	2014	Steinthorsdottir et al. (2014), Nat Genet
6	7,258,617	*SSR1/ RREB1*	UC	rs9502570	1.06 (1.04–1.08)	A/G	NA	2014	DIAGRAM Consortium et al. (2014), Nat Genet
6	20,679,709	*CDKAL1*	BC	rs7756992	1.17 (1.14–1.20)	G/A	0.29	2007	Morris et al. (2012), Nat Genet
6	31,136,714	*POU5F1/ TCF19*	UC	rs3132524	1.07 (1.04–1.09)	G/A	NA	2014	DIAGRAM Consortium et al. (2014), Nat Genet
6	38,177,667	*ZFAND3*	UC	rs4299828	1.04 (1.01–1.07)	A/G	0.79	2012	Morris et al. (2012), Nat Genet
6	39,304,211	*KCNK16*	BC	rs3734621	1.07 (1.00–1.15)	C/A	0.03	2012	Morris et al. (2012), Nat Genet
7	14,898,282	***DGKB***	BC	rs17168486	1.15 (1.11–1.19)	T/C	0.19	2012	Morris et al. (2012), Nat Genet
7	15,052,860	***DGKB, secondary***	BC	rs6960043	1.06 (1.04–1.09)	C/T	0.47	2010	Morris et al. (2012), Nat Genet

(continued)

Table 2.2 (continued)

Chromosome	Position (HG19)	Locus (nearest genes)	Effect on glucose homeostasis[a]	Index variant	Effect size (95 %CI)	Risk allele/other alleles	Risk allele frequency	Year first association report	Locus lead SNP, effect size, and allele frequency reference
7	28,196,413	*JAZF1*	UC	rs849135	1.11 (1.08–1.13)	G/A	0.52	2008	Morris et al. (2012), Nat Genet
7	44,245,363	*GCK*	Hyperglycemic[d]	rs10278336	1.07 (1.04–1.10)	A/G	0.50	2010	Morris et al. (2012), Nat Genet
7	126,996,837	***GCC1***	UC	rs17867832	1.09 (1.03–1.15)	T/G	0.91	2012	Morris et al. (2012), Nat Genet
7	127,246,903	***PAX4***	BC[b]	rs10229583	1.15 (1.08–1.22)	G/A	0.85	2013	Ma et al. (2013), Diabetes[c]
7	127,862,802	*MIR129-LEP*	IR	rs791595	1.17 (1.12–1.22)	A/G	0.08	2014	Hara et al. (2014), Hum Mol Gen[c]
7	130,437,689	*KLF14*	IR[d]	rs13233731	1.05 (1.02–1.07)	G/A	0.51	2010	Morris et al. (2012), Nat Genet
8	41,519,248	*ANK1*	BC[b]	rs516946	1.08 (1.05–1.11)	C/T	0.76	2012	Morris et al. (2012), Nat Genet
8	95,937,502	*TP53INP1*	UC	rs7845219	1.06 (1.03–1.08)	T/C	0.52	2010	Morris et al. (2012), Nat Genet
8	118,185,025	*SLC30A8*	BC	rs3802177	1.14 (1.11–1.17)	G/A	0.66	2007	Morris et al. (2012), Nat Genet
9	4,292,083	*GLIS3*	BC[b]	rs10758593	1.06 (1.04–1.09)	A/G	0.42	2012	Morris et al. (2012), Nat Genet
9	8,369,533	*PTPRD*	UC	rs16927668	1.04 (1.01–1.07)	T/C	0.24	2010	Morris et al. (2012), Nat Genet
9	22,051,670	***CDKN2A/B, secondary***	BC	rs944801	1.08 (1.05–1.10)	C/G	0.53	2012	Morris et al. (2012), Nat Genet
9	22,134,094	***CDKN2A/B***	BC	rs10811661	1.18 (1.15–1.22)	T/C	0.82	2007	Morris et al. (2012), Nat Genet
9	81,905,590	*TLE4 [CHCHD9]*[e]	UC	rs17791513	1.12 (1.07–1.17)	A/G	0.91	2010	Morris et al. (2012), Nat Genet

9	84,308,948	*TLE1*	UC	rs2796441	1.07 (1.05–1.10)	G/A	0.70	2012	Morris et al. (2012), Nat Genet
9	139,252,148	*GPSM1*	UC	rs11787792	1.15 (1.10–1.20)	A/G	0.87	2014	Hara et al. (2014), Hum Mol Gen[c]
10	12,307,894	*CDC123/ CAMK1D*	BC	rs11257655	1.07 (1.04–1.10)	T/C	0.23	2008	Morris et al. (2012), Nat Genet
10	70,865,342	*VPS26A*	UC	rs12242953	1.07 (1.02–1.12)	G/A	0.93	2011	Morris et al. (2012), Nat Genet
10	80,942,631	*ZMIZ1*	UC	rs12571751	1.08 (1.05–1.10)	A/G	0.52	2012	Morris et al. (2012), Nat Genet
10	94,462,882	*HHEX/IDE*	BC	rs1111875	1.11 (1.09–1.14)	C/T	0.58	2007	Morris et al. (2012), Nat Genet
10	114,758,349	*TCF7L2*	BC	rs7903146	1.40 (1.34–1.46)	T/C	NA	2006	Morris et al. (2012), Nat Genet
10	121,149,403	*GRK5*	IR	rs10886471	1.12 (1.08–1.16)	C/T	0.78	2013	Li et al. (2013), Diabetes[c]
11	1,696,849	*DUSP8*	UC	rs2334499	1.04 (1.02–1.06)	T/C	0.43	2009	Morris et al. (2012), Nat Genet
11	2,150,895	*INS-IGF2*	UC	rs11564732	1.30 (1.19–1.43)	G/A	NA	2014	SIGMA Type 2 Diabetes Consortium 2014b, Nature[c]
11	2,691,500	***KCNQ1*, *secondary***	BC	rs231361	1.09 (1.06–1.12)	A/G	0.29	2010	Morris et al. (2012), Nat Genet
11	2,847,069	***KCNQ1***	BC	rs163184	1.12 (1.09–1.16)	G/T	0.50	2012	Morris et al. (2012), Nat Genet
11	17,408,630	*KCNJ11*	UC	rs5215	1.07 (1.05–1.10)	C/T	0.41	2003	Morris et al. (2012), Nat Genet
11	72,433,098	*ARAP1 (CENTD2)*	BC; decreased fasting proinsulin[d]	rs1552224	1.11 (1.07–1.14)	A/C	0.88	2010	Morris et al. (2012), Nat Genet
11	92,708,710	*MTNR1B*	BC; hyperglycemic[d]	rs10830963	1.09 (1.06–1.12)	G/C	NA	2009	Morris et al. (2012), Nat Genet

(continued)

Table 2.2 (continued)

Chromosome	Position (HG19)	Locus (nearest genes)	Effect on glucose homeostasis[a]	Index variant	Effect size (95 %CI)	Risk allele/ other alleles	Risk allele frequency	Year first association report	Locus lead SNP, effect size, and allele frequency reference
12	4,305,972	***CCND2, secondary***	Insulin secretion	rs75615236	1.16	G/C	0.07	2014	Steinthorsdottir et al. (2014), Nat Genet
12	4,374,373	***CCND2***	UC	rs11063069	1.08 (1.05–1.11)	G/A	0.21	2012	Morris et al. (2012), Nat Genet
12	4,384,844	***CCND2***	Insulin secretion	rs76895963	0.53	G/T	0.02	2014	Steinthorsdottir et al. (2014), Nat Genet
12	27,965,150	*KLHDC5*	UC	rs10842994	1.10 (1.06–1.13)	C/T	0.80	2012	Morris et al. (2012), Nat Genet
12	66,212,318	*HMGA2*	IR	rs2261181	1.13 (1.08–1.17)	T/C	0.10	2010	Morris et al. (2012), Nat Genet
12	71,433,293	*TSPAN8/ LGR5*	UC	rs7955901	1.07 (1.05–1.10)	C/T	0.45	2008	Morris et al. (2012), Nat Genet
12	121,426,901	*HNF1A (TCF1)*	UC	rs12427353	1.08 (1.05–1.12)	G/A	0.79	2010	Morris et al. (2012), Nat Genet
12	123,640,853	*MPHOSPH9*	UC	rs1727313	1.06 (1.04–1.08)	C/T	NA	2014	DIAGRAM Consortium et al. (2014), Nat Genet
13	23,864,657	*SGCG*	IR	rs9552911	1.49 (1.29–1.72)	G/A	0.82	2013	Saxena et al. (2013), Diabetes[c]
13	27,396,636	*PDX1*	BC[b]	Chr. 13: g.27396636delT	2.27	T/-	0.00	2014	Steinthorsdottir et al. (2014), Nat Genet
13	80,717,156	*SPRY2*	BC	rs1359790	1.08 (1.05–1.10)	G/A	NA	2010	Morris et al. (2012), Nat Genet

15	38,822,905	*RASGRP1*	BC[b]	rs7403531	1.10 (1.06–1.13)	T/C	0.35	2013	Li et al. (2013), Diabetes[c]
15	62,383,155	*C2CD4A*	BC	rs4502156	1.06 (1.03–1.08)	T/C	0.52	2010	Morris et al. (2012), Nat Genet
15	77,832,762	*HMG20A*	UC	rs7177055	1.08 (1.05–1.10)	A/G	0.68	2011	Morris et al. (2012), Nat Genet
15	80,432,222	*ZFAND6*	UC	rs11634397	1.05 (1.02–1.07)	G/A	0.60	2010	Morris et al. (2012), Nat Genet
15	90,345,335	*AP3S2*	UC	rs2007084	1.02 (0.98–1.07)	G/A	0.92	2011	Morris et al. (2012), Nat Genet
15	91,544,076	*PRC1*	UC	rs12899811	1.08 (1.05–1.10)	G/A	0.31	2010	Morris et al. (2012), Nat Genet
16	53,819,169	*FTO*	IR	rs9936385	1.13 (1.10–1.16)	C/T	0.41	2007	Morris et al. (2012), Nat Genet
16	75,247,245	*BCAR1*	BC[b]	rs7202877	1.12 (1.07–1.16)	T/G	0.89	2012	Morris et al. (2012), Nat Genet
17	2,298,974	*SRR*	UC	rs2447090	1.04 (1.01–1.06)	A/G	0.62	2010	Morris et al. (2012), Nat Genet
17	6,940,393	***SLC16A13***	UC	rs312457	1.20 (1.14–1.26)	G/A	0.08	2014	Hara et al. (2014), Hum Mol Gen[c]
17	6,945,940	***SLC16A11***	Triacylglycerol metabolism	rs13342232	-	G/A	NA	2014	SIGMA Type 2 Diabetes Consortium 2014b, Nature[c]
17	6,946,287	***SLC16A11***	Triacylglycerol metabolism	rs13342692	1.13 (1.06–1.20)	C/T	NA	2014	SIGMA Type 2 Diabetes Consortium, Consortium et al. 2014b, Nature[c]
17	36,102,381	*HNF1B (TCF2)*	UC	rs11651052	1.10 (1.07–1.14)	A/G	0.44	2007	Morris et al. (2012), Nat Genet
18	7,068,462	*LAMA1*	BC[b]	rs8090011	1.13 (1.09–1.189	G/C	0.38	2012	Perry et al. (2012), Plos Genet
18	57,884,750	***MC4R***	IR[b]	rs12970134	1.08 (1.05–1.11)	A/G	0.27	2012	Morris et al. (2012), Nat Genet

(continued)

Table 2.2 (continued)

Chromosome	Position (HG19)	Locus (nearest genes)	Effect on glucose homeostasis[a]	Index variant	Effect size (95 %CI)	Risk allele/ other alleles	Risk allele frequency	Year first association report	Locus lead SNP, effect size, and allele frequency reference
18	58,049,192	***MC4R, secondary***	IR[b]	rs11873305	1.18 (1.11–1.26)	A/C	0.96	2012	Morris et al. (2012), Nat Genet
18	60,845,884	*BCL2*	UC	rs12454712	1.08 (1.04–1.11)	T/C	0.63	2012	Saxena et al. (2012), Am J Hum Genet
19	19,407,718	*CILP2*	UC	rs10401969	1.13 (1.09–1.18)	C/T	0.08	2012	Morris et al. (2012), Nat Genet
19	33,909,710	*PEPD*	UC	rs8182584	1.04 (1.01–1.07)	T/G	0.38	2012	Morris et al. (2012), Nat Genet
19	46,158,513	*GIPR*	BC[b]	rs8108269	1.10 (1.06–1.14)	G/T	0.31	2012	Morris et al. (2012), Nat Genet
20	42,946,966	***HNF4A/ FITM2/ R3HDML***	UC	rs6017317	1.09 (1.07–1.12)	G/T	NA	2012	Cho et al. (2012), Nat Genet
20	42,989,267	***HNF4A***	UC	rs4812829	1.06 (1.03–1.09)	G/A	0.19	2011	Morris et al. (2012), Nat Genet
23	152,899,922	*DUSP9*	UC	rs5945326	1.27 (1.18–1.37)	A/G	0.79	2010	Voight et al. (2010), Nat Genet

[a]Effect on glucose homeostasis: insulin resistance (IR) or reduced β-cell function (BC), unclassified (UC). Classification reported in Ayub et al. (2014), Am J Hum Genet for Morris et al. set of loci
[b]Classification from original publication or other published literatures
[c]Discovery in non-Europeans, risk allele frequency, and effect size are reported for the ethnicity of discovery
[d]Classification from Dimas et al. (2014), Diabetes
[e]Locus name reported in discovery study
Loci in bold contain multiple independent signals; for secondary signals of association threshold of $P < 10^{-5}$ significance is used; loci with different names and located within small distance are considered as one locus, e.g., *GCC1* and *PAX4* are considered as one locus

(Scott et al. 2007b; Zeggini et al. 2007; Diabetes Genetics Initiative of Broad Institute of H et al. 2007). Following a 2-stage replication with genotyping of selected SNPs in ~75,500 individuals, the DIAGRAM study identified six novel loci (Tables 2.1 and 2.2), including only one reasonable biological candidate gene (*NOTCH2*, Notch homologue 2, *Drosophila*), which is involved in pancreatic development.

The DIAGRAM consortium published two further meta-analyses, each based on increasingly larger case-control samples from European populations. The first combined discovery data from 21 GWA studies in up to 8130 individuals with T2D and 38,987 controls all imputed to a HapMap 2 reference panel, followed by large-scale replication in 34,412 cases and 59,925 controls where 13 (11 novel) out of 23 autosomal signals were confirmed (Tables 2.1 and 2.2) (Voight et al. 2010). This meta-analysis was the first to examine T2D associations on chromosome X (taking X-inactivation into account) and identified an association at *DUSP9* with a large effect on T2D risk (OR = 1.27, Table 2.2; Fig. 2.1) (Voight et al. 2010). The second meta-analysis, in addition to dramatically increasing the sample size (34,840 cases and 114,981 controls), implemented a novel cost-effective strategy for large-scale replication based on the CardioMetabochip (Metabochip), an Illumina iSelect genotyping array. Metabochip, which was designed through collaboration between six GWA consortia studying metabolic and atherosclerotic/cardiovascular diseases and traits (Voight et al. 2012), permitted follow-up of ~66,000 putative signals for cardiometabolic phenotypes (~5000 of which were selected for T2D) (Morris et al. 2012). The Metabochip array also contained approximately 120,000 SNP probes to fine map 257 established loci in an attempt to identify causal T2D susceptibility variants. The DIAGRAM meta-analysis with Metabochip follow-up established T2D associations at 10 loci (Tables 2.1 and 2.2), including two at *CCND2* and *GIPR* with larger effects on T2D risk in males and females, respectively (Morris et al. 2012). Among previously established T2D loci, sex differentiation in effect size has been shown for *KCNQ1*, *DGKB*, and *BCL11A* (larger effects in males) and *GRB14* (larger effects in females).

A separate DIAGRAM GWA meta-analysis of the effects of obesity on T2D risk, performed in Europeans through GWA meta-analysis of lean ($BMI < 25\ kg/m^2$) and obese ($BMI \geq 30\ kg/m^2$) T2D diabetics with ~54,000 controls, identified associations with lean diabetic participants at *LAMA1* and with obese subjects at *HMG20A* (Perry et al. 2012). A GWA meta-analysis in >8000 T2D cases and >10,870 controls in Europeans with large replication, including several additional datasets with de novo genotyping and the DIAGRAM discovery meta-analysis data in silico, reported association at *RBMS1* (Tables 2.1 and 2.2) (Qi et al. 2010).

In parallel to studies in European populations, T2D GWA studies in Asian ethnic groups (representing Japanese, Chinese, Punjabi Sikhs, Indians, South Asian, and East Asian subjects) have established T2D associations at 27 loci (Table 2.1). These studies have generally followed a design based on a GWA study with large-scale replication in an individual ethnic group, frequently undertaken in multistage fashion. In addition, several groups have combined efforts to complete a recent

East Asian GWA meta-analysis in up to 6952 T2D cases and 11,865 controls (with imputation based on the East Asian HapMap 2 reference panel) and identified eight novel loci, including *GLIS3*, *PEPD*, *FITM2-R3HDML-HNF4A*, *KCNK16*, *MAEA*, *GCC1-PAX4*, *PSMD6*, and *ZFAND3* (Cho et al. 2012). A second meta-analysis of Chinese samples (with imputation based on the 1000 Genomes Project JPT (Japanese in Tokyo) and CHB (Han Chinese in Beijing) reference panels) has described T2D association with a common variant in the *PAX4* gene, which is expressed in early pancreatic endocrine cells. The association, which was confirmed in a multiethnic analysis including European and five East Asian populations (Ma et al. 2013), adds another example of common variant associations with T2D at a MODY locus as heterozygous mutations in *PAX4* have been identified as a cause of MODY9 (omim.org/entry/606391). Therefore, while rare coding mutations severely impair islet function and cause rare monogenic forms of diabetes, common variants can act through the same genes, but with smaller effects, to increase an individual's risk of developing a more common form of diabetes.

A small number of GWA studies have been reported for other ethnic groups. Studies in Mexican individuals reported associations at several established loci (Parra et al. 2011) and a novel association at *SLC16A11/SLC16A13* where the haplotype carriers had amino acid substitutions in *SLC16A11* (Consortium et al. 2014b). The locus is thought to affect triacylglycerol metabolism and shows stronger association in leaner and younger people. While common in Native Americans and Asians, risk variants at this locus are rare in European and African individuals and have introgressed into modern humans through admixture with Neanderthals. A second study in American Pima Indians confirmed associations for a set of previously established loci while reaching study-wise significance (P-value $= 6.6 \times 10^{-8}$) at the *DNER* gene (Hanson et al. 2014). Finally, an African American GWA study has provided evidence for association at *RND3/RBM43* (Palmer et al. 2012).

Methodological development in to combine data from multiple ancestry groups by accounting for heterogeneous allelic effects (Morris 2011) has enabled performing meta-analysis across different ethnicities. For example, combining European, East Asian, South Asian, and Mexican and Mexican-American GWA meta-analyses in up to 26,488 T2D cases and 83,964 controls has identified seven novel T2D susceptibility loci *TMEM154*, *SSR1-RREB1*, *FAF1*, *POU5F1-TCF19*, *LPP*, *ARL15*, and *MPHOSPH9* (Replication et al. 2014). Importantly, the study demonstrated an overwhelming concordance of allelic effects across ethnicities, even at loci with only weak evidence of association, supporting the hypothesis that T2D risk variants predate migration of humans out of Africa and arguing against the "synthetic association" hypothesis, which predicts that associations at common variants are driven by unobserved lower frequency causal alleles with large effects (Dickson et al. 2010).

It has long been suggested that the high prevalence of metabolic disorders related to impaired glucose homeostasis may be a result of selective evolutionary advantage of T2D and obesity-risk variants during periods of scarce food resources, which resulted in an increase in their frequency at the population level (thrifty gene hypothesis) (Neel 1962, 1999). Given that food intake is known to act as a trigger for insulin release, it has also been hypothesized that a positive selection may have operated in particular on those loci associated with T2D through an influence on β-cell function (Ayub et al. 2014). Some evidence of directional population differentiation and nominal positive selection at individual T2D risk loci, including *TCF7L2*, *THADA*, and *NOTCH2*, has been reported (Chen et al. 2012; Corona et al. 2013; Klimentidis et al. 2011). The collective analysis of all T2D-associated variants along with stratified by their impact on β-cell function or insulin resistance has to date found no support for global or differential positive selection at T2D loci, thus offering little support for the thrifty gene hypothesis (Ayub et al. 2014; Southam et al. 2009).

2.4 Common Variants with Modest Effect Sizes

Most GWA study designs are based on common variant genotyping arrays, which have determined the allele spectrum of the resulting T2D-associated variants (Table 2.2). The 88 known T2D risk loci (Table 2.2) show only modest effects (OR = 1.1–1.2), with *TCF7L2* being the only locus showing larger effects in European populations (OR ~1.40, Fig. 2.1) (Morris et al. 2012). While this has led to an intense search for additional rare and common variants (particularly for causal variants which are expected to have larger effects), the early search for rare coding variants has had limited success (Table 2.2) (Steinthorsdottir et al. 2014; Albrechtsen et al. 2013). Additionally, studies in non-Europeans have recently provided support for a number of novel T2D susceptibility loci that show low allele frequencies in European populations (Unoki et al. 2008; Hanson et al. 2014; Consortium et al. 2014b). While this provides a challenge to validating these loci in European populations, the high concordance of the direction of effects across ethnicities for T2D risk variants (Replication et al. 2014) suggests that additional common T2D risk variants with consistent and modest effects across ethnic groups remain to be described. Their identification will require larger sample sizes and combined efforts of many studies and research centers (Morris et al. 2012).

The discriminatory capacity of genetic variants for T2D risk prediction and patient stratification has been assessed in longitudinal studies by examining whether inclusion of genetic risk scores (GRS) in predictive models increases the area under the receiver-operating-characteristic curve compared to predictive models including only clinical parameters. Early studies suggested that inclusion of GRS provided little improvement in T2D risk prediction compared to clinical risk factors and family history alone (Lyssenko et al. 2008; Meigs et al. 2008; Balkau et al. 2008; Talmud et al. 2010; de Miguel-Yanes et al. 2011). More recent studies, incorporating increasing numbers of T2D risk variants into the GRS, have also had mixed results (Hivert et al. 2011; Muhlenbruch et al. 2013; Vaxillaire et al. 2014). For example, while a recent study incorporating 43 T2D associated variants showed little improvement in T2D prediction, inclusion of the GRS in predictive models improved the receiver-operating-characteristic curve for subgroups of subjects at increased risk of T2D, including obese subjects, older participants, and those with a family history of diabetes (Muhlenbruch et al. 2013). Similarly, Hivert et al. have shown that a GRS with 34 variants was significantly associated with increased risk of progression to T2D in high-risk individuals, as well as a reduced effect of lifestyle interventions on genetic risk (Hivert et al. 2011).

A recent study comparing the discriminative capacity of GRSs including 65 - T2D-associated loci and 36 FG-associated loci FG showed modest but significant improvement in T2D reclassification rates in models including a GRS incorporating T2D risk loci and modestly improved reclassification rates of incident and non-incident T2D and impaired fasting glucose (IFG) using the GRS incorporating both T2D risk and FG loci, suggesting that inclusion of risk loci associated with glycemic traits may be beneficial for intermediate phenotypes such as IFG (Vaxillaire et al. 2014). Further studies using GRS based on new loci and causative variants will help to improve insight into the longitudinal impact of genetic variants associated with glycemic traits on T2D risk of and disease trajectories.

2.5 Understanding Relationship with Other Phenotypes

Two critical processes leading to T2D development are β-cell dysfunction and insulin resistance in peripheral tissues including fat, muscle, liver, and elsewhere (Prokopenko et al. 2008). Beginning long before the clinical diagnosis of T2D, these processes are hallmarks of prediabetes; following which, progressive deterioration of β-cell function reaches a point when they are no longer able to meet the increased insulin demands from peripheral tissues, leading to the development of diabetes. In parallel to T2D GWA meta-analyses, a number of large-scale association studies have been successful in identifying genetic loci that influence quantitative glycemic traits, including fasting and postprandial glucose and serum insulin levels. These studies take advantage of the increased power that can be obtained when similarly sized cohorts studied for continuous traits compared to dichotomous outcomes; their success relies on the hypothesis that genes influencing blood levels in normal subjects will also increase diabetes risk. Significantly, while the genetic risk loci identified for T2D overlap to some degree with quantitative trait loci for blood glucose and insulin, several genes have shown association only with glycemic traits or only with increased T2D risk (Fig. 2.1). While it's possible that this discordance may reflect the statistical power of the studies completed to date, the milder phenotypes observed in patients with glucokinase mutations compared to patients with other forms of MODY (McDonald and Ellard 2013) suggest that it is important to distinguish two overlapping but distinct groups of GWAs loci that are associated with altered glucose homeostasis on the one hand and the progressive of metabolic decompensation that leads to T2D on the other.

The only association with FG established before the GWA study era was at the glucokinase (*GCK*) locus (Weedon et al. 2005, 2006), a gene in which rare mutations cause MODY2 (Froguel et al. 1992). *GCK* catalyzes the first step in glycolysis and is one of the principal regulators of FG concentration and of β-cell secretory activity. An indicative association at the glucokinase regulator (*GCKR*) locus (rs780094) with FG, as well as an association at the same variant with serum triglyceride levels, was described by the DGI T2D GWA study, which however was not powered enough to detect an effect on T2D (Diabetes Genetics Initiative of Broad Institute of H et al. 2007). The product of *GCKR* regulates GCK activity and is a highly plausible candidate involved in T2D pathogenesis (see Chap. 16). The *GCK* and *GCKR* loci have since been associated with FG/HOMA-B (homeostasis model assessment of β-cell function) and FG/FI/HOMA-IR (homeostasis model assessment of insulin resistance), respectively, and with T2D (Dupuis et al. 2010; Manning et al. 2012). These findings prompted further interest in well-powered GWA studies for glycemic traits to detect reliable genetic associations which may be relevant to T2D pathogenesis (see Chap. 3).

In 2009, the collaborative Meta-Analyses of Glucose and Insulin-related traits Consortium (MAGIC, http://www.magicinvestigators.org/) was established to consolidate the efforts of many groups working on glycemic trait genetics, in order to understand the variation of these traits within the physiological range and investigate their impact on T2D risk and other cardiometabolic traits (Prokopenko et al. 2009a). The first effort of MAGIC confirmed the association at *GCK* and *G6PC2* loci and identified a novel signal at the melatonin receptor 1B (*MTNR1B*) locus for higher FG and lower insulin secretion. The inverse correlation between the levels of the neurohormone melatonin, secreted by the pineal gland, and insulin has long been known. However, few studies had investigated the relationship between melatonin signaling in pancreatic islets and metabolic disease (Peschke et al. 2007), prior to publication of large-scale association studies (Prokopenko et al. 2009a; Bouatia-Naji et al. 2009; Chambers et al. 2009; Go et al. 2013; Lyssenko et al. 2009). Association with T2D at *MTNR1B* locus was subsequently confirmed at genome-wide significance (Prokopenko et al. 2009a; Dupuis et al. 2010; Voight et al. 2010; Lyssenko et al. 2009).

To extend the first MAGIC study, a new, larger, whole GWA meta-analysis (21 studies, up to 46,186 nondiabetic individuals) was performed (Dupuis et al. 2010). It increased the number of glycemic trait loci to 16 and reported novel effects on T2D from a large-scale analysis at five of the FG/FI-associated loci (*ADCY5*, *GCK*, *GCKR*, *DGKB*, *PROX1*), thus highlighting that the overlap between the genetic variation influencing glucose homeostasis and risk of T2D is only partial (Fig. 2.1). Four of these loci contributed to impaired β-cell function as measured by HOMA-B and one (*GCKR*) was associated with insulin resistance.

Three MAGIC GWA meta-analyses for additional glycemic traits provided further insights into pathophysiology of T2D. A study focusing on 2-hour postprandial glucose (2hGlu) levels (15,234 nondiabetic individuals in discovery and up to 30,620 in replication) identified five associated loci (*GIPR*, *VPS13C*, *ADCY5*, *GCKR*, *TCF7L2*), including the novel locus *GIPR* (rs10423928) containing the gene encoding the GIP receptor for the insulin-response stimulating hormone GIP (glucose-dependent insulinotropic polypeptide) in pancreatic islet β-cells

(Saxena et al. 2010) and in linkage disequilibrium (LD, HapMap CEU $r^2 = 0.78$) with BMI-associated rs2287019 (Speliotes et al. 2010). Genome-wide meta-analysis of HbA_{1c}, including study in 46,368 in nondiabetic individuals by the MAGIC investigators, identified 10 genetic loci, of which *MTNR1B* and *GCK* also increase T2D risk, suggesting that their effect on hyperglycemia (as measured by FG) extends an effect on average glycemia over a 2- to 3-month period (as detected through HbA_{1c}) and is related to T2D pathogenesis, while *ANK1* maps close to T2D risk variant (Soranzo et al. 2010). Variants at *VPS13C/C2CD4A/B* and *GIPR* were subsequently associated with T2D, the latter showing larger effects in women, but both were in weak LD with glycemic trait variants (Yamauchi et al. 2010; Morris et al. 2012). Similarly, the *ANK1* HbA_{1c} locus variant rs4737009 identified by Soranzo et al. is not in LD with the T2D risk variant (rs516946, HapMap CEU $r^2 < 0.01$) (Morris et al. 2012). The genetic architecture at these three loci is complex and requires further investigation to dissect the relationships between genetic effects on the associated glycemic phenotypes.

Large-scale studies of glycemic traits using Metabochip have discovered additional common variant loci with small effects on FG/FI/2hGlu loci trait variability and further increased the overlap with T2D risk loci. In this study, 39 FG-raising alleles were related to increased T2D risk, although only 20 (>60 %) of them showed at least nominal significance ($P < 0.05$) for T2D. Similarly, 13 of the 19 FI loci were nominally associated with T2D and all but *TCF7L2*. Similarly, 13 of the 19 FI loci were nominal association with T2D and all, but *TCF7L2*, FI/insulin resistance-increasing alleles were associated with higher T2D risk and showed an impaired lipid profile (Fig. 2.2) (Scott et al. 2012).

FG-associated loci from GWAS studies have also helped define the relationship between T2D and abnormal insulin processing and secretion in β-cells. Among other glycemic trait analyses by the MAGIC, nine genome-wide significant loci were described for corrected insulin response (CIR), seven of which were previously associated with both T2D and other glycemic traits (*MTNR1B*, *GCK*, *HHEX/IDE*, *CDKAL1*, *CDKN2A/2B*, *ANK1*, *C2CD4A/B*) (Prokopenko et al. 2014). Two other loci included *G6PC2* associated with glycemic trait variability in nondiabetic individuals and the novel *GRB10* association, which showed potential tissue-specific methylation and parental imprinting that might mask its association with T2D). Meta-analysis of GWA studies by MAGIC for fasting proinsulin levels adjusted for FI identified eight loci, of which four demonstrated that both proinsulin-raising (for *TCF7L2*, *SLC30A8*, and *VPS13C/C2CD4A/B*) and proinsulin-lowering alleles (for *ARAP1*) influenced T2D risk through a decrease in insulin secretion caused by distal or proximal impairment of proinsulin conversion, respectively (Strawbridge et al. 2011). Similarly, Dimas and colleagues described associations at the *HHEX/IDE* and *MTNR1B* loci with defects in early insulin secretion through reduced insulinogenic index for the T2D risk allele and showed that the T2D risk allele at *ARAP1* was related to defects in the first steps of insulin production, through association with 32,33 split proinsulin (Dimas et al. 2014).

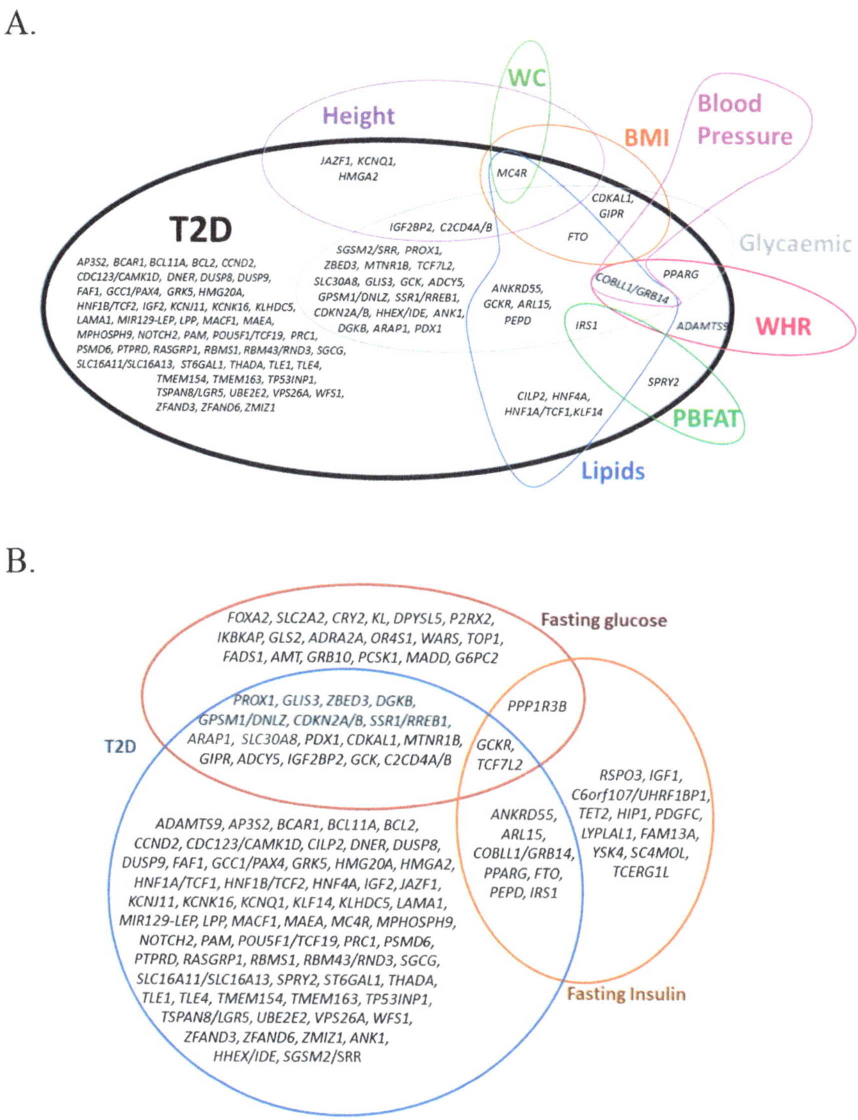

Fig. 2.2 Effects of established T2D loci: (**a**) on cardiometabolic phenotypes and (**b**) on glycemic traits. (**a**) Among a total 88 T2D loci, only 49 do not overlap with other cardiometabolic traits. The independent loci were defined by physical distance more than 500 kb from each other and by CEU LD $r^2 > 0.01$. (**b**) Among a total 88 T2D loci, only 27 overlap with fasting glucose or fasting insulin levels. The independent loci were defined by physical distance more than 500 kb from each other and by CEU LD $r^2 > 0.01$

These large-scale discovery efforts in nondiabetic individuals have provided genetic markers that may provide mechanistic insights into the pathogenesis of T2D and possibly to classify disease mechanisms that are active in individual patients.

For example, physiological characterization of the effects of glycemic and T2D loci on quantitative glycemic traits has revealed a clear separation of hyperglycemic loci (*MTNR1B* and *GCK*) which are associated with reduced basal and stimulated β-cell secretion and consequent fasting hyperglycemia without large effects on T2D risk from β-cell loci that show an effect on insulin processing and secretion that only modestly change FG but exert much stronger effects on T2D risk (*TCF7L2*, *SLC30A8*, *HHEX/IDE*, *CDKAL1*, *CDKN2A/2B*, *THADA*, *DGKB*, *PROX1*, *ADCY5*) (Dimas et al. 2014).

Loci with effects on insulin sensitivity represent a much smaller proportion of T2D variants. Physiological characterization of T2D loci grouped variants with primary effects on insulin sensitivity in basal and stimulated state (*IRS1*, *GCKR*, *PPARG*, *KLF14*); in addition, weak effects on insulin sensitivity have also been suggested for *HMGA2* (Voight et al. 2010). Insulin sensitivity indices showed consistently decreased effects for T2D risk alleles only for loci with known effects on insulin resistance at basal measures (HOMA-IR) (Dimas et al. 2014). In many cases, these loci may exert widespread biochemical changes affecting cardiometabolic risk (Fig. 2.2): some FI-associated loci can alter BMI and body fat distribution, while most loci associated with higher insulin levels are also associated with lower HDL cholesterol and higher triglyceride levels (Manning et al. 2012). For example, variants within the fat mass and obesity-associated (*FTO*) gene and at melanocortin-4 receptor (*MC4R*) exert their T2D effect through a primary impact on BMI (Frayling et al. 2007; Loos et al. 2008; Morris et al. 2012). In contrast, effects of *IRS1* and *PPARG* on insulin resistance and T2D are independent from obesity (Scott et al. 2012; Kilpelainen et al. 2011; Rung et al. 2009). For a number of loci, the association with lipids and T2D (*HNF4A*, *CILP2*, *KLF14*, *HNF1A/TCF1*, *MC4R*) and additionally with FI (*GRB14*, *GCKR*, *FTO*, *PEPD*, *ANKRD55*, *IRS1*, *ARL15*) has been reported independently for each phenotype, underlying the close relationship between increased lipids/adiposity and increased insulin (Fig. 2.2) (Scott et al. 2012). This picture is consistent with the first stages of diabetes, where high adiposity in peripheral tissues causes insulin resistance, which is complemented by an increase in β-cell insulin production.

Several T2D loci appear to have an effect on complex diseases whose pathogenesis is not commonly associated with changes in metabolic fitness: pleiotropy could be a probable mechanism for these effects, since the correlation between the associated disease outcomes is low for them to be considered as comorbidities. Thus, variants at ~20 T2D loci, including *CDKN2A/2B*, *JAZF1*, *HNF1B*, *THADA*, *CCND2*, *ZMIZ1*, and *IGF2*, have a role in cancer susceptibility (Gudmundsson et al. 2007; Thomas et al. 2008; Finkel et al. 2007). Interestingly, T2D risk alleles at *THADA*, *TSPAN8*, and *HNF1B* are protective against prostate cancer, an inverse relationship that supports epidemiological observations. The genetic links between diabetes and cancer point to a set of shared biological pathways, including opposing roles in regulation of cell cycle and common signaling pathways.

2.6 What Is Next in T2D GWA Studies?

Despite the success of GWA studies in identification of common variant associations, the largest heritable component of T2D susceptibility remains unexplained. Rapid development and reduced costs of exome sequencing approaches has opened wide opportunities in both sequencing of large numbers of individuals and generation of large reference panels for imputation of rare variants from resequencing [e.g., those from the 1000 Genomes Project (Genomes Project et al. 2010)]. Population-based studies have also benefited from sequencing through implementation of population-specific next-generation sequencing-based reference panels, including deCODE Icelandic and Genome of the Netherlands (GoNL) reference panels (Boomsma et al. 2014; Steinthorsdottir et al. 2014). To date, these sequencing studies have not succeeded in identifying a large number of novel risk loci. For example, whole-exome sequencing at 8× depth in a Danish sample of 1000 T2D cases and 1000 controls hasn't produced evidence of association with T2D at rare exomic variants, but has confirmed associations with T2D at common variants in *COBLL1* and *MACF1* (Tables 2.1 and 2.2) (Albrechtsen et al. 2013). While the sample size used in the study was small and the variant calling accuracy was not optimal for detecting small indels or changes in copy number, the results are consistent with previous regional resequencing studies which suggest that most causative variants linked to the GWA risk loci will not alter protein coding sequences. A recent whole-exome sequencing study in 3756 Latinos with an average depth 67.17× has identified a rare missense variant in *HNF1A* (c.1522G>A [p.E508K], odds ratio [OR] = 5.48) (Consortium et al. 2014a). As a result, there is considerable interest in pursuing whole-genome and whole-exome sequencing studies, particularly in cohorts that have sufficient statistical power to detect epistatic interactions that may confer additional T2D risk. Several international T2D collaborations have recently focused their efforts on large-scale sequencing projects, including the GoT2D (genomics of T2D) consortium that has undertook whole-genome (low-pass 4–6×) and deep whole-exome sequencing for ~2800 T2D case and control individuals from Northern Europe and the T2D Genetic Exploration by Next-generation sequencing in multi-Ethnic Samples (T2D-GENES) consortium that has undertook trans-ethnic deep whole-exome sequencing in ~10,000 individuals distributed equally between five ethnic groups (McCarthy 2010).

GWA studies have provided an excellent springboard for large-scale T2D studies through international collaborative efforts focused on Europeans and being widely extended to other ethnic groups. Improved sequencing technologies and variant calling algorithms will extend the variant set to other types of genetic variability, including copy number variation, which may have significant impact on the dissection of T2D susceptibility. These collaborations will enable well-powered fine-mapping studies and identification and functional characterization of disease-causing variants. Overall, identifying causative genetic variants and discovering the molecular mechanisms linking them to the development of prediabetic changes will be essential in understanding the pathophysiology of T2D. This in turn may lead to rational drug development and suggest therapies that can be applied appropriately and early to those most at risk of developing T2D (Tuomilehto and Lindstrom 2003). This outcome is potentially feasible as genes that have already been associated with diabetes have also acted as targets for its treatment: while this is best demonstrated by the use of sulfonylureas to treat neonatal diabetes associated with inactivating mutations of the Sur1 protein (Gloyn et al. 2004), the same family of drugs have also provided a mainstay for treating adults with polygenic T2D for many years.

Acknowledgments We would like to thank Dr. Letizia Marullo (University of Ferrara, Italy) and Dr. Julia Sarah El-Sayed Moustafa (Imperial College London, UK) for their help in the context of manuscript preparation.

References

Albrechtsen A, Grarup N, Li Y, Sparso T, Tian G, Cao H, Jiang T, Kim SY, Korneliussen T, Li Q, Nie C, Wu R, Skotte L, Morris AP, Ladenvall C, Cauchi S, Stancakova A, Andersen G, Astrup A, Banasik K, Bennett AJ, Bolund L, Charpentier G, Chen Y, Dekker JM, Doney AS, Dorkhan M, Forsen T, Frayling TM, Groves CJ, Gui Y, Hallmans G, Hattersley AT, He K, Hitman GA, Holmkvist J, Huang S, Jiang H, Jin X, Justesen JM, Kristiansen K, Kuusisto J, Lajer M, Lantieri O, Li W, Liang H, Liao Q, Liu X, Ma T, Ma X, Manijak MP, Marre M, Mokrosinski J, Morris AD, Mu B, Nielsen AA, Nijpels G, Nilsson P, Palmer CN, Rayner NW, Renstrom F, Ribel-Madsen R, Robertson N, Rolandsson O, Rossing P, Schwartz TW, Group DESIRS, Slagboom PE, Sterner M, Consortium D, Tang M, Tarnow L, Tuomi T, Van't Riet E, van Leeuwen N, Varga TV, Vestmar MA, Walker M, Wang B, Wang Y, Wu H, Xi F, Yengo L, Yu C, Zhang X, Zhang J, Zhang Q, Zhang W, Zheng H, Zhou Y, Altshuler D, Thart LM, Franks PW, Balkau B, Froguel P, McCarthy MI, Laakso M, Groop L, Christensen C, Brandslund I, Lauritzen T, Witte DR, Linneberg A, Jorgensen T, Hansen T, Wang J, Nielsen R, Pedersen O (2013) Exome sequencing-driven discovery of coding polymorphisms associated with common metabolic phenotypes. Diabetologia 56(2):298–310. doi:10.1007/s00125-012-2756-1

Altshuler D, Hirschhorn JN, Klannemark M, Lindgren CM, Vohl MC, Nemesh J, Lane CR, Schaffner SF, Bolk S, Brewer C, Tuomi T, Gaudet D, Hudson TJ, Daly M, Groop L, Lander ES (2000) The common PPARgamma Pro12Ala polymorphism is associated with decreased risk of type 2 diabetes. Nat Genet 26(1):76–80. doi:10.1038/79216

Ayub Q, Moutsianas L, Chen Y, Panoutsopoulou K, Colonna V, Pagani L, Prokopenko I, Ritchie GR, Tyler-Smith C, McCarthy MI, Zeggini E, Xue Y (2014) Revisiting the thrifty gene hypothesis via 65 loci associated with susceptibility to type 2 diabetes. Am J Hum Genet 94 (2):176–185. doi:10.1016/j.ajhg.2013.12.010

Balkau B, Lange C, Fezeu L, Tichet J, de Lauzon-Guillain B, Czernichow S, Fumeron F, Froguel P, Vaxillaire M, Cauchi S, Ducimetiere P, Eschwege E (2008) Predicting diabetes: clinical, biological, and genetic approaches: data from the Epidemiological Study on the Insulin Resistance Syndrome (DESIR). Diabet Care 31(10):2056–2061. doi:10.2337/dc08-0368

Boj SF, van Es JH, Huch M, Li VS, Jose A, Hatzis P, Mokry M, Haegebarth A, van den Born M, Chambon P, Voshol P, Dor Y, Cuppen E, Fillat C, Clevers H (2012) Diabetes risk gene and Wnt effector Tcf7l2/TCF4 controls hepatic response to perinatal and adult metabolic demand. Cell 151(7):1595–1607. doi:10.1016/j.cell.2012.10.053

Boomsma DI, Wijmenga C, Slagboom EP, Swertz MA, Karssen LC, Abdellaoui A, Ye K, Guryev V, Vermaat M, van Dijk F, Francioli LC, Hottenga JJ, Laros JF, Li Q, Li Y, Cao H, Chen R, Du Y, Li N, Cao S, van Setten J, Menelaou A, Pulit SL, Hehir-Kwa JY, Beekman M, Elbers CC, Byelas H, de Craen AJ, Deelen P, Dijkstra M, den Dunnen JT, de Knijff P, Houwing-Duistermaat J, Koval V, Estrada K, Hofman A, Kanterakis A, Enckevort D, Mai H, Kattenberg M, van Leeuwen EM, Neerincx PB, Oostra B, Rivadeneira F, Suchiman EH, Uitterlinden AG, Willemsen G, Wolffenbuttel BH, Wang J, de Bakker PI, van Ommen GJ, van Duijn CM (2014) The genome of the Netherlands: design, and project goals. Eur J Hum Genet 22(2):221–227. doi:10.1038/ejhg.2013.118

Bouatia-Naji N, Bonnefond A, Cavalcanti-Proenca C, Sparso T, Holmkvist J, Marchand M, Delplanque J, Lobbens S, Rocheleau G, Durand E, De Graeve F, Chevre JC, Borch-Johnsen K, Hartikainen AL, Ruokonen A, Tichet J, Marre M, Weill J, Heude B, Tauber M, Lemaire K, Schuit F, Elliott P, Jorgensen T, Charpentier G, Hadjadj S, Cauchi S, Vaxillaire M, Sladek R, Visvikis-Siest S, Balkau B, Levy-Marchal C, Pattou F, Meyre D, Blakemore AI, Jarvelin MR, Walley AJ, Hansen T, Dina C, Pedersen O, Froguel P (2009) A variant near MTNR1B is associated with increased fasting plasma glucose levels and type 2 diabetes risk. Nat Genet 41 (1):89–94. doi:10.1038/ng.277

Chambers JC, Zhang W, Zabaneh D, Sehmi J, Jain P, McCarthy MI, Froguel P, Ruokonen A, Balding D, Jarvelin MR, Scott J, Elliott P, Kooner JS (2009) Common genetic variation near melatonin receptor MTNR1B contributes to raised plasma glucose and increased risk of type 2 diabetes among Indian Asians and European Caucasians. Diabetes 58(11):2703–2708. doi:10.2337/db08-1805

Chen R, Corona E, Sikora M, Dudley JT, Morgan AA, Moreno-Estrada A, Nilsen GB, Ruau D, Lincoln SE, Bustamante CD, Butte AJ (2012) Type 2 diabetes risk alleles demonstrate extreme directional differentiation among human populations, compared to other diseases. PLoS Genet 8(4):e1002621. doi:10.1371/journal.pgen.1002621

Cho YS, Chen CH, Hu C, Long J, Ong RT, Sim X, Takeuchi F, Wu Y, Go MJ, Yamauchi T, Chang YC, Kwak SH, Ma RC, Yamamoto K, Adair LS, Aung T, Cai Q, Chang LC, Chen YT, Gao Y, Hu FB, Kim HL, Kim S, Kim YJ, Lee JJ, Lee NR, Li Y, Liu JJ, Lu W, Nakamura J, Nakashima E, Ng DP, Tay WT, Tsai FJ, Wong TY, Yokota M, Zheng W, Zhang R, Wang C, So WY, Ohnaka K, Ikegami H, Hara K, Cho YM, Cho NH, Chang TJ, Bao Y, Hedman AK, Morris AP, McCarthy MI, Consortium D, Mu TC, Takayanagi R, Park KS, Jia W, Chuang LM, Chan JC, Maeda S, Kadowaki T, Lee JY, Wu JY, Teo YY, Tai ES, Shu XO, Mohlke KL, Kato N, Han BG, Seielstad M (2012) Meta-analysis of genome-wide association studies identifies eight new loci for type 2 diabetes in east Asians. Nat Genet 44(1):67–72. doi:10.1038/ng.1019

Consortium IH (2005) A haplotype map of the human genome. Nature 437(7063):1299–1320. doi:10.1038/nature04226

Consortium STD, Estrada K, Aukrust I, Bjorkhaug L, Burtt NP, Mercader JM, Garcia-Ortiz H, Huerta-Chagoya A, Moreno-Macias H, Walford G, Flannick J, Williams AL, Gomez-Vazquez MJ, Fernandez-Lopez JC, Martinez-Hernandez A, Centeno-Cruz F, Mendoza-Caamal E, Revilla-Monsalve C, Islas-Andrade S, Cordova EJ, Soberon X, Gonzalez-Villalpando ME, Henderson E, Wilkens LR, Le Marchand L, Arellano-Campos O, Ordonez-Sanchez ML, Rodriguez-Torres M, Rodriguez-Guillen R, Riba L, Najmi LA, Jacobs SB, Fennell T, Gabriel S, Fontanillas P, Hanis CL, Lehman DM, Jenkinson CP, Abboud HE, Bell GI, Cortes ML, Boehnke M, Gonzalez-Villalpando C, Orozco L, Haiman CA, Tusie-Luna T, Aguilar-Salinas CA, Altshuler D, Njolstad PR, Florez JC, MacArthur DG (2014a) Association of a low-frequency variant in HNF1A with type 2 diabetes in a Latino population. JAMA 311 (22):2305–2314. doi:10.1001/jama.2014.6511

Consortium STD, Williams AL, Jacobs SB, Moreno-Macias H, Huerta-Chagoya A, Churchhouse C, Marquez-Luna C, Garcia-Ortiz H, Gomez-Vazquez MJ, Burtt NP, Aguilar-Salinas CA, Gonzalez-Villalpando C, Florez JC, Orozco L, Haiman CA, Tusie-Luna T,

Altshuler D (2014b) Sequence variants in SLC16A11 are a common risk factor for type 2 diabetes in Mexico. Nature 506(7486):97–101. doi:10.1038/nature12828

Corona E, Chen R, Sikora M, Morgan AA, Patel CJ, Ramesh A, Bustamante CD, Butte AJ (2013) Analysis of the genetic basis of disease in the context of worldwide human relationships and migration. PLoS Genet 9(5):e1003447. doi:10.1371/journal.pgen.1003447

da Silva XG, Loder MK, McDonald A, Tarasov AI, Carzaniga R, Kronenberger K, Barg S, Rutter GA (2009) TCF7L2 regulates late events in insulin secretion from pancreatic islet beta-cells. Diabetes 58(4):894–905. doi:10.2337/db08-1187

da Silva XG, Mondragon A, Sun G, Chen L, McGinty JA, French PM, Rutter GA (2012) Abnormal glucose tolerance and insulin secretion in pancreas-specific Tcf7l2-null mice. Diabetologia 55 (10):2667–2676. doi:10.1007/s00125-012-2600-7

de Miguel-Yanes JM, Shrader P, Pencina MJ, Fox CS, Manning AK, Grant RW, Dupuis J, Florez JC, D'Agostino RB Sr, Cupples LA, Meigs JB, Investigators M, Investigators D (2011) Genetic risk reclassification for type 2 diabetes by age below or above 50 years using 40 type 2 diabetes risk single nucleotide polymorphisms. Diabet Care 34(1):121–125. doi:10.2337/dc10-1265

Deeb SS, Fajas L, Nemoto M, Pihlajamaki J, Mykkanen L, Kuusisto J, Laakso M, Fujimoto W, Auwerx J (1998) A Pro12Ala substitution in PPARgamma2 associated with decreased receptor activity, lower body mass index and improved insulin sensitivity. Nat Genet 20(3):284–287. doi:10.1038/3099

Diabetes Genetics Initiative of Broad Institute of H, Mit LU, Novartis Institutes of BioMedical R, Saxena R, Voight BF, Lyssenko V, Burtt NP, de Bakker PI, Chen H, Roix JJ, Kathiresan S, Hirschhorn JN, Daly MJ, Hughes TE, Groop L, Altshuler D, Almgren P, Florez JC, Meyer J, Ardlie K, Bengtsson Bostrom K, Isomaa B, Lettre G, Lindblad U, Lyon HN, Melander O, Newton-Cheh C, Nilsson P, Orho-Melander M, Rastam L, Speliotes EK, Taskinen MR, Tuomi T, Guiducci C, Berglund A, Carlson J, Gianniny L, Hackett R, Hall L, Holmkvist J, Laurila E, Sjogren M, Sterner M, Surti A, Svensson M, Svensson M, Tewhey R, Blumenstiel B, Parkin M, Defelice M, Barry R, Brodeur W, Camarata J, Chia N, Fava M, Gibbons J, Handsaker B, Healy C, Nguyen K, Gates C, Sougnez C, Gage D, Nizzari M, Gabriel SB, Chirn GW, Ma Q, Parikh H, Richardson D, Ricke D, Purcell S (2007) Genome-wide association analysis identifies loci for type 2 diabetes and triglyceride levels. Science 316 (5829):1331–1336. doi:10.1126/science.1142358

DIAbetes Genetics Replication And Meta-analysis (DIAGRAM) Consortium, Asian Genetic Epidemiology Network Type 2 Diabetes (AGEN-T2D) Consortium, South Asian Type 2 Diabetes (SAT2D) Consortium, Mexican American Type 2 Diabetes (MAT2D) Consortium, Type 2 Diabetes Genetic Exploration by Nex-generation sequencing in multi-Ethnic Samples (T2D-GENES) Consortium, Mahajan A et al (2014) Genome-wide trans-ancestry meta-analysis provides insight into the genetic architecture of type 2 diabetes susceptibility. Nat Genet 46(3):234–244

Dickson SP, Wang K, Krantz I, Hakonarson H, Goldstein DB (2010) Rare variants create synthetic genome-wide associations. PLoS Biol 8(1):e1000294. doi:10.1371/journal.pbio.1000294

Dimas AS, Lagou V, Barker A, Knowles JW, Magi R, Hivert MF, Benazzo A, Rybin D, Jackson AU, Stringham HM, Song C, Fischer-Rosinsky A, Boesgaard TW, Grarup N, Abbasi FA, Assimes TL, Hao K, Yang X, Lecoeur C, Barroso I, Bonnycastle LL, Bottcher Y, Bumpstead S, Chines PS, Erdos MR, Graessler J, Kovacs P, Morken MA, Narisu N, Payne F, Stancakova A, Swift AJ, Tonjes A, Bornstein SR, Cauchi S, Froguel P, Meyre D, Schwarz PE, Haring HU, Smith U, Boehnke M, Bergman RN, Collins FS, Mohlke KL, Tuomilehto J, Quertemous T, Lind L, Hansen T, Pedersen O, Walker M, Pfeiffer AF, Spranger J, Stumvoll M, Meigs JB, Wareham NJ, Kuusisto J, Laakso M, Langenberg C, Dupuis J, Watanabe RM, Florez JC, Ingelsson E, McCarthy MI, Prokopenko I, Investigators M (2014) Impact of type 2 diabetes susceptibility variants on quantitative glycemic traits reveals mechanistic heterogeneity. Diabetes 63(6):2158–2171. doi:10.2337/db13-0949

Dupuis J, Langenberg C, Prokopenko I, Saxena R, Soranzo N, Jackson AU, Wheeler E, Glazer NL, Bouatia-Naji N, Gloyn AL, Lindgren CM, Magi R, Morris AP, Randall J, Johnson T, Elliott P,

Rybin D, Thorleifsson G, Steinthorsdottir V, Henneman P, Grallert H, Dehghan A, Hottenga JJ, Franklin CS, Navarro P, Song K, Goel A, Perry JR, Egan JM, Lajunen T, Grarup N, Sparso T, Doney A, Voight BF, Stringham HM, Li M, Kanoni S, Shrader P, Cavalcanti-Proenca C, Kumari M, Qi L, Timpson NJ, Gieger C, Zabena C, Rocheleau G, Ingelsson E, An P, O'Connell J, Luan J, Elliott A, McCarroll SA, Payne F, Roccasecca RM, Pattou F, Sethupathy P, Ardlie K, Ariyurek Y, Balkau B, Barter P, Beilby JP, Ben-Shlomo Y, Benediktsson R, Bennett AJ, Bergmann S, Bochud M, Boerwinkle E, Bonnefond A, Bonnycastle LL, Borch-Johnsen K, Bottcher Y, Brunner E, Bumpstead SJ, Charpentier G, Chen YD, Chines P, Clarke R, Coin LJ, Cooper MN, Cornelis M, Crawford G, Crisponi L, Day IN, de Geus EJ, Delplanque J, Dina C, Erdos MR, Fedson AC, Fischer-Rosinsky A, Forouhi NG, Fox CS, Frants R, Franzosi MG, Galan P, Goodarzi MO, Graessler J, Groves CJ, Grundy S, Gwilliam R, Gyllensten U, Hadjadj S, Hallmans G, Hammond N, Han X, Hartikainen AL, Hassanali N, Hayward C, Heath SC, Hercberg S, Herder C, Hicks AA, Hillman DR, Hingorani AD, Hofman A, Hui J, Hung J, Isomaa B, Johnson PR, Jorgensen T, Jula A, Kaakinen M, Kaprio J, Kesaniemi YA, Kivimaki M, Knight B, Koskinen S, Kovacs P, Kyvik KO, Lathrop GM, Lawlor DA, Le Bacquer O, Lecoeur C, Li Y, Lyssenko V, Mahley R, Mangino M, Manning AK, Martinez-Larrad MT, McAteer JB, McCulloch LJ, McPherson R, Meisinger C, Melzer D, Meyre D, Mitchell BD, Morken MA, Mukherjee S, Naitza S, Narisu N, Neville MJ, Oostra BA, Orru M, Pakyz R, Palmer CN, Paolisso G, Pattaro C, Pearson D, Peden JF, Pedersen NL, Perola M, Pfeiffer AF, Pichler I, Polasek O, Posthuma D, Potter SC, Pouta A, Province MA, Psaty BM, Rathmann W, Rayner NW, Rice K, Ripatti S, Rivadeneira F, Roden M, Rolandsson O, Sandbaek A, Sandhu M, Sanna S, Sayer AA, Scheet P, Scott LJ, Seedorf U, Sharp SJ, Shields B, Sigurethsson G, Sijbrands EJ, Silveira A, Simpson L, Singleton A, Smith NL, Sovio U, Swift A, Syddall H, Syvanen AC, Tanaka T, Thorand B, Tichet J, Tonjes A, Tuomi T, Uitterlinden AG, van Dijk KW, van Hoek M, Varma D, Visvikis-Siest S, Vitart V, Vogelzangs N, Waeber G, Wagner PJ, Walley A, Walters GB, Ward KL, Watkins H, Weedon MN, Wild SH, Willemsen G, Witteman JC, Yarnell JW, Zeggini E, Zelenika D, Zethelius B, Zhai G, Zhao JH, Zillikens MC, Consortium D, Consortium G, Global BC, Borecki IB, Loos RJ, Meneton P, Magnusson PK, Nathan DM, Williams GH, Hattersley AT, Silander K, Salomaa V, Smith GD, Bornstein SR, Schwarz P, Spranger J, Karpe F, Shuldiner AR, Cooper C, Dedoussis GV, Serrano-Rios M, Morris AD, Lind L, Palmer LJ, Hu FB, Franks PW, Ebrahim S, Marmot M, Kao WH, Pankow JS, Sampson MJ, Kuusisto J, Laakso M, Hansen T, Pedersen O, Pramstaller PP, Wichmann HE, Illig T, Rudan I, Wright AF, Stumvoll M, Campbell H, Wilson JF, Anders Hamsten on behalf of Procardis C, Investigators M, Bergman RN, Buchanan TA, Collins FS, Mohlke KL, Tuomilehto J, Valle TT, Altshuler D, Rotter JI, Siscovick DS, Penninx BW, Boomsma DI, Deloukas P, Spector TD, Frayling TM, Ferrucci L, Kong A, Thorsteinsdottir U, Stefansson K, van Duijn CM, Aulchenko YS, Cao A, Scuteri A, Schlessinger D, Uda M, Ruokonen A, Jarvelin MR, Waterworth DM, Vollenweider P, Peltonen L, Mooser V, Abecasis GR, Wareham NJ, Sladek R, Froguel P, Watanabe RM, Meigs JB, Groop L, Boehnke M, McCarthy MI, Florez JC, Barroso I (2010) New genetic loci implicated in fasting glucose homeostasis and their impact on type 2 diabetes risk. Nat Genet 42(2):105–116. doi:10.1038/ng.520

Fall T, Hagg S, Magi R, Ploner A, Fischer K, Horikoshi M, Sarin AP, Thorleifsson G, Ladenvall C, Kals M, Kuningas M, Draisma HH, Ried JS, van Zuydam NR, Huikari V, Mangino M, Sonestedt E, Benyamin B, Nelson CP, Rivera NV, Kristiansson K, Shen HY, Havulinna AS, Dehghan A, Donnelly LA, Kaakinen M, Nuotio ML, Robertson N, de Bruijn RF, Ikram MA, Amin N, Balmforth AJ, Braund PS, Doney AS, Doring A, Elliott P, Esko T, Franco OH, Gretarsdottir S, Hartikainen AL, Heikkila K, Herzig KH, Holm H, Hottenga JJ, Hypponen E, Illig T, Isaacs A, Isomaa B, Karssen LC, Kettunen J, Koenig W, Kuulasmaa K, Laatikainen T, Laitinen J, Lindgren C, Lyssenko V, Laara E, Rayner NW, Mannisto S, Pouta A, Rathmann W, Rivadeneira F, Ruokonen A, Savolainen MJ, Sijbrands EJ, Small KS, Smit JH, Steinthorsdottir V, Syvanen AC, Taanila A, Tobin MD, Uitterlinden AG, Willems SM, Willemsen G, Witteman J, Perola M, Evans A, Ferrieres J, Virtamo J, Kee F, Tregouet DA,

Arveiler D, Amouyel P, Ferrario MM, Brambilla P, Hall AS, Heath AC, Madden PA, Martin NG, Montgomery GW, Whitfield JB, Jula A, Knekt P, Oostra B, van Duijn CM, Penninx BW, Davey Smith G, Kaprio J, Samani NJ, Gieger C, Peters A, Wichmann HE, Boomsma DI, de Geus EJ, Tuomi T, Power C, Hammond CJ, Spector TD, Lind L, Orho-Melander M, Palmer CN, Morris AD, Groop L, Jarvelin MR, Salomaa V, Vartiainen E, Hofman A, Ripatti S, Metspalu A, Thorsteinsdottir U, Stefansson K, Pedersen NL, McCarthy MI, Ingelsson E, Prokopenko I, European Network for G, Genomic Epidemiology C (2013) The role of adiposity in cardiometabolic traits: a Mendelian randomization analysis. PLoS Med 10(6): e1001474. doi:10.1371/journal.pmed.1001474

Federation ID (2013) IDF diabetes atlas, 6th edn. International Diabetes Federation, Brussels

Finkel T, Serrano M, Blasco MA (2007) The common biology of cancer and ageing. Nature 448 (7155):767–774. doi:10.1038/nature05985

Franks PW, Rolandsson O, Debenham SL, Fawcett KA, Payne F, Dina C, Froguel P, Mohlke KL, Willer C, Olsson T, Wareham NJ, Hallmans G, Barroso I, Sandhu MS (2008) Replication of the association between variants in WFS1 and risk of type 2 diabetes in European populations. Diabetologia 51(3):458–463. doi:10.1007/s00125-007-0887-6

Frayling TM, Timpson NJ, Weedon MN, Zeggini E, Freathy RM, Lindgren CM, Perry JR, Elliott KS, Lango H, Rayner NW, Shields B, Harries LW, Barrett JC, Ellard S, Groves CJ, Knight B, Patch AM, Ness AR, Ebrahim S, Lawlor DA, Ring SM, Ben-Shlomo Y, Jarvelin MR, Sovio U, Bennett AJ, Melzer D, Ferrucci L, Loos RJ, Barroso I, Wareham NJ, Karpe F, Owen KR, Cardon LR, Walker M, Hitman GA, Palmer CN, Doney AS, Morris AD, Smith GD, Hattersley AT, McCarthy MI (2007) A common variant in the FTO gene is associated with body mass index and predisposes to childhood and adult obesity. Science 316(5826):889–894. doi:10.1126/science.1141634, 1141634 [pii]

Freathy RM, Timpson NJ, Lawlor DA, Pouta A, Ben-Shlomo Y, Ruokonen A, Ebrahim S, Shields B, Zeggini E, Weedon MN, Lindgren CM, Lango H, Melzer D, Ferrucci L, Paolisso G, Neville MJ, Karpe F, Palmer CN, Morris AD, Elliott P, Jarvelin MR, Smith GD, McCarthy MI, Hattersley AT, Frayling TM (2008) Common variation in the FTO gene alters diabetes-related metabolic traits to the extent expected given its effect on BMI. Diabetes 57 (5):1419–1426. doi:10.2337/db07-1466, db07-1466 [pii]

Froguel P, Vaxillaire M, Sun F, Velho G, Zouali H, Butel MO, Lesage S, Vionnet N, Clement K, Fougerousse F et al (1992) Close linkage of glucokinase locus on chromosome 7p to early-onset non-insulin-dependent diabetes mellitus. Nature 356(6365):162–164. doi:10.1038/356162a0

Genomes Project C, Abecasis GR, Altshuler D, Auton A, Brooks LD, Durbin RM, Gibbs RA, Hurles ME, McVean GA (2010) A map of human genome variation from population-scale sequencing. Nature 467(7319):1061–1073. doi:10.1038/nature09534

Gloyn AL, McCarthy MI (2001) The genetics of type 2 diabetes. Best Pract Res Clin Endocrinol Metab 15(3):293–308. doi:10.1053/beem.2001.0147

Gloyn AL, Weedon MN, Owen KR, Turner MJ, Knight BA, Hitman G, Walker M, Levy JC, Sampson M, Halford S, McCarthy MI, Hattersley AT, Frayling TM (2003) Large-scale association studies of variants in genes encoding the pancreatic beta-cell KATP channel subunits Kir6.2 (KCNJ11) and SUR1 (ABCC8) confirm that the KCNJ11 E23K variant is associated with type 2 diabetes. Diabetes 52(2):568–572

Gloyn AL, Pearson ER, Antcliff JF, Proks P, Bruining GJ, Slingerland AS, Howard N, Srinivasan S, Silva JM, Molnes J, Edghill EL, Frayling TM, Temple IK, Mackay D, Shield JP, Sumnik Z, van Rhijn A, Wales JK, Clark P, Gorman S, Aisenberg J, Ellard S, Njolstad PR, Ashcroft FM, Hattersley AT (2004) Activating mutations in the gene encoding the ATP-sensitive potassium-channel subunit Kir6.2 and permanent neonatal diabetes. N Engl J Med 350(18):1838–1849. doi:10.1056/NEJMoa032922

Go MJ, Hwang JY, Kim YJ, Hee Oh J, Kim YJ, Heon Kwak S, Soo Park K, Lee J, Kim BJ, Han BG, Cho MC, Cho YS, Lee JY (2013) New susceptibility loci in MYL2, C12orf51 and OAS1 associated with 1-h plasma glucose as predisposing risk factors for type 2 diabetes in the Korean population. J Hum Genet 58(6):362–365. doi:10.1038/jhg.2013.14

Grant SF, Thorleifsson G, Reynisdottir I, Benediktsson R, Manolescu A, Sainz J, Helgason A, Stefansson H, Emilsson V, Helgadottir A, Styrkarsdottir U, Magnusson KP, Walters GB, Palsdottir E, Jonsdottir T, Gudmundsdottir T, Gylfason A, Saemundsdottir J, Wilensky RL, Reilly MP, Rader DJ, Bagger Y, Christiansen C, Gudnason V, Sigurdsson G, Thorsteinsdottir U, Gulcher JR, Kong A, Stefansson K (2006) Variant of transcription factor 7-like 2 (TCF7L2) gene confers risk of type 2 diabetes. Nat Genet 38(3):320–323. doi:10.1038/ng1732, ng1732 [pii]

Gudmundsson J, Sulem P, Steinthorsdottir V, Bergthorsson JT, Thorleifsson G, Manolescu A, Rafnar T, Gudbjartsson D, Agnarsson BA, Baker A, Sigurdsson A, Benediktsdottir KR, Jakobsdottir M, Blondal T, Stacey SN, Helgason A, Gunnarsdottir S, Olafsdottir A, Kristinsson KT, Birgisdottir B, Ghosh S, Thorlacius S, Magnusdottir D, Stefansdottir G, Kristjansson K, Bagger Y, Wilensky RL, Reilly MP, Morris AD, Kimber CH, Adeyemo A, Chen Y, Zhou J, So WY, Tong PC, Ng MC, Hansen T, Andersen G, Borch-Johnsen K, Jorgensen T, Tres A, Fuertes F, Ruiz-Echarri M, Asin L, Saez B, van Boven E, Klaver S, Swinkels DW, Aben KK, Graif T, Cashy J, Suarez BK, van Vierssen TO, Frigge ML, Ober C, Hofker MH, Wijmenga C, Christiansen C, Rader DJ, Palmer CN, Rotimi C, Chan JC, Pedersen O, Sigurdsson G, Benediktsson R, Jonsson E, Einarsson GV, Mayordomo JI, Catalona WJ, Kiemeney LA, Barkardottir RB, Gulcher JR, Thorsteinsdottir U, Kong A, Stefansson K (2007) Two variants on chromosome 17 confer prostate cancer risk, and the one in TCF2 protects against type 2 diabetes. Nat Genet 39(8):977–983. doi:10.1038/ng2062

Hanson RL, Muller YL, Kobes S, Guo T, Bian L, Ossowski V, Wiedrich K, Sutherland J, Wiedrich C, Mahkee D, Huang K, Abdussamad M, Traurig M, Weil EJ, Nelson RG, Bennett PH, Knowler WC, Bogardus C, Baier LJ (2014) A genome-wide association study in American Indians implicates DNER as a susceptibility locus for type 2 diabetes. Diabetes 63(1):369–376. doi:10.2337/db13-0416

Hara K, Boutin P, Mori Y, Tobe K, Dina C, Yasuda K, Yamauchi T, Otabe S, Okada T, Eto K, Kadowaki H, Hagura R, Akanuma Y, Yazaki Y, Nagai R, Taniyama M, Matsubara K, Yoda M, Nakano Y, Tomita M, Kimura S, Ito C, Froguel P, Kadowaki T (2002) Genetic variation in the gene encoding adiponectin is associated with an increased risk of type 2 diabetes in the Japanese population. Diabetes 51(2):536–540

Hara K, Fujita H, Johnson TA, Yamauchi T, Yasuda K, Horikoshi M, Peng C, Hu C, Ma RC, Imamura M, Iwata M, Tsunoda T, Morizono T, Shojima N, So WY, Leung TF, Kwan P, Zhang R, Wang J, Yu W, Maegawa H, Hirose H, Consortium D, Kaku K, Ito C, Watada H, Tanaka Y, Tobe K, Kashiwagi A, Kawamori R, Jia W, Chan JC, Teo YY, Shyong TE, Kamatani N, Kubo M, Maeda S, Kadowaki T (2014) Genome-wide association study identifies three novel loci for type 2 diabetes. Hum Mol Genet 23(1):239–246. doi:10.1093/hmg/ddt399

Hivert MF, Jablonski KA, Perreault L, Saxena R, McAteer JB, Franks PW, Hamman RF, Kahn SE, Haffner S, Consortium D, Meigs JB, Altshuler D, Knowler WC, Florez JC, Diabetes Prevention Program Research G (2011) Updated genetic score based on 34 confirmed type 2 diabetes loci is associated with diabetes incidence and regression to normoglycemia in the diabetes prevention program. Diabetes 60(4):1340–1348. doi:10.2337/db10-1119

Horikawa Y, Oda N, Cox NJ, Li X, Orho-Melander M, Hara M, Hinokio Y, Lindner TH, Mashima H, Schwarz PE, del Bosque-Plata L, Horikawa Y, Oda Y, Yoshiuchi I, Colilla S, Polonsky KS, Wei S, Concannon P, Iwasaki N, Schulze J, Baier LJ, Bogardus C, Groop L, Boerwinkle E, Hanis CL, Bell GI (2000) Genetic variation in the gene encoding calpain-10 is associated with type 2 diabetes mellitus. Nat Genet 26(2):163–175. doi:10.1038/79876

Hu FB (2011) Globalization of diabetes: the role of diet, lifestyle, and genes. Diabet Care 34 (6):1249–1257. doi:10.2337/dc11-0442

Kilpelainen TO, Zillikens MC, Stancakova A, Finucane FM, Ried JS, Langenberg C, Zhang W, Beckmann JS, Luan J, Vandenput L, Styrkarsdottir U, Zhou Y, Smith AV, Zhao JH, Amin N, Vedantam S, Shin SY, Haritunians T, Fu M, Feitosa MF, Kumari M, Halldorsson BV,

Tikkanen E, Mangino M, Hayward C, Song C, Arnold AM, Aulchenko YS, Oostra BA, Campbell H, Cupples LA, Davis KE, Doring A, Eiriksdottir G, Estrada K, Fernandez-Real JM, Garcia M, Gieger C, Glazer NL, Guiducci C, Hofman A, Humphries SE, Isomaa B, Jacobs LC, Jula A, Karasik D, Karlsson MK, Khaw KT, Kim LJ, Kivimaki M, Klopp N, Kuhnel B, Kuusisto J, Liu Y, Ljunggren O, Lorentzon M, Luben RN, McKnight B, Mellstrom D, Mitchell BD, Mooser V, Moreno JM, Mannisto S, O'Connell JR, Pascoe L, Peltonen L, Peral B, Perola M, Psaty BM, Salomaa V, Savage DB, Semple RK, Skaric-Juric T, Sigurdsson G, Song KS, Spector TD, Syvanen AC, Talmud PJ, Thorleifsson G, Thorsteinsdottir U, Uitterlinden AG, van Duijn CM, Vidal-Puig A, Wild SH, Wright AF, Clegg DJ, Schadt E, Wilson JF, Rudan I, Ripatti S, Borecki IB, Shuldiner AR, Ingelsson E, Jansson JO, Kaplan RC, Gudnason V, Harris TB, Groop L, Kiel DP, Rivadeneira F, Walker M, Barroso I, Vollenweider P, Waeber G, Chambers JC, Kooner JS, Soranzo N, Hirschhorn JN, Stefansson K, Wichmann HE, Ohlsson C, O'Rahilly S, Wareham NJ, Speliotes EK, Fox CS, Laakso M, Loos RJ (2011) Genetic variation near IRS1 associates with reduced adiposity and an impaired metabolic profile. Nat Genet 43(8):753–760. doi:10.1038/ng.866

Klimentidis YC, Abrams M, Wang J, Fernandez JR, Allison DB (2011) Natural selection at genomic regions associated with obesity and type-2 diabetes: East Asians and sub-Saharan Africans exhibit high levels of differentiation at type-2 diabetes regions. Hum Genet 129 (4):407–418. doi:10.1007/s00439-010-0935-z

Kooner JS, Saleheen D, Sim X, Sehmi J, Zhang W, Frossard P, Been LF, Chia KS, Dimas AS, Hassanali N, Jafar T, Jowett JB, Li X, Radha V, Rees SD, Takeuchi F, Young R, Aung T, Basit A, Chidambaram M, Das D, Grundberg E, Hedman AK, Hydrie ZI, Islam M, Khor CC, Kowlessur S, Kristensen MM, Liju S, Lim WY, Matthews DR, Liu J, Morris AP, Nica AC, Pinidiyapathirage JM, Prokopenko I, Rasheed A, Samuel M, Shah N, Shera AS, Small KS, Suo C, Wickremasinghe AR, Wong TY, Yang M, Zhang F, Diagram MT, Abecasis GR, Barnett AH, Caulfield M, Deloukas P, Frayling TM, Froguel P, Kato N, Katulanda P, Kelly MA, Liang J, Mohan V, Sanghera DK, Scott J, Seielstad M, Zimmet PZ, Elliott P, Teo YY, McCarthy MI, Danesh J, Tai ES, Chambers JC (2011) Genome-wide association study in individuals of South Asian ancestry identifies six new type 2 diabetes susceptibility loci. Nat Genet 43(10):984–989. doi:10.1038/ng.921

Laukkanen O, Pihlajamaki J, Lindstrom J, Eriksson J, Valle TT, Hamalainen H, Ilanne-Parikka P, Keinanen-Kiukaanniemi S, Tuomilehto J, Uusitupa M, Laakso M, Finnish Diabetes Prevention Study G (2004) Polymorphisms of the SUR1 (ABCC8) and Kir6.2 (KCNJ11) genes predict the conversion from impaired glucose tolerance to type 2 diabetes. The Finnish Diabetes Prevention Study. J Clin Endocrinol Metab 89(12):6286–6290. doi:10.1210/jc.2004-1204

Li H, Gan W, Lu L, Dong X, Han X, Hu C, Yang Z, Sun L, Bao W, Li P, He M, Sun L, Wang Y, Zhu J, Ning Q, Tang Y, Zhang R, Wen J, Wang D, Zhu X, Guo K, Zuo X, Guo X, Yang H, Zhou X, Consortium D, Consortium A-TD, Zhang X, Qi L, Loos RJ, Hu FB, Wu T, Liu Y, Liu L, Yang Z, Hu R, Jia W, Ji L, Li Y, Lin X (2013) A genome-wide association study identifies GRK5 and RASGRP1 as type 2 diabetes loci in Chinese Hans. Diabetes 62 (1):291–298. doi:10.2337/db12-0454

Lohmueller KE, Pearce CL, Pike M, Lander ES, Hirschhorn JN (2003) Meta-analysis of genetic association studies supports a contribution of common variants to susceptibility to common disease. Nat Genet 33(2):177–182. doi:10.1038/ng1071

Loos RJ, Lindgren CM, Li S, Wheeler E, Zhao JH, Prokopenko I, Inouye M, Freathy RM, Attwood AP, Beckmann JS, Berndt SI, Jacobs KB, Chanock SJ, Hayes RB, Bergmann S, Bennett AJ, Bingham SA, Bochud M, Brown M, Cauchi S, Connell JM, Cooper C, Smith GD, Day I, Dina C, De S, Dermitzakis ET, Doney AS, Elliott KS, Elliott P, Evans DM, Sadaf Farooqi I, Froguel P, Ghori J, Groves CJ, Gwilliam R, Hadley D, Hall AS, Hattersley AT, Hebebrand J, Heid IM, Lamina C, Gieger C, Illig T, Meitinger T, Wichmann HE, Herrera B, Hinney A, Hunt SE, Jarvelin MR, Johnson T, Jolley JD, Karpe F, Keniry A, Khaw KT, Luben RN, Mangino M, Marchini J, McArdle WL, McGinnis R, Meyre D, Munroe PB, Morris AD, Ness AR, Neville MJ, Nica AC, Ong KK, O'Rahilly S, Owen KR, Palmer CN, Papadakis K, Potter S, Pouta A,

Qi L, Randall JC, Rayner NW, Ring SM, Sandhu MS, Scherag A, Sims MA, Song K, Soranzo N, Speliotes EK, Syddall HE, Teichmann SA, Timpson NJ, Tobias JH, Uda M, Vogel CI, Wallace C, Waterworth DM, Weedon MN, Willer CJ, Wraight YX, Zeggini E, Hirschhorn JN, Strachan DP, Ouwehand WH, Caulfield MJ, Samani NJ, Frayling TM, Vollenweider P, Waeber G, Mooser V, Deloukas P, McCarthy MI, Wareham NJ, Barroso I, Kraft P, Hankinson SE, Hunter DJ, Hu FB, Lyon HN, Voight BF, Ridderstrale M, Groop L, Scheet P, Sanna S, Abecasis GR, Albai G, Nagaraja R, Schlessinger D, Jackson AU, Tuomilehto J, Collins FS, Boehnke M, Mohlke KL (2008) Common variants near MC4R are associated with fat mass, weight and risk of obesity. Nat Genet 40(6):768–775. doi:10.1038/ng.140

Lyssenko V, Jonsson A, Almgren P, Pulizzi N, Isomaa B, Tuomi T, Berglund G, Altshuler D, Nilsson P, Groop L (2008) Clinical risk factors, DNA variants, and the development of type 2 diabetes. N Engl J Med 359(21):2220–2232. doi:10.1056/NEJMoa0801869

Lyssenko V, Nagorny CL, Erdos MR, Wierup N, Jonsson A, Spegel P, Bugliani M, Saxena R, Fex M, Pulizzi N, Isomaa B, Tuomi T, Nilsson P, Kuusisto J, Tuomilehto J, Boehnke M, Altshuler D, Sundler F, Eriksson JG, Jackson AU, Laakso M, Marchetti P, Watanabe RM, Mulder H, Groop L (2009) Common variant in MTNR1B associated with increased risk of type 2 diabetes and impaired early insulin secretion. Nat Genet 41(1):82–88. doi:10.1038/ng.288

Ma RC, Hu C, Tam CH, Zhang R, Kwan P, Leung TF, Thomas GN, Go MJ, Hara K, Sim X, Ho JS, Wang C, Li H, Lu L, Wang Y, Li JW, Wang Y, Lam VK, Wang J, Yu W, Kim YJ, Ng DP, Fujita H, Panoutsopoulou K, Day-Williams AG, Lee HM, Ng AC, Fang YJ, Kong AP, Jiang F, Ma X, Hou X, Tang S, Lu J, Yamauchi T, Tsui SK, Woo J, Leung PC, Zhang X, Tang NL, Sy HY, Liu J, Wong TY, Lee JY, Maeda S, Xu G, Cherny SS, Chan TF, Ng MC, Xiang K, Morris AP, Consortium D, Keildson S, Mu TC, Hu R, Ji L, Lin X, Cho YS, Kadowaki T, Tai ES, Zeggini E, McCarthy MI, Hon KL, Baum L, Tomlinson B, So WY, Bao Y, Chan JC, Jia W (2013) Genome-wide association study in a Chinese population identifies a susceptibility locus for type 2 diabetes at 7q32 near PAX4. Diabetologia 56(6):1291–1305. doi:10.1007/s00125-013-2874-4

Majithia AR, Florez JC (2009) Clinical translation of genetic predictors for type 2 diabetes. Curr Opin Endocrinol Diabetes Obes 16(2):100–106

Manning AK, Hivert MF, Scott RA, Grimsby JL, Bouatia-Naji N, Chen H, Rybin D, Liu CT, Bielak LF, Prokopenko I, Amin N, Barnes D, Cadby G, Hottenga JJ, Ingelsson E, Jackson AU, Johnson T, Kanoni S, Ladenvall C, Lagou V, Lahti J, Lecoeur C, Liu Y, Martinez-Larrad MT, Montasser ME, Navarro P, Perry JR, Rasmussen-Torvik LJ, Salo P, Sattar N, Shungin D, Strawbridge RJ, Tanaka T, van Duijn CM, An P, de Andrade M, Andrews JS, Aspelund T, Atalay M, Aulchenko Y, Balkau B, Bandinelli S, Beckmann JS, Beilby JP, Bellis C, Bergman RN, Blangero J, Boban M, Boehnke M, Boerwinkle E, Bonnycastle LL, Boomsma DI, Borecki IB, Bottcher Y, Bouchard C, Brunner E, Budimir D, Campbell H, Carlson O, Chines PS, Clarke R, Collins FS, Corbaton-Anchuelo A, Couper D, de Faire U, Dedoussis GV, Deloukas P, Dimitriou M, Egan JM, Eiriksdottir G, Erdos MR, Eriksson JG, Eury E, Ferrucci L, Ford I, Forouhi NG, Fox CS, Franzosi MG, Franks PW, Frayling TM, Froguel P, Galan P, de Geus E, Gigante B, Glazer NL, Goel A, Groop L, Gudnason V, Hallmans G, Hamsten A, Hansson O, Harris TB, Hayward C, Heath S, Hercberg S, Hicks AA, Hingorani A, Hofman A, Hui J, Hung J, Jarvelin MR, Jhun MA, Johnson PC, Jukema JW, Jula A, Kao WH, Kaprio J, Kardia SL, Keinanen-Kiukaanniemi S, Kivimaki M, Kolcic I, Kovacs P, Kumari M, Kuusisto J, Kyvik KO, Laakso M, Lakka T, Lannfelt L, Lathrop GM, Launer LJ, Leander K, Li G, Lind L, Lindstrom J, Lobbens S, Loos RJ, Luan J, Lyssenko V, Magi R, Magnusson PK, Marmot M, Meneton P, Mohlke KL, Mooser V, Morken MA, Miljkovic I, Narisu N, O'Connell J, Ong KK, Oostra BA, Palmer LJ, Palotie A, Pankow JS, Peden JF, Pedersen NL, Pehlic M, Peltonen L, Penninx B, Pericic M, Perola M, Perusse L, Peyser PA, Polasek O, Pramstaller PP, Province MA, Raikkonen K, Rauramaa R, Rehnberg E, Rice K, Rotter JI, Rudan I, Ruokonen A, Saaristo T, Sabater-Lleal M, Salomaa V, Savage DB, Saxena R, Schwarz P, Seedorf U, Sennblad B, Serrano-Rios M, Shuldiner AR, Sijbrands EJ, Siscovick

DS, Smit JH, Small KS, Smith NL, Smith AV, Stancakova A, Stirrups K, Stumvoll M, Sun YV, Swift AJ, Tonjes A, Tuomilehto J, Trompet S, Uitterlinden AG, Uusitupa M, Vikstrom M, Vitart V, Vohl MC, Voight BF, Vollenweider P, Waeber G, Waterworth DM, Watkins H, Wheeler E, Widen E, Wild SH, Willems SM, Willemsen G, Wilson JF, Witteman JC, Wright AF, Yaghootkar H, Zelenika D, Zemunik T, Zgaga L, Replication DIG, Meta-analysis C, Multiple Tissue Human Expression Resource C, Wareham NJ, McCarthy MI, Barroso I, Watanabe RM, Florez JC, Dupuis J, Meigs JB, Langenberg C (2012) A genome-wide approach accounting for body mass index identifies genetic variants influencing fasting glycemic traits and insulin resistance. Nat Genet 44(6):659–669. doi:10.1038/ng.2274

Manolio TA, Collins FS, Cox NJ, Goldstein DB, Hindorff LA, Hunter DJ, McCarthy MI, Ramos EM, Cardon LR, Chakravarti A, Cho JH, Guttmacher AE, Kong A, Kruglyak L, Mardis E, Rotimi CN, Slatkin M, Valle D, Whittemore AS, Boehnke M, Clark AG, Eichler EE, Gibson G, Haines JL, Mackay TF, McCarroll SA, Visscher PM (2009) Finding the missing heritability of complex diseases. Nature 461(7265):747–753. doi:10.1038/nature08494, nature08494 [pii]

McCarthy MI (2003) Growing evidence for diabetes susceptibility genes from genome scan data. Curr Diab Rep 3(2):159–167

McCarthy MI (2008) Casting a wider net for diabetes susceptibility genes. Nat Genet 40 (9):1039–1040. doi:10.1038/ng0908-1039

McCarthy MI (2010) Genomics, type 2 diabetes, and obesity. N Engl J Med 363(24):2339–2350. doi:10.1056/NEJMra0906948

McCarthy MI, Hattersley AT (2008) Learning from molecular genetics: novel insights arising from the definition of genes for monogenic and type 2 diabetes. Diabetes 57(11):2889–2898. doi:10.2337/db08-0343

McDonald TJ, Ellard S (2013) Maturity onset diabetes of the young: identification and diagnosis. Ann Clin Biochem 50(Pt 5):403–415. doi:10.1177/0004563213483458

Meigs JB, Shrader P, Sullivan LM, McAteer JB, Fox CS, Dupuis J, Manning AK, Florez JC, Wilson PW, D'Agostino RB Sr, Cupples LA (2008) Genotype score in addition to common risk factors for prediction of type 2 diabetes. N Engl J Med 359(21):2208–2219. doi:10.1056/NEJMoa0804742

Meyre D, Bouatia-Naji N, Tounian A, Samson C, Lecoeur C, Vatin V, Ghoussaini M, Wachter C, Hercberg S, Charpentier G, Patsch W, Pattou F, Charles MA, Tounian P, Clement K, Jouret B, Weill J, Maddux BA, Goldfine ID, Walley A, Boutin P, Dina C, Froguel P (2005) Variants of ENPP1 are associated with childhood and adult obesity and increase the risk of glucose intolerance and type 2 diabetes. Nat Genet 37(8):863–867. doi:10.1038/ng1604

Morris AP (2011) Transethnic meta-analysis of genomewide association studies. Genet Epidemiol 35(8):809–822. doi:10.1002/gepi.20630

Morris AP, Voight BF, Teslovich TM, Ferreira T, Segre AV, Steinthorsdottir V, Strawbridge RJ, Khan H, Grallert H, Mahajan A, Prokopenko I, Kang HM, Dina C, Esko T, Fraser RM, Kanoni S, Kumar A, Lagou V, Langenberg C, Luan J, Lindgren CM, Muller-Nurasyid M, Pechlivanis S, Rayner NW, Scott LJ, Wiltshire S, Yengo L, Kinnunen L, Rossin EJ, Raychaudhuri S, Johnson AD, Dimas AS, Loos RJ, Vedantam S, Chen H, Florez JC, Fox C, Liu CT, Rybin D, Couper DJ, Kao WH, Li M, Cornelis MC, Kraft P, Sun Q, van Dam RM, Stringham HM, Chines PS, Fischer K, Fontanillas P, Holmen OL, Hunt SE, Jackson AU, Kong A, Lawrence R, Meyer J, Perry JR, Platou CG, Potter S, Rehnberg E, Robertson N, Sivapalaratnam S, Stancakova A, Stirrups K, Thorleifsson G, Tikkanen E, Wood AR, Almgren P, Atalay M, Benediktsson R, Bonnycastle LL, Burtt N, Carey J, Charpentier G, Crenshaw AT, Doney AS, Dorkhan M, Edkins S, Emilsson V, Eury E, Forsen T, Gertow K, Gigante B, Grant GB, Groves CJ, Guiducci C, Herder C, Hreidarsson AB, Hui J, James A, Jonsson A, Rathmann W, Klopp N, Kravic J, Krjutskov K, Langford C, Leander K, Lindholm E, Lobbens S, Mannisto S, Mirza G, Muhleisen TW, Musk B, Parkin M, Rallidis L, Saramies J, Sennblad B, Shah S, Sigurethsson G, Silveira A, Steinbach G, Thorand B, Trakalo J, Veglia F, Wennauer R, Winckler W, Zabaneh D, Campbell H, van Duijn C, Uitterlinden AG, Hofman A, Sijbrands E, Abecasis GR, Owen KR, Zeggini E, Trip

MD, Forouhi NG, Syvanen AC, Eriksson JG, Peltonen L, Nothen MM, Balkau B, Palmer CN, Lyssenko V, Tuomi T, Isomaa B, Hunter DJ, Qi L, Wellcome Trust Case Control C, Meta-Analyses of G, Insulin-Related Traits Consortium I, Genetic Investigation of ATC, Asian Genetic Epidemiology Network-Type 2 Diabetes C, South Asian Type 2 Diabetes C, Shuldiner AR, Roden M, Barroso I, Wilsgaard T, Beilby J, Hovingh K, Price JF, Wilson JF, Rauramaa R, Lakka TA, Lind L, Dedoussis G, Njolstad I, Pedersen NL, Khaw KT, Wareham NJ, Keinanen-Kiukaanniemi SM, Saaristo TE, Korpi-Hyovalti E, Saltevo J, Laakso M, Kuusisto J, Metspalu A, Collins FS, Mohlke KL, Bergman RN, Tuomilehto J, Boehm BO, Gieger C, Hveem K, Cauchi S, Froguel P, Baldassarre D, Tremoli E, Humphries SE, Saleheen D, Danesh J, Ingelsson E, Ripatti S, Salomaa V, Erbel R, Jockel KH, Moebus S, Peters A, Illig T, de Faire U, Hamsten A, Morris AD, Donnelly PJ, Frayling TM, Hattersley AT, Boerwinkle E, Melander O, Kathiresan S, Nilsson PM, Deloukas P, Thorsteinsdottir U, Groop LC, Stefansson K, Hu F, Pankow JS, Dupuis J, Meigs JB, Altshuler D, Boehnke M, McCarthy MI, Replication DIG, Meta-analysis C (2012) Large-scale association analysis provides insights into the genetic architecture and pathophysiology of type 2 diabetes. Nat Genet 44(9):981–990. doi:10.1038/ng.2383

Muhlenbruch K, Jeppesen C, Joost HG, Boeing H, Schulze MB (2013) The value of genetic information for diabetes risk prediction - differences according to sex, age, family history and obesity. PloS One 8(5):e64307. doi:10.1371/journal.pone.0064307

Neel JV (1962) Diabetes mellitus: a "thrifty" genotype rendered detrimental by "progress"? Am J Hum Genet 14:353–362

Neel JV (1999) The "thrifty genotype" in 1998. Nutr Rev 57(5 Pt 2):S2–S9

Palmer ND, McDonough CW, Hicks PJ, Roh BH, Wing MR, An SS, Hester JM, Cooke JN, Bostrom MA, Rudock ME, Talbert ME, Lewis JP, Consortium D, Investigators M, Ferrara A, Lu L, Ziegler JT, Sale MM, Divers J, Shriner D, Adeyemo A, Rotimi CN, Ng MC, Langefeld CD, Freedman BI, Bowden DW, Voight BF, Scott LJ, Steinthorsdottir V, Morris AP, Dina C, Welch RP, Zeggini E, Huth C, Aulchenko YS, Thorleifsson G, McCulloch LJ, Ferreira T, Grallert H, Amin N, Wu G, Willer CJ, Raychaudhuri S, McCarroll SA, Langenberg C, Hofmann OM, Dupuis J, Qi L, Segre AV, van Hoek M, Navarro P, Ardlie K, Balkau B, Benediktsson R, Bennett AJ, Blagieva R, Boerwinkle E, Bonnycastle LL, Bostrom KB, Bravenboer B, Bumpstead S, Burtt NP, Charpentier G, Chines PS, Cornelis M, Couper DJ, Crawford G, Doney AS, Elliott KS, Elliott AL, Erdos MR, Fox CS, Franklin CS, Ganser M, Gieger C, Grarup N, Green T, Griffin S, Groves CJ, Guiducci C, Hadjadj S, Hassanali N, Herder C, Isomaa B, Jackson AU, Johnson PR, Jorgensen T, Kao WH, Klopp N, Kong A, Kraft P, Kuusisto J, Lauritzen T, Li M, Lieverse A, Lindgren CM, Lyssenko V, Marre M, Meitinger T, Midthjell K, Morken MA, Narisu N, Nilsson P, Owen KR, Payne F, Perry JR, Petersen AK, Platou C, Proenca C, Prokopenko I, Rathmann W, Rayner NW, Robertson NR, Rocheleau G, Roden M, Sampson MJ, Saxena R, Shields BM, Shrader P, Sigurdsson G, Sparso T, Strassburger K, Stringham HM, Sun Q, Swift AJ, Thorand B, Tichet J, Tuomi T, van Dam RM, van Haeften TW, van Herpt T, van Vliet-Ostaptchouk JV, Walters GB, Weedon MN, Wijmenga C, Witteman J, Bergman RN, Cauchi S, Collins FS, Gloyn AL, Gyllensten U, Hansen T, Hide WA, Hitman GA, Hofman A, Hunter DJ, Hveem K, Laakso M, Mohlke KL, Morris AD, Palmer CN, Pramstaller PP, Rudan I, Sijbrands E, Stein LD, Tuomilehto J, Uitterlinden A, Walker M, Wareham NJ, Watanabe RM, Abecasis GR, Boehm BO, Campbell H, Daly MJ, Hattersley AT, Hu FB, Meigs JB, Pankow JS, Pedersen O, Wichmann HE, Barroso I, Florez JC, Frayling TM, Groop L, Sladek R, Thorsteinsdottir U, Wilson JF, Illig T, Froguel P, van Duijn CM, Stefansson K, Altshuler D, Boehnke M, McCarthy MI, Soranzo N, Wheeler E, Glazer NL, Bouatia-Naji N, Magi R, Randall J, Johnson T, Elliott P, Rybin D, Henneman P, Dehghan A, Hottenga JJ, Song K, Goel A, Egan JM, Lajunen T, Doney A, Kanoni S, Cavalcanti-Proenca C, Kumari M, Timpson NJ, Zabena C, Ingelsson E, An P, O'Connell J, Luan J, Elliott A, McCarroll SA, Roccasecca RM, Pattou F, Sethupathy P, Ariyurek Y, Barter P, Beilby JP, Ben-Shlomo Y, Bergmann S, Bochud M, Bonnefond A, Borch-Johnsen K, Bottcher Y, Brunner E, Bumpstead SJ, Chen YD, Chines P, Clarke R, Coin

LJ, Cooper MN, Crisponi L, Day IN, de Geus EJ, Delplanque J, Fedson AC, Fischer-Rosinsky-A, Forouhi NG, Frants R, Franzosi MG, Galan P, Goodarzi MO, Graessler J, Grundy S, Gwilliam R, Hallmans G, Hammond N, Han X, Hartikainen AL, Hayward C, Heath SC, Hercberg S, Hicks AA, Hillman DR, Hingorani AD, Hui J, Hung J, Jula A, Kaakinen M, Kaprio J, Kesaniemi YA, Kivimaki M, Knight B, Koskinen S, Kovacs P, Kyvik KO, Lathrop GM, Lawlor DA, Le Bacquer O, Lecoeur C, Li Y, Mahley R, Mangino M, Manning AK, Martinez-Larrad MT, McAteer JB, McPherson R, Meisinger C, Melzer D, Meyre D, Mitchell BD, Mukherjee S, Naitza S, Neville MJ, Oostra BA, Orru M, Pakyz R, Paolisso G, Pattaro C, Pearson D, Peden JF, Pedersen NL, Perola M, Pfeiffer AF, Pichler I, Polasek O, Posthuma D, Potter SC, Pouta A, Province MA, Psaty BM, Rayner NW, Rice K, Ripatti S, Rivadeneira F, Rolandsson O, Sandbaek A, Sandhu M, Sanna S, Sayer AA, Scheet P, Seedorf U, Sharp SJ, Shields B, Sijbrands EJ, Silveira A, Simpson L, Singleton A, Smith NL, Sovio U, Swift A, Syddall H, Syvanen AC, Tanaka T, Tonjes A, Uitterlinden AG, van Dijk KW, Varma D, Visvikis-Siest S, Vitart V, Vogelzangs N, Waeber G, Wagner PJ, Walley A, Ward KL, Watkins H, Wild SH, Willemsen G, Witteman JC, Yarnell JW, Zelenika D, Zethelius B, Zhai G, Zhao JH, Zillikens MC, Borecki IB, Loos RJ, Meneton P, Magnusson PK, Nathan DM, Williams GH, Silander K, Salomaa V, Smith GD, Bornstein SR, Schwarz P, Spranger J, Karpe F, Shuldiner AR, Cooper C, Dedoussis GV, Serrano-Rios M, Lind L, Palmer LJ, Franks PW, Ebrahim S, Marmot M, Kao WH, Pramstaller PP, Wright AF, Stumvoll M, Hamsten A, Buchanan TA, Valle TT, Rotter JI, Siscovick DS, Penninx BW, Boomsma DI, Deloukas P, Spector TD, Ferrucci L, Cao A, Scuteri A, Schlessinger D, Uda M, Ruokonen A, Jarvelin MR, Waterworth DM, Vollenweider P, Peltonen L, Mooser V, Sladek R (2012) A genome-wide association search for type 2 diabetes genes in African Americans. PloS One 7(1):e29202. doi:10.1371/journal.pone.0029202

Parikh H, Groop L (2004) Candidate genes for type 2 diabetes. Rev Endocr Metab Disord 5 (2):151–176. doi:10.1023/B:REMD.0000021437.46773.26

Parra EJ, Below JE, Krithika S, Valladares A, Barta JL, Cox NJ, Hanis CL, Wacher N, Garcia-Mena J, Hu P, Shriver MD, Diabetes Genetics R, Meta-analysis C, Kumate J, McKeigue PM, Escobedo J, Cruz M (2011) Genome-wide association study of type 2 diabetes in a sample from Mexico City and a meta-analysis of a Mexican-American sample from Starr County, Texas. Diabetologia 54(8):2038–2046. doi:10.1007/s00125-011-2172-y

Permutt MA, Wasson J, Cox N (2005) Genetic epidemiology of diabetes. J Clin Invest 115 (6):1431–1439. doi:10.1172/JCI24758

Perry JR, Voight BF, Yengo L, Amin N, Dupuis J, Ganser M, Grallert H, Navarro P, Li M, Qi L, Steinthorsdottir V, Scott RA, Almgren P, Arking DE, Aulchenko Y, Balkau B, Benediktsson R, Bergman RN, Boerwinkle E, Bonnycastle L, Burtt NP, Campbell H, Charpentier G, Collins FS, Gieger C, Green T, Hadjadj S, Hattersley AT, Herder C, Hofman A, Johnson AD, Kottgen A, Kraft P, Labrune Y, Langenberg C, Manning AK, Mohlke KL, Morris AP, Oostra B, Pankow J, Petersen AK, Pramstaller PP, Prokopenko I, Rathmann W, Rayner W, Roden M, Rudan I, Rybin D, Scott LJ, Sigurdsson G, Sladek R, Thorleifsson G, Thorsteinsdottir U, Tuomilehto J, Uitterlinden AG, Vivequin S, Weedon MN, Wright AF, Magic CD, Consortium G, Hu FB, Illig T, Kao L, Meigs JB, Wilson JF, Stefansson K, van Duijn C, Altschuler D, Morris AD, Boehnke M, McCarthy MI, Froguel P, Palmer CN, Wareham NJ, Groop L, Frayling TM, Cauchi S (2012) Stratifying type 2 diabetes cases by BMI identifies genetic risk variants in LAMA1 and enrichment for risk variants in lean compared to obese cases. PLoS Genet 8(5): e1002741. doi:10.1371/journal.pgen.1002741

Peschke E, Stumpf I, Bazwinsky I, Litvak L, Dralle H, Muhlbauer E (2007) Melatonin and type 2 diabetes - a possible link? J Pineal Res 42(4):350–358. doi:10.1111/j.1600-079X.2007.00426.x

Prokopenko I, McCarthy MI, Lindgren CM (2008) Type 2 diabetes: new genes, new understanding. Trends Genet 24(12):613–621. doi:10.1016/j.tig.2008.09.004

Prokopenko I, Langenberg C, Florez JC, Saxena R, Soranzo N, Thorleifsson G, Loos RJ, Manning AK, Jackson AU, Aulchenko Y, Potter SC, Erdos MR, Sanna S, Hottenga JJ, Wheeler E,

Kaakinen M, Lyssenko V, Chen WM, Ahmadi K, Beckmann JS, Bergman RN, Bochud M, Bonnycastle LL, Buchanan TA, Cao A, Cervino A, Coin L, Collins FS, Crisponi L, de Geus EJ, Dehghan A, Deloukas P, Doney AS, Elliott P, Freimer N, Gateva V, Herder C, Hofman A, Hughes TE, Hunt S, Illig T, Inouye M, Isomaa B, Johnson T, Kong A, Krestyaninova M, Kuusisto J, Laakso M, Lim N, Lindblad U, Lindgren CM, McCann OT, Mohlke KL, Morris AD, Naitza S, Orru M, Palmer CN, Pouta A, Randall J, Rathmann W, Saramies J, Scheet P, Scott LJ, Scuteri A, Sharp S, Sijbrands E, Smit JH, Song K, Steinthorsdottir V, Stringham HM, Tuomi T, Tuomilehto J, Uitterlinden AG, Voight BF, Waterworth D, Wichmann HE, Willemsen G, Witteman JC, Yuan X, Zhao JH, Zeggini E, Schlessinger D, Sandhu M, Boomsma DI, Uda M, Spector TD, Penninx BW, Altshuler D, Vollenweider P, Jarvelin MR, Lakatta E, Waeber G, Fox CS, Peltonen L, Groop LC, Mooser V, Cupples LA, Thorsteinsdottir U, Boehnke M, Barroso I, Van Duijn C, Dupuis J, Watanabe RM, Stefansson K, McCarthy MI, Wareham NJ, Meigs JB, Abecasis GR (2009a) Variants in MTNR1B influence fasting glucose levels. Nat Genet 41(1):77–81. doi:10.1038/ng.290

Prokopenko I, Zeggini E, Hanson RL, Mitchell BD, Rayner NW, Akan P, Baier L, Das SK, Elliott KS, Fu M, Frayling TM, Groves CJ, Gwilliam R, Scott LJ, Voight BF, Hattersley AT, Hu C, Morris AD, Ng M, Palmer CN, Tello-Ruiz M, Vaxillaire M, Wang CR, Stein L, Chan J, Jia W, Froguel P, Elbein SC, Deloukas P, Bogardus C, Shuldiner AR, McCarthy MI, International Type 2 Diabetes 1q C (2009b) Linkage disequilibrium mapping of the replicated type 2 diabetes linkage signal on chromosome 1q. Diabetes 58(7):1704–1709. doi:10.2337/db09-0081

Prokopenko I, Poon W, Magi R, Prasad BR, Salehi SA, Almgren P, Osmark P, Bouatia-Naji N, Wierup N, Fall T, Stancakova A, Barker A, Lagou V, Osmond C, Xie W, Lahti J, Jackson AU, Cheng YC, Liu J, O'Connell JR, Blomstedt PA, Fadista J, Alkayyali S, Dayeh T, Ahlqvist E, Taneera J, Lecoeur C, Kumar A, Hansson O, Hansson K, Voight BF, Kang HM, Levy-Marchal C, Vatin V, Palotie A, Syvanen AC, Mari A, Weedon MN, Loos RJ, Ong KK, Nilsson P, Isomaa B, Tuomi T, Wareham NJ, Stumvoll M, Widen E, Lakka TA, Langenberg C, Tonjes A, Rauramaa R, Kuusisto J, Frayling TM, Froguel P, Walker M, Eriksson JG, Ling C, Kovacs P, Ingelsson E, McCarthy MI, Shuldiner AR, Silver KD, Laakso M, Groop L, Lyssenko V (2014) A central role for GRB10 in regulation of islet function in man. PLoS Genet 10(4):e1004235. doi:10.1371/journal.pgen.1004235

Qi L, Cornelis MC, Kraft P, Stanya KJ, Linda Kao WH, Pankow JS, Dupuis J, Florez JC, Fox CS, Pare G, Sun Q, Girman CJ, Laurie CC, Mirel DB, Manolio TA, Chasman DI, Boerwinkle E, Ridker PM, Hunter DJ, Meigs JB, Lee CH, Hu FB, van Dam RM, Meta-Analysis of G, Insulin-Related Traits C, Diabetes Genetics R, Meta-Analysis C (2010) Genetic variants at 2q24 are associated with susceptibility to type 2 diabetes. Hum Mol Genet 19(13):2706–2715. doi:10.1093/hmg/ddq156

Reich DE, Lander ES (2001) On the allelic spectrum of human disease. Trends Genetics 17 (9):502–510

Replication DIG, Meta-analysis C (2014) Genome-wide trans-ancestry meta-analysis provides insight into the genetic architecture of type 2 diabetes susceptibility. Nat Genet 46(3):234–244. doi:10.1038/ng.2897

Replication DIG, Meta-analysis C, Asian Genetic Epidemiology Network Type 2 Diabetes C, South Asian Type 2 Diabetes C, Mexican American Type 2 Diabetes C, Type 2 Diabetes Genetic Exploration by Next-generation sequencing in multi-Ethnic Samples C, Mahajan A, Go MJ, Zhang W, Below JE, Gaulton KJ, Ferreira T, Horikoshi M, Johnson AD, Ng MCY, Prokopenko I, Saleheen D, Wang X, Zeggini E, Abecasis GR, Adair LS, Almgren P, Atalay M, Aung T, Baldassarre D, Balkau B, Bao Y, Barnett AH, Barroso I, Basit A, Been LF, Beilby J, Bell GI, Benediktsson R, Bergman RN, Boehm BO, Boerwinkle E, Bonnycastle LL, Burtt N, Cai Q, Campbell H, Carey J, Cauchi S, Caulfield M, Chan JCN, Chang L-C, Chang T-J, Chang Y-C, Charpentier G, Chen C-H, Chen H, Chen Y-T, Chia K-S, Chidambaram M, Chines PS, Cho NH, Cho YM, Chuang L-M, Collins FS, Cornelis MC, Couper DJ, Crenshaw AT, van Dam RM, Danesh J, Das D, de Faire U, Dedoussis G, Deloukas P, Dimas AS, Dina C, Doney ASF, Donnelly PJ, Dorkhan M, van Duijn C, Dupuis J, Edkins S, Elliott P, Emilsson V,

Erbel R, Eriksson JG, Escobedo J, Esko T, Eury E, Florez JC, Fontanillas P, Forouhi NG, Forsen T, Fox C, Fraser RM, Frayling TM, Froguel P, Frossard P, Gao Y, Gertow K, Gieger C, Gigante B, Grallert H, Grant GB, Groop LC, Groves CJ, Grundberg E, Guiducci C, Hamsten A, Han B-G, Hara K, Hassanali N, Hattersley AT, Hayward C, Hedman AK, Herder C, Hofman A, Holmen OL, Hovingh K, Hreidarsson AB, Hu C, Hu FB, Hui J, Humphries SE, Hunt SE, Hunter DJ, Hveem K, Hydrie ZI, Ikegami H, Illig T, Ingelsson E, Islam M, Isomaa B, Jackson AU, Jafar T, James A, Jia W, Jockel K-H, Jonsson A, Jowett JBM, Kadowaki T, Kang HM, Kanoni S, Kao WHL, Kathiresan S, Kato N, Katulanda P, Keinanen-Kiukaanniemi SM, Kelly AM, Khan H, Khaw K-T, Khor C-C, Kim H-L, Kim S, Kim YJ, Kinnunen L, Klopp N, Kong A, Korpi-Hyovalti E, Kowlessur S, Kraft P, Kravic J, Kristensen MM, Krithika S, Kumar A, Kumate J, Kuusisto J, Kwak SH, Laakso M, Lagou V, Lakka TA, Langenberg C, Langford C, Lawrence R, Leander K, Lee J-M, Lee NR, Li M, Li X, Li Y, Liang J, Liju S, Lim W-Y, Lind L, Lindgren CM, Lindholm E, Liu C-T, Liu JJ, Lobbens S, Long J, Loos RJF, Lu W, Luan J, Lyssenko V, Ma RCW, Maeda S, Magi R, Mannisto S, Matthews DR, Meigs JB, Melander O, Metspalu A, Meyer J, Mirza G, Mihailov E, Moebus S, Mohan V, Mohlke KL, Morris AD, Muhleisen TW, Muller-Nurasyid M, Musk B, Nakamura J, Nakashima E, Navarro P, Ng P-K, Nica AC, Nilsson PM, Njolstad I, Nothen MM, Ohnaka K, Ong TH, Owen KR, Palmer CNA, Pankow JS, Park KS, Parkin M, Pechlivanis S, Pedersen NL, Peltonen L, Perry JRB, Peters A, Pinidiyapathirage JM, Platou CGP, Potter S, Price JF, Qi L, Radha V, Rallidis L, Rasheed A, Rathmann W, Rauramaa R, Raychaudhuri S, Rayner NW, Rees SD, Rehnberg E, Ripatti S, Robertson N, Roden M, Rossin EJ, Rudan I, Rybin D, Saaristo TE, Salomaa V, Saltevo J, Samuel M, Sanghera DK, Saramies J, Scott J, Scott LJ, Scott RA, Segre AV, Sehmi J, Sennblad B, Shah N, Shah S, Shera AS, Shu XO, Shuldiner AR, Sigursson G, Sijbrands E, Silveira A, Sim X, Sivapalaratnam S, Small KS, So WY, Stancakova A, Stefansson K, Steinbach G, Steinthorsdottir V, Stirrups K, Strawbridge RJ, Stringham HM, Sun Q, Suo C, Syvanen A-C, Takayanagi R, Takeuchi F, Tay WT, Teslovich TM, Thorand B, Thorleifsson G, Thorsteinsdottir U, Tikkanen E, Trakalo J, Tremoli E, Trip MD, Tsai FJ, Tuomi T, Tuomilehto J, Uitterlinden AG, Valladares-Salgado A, Vedantam S, Veglia F, Voight BF, Wang C, Wareham NJ, Wennauer R, Wickremasinghe AR, Wilsgaard T, Wilson JF, Wiltshire S, Winckler W, Wong TY, Wood AR, Wu J-Y, Wu Y, Yamamoto K, Yamauchi T, Yang M, Yengo L, Yokota M, Young R, Zabaneh D, Zhang F, Zhang R, Zheng W, Zimmet PZ, Altshuler D, Bowden DW, Cho YS, Cox NJ, Cruz M, Hanis CL, Kooner J, Lee J-Y, Seielstad M, Teo YY, Boehnke M, Parra EJ, Chambers JC, Tai ES, McCarthy MI, Morris AP (2014) Genome-wide trans-ancestry meta-analysis provides insight into the genetic architecture of type 2 diabetes susceptibility. Nat Genet 46(3):234–244. doi:10.1038/ng.2897, http://www.nature.com/ng/journal/v46/n3/abs/ng.2897.html#supplementary-information

Reynisdottir I, Thorleifsson G, Benediktsson R, Sigurdsson G, Emilsson V, Einarsdottir AS, Hjorleifsdottir EE, Orlygsdottir GT, Bjornsdottir GT, Saemundsdottir J, Halldorsson S, Hrafnkelsdottir S, Sigurjonsdottir SB, Steinsdottir S, Martin M, Kochan JP, Rhees BK, Grant SF, Frigge ML, Kong A, Gudnason V, Stefansson K, Gulcher JR (2003) Localization of a susceptibility gene for type 2 diabetes to chromosome 5q34-q35.2. Am J Hum Genet 73(2):323–335. doi:10.1086/377139

Risch N, Merikangas K (1996) The future of genetic studies of complex human diseases. Science 273(5281):1516–1517

Rung J, Cauchi S, Albrechtsen A, Shen L, Rocheleau G, Cavalcanti-Proenca C, Bacot F, Balkau B, Belisle A, Borch-Johnsen K, Charpentier G, Dina C, Durand E, Elliott P, Hadjadj S, Jarvelin MR, Laitinen J, Lauritzen T, Marre M, Mazur A, Meyre D, Montpetit A, Pisinger C, Posner B, Poulsen P, Pouta A, Prentki M, Ribel-Madsen R, Ruokonen A, Sandbaek A, Serre D, Tichet J, Vaxillaire M, Wojtaszewski JF, Vaag A, Hansen T, Polychronakos C, Pedersen O, Froguel P, Sladek R (2009) Genetic variant near IRS1 is associated with type 2 diabetes, insulin resistance and hyperinsulinemia. Nat Genet 41(10):1110–1115. doi:10.1038/ng.443

Sandhu MS, Weedon MN, Fawcett KA, Wasson J, Debenham SL, Daly A, Lango H, Frayling TM, Neumann RJ, Sherva R, Blech I, Pharoah PD, Palmer CN, Kimber C, Tavendale R, Morris AD, McCarthy MI, Walker M, Hitman G, Glaser B, Permutt MA, Hattersley AT, Wareham NJ, Barroso I (2007) Common variants in WFS1 confer risk of type 2 diabetes. Nat Genet 39 (8):951–953. doi:10.1038/ng2067, ng2067 [pii]

Saxena R, Hivert MF, Langenberg C, Tanaka T, Pankow JS, Vollenweider P, Lyssenko V, Bouatia-Naji N, Dupuis J, Jackson AU, Kao WH, Li M, Glazer NL, Manning AK, Luan J, Stringham HM, Prokopenko I, Johnson T, Grarup N, Boesgaard TW, Lecoeur C, Shrader P, O'Connell J, Ingelsson E, Couper DJ, Rice K, Song K, Andreasen CH, Dina C, Kottgen A, Le Bacquer O, Pattou F, Taneera J, Steinthorsdottir V, Rybin D, Ardlie K, Sampson M, Qi L, van Hoek M, Weedon MN, Aulchenko YS, Voight BF, Grallert H, Balkau B, Bergman RN, Bielinski SJ, Bonnefond A, Bonnycastle LL, Borch-Johnsen K, Bottcher Y, Brunner E, Buchanan TA, Bumpstead SJ, Cavalcanti-Proenca C, Charpentier G, Chen YD, Chines PS, Collins FS, Cornelis M, JC G, Delplanque J, Doney A, Egan JM, Erdos MR, Firmann M, Forouhi NG, Fox CS, Goodarzi MO, Graessler J, Hingorani A, Isomaa B, Jorgensen T, Kivimaki M, Kovacs P, Krohn K, Kumari M, Lauritzen T, Levy-Marchal C, Mayor V, McAteer JB, Meyre D, Mitchell BD, Mohlke KL, Morken MA, Narisu N, Palmer CN, Pakyz R, Pascoe L, Payne F, Pearson D, Rathmann W, Sandbaek A, Sayer AA, Scott LJ, Sharp SJ, Sijbrands E, Singleton A, Siscovick DS, Smith NL, Sparso T, Swift AJ, Syddall H, Thorleifsson G, Tonjes A, Tuomi T, Tuomilehto J, Valle TT, Waeber G, Walley A, Waterworth DM, Zeggini E, Zhao JH, Consortium G, Investigators M, Illig T, Wichmann HE, Wilson JF, van Duijn C, Hu FB, Morris AD, Frayling TM, Hattersley AT, Thorsteinsdottir U, Stefansson K, Nilsson P, Syvanen AC, Shuldiner AR, Walker M, Bornstein SR, Schwarz P, Williams GH, Nathan DM, Kuusisto J, Laakso M, Cooper C, Marmot M, Ferrucci L, Mooser V, Stumvoll M, Loos RJ, Altshuler D, Psaty BM, Rotter JI, Boerwinkle E, Hansen T, Pedersen O, Florez JC, McCarthy MI, Boehnke M, Barroso I, Sladek R, Froguel P, Meigs JB, Groop L, Wareham NJ, Watanabe RM (2010) Genetic variation in GIPR influences the glucose and insulin responses to an oral glucose challenge. Nat Genet 42(2):142–148. doi:10.1038/ng.521

Saxena R, Elbers CC, Guo Y, Peter I, Gaunt TR, Mega JL, Lanktree MB, Tare A, Castillo BA, Li YR, Johnson T, Bruinenberg M, Gilbert-Diamond D, Rajagopalan R, Voight BF, Balasubramanyam A, Barnard J, Bauer F, Baumert J, Bhangale T, Bohm BO, Braund PS, Burton PR, Chandrupatla HR, Clarke R, Cooper-DeHoff RM, Crook ED, Davey-Smith G, Day IN, de Boer A, de Groot MC, Drenos F, Ferguson J, Fox CS, Furlong CE, Gibson Q, Gieger C, Gilhuijs-Pederson LA, Glessner JT, Goel A, Gong Y, Grant SF, Grobbee DE, Hastie C, Humphries SE, Kim CE, Kivimaki M, Kleber M, Meisinger C, Kumari M, Langaee TY, Lawlor DA, Li M, Lobmeyer MT, Maitland-van der Zee AH, Meijs MF, Molony CM, Morrow DA, Murugesan G, Musani SK, Nelson CP, Newhouse SJ, O'Connell JR, Padmanabhan S, Palmen J, Patel SR, Pepine CJ, Pettinger M, Price TS, Rafelt S, Ranchalis J, Rasheed A, Rosenthal E, Ruczinski I, Shah S, Shen H, Silbernagel G, Smith EN, Spijkerman AW, Stanton A, Steffes MW, Thorand B, Trip M, van der Harst P, van Der AD, van Iperen EP, van Setten J, van Vliet-Ostaptchouk JV, Verweij N, Wolffenbuttel BH, Young T, Zafarmand MH, Zmuda JM, Look ARG, Consortium D, Boehnke M, Altshuler D, McCarthy M, Kao WH, Pankow JS, Cappola TP, Sever P, Poulter N, Caulfield M, Dominiczak A, Shields DC, Bhatt DL, Zhang L, Curtis SP, Danesh J, Casas JP, van der Schouw YT, Onland-Moret NC, Doevendans PA, Dorn GW 2nd, Farrall M, FitzGerald GA, Hamsten A, Hegele R, Hingorani AD, Hofker MH, Huggins GS, Illig T, Jarvik GP, Johnson JA, Klungel OH, Knowler WC, Koenig W, Marz W, Meigs JB, Melander O, Munroe PB, Mitchell BD, Bielinski SJ, Rader DJ, Reilly MP, Rich SS, Rotter JI, Saleheen D, Samani NJ, Schadt EE, Shuldiner AR, Silverstein R, Kottke-Marchant K, Talmud PJ, Watkins H, Asselbergs FW, de Bakker PI, McCaffery J, Wijmenga C, Sabatine MS, Wilson JG, Reiner A, Bowden DW, Hakonarson H, Siscovick DS, Keating BJ (2012) Large-scale gene-centric meta-analysis across 39 studies identifies type 2 diabetes loci. Am J Hum Genet 90(3):410–425. doi:10.1016/j.ajhg.2011.12.022

Saxena R, Saleheen D, Been LF, Garavito ML, Braun T, Bjonnes A, Young R, Ho WK, Rasheed A, Frossard P, Sim X, Hassanali N, Radha V, Chidambaram M, Liju S, Rees SD, Ng DP, Wong TY, Yamauchi T, Hara K, Tanaka Y, Hirose H, McCarthy MI, Morris AP, Diagram MT, Agen BA, Barnett AH, Katulanda P, Matthews D, Mohan V, Wander GS, Singh JR, Mehra NK, Ralhan S, Kamboh MI, Mulvihill JJ, Maegawa H, Tobe K, Maeda S, Cho YS, Tai ES, Kelly MA, Chambers JC, Kooner JS, Kadowaki T, Deloukas P, Rader DJ, Danesh J, Sanghera DK (2013) Genome-wide association study identifies a novel locus contributing to type 2 diabetes susceptibility in Sikhs of Punjabi origin from India. Diabetes 62(5):1746–1755. doi:10.2337/db12-1077

Scott LJ, Mohlke KL, Bonnycastle LL, Willer CJ, Li Y, Duren WL, Erdos MR, Stringham HM, Chines PS, Jackson AU, Prokunina-Olsson L, Ding CJ, Swift AJ, Narisu N, Hu T, Pruim R, Xiao R, Li XY, Conneely KN, Riebow NL, Sprau AG, Tong M, White PP, Hetrick KN, Barnhart MW, Bark CW, Goldstein JL, Watkins L, Xiang F, Saramies J, Buchanan TA, Watanabe RM, Valle TT, Kinnunen L, Abecasis GR, Pugh EW, Doheny KF, Bergman RN, Tuomilehto J, Collins FS, Boehnke M (2007a) A genome-wide association study of type 2 diabetes in Finns detects multiple susceptibility variants. Science 316(5829):1341–1345

Scott LJ, Mohlke KL, Bonnycastle LL, Willer CJ, Li Y, Duren WL, Erdos MR, Stringham HM, Chines PS, Jackson AU, Prokunina-Olsson L, Ding CJ, Swift AJ, Narisu N, Hu T, Pruim R, Xiao R, Li XY, Conneely KN, Riebow NL, Sprau AG, Tong M, White PP, Hetrick KN, Barnhart MW, Bark CW, Goldstein JL, Watkins L, Xiang F, Saramies J, Buchanan TA, Watanabe RM, Valle TT, Kinnunen L, Abecasis GR, Pugh EW, Doheny KF, Bergman RN, Tuomilehto J, Collins FS, Boehnke M (2007b) A genome-wide association study of type 2 diabetes in Finns detects multiple susceptibility variants. Science 316(5829):1341–1345. doi:10.1126/science.1142382

Scott RA, Lagou V, Welch RP, Wheeler E, Montasser ME, Luan J, Magi R, Strawbridge RJ, Rehnberg E, Gustafsson S, Kanoni S, Rasmussen-Torvik LJ, Yengo L, Lecoeur C, Shungin D, Sanna S, Sidore C, Johnson PC, Jukema JW, Johnson T, Mahajan A, Verweij N, Thorleifsson G, Hottenga JJ, Shah S, Smith AV, Sennblad B, Gieger C, Salo P, Perola M, Timpson NJ, Evans DM, Pourcain BS, Wu Y, Andrews JS, Hui J, Bielak LF, Zhao W, Horikoshi M, Navarro P, Isaacs A, O'Connell JR, Stirrups K, Vitart V, Hayward C, Esko T, Mihailov E, Fraser RM, Fall T, Voight BF, Raychaudhuri S, Chen H, Lindgren CM, Morris AP, Rayner NW, Robertson N, Rybin D, Liu CT, Beckmann JS, Willems SM, Chines PS, Jackson AU, Kang HM, Stringham HM, Song K, Tanaka T, Peden JF, Goel A, Hicks AA, An P, Muller-Nurasyid M, Franco-Cereceda A, Folkersen L, Marullo L, Jansen H, Oldehinkel AJ, Bruinenberg M, Pankow JS, North KE, Forouhi NG, Loos RJ, Edkins S, Varga TV, Hallmans G, Oksa H, Antonella M, Nagaraja R, Trompet S, Ford I, Bakker SJ, Kong A, Kumari M, Gigante B, Herder C, Munroe PB, Caulfield M, Antti J, Mangino M, Small K, Miljkovic I, Liu Y, Atalay M, Kiess W, James AL, Rivadeneira F, Uitterlinden AG, Palmer CN, Doney AS, Willemsen G, Smit JH, Campbell S, Polasek O, Bonnycastle LL, Hercberg S, Dimitriou M, Bolton JL, Fowkes GR, Kovacs P, Lindstrom J, Zemunik T, Bandinelli S, Wild SH, Basart HV, Rathmann W, Grallert H, Replication DIG, Meta-analysis C, Maerz W, Kleber ME, Boehm BO, Peters A, Pramstaller PP, Province MA, Borecki IB, Hastie ND, Rudan I, Campbell H, Watkins H, Farrall M, Stumvoll M, Ferrucci L, Waterworth DM, Bergman RN, Collins FS, Tuomilehto J, Watanabe RM, de Geus EJ, Penninx BW, Hofman A, Oostra BA, Psaty BM, Vollenweider P, Wilson JF, Wright AF, Hovingh GK, Metspalu A, Uusitupa M, Magnusson PK, Kyvik KO, Kaprio J, Price JF, Dedoussis GV, Deloukas P, Meneton P, Lind L, Boehnke M, Shuldiner AR, van Duijn CM, Morris AD, Toenjes A, Peyser PA, Beilby JP, Korner A, Kuusisto J, Laakso M, Bornstein SR, Schwarz PE, Lakka TA, Rauramaa R, Adair LS, Smith GD, Spector TD, Illig T, de Faire U, Hamsten A, Gudnason V, Kivimaki M, Hingorani A, Keinanen-Kiukaanniemi SM, Saaristo TE, Boomsma DI, Stefansson K, van der Harst P, Dupuis J, Pedersen NL, Sattar N, Harris TB, Cucca F, Ripatti S, Salomaa V,

Mohlke KL, Balkau B, Froguel P, Pouta A, Jarvelin MR, Wareham NJ, Bouatia-Naji N, McCarthy MI, Franks PW, Meigs JB, Teslovich TM, Florez JC, Langenberg C, Ingelsson E, Prokopenko I, Barroso I (2012) Large-scale association analyses identify new loci influencing glycemic traits and provide insight into the underlying biological pathways. Nat Genet 44 (9):991–1005. doi:10.1038/ng.2385

Shu XO, Long J, Cai Q, Qi L, Xiang YB, Cho YS, Tai ES, Li X, Lin X, Chow WH, Go MJ, Seielstad M, Bao W, Li H, Cornelis MC, Yu K, Wen W, Shi J, Han BG, Sim XL, Liu L, Qi Q, Kim HL, Ng DP, Lee JY, Kim YJ, Li C, Gao YT, Zheng W, Hu FB (2010) Identification of new genetic risk variants for type 2 diabetes. PLoS Genet 6(9):e1001127. doi:10.1371/journal.pgen.1001127

Silander K, Scott LJ, Valle TT, Mohlke KL, Stringham HM, Wiles KR, Duren WL, Doheny KF, Pugh EW, Chines P, Narisu N, White PP, Fingerlin TE, Jackson AU, Li C, Ghosh S, Magnuson VL, Colby K, Erdos MR, Hill JE, Hollstein P, Humphreys KM, Kasad RA, Lambert J, Lazaridis KN, Lin G, Morales-Mena A, Patzkowski K, Pfahl C, Porter R, Rha D, Segal L, Suh YD, Tovar J, Unni A, Welch C, Douglas JA, Epstein MP, Hauser ER, Hagopian W, Buchanan TA, Watanabe RM, Bergman RN, Tuomilehto J, Collins FS, Boehnke M (2004) A large set of Finnish affected sibling pair families with type 2 diabetes suggests susceptibility loci on chromosomes 6, 11, and 14. Diabetes 53(3):821–829

Sladek R, Rocheleau G, Rung J, Dina C, Shen L, Serre D, Boutin P, Vincent D, Belisle A, Hadjadj S, Balkau B, Heude B, Charpentier G, Hudson TJ, Montpetit A, Pshezhetsky AV, Prentki M, Posner BI, Balding DJ, Meyre D, Polychronakos C, Froguel P (2007) A genome-wide association study identifies novel risk loci for type 2 diabetes. Nature 445 (7130):881–885. doi:10.1038/nature05616, nature05616 [pii]

Soranzo N, Sanna S, Wheeler E, Gieger C, Radke D, Dupuis J, Bouatia-Naji N, Langenberg C, Prokopenko I, Stolerman E, Sandhu MS, Heeney MM, Devaney JM, Reilly MP, Ricketts SL, Stewart AF, Voight BF, Willenborg C, Wright B, Altshuler D, Arking D, Balkau B, Barnes D, Boerwinkle E, Bohm B, Bonnefond A, Bonnycastle LL, Boomsma DI, Bornstein SR, Bottcher Y, Bumpstead S, Burnett-Miller MS, Campbell H, Cao A, Chambers J, Clark R, Collins FS, Coresh J, de Geus EJ, Dei M, Deloukas P, Doring A, Egan JM, Elosua R, Ferrucci L, Forouhi N, Fox CS, Franklin C, Franzosi MG, Gallina S, Goel A, Graessler J, Grallert H, Greinacher A, Hadley D, Hall A, Hamsten A, Hayward C, Heath S, Herder C, Homuth G, Hottenga JJ, Hunter-Merrill R, Illig T, Jackson AU, Jula A, Kleber M, Knouff CW, Kong A, Kooner J, Kottgen A, Kovacs P, Krohn K, Kuhnel B, Kuusisto J, Laakso M, Lathrop M, Lecoeur C, Li M, Li M, Loos RJ, Luan J, Lyssenko V, Magi R, Magnusson PK, Malarstig A, Mangino M, Martinez-Larrad MT, Marz W, McArdle WL, McPherson R, Meisinger C, Meitinger T, Melander O, Mohlke KL, Mooser VE, Morken MA, Narisu N, Nathan DM, Nauck M, O'Donnell C, Oexle K, Olla N, Pankow JS, Payne F, Peden JF, Pedersen NL, Peltonen L, Perola M, Polasek O, Porcu E, Rader DJ, Rathmann W, Ripatti S, Rocheleau G, Roden M, Rudan I, Salomaa V, Saxena R, Schlessinger D, Schunkert H, Schwarz P, Seedorf U, Selvin E, Serrano-Rios M, Shrader P, Silveira A, Siscovick D, Song K, Spector TD, Stefansson K, Steinthorsdottir V, Strachan DP, Strawbridge R, Stumvoll M, Surakka I, Swift AJ, Tanaka T, Teumer A, Thorleifsson G, Thorsteinsdottir U, Tonjes A, Usala G, Vitart V, Volzke H, Wallaschofski H, Waterworth DM, Watkins H, Wichmann HE, Wild SH, Willemsen G, Williams GH, Wilson JF, Winkelmann J, Wright AF, Wtccc, Zabena C, Zhao JH, Epstein SE, Erdmann J, Hakonarson HH, Kathiresan S, Khaw KT, Roberts R, Samani NJ, Fleming MD, Sladek R, Abecasis G, Boehnke M, Froguel P, Groop L, McCarthy MI, Kao WH, Florez JC, Uda M, Wareham NJ, Barroso I, Meigs JB (2010) Common variants at 10 genomic loci influence hemoglobin A(1)(C) levels via glycemic and nonglycemic pathways. Diabetes 59(12):3229–3239. doi:10.2337/db10-0502

Southam L, Soranzo N, Montgomery SB, Frayling TM, McCarthy MI, Barroso I, Zeggini E (2009) Is the thrifty genotype hypothesis supported by evidence based on confirmed type 2 diabetes- and obesity-susceptibility variants? Diabetologia 52(9):1846–1851. doi:10.1007/s00125-009-1419-3

Speliotes EK, Willer CJ, Berndt SI, Monda KL, Thorleifsson G, Jackson AU, Lango Allen H, Lindgren CM, Luan J, Magi R, Randall JC, Vedantam S, Winkler TW, Qi L, Workalemahu T, Heid IM, Steinthorsdottir V, Stringham HM, Weedon MN, Wheeler E, Wood AR, Ferreira T, Weyant RJ, Segre AV, Estrada K, Liang L, Nemesh J, Park JH, Gustafsson S, Kilpelainen TO, Yang J, Bouatia-Naji N, Esko T, Feitosa MF, Kutalik Z, Mangino M, Raychaudhuri S, Scherag A, Smith AV, Welch R, Zhao JH, Aben KK, Absher DM, Amin N, Dixon AL, Fisher E, Glazer NL, Goddard ME, Heard-Costa NL, Hoesel V, Hottenga JJ, Johansson A, Johnson T, Ketkar S, Lamina C, Li S, Moffatt MF, Myers RH, Narisu N, Perry JR, Peters MJ, Preuss M, Ripatti S, Rivadeneira F, Sandholt C, Scott LJ, Timpson NJ, Tyrer JP, van Wingerden S, Watanabe RM, White CC, Wiklund F, Barlassina C, Chasman DI, Cooper MN, Jansson JO, Lawrence RW, Pellikka N, Prokopenko I, Shi J, Thiering E, Alavere H, Alibrandi MT, Almgren P, Arnold AM, Aspelund T, Atwood LD, Balkau B, Balmforth AJ, Bennett AJ, Ben-Shlomo Y, Bergman RN, Bergmann S, Biebermann H, Blakemore AI, Boes T, Bonnycastle LL, Bornstein SR, Brown MJ, Buchanan TA, Busonero F, Campbell H, Cappuccio FP, Cavalcanti-Proenca C, Chen YD, Chen CM, Chines PS, Clarke R, Coin L, Connell J, Day IN, den Heijer M, Duan J, Ebrahim S, Elliott P, Elosua R, Eiriksdottir G, Erdos MR, Eriksson JG, Facheris MF, Felix SB, Fischer-Posovszky P, Folsom AR, Friedrich N, Freimer NB, Fu M, Gaget S, Gejman PV, Geus EJ, Gieger C, Gjesing AP, Goel A, Goyette P, Grallert H, Grassler J, Greenawalt DM, Groves CJ, Gudnason V, Guiducci C, Hartikainen AL, Hassanali N, Hall AS, Havulinna AS, Hayward C, Heath AC, Hengstenberg C, Hicks AA, Hinney A, Hofman A, Homuth G, Hui J, Igl W, Iribarren C, Isomaa B, Jacobs KB, Jarick I, Jewell E, John U, Jorgensen T, Jousilahti P, Jula A, Kaakinen M, Kajantie E, Kaplan LM, Kathiresan S, Kettunen J, Kinnunen L, Knowles JW, Kolcic I, Konig IR, Koskinen S, Kovacs P, Kuusisto J, Kraft P, Kvaloy K, Laitinen J, Lantieri O, Lanzani C, Launer LJ, Lecoeur C, Lehtimaki T, Lettre G, Liu J, Lokki ML, Lorentzon M, Luben RN, Ludwig B, Manunta P, Marek D, Marre M, Martin NG, McArdle WL, McCarthy A, McKnight B, Meitinger T, Melander O, Meyre D, Midthjell K, Montgomery GW, Morken MA, Morris AP, Mulic R, Ngwa JS, Nelis M, Neville MJ, Nyholt DR, O'Donnell CJ, O'Rahilly S, Ong KK, Oostra B, Pare G, Parker AN, Perola M, Pichler I, Pietilainen KH, Platou CG, Polasek O, Pouta A, Rafelt S, Raitakari O, Rayner NW, Ridderstrale M, Rief W, Ruokonen A, Robertson NR, Rzehak P, Salomaa V, Sanders AR, Sandhu MS, Sanna S, Saramies J, Savolainen MJ, Scherag S, Schipf S, Schreiber S, Schunkert H, Silander K, Sinisalo J, Siscovick DS, Smit JH, Soranzo N, Sovio U, Stephens J, Surakka I, Swift AJ, Tammesoo ML, Tardif JC, Teder-Laving M, Teslovich TM, Thompson JR, Thomson B, Tonjes A, Tuomi T, van Meurs JB, van Ommen GJ, Vatin V, Viikari J, Visvikis-Siest S, Vitart V, Vogel CI, Voight BF, Waite LL, Wallaschofski H, Walters GB, Widen E, Wiegand S, Wild SH, Willemsen G, Witte DR, Witteman JC, Xu J, Zhang Q, Zgaga L, Ziegler A, Zitting P, Beilby JP, Farooqi IS, Hebebrand J, Huikuri HV, James AL, Kahonen M, Levinson DF, Macciardi F, Nieminen MS, Ohlsson C, Palmer LJ, Ridker PM, Stumvoll M, Beckmann JS, Boeing H, Boerwinkle E, Boomsma DI, Caulfield MJ, Chanock SJ, Collins FS, Cupples LA, Smith GD, Erdmann J, Froguel P, Gronberg H, Gyllensten U, Hall P, Hansen T, Harris TB, Hattersley AT, Hayes RB, Heinrich J, Hu FB, Hveem K, Illig T, Jarvelin MR, Kaprio J, Karpe F, Khaw KT, Kiemeney LA, Krude H, Laakso M, Lawlor DA, Metspalu A, Munroe PB, Ouwehand WH, Pedersen O, Penninx BW, Peters A, Pramstaller PP, Quertermous T, Reinehr T, Rissanen A, Rudan I, Samani NJ, Schwarz PE, Shuldiner AR, Spector TD, Tuomilehto J, Uda M, Uitterlinden A, Valle TT, Wabitsch M, Waeber G, Wareham NJ, Watkins H, Wilson JF, Wright AF, Zillikens MC, Chatterjee N, McCarroll SA, Purcell S, Schadt EE, Visscher PM, Assimes TL, Borecki IB, Deloukas P, Fox CS, Groop LC, Haritunians T, Hunter DJ, Kaplan RC, Mohlke KL, O'Connell JR, Peltonen L, Schlessinger D, Strachan DP, van Duijn CM, Wichmann HE, Frayling TM, Thorsteinsdottir U, Abecasis GR, Barroso I, Boehnke M, Stefansson K, North KE, McCarthy MI, Hirschhorn JN, Ingelsson E, Loos RJ (2010) Association analyses of 249,796 individuals reveal 18 new loci associated with body mass index. Nat Genet 42(11):937–948. doi:10.1038/ng.686

Steinthorsdottir V, Thorleifsson G, Reynisdottir I, Benediktsson R, Jonsdottir T, Walters GB, Styrkarsdottir U, Gretarsdottir S, Emilsson V, Ghosh S, Baker A, Snorradottir S, Bjarnason H, Ng MC, Hansen T, Bagger Y, Wilensky RL, Reilly MP, Adeyemo A, Chen Y, Zhou J, Gudnason V, Chen G, Huang H, Lashley K, Doumatey A, So WY, Ma RC, Andersen G, Borch-Johnsen K, Jorgensen T, van Vliet-Ostaptchouk JV, Hofker MH, Wijmenga C, Christiansen C, Rader DJ, Rotimi C, Gurney M, Chan JC, Pedersen O, Sigurdsson G, Gulcher JR, Thorsteinsdottir U, Kong A, Stefansson K (2007) A variant in CDKAL1 influences insulin response and risk of type 2 diabetes. Nat Genet 39(6):770–775

Steinthorsdottir V, Thorleifsson G, Sulem P, Helgason H, Grarup N, Sigurdsson A, Helgadottir HT, Johannsdottir H, Magnusson OT, Gudjonsson SA, Justesen JM, Harder MN, Jorgensen ME, Christensen C, Brandslund I, Sandbaek A, Lauritzen T, Vestergaard H, Linneberg A, Jorgensen T, Hansen T, Daneshpour MS, Fallah MS, Hreidarsson AB, Sigurdsson G, Azizi F, Benediktsson R, Masson G, Helgason A, Kong A, Gudbjartsson DF, Pedersen O, Thorsteinsdottir U, Stefansson K (2014) Identification of low-frequency and rare sequence variants associated with elevated or reduced risk of type 2 diabetes. Nat Genet 46(3):294–298. doi:10.1038/ng.2882

Stolerman ES, Florez JC (2009) Genomics of type 2 diabetes mellitus: implications for the clinician. Nat Rev Endocrinol 5(8):429–436. doi:10.1038/nrendo.2009.129

Strawbridge RJ, Dupuis J, Prokopenko I, Barker A, Ahlqvist E, Rybin D, Petrie JR, Travers ME, Bouatia-Naji N, Dimas AS, Nica A, Wheeler E, Chen H, Voight BF, Taneera J, Kanoni S, Peden JF, Turrini F, Gustafsson S, Zabena C, Almgren P, Barker DJ, Barnes D, Dennison EM, Eriksson JG, Eriksson P, Eury E, Folkersen L, Fox CS, Frayling TM, Goel A, Gu HF, Horikoshi M, Isomaa B, Jackson AU, Jameson KA, Kajantie E, Kerr-Conte J, Kuulasmaa T, Kuusisto J, Loos RJ, Luan J, Makrilakis K, Manning AK, Martinez-Larrad MT, Narisu N, Nastase Mannila M, Ohrvik J, Osmond C, Pascoe L, Payne F, Sayer AA, Sennblad B, Silveira A, Stancakova A, Stirrups K, Swift AJ, Syvanen AC, Tuomi T, Hooft FM, Walker M, Weedon MN, Xie W, Zethelius B, Consortium D, Consortium G, Mu TC, Consortium CA, Consortium CD, Ongen H, Malarstig A, Hopewell JC, Saleheen D, Chambers J, Parish S, Danesh J, Kooner J, Ostenson CG, Lind L, Cooper CC, Serrano-Rios M, Ferrannini E, Forsen TJ, Clarke R, Franzosi MG, Seedorf U, Watkins H, Froguel P, Johnson P, Deloukas P, Collins FS, Laakso M, Dermitzakis ET, Boehnke M, McCarthy MI, Wareham NJ, Groop L, Pattou F, Gloyn AL, Dedoussis GV, Lyssenko V, Meigs JB, Barroso I, Watanabe RM, Ingelsson E, Langenberg C, Hamsten A, Florez JC (2011) Genome-wide association identifies nine common variants associated with fasting proinsulin levels and provides new insights into the pathophysiology of type 2 diabetes. Diabetes 60 (10):2624–2634. doi:10.2337/db11-0415

Stumvoll M, Goldstein BJ, van Haeften TW (2005) Type 2 diabetes: principles of pathogenesis and therapy. Lancet 365(9467):1333–1346. doi:10.1016/S0140-6736(05)61032-X, S0140-6736(05)61032-X [pii]

Tabassum R, Chauhan G, Dwivedi OP, Mahajan A, Jaiswal A, Kaur I, Bandesh K, Singh T, Mathai BJ, Pandey Y, Chidambaram M, Sharma A, Chavali S, Sengupta S, Ramakrishnan L, Venkatesh P, Aggarwal SK, Ghosh S, Prabhakaran D, Srinath RK, Saxena M, Banerjee M, Mathur S, Bhansali A, Shah VN, Madhu SV, Marwaha RK, Basu A, Scaria V, McCarthy MI, Diagram, Indico, Venkatesan R, Mohan V, Tandon N, Bharadwaj D (2013) Genome-wide association study for type 2 diabetes in Indians identifies a new susceptibility locus at 2q21. Diabetes 62(3):977–986. doi:10.2337/db12-0406

Takamoto I, Kubota N, Nakaya K, Kumagai K, Hashimoto S, Kubota T, Inoue M, Kajiwara E, Katsuyama H, Obata A, Sakurai Y, Iwamoto M, Kitamura T, Ueki K, Kadowaki T (2014) TCF7L2 in mouse pancreatic beta cells plays a crucial role in glucose homeostasis by regulating beta cell mass. Diabetologia 57(3):542–553. doi:10.1007/s00125-013-3131-6

Talmud PJ, Hingorani AD, Cooper JA, Marmot MG, Brunner EJ, Kumari M, Kivimaki M, Humphries SE (2010) Utility of genetic and non-genetic risk factors in prediction of type 2 diabetes: Whitehall II prospective cohort study. Br Med J 340:b4838. doi:10.1136/bmj.b4838

Thomas G, Jacobs KB, Yeager M, Kraft P, Wacholder S, Orr N, Yu K, Chatterjee N, Welch R, Hutchinson A, Crenshaw A, Cancel-Tassin G, Staats BJ, Wang Z, Gonzalez-Bosquet J, Fang J, Deng X, Berndt SI, Calle EE, Feigelson HS, Thun MJ, Rodriguez C, Albanes D, Virtamo J, Weinstein S, Schumacher FR, Giovannucci E, Willett WC, Cussenot O, Valeri A, Andriole GL, Crawford ED, Tucker M, Gerhard DS, Fraumeni JF Jr, Hoover R, Hayes RB, Hunter DJ, Chanock SJ (2008) Multiple loci identified in a genome-wide association study of prostate cancer. Nat Genet 40(3):310–315. doi:10.1038/ng.91

Tsai FJ, Yang CF, Chen CC, Chuang LM, Lu CH, Chang CT, Wang TY, Chen RH, Shiu CF, Liu YM, Chang CC, Chen P, Chen CH, Fann CS, Chen YT, Wu JY (2010) A genome-wide association study identifies susceptibility variants for type 2 diabetes in Han Chinese. PLoS Genet 6(2):e1000847. doi:10.1371/journal.pgen.1000847

Tuomilehto J, Lindstrom J (2003) The major diabetes prevention trials. Curr Diab Rep 3 (2):115–122

Unoki H, Takahashi A, Kawaguchi T, Hara K, Horikoshi M, Andersen G, Ng DP, Holmkvist J, Borch-Johnsen K, Jorgensen T, Sandbaek A, Lauritzen T, Hansen T, Nurbaya S, Tsunoda T, Kubo M, Babazono T, Hirose H, Hayashi M, Iwamoto Y, Kashiwagi A, Kaku K, Kawamori R, Tai ES, Pedersen O, Kamatani N, Kadowaki T, Kikkawa R, Nakamura Y, Maeda S (2008) SNPs in KCNQ1 are associated with susceptibility to type 2 diabetes in East Asian and European populations. Nat Genet 40(9):1098–1102. doi:10.1038/ng.208

Vaxillaire M, Yengo L, Lobbens S, Rocheleau G, Eury E, Lantieri O, Marre M, Balkau B, Bonnefond A, Froguel P (2014) Type 2 diabetes-related genetic risk scores associated with variations in fasting plasma glucose and development of impaired glucose homeostasis in the prospective DESIR study. Diabetologia. doi:10.1007/s00125-014-3277-x

Voight BF, Scott LJ, Steinthorsdottir V, Morris AP, Dina C, Welch RP, Zeggini E, Huth C, Aulchenko YS, Thorleifsson G, McCulloch LJ, Ferreira T, Grallert H, Amin N, Wu G, Willer CJ, Raychaudhuri S, McCarroll SA, Langenberg C, Hofmann OM, Dupuis J, Qi L, Segre AV, van Hoek M, Navarro P, Ardlie K, Balkau B, Benediktsson R, Bennett AJ, Blagieva R, Boerwinkle E, Bonnycastle LL, Bengtsson Bostrom K, Bravenboer B, Bumpstead S, Burtt NP, Charpentier G, Chines PS, Cornelis M, Couper DJ, Crawford G, Doney AS, Elliott KS, Elliott AL, Erdos MR, Fox CS, Franklin CS, Ganser M, Gieger C, Grarup N, Green T, Griffin S, Groves CJ, Guiducci C, Hadjadj S, Hassanali N, Herder C, Isomaa B, Jackson AU, Johnson PR, Jorgensen T, Kao WH, Klopp N, Kong A, Kraft P, Kuusisto J, Lauritzen T, Li M, Lieverse A, Lindgren CM, Lyssenko V, Marre M, Meitinger T, Midthjell K, Morken MA, Narisu N, Nilsson P, Owen KR, Payne F, Perry JR, Petersen AK, Platou C, Proenca C, Prokopenko I, Rathmann W, Rayner NW, Robertson NR, Rocheleau G, Roden M, Sampson MJ, Saxena R, Shields BM, Shrader P, Sigurdsson G, Sparso T, Strassburger K, Stringham HM, Sun Q, Swift AJ, Thorand B, Tichet J, Tuomi T, van Dam RM, van Haeften TW, van Herpt T, van Vliet-Ostaptchouk JV, Walters GB, Weedon MN, Wijmenga C, Witteman J, Bergman RN, Cauchi S, Collins FS, Gloyn AL, Gyllensten U, Hansen T, Hide WA, Hitman GA, Hofman A, Hunter DJ, Hveem K, Laakso M, Mohlke KL, Morris AD, Palmer CN, Pramstaller PP, Rudan I, Sijbrands E, Stein LD, Tuomilehto J, Uitterlinden A, Walker M, Wareham NJ, Watanabe RM, Abecasis GR, Boehm BO, Campbell H, Daly MJ, Hattersley AT, Hu FB, Meigs JB, Pankow JS, Pedersen O, Wichmann HE, Barroso I, Florez JC, Frayling TM, Groop L, Sladek R, Thorsteinsdottir U, Wilson JF, Illig T, Froguel P, van Duijn CM, Stefansson K, Altshuler D, Boehnke M, McCarthy MI, Investigators M, Consortium G (2010) Twelve type 2 diabetes susceptibility loci identified through large-scale association analysis. Nat Genet 42(7):579–589. doi:10.1038/ng.609

Voight BF, Kang HM, Ding J, Palmer CD, Sidore C, Chines PS, Burtt NP, Fuchsberger C, Li Y, Erdmann J, Frayling TM, Heid IM, Jackson AU, Johnson T, Kilpelainen TO, Lindgren CM, Morris AP, Prokopenko I, Randall JC, Saxena R, Soranzo N, Speliotes EK, Teslovich TM, Wheeler E, Maguire J, Parkin M, Potter S, Rayner NW, Robertson N, Stirrups K, Winckler W, Sanna S, Mulas A, Nagaraja R, Cucca F, Barroso I, Deloukas P, Loos RJ, Kathiresan S, Munroe PB, Newton-Cheh C, Pfeufer A, Samani NJ, Schunkert H, Hirschhorn JN, Altshuler D,

McCarthy MI, Abecasis GR, Boehnke M (2012) The metabochip, a custom genotyping array for genetic studies of metabolic, cardiovascular, and anthropometric traits. PLoS Genet 8(8): e1002793. doi:10.1371/journal.pgen.1002793

Weedon MN (2007) The importance of TCF7L2. Diabet Med 24(10):1062–1066. doi:10.1111/j.1464-5491.2007.02258.x

Weedon MN, Frayling TM, Shields B, Knight B, Turner T, Metcalf BS, Voss L, Wilkin TJ, McCarthy A, Ben-Shlomo Y, Davey Smith G, Ring S, Jones R, Golding J, Byberg L, Mann V, Axelsson T, Syvanen AC, Leon D, Hattersley AT (2005) Genetic regulation of birth weight and fasting glucose by a common polymorphism in the islet cell promoter of the glucokinase gene. Diabetes 54(2):576–581

Weedon MN, Clark VJ, Qian Y, Ben-Shlomo Y, Timpson N, Ebrahim S, Lawlor DA, Pembrey ME, Ring S, Wilkin TJ, Voss LD, Jeffery AN, Metcalf B, Ferrucci L, Corsi AM, Murray A, Melzer D, Knight B, Shields B, Smith GD, Hattersley AT, Di Rienzo A, Frayling TM (2006) A common haplotype of the glucokinase gene alters fasting glucose and birth weight: association in six studies and population-genetics analyses. Am J Hum Genet 79(6):991–1001. doi:10.1086/509517

Winckler W, Weedon MN, Graham RR, McCarroll SA, Purcell S, Almgren P, Tuomi T, Gaudet D, Bostrom KB, Walker M, Hitman G, Hattersley AT, McCarthy MI, Ardlie KG, Hirschhorn JN, Daly MJ, Frayling TM, Groop L, Altshuler D (2007) Evaluation of common variants in the six known maturity-onset diabetes of the young (MODY) genes for association with type 2 diabetes. Diabetes 56(3):685–693. doi:10.2337/db06-0202

Yamauchi T, Hara K, Maeda S, Yasuda K, Takahashi A, Horikoshi M, Nakamura M, Fujita H, Grarup N, Cauchi S, Ng DP, Ma RC, Tsunoda T, Kubo M, Watada H, Maegawa H, Okada-Iwabu M, Iwabu M, Shojima N, Shin HD, Andersen G, Witte DR, Jorgensen T, Lauritzen T, Sandbaek A, Hansen T, Ohshige T, Omori S, Saito I, Kaku K, Hirose H, So WY, Beury D, Chan JC, Park KS, Tai ES, Ito C, Tanaka Y, Kashiwagi A, Kawamori R, Kasuga M, Froguel P, Pedersen O, Kamatani N, Nakamura Y, Kadowaki T (2010) A genome-wide association study in the Japanese population identifies susceptibility loci for type 2 diabetes at UBE2E2 and C2CD4A-C2CD4B. Nat Genet 42(10):864–868. doi:10.1038/ng.660

Zeggini E, Weedon MN, Lindgren CM, Frayling TM, Elliott KS, Lango H, Timpson NJ, Perry JR, Rayner NW, Freathy RM, Barrett JC, Shields B, Morris AP, Ellard S, Groves CJ, Harries LW, Marchini JL, Owen KR, Knight B, Cardon LR, Walker M, Hitman GA, Morris AD, Doney AS, McCarthy MI, Hattersley AT (2007) Replication of genome-wide association signals in UK samples reveals risk loci for type 2 diabetes. Science 316(5829):1336–1341. doi:10.1126/science.1142364, 1142364 [pii]

Zeggini E, Scott LJ, Saxena R, Voight BF, Marchini JL, Hu T, de Bakker PI, Abecasis GR, Almgren P, Andersen G, Ardlie K, Bostrom KB, Bergman RN, Bonnycastle LL, Borch-Johnsen K, Burtt NP, Chen H, Chines PS, Daly MJ, Deodhar P, Ding CJ, Doney AS, Duren WL, Elliott KS, Erdos MR, Frayling TM, Freathy RM, Gianniny L, Grallert H, Grarup N, Groves CJ, Guiducci C, Hansen T, Herder C, Hitman GA, Hughes TE, Isomaa B, Jackson AU, Jorgensen T, Kong A, Kubalanza K, Kuruvilla FG, Kuusisto J, Langenberg C, Lango H, Lauritzen T, Li Y, Lindgren CM, Lyssenko V, Marvelle AF, Meisinger C, Midthjell K, Mohlke KL, Morken MA, Morris AD, Narisu N, Nilsson P, Owen KR, Palmer CN, Payne F, Perry JR, Pettersen E, Platou C, Prokopenko I, Qi L, Qin L, Rayner NW, Rees M, Roix JJ, Sandbaek A, Shields B, Sjogren M, Steinthorsdottir V, Stringham HM, Swift AJ, Thorleifsson G, Thorsteinsdottir U, Timpson NJ, Tuomi T, Tuomilehto J, Walker M, Watanabe RM, Weedon MN, Willer CJ, Wellcome Trust Case Control C, Illig T, Hveem K, Hu FB, Laakso M, Stefansson K, Pedersen O, Wareham NJ, Barroso I, Hattersley AT, Collins FS, Groop L, McCarthy MI, Boehnke M, Altshuler D (2008) Meta-analysis of genome-wide association data and large-scale replication identifies additional susceptibility loci for type 2 diabetes. Nat Genet 40(5):638–645. doi:10.1038/ng.120

Chapter 3
Genome-Wide Association Studies of Quantitative Glycaemic Traits

Inês Barroso and Robert Scott

Abstract Genome-wide association studies (GWASs) of patients with type 2 diabetes (T2D) and unaffected control participants (case-control studies) are designed to identify genetic variants that predispose to T2D, although the mechanisms by which these variants predispose to T2D are unclear. In 2008, the *M*eta-*A*nalysis of *G*lucose and *I*nsulin-related traits *C*onsortium (MAGIC) was established to facilitate meta-analysis of GWAS data of quantitative glycaemic traits (including fasting and post-challenge glycaemic measures) from persons without diabetes. These traits are associated with cardiovascular outcomes even below the diabetic threshold and are important in and of themselves. Our aims were threefold: (a) to identify loci influencing glycaemic traits as a way to understand the similarities and differences between loci influencing glucose regulation within the normal physiological range and those affecting pathophysiological states (T2D); (b) to identify new loci impacting T2D risk using an alternative approach; and (c) to use glycaemic traits to begin to elucidate disease mechanism, i.e. how loci impact biological pathways to promote disease. Here, we describe the approaches used in MAGIC, what we have learned about the genetic architecture of glycaemic traits and T2D itself, and how we see future genomic studies further refining disease aetiology.

3.1 Genetic Association Studies: From Diabetes to Quantitative Glycaemic Traits

The advent of genome-wide association studies (GWASs) in 2007 quickly transformed the landscape of complex disease genetics. Prior to 2007, genetic variation at three loci [*KCNJ11*, *PPARG*, and *TCF7L2* (Hani et al. 1998; Altshuler

I. Barroso (✉)
Wellcome Trust Sanger Institute, Wellcome Genome Campus, Cambridge, UK
e-mail: ib1@sanger.ac.uk

R. Scott
MRC Epidemiology Unit, University of Cambridge School of Clinical Medicine, Institute of Metabolic Science, Cambridge Biomedical Campus, Cambridge, UK

J.C. Florez (ed.), *The Genetics of Type 2 Diabetes and Related Traits*,
DOI 10.1007/978-3-319-01574-3_3

et al. 2000; Grant et al. 2006)] had been robustly associated with type 2 diabetes (T2D). Within the first year of GWAS, this number had more than quadrupled to 13 (Zeggini et al. 2007; Sladek et al. 2007; Saxena et al. 2007; Steinthorsdottir et al. 2007; WTCCC, 2007; Scott et al. 2007), although these variants were of relatively small effect (odds ratios of ~1.1–1.4 per risk allele) and explained only a small proportion of the genetic contribution to T2D risk. It was recognised that these small effect sizes would necessitate large sample sizes for their identification, and large, international consortia were formed to investigate the genetic basis of complex disease. Furthermore, it was identified that some loci associated with T2D increased risk via diverse pathways. For example, variants in the *FTO* gene were associated with T2D (WTCCC, 2007), although this association was identified to be secondary to the primary association of these variants with adiposity (Frayling et al. 2007). This suggested that a complementary approach of studying the genetic basis of T2D-related quantitative glycaemic traits, such as fasting glucose (FG), would be informative. Such phenotypes were felt to relate to in vivo physiology more closely, had been precisely measured, and it was hypothesised that their genetic dissection would lead to clearer insights into disease pathophysiology. Fasting glucose had been shown to be heritable in family studies (Snieder et al. 1999), and prior to GWAS approaches, specific genetic variants in the glucokinase gene (*GCK*) had been shown to influence FG levels (Weedon et al. 2006). Early GWAS efforts replicated these findings (Bouatia-Naji et al. 2008; Chen et al. 2008) and identified other highly biologically plausible genes including *G6PC2* (see Chap. 17) and *GCKR* (see Chap. 16) associated with FG levels (Scott et al. 2007; Bouatia-Naji et al. 2008; Vaxillaire et al. 2008; Orho-Melander et al. 2008) highlighting the utility of the GWAS approach in elucidating the genetic aetiology of glycaemia. Thus, the *M*eta-*A*nalysis of *G*lucose and *I*nsulin-related traits *C*onsortium (MAGIC), a large-scale international collaboration, was formed to investigate the genetic bases of quantitative glycaemic traits including fasting glucose and insulin as well as post-challenge phenotypes.

Although the diagnosis of T2D is based on either fasting or post-challenge hyperglycaemia, it likely encompasses individuals with many different types of disease. T2D is a complex disease that typically develops as a result of impaired beta-cell function, insulin resistance, or a combination of both. Indeed, T2D often develops when beta-cell function is insufficient to maintain normoglycaemia in the face of obesity-induced insulin resistance. Thus, improved understanding of the underlying pathophysiology of T2D in each affected person might lead to better disease sub-classification and potentially improved or more targeted treatments. Identifying genetic loci that influence glycaemic traits, which are tightly regulated within each individual, presented a potential avenue for disease sub-classification and exploration of disease pathophysiology. Fasting or 2h glucose levels after an oral glucose tolerance test are also associated with future cardiovascular disease, even below the diabetic threshold, further highlighting the independent utility in understanding their regulation in nondiabetic individuals. Indeed, it has been suggested that there are distinct aetiologies among individuals diagnosed with T2D on the basis of either fasting glucose or 2h glucose (Færch et al. 2013), with 2h glucose concentrations more strongly associated with cardiovascular risk than fasting glucose levels (DECODE Study Group, the European Diabetes Epidemiology Group, 2001). Recently, glycated haemoglobin (HbA1c) has been proposed as an additional diagnostic measure of T2D (Farmer 2012), as it reflects average glycaemia over the preceding 120 days and also predicts future vascular complications (Khaw et al. 2004). Further, insulin resistance, often closely associated with obesity, is an important risk factor for T2D, and understanding its genetic basis, and that of its more widely available surrogate measures such as fasting insulin (FI) and insulin resistance by homeostasis model assessment [HOMA-IR (Matthews et al. 1985)], may also provide clues of how various loci impact T2D risk. MAGIC was therefore established to garner power from large-scale meta-analyses of GWAS of glycaemic traits in persons without T2D to improve our understanding of glucose homeostasis. While a diagnosis of T2D is likely to result in a number of changes: lifestyle, anthropometric, and metabolic, by studying the genetic basis of quantitative traits in individuals free from T2D, we can minimise the effect of these potential confounding factors resulting from disease and its treatment and begin to dissect individual pathophysiological pathways to T2D. More sophisticated phenotypes, such as measures of proinsulin, data from insulin or glucose clamps, or longitudinal follow-up, are generally harder to obtain in large epidemiological settings; therefore, GWASs of these phenotypes have been more limited. However, they do provide significant additional utility to dissect disease pathways, as discussed further below (Sect. 3.2.5).

3.2 Genetic Associations with Glycaemic Traits

3.2.1 Early GWAS Approaches to Fasting Glucose and Fasting Insulin

The first effort from MAGIC focused on FG levels and described a novel association with the *MTNR1B* locus (encoding the melatonin receptor 1B), where the glucose-raising allele also increased T2D risk (Prokopenko et al. 2009). The same locus was independently identified in French individuals without diabetes (Bouatia-Naji et al. 2009) and soon after validated in individuals of Finnish (Sabatti et al. 2009) and Indian Asian descent (Chambers et al. 2009). More recently, loss of function mutations in *MTNR1B* were associated with T2D, implicating *MTNR1B* as the causal gene in the association region and suggesting that GWAS may identify loci in which independent rare variants (not tagged by common alleles) further contribute to disease heritability (Bonnefond et al. 2012). Functional follow-up work on this locus is described in more detail in Chap. 21. Among these initial GWASs, variants near *DGKB-TMEM195* (Sabatti et al. 2009) were also associated with fasting glucose. In independent work, a variant in *IRS1* (encoding insulin receptor substrate 1) was associated with T2D risk, hyperglycaemia, and also fasting insulin (FI) and HOMA-IR (Rung et al. 2009).

3.2.2 Large-Scale Genome-Wide Collaboration in MAGIC

Soon after, MAGIC efforts were expanded to include a genome-wide discovery sample size of over 45,000 nondiabetic participants for fasting glucose and to include surrogate measures of beta-cell function including homeostasis model assessment (HOMA-B). This aided identification of a further nine loci (in or near *ADCY5*, *MADD*, *ADRA2A*, *CRY2*, *FADS1*, *GLIS3*, *SLC2A2*, *PROX1*, and *C2CD4B*) associated with FG, increasing the total number of loci to 16 (Dupuis et al. 2010) (Table 3.1). This effort also identified the association of known T2D genes *TCF7L2* and *SLC30A8* with FG. This expanded effort also included fasting insulin (FI) and HOMA-IR. While the discovery sample size for FI was broadly comparable ($N = 38{,}238$) to that for FG ($N = 46{,}186$), only two genome-wide significant associations were identified: a novel locus near *IGF1* and *GCKR*, where the FG-raising allele was also associated with higher fasting insulin. In parallel, analyses of 2h glucose were also performed (Saxena et al. 2010). However, fewer studies had measured 2h glucose, and as such, the sample size was smaller ($N = 15{,}224$) and subsequently yielded fewer associations, namely, those with *TCF7L2*, *GCKR*, *GIPR*, *ADCY5*, and *VPS13C* (Table 3.1). While four of these loci were associated with fasting traits, the variants in *GIPR* were identified to have an association with glycaemic traits for the first time. *GIPR* encodes the receptor for gastric inhibitory polypeptide (GIP), an incretin hormone secreted from the gut, which potentiates insulin secretion in response to oral glucose challenge above and beyond that in response to an intravenous glucose challenge. This association with 2h glucose highlighted the role of genetic variation in the incretin pathway in impaired glucose tolerance.

3.2.3 Alternative Approaches to Unravel the Genetic Control of Glycaemic Traits

A surprising finding of the MAGIC effort was the discrepancy in the number of loci associated with FG compared to only two identified for FI. While it has been suggested that the heritability of insulin resistance, a trait for which FI is a commonly used proxy, is lower than that for insulin secretion (Prudente et al. 2009), estimates typically suggest that up to half of the variation in insulin resistance is attributable to genetic factors (Poulsen et al. 2005). It was speculated that the fewer FI associations detected might be due to a different genetic architecture of this trait (e.g. perhaps the variants influencing FI had lower allele frequencies or smaller effect sizes). However, another possibility was considered, namely, that the effect of body mass index (BMI) on each of these traits might influence the results obtained. While FG is weakly associated with BMI (Scott et al. 2012a), insulin levels are strongly associated with BMI (Reaven 1988), therefore a MAGIC effort was undertaken to account for differences in BMI in these genetic

Table 3.1 Loci influencing glycaemic traits and their associated T2D effect

Locus	Index SNP	Alleles [E/O]	Type of variant	EAF %	Effect size (SE)	Trait	T2D OR	*p*-value
ABO	rs505922 (Huyghe et al. 2013)	C/T	Intronic	0.471	−0.038 (0.006)	Disposition index	Unknown	
ADCY5	rs11708067 (Dupuis et al. 2010)	A/G	Intronic	0.77	0.027 (0.003)	FG	1.11 (1.08–1.14)	7.19×10^{-14}
					−0.023 (0.004)	HOMA-B		
	rs2877716 (Saxena et al. 2010)	C/T	Intronic	0.75	0.07 (0.01)	2hGlu	1.10 (1.06–1.13)	1.04×10^{-9}
ADRA2A	rs10885122 (Dupuis et al. 2010)	G/T	210 kb D	0.90	0.022 (0.004)	FG	1.04 (1.01–1.08)	2.19×10^{-2}
AMT	rs11715915 (Scott et al. 2012b)	C/T	R318R	0.73	0.012 (0.002)	FG	1.00 (0.98–1.03)	8.17×10^{-1}
ANK1	rs6474359 (Soranzo et al. 2010)	T/C	Intronic	0.97	0.058 (0.011)	HbA1c	0.93 (0.87–0.99)	2.68×10^{-2}
	rs4737009 (Soranzo et al. 2010)	A/G	Intronic	0.27	0.027 (0.004)	HbA1c	1.01 (0.98–1.03)	6.72×10^{-1}
ANKRD55/ MAP3K1	rs459193 (Scott et al. 2012b)	G/A	Intergenic	0.78	0.015 (0.002)	FI(BMI)	1.08 (1.05–1.11)	5.99×10^{-9}
ARAP1	rs11603334 (Scott et al. 2012b)	G/A	Intronic	0.87	0.019 (0.003)	FG	1.10 (1.07–1.14)	4.07×10^{-10}
					−0.093 (0.005)	Proinsulin		
ARL15	rs4865796 (Scott et al. 2012b)	A/G	Intronic	0.71	0.015 (0.003)	FI	1.06 (1.03–1.09)	8.29×10^{-6}
					0.015 (0.002)	FI(BMI)		

(continued)

Table 3.1 (continued)

Locus	Index SNP	Alleles [E/O]	Type of variant	EAF %	Effect size (SE)	Trait	T2D OR	*p*-value
ATP11A	rs7998202 (Soranzo et al. 2010)	G/A	13 kb Up	0.13	0.031 (0.005)	HbA1c	1.00 (0.97–1.05)[a]	6.00×10^{-1}
CDKAL1	rs9368222 (Scott et al. 2012b)	A/C	Intronic	0.28	0.014 (0.002)	FG	1.17 (1.14–1.20)	7.04×10^{-34}
CDKN2B	rs10811661 (Scott et al. 2012b)	T/C	125 kb Up	0.80	0.024 (0.003)	FG	1.18 (1.15–1.22)	3.72×10^{-27}
CRY2	rs11605924 (Dupuis et al. 2010)	A/C	Intronic	0.54	0.015 (0.003)	FG	1.03 (1.01–1.06)	9.25×10^{-3}
DGKB/ TMEM195	rs2191349 (Dupuis et al. 2010)	T/G	Intergenic	0.47	0.030 (0.003)	FG	1.05 (1.03–1.08)	2.99×10^{-5}
DNLZ	rs3829109 (Scott et al. 2012b)	G/A	Intronic	0.66	0.017 (0.003)	FG	1.06 (1.03–1.09)	1.13×10^{-4}
DPYSL5	rs1371614 (Manning et al. 2012)	T/C	Intronic	0.29	0.015 (0.006)[b]	FG(BMI)	1.01 (0.98–1.03)	5.82×10^{-3}
ERAP2	rs1019503 (Scott et al. 2012b)	A/G	3′UTR	0.53	0.063 (0.011)	2hGlu	1.02 (0.99–1.04)	1.54×10^{-1}
FADS1	rs174550 (Dupuis et al. 2010)	T/C	Intronic	0.63	0.017 (0.003)	FG	1.02 (1.00–1.05)	5.27×10^{-2}
					−0.020 (0.003)	HOMA-B		
FAM13A	rs3822072 (Scott et al. 2012b)	A/G	Intronic	0.46	0.012 (0.002)	FI(BMI)	1.05 (1.02–1.07)	2.09×10^{-4}
FN3K	rs1046896 (Soranzo et al. 2010)	T/C	8 kb Up	0.31	0.035 (0.003)	HbA1c	1.01 (0.99–1.04)	3.42×10^{-1}

FOXA2	rs6048205 (Manning et al. 2012)	A/G	2 kb D	0.95	0.023 (0.012)	FG(BMI)	1.06 (1.00–1.13)	5.00×10^{-2}
	rs6113722 (Scott et al. 2012b)	G/A	4.5 kb D	0.96	0.035 (0.005)	FG	1.05 (0.99–1.11)	1.36×10^{-1}
FTO	rs1421085 (Scott et al. 2012b)	C/T	Intronic	0.46	0.020 (0.003)	FI	1.12 (1.09–1.14)	5.42×10^{-22}
G6PC2/	rs560887 (Dupuis et al. 2010)	C/T	Intronic	0.67	0.075 (0.004)	FG	0.99 (0.96–1.02)	3.48×10^{-1}
					−0.042 (0.004)	HOMA-B		
					0.032 (0.004)	HbA1c		
ABCB11	rs552976 (Soranzo et al. 2010)	G/A	Intronic	0.62	0.047 (0.003)	HbA1c	0.98 (0.96–1.00)	1.02×10^{-1}
GCK	rs4607517 (Dupuis et al. 2010)	A/G	36 kb Up	0.20	0.062 (0.004)	FG	1.08 (1.04–1.11)	1.03×10^{-5}
	rs1799884 (Soranzo et al. 2010)	T/C	46bp Up	0.20	0.038 (0.004)	HbA1c	1.07 (1.04–1.11)	1.13×10^{-5}
	rs6975024 (Scott et al. 2012b)	C/T	2.8 kb Up	0.20	0.1026 (0.016)	2hGlu	1.07 (1.04–1.11)	1.30×10^{-5}
GCKR	rs780094 (Dupuis et al. 2010)	C/T	Intronic	0.61	0.029 (0.003)	FG	1.06 (1.04–1.09)	5.37×10^{-7}
					0.032 (0.004)	FI		
					0.035 (0.004)	HOMA-IR		
	rs1260326 (Saxena et al. 2007)	T/C	L446P	0.42	0.100 (0.01)	2hGlu	0.94 (0.92–0.97)	1.63×10^{-6}

(continued)

Table 3.1 (continued)

Locus	Index SNP	Alleles [E/O]	Type of variant	EAF %	Effect size (SE)	Trait	T2D OR	*p*-value
GIPR	rs10423928 (Saxena et al. 2010)	A/T	Intronic	0.18	0.110 (0.01)	2hGlu	1.05 (1.02–1.08)[a]	1.00×10^{-3}
	rs2302593 (Scott et al. 2012b)	C/G	9.7 kb D	0.53	0.014 (0.002)	FG	0.99 (0.96–1.02)	4.19×10^{-1}
GLIS3	rs7034200 (Dupuis et al. 2010)	A/C	Intronic	0.53	0.018 (0.003)	FG	1.05 (1.02–1.07)	1.27×10^{-4}
					−0.020 (0.004)	HOMA-B		
GLS2	rs2657879 (Scott et al. 2012b)	G/A	L581P	0.19	0.016 (0.003)	FG(BMI)	1.01 (0.98–1.05)	3.40×10^{-1}
GPSM1	rs60980157 (Huyghe et al. 2013)	T/C	S391L	0.3	0.072 (0.013)	Insulinogenic index	Unknown	
GRB10	rs6943153 (Scott et al. 2012b)	T/C	Intronic	0.28	0.015 (0.002)	FG	0.99 (0.97–1.02)	5.55×10^{-1}
GRB14/	rs10195252 (Scott et al. 2012b)	T/C	34 kb Up	0.56	0.016 (0.003)	FI	1.07 (1.04–1.1)	3.01×10^{-8}
COBLL1					0.017 (0.002)	FI(BMI)		
	rs7607980 (Manning et al. 2012)	T/C	N939D	0.86	0.039 (0.008)[b]	FI(BMI)	1.10 (1.06–1.13)	2.93×10^{-7}
HFE	rs1800562 (Soranzo et al. 2010)	G/A	C282Y	0.95	0.063 (0.007)	HbA1c	0.96 (0.91–1.01)	9.00×10^{-2}
HIP1	rs1167800 (Scott et al. 2012b)	A/G	Intron–exon junction	0.55	0.016 (0.003)	FI	0.99 (0.97–1.02)	4.80×10^{-1}
HK1	rs16926246 (Soranzo et al. 2010)	C/T	Intronic	0.89	0.089 (0.004)	HbA1c	1.04 (1.00–1.09)	4.13×10^{-2}

HNF1A	rs2650000 (Huyghe et al. 2013)	A/C	Intergenic	0.46	−0.076 (0.012)	Insulinogenic index	Unknown	
IFG1	rs35767 (Dupuis et al. 2010)	G/A	1.2 kb Up	0.89	0.010 (0.006)	FI	1.02 (0.99–1.06)	1.46×10^{-1}
					0.013 (0.006)	HOMA-IR		
IGF2BP2	rs7651090 (Scott et al. 2012b)	G/A	Intronic	0.30	0.013 (0.002)	FG	1.13 (1.10–1.16)	3.41×10^{-23}
					0.064 (0.012)	2hGlu (BMI)		
IKBKAP	rs16913693 (Scott et al. 2012b)	T/G	Intronic	0.98	0.043 (0.007)	FG	1.02 (0.95–1.09)	6.68×10^{-1}
IRS1	rs2943634 (Manning et al. 2012)	C/A	528 kb D	0.66	0.021 (0.010)[b]	FI(BMI)	1.09 (1.06–1.11)	8.41×10^{-12}
	rs2972143 (Scott et al. 2012b)	G/A	480 kb D	0.63	0.014 (0.003)	FI	1.09 (1.06–1.12)	5.03×10^{-12}
	rs2943645 (Scott et al. 2012b)	T/C	497 kb D	0.63	0.019 (0.002)	FI(BMI)	1.09 (1.07–1.12)	1.15×10^{-12}
KANK1	rs3824420 (Huyghe et al. 2013)	A/G	R279C	0.029	0.107 (0.018)	Proinsulin AUC^{0-30}	Unknown	
KL	rs576674 (Scott et al. 2012b)	G/A	36 kb Up	0.13	0.017 (0.003)	FG	1.08 (1.04–1.11)	1.05×10^{-5}
LARP6	rs1549318 (Strawbridge et al. 2011)	T/C	15 kb D	0.60	0.019 (0.005)	Proinsulin	1.01 (0.99–1.04)[a]	3.00×10^{-1}
LYPLAL1	rs2820436 (Scott et al. 2012b)	C/A	254 kb D	0.68	0.015 (0.003)	FI	1.04 (1.01–1.06)	5.82×10^{-3}
	rs4846565 (Scott et al. 2012b)	G/A	336 kb D	0.69	0.013 (0.002)	FI(BMI)	1.04 (1.01–1.06)	5.82×10^{-3}
	rs2785980 (Manning et al. 2012)	T/C	314 kb D	0.69	0.018 (0.010)[b]	FI(BMI)	1.06 (1.03–1.10)[c]	1.30×10^{-3}

(continued)

Table 3.1 (continued)

Locus	Index SNP	Alleles [E/O]	Type of variant	EAF %	Effect size (SE)	Trait	T2D OR	*p*-value
MADD	rs7944584 (Dupuis et al. 2010)	A/T	Intronic	0.71	0.021 (0.003)	FG	1.01 (0.99–1.04)	2.68×10^{-1}
	rs10501320 (Strawbridge et al. 2011)	G/C	Intronic	0.71	0.081 (0.006)	Proinsulin	0.98 (0.96–1.01)[a]	2.40×10^{-1}
	rs10838687 (Strawbridge et al. 2011)	T/G	Intronic	0.81	0.025 (0.005)	Proinsulin	1.00 (0.97–1.03)	8.39×10^{-1}
	rs35233100 (Huyghe et al. 2013)	T/C	R766X	0.037	−0.32 (0.04)	Proinsulin	Unknown	
MTNR1B	rs10830963 (Dupuis et al. 2010)	G/C	Intronic	0.30	0.067 (0.003)	FG	1.10 (1.07–1.13)	5.32×10^{-13}
					−0.034 (0.004)	HOMA-B		
					0.024 (0.004)	HbA1c		
	rs1387153 (Soranzo et al. 2010)	T/C	29 kb Up	0.27	0.028 (0.004)	HbA1c	1.09 (1.06–1.12)	1.59×10^{-11}
OR4S1	rs1483121 (Manning et al. 2012)	G/A	4.7 kb D	0.85	0.006 (0.008)[b]	FG(BMI)	1.00 (0.97–1.03)	9.75×10^{-1}
PAM	rs35658696 (Huyghe et al. 2013)	G/A	N563G	0.053	−0.152 (0.027)	Insulinogenic index	Unknown	
P2RX2	rs10747083 (Scott et al. 2012b)	A/G	154 kb Up	0.75	0.013 (0.002)	FG	1.01 (0.99–1.04)	3.78×10^{-1}
PCSK1	rs13179048 (Manning et al. 2012)	C/A	183 kb D	0.72	0.027 (0.013)[b]	FG(BMI)	1.02 (0.98–1.06)[c]	2.9×10^{-1}
	rs4869272 (Scott et al. 2012b)	T/C	187 kb D	0.68	0.018 (0.002)	FG	1.00 (0.98–1.03)	7.59×10^{-1}
	rs6235 (Strawbridge et al. 2011)	G/C	S690T	0.27	0.039 (0.005)	Proinsulin	1.01 (0.99–104)[a]	2.9×10^{-1}

PDGFC	rs4691380 (Manning et al. 2012)	C/T	Intronic	0.66	0.020 (0.010)[b]	FI(BMI)	1.04 (1.00–1.07)[c]	4.30×10^{-2}
	rs6822892 (Scott et al. 2012b)	A/G	Intronic	0.65	0.014 (0.002)	FI(BMI)	1.04 (1.02–1.07)	6.83×10^{-4}
PDX1	rs2293941 (Manning et al. 2012)	A/G	3 kb Up	0.21	0.016 (0.006)[b]	FG(BMI)	1.01 (0.98–1.04)	5.25×10^{-1}
	rs11619319 (Scott et al. 2012b)	G/A	6.6 kb Up	0.21	0.019 (0.002)	FG	1.01 (0.98–1.04)	5.45×10^{-1}
PEPD	rs731839 (Scott et al. 2012b)	G/A	Intronic	0.34	0.015 (0.003)	FI	1.03 (1–1.05)	2.80×10^{-2}
					0.015 (0.002)	FI(BMI)		
PPARG	rs17036328 (Scott et al. 2012b)	T/C	Intronic	0.90	0.021 (0.003)	FI(BMI)	1.12 (1.08–1.16)	1.28×10^{-10}
PPIP5K2	rs36046591 (Huyghe et al. 2013)	G/A	S1228G	0.053	−0.152 (0.027)	Insulinogenic index	Unknown	
PPP1R3B	rs4841132 (Manning et al. 2012)	A/G	175 kb Up	0.08	0.032 (0.016)[b]	FI(BMI)	1.07 (1.03–1.11)	4.56×10^{-4}
					0.054 (0.021)[b]	FG (adjBMI)		
	rs983309 (Scott et al. 2012b)	T/G	170 kb Up	0.10	0.026 (0.003)	FG	1.06 (1.02–1.1)	1.48×10^{-3}
					0.029 (0.004)	FI		
	rs2126259 (Scott et al. 2012b)	T/C	176 kb Up	0.09	0.024 (0.003)	FI(BMI)	1.06 (1.02–1.1)	4.33×10^{-3}
	rs11782386 (Scott et al. 2012b)	C/T	194 kb Up	0.97	0.098 (0.017)	2hGlu	0.95 (0.92–0.99)	6.33×10^{-3}

(continued)

Table 3.1 (continued)

Locus	Index SNP	Alleles [E/O]	Type of variant	EAF %	Effect size (SE)	Trait	T2D OR	*p*-value
PROX1	rs340874 (Dupuis et al. 2010)	C/T	2 kb Up	0.56	0.013 (0.003)	FG	1.07 (1.04–1.09)	1.05×10^{-7}
RREB1	rs17762454 (Scott et al. 2012b)	T/C	Intronic	0.22	0.014 (0.002)	FG	1.04 (1.01–1.07)	5.20×10^{-3}
RSPO3	rs2745353 (Scott et al. 2012b)	T/C	Intronic	0.55	0.014 (0.002)	FI	0.99 (0.96–1.01)	2.26×10^{-1}
SGSM2	rs4790333 (Strawbridge et al. 2011)	T/C	Intronic	0.40	0.015 (0.004)	Proinsulin	0.97 (0.95–0.99)[a]	8.20×10^{-3}
	rs61741902 (Huyghe et al. 2013)	A/G	V996I	0.014	0.126 (0.021)	Proinsulin	Unknown	
SLC2A2	rs11920090 (Dupuis et al. 2010)	T/A	Intronic	0.85	0.020 (0.004)	FG	1.02 (0.99–1.06)	1.97×10^{-1}
SLC30A8	rs13266634 (Dupuis et al. 2010)	C/T	R325W	0.76	0.027 (0.004)	FG	1.13 (1.10–1.16)	4.97×10^{-21}
					0.029 (0.002)	FG	1.13 (1.10–1.16)	1.06×10^{-20}
	rs11558471 (Scott et al. 2012b; Strawbridge et al. 2011)	A/G	3′UTR	0.75	0.028 (0.005)	Proinsulin		
SNX7	rs9727115 (Strawbridge et al. 2011)	G/A	Intronic	0.54	0.013 (0.005)	Proinsulin	1.01 (0.99–1.03)[a]	4.0×10^{-1}
SPTA1	rs2779116 (Soranzo et al. 2010)	T/C	Intronic	0.28	0.024 (0.004)	HbA1c	1.00 (0.98–1.03)[a]	8.1×10^{-1}
TBC1D30	rs150781447 (Huyghe et al. 2013)	T/C	R279C	0.02	0.204 (0.025)	Proinsulin $AUC^{30–120}$	Unknown	

TCF7L2	rs7903146 (Scott et al. 2012b; Franklin et al. 2010; Strawbridge et al. 2011)	T/C	Intronic	0.28	0.022 (0.002)	FG	1.39 (1.35–1.42)	1.2×10^{-139}
					−0.018 (0.003)	FI		
					0.032 (0.007)	Proinsulin		
					0.05 (0.03)	HbA1c		
	rs12243326 (Saxena et al. 2010)	C/T	Intronic	0.24	0.07 (0.01)	2hGlu	1.33 (1.30–1.37)[a]	2.9×10^{-112}
TET2	rs9884482 (Scott et al. 2012b)	C/T	Intronic	0.35	0.017 (0.002)	FI	0.99 (0.97–1.02)	5.42×10^{-1}
	rs974801 (Scott et al. 2012b)	G/A	Intronic	0.39	0.014 (0.002)	FI(BMI)	0.99 (0.97–1.01)	4.35×10^{-1}
TMPRSS6	rs855791 (Soranzo et al. 2010)	A/G	V736A	0.41	0.027 (0.004)	HbA1c	0.99 (0.97–1.02)	4.56×10^{-1}
TOP1	rs6072275 (Scott et al. 2012b)	A/G	Intronic	0.14	0.016 (0.003)	FG	1.07 (1.04–1.10)	1.92×10^{-5}
UHRF1BP1	rs4646949 (Manning et al. 2012)	T/G	Intronic	0.73	0.009 (0.010)[b]	FI(BMI)	1.04 (1.00–1.08)[c]	3.90×10^{-2}
	rs6912327 (Scott et al. 2012b)	T/C	Intronic	0.76	0.017 (0.003)	FI(BMI)	1.03 (1–1.07)	8.67×10^{-2}
VPS13C/	rs17271305 (Saxena et al. 2010)	G/A	Intronic	0.42	0.070 (0.01)	2hGlu	0.96 (0.94–0.99)[a]	1.60×10^{-3}
C2CD4A/B	rs11071657 (Dupuis et al. 2010)	A/G	21 kb D	0.59	0.008 (0.003)	FG	1.03 (1.00–1.05)[a]	7.60×10^{-2}
	rs4502156 (Strawbridge et al. 2011)	T/C	73 kb D	0.58	0.029 (0.004)	Proinsulin	1.06 (1.03–1.08)	2.25×10^{-6}

(continued)

Table 3.1 (continued)

Locus	Index SNP	Alleles [E/O]	Type of variant	EAF %	Effect size (SE)	Trait	T2D OR	*p*-value
WARS	rs3783347 (Scott et al. 2012b)	G/T	Intronic	0.78	0.017 (0.003)	FG	1.03 (1.00–1.06)	7.72×10^{-2}
YSK4	rs1530559 (Scott et al. 2012b)	A/G	Intronic	0.60	0.015 (0.003)	FI	1.00 (0.97–1.02)	9.57×10^{-1}
ZBED3	rs7708285 (Scott et al. 2012b)	G/A	Intronic	0.26	0.015 (0.003)	FG(BMI)	1.10 (1.07–1.13)	1.45×10^{-10}

FG, fasting glucose (mmol/L); FG(BMI), fasting glucose BMI adjusted; FI, fasting insulin (pmol/L); FI(BMI), fasting insulin BMI adjusted; 2hGlu, 2hr glucose (FG adjusted, mmol/L); HbA1c, glycated haemoglobin (%); HOMA-B, β-cell function by homeostasis model assessment; HOMA-IR, insulin resistance by homeostasis model assessment; proinsulin (pmol/L); Proinsulin AUC^{0-30}, proinsulin area under the curve (pM × min) 0 to 30 minute of an OGTT; Proinsulin AUC^{30-120}, proinsulin area under the curve (pM × min) 30 to 120 minute of an OGTT. Effect estimates are taken from original references where SNPs attained genome-wide significant *p*-values for the trait unless otherwise noted and are all rounded to three decimal points. The closest gene to each index SNP (or most likely biological candidate) is listed; in most cases, the underlying causal variant is still unknown

EAF effect allele frequency (HapMap CEU); alleles [E/O], effect/other. Type of variant: up, upstream, D, downstream; missense and silent variants are coded as single letter amino acid code and residue number; T2D OR, type 2 diabetes odds ratio (95 % CI) taken from recent DIAGRAM analysis (Morris et al. 2012)

[a]SNPs not on metabochip but with a new lead SNP on this array (lead SNP, metabochip lead, new effect allele, r2): rs9727115, rs1358431, A,0.18; rs2779116, rs2246434, A, 0.85; rs6235, rs7713317, G, 1; rs12243326, rs12255372, T, 0.95; rs10501320, rs11039182, C, 0.92; rs7998202, rs282606, A, 1; rs17271305, rs1436958, T, 0.84; rs11071657, rs12440695, T, 1; rs1549318, rs11072221, C, 0.43; rs4790333, rs2429909, A, 0.79; rs10423928, rs11672660, T, 0.95

[b]For individuals with BMI = 30 kg/m^2

[c]SNPs not present on metabochip and no alternative lead, results are presented for DIAGRAMv3 meta-analysis

associations (Manning et al. 2012). An approach was applied to jointly meta-analyse the association of genetic variants with glycaemic traits, adjusted for BMI, as well as their interaction with BMI. This approach identified six previously unidentified associations with FI, including variants in or near *LYPLAL1*, *COBLL1-GRB14*, *IRS1*, *PDGFC*, *UHRF1BP1*, and *PPP1R3B* (Table 3.1). This effort also identified another seven loci to be associated with FG (Table 3.1) (Manning et al. 2012).

One of the major challenges of GWAS is the stringent levels of statistical significance that have to be reached in order to overcome the burden of multiple testing arising from the number of single nucleotide polymorphisms (SNPs) tested. Considering one million independent tests, it follows that a SNP must reach $p < 5 \times 10^{-8}$ (0.05/1,000,000) to be considered a significant association. The conventional approach is that the most promising SNPs in the discovery stage are genotyped de novo in additional replication samples to maximise the opportunity to reach this stringent level of association. The limited capacity and high per-unit cost for de novo genotyping meant that only the top loci were routinely followed up and, as such, it was likely that some "real" associations remained among SNPs not previously selected for replication. In order to address this challenge and leverage the large amount of data generated in GWAS, an ambitious update to the experimental design was undertaken. MAGIC and other consortia collaborated with Illumina to design the Illumina Cardio-MetaboChip (Voight et al. 2012), an iPLEX custom-array with approximately 200,000 markers designed to facilitate cost-effective and large-scale replication and fine-mapping of loci influencing cardiometabolic traits. The chip comprised the most promising loci from the discovery phase association analyses from a range of cardiometabolic traits and disease outcomes including anthropometric (Speliotes et al. 2010; Heid et al. 2010), glycaemic (Dupuis et al. 2010; Saxena et al. 2010), and lipid (Teslovich et al. 2010) traits, as well as T2D (Voight et al. 2010). This experimental design allowed an expanded analysis within MAGIC, including up to 133,010 individuals and 66,000 variants that were suggestively associated with cardiometabolic traits. This approach identified or confirmed 20 additional loci for FG and 17 for FI relative to the initial MAGIC discovery effort from which the follow-up list was compiled (Dupuis et al. 2010), as well as another 4 loci for 2h glucose (Scott et al. 2012b) (Table 3.1). Of these 53 nonoverlapping loci, 33 were also associated with T2D (Morris et al. 2012).

3.2.4 HbA1c

HbA1c has recently been proposed as a useful diagnostic marker of diabetes (Farmer 2012), as it reflects average glycaemia over the life of the erythrocyte (~3 months). This means that day-to-day variability of the measure is lower than that for fasting or 2h glucose. An additional advantage of HbA1c is that it does not require the participant to be fasted before measurement and may, therefore, be more feasible in large population-based studies. The potential for larger sample sizes, required for successful GWAS, means this is a promising alternative to unravel the genetic basis of T2D and related traits. Initial loci associated with HbA1c levels were first identified by earlier approaches focused on other glycaemic traits (*G6PC2* and *MTNR1B*) (Bouatia-Naji et al. 2008, 2009), and the first HbA1c GWAS, performed in over 14,000 women, confirmed previously established glycaemic loci (*GCK*, *G6PC2*, and *SLC30A8*) and identified a novel association at *HK1* (Paré et al. 2008). Subsequent studies by MAGIC in up to 46,368 individuals identified six novel associations with HbA1c and extended the number of genome-wide significant associations to ten (Soranzo et al. 2010) (Table 3.1). As expected, many previously identified glycaemic and T2D risk loci also have effects on HbA1c, though not all reach genome-wide significance levels (Paré et al. 2008; Franklin et al. 2010).

While HbA1c is a valid marker of glycaemia, it is known that a number of medical conditions, including hereditary anaemias and iron storage disease, change erythrocyte number and turnover and affect glycated haemoglobin levels (Coban et al. 2004; Roberts et al. 2005). By leveraging information gleaned from GWAS for blood traits (Soranzo et al. 2009), the role of HbA1c-associated loci in both glycaemic and haematological parameters could be investigated. Of the ten loci associated with HbA1c, seven harbour rare mutations that cause hereditary anaemias or iron storage disease. Analyses of HbA1c associations adjusted for FG and haematological traits suggested that these seven loci were likely to affect HbA1c levels via non-glucose-mediated pathways (Soranzo et al. 2010). Further, the other three loci (*G6PC2*, *GCK*, and *MTNR1B*) are the only ones to have shown association with other glycaemic traits in recent meta-analyses (Scott et al. 2012b). This posed the question of whether loci influencing HbA1c through their effect on haematological traits might lead to diabetes misdiagnosis, if HbA1c were to be used as a diagnostic measure. However, the degree of diabetes misclassification due to variation in the levels of HbA1c influenced by common variants mediating their effect via non-glycaemic pathways was estimated to be low (Soranzo et al. 2010).

3.2.5 Detailed Phenotypes and Indices of Insulin Secretion

While isolated fasting and 2h measures of glycaemia represent widely available and useful approximations to the major components of glucose homeostasis, their correlation with "gold-standard" estimates of insulin secretion or sensitivity can be modest (Muniyappa et al. 2008; Hanson et al. 2000). The ratio of circulating proinsulin relative to circulating insulin is elevated in T2D and reflects reduced beta-cell secretory capacity (Røder et al. 1998) or impaired early insulin processing. Indeed, several of the T2D-associated loci known to impact beta-cell function show associations with proinsulin/insulin ratio (Ingelsson et al. 2010; González-Sánchez et al. 2008; Dimas et al. 2013). Proinsulin levels are available in some epidemiological cohorts, which enabled GWAS of proinsulin adjusted for fasting insulin (essentially equivalent to the ratio, but more amenable to statistical analyses) (Strawbridge et al. 2011). Nine variants at eight loci were associated with fasting proinsulin including highly plausible biological candidates, such as a missense variant in *PCSK1* (which encodes prohormone convertase 1, the key enzyme in the early processing of proinsulin to insulin). Previously identified variants associated with T2D and thought to influence beta-cell function also showed associations with proinsulin levels (Strawbridge et al. 2011). A known variant at *ARAP1* showed an interesting pattern of association, being associated with lower proinsulin levels as well as *higher* fasting glucose, *lower* beta-cell function, and *higher* T2D risk. This result runs counter to the observed epidemiological associations, is suggestive of an early impairment in the proinsulin processing pathway, and illustrates that both elevated and reduced proinsulin levels relative to FI can signal beta-cell stress. A risk score comprising these nine loci was not associated with coronary artery disease (CAD), arguing against a direct role for proinsulin in the aetiology of CAD, despite the observational epidemiological associations (Zethelius et al. 2002), and further demonstrating the utility of studying intermediary quantitative traits to aid causal inference (Chap. 26).

A recent effort focusing on rare and low-frequency coding variation associated with proinsulin levels (adjusted for insulin) identified low-frequency, nonsynonymous variants in *SGSM2* and *MADD* to be associated with insulin processing (Huyghe et al. 2013). Notably, these variants had a larger effect size than those reported previously for common variants: each allele being associated with a difference of >0.3 standard deviations. Further, the identification of coding variants in these genes lends support to these being the causal genes underlying previously identified associations at these loci. This effort also identified three novel associations with low-frequency variants for proinsulin processing phenotypes and insulinogenic index (a measure of early insulin secretion) (Table 3.1) (Huyghe et al. 2013), as well as three loci which showed aggregate gene-level associations of low-frequency nonsynonymous variants with proinsulin, including one at *ATG13*, which had not previously been identified.

A recent GWAS investigating the determinants of early insulin secretion identified variants in the *GRB10* gene (encoding growth factor receptor-bound protein 10) and seven other previously reported glycaemic loci to be associated with reduced glucose-stimulated insulin secretion (GSIS) (Prokopenko et al. 2014). While variants in *GRB10* had previously been associated with FG, follow-up analyses in this effort suggested a parent-of-origin effect, where the maternally inherited alleles showed association with GSIS, while the paternally inherited alleles did not. Furthermore, results actually suggested opposite directions of effect on FG for alleles dependent on their parental origin which, while difficult to reconcile with previous findings, may explain the absence of association of these variants with T2D and highlight additional complexity worth consideration when investigating the genetic basis of glycaemic traits.

3.3 What Insights Have We Gained from These Studies?

3.3.1 *Biological Insights*

Most loci associated with glycaemic traits are represented by a lead SNP with no known functional consequence, often being intergenic or intronic. Recent analyses focusing on coding variants have identified some associations with missense variants (Huyghe et al. 2013), but these remain in the minority. This makes immediate functional interpretation of these findings less tractable, but recent advances in understanding the tissue-specific regulatory function of the non-coding genome (Dunham et al. 2012) show promise in interrogating genetic associations. A recent study identifying genomic sequences with regulatory function in pancreatic islet cells showed that loci previously associated with FG were enriched for these regulatory sequences (Pasquali et al. 2014). Indeed, loci associated with FG showed more enrichment than those associated with T2D, again highlighting the value of studying quantitative traits to minimise heterogeneity in the observed associations.

From these genetic studies, specific pathways and processes linked to glycaemic traits are beginning to emerge. Observational epidemiological data previously demonstrated circadian rhythms in metabolic parameters (Boden et al. 1996a, b), while this rhythmicity is impaired in T2D (Polonsky et al. 1988). In addition, *Clock* and *Bmal1* mutant mice show impaired glycaemic control (Turek et al. 2005; Bunger et al. 2000), while ablation of pancreatic rhythmicity in these genes leads to diabetes (Marcheva et al. 2010). However, the association of *MTNR1B* and *CRY2* with FG provided the first human genetic evidence for this link, and subsequent studies in human islets highlighted the differential expression of CRY2 in islets from individuals with T2D (Stamenkovic et al. 2012). Also, and while the causal variants are still unknown, a number of loci influencing glycaemic traits are very proximal to classical candidate genes with a role in glucose metabolism or insulin secretion and processing, such as *SLC2A2* (encoding the glucose transporter two), *GCK, GCKR,* and *PCSK1,* as well as transcription factors with a role in pancreas development including *PDX1* and *FOXA2* (Jonsson et al. 1994; Gao et al. 2008). While these associations highlight genes known to be associated with glycaemic control, such findings also highlight the utility of glycaemic traits to identify highly plausible genes and suggest that many of the genes identified with unknown function are likely to have similarly important roles in glycaemic control, which are as yet uncharacterised. The identification of a number of "classical" genes with a plausible role in glycaemia contrasts with T2D loci where many of the underlying genes and pathways were previously unanticipated and of broad importance (e.g. cell cycle regulation) (Morris et al. 2012).

While insulin resistance is a major risk factor for T2D (Reaven 1988), few loci associated with T2D are implicated in mediating their effect on T2D via insulin resistance (Ingelsson et al. 2010; Dimas et al. 2013). As discussed, loci associated with fasting insulin were less tractable to GWAS (Dupuis et al. 2010). Adjustment for BMI aided in the identification of more SNPs associated with insulin resistance

(Manning et al. 2012; Scott et al. 2012b). This was at least in part attributable to removing the variance in fasting insulin attributable to BMI, as evidenced by the reduced variance estimates for the effect sizes (Scott et al. 2012b). While fasting insulin is an imperfect measure of insulin resistance, we noted that of the 19 loci associated with fasting insulin, 10 were associated with a dyslipidaemic profile of elevated triglycerides and lower HDL, a hallmark of insulin resistance. The identification of *FTO* as being associated with fasting insulin, an association entirely mediated by BMI, highlights the positive epidemiological association between BMI and fasting insulin. Also, several of these loci were associated with a higher waist-hip ratio, indicative of a tendency to store fat centrally, another trait strongly associated with insulin resistance (Kahn et al. 2006). However, we also noted that some of these loci were associated with lower BMI (Scott et al. 2012b), which runs counter to the observed epidemiological evidence. It is likely, therefore, that these associations were masked in previous analyses, at least in part due to negative confounding by BMI. One such locus was *IRS1*, which had previously been identified to be associated with body fat percentage although the body-fat-lowering allele was associated with increased risk of T2D and CAD (Kilpeläinen et al. 2011). These intriguing, and apparently paradoxical, associations highlight the novel insights that the study of T2D-related quantitative traits can bring. For example, loci such as *IRS1* and *GRB14* have no demonstrable association with fasting glucose, yet are associated with fasting insulin and dyslipidaemia, indicative of an effect on insulin resistance (Fig. 3.1).

3.3.2 From Biology to Disease and Vice Versa

Although few of the loci associated with T2D were characterised as influencing insulin resistance (Dimas et al. 2013). Among the 19 loci with genome-wide significant associations with FI, 13 were associated with T2D ($p < 0.05$) (Scott et al. 2012b). Thus the study of quantitative traits can bring additional, and more precisely focused, insights into the aetiology of T2D. Indeed, even among those loci now associated with T2D at genome-wide significance, many were first uncovered through studies focused on glycaemic traits (e.g. *MTNR1B, ADCY5, PROX1, GCK, GCKR, GIPR,* and *DGKB-TMEM195*). Of 53 loci associated with FG, FI, or 2h glucose, 33 are associated with risk of T2D ($p < 0.05$) (Scott et al. 2012b), although only 14 are at genome-wide levels of significance for both glycaemic traits and T2D. Another utility of identifying genetic variants associated with quantitative traits is the opportunity to use these as instrumental variables to understand the causal role of a range of risk factors in T2D (De Silva et al. 2011; Li et al. 2011), as described further in Chap. 26. While the major route to diagnosis of T2D is a measurement of fasting glucose, there is a poor correlation between the effect size on FG and the magnitude of association with T2D. Indeed some of the loci associated with FG at genome-wide significance have no discernible association with T2D (e.g. *G6PC2, FADS1, MADD*) (Scott et al. 2012b; Morris et al. 2012)

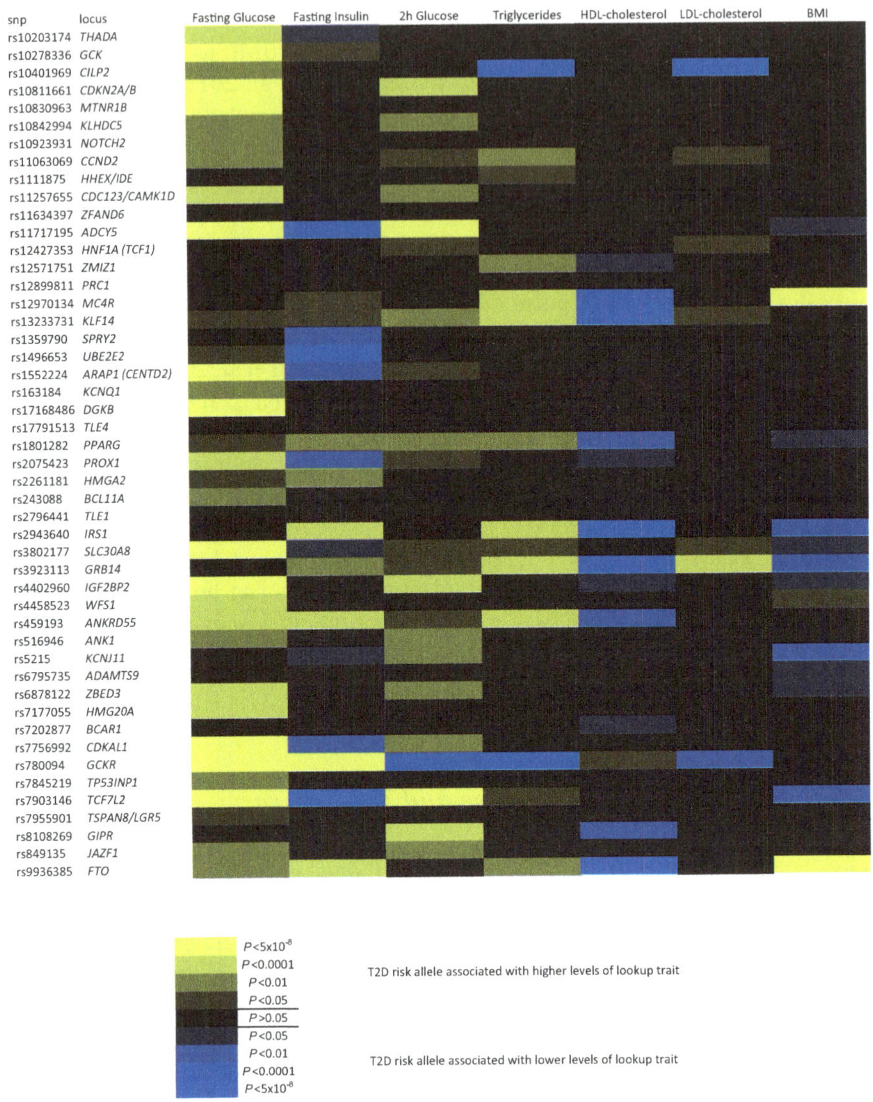

Fig. 3.1 T2D loci and their association with FG, FI, 2hGlu, and other metabolic traits

(Fig. 3.1). Given the vast sample sizes in these analyses, it is unlikely that these discordant associations are driven completely by lack of power to detect them; rather that those loci which influence variation in FG among healthy individuals are not inevitably associated with disease. Alternatively, they may highlight unknown pleiotropic associations where the associations with FG are counteracted by other associations which reduce risk of T2D through some other pathway, as has been suggested for *GIPR*, where the 2h glucose-raising allele is associated with lower BMI, largely mitigating the increased risk of T2D.

An advantage of studying quantitative traits is the ability to highlight particular aetiological pathways to disease through follow-up of loci associated with T2D via close collaboration between MAGIC and the T2D consortium (DIAGRAM). Initial follow-up of those loci involved testing the association of SNPs including those associated with T2D with a range of detailed quantitative traits and highlighted the considerable heterogeneity in the associations they displayed with traits including insulin secretion, processing, and sensitivity (Ingelsson et al. 2010). A more recent approach sought to take an objective hierarchical clustering approach on loci associated with T2D at genome-wide levels of significance (Voight et al. 2010) by identifying clusters of SNPs displaying distinct patterns of association with quantitative traits (Dimas et al. 2013). This approach identified four main clusters: insulin resistance, hyperglycaemic, proinsulin, and beta cell. Interestingly, fewer than half of the 37 SNPs tested were classified into one of the four defined clusters, highlighting that a diverse range of aetiological pathways can lead to the hyperglycaemia common to a diagnosis of T2D (Ingelsson et al. 2010; Dimas et al. 2013). This work highlights the diverse range of pathways that can lead to T2D and also leaves the question of why many SNPs were not associated with any of these traits. It is intriguing that the majority of variants did not cluster into a particular category; yet whether this reflects type II error or these SNPs are associated with T2D through other, as yet unknown mechanisms, is unclear.

3.4 Where Do We Go from Here?

Genetic approaches have allowed major advances in understanding the genetic aetiology of T2D and variance in glycaemic traits since the advent of genome-wide technologies. Indeed, while some of the loci identified are proximal to prominent drug targets for T2D (Plenge et al. 2013), the identification of ~100 further loci causally implicated in the aetiology of T2D suggests that they may have the potential to uncover the drug targets of tomorrow. Recent work on rheumatoid arthritis has employed a range of bioinformatics techniques including characterising eQTLs and other functional annotation to identify candidate genes in associated loci. Their approach also showed evidence of enrichment of association in genes encoding drug targets and identified treatments for other diseases which may represent candidates for repositioning (Okada et al. 2013). However, a number of steps are required for this goal to become a plausible reality for glycaemic associations. For example, while most genetic associations map to loci containing multiple potentially functional genes, a major step in understanding the nature of the associations is to identify the causal genes. While compelling evidence exists for some of these loci (*KCNJ11, TCF7L2, SLC30A8, GCK, GCKR*), the majority are uncertain. This is further complicated by the observation that the most of these associations are in non-coding elements of the genome. The recent report of a variety of previously unrecognised functional roles for these non-coding elements via the ENCODE project (Dunham et al. 2012), and the observation that the downstream functional mediators of these associations may be quite distal to the associated SNP (Sanyal et al. 2012), offers both promise to understand the nature of these associations (Trynka and Raychaudhuri 2013) and caution that understanding these associations will be challenging (Smemo et al. 2014) (Chaps. 15–21). Regardless, a useful first step in this process is fine-mapping association signals to a high degree of resolution to allow more refined functional follow-up. These efforts are described in more detail in Chap. 6.

While over 100 individual loci are associated with T2D and related quantitative traits to date, the SNPs typically explain less than 5 % of the variance in quantitative glycaemic traits (Scott et al. 2012b) or T2D risk (Morris et al. 2012). This suggests that a large number of loci are yet to be uncovered. A range of approaches is being employed to uncover these loci, including imputation to higher density reference panels to cover more of the variation in the genome (Abecasis et al. 2012; Marchini J on behalf of the HC 2013), newer generation arrays designed to target low-frequency and rare coding variation (Huyghe et al. 2013), and targeted (Flannick et al. 2014), whole exome (Albrechtsen et al. 2013), and genome-sequencing efforts (Steinthorsdottir et al. 2014). Recent work (Scott et al. 2012b; Morris et al. 2012) suggests that a large number of individual loci remain to be discovered among those not currently reaching genome-wide significance, and with increasing sample sizes, further loci will be uncovered. However, it is likely that the development of alternative modelling approaches accounting for environmental factors, epistasis, or disease stratification will help to exploit the currently available data and aid future discovery efforts. In addition, with newer sequencing technology that brings the aspiration for the $1000 genome closer to reality (Check Hayden 2014), the large sample sizes needed to uncover additional lower-frequency variants with modest allele frequency and effect size (Agarwala et al. 2013; Zuk et al. 2014) will be more attainable. Indeed large-scale sequencing efforts are now planned within existing healthcare systems bringing these approaches directly to the clinic (http://www.genomicsengland.co.uk/), which are likely to yield additional knowledge on the allelic architecture of disease in the forthcoming years.

References

Abecasis GR, Auton A, Brooks LD, DePristo M a, Durbin RM, Handsaker RE et al (2012) An integrated map of genetic variation from 1,092 human genomes. Nature 491(7422):56–65

Agarwala V, Flannick J, Sunyaev S, Altshuler D (2013) Evaluating empirical bounds on complex disease genetic architecture. Nat Genet 45(12):1418–1427

Albrechtsen A, Grarup N, Li Y, Sparsø T, Tian G, Cao H et al (2013) Exome sequencing-driven discovery of coding polymorphisms associated with common metabolic phenotypes. Diabetologia 56(2):298–310

Altshuler D, Hirschhorn JN, Klannemark M, Lindgren CM, Vohl MC, Nemesh J et al (2000) The common PPARgamma Pro12Ala polymorphism is associated with decreased risk of type 2 diabetes. Nat Genet 26(1):76–80

Boden G, Ruiz J, Urbain JL, Chen X (1996a) Evidence for a circadian rhythm of insulin secretion. Am J Physiol 271(2 Pt 1):E246–E252

Boden G, Chen X, Urbain JL (1996b) Evidence for a circadian rhythm of insulin sensitivity in patients with NIDDM caused by cyclic changes in hepatic glucose production. Diabetes 45 (8):1044–1050

Bonnefond A, Clément N, Fawcett K, Yengo L, Vaillant E, Guillaume J-L et al (2012) Rare MTNR1B variants impairing melatonin receptor 1B function contribute to type 2 diabetes. Nat Genet 44(3):297–301

Bouatia-Naji N, Rocheleau G, Van Lommel L, Lemaire K, Schuit F, Cavalcanti-Proença C et al (2008) A polymorphism within the G6PC2 gene is associated with fasting plasma glucose levels. Science 320(5879):1085–1088

Bouatia-Naji N, Bonnefond A, Cavalcanti-Proença C, Sparsø T, Holmkvist J, Marchand M et al (2009) A variant near MTNR1B is associated with increased fasting plasma glucose levels and type 2 diabetes risk. Nat Genet 41(1):89–94

Bunger MK, Wilsbacher LD, Moran SM, Clendenin C, Radcliffe L a, Hogenesch JB et al (2000) Mop3 is an essential component of the master circadian pacemaker in mammals. Cell 103 (7):1009–1017

Chambers JC, Zhang W, Zabaneh D, Sehmi J, Jain P, Mccarthy MI et al (2009) Common genetic variation near melatonin receptor MTNR1B contributes to raised plasma glucose and increased risk of type 2 diabetes among Indian Asians and European Caucasians. Diabetes; 58 (11):2703–2708

Check Hayden E (2014) Is the $1,000 genome for real? Nature. doi:10.1038/nature.2014.14530

Chen W, Erdos MR, Jackson AU, Saxena R, Sanna S, Silver KD et al (2008) Variations in the G6PC2/ABCB11 genomic region are associated with fasting glucose levels. J Clin Invest 118 (7):2620–2628

Coban E, Ozdogan M, Timuragaoglu A (2004) Effect of iron deficiency anemia on the levels of hemoglobin A1c in nondiabetic patients. Acta Haematol 112(3):126–128

De Silva NMG, Freathy RM, Palmer TM, Donnelly L a, Luan J, Guant T et al (2011) Mendelian randomization studies do not support a role for raised circulating triglyceride levels influencing type 2 diabetes, glucose levels, or insulin resistance. Diabetes 60(3):1008–1018

Dimas AS, Lagou V, Barker A, Knowles JW, Mägi R, Hivert M-F et al (2013) Impact of type 2 diabetes susceptibility variants on quantitative glycemic traits reveals mechanistic heterogeneity. Diabetes 2013:3–75

Dunham I, Kundaje A, Aldred SF, Collins PJ, Davis C a, Doyle F et al (2012) An integrated encyclopedia of DNA elements in the human genome. Nature 489(7414):57–74

Dupuis J, Langenberg C, Prokopenko I, Saxena R, Soranzo N, Jackson AU et al (2010) New genetic loci implicated in fasting glucose homeostasis and their impact on type 2 diabetes risk. Nat Genet 42(2):105–116

Færch K, Witte DR, Tabák AG, Perreault L, Herder C, Brunner EJ et al (2013) Trajectories of cardiometabolic risk factors before diagnosis of three subtypes of type 2 diabetes: a post-hoc analysis of the longitudinal Whitehall II cohort study. Lancet Diabetes Endocrinol 1(1):43–51

Farmer A (2012) Use of HbA1c in the diagnosis of diabetes. BMJ 345:e7293

Flannick J, Thorleifsson G, Beer NL, Jacobs SBR, Grarup N, Burtt NP et al (2014) Loss-of-function mutations in SLC30A8 protect against type 2 diabetes. Nat Genet 46(4):357–363

Franklin CS, Aulchenko YS, Huffman JE, Vitart V, Hayward C, Polašek O et al (2010) The TCF7L2 diabetes risk variant is associated with $HbA_1(C)$ levels: a genome-wide association meta-analysis. Ann Hum Genet 74(6):471–478

Frayling TM, Timpson NJ, Weedon MN, Zeggini E, Freathy RM, Lindgren CM et al (2007) A common variant in the FTO gene is associated with body mass index and predisposes to childhood and adult obesity. Science 316(5826):889–894

Gao N, LeLay J, Vatamaniuk MZ, Rieck S, Friedman JR, Kaestner KH (2008) Dynamic regulation of Pdx1 enhancers by Foxa1 and Foxa2 is essential for pancreas development. Genes Dev 22 (24):3435–3448

González-Sánchez JL, Martínez-Larrad MT, Zabena C, Pérez-Barba M, Serrano-Ríos M (2008) Association of variants of the TCF7L2 gene with increases in the risk of type 2 diabetes and the proinsulin:insulin ratio in the Spanish population. Diabetologia 51(11):1993–1997

Grant SF, Thorleifsson G, Reynisdottir I, Benediktsson R, Manolescu A, Sainz J et al (2006) Variant of transcription factor 7-like 2 (TCF7L2) gene confers risk of type 2 diabetes. Nat Genet 38(3):320–323

Hani E, Boutin P, Durand E, Inoue H, Permutt M a, Velho G et al (1998) Missense mutations in the pancreatic islet beta cell inwardly rectifying K+ channel gene (KIR6.2/BIR): a meta-analysis suggests a role in the polygenic basis of Type II diabetes mellitus in Caucasians. Diabetologia 41(12):1511–1515

Hanson RL, Pratley RE, Bogardus C, Narayan KM, Roumain JM, Imperatore G et al (2000) Evaluation of simple indices of insulin sensitivity and insulin secretion for use in epidemiologic studies. Am J Epidemiol 151(2):190–198

Heid IM, Jackson AU, Randall JC, Winkler TW, Qi L, Steinthorsdottir V et al (2010) Meta-analysis identifies 13 new loci associated with waist-hip ratio and reveals sexual dimorphism in the genetic basis of fat distribution. Nat Genet 42(11):949–960

Huyghe JR, Jackson AU, Fogarty MP, Buchkovich ML, Stančáková A, Stringham HM et al (2013) Exome array analysis identifies new loci and low-frequency variants influencing insulin processing and secretion. Nat Genet 45(2):197–201

Ingelsson E, Langenberg C, Hivert M, Prokopenko I, Lyssenko V, Dupuis J et al (2010) Detailed physiologic characterization reveals diverse mechanisms for novel genetic Loci regulating glucose and insulin metabolism in humans. Diabetes 59(5):1266–1275

DECODE Study Group, the European Diabetes Epidemiology Group. (2001) Glucose tolerance and cardiovascular mortality: comparison of fasting and 2 hour diagnostic criteria. Arch Intern Med 161(3):397–405

Jonsson J, Carlsson L, Edlund T, Edlund H (1994) Insulin-promoter-factor 1 is required for pancreas development in mice. Nature 371(6498):606–609

Kahn SE, Hull RL, Utzschneider KM (2006) Mechanisms linking obesity to insulin resistance and type 2 diabetes. Nature 444(7121):840–846

Khaw K-T, Wareham N, Bingham S, Luben R, Welch A, Day N (2004) Association of hemoglobin A1c with cardiovascular disease and mortality in adults: the European prospective investigation into cancer in Norfolk. Ann Intern Med 141(6):413–420

Kilpeläinen TO, Zillikens MC, Stančákova A, Finucane FM, Ried JS, Langenberg C et al (2011) Genetic variation near IRS1 associates with reduced adiposity and an impaired metabolic profile. Nat Genet 43(8):753–760

Li S, Zhao JH, Luan J, Langenberg C, Luben RN, Khaw KT et al (2011) Genetic predisposition to obesity leads to increased risk of type 2 diabetes. Diabetologia 54(4):776–782

Manning AK, Hivert M-F, Scott R a, Grimsby JL, Bouatia-Naji N, Chen H et al (2012) A genome-wide approach accounting for body mass index identifies genetic variants influencing fasting glycemic traits and insulin resistance. Nat Genet 44(6):659–669

Marcheva B, Ramsey KM, Buhr ED, Kobayashi Y, Su H, Ko CH et al (2010) Disruption of the clock components CLOCK and BMAL1 leads to hypoinsulinaemia and diabetes. Nature 466 (7306):627–631

Marchini J on behalf of the HC (2013) A haplotype map derived from whole genome low-coverage sequencing of over 25,000 individuals. In: Paper presented at the 63rd annual meeting of the American Society of Human Genetics, Abstract 349

Matthews DR, Hosker JPR, Rudenski AS, Naylor BA, Treacher DF, Turner RC et al (1985) Homeostasis model assessment: insulin resistance and beta-cell function from fasting plasma glucose and insulin concentrations in man. Diabetologia 28(7):412–419

Morris ADP, Voight BF, Teslovich TM, Ferreira T, Segrè AV, Steinthorsdottir V et al (2012) Large-scale association analysis provides insights into the genetic architecture and pathophysiology of type 2 diabetes. Nat Genet 44(9):981–990

Muniyappa R, Lee S, Chen H, Quon MJ (2008) Current approaches for assessing insulin sensitivity and resistance in vivo: advantages, limitations, and appropriate usage. Am J Physiol Endocrinol Metab 294(1):E15–E26

Okada Y, Wu D, Trynka G, Raj T, Terao C, Ikari K et al (2013) Genetics of rheumatoid arthritis contributes to biology and drug discovery. Nature 506(7488):376–381

Orho-Melander M, Melander O, Guiducci C, Perez-Martinez P, Corella D, Roos C et al (2008) Common missense variant in the glucokinase regulatory protein gene is associated with increased plasma triglyceride and C-reactive protein but lower fasting glucose concentrations. Diabetes 57(11):3112–3121

Paré G, Chasman DI, Parker AN, Nathan DM, Miletich JP, Zee RY et al (2008) Novel association of HK1 with glycated hemoglobin in a non-diabetic population: a genome-wide evaluation of 14,618 participants in the Women's Genome Health Study. PLoS Genet 4(12):e1000312

Pasquali L, Gaulton KJ, Rodríguez-Seguí S a, Mularoni L, Miguel-Escalada I, Akerman I et al (2014) Pancreatic islet enhancer clusters enriched in type 2 diabetes risk-associated variants. Nat Genet 46:136–143

Plenge RM, Scolnick EM, Altshuler D (2013) Validating therapeutic targets through human genetics. Nat Rev Drug Discov 12(8):581–594

Polonsky KS, Given BD, Hirsch LJ, Tillil H, Shapiro ET, Beebe C et al (1988) Abnormal patterns of insulin secretion in non-insulin-dependent diabetes mellitus. N Engl J Med 318 (19):1231–1239

Poulsen P, Levin K, Petersen I, Christensen K, Beck-Nielsen H, Vaag A (2005) Heritability of insulin secretion, peripheral and hepatic insulin action, and intracellular glucose partitioning in young and old Danish twins. Diabetes 54(1):275–283

Prokopenko I, Langenberg C, Florez JC, Saxena R, Soranzo N, Thorleifsson G et al (2009) Variants in MTNR1B influence fasting glucose levels. Nat Genet 41(1):77–81

Prokopenko I, Poon W, Mägi R, Prasad BR, Salehi SA, Almgren P et al (2014) A central role for GRB10 in regulation of islet function in man. PLoS Genet 10(4):e1004235, Williams SM (editor)

Prudente S, Morini E, Trischitta V (2009) Insulin signaling regulating genes: effect on T2DM and cardiovascular risk. Nat Rev Endocrinol 5(12):682–693

Reaven GM (1988) Banting lecture 1988. Role of insulin resistance in human disease. Diabetes 37 (12):1595–1607

Roberts WL, Safar-Pour S, De BK, Rohlfing CL, Weykamp CW, Little RR (2005) Effects of hemoglobin C and S traits on glycohemoglobin measurements by eleven methods. Clin Chem 51(4):776–778

Røder ME, Porte D, Schwartz RS, Kahn SE (1998) Disproportionately elevated proinsulin levels reflect the degree of impaired B cell secretory capacity in patients with noninsulin-dependent diabetes mellitus. J Clin Endocrinol Metab 83(2):604–608

Rung J, Cauchi S, Albrechtsen A, Shen L, Rocheleau G, Cavalcanti-Proença C et al (2009) Genetic variant near IRS1 is associated with type 2 diabetes, insulin resistance and hyperinsulinemia. Nat Genet 41(10):1110–1115

Sabatti C, Service SK, Hartikainen A-L, Pouta A, Ripatti S, Brodsky J et al (2009) Genome-wide association analysis of metabolic traits in a birth cohort from a founder population. Nat Genet 41(1):35–46

Sanyal A, Lajoie BR, Jain G, Dekker J (2012) The long-range interaction landscape of gene promoters. Nature 489(7414):109–113

Saxena R, Voight BF, Lyssenko V, Burtt NP, de Bakker PIW, Chen H et al (2007) Genome-wide association analysis identifies loci for type 2 diabetes and triglyceride levels. Science 316 (5829):1331–1336

Saxena R, Hivert M-F, Langenberg C, Tanaka T, Pankow JS, Vollenweider P et al (2010) Genetic variation in GIPR influences the glucose and insulin responses to an oral glucose challenge. Nat Genet 42(2):142–148
Scott LJ, Mohlke KL, Bonnycastle LL, Willer CJ, Li Y, Duren WL et al (2007) A genome-wide association study of type 2 diabetes in Finns detects multiple susceptibility variants. Science 316(5829):1341–1345
Scott RA, Chu AY, Grarup N, Manning AK, Hivert M-F, Shungin D et al (2012a) No interactions between previously associated 2hour glucose gene variants and physical activity or BMI on 2hour glucose levels. Diabetes 61(5):1291–1296
Scott RA, Lagou V, Welch RP, Wheeler E, Montasser ME, Luan J et al (2012b) Large-scale association analyses identify new loci influencing glycemic traits and provide insight into the underlying biological pathways. Nat Genet 44(9):991–1005
Sladek R, Rocheleau G, Rung J, Dina C, Shen L, Serre D et al (2007) A genome-wide association study identifies novel risk loci for type 2 diabetes. Nature 445(7130):881–885
Smemo S, Tena JJ, Kim K-H, Gamazon ER, Sakabe NJ, Gómez-Marín C et al (2014) Obesity-associated variants within FTO form long-range functional connections with IRX3. Nature 507 (7492):371–375
Snieder H, Boomsma DI, van Doornen LJ, Neale MC (1999) Bivariate genetic analysis of fasting insulin and glucose levels. Genet Epidemiol 16(4):426–446
Soranzo N, Spector TD, Mangino M, Kühnel B, Rendon A, Teumer A et al (2009) A genome-wide meta-analysis identifies 22 loci associated with eight hematological parameters in the HaemGen consortium. Nat Genet 41(11):1182–1190
Soranzo N, Sanna S, Wheeler E, Gieger C, Radke D, Dupuis J et al (2010) Common variants at 10 genomic loci influence hemoglobin A_1(C) levels via glycemic and nonglycemic pathways. Diabetes 59(12):3229–3239
Speliotes EK, Willer CJ, Berndt SI, Monda KL, Thorleifsson G, Jackson AU et al (2010) Association analyses of 249,796 individuals reveal 18 new loci associated with body mass index. Nat Genet 42(11):937–948
Stamenkovic JA, Olsson AH, Nagorny CL, Malmgren S, Dekker-Nitert M, Ling C et al (2012) Regulation of core clock genes in human islets. Metabolism 61(7):978–985
Steinthorsdottir V, Thorleifsson G, Reynisdottir I, Benediktsson R, Jonsdottir T, Walters GB et al (2007) A variant in CDKAL1 influences insulin response and risk of type 2 diabetes. Nat Genet 39(6):770–775
Steinthorsdottir V, Thorleifsson G, Sulem P, Helgason H, Grarup N, Sigurdsson A et al (2014) Identification of low-frequency and rare sequence variants associated with elevated or reduced risk of type 2 diabetes. Nat Genet 46(3):294–298
Strawbridge RJ, Dupuis J, Prokopenko I, Barker A, Ahlqvist E, Rybin D et al (2011) Genome-wide association identifies nine common variants associated with fasting proinsulin levels and provides new insights into the pathophysiology of type 2 diabetes. Diabetes 60(10):2624–2634
Teslovich TM, Musunuru K, Smith AV, Edmondson AC, Stylianou IM, Koseki M et al (2010) Biological, clinical and population relevance of 95 loci for blood lipids. Nature 466 (7307):707–713
Trynka G, Raychaudhuri S (2013) Using chromatin marks to interpret and localize genetic associations to complex human traits and diseases. Curr Opin Genet Dev 23(6):635–641
Turek FW, Joshu C, Kohsaka A, Lin E, Ivanova G, McDearmon E et al (2005) Obesity and metabolic syndrome in circadian Clock mutant mice. Science 308(5724):1043–1045
Vaxillaire M, Cavalcanti-proenc C, Tichet J, Marre M, Balkau B, Froguel P et al (2008) The common P446L polymorphism in GCKR inversely modulates fasting glucose and triglyceride levels and reduces type 2 diabetes risk in the DESIR prospective general French population. Diabetes 2008:2253–2257
Voight BF, Scott LJ, Steinthorsdottir V, Morris AP, Dina C, Welch RP et al (2010) Twelve type 2 diabetes susceptibility loci identified through large-scale association analysis. Nat Genet 42 (7):579–589

Voight BF, Kang HM, Ding J, Palmer CD, Sidore C, Chines PS et al (2012) The metabochip, a custom genotyping array for genetic studies of metabolic, cardiovascular, and anthropometric traits. PLoS Genet 8(8):e1002793

Weedon MN, Clark VJ, Qian Y, Ben-Shlomo Y, Timpson N, Ebrahim S et al (2006) A common haplotype of the glucokinase gene alters fasting glucose and birth weight: association in six studies and population-genetics analyses. Am J Hum Genet 79(6):991–1001

WTCCC (2007) Genome-wide association study of 14,000 cases of seven common diseases and 3,000 shared controls. Nature 447(7145):661–678

Zeggini E, Weedon MN, Lindgren CM, Frayling TM, Elliott KS, Lango H et al (2007) Replication of genome-wide association signals in UK samples reveals risk loci for type 2 diabetes. Science 316(5829):1336–1341

Zethelius B, Byberg L, Hales CN, Lithell H, Berne C (2002) Proinsulin is an independent predictor of coronary heart disease: report from a 27-year follow-up study. Circulation 105 (18):2153–2158

Zuk O, Schaffner SF, Samocha K, Do R, Hechter E, Kathiresan S et al (2014) Searching for missing heritability: designing rare variant association studies. Proc Natl Acad Sci USA 111 (4):E455–E464

Chapter 4
Genome-Wide Association Studies (GWAS) of Adiposity

Erik Ingelsson and Tuomas O. Kilpeläinen

Abstract Adiposity is strongly heritable and one of the leading risk factors for type 2 diabetes, cardiovascular disease, cancer, and premature death. In the past 8 years, genome-wide association studies (GWAS) have greatly increased our understanding of the genes and biological pathways that regulate adiposity by identifying more than 100 novel susceptibility loci for overall adiposity and more than 70 loci for body fat distribution. The results for overall adiposity highlight a significant neuronal component, whereas loci regulating body fat distribution demonstrate a central role for adipocyte biology and insulin resistance in the pathophysiology. The effect sizes of all identified loci are small, and even in aggregate, they explain <3 % of the variance in each adiposity trait. This and other evidence suggest that numerous new loci will be identified in extended meta-analyses in the future. The translation of the new discoveries into clinical care remains a major challenge. As the first step, further studies are required to establish the causal genes and variants and to disentangle the exact physiological mechanisms underlying each genotype-phenotype association.

4.1 Introduction

Since the advent of the genome-wide association study (GWAS) approach in complex disease genetics, some of the most extensive collaborations utilizing the method have focused on adiposity traits. Most recently, meta-analyses of adiposity

E. Ingelsson
Department of Medical Sciences, Molecular Epidemiology and Science for Life Laboratory, Uppsala University, Uppsala, Sweden

Department of Medicine, Division of Cardiovascular Medicine, Stanford University School of Medicine, Stanford, California, USA

T.O. Kilpeläinen (✉)
The Novo Nordisk Foundation Center for Basic Metabolic Research, Section of Metabolic Genetics, Faculty of Health and Medical Sciences, University of Copenhagen, Copenhagen, Denmark
e-mail: tuomas.kilpelainen@sund.ku.dk

J.C. Florez (ed.), *The Genetics of Type 2 Diabetes and Related Traits*,
DOI 10.1007/978-3-319-01574-3_4

GWAS have amassed >300,000 samples from up to 125 studies (Locke et al. 2015; Shungin et al. 2015). Such large-scale efforts have been motivated by (1) the major impact of adiposity on public health and in particular its strong link to type 2 diabetes, the incidence of which is currently increasing in a pandemic manner (WHO 2000; Zimmet et al. 2001); (2) the evident heritability of obesity of ~40 % (Hemani et al. 2013) and the poor success in identifying genetic susceptibility loci before the GWAS era (Li and Loos 2008); and (3) the availability of the simple, noninvasive adiposity measurements in most epidemiological studies, which has enabled GWAS investigators to collect very large sample sizes for meta-analyses.

Most GWAS of adiposity traits have focused on overall adiposity (assessed as BMI, weight divided by height squared) that has a strong link to morbidity and mortality (WHO 2000; Pischon et al. 2008). However, the genetic basis of body fat distribution (often assessed by measuring waist and hip circumferences and their ratio) has raised additional interest because of the well-documented association between intra-abdominal fat and elevated risk of metabolic and cardiovascular diseases, beyond that of overall adiposity (Pischon et al. 2008).

Since 2006, meta-analyses of GWAS have identified more than 100 loci for overall adiposity and more than 70 loci for fat distribution. These discoveries have given valuable novel insights into the genetic architecture of adiposity, which may ultimately open up new avenues for the prevention of obesity and its related comorbidities. In this chapter, we will review these recent discoveries, discuss their biological and public health significance, and reflect on the prospects for the next few years.

4.2 GWAS of Overall Adiposity

4.2.1 The First Discoveries: FTO *and* MC4R

The first discovery of a locus regulating overall adiposity was made in 2007 when three separate GWAS identified variants in the first intron of the *FTO* gene as being unequivocally associated with BMI (Frayling et al. 2007; Scuteri et al. 2007; Hinney et al. 2007). The initial discovery of *FTO* did not, however, occur in a GWAS of an adiposity trait, but in a GWAS of type 2 diabetes. This study identified a common variant in the first intron of the *FTO* gene to be highly significantly associated with type 2 diabetes, but adjustment for BMI abolished the association, indicating that the association was mediated through increased adiposity (Frayling et al. 2007). The association of *FTO* with BMI was replicated in 38,759 adults and children, where each risk allele of the *FTO* variant increased adult BMI by 0.34–0.46 kg/m^2 (~1 kg of body weight) and the variant explained 0.34 % of the interindividual variance in BMI (Frayling et al. 2007). Two other studies, including a GWAS of BMI among 4741 Sardinians (Scuteri et al. 2007) and a GWAS of early-onset extreme obesity in 487 cases and 442 healthy lean controls (Hinney et al. 2007) confirmed *FTO* as the first locus robustly associated with common adiposity.

In addition to *FTO*, the three initial GWAS of adiposity traits took forward several other variants, but none of these initial discoveries were replicated in subsequent analyses suggesting that larger sample sizes were required to identify additional adiposity loci. Consequently, an extensive collaboration between scientists from Europe and the USA was initiated, and the first meta-analysis of the Genomic Investigation of ANthropometric Traits (GIANT) consortium was published in 2008. By combining GWAS data from 16,876 adults of European descent and with a replication in additional 60,000 adults, the consortium reported a novel BMI signal 188 kb downstream of *MC4R*, a very obvious candidate gene due to its role in the hypothalamic regulation of food intake and whose mutations were known to be the most common source of monogenic obesity (Loos et al. 2008; Huszar et al. 1997; Farooqi et al. 2003).

4.2.2 Extended Meta-Analyses Identify 100 New BMI Loci

The number of known BMI loci increased drastically in 2009 when two large-scale GWAS meta-analyses were carried out by the GIANT consortium and deCODE genetics. GIANT used data from 32,387 adults of European descent to identify six novel BMI loci in or near the *NEGR1*, *TMEM18*, *SH2B1*, *KCTD15*, *GNPDA2*, and *MTCH2* genes (Willer et al. 2009). At the same time, deCODE genetics carried out an independent GWAS meta-analysis of 31,392 adults of mainly Icelandic origin, which identified four of the loci discovered by the GIANT consortium while also identifying four additional new loci in or near the *SEC16B*, *ETV5*, *BDNF*, and *BCDIN3D* genes (Thorleifsson et al. 2009).

In 2010, the third meta-analysis of the GIANT consortium was published, this time involving 123,865 individuals of European ancestry at the genome-wide stage with a follow-up of the strongest signals in 125,931 additional adults (Speliotes et al. 2010). This meta-analysis confirmed all 12 loci identified in previous studies and enabled the identification of 20 novel BMI-associated loci (Fig. 4.1). In aggregate, the 32 confirmed BMI loci explained 1.45 % of the interindividual variation in BMI (Speliotes et al. 2010).

The discoveries of BMI loci in GWAS of individuals of European ancestry were followed by large-scale meta-analyses in populations of non-European ancestry, in whom the differing allele frequencies and effect sizes may facilitate the discovery of novel adiposity loci. Indeed, in 2012, two large GWAS meta-analyses of South-Asian populations with genome-wide sample sizes of 26,620 and 27,715 individuals discovered four novel loci in or near the *CDKAL1*, *KLF9*, *PCSK1*, and *GP2* genes (Wen et al. 2012; Okada et al. 2012). A year later, a meta-analysis of 39,144 men and women of African ancestry with replication in 32,268 Africans identified a new locus in *GALNT10*. Furthermore, when the African-ancestry data were meta-analyzed with the data of 123,865 white European adults from the meta-analysis of

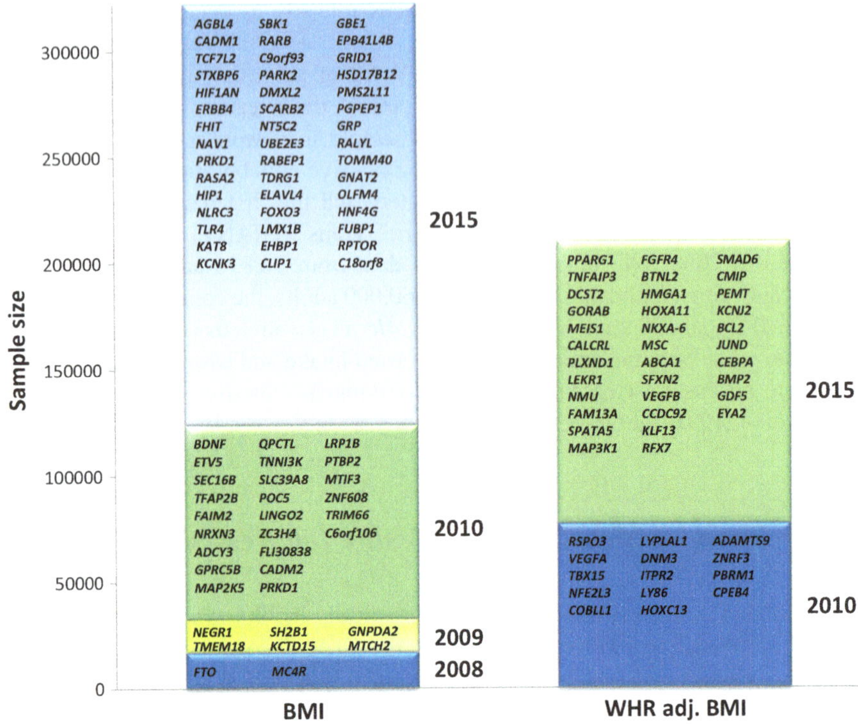

Fig. 4.1 Sample size of the discovery stage (number of individuals with genome-wide data) and the new loci identified in the GIANT consortium meta-analyses of individuals of European descent for BMI in 2008 (Loos et al. 2008), 2009 (Willer et al. 2009), 2010 (Speliotes et al. 2010), and 2015 (Locke et al. 2015) and for BMI-adjusted waist-hip ratio (WHR adj. BMI) in 2010 (Heid et al. 2010) and 2015 (Shungin et al. 2015)

the GIANT consortium (Speliotes et al. 2010), another novel locus was identified near the *NFE2L3* gene (Monda et al. 2013) thus raising the total number of identified BMI loci to 38.

In 2013, yet another novel adiposity locus, *TOMM40*, was identified in a meta-analysis utilizing a gene-centric array of 49,320 SNPs across ~2100 metabolic and cardiovascular-related loci in 108,912 adults of European, African-American, Hispanic, or East Asian ancestry (Guo et al. 2013).

The most recent meta-analysis from the GIANT consortium, published in 2014, included 236,231 individuals with genome-wide data and 103,047 individuals genotyped with the Metabochip—a custom-selected genotyping array designed to cover the loci that fell just below genome-wide significant thresholds in previous GWAS of anthropometric and metabolic traits (Locke et al. 2015; Voight et al. 2012). Of the 339,226 individuals that were included in this meta-analysis, 322,154 were of European descent and 17,072 of non-European descent. A meta-analysis of the European-ancestry individuals identified 77 loci reaching genome-wide significance (Fig. 4.1). Ten additional loci were identified when including the

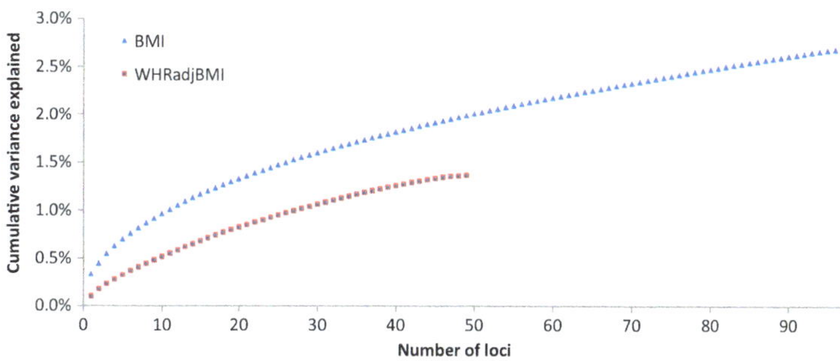

Fig. 4.2 Cumulative variance explained by the 97 BMI loci identified in Locke et al. (2015) and 49 BMI-adjusted waist-hip ratio loci in Shungin et al. (2015). The loci are ordered by the proportion of the phenotypic variance explained with the loci explaining largest proportion of variance on the *left* and the loci explaining smallest proportion of variance on the *right*

non-European-descent individuals and 10 further loci in stratified analyses (3 loci in men, $n = 152{,}893$; 3 loci in women, $n = 171{,}799$; and 4 loci in population-based studies, $n = 209{,}521$). The total number of loci with a genome-wide significant association with BMI in this meta-analysis thus reached 97, of which 56 were novel and 41 had been identified in previous meta-analyses of adiposity traits (Monda et al. 2013; Berndt et al. 2013; Kilpelainen et al. 2011; Guo et al. 2013; Speliotes et al. 2010; Bradfield et al. 2012).

In summary, GWAS meta-analyses of adult populations have discovered 102 loci to be associated with BMI. These findings have been highly significant and reproducible in large populations, but the effect sizes of all loci are tiny. The more recently identified loci generally have lower minor allele frequency and/or smaller effect sizes than the loci identified in the early meta-analyses and explain a smaller proportion of variance in BMI. In total, the 97 loci identified in the latest GIANT paper (Locke et al. 2015) explain 2.7 % of the variance in BMI in adults of white European descent (Fig. 4.2).

4.2.3 GWAS of Body Fat Mass and Fat Percentage

Although BMI is a good indicator of overall adiposity and disease risk, it cannot distinguish between lean and fat body mass. The identified BMI loci may thus increase BMI by regulating either the level of body fat mass, body lean mass, or both. Studies of more specific measures of adiposity, including body fat mass or body fat percentage (fat mass/weight), assessed with dual-emission X-ray absorptiometry (DEXA) or bioelectrical impedance analysis, may therefore be helpful in disentangling genetics of adiposity.

In 2011, a GWAS meta-analysis of body fat percentage, measured with DEXA or bioimpedance, in 36,626 individuals of European ($n = 29{,}069$) or Indian-Asian ($n = 7557$) descent and a follow-up in 39,576 additional individuals of European ancestry confirmed *FTO* as an adiposity locus but also reported two novel loci associated with body fat percentage near the *IRS1* and *SPRY2* genes (Kilpelainen et al. 2011).

In 2013, a meta-analysis of 10,196 individuals of mainly European ethnicity and a follow-up in 3923 individuals of European ancestry and 2740 Hans Chinese identified a novel locus in the *CTSS* gene associated with DEXA-measured fat body mass while adjusting for lean body mass (Pei et al. 2013).

4.3 GWAS of Extreme Obesity

The reasons for using samples of extremely obese individuals when searching for new adiposity loci include the assumptions of larger effect sizes with resulting higher statistical power per individual included, greater genetic contribution to trait variance, enrichment of highly penetrant variants, and lower locus heterogeneity. The definitions of extreme obesity have typically been based on fixed levels of BMI, such as clinically established cut points for obesity class II (BMI > =35) or class III (BMI > =40) or on relative cut points based on the BMI distribution, such as >99th percentile of BMI. Studies have included children and adolescents (Hinney et al. 2007; Scherag et al. 2010; Bradfield et al. 2012; Wheeler et al. 2013), adults (Cotsapas et al. 2009; Jiao et al. 2011; Paternoster et al. 2011; Wang et al. 2011; Berndt et al. 2013), or both children and adults (Meyre et al. 2009) in their discovery stages.

The earliest GWAS of extreme obesity, published in 2007–2011, were based on a relatively small number of extremely obese cases (less than 1500 individuals in the discovery stage) and normal-weight controls. As a result, these studies had limited statistical power and most either failed to report any statistically significant loci (Hinney et al. 2007); reported only known adiposity loci, such as *FTO* or *MC4R* (Cotsapas et al. 2009; Scherag et al. 2010; Wang et al. 2011); or reported loci that have proven difficult to replicate in further studies, such as *KCNMA1* (Jiao et al. 2011). Meyre and colleagues included 1380 individuals with early-onset and morbid adult obesity and 1416 age-matched normal-weight controls with follow-up in 14,186 additional individuals. In addition to *FTO* and *MC4R*, they reported three new risk loci: *NPC1*, *PTER*, and *MAF* (Meyre et al. 2009). *NPC1* and *PTER* have been replicated in later studies (Berndt et al. 2013), while it still remains uncertain whether *MAF* represents a false-positive or a population-specific finding.

In 2011, a study of 2633 individuals drawn from the extremes of the BMI distribution and 2740 controls from two very large Danish cohorts reported genome-wide significant associations with *FTO*, *MC4R*, *FAIM2*, and *TFAP2B*, all previously known adiposity loci (Paternoster et al. 2011). Bradfield and colleagues used a slightly less extreme definition of childhood obesity (>95th percentile of BMI) and increased the sample size to 5530 cases and 8318 controls (<50th percentile of BMI), which resulted in two new loci that reached genome-wide significance when combined with the replication stage, *OLFM4* and *HOXB5*

(Bradfield et al. 2012). Both loci showed evidence of association in two extreme childhood obesity cohorts, as well as in the GIANT meta-analysis of adult BMI published in 2010 (Speliotes et al. 2010).

In 2013, a study of 1509 children with severe early-onset obesity (BMI standard deviation score (SDS) > 3 and onset of obesity before the age of 10 years) and 5380 controls with follow-up in additional 971 severely obese children and 1990 controls identified 4 new loci associated with severe obesity (*LEPR*, *PRKCH*, *PACS1*, and *RMST*) (Wheeler et al. 2013). They also reported a significant burden of rare, single CNVs in severely obese cases. Integrative gene network pathway analysis of rare deletions indicated enrichment of genes affecting G protein-coupled receptors involved in the neuronal regulation of energy homeostasis.

In the same issue of *Nature Genetics*, the GIANT consortium presented their study of extreme obesity conducted in a sampling frame of up to 168,267 individuals with follow-up in up to 109,703 additional individuals (Berndt et al. 2013). Using relative cut points (upper vs. lower 5th percentiles of body mass index), as well as clinical classes of obesity, they found seven new loci (*HNF4G*, *RPTOR*, *GNAT2*, *MRPS33P4*, *ADCY9*, *HS6ST3*, *ZZZ3*) associated with clinical classes of obesity. Further, using various methods, they showed that there is a large overlap in terms of genetic structure and distribution of variants between traits based on the extremes and the general population and little etiologic heterogeneity between obesity subgroups. The seven novel loci showed a consistent direction of effect and a similar effect size in five studies that applied other ascertainment strategies for defining extremely obese; and conversely, out of the 13 loci previously associated with extreme obesity, only *MAF* and *KCNMA1* failed to show evidence of association in the GIANT study. They also performed systematic comparisons of the genetic inheritance and distribution of SNPs between the tails and full distribution, which indicated that (a) effect sizes observed in tails and those expected based on the overall distribution were similar and (b) genetic determinants for the tails were similar to those for the full distribution and that common variant loci contribute to extreme phenotypes.

In summary, current knowledge implies that common genetic variation associated with extreme forms of obesity is overlapping with that of overall BMI to a large degree. Also, while some common variants can have larger effects in the extremes, these effects as a whole are not larger than expected based on the effects in the overall distribution. A strategy with selection of individuals from the extremes for genetic analyses is a cost-effective approach that will reveal loci that are likely to be relevant and largely generalizable to the full population. That said, the sample sizes needed to reveal true novel loci are still substantial, as evidenced by the earliest GWAS of extreme obesity, which were underpowered to robustly detect novel loci.

4.4 GWAS of Body Fat Distribution

Increased abdominal fat is a risk factor for cardiovascular disease and type 2 diabetes, independent of overall adiposity (Wang et al. 2005; Canoy 2008), whereas the accumulation of gluteal subcutaneous fat may be protective (Snijder et al. 2003; Yusuf et al. 2005). Therefore, in addition to studies of overall adiposity, it is important to unravel the genetic basis of body fat distribution. Most commonly, fat distribution is assessed by waist and hip circumference and their ratio. A larger waist circumference or waist-hip ratio indicates relatively more abdominal fat whereas a larger hip circumference or smaller waist-hip ratio indicates relatively greater gluteal fat accumulation. However, as these measures are strongly correlated with overall adiposity, recent studies have adjusted circumference-based measures for BMI. Other GWAS have focused on computer tomography-based measures of fat distribution to give a more direct and precise measurement of the visceral and subcutaneous fat tissue compartments (Fox et al. 2007).

4.4.1 Discovery of the First Fat Distribution Loci

In 2009, the GIANT consortium carried out its first GWAS meta-analysis of waist circumference and waist-hip ratio, including 38,580 adults of European ancestry with follow-up in 70,689 additional adults (Lindgren et al. 2009). Two loci were found to be robustly associated with waist circumference, *TFAP2B* and *MSRA*. In addition, a locus near the *LYPLAL1* gene was associated with waist-hip ratio, but only in women. At the same time, a meta-analysis of the Cohorts for Heart and Aging Research in Genomic Epidemiology (CHARGE) consortium for waist circumference, including 31,373 individuals of European ancestry and a replication in the GIANT consortium meta-analysis, identified a novel locus in the *NRXN3* gene (Heard-Costa et al. 2009). Each of the four loci identified in these first meta-analyses were, however, also highly significantly associated with BMI. This suggested that the loci do not only affect fat distribution but also overall adiposity, which was expected due to the strong phenotypic correlation of waist and hip circumferences with BMI (Lindgren et al. 2009; Heard-Costa et al. 2009).

To identify loci that are more purely associated with fat distribution rather than overall adiposity, GWAS meta-analyses published since 2010 have mainly focused on BMI-adjusted waist-hip ratio. This measure provides a relative comparison of central and subcutaneous fat compartments after accounting for overall adiposity. The 2010 meta-analysis from the GIANT consortium in 77,167 adults of European ancestry with follow-up of the strongest findings in 113,636 adults identified 13 novel loci associated with BMI-adjusted waist-hip ratio, along with the previously identified signal in the *LYPLAL1* gene (Heid et al. 2010) (Fig. 4.1). Interestingly, seven of the 14 loci showed a stronger association in women than in men, whereas the other 7 loci showed a similar magnitude of association in both sexes.

Waist circumference is a good proxy for visceral adiposity and increased disease risk, but it is not able to discriminate between visceral and subcutaneous adipose tissue. More refined phenotypes for fat distribution may help to identify novel loci associated with specific adipose tissue compartments. In 2012, a GWAS of abdominal adipose tissue depots, assessed with computed tomography in 5560 women and 4997 men, identified the known waist-hip ratio locus in the *LYPLAL1* gene to be associated with the ratio of visceral fat to abdominal subcutaneous fat (Fox et al. 2012). Furthermore, a novel locus near the *THNSL2* gene was associated with visceral fat in women, but not in men.

In 2013, a locus in the *TMCC1* gene was identified to be associated with waist-hip ratio adjusted for BMI in a meta-analysis utilizing a gene-centric array of 49,320 SNPs across ~2100 metabolic- and cardiovascular-related loci (Guo et al. 2013).

4.4.2 Sex and Ethnicity-Specific Meta-Analyses Identify Additional Loci

The findings of pronounced sexual dimorphism for many of the loci regulating fat distribution led to a focused meta-analysis of sex differences in anthropometric traits within the GIANT consortium (Randall et al. 2013). In 2012, this sex-stratified GWAS meta-analysis of up to 60,586 men and 73,137 women with follow-up in 62,212 men and 74,657 women identified seven loci with sex-specific effects, including four of the previously established loci for waist-hip ratio adjusted for BMI (*COBLL1*, *LYPLAL1*, *VEGFA*, *ADAMTS9*) but also two novel loci near the *HSD17B4* and *PPARG* genes and a novel locus for BMI-adjusted waist circumference near *MAP3K1*. All seven loci were genome-wide significant in women but not in men. Interestingly, no loci showed strong sex differences in body weight, BMI, or hip circumference.

There is also wide variation in body fat distribution between ethnic groups (Lear et al. 2010), encouraging genetic studies of fat distribution traits in individuals of non-European ancestry. In 2008, a GWAS of 2684 Indian-Asians with a replication in 11,955 individuals of Indian-Asian or European ancestry identified the same *MC4R* locus that has been associated with BMI in European-ancestry individuals, to be associated with waist circumference (Chambers et al. 2008). In 2009, a GWAS meta-analysis of 8842 Korean individuals and a replication in 7861 additional Koreans identified a locus near the *C12orf51* gene to be associated with waist-hip ratio (Cho et al. 2009). In these two studies, the associations were not adjusted for BMI and the findings could thus be explained by association with overall adiposity. More recently, GWAS of BMI-adjusted traits in up to 33,591 individuals of African ancestry identified a locus in *LHX2* for waist circumference adjusted for BMI and a locus in *RREB1* for waist-hip ratio adjusted for BMI (Liu et al. 2013).

4.4.3 Extended Meta-Analyses Identify 51 New Loci for Waist- and Hip-Related Traits

The most recent meta-analysis of the GIANT consortium for waist- and hip-related traits, published in 2015, pooled data from 224,459 individuals, of whom 142,762 had genome-wide data and 81,697 had been genotyped with the Metabochip (Shungin et al. 2015; Voight et al. 2012). Of the Metabochip-genotyped individuals, 14,371 were of non-European ancestry. Meta-analysis of waist-hip ratio adjusted for BMI identified 49 loci, of which 32 were new and 17 were already known (Fig. 4.1). Again, the results emphasized strong sexual dimorphism in the genetic regulation of fat distribution, with 19 loci showing significantly stronger effects in women and only one locus showing a stronger effect in men, and supported by higher heritability of these traits in women (Shungin et al. 2015). The 49 SNPs explained 1.36 % of the variance in WHR adjusted for BMI overall (Fig. 4.2). In meta-analyses of other BMI-adjusted and unadjusted waist and hip measures, 19 additional loci were identified, of which seven for BMI-adjusted waist circumference, six for BMI-adjusted hip circumference, three for unadjusted hip circumference, and three for unadjusted waist-hip ratio.

In summary, the GWAS of fat distribution traits have identified 75 independent loci in recent years. These results have demonstrated that the genetic regulation of fat distribution is largely distinct from the mechanisms regulating overall adiposity. Furthermore, the strong pattern of sexual dimorphism that is seen for the genetic loci regulating body fat distribution has not been observed for loci regulating overall adiposity.

4.5 Many Obesity Loci from GWAS: What to Make Out of It?

4.5.1 Pathways Implicated by Adiposity Loci

It is becoming clear that overall obesity as assessed by BMI and body fat distribution as assessed by BMI-adjusted waist-hip ratio are driven by different and largely non-overlapping biological pathways. As highlighted already in the first study of BMI by the GIANT consortium in 2009, neuronal pathways are central to the development of overall obesity (Willer et al. 2009). Several of the first loci found to be associated with BMI were indeed suggested to act through the central nervous system (CNS), as well as to be highly expressed in the brain and often in the hypothalamus. The important role of the CNS for obesity development was further detailed in the subsequent GIANT paper on BMI in 2010 (Speliotes et al. 2010), where several lead variants mapped near key hypothalamic regulators of energy balance. In the most recent study on BMI by the GIANT consortium (Locke et al. 2015), an enrichment in gene expression of BMI-associated loci was seen not only in the hypothalamus and pituitary gland, important for appetite regulation and implicated in earlier work, but even more clearly in the hippocampus and limbic system—tissues that are important for functions such as learning, cognition, and emotion. This study also reported strong enrichment for gene sets associated with CNS, such as synaptic function, long-term potentiation, and neurotransmitter signaling, including glutamate signaling, norepinephrine, dopamine, and serotonin release cycles. Examples of CNS-related genes within BMI loci include *ELAVL4*, *GRID1*, *CADM2*, *NRXN3*, *NEGR1*, *SCG3*, *PCDH9*, *TAOK2*, and *STX1B* that are implicated in synaptic function, cell-cell adhesion, or glutamate signaling (Locke et al. 2015). There is evidence also for other mechanisms being involved in overall obesity coming from GWAS on BMI, as exemplified by the intriguing *GIPR* locus—GIPR is the receptor of GIP, an incretin that stimulates insulin release potently in the presence of orally ingested glucose (Speliotes et al. 2010). Also, gene set analyses have shown enrichment for mechanisms such as energy metabolism, polyphagia, secretion and action of insulin and related hormones, and MTOR signaling (involved in cell growth after nutrient intake via insulin and growth factors) (Locke et al. 2015). That said, there is now strong evidence that genetic determinants of overall obesity primarily derive from neuronal tissues and pathways.

This is in contrast to the genetics underlying body fat distribution, which seems to be driven by processes related to insulin and adipocyte biology. Gene set enrichment analyses of BMI-adjusted waist-hip ratio have highlighted gene sets involved in body fat regulation (including adiponectin signaling, insulin sensitivity, and regulation of glucose levels), skeletal growth, transcriptional regulation, and development (Shungin et al. 2015). Among these gene sets, there are several that are specific for metabolically active tissues including the adipose, heart, liver, and muscle—tissues that also showed higher expression of genes within loci associated with BMI-adjusted waist-hip ratio. Examples of specific pathways that are associated with body fat distribution include the VEGF and PTEN signaling pathways. VEGF signaling has a central role in angiogenesis, insulin resistance, and obesity (Elias et al. 2013), while phosphatase and tensin homolog (PTEN) signaling has been shown to promote insulin resistance (Pal et al. 2012). In the recent paper from the GIANT consortium (Shungin et al. 2015), the 49 BMI-adjusted waist-hip ratio loci were examined for overlap with regulatory elements from the ENCODE and RoadMap Epigenomic data. The strongest enrichments were seen for enhancer activity in the adipose, muscle, endothelial cells, and bone, suggesting that variation in loci involved in body fat regulates transcription in these tissues and cells. Taken together, current evidence highlights the central role of genes, pathways, and tissues involved in adipocyte metabolism and insulin resistance as being central for body fat distribution.

4.5.2 Associations of Adiposity Loci with Other Cardiovascular and Metabolic Traits

As expected, loci that have been discovered for their association with BMI or BMI-adjusted waist-hip ratio are also associated with other cardiovascular and metabolic traits more than expected by chance. Specifically, both BMI and BMI-adjusted waist-hip ratio loci are often associated with high-density lipoprotein cholesterol, triglycerides, type 2 diabetes, and insulin (Locke et al. 2015; Shungin et al. 2015). BMI loci also showed directionally consistent effects on age at menarche, while several BMI-adjusted waist-hip ratio variants were strongly associated with low-density lipoprotein cholesterol or adiponectin. These observations emphasize the common pathophysiology between adiposity and cardiometabolic traits; however, based on GWAS data, it is not possible to disentangle whether this represents pleiotropy or underlying causal relations between adiposity and other traits.

There are also several adiposity loci which show effects in directions opposite to what would be expected based on known observational trait correlations. One such locus contains *GIPR*, the incretin receptor, where the BMI-increasing allele is associated with higher fasting glucose levels and *lower* 2-h glucose levels after an oral glucose challenge (Speliotes et al. 2010). GIP was originally called gastric inhibitory polypeptide, but more recently it has been referred to as the glucose-dependent insulinotropic polypeptide, since it has been shown to be an incretin hormone with similar effects as the other main incretin hormone, GLP-1, i.e., increasing release of insulin from the pancreas in response to high levels of glucose or fat in the duodenum (Speakman 2013). Mice with disruption of Gipr show higher glucose and an impaired early insulin response, which is consistent with the incretin function, as well as resistance to diet-induced obesity. These data suggest that GIP and GIPR have an important role in the utilization of ingested nutrients by increasing insulin secretion, stimulation of fat uptake in adipocytes, and adipogenesis. Consequently, variants in *GIPR* that are associated with increased incretin effect should increase adiposity as well as insulin secretion, with resulting lower 2-h glucose.

Another intriguing locus is *IRS1*, where the BMI-increasing allele is associated with a beneficial cardiovascular risk profile (less dyslipidemia, lower fasting insulin, and higher adiponectin) and lower risk of coronary heart disease, type 2 diabetes, and diabetic nephropathy (Locke et al. 2015). The BMI-increasing variant is in perfect LD with a variant that has been associated with increased body fat percentage (and improved cardiovascular risk profile), but also with a higher ratio of subcutaneous adipose tissue to visceral adipose tissue in men (Kilpelainen et al. 2011). This suggests that the BMI-increasing and body fat percentage-increasing alleles increase subcutaneous deposition of fat, leading to lower visceral and ectopic fat accumulation and more beneficial cardiometabolic profile.

Other examples of loci with unexpected effect directions include *HHIP* (where the BMI-increasing allele is associated with decreased risk of type 2 diabetes and higher high-density lipoprotein cholesterol) and *TCF7L2* (where the BMI-increasing allele is associated with reduced risk of coronary heart disease and type 2 diabetes and decreased fasting glucose, 2-h glucose, and total cholesterol levels) (Locke et al. 2015). However, further studies are needed in order to disentangle these associations, especially in light that there is quite convincing data indicating that the association of the *TCF7L2* diabetes risk allele with adiposity is driven by ascertainment bias (Stolerman et al. 2009).

4.5.3 Functional Follow-Up of Adiposity Loci

To be able to fully appreciate the new adiposity loci arising from GWAS, there is a need for mechanistic studies to understand the pathophysiology in detail. However, before any such studies can be undertaken, the first step is to establish the causal genes and variants from the GWAS loci as they usually contain many genes, several of which are good candidates. One approach to search for the causal variants is fine-mapping of GWAS loci. Using data from both European and non-European populations genotyped on the Metabochip (Voight et al. 2012), the GIANT consortium managed to narrow the genomic regions and decrease the number of SNPs in the regions that were likely to contain the causal variant in 10 BMI and 17 - BMI-adjusted waist-hip ratio loci that were fine-mapped on the Metabochip (Locke et al. 2015; Shungin et al. 2015). For example, these analyses narrowed the *SEC16B* and *FTO* loci (BMI) and the *HOXC13* locus (BMI-adjusted waist-hip ratio) to include a single SNP (see Chap. 20 for more details on *FTO* and fine-mapping). Future fine-mapping efforts using more ethnically diverse study samples and more complete panels of variation are likely to narrow GWAS signals further, as will other fine-mapping approaches including resequencing projects. Other approaches attempting to establish causal variants and genes include analyses of functional variants, as well as eQTL analyses. Such analyses have been employed in large consortium papers (Speliotes et al. 2010; Heid et al. 2010; Locke et al. 2015; Shungin et al. 2015), resulting in lists of genes that are more likely to be causal within the GWAS loci. As an example, there are two BMI-associated loci which harbor copy number variants (CNVs): a 45-kb deletion near *NEGR1* (Willer et al. 2009) and a 21-kb deletion 50 kb upstream of *GPRC5B* (Speliotes et al. 2010). Even if correlations of lead variants from GWAS loci with plausible functional variants or eQTLs cannot ultimately prove that these are the causal variants and genes, they provide good clues and form a basis for functional follow-up studies.

Since the adiposity loci uncovered by GWAS were established only in the past few years and since the causal genes are not unequivocally established for most loci, the functional aspects of the absolute majority of adiposity loci are still not well understood. The first adiposity locus that was established from GWAS was the *FTO* locus, and although the exact function of the protein encoded by the fat mass and obesity-associated gene (*FTO*) is still unknown, there has been some progress in the past years (see Chap. 20).

Also the second locus to be established for adiposity, the melanocortin-4 receptor (*MC4R*) locus, is well characterized—primarily since loss-of-function variants in *MC4R* were already known to cause monogenic obesity (Yeo et al. 1998; Vaisse et al. 1998). Mc4r-deficient mice are characterized by hyperphagia, hyperinsulinemia, hyperglycemia, and mature-onset obesity (Huszar et al. 1997). The MC4R is expressed on the surface of neurons in the arcuate nucleus of the hypothalamus and is activated by the alpha-melanocyte-stimulating hormone [a cleavage product of proopiomelanokortin (POMC)] leading to appetite-decreasing effects (De Jonghe et al. 2011). For other adiposity loci identified in GWAS, the level of mechanistic understanding is lower. Some of the loci that were established in 2009 (De Jonghe et al. 2011) are better understood, such as *SH2B1*, *NEGR1*, and *ETV5* (Speakman 2013), but for most loci that have been discovered thereafter, there is still much research to be done before we can harvest the fruits of GWAS of adiposity.

4.6 Future Avenues in Obesity Genetics

The GWAS methodology has been extremely successful in identifying novel loci associated with adiposity and other complex traits. This means that we now have a large number of loci robustly associated with adiposity traits that need further investigation using a range of methodologies, as well as a large number of samples that have been genotyped on GWAS microarrays that can be utilized for additional studies. We have summarized some of our suggestions of where the field may be moving in the next few years in Fig. 4.3.

In terms of follow-up of the already established adiposity loci, the first step is to establish the causal genes and variants, as discussed above. Apart from fine-mapping methods using GWAS data with denser catalogues of variation [e.g., the 1000 Genomes Project (Abecasis et al. 2012) or UK10K (Muddyman et al. 2013)], along with more ethnically diverse samples and improved imputation methods, resequencing approaches can also provide evidence for causal variants and genes. Such projects can be targeted at the known adiposity loci or can use whole-exome sequencing data to assess rarer variation that may not be well-tagged by GWAS arrays. Another approach addressing rare variation in the exomes is to use the HumanExome BeadChip (Grove et al. 2013), a microarray that has been designed based on exome and whole-genome sequencing projects to allow for cost-effective mapping of exonic variation in large study samples. Other methods to identify causal variants include eQTL studies (where associations of gene variants from GWAS are studied in relation to gene transcripts from the region), targeted proteomics (where proteins encoded by genes in the GWAS loci are studied in relation to outcome), and large-scale screening in model systems (such as yeast, *Drosophila melanogaster* (fruit fly), or human cells).

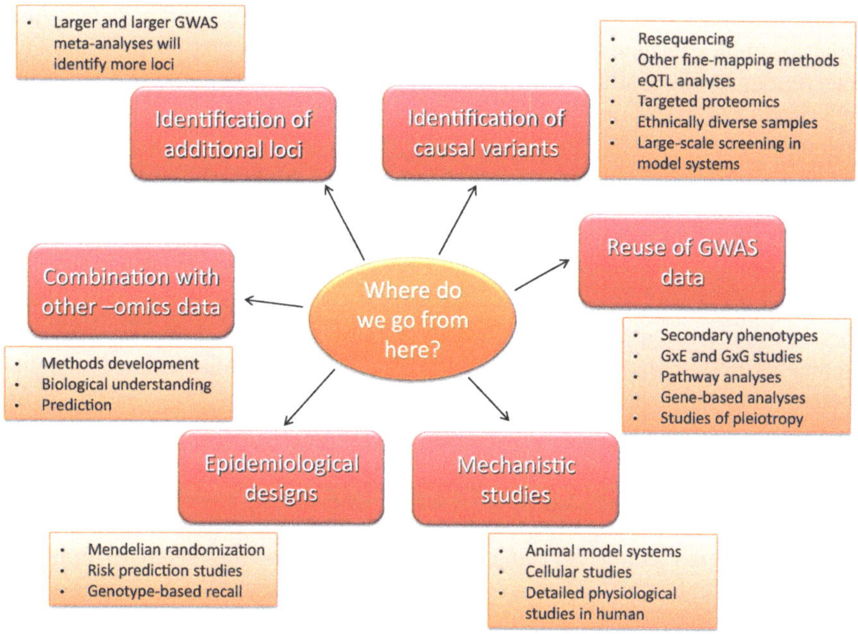

Fig. 4.3 Suggestions of future avenues of research in obesity genetics

Once the causal genes have been identified, the next step should be to undertake mechanistic studies to understand the pathophysiology underlying the genotype-phenotype association. Here, we refer to mechanistic studies in its broadest sense, and the methods can vary from in vitro model systems, to in vivo model systems, to further studies in humans. Often, it will be beneficial to combine several of these approaches to achieve a more complete understanding, as the methods are complementary. Genomic editing in cell cultures as well as in animal model systems is a very useful tool to study the effect of overexpression or knockdown of a certain gene and allows for detailed studies of specific gene variants, as well as tissue-specific effects. Animals used for in vivo systems have traditionally often been rodents, but given the large number of loci to follow-up and the advent of high-throughput techniques in new fields, more and more groups are turning to simpler model systems such as *Caenorhabditis elegans* (roundworm), *Drosophila melanogaster*, and *Danio rerio* (zebrafish), sometimes as an intermediate step before proceeding to rodent models. Overall, since the targets have first been identified in human studies, this represents a translation-back translation approach, which may be more successful in informing human medicine than studies that are based entirely on animal model systems where findings sometimes have been proven difficult to translate to humans. Mechanistic studies in humans often utilize more detailed phenotypes than in the original GWAS, for example, computer tomography-determined fat distribution or longitudinal studies with repeated phenotype assessments, but can also involve combinations with other -omics data, such as transcriptomics, epigenomics, proteomics, metabolomics, or metagenomics. Other examples of human studies that can further characterize gene function are interventions with pharmaceutical agents or genotype-based recall studies, where individuals with certain genotypes are selected from a large population and re-invited for detailed physiological examinations and/or functional studies.

As noted, there are a large number of samples that have already been genotyped on GWAS microarrays that can be utilized for additional studies, for example, gene-environment or gene-gene interaction studies, pathway analyses, gene-based analyses, studies of pleiotropy, or studies of other phenotypes than the phenotype for which the study was originally initiated. The large number of new GWAS loci together with genotyped samples also provides us with new and improved tools for studies of risk prediction (see Chap. 23) as well as causality using Mendelian randomization (see Chap. 26). Finally, even if the number of loci robustly associated with adiposity and other complex traits is already large, there is overwhelming evidence that there are even more loci out there to find (Locke et al. 2015; Shungin et al. 2015; Wood et al. 2014). Since the potential biological value is not likely to be lower with the decreasing effect sizes or minor allele frequencies of variants that can be discovered with larger sample sizes, there are good reasons to perform even larger GWAS meta-analyses than those that have already been performed. Since there are several very large genotyping efforts ongoing, for example, within the UK Biobank (Manolio et al. 2012) where 500,000 participants are being genotyped with a GWAS microarray, we can rest assured that there will be additional large GWAS meta-analyses which most likely will report a substantial number of new loci in the next few years to come.

4.7 Conclusions

In the past 8 years, GWAS have been tremendously successful in increasing our understanding of the genetic basis of adiposity with more than 100 novel susceptibility loci identified for overall adiposity and more than 50 loci for body fat distribution. Each susceptibility locus, however, confers only a small effect, and even in aggregate the loci explain $<3\%$ of the variability in each adiposity trait. Nevertheless, these genetic discoveries have already provided invaluable new insights into the biological mechanisms and pathways that underlie adiposity, and numerous new loci are likely to be identified in extended GWAS meta-analyses in the future. However, it is now crucial to increasingly focus on the translation of novel discoveries into clinical care. The presence of hundreds of loci robustly associated with adiposity presents the scientific community with unparalleled opportunities for groundbreaking discoveries in the next decade to come.

References

Abecasis GR, Auton A, Brooks LD et al (2012) An integrated map of genetic variation from 1,092 human genomes. Nature 491(7422):56–65. doi:10.1038/nature11632

Berndt SI, Gustafsson S, Magi R et al (2013) Genome-wide meta-analysis identifies 11 new loci for anthropometric traits and provides insights into genetic architecture. Nat Genet 45 (5):501–512. doi:10.1038/ng.2606

Bradfield JP, Taal HR, Timpson NJ et al (2012) A genome-wide association meta-analysis identifies new childhood obesity loci. Nat Genet 44(5):526–531. doi:10.1038/ng.2247

Canoy D (2008) Distribution of body fat and risk of coronary heart disease in men and women. Curr Opin Cardiol 23(6):591–598. doi:10.1097/HCO.0b013e328313133a

Chambers JC, Elliott P, Zabaneh D et al (2008) Common genetic variation near MC4R is associated with waist circumference and insulin resistance. Nat Genet 40(6):716–718. doi:10.1038/ng.156

Cho YS, Go MJ, Kim YJ et al (2009) A large-scale genome-wide association study of Asian populations uncovers genetic factors influencing eight quantitative traits. Nat Genet 41 (5):527–534. doi:10.1038/ng.357

Cotsapas C, Speliotes EK, Hatoum IJ et al (2009) Common body mass index-associated variants confer risk of extreme obesity. Hum Mol Genet 18(18):3502–3507. doi:10.1093/hmg/ddp292

De Jonghe BC, Hayes MR, Bence KK (2011) Melanocortin control of energy balance: evidence from rodent models. Cell Mol Life Sci 68(15):2569–2588. doi:10.1007/s00018-011-0707-5

Elias I, Franckhauser S, Bosch F (2013) New insights into adipose tissue VEGF-A actions in the control of obesity and insulin resistance. Adipocyte 2(2):109–112. doi:10.4161/adip.22880

Farooqi IS, Keogh JM, Yeo GS et al (2003) Clinical spectrum of obesity and mutations in the melanocortin 4 receptor gene. N Engl J Med 348(12):1085–1095. doi:10.1056/NEJMoa022050

Fox CS, Massaro JM, Hoffmann U et al (2007) Abdominal visceral and subcutaneous adipose tissue compartments: association with metabolic risk factors in the Framingham Heart Study. Circulation 116(1):39–48. doi:10.1161/circulationaha.106.675355

Fox CS, Liu Y, White CC et al (2012) Genome-wide association for abdominal subcutaneous and visceral adipose reveals a novel locus for visceral fat in women. PLoS Genet 8(5), e1002695. doi:10.1371/journal.pgen.1002695

Frayling TM, Timpson NJ, Weedon MN et al (2007) A common variant in the FTO gene is associated with body mass index and predisposes to childhood and adult obesity. Science (New York, NY) 316(5826):889–894. doi:10.1126/science.1141634

Grove ML, Yu B, Cochran BJ et al (2013) Best practices and joint calling of the HumanExome BeadChip: the CHARGE consortium. PLoS One 8(7), e68095. doi:10.1371/journal.pone.0068095

Guo Y, Lanktree MB, Taylor KC et al (2013) Gene-centric meta-analyses of 108 912 individuals confirm known body mass index loci and reveal three novel signals. Hum Mol Genet 22 (1):184–201. doi:10.1093/hmg/dds396

Heard-Costa NL, Zillikens MC, Monda KL et al (2009) NRXN3 is a novel locus for waist circumference: a genome-wide association study from the CHARGE Consortium. PLoS Genet 5(6), e1000539. doi:10.1371/journal.pgen.1000539

Heid IM, Jackson AU, Randall JC et al (2010) Meta-analysis identifies 13 new loci associated with waist-hip ratio and reveals sexual dimorphism in the genetic basis of fat distribution. Nat Genet 42(11):949–960. doi:10.1038/ng.685

Hemani G, Yang J, Vinkhuyzen A et al (2013) Inference of the genetic architecture underlying BMI and height with the use of 20,240 sibling pairs. Am J Hum Genet 93(5):865–875. doi:10.1016/j.ajhg.2013.10.005

Hinney A, Nguyen TT, Scherag A et al (2007) Genome wide association (GWA) study for early onset extreme obesity supports the role of fat mass and obesity associated gene (FTO) variants. PLoS One 2(12), e1361. doi:10.1371/journal.pone.0001361

Huszar D, Lynch CA, Fairchild-Huntress V et al (1997) Targeted disruption of the melanocortin-4 receptor results in obesity in mice. Cell 88(1):131–141

Jiao H, Arner P, Hoffstedt J et al (2011) Genome wide association study identifies KCNMA1 contributing to human obesity. BMC Med Genomics 4:51. doi:10.1186/1755-8794-4-51

Kilpelainen TO, Zillikens MC, Stancakova A et al (2011) Genetic variation near IRS1 associates with reduced adiposity and an impaired metabolic profile. Nat Genet 43(8):753–760. doi:10.1038/ng.866

Lear SA, James PT, Ko GT et al (2010) Appropriateness of waist circumference and waist-to-hip ratio cutoffs for different ethnic groups. Eur J Clin Nutr 64(1):42–61. doi:10.1038/ejcn.2009.70

Li S, Loos RJ (2008) Progress in the genetics of common obesity: size matters. Curr Opin Lipidol 19(2):113–121. doi:10.1097/MOL.0b013e3282f6a7f3

Lindgren CM, Heid IM, Randall JC et al (2009) Genome-wide association scan meta-analysis identifies three Loci influencing adiposity and fat distribution. PLoS Genet 5(6), e1000508. doi:10.1371/journal.pgen.1000508

Liu CT, Monda KL, Taylor KC et al (2013) Genome-wide association of body fat distribution in African ancestry populations suggests new loci. PLoS Genet 9(8), e1003681. doi:10.1371/journal.pgen.1003681

Locke AE, Kahali B, Berndt SI et al. (2015) Genetic studies of body mass index yield new insights for obesity biology. Nature 518(7538):197–206. doi: 10.1038/nature14177

Loos RJ, Lindgren CM, Li S et al (2008) Common variants near MC4R are associated with fat mass, weight and risk of obesity. Nat Genet 40(6):768–775. doi:10.1038/ng.140

Manolio TA, Weis BK, Cowie CC et al (2012) New models for large prospective studies: is there a better way? Am J Epidemiol 175(9):859–866. doi:10.1093/aje/kwr453

Meyre D, Delplanque J, Chevre JC et al (2009) Genome-wide association study for early-onset and morbid adult obesity identifies three new risk loci in European populations. Nat Genet 41 (2):157–159. doi:10.1038/ng.301

Monda KL, Chen GK, Taylor KC et al (2013) A meta-analysis identifies new loci associated with body mass index in individuals of African ancestry. Nat Genet 45(6):690–696. doi:10.1038/ng.2608

Muddyman D, Smee C, Griffin H et al (2013) Implementing a successful data-management framework: the UK10K managed access model. Genome Med 5(11):100. doi:10.1186/gm504

Okada Y, Kubo M, Ohmiya H et al (2012) Common variants at CDKAL1 and KLF9 are associated with body mass index in east Asian populations. Nat Genet 44(3):302–306. doi:10.1038/ng.1086

Pal A, Barber TM, Van de Bunt M et al (2012) PTEN mutations as a cause of constitutive insulin sensitivity and obesity. N Engl J Med 367(11):1002–1011. doi:10.1056/NEJMoa1113966

Paternoster L, Evans DM, Nohr EA et al (2011) Genome-wide population-based association study of extremely overweight young adults—the GOYA study. PLoS One 6(9), e24303. doi:10.1371/journal.pone.0024303

Pei YF, Zhang L, Liu Y et al (2013) Meta-analysis of genome-wide association data identifies novel susceptibility loci for obesity. Hum Mol Genet. doi:10.1093/hmg/ddt464

Pischon T, Boeing H, Hoffmann K et al (2008) General and abdominal adiposity and risk of death in Europe. N Engl J Med 359(20):2105–2120. doi:10.1056/NEJMoa0801891

Randall JC, Winkler TW, Kutalik Z et al (2013) Sex-stratified genome-wide association studies including 270,000 individuals show sexual dimorphism in genetic loci for anthropometric traits. PLoS Genet 9(6), e1003500. doi:10.1371/journal.pgen.1003500

Scherag A, Dina C, Hinney A et al (2010) Two new loci for body-weight regulation identified in a joint analysis of genome-wide association studies for early-onset extreme obesity in French and German study groups. PLoS Genet 6(4), e1000916. doi:10.1371/journal.pgen.1000916

Scuteri A, Sanna S, Chen WM et al (2007) Genome-wide association scan shows genetic variants in the FTO gene are associated with obesity-related traits. PLoS Genet 3(7), e115. doi:10.1371/journal.pgen.0030115

Shungin D, Winkler TW, Croteau-Chonka DC et al. (2015) New genetic loci link adipose and insulin biology to body fat distribution. Nature 518(7538):187–196. doi: 10.1038/nature14132
Snijder MB, Dekker JM, Visser M et al (2003) Associations of hip and thigh circumferences independent of waist circumference with the incidence of type 2 diabetes: the Hoorn Study. Am J Clin Nutr 77(5):1192–1197
Speakman JR (2013) Functional analysis of seven genes linked to body mass index and adiposity by genome-wide association studies: a review. Hum Hered 75(2–4):57–79. doi:10.1159/000353585
Speliotes EK, Willer CJ, Berndt SI et al (2010) Association analyses of 249,796 individuals reveal 18 new loci associated with body mass index. Nat Genet 42(11):937–948. doi:10.1038/ng.686
Stolerman ES, Manning AK, McAteer JB et al (2009) TCF7L2 variants are associated with increased proinsulin/insulin ratios but not obesity traits in the Framingham Heart Study. Diabetologia 52(4):614–620. doi:10.1007/s00125-009-1266-2
Thorleifsson G, Walters GB, Gudbjartsson DF et al (2009) Genome-wide association yields new sequence variants at seven loci that associate with measures of obesity. Nat Genet 41(1):18–24. doi:10.1038/ng.274
Vaisse C, Clement K, Guy-Grand B et al (1998) A frameshift mutation in human MC4R is associated with a dominant form of obesity. Nat Genet 20(2):113–114. doi:10.1038/2407
Voight BF, Kang HM, Ding J et al (2012) The metabochip, a custom genotyping array for genetic studies of metabolic, cardiovascular, and anthropometric traits. PLoS Genet 8(8), e1002793. doi:10.1371/journal.pgen.1002793
Wang Y, Rimm EB, Stampfer MJ et al (2005) Comparison of abdominal adiposity and overall obesity in predicting risk of type 2 diabetes among men. Am J Clin Nutr 81(3):555–563
Wang K, Li WD, Zhang CK et al (2011) A genome-wide association study on obesity and obesity-related traits. PLoS One 6(4), e18939. doi:10.1371/journal.pone.0018939
Wen W, Cho YS, Zheng W et al (2012) Meta-analysis identifies common variants associated with body mass index in east Asians. Nat Genet 44(3):307–311. doi:10.1038/ng.1087
Wheeler E, Huang N, Bochukova EG et al (2013) Genome-wide SNP and CNV analysis identifies common and low-frequency variants associated with severe early-onset obesity. Nat Genet 45 (5):513–517. doi:10.1038/ng.2607
WHO (2000) Obesity: preventing and managing the global epidemic. Report of a WHO consultation. World Health Organ Tech Rep Ser. 894:i–xii, 1–253
Willer CJ, Speliotes EK, Loos RJ et al (2009) Six new loci associated with body mass index highlight a neuronal influence on body weight regulation. Nat Genet 41(1):25–34. doi:10.1038/ng.287
Wood AR, Esko T, Yang J et al (2014) Defining the role of common variation in the genomic and biological architecture of adult human height. Nat Genet 46(11):1173–1186
Yeo GS, Farooqi IS, Aminian S et al (1998) A frameshift mutation in MC4R associated with dominantly inherited human obesity. Nat Genet 20(2):111–112. doi:10.1038/2404
Yusuf S, Hawken S, Ounpuu S et al (2005) Obesity and the risk of myocardial infarction in 27,000 participants from 52 countries: a case–control study. Lancet 366(9497):1640–1649. doi:10.1016/s0140-6736(05)67663-5
Zimmet P, Alberti KG, Shaw J (2001) Global and societal implications of the diabetes epidemic. Nature 414(6865):782–787. doi:10.1038/414782a

Chapter 5
Sequencing Approaches to Type 2 Diabetes

Jeffrey W. Kleinberger and Toni I. Pollin

Abstract Rapid advancement and decreasing costs of DNA sequencing technologies have yielded great strides in improving our understanding of the genetic etiology of human disease, including type 2 diabetes (T2D). The state of the science of sequencing in human disease has progressed from being largely restricted to single-gene disorders, to elucidating common variants conferring susceptibility to common diseases, and most recently to the cataloging of rare variants involved in common diseases. Sanger sequencing, DNA amplification, and microarrays provided early insights into the genetic architecture of type 2 diabetes, but the establishment of massively parallel sequencing platforms has accelerated the process. There are several different platforms, including Illumina, Pacific Biosciences, and Ion Torrent technologies. Each platform has its own specific strengths and weakness, and proper quality control and variant-calling techniques are crucial for accurate sequencing data. The high-throughput capabilities of these technologies have allowed population-based sequencing projects, such as the 1000 Genomes Project, to create repositories of human genomic variation. Genetic variants first associated with T2D were discovered through sequencing a number of candidate genes, with *PPARG* and *KCNJ11* yielding variants with consistent enough association to be established universally as the first two T2D susceptibility genes. Variants in the *HNF4A* pancreatic-specific P2 promoter and *TCF7L2* gene were both discovered through sequencing of linkage signals. Genome-wide association studies have found many T2D-associated regions, but a limited amount of sequencing has been performed to follow up these studies with limited numbers of causal variants being discovered. The growth of massively parallel sequencing has led to

J.W. Kleinberger
Program for Personalized and Genomic Medicine, Department of Medicine, Division of Endocrinology, Diabetes, and Nutrition, University of Maryland School of Medicine, Baltimore, MD, USA

T.I. Pollin, MS, PhD (✉)
Program for Personalized and Genomic Medicine, Department of Medicine, Division of Endocrinology, Diabetes, and Nutrition, University of Maryland School of Medicine, Baltimore, MD, USA

Department of Epidemiology and Public Health, University of Maryland School of Medicine, Baltimore, MD, USA
e-mail: tpollin@medicine.umaryland.edu

J.C. Florez (ed.), *The Genetics of Type 2 Diabetes and Related Traits*,
DOI 10.1007/978-3-319-01574-3_5

the generation of comprehensive sequencing across the exome and genome of large numbers of individuals with T2D in projects such as the T2D-GENES consortium. These data will provide the enhanced understanding of the genetic architecture of T2D necessary for improving approaches to the prevention and treatment of T2D.

5.1 Introduction

All discovery of genetic variants begins with DNA sequencing. In the pre-Human Genome Project era, the focus of human disease genetics was on finding the genetic cause of Mendelian diseases or diseases where there is a clear and near-perfect relationship between genetic variation and phenotypic variation. Then the gene would be cloned and sequenced in affected individuals to find disease-causing variants. Next, a number of healthy individuals including relatives of affected individuals would be studied to make sure those variants were specific to the disease.

In the common disease era, a group with a particular interest in the role of a candidate gene in a common disease, based on prior knowledge of that gene's function, would sequence a small number of individuals to find what variant(s) occurred naturally in the coding region of the gene. The investigators would then genotype that variant(s) in a number of individuals with and without the disease or in group of individuals with appropriate quantitative phenotypes. They would then look for genotype/phenotype associations. Once the variants in a particular gene were published, other groups would seek to replicate the association of the common variant with the same disease after genotyping specific variants only. Such non-sequencing-based approaches became increasingly feasible after the completion of the Human Genome Project and with the dawning of dbSNP and other related resources that followed, including HapMap. HapMap revealed the correlation between SNPs and thus made it possible to reduce the number of SNPs that needed to be genotyped in order to get equivalent information.

More recently, 1000 Genomes and other publicly available datasets are beginning to provide comprehensive information on genetic architecture. The latter was enabled by the evolution of next-generation or massively parallel sequencing technologies. However, the databases still do not obviate the interest in sequencing within individual populations, which allows for the possibility of discovery of novel, rare variants in single individuals or small numbers of individuals.

5.2 Overview of Sequencing Technologies

5.2.1 *Sanger Sequencing*

DNA sequencing was first performed in 1977 and has created a scientific revolution as a result. The first method of sequencing, called Sanger sequencing after discoverer Frederick Sanger, utilized primer-based replication of template DNA with both natural nucleotides and chain-terminating dideoxynucleotides to create differently-sized fragments (Sanger et al. 1977). By tagging the dideoxynucleotides and aligning the fragments by size through electrophoresis, the sequence of the template DNA could be uncovered. This process could initially sequence DNA fragments of up to 200 base pairs (bp), but has been optimized to accurately sequence 1000 bp with current technology (e.g., Applied Biosystems 3730xl DNA Analyzer: http://www6.appliedbiosystems.com/products/abi3730xlspecs.cfm). Sanger sequencing is still considered the most accurate method to obtain and validate DNA sequence.

5.2.2 *PCR/YAC/Genome Project*

The invention of the polymerase chain reaction (PCR) and discovery of yeast artificial chromosomes (YACs) in the 1980s paved the way for larger scale sequencing studies (Murray and Szostak 1983; Mullis et al. 1986). PCR uses DNA polymerase and thermal cycling to exponentially amplify DNA. This process greatly improved the speed and output of previous sequencing techniques. The discovery of YACs allowed larger fragments of DNA (up to 1 Mbp) to be carried and transported into cells. The YACs paved the way for bacterial artificial chromosomes (BACs), which can more stably carry about 200 Kbp DNA (O'Connor et al. 1989). These techniques allowed the Human Genome Project to sequence the entire human genome using a technique called "chromosome walking" (Kere et al. 1992). This technique used Sanger sequencing to perform sequential DNA analysis along a fragment of human DNA that had been incorporated into a BAC. This process was repeated for thousands of BACs that together contained the entirety of the human genome. This painstaking and expensive process produced the first reference human genome in 2003 (Schmutz et al. 2004; International Human Genome Sequencing Consortium 2004).

5.2.3 *Arrays*

While the Human Genome project was in its early days, microarray technology was beginning to emerge as a potential tool for genetic studies (Fodor et al. 1991). DNA microarrays are single chips that can carry hundreds to millions of genetic markers across the genome. An individual's DNA can be fragmented and bound to the genetic markers of a DNA microarray to determine the individual's DNA sequence at those markers. It provides a low-resolution framework of the identity of regions of the genome (Dupuis et al. 1995). This greatly pushed forward the capabilities of position-dependent cloning of genes and provided a general roadmap for the entire genome.

5.2.4 Massively Parallel Sequencing

In the mid-1990s, massively parallel sequencing, or next-generation sequencing, first became a possibility with the inception of the pyrosequencing method of sequencing (Nyren 2007). Massively parallel sequencing uses many independent sequencing operations on amplified input DNA to simultaneously determine the DNA sequence through the use of a flow cell. Pyrosequencing uses emulsion PCR to amplify the target DNA and bind the amplified region to many beads. In the presence of primers and polymerase, each of the four nucleotides is flowed over the beads. If the nucleotide matches the target DNA and is incorporated, it will release a pyrophosphate molecule that can be detected through a luciferase reaction. The output of the pyrosequencing reaction is the sequence of the amplified DNA, as aggregated from the parallel sequencing events of each of the beads (Gharizadeh et al. 2006). This platform became commercially available in 2005, popularized by Roche's 454 platform. In 2011, Ion Torrent released a sequencing platform that uses a similar sequencing method. Ion Torrent uses the same emulsion PCR, bead attachment, and nucleotide flow system, but the proton released from the polymerase incorporation of the flowed nucleotide is the molecule that is detected through a pH monitoring system (Rothberg et al. 2011). However, the most prominent sequencing platform in the field is Illumina and their sequencing-by-synthesis method (Mardis 2013). This platform uses short-read shotgun sequencing. Short-read sequencing allows Illumina to use paired-end sequencing that amplifies input DNA using a bridge-amplification technique that creates fragments of equal length that represent a "scaffold" of a known distance separating two sequencing reads. Each end of the fragment is sequenced using a polymerase with cyclic reversible terminators and fluorescently labeled nucleotides. As each nucleotide is polymerized to the growing strand, the elongation process is stopped as the fluorescence of the nucleotide is detected by a laser before the process may proceed for addition of the subsequent nucleotide (Mardis 2013). This process allows much greater throughput and the paired-end sequencing data can be more easily assembled. Another recently introduced technology is single molecule, real-time sequencing (SMRT). This system has a suspended polymerase molecule that circularized input DNA is read through in the presence of fluorescent nucleotides. As the nucleotides are added, the fluorescence is detected before being cleaved off the molecule (Eid et al. 2009). This technology, found in Pacific Bioscience's sequencing platform, allows much longer reads than any other system currently available (Quail et al.

2012). There is also much anticipation with the future of sequencing technology moving into nanopore sequencing techniques. These methods push input DNA through a single pore with a voltage bias. As the different nucleotides are passed, the residual pore current is altered to different levels, which provide the identity of the nucleotides (Buermans and den Dunnen 2014). The continued evolution of DNA sequencing technology promises to continue to explode the scale of human genetic sequencing.

5.2.5 *Advantages and Disadvantages of Each Platform*

With the growing types of sequencing platforms available, it is important to take into account the strengths and weaknesses of each. Illumina is the most popular platform, mainly due to its high throughput and lower overall cost per data output (Liu et al. 2012). Illumina also boasts the first FDA-approved next-generation sequencing platform for clinical use in the lower throughput MiSeqDx (FDA 2015). Additionally, they offer the HiSeq X Ten, a platform that is advertised as having the capacity to sequence a human genome for under $1000 (Check 2014). However, Illumina platforms have difficulty with substitution errors and the short sequencing reads can have difficulties in long repetitive elements of DNA (Liu et al. 2012). The Pacific Bioscience's platform solves the problems associated with short reads by having a read length of several thousand Kbp; however, it has the most common error rate and difficulty calling indels, especially in homopolymer regions (Quail et al. 2012). An interesting strength of Pacific Bioscience is that it does not need PCR amplification prior to sequencing (Liu et al. 2012). Pyrosequencing platforms like 454 or Ion Torrent also have difficulty calling indels and homopolymer regions. Each platform has its strengths though, with 454's low error rate and Ion Torrent's rapid runtime (Frey et al. 2014).

5.2.6 *Quality Control and Variant Calling*

After sequencing has been performed, it is important for proper quality control procedures to remove all low-quality data. This process carries a great deal of importance for massively parallel sequencing because the large amount of data produced ensures that there will also be a larger amount of low-quality data that needs to be removed. Sequencing data provides quality scores for each base, and this information is stored in the raw sequencing data. If the reads' average quality score is below the QC threshold, those reads will be removed completely through filtering. Each individual read is also assessed for low-quality bases that need to be trimmed from the ends of the read (Pabinger et al. 2014). Factors such as total reads, filtered reads, trimmed reads, and average read length all constitute important components that can describe the quality of the sequencing run. After QC procedures, the sequences are mapped to the reference genome. Metrics such as average coverage, proportion covered at >20X, and proportion of bases with zero coverage can all lend information about the quality of the sequencing data post-alignment (Allcock 2014). Finally, variant-calling software is necessary to analyze the number of base calls that match the reference genome compared to alternate base calls. The variant caller must also detect indels by detecting where gaps or insertions between the reference genome and the sequenced reads are. By correlating this information with the quality of the bases in the read, highly accurate variant calls can be made. In any case, the QC, alignment, and variant caller should be optimized for both the sequencing platform and the DNA region being sequenced for the most accurate data.

5.2.7 HapMap/dbSNP/1000Genomes

The explosion of data coming from advances in sequencing technology has provided a number of resources that can fuel future study. Based initially on DNA microarray data, the International HapMap project was performed in the mid-2000s. This project developed a haplotype map of the human genome by genotyping over 1.6 million SNPs across what turned out to be 11 populations. This provided a great deal of information about blocks of linkage disequilibrium in each population (International HapMap 3 Consortium et al. 2010). The HapMap project has been an important resource for discovery of genetic variants associated with disease. The HapMap project has also provided many entries to the NCBI's dbSNP database that is a collection of SNPs, indels, and other short sequence variants submitted by individuals or groups that are then validated before becoming entries. In addition to these tools, massively parallel sequencing has been utilized to create multiple databases of variants across the genome. The 1000 Genomes Project is a multi-institute project that has already performed whole-genome sequencing at an average of 4X depth for 1092 individuals across 4 different populations, which will expand to over 2500 individuals for the next phase of the project (1000 Genomes Project Consortium et al. 2012). This study has become the primary resource for determining the prevalence of variants across the genome. Other large-scale sequencing projects include the NHLBI Grand Opportunity Exome Sequencing Project and the UK 10,000 Genomes Project (Tennessen et al. 2012; Kaye et al. 2014). These projects, in addition to future large-scale collaborations, help characterize the variation across the human genome, including common variants, rare variants, and structural variants.

5.3 Candidate Gene Approaches Pre-linkage

Prior to the GWAS era, genetic discoveries in type 2 diabetes (T2D) relied on sequencing candidate genes in small numbers of individuals to discover common variants, followed by genotyping identified variants in cases and controls. Two variants from that era withstood replication efforts and are considered to be T2D susceptibility genes.

In 1997, *PPARG* was mapped and cloned (Beamer et al. 1997) and was chosen as a candidate gene for T2D because of its respective proteins role in adipocyte differentiation and as a target for thiazolidinediones. The Pro12Ala variant was discovered by molecular scanning (Yen et al. 1997). In this case, molecular scanning meant running single-strand conformational polymorphism analysis (SSCP) followed by sequencing of subjects with aberrant SSCP patterns. This variant was subsequently associated with insulin resistance (Deeb et al. 1998; Jacob et al. 2000; Stumvoll et al. 2001) and then confirmed in a meta-analysis to be associated with diabetes (Altshuler et al. 2000).

The *KCNJ11* E23K variant was identified in a similar manner: Samples from 35 individuals with T2D were subjected to SSCP followed by sequencing of variants identified (Inoue et al. 1997). Initially the E23K was not associated with diabetes in an expanded set of 306 diabetic cases and 175 nondiabetic controls. However, a subsequent meta-analyses provided evidence of association, particularly for a recessive model, (Hani et al. 1998) and a larger meta-analysis of 2486 subjects confirmed the association (Gloyn et al. 2003).

5.4 Sequencing Post-linkage

Following up on a linkage for T2D signal on chromosome 2q37 in Mexican American families which was denoted as NIDDM1, the gene *CAPN10* encoding Calpain-10 (Horikawa et al. 2000) was proposed as a susceptibility gene for T2D. The investigators used information from revised genetic maps (Broman et al. 1998) and additional analyses showing an interaction with chromosome 15 (Cox et al. 1999) to a 7 cM region on chromosome 2, 259–266 cM. They needed to first construct a physical map of the region using yeast, bacterial, and plasmid-derived artificial chromosomes. They then sequenced regions that mapped to the GenBank database in 10–20 diabetic cases to look for polymorphisms. After finding each polymorphism, they would genotype it in 112 randomly selected individuals and 110 patients. They evaluated the variants both individually and as haplotypes. In a 66 kb interval containing three genes, a variant then labeled SNP-UC43 was found to be increased in frequency in the patient sample vs. the random sample (80 % vs. 75 %); the case frequency increased to 88 % when restricted to those contributing to the linkage and 95 % when further restricted to those contributing to the joint linkage on chromosome 15. They also showed that the variant had strong evidence for linkage in the families. Further sequencing and genotyping resulted in the identification of a three-SNP haplotype explaining more of the association (Horikawa et al. 2000). It was not concluded that NIDDM1 had been positionally cloned. Results of replication attempts were inconsistent, but a subsequent meta-analysis was consistent with an association (Tsuchiya et al. 2006).

HNF4A was considered a strong positional candidate gene for T2D because rare, highly penetrant mutations cause maturity-onset diabetes of the young type 1 (MODY1) and several groups found linkage to the region of chromosome 20 containing it (Bowden et al. 1997; Ji et al. 1997; Zouali et al. 1997; Ghosh et al. 1999; Permutt et al. 2001; Luo et al. 2001; Mori et al. 2002). However, sequencing of the coding region and proximal putative promoter region did not initially reveal any variants associated with T2D or explaining the linkage (Ghosh et al. 1999). Subsequently, a pancreatic-specific promoter P2 was discovered 46 kb upstream of the hepatic promoter P1 (Thomas et al. 2001). A comprehensive analysis of the linkage region identified SNPs near this promoter that were associated, including the sequencing of 12 individuals contributing to the linkage signal (Silander et al. 2004).

Another illustrative example of post-linkage sequencing is that of *TCF7L2*. Following up on a suggestive linkage signal on chromosome 10q, (Reynisdottir et al. 2003; Duggirala et al. 1999) Grant et al. conducted a fine mapping study using densely placed microsatellite markers that uncovered a strong association with a microsatellite marker located in an intron of the gene *TCF7L2* (Grant et al. 2006). Because of the availability of HapMap, no sequencing was required to find SNPs in strong linkage disequilibrium with the microsatellite; they simply genotyped the microsatellite in the HapMap CEU samples and initially found one strongly correlated intronic SNP variant. In this case sequencing was used to try to find the causal variant that the microsatellite and SNPs might be tagging. Since the effects of coding variants are better understood and because only exon 4 appeared within the same linkage disequilibrium block as the association signals, they first sequenced exon 4 of *TCF7L2* in a subset of 557 diabetic cases and 769 controls and found no exonic variants to explain the association. They then sequenced all exons of the gene in 93 cases and 91 controls with a similar result. They first searched the entire LD block in HapMap and identified the five most highly correlated SNPs. The final sequencing effort was to sequence the entire LD block in 538 cases and 462 controls. In this case the goal of the sequencing was to fine map the association to find a better candidate for a causal variant; the effect was to rule out anything else (Grant et al. 2006). Therefore as the authors recommended, replication efforts focused on the two SNPs in highest LD with the initial signal, and little additional sequencing was done. These replication efforts were highly successful (Cauchi et al. 2007) and placed *TCF7L2* on the map as the strongest T2D susceptibility gene to date (at that time and now). Understanding the specific manner in which variation in the gene impacts disease has been complex but has provided new insights into the etiology of the role of the Wnt signaling pathway in diabetes (Cauchi and Froguel 2008).

5.5 Sequencing Post-GWAS

Limited sequencing follow-up was done with GWAS genes prior to the convening of the T2D-GENES consortium described below. A resequencing effort was described genes for quantitative traits including glucose and insulin (Service et al. 2014). The authors selected loci previously associated with one or more of six quantitative traits: triglycerides (TGs), high-density lipoprotein cholesterol (HDL-C), low-density lipoprotein cholesterol (LDL-C), total cholesterol (TC), glucose, and insulin. They resequenced 78 genes from 17 loci in nearly 8000 individuals. Most variants identified were for lipids, but they identified a probably damaging missense variant in *G6PC2* associated with fasting glucose. They also evaluated association of the variants with T2D status but found no associations (Service et al. 2014).

Few causal variants have been identified for T2D GWAS genes. To find loss-of-function variants in T2D genes that could identify targets for downregulation as novel treatment approaches, Flannick et al. (2014) sequenced exonic regions of 115 T2D GWAS genes from the 65 replicated loci in 758 Finnish or Swedish individuals (352 cases and 406 controls). They were unable to identify any SNP or gene with study-wide significant association with T2D. However, they did find a rare truncating variant in *SLC30A8* with borderline nominal evidence toward protection from diabetes. The ability to identify even this variant was due to a founder effect, as subsequent studies to expand the sample size to examine this variant found it to be virtually nonexistent outside of western Finland, (Flannick et al. 2014) though other rare variants were found. An additional founder variant was identified by taking advantage of existing data from the DeCODE project, in which 2500 Icelanders had undergone whole-genome sequencing (Gudmundsson et al. 2012). Additional rare, mostly singleton variants were found in the 13,000 exomes from the T2D-Genes/GoT2D consortium. These data strengthened the evidence for loss-of-function variants in *SLC30A8* having a protective effect against T2D. This example illustrates how the decreased cost of sequencing changes what is possible. In contrast to the candidate gene studies of the 1990s discussed above, in which a small number of individuals were sequenced after first using SSCP to select regions to sequence, here a large number of people were sequenced, necessary in order to discover rare variants with large effect size. It is important to note the value of founder populations for having any power to associate these variants with disease.

5.6 Emergence of Next-Generation Sequencing and Effects on Diabetes Gene Discovery

5.6.1 Effects on Scope

The massive amount of information that results from next-generation sequencing has altered the methods of genetic analysis moving forward. In the past, it was necessary to have either common genetic variants, large effect sizes, or large numbers of individuals to properly power an association between a genetic variant and a phenotype. Massively parallel sequencing, in conjunction with the genetic variant databases that have been produced by the sequencing technology, allows associations to be made with rare variants in smaller populations. These associations are possible due to the increased amount of data, but it becomes more vital to properly filter true associations from background noise. This can be done through proper QC and variant calling, in addition to appropriate predesigned variant filtering based on variant annotation tools. The implementation of massively parallel sequencing has shifted the target of genetic analysis from *finding* the causative variants to *filtering down to* the causative variants.

5.6.2 Targeted Sequencing

One common approach to filter out background data uses sequencing targeted to only select regions of the genome. This approach differs from Sanger sequencing because it has the capacity to sequence up to hundreds of different genomic regions simultaneously, while Sanger sequencing only identifies a single region at a time. Targeted sequencing uses primers that only amplify regions of interest, such as the coding sequence of specific genes (Summerer 2009). Panels for targeted sequencing therefore only sequence a fraction of the genome, which can be done more quickly and less expensively than whole-exome or whole-genome sequencing. Lower amounts of target sequence also provide the opportunity for sample multiplexing, which can greatly increase the efficiency of each sequencing run (Fokstuen et al. 2014). Platforms like the Illumina MiSeq or the Ion Torrent Personal Genome Machine were created specifically for the lower throughput needs of targeted sequencing panels (Koshimizu et al. 2013). These techniques are useful for identifying previously discovered variants or discovering novel variants in genes or loci previously associated with the phenotype. Targeted gene sequencing is already being incorporated into the analysis of highly penetrant, genetic forms of diabetes, like maturity-onset diabetes of the young (MODY) and transient or permanent neonatal diabetes mellitus (NDM) (Ellard et al. 2013; Gao et al. 2014; Chapla et al. 2014).

5.6.3 Whole-Exome Sequencing

The human exome, making up only 1–2 % of the human genome, is the region of DNA that codes for the mature mRNA translated into peptides. Sequencing of the exome allows a snapshot of the most well-understood genetic material without the flood of information that makes up the whole genome. The exome is the region in which our current understanding of genomic architecture and function best allows us to predict whether a mutation is damaging. Exome sequencing, like targeted sequencing, uses specific primers to amplify all of the coding regions of the human genome. The whole human exome can be analyzed singularly on the Illumina MiSeq or the Ion Torrent Proton, or multiple samples can be multiplexed in a single run using the Illumina HiSeq (Boland et al. 2013). The data from exome sequencing can also be analyzed in a targeted manner. Using that method, only the targeted genes and loci are analyzed for phenotype-causing variants, with the option of expanding the search for phenotype-causing variants post hoc. Whole-exome sequencing is already being implemented as a tool for discovering causative variants in cases of families with monogenic diabetes that lack a genetic diagnosis (Bonnefond et al. 2012; Dusatkova et al. 2014; Johansson et al. 2012).

5.6.4 Whole-Genome Sequencing

Whole-genome sequencing represents the most comprehensive method to discover a phenotype-causing variant. This method uses random primers to amplify the input DNA, which allows an even distribution of amplification across the genome. As a result, whole-genome sequencing can more efficiently detect structural variants such as copy number variants, translocations, and mobile element insertions, in addition to SNPs and indels (Tan et al. 2014; Suzuki et al. 2014; Wu et al. 2014). However, the sheer amount of data can be overwhelming, with a single-genome sequencing file taking up several dozens of gigabytes of storage. This amount of data is not easily manipulated, and analysis of genome can take hours to days of computing time (O'Driscoll et al. 2013). Additionally, sequencing of a single genome still generally costs over $1000, which can be prohibitive when added to the costs of data storage and management for large studies (Check 2014). Despite these difficulties, the wealth of information from the genome may outweigh the costs and difficulties, especially as technology and computing capabilities improve in the future, and our understanding of the function of noncoding DNA through efforts such as the ENCODE Project (ENCODE Project Consortium. An integrated encyclopedia of DNA elements in the human genome. Nature 2012) progresses.

5.6.5 T2D-GENES Consortium Goals

The Type 2 Diabetes Genetic Exploration by Next-generation sequencing in multi-Ethnic Samples (T2D-GENES) consortium is incorporating many next-generation sequencing techniques to further understand the pathophysiology of T2D. The first of three projects in T2D-GENES uses whole-exome sequencing of 10,000 individuals across five different ethnicities, of which half of the participants are controls (Teslovich et al. 2012). This goal of this project is to discover the genes related to T2D pathogenesis in addition to studying genes associated with monogenic and syndromic T2D cases. Whole-exome sequencing data from this study was used by Flannick et al. to find additional apparently protective loss-of-function mutations in *SLC30A8* (Flannick et al. 2014). This is an example of filtering whole-exome sequencing down to only the genes of interest, and the T2D-GENES consortium exome data has also been used this way for the *PPARG* and *NPHS1* genes (Majithia et al. 2014; Bonomo et al. 2014). The second project of T2D-GENES is using whole-genome sequencing of 600 individuals from Mexican American families (dbGaP|phs000462.v1.p1|T2D-GENES Project 2 2015). This study intends to discover rare variants found in less than 1 % of the population that can alter risk for T2D. The third project used over 100,000 individuals from five consortia of different ancestries in order to discover the causative variants linked to GWAS hits for T2D. The results from this study have been published as a meta-analysis, demonstrating an improved resolution of common variant associations for T2D in addition to discovering new susceptibility loci through trans-ethnic analysis (Consortium et al. 2014). Overall this project utilizes multiple sequencing techniques (whole-exome sequencing, whole-genome sequencing, and GWAS) to thoroughly analyze the human genome and its role in T2D pathophysiology. Publication of the results of these efforts and further analysis are ongoing and ultimately provide a greatly enhanced and importantly comprehensive understanding of the underlying genetic architecture of T2D.

5.7 Concluding Remarks

In 1974, R.B. Tattersall wrote:

> The genetics of diabetes is a confused subject in which almost every possible mode of inheritance has been proposed. It is apparent, however, that no simple genetic hypothesis is compatible with the data gathered from large groups of diabetics. One reason . . . may be that diabetes is not homogeneous but rather consists of a number of different diseases in which carbohydrate intolerance is the common factor (Tattersall 1974).

Today we know that even T2D alone is not one disease but many. Over the past couple of decades, the genetic heterogeneity of T2D has begun to be unraveled, enabled by the discovery of increasingly efficient and powerful sequencing methods, analytical tools, and the work of multiple determined individuals. The journey is only really at the beginning, and now that we are on the cusp of the $1000 genome, technology and cost will cease to be a limit of the goal of elucidating the etiology and therefore hope for prevention, treatment, and cure of every case of T2D.

References

1000 Genomes Project Consortium, Abecasis GR, Auton A et al (2012) An integrated map of genetic variation from 1,092 human genomes. Nature 491(7422):56–65. doi:10.1038/nature11632

Allcock RJ (2014) Production and analytic bioinformatics for next-generation DNA sequencing. Methods Mol Biol 1168:17–29. doi:10.1007/978-1-4939-0847-9_2

Altshuler D, Hirschhorn JN, Klannemark M et al (2000) The common PPARgamma Pro12Ala polymorphism is associated with decreased risk of type 2 diabetes. Nat Genet 26(1):76–80

Beamer BA, Negri C, Yen CJ et al (1997) Chromosomal localization and partial genomic structure of the human peroxisome proliferator activated receptor-gamma (hPPAR gamma) gene. Biochem Biophys Res Commun 233:756–759

Boland JF, Chung CC, Roberson D et al (2013) The new sequencer on the block: comparison of Life Technology's Proton sequencer to an Illumina HiSeq for whole-exome sequencing. Hum Genet 132(10):1153–1163. doi:10.1007/s00439-013-1321-4

Bonnefond A, Philippe J, Durand E et al (2012) Whole-exome sequencing and high throughput genotyping identified KCNJ11 as the thirteenth MODY gene. PLoS One 7(6), e37423. doi:10.1371/journal.pone.0037423

Bonomo JA, Ng MC, Palmer ND et al (2014) Coding variants in nephrin (NPHS1) and susceptibility to nephropathy in African Americans. Clin J Am Soc Nephrol 9(8):1434–1440. doi:10.2215/CJN.00290114

Bowden DW, Sale M, Howard TD et al (1997) Linkage of genetic markers on human chromosomes 20 and 12 to NIDDM in Caucasian sib pairs with a history of diabetic nephropathy. Diabetes 46(5):882–886

Broman KW, Murray JC, Sheffield VC, White RL, Weber JL (1998) Comprehensive human genetic maps: individual and sex-specific variation in recombination. Am J Hum Genet 63(3):861–869, doi:S0002-9297(07)61389-5 [pii]

Buermans HP, den Dunnen JT (2014) Next generation sequencing technology: advances and applications. Biochim Biophys Acta 1842(10):1932–1941, doi:S0925-4439(14)00180-X [pii]

Cauchi S, Froguel P (2008) TCF7L2 genetic defect and type 2 diabetes. Curr Diab Rep 8(2):149–155

Cauchi S, El Achhab Y, Choquet H et al (2007) TCF7L2 is reproducibly associated with type 2 diabetes in various ethnic groups: a global meta-analysis. J Mol Med (Berl) 85(7):777–782. doi:10.1007/s00109-007-0203-4

Chapla A, Mruthyunjaya MD, Asha HS et al (2014) Maturity onset diabetes of the young in India—a distinctive mutation pattern identified through targeted next-generation sequencing. Clin Endocrinol (Oxf). doi:10.1111/cen.12541

Check HE (2014) Is the $1,000 genome for real? Nature. doi:10.1038/nature.2014.14530

Cox NJ, Frigge M, Nicolae DL et al (1999) Loci on chromosomes 2 (NIDDM1) and 15 interact to increase susceptibility to diabetes in Mexican Americans. Nat Genet 21(2):213–215. doi:10.1038/6002

dbGaP|phs000462.v1.p1|T2D-GENES Project 2: San Antonio Mexican American Family Studies. (2015) http://www.ncbi.nlm.nih.gov/projects/gap/cgi-bin/study.cgi?study_id=phs000462.v1.p1. Accessed 21 Jan 2015

Deeb SS, Fajas L, Nemoto M et al (1998) A Pro12Ala substitution in PPARgamma2 associated with decreased receptor activity, lower body mass index and improved insulin sensitivity. Nat Genet 20(3):284–287. doi:10.1038/3099

DIAbetes Genetics Replication and Meta-analysis (DIAGRAM) Consortium, Asian Genetic Epidemiology Network Type 2 Diabetes (AGEN-T2D) Consortium, South Asian Type 2 Diabetes (SAT2D) Consortium et al (2014) Genome-wide trans-ancestry meta-analysis provides insight into the genetic architecture of type 2 diabetes susceptibility. Nat Genet 46(3):234–244. doi:10.1038/ng.2897

Duggirala R, Blangero J, Almasy L et al (1999) Linkage of type 2 diabetes mellitus and of age at onset to a genetic location on chromosome 10q in Mexican Americans. Am J Hum Genet 64(4):1127–1140, doi:AJHG980631 [pii]

Dupuis J, Brown PO, Siegmund D (1995) Statistical methods for linkage analysis of complex traits from high-resolution maps of identity by descent. Genetics 140(2):843–856

Dusatkova P, Fang M, Pruhova S et al (2014) Lessons from whole-exome sequencing in MODYX families. Diabetes Res Clin Pract 104(3):e72–e74. doi:10.1016/j.diabres.2014.03.008

Eid J, Fehr A, Gray J et al (2009) Real-time DNA sequencing from single polymerase molecules. Science 323(5910):133–138. doi:10.1126/science.1162986

Ellard S, Lango Allen H, De Franco E, Flanagan SE, Hysenaj G, Colclough K, Houghton JAL, Shepherd M, Hattersley AT, Weedon MN, Caswell R (2013) Improved genetic testing for monogenic diabetes using targeted next-generation sequencing. Diabetologia. doi:10.1007/s00125-013-2962-5

ENCODE Project Consortium (2012) An integrated encyclopedia of DNA elements in the human genome. Nature 489(7414):57–74. doi:10.1038/nature11247

FDA (2015) Press Announcements>FDA allows marketing of four "next generation" gene sequencing devices. http://www.fda.gov/NewsEvents/Newsroom/PressAnnouncements/ucm375742.htm. Accessed 21 Jan 2015

Flannick J, Thorleifsson G, Beer NL et al (2014) Loss-of-function mutations in SLC30A8 protect against type 2 diabetes. Nat Genet 46(4):357–363. doi:10.1038/ng.2915

Fodor SP, Read JL, Pirrung MC, Stryer L, Lu AT, Solas D (1991) Light-directed, spatially addressable parallel chemical synthesis. Science 251(4995):767–773

Fokstuen S, Makrythanasis P, Nikolaev S et al (2014) Multiplex targeted high-throughput sequencing for Mendelian cardiac disorders. Clin Genet 85(4):365–370. doi:10.1111/cge.12168

Frey KG, Herrera-Galeano JE, Redden CL et al (2014) Comparison of three next-generation sequencing platforms for metagenomic sequencing and identification of pathogens in blood. BMC Genomics 15:96. doi:10.1186/1471-2164-15-96

Gao R, Liu Y, Gjesing AP et al (2014) Evaluation of a target region capture sequencing platform using monogenic diabetes as a study-model. BMC Genet 15(1):13. doi:10.1186/1471-2156-15-13

Gharizadeh B, Akhras M, Nourizad N et al (2006) Methodological improvements of pyrosequencing technology. J Biotechnol 124(3):504–511, doi:S0168-1656(06)00077-0 [pii]

Ghosh S, Watanabe RM, Hauser ER et al (1999) Type 2 diabetes: evidence for linkage on chromosome 20 in 716 Finnish affected sib pairs. Proc Natl Acad Sci USA 96(5):2198–2203

Gloyn AL, Weedon MN, Owen KR et al (2003) Large-scale association studies of variants in genes encoding the pancreatic beta-cell KATP channel subunits Kir6.2 (KCNJ11) and SUR1 (ABCC8) confirm that the KCNJ11 E23K variant is associated with type 2 diabetes. Diabetes 52(2):568–572

Grant SF, Thorleifsson G, Reynisdottir I et al (2006) Variant of transcription factor 7-like 2 (TCF7L2) gene confers risk of type 2 diabetes. Nat Genet 38(3):320–323, doi:ng1732 [pii]

Gudmundsson J, Sulem P, Gudbjartsson DF et al (2012) A study based on whole-genome sequencing yields a rare variant at 8q24 associated with prostate cancer. Nat Genet 44(12):1326–1329. doi:10.1038/ng.2437

Hani EH, Boutin P, Durand E et al (1998) Missense mutations in the pancreatic islet beta cell inwardly rectifying K+ channel gene (KIR6.2/BIR): a meta-analysis suggests a role in the polygenic basis of Type II diabetes mellitus in Caucasians. Diabetologia 41(12):1511–1515. doi:10.1007/s001250051098

Horikawa Y, Oda N, Cox NJ et al (2000) Genetic variation in the gene encoding calpain-10 is associated with type 2 diabetes mellitus. Nat Genet 26(2):163–175. doi:10.1038/79876

Inoue H, Ferrer J, Warren-Perry M et al (1997) Sequence variants in the pancreatic islet beta-cell inwardly rectifying K+ channel Kir6.2 (Bir) gene: identification and lack of role in Caucasian patients with NIDDM. Diabetes 46(3):502–507

International HapMap 3 Consortium, Altshuler DM, Gibbs RA et al (2010) Integrating common and rare genetic variation in diverse human populations. Nature 467(7311):52–58. doi:10.1038/nature09298

International Human Genome Sequencing Consortium (2004) Finishing the euchromatic sequence of the human genome. Nature 431(7011):931–945, doi:nature03001 [pii]

Jacob S, Stumvoll M, Becker R et al (2000) The PPARgamma2 polymorphism pro12Ala is associated with better insulin sensitivity in the offspring of type 2 diabetic patients. Horm Metab Res 32(10):413–416. doi:10.1055/s-2007-978663

Ji L, Malecki M, Warram JH, Yang Y, Rich SS, Krolewski AS (1997) New susceptibility locus for NIDDM is localized to human chromosome 20q. Diabetes 46(5):876–881

Johansson S, Irgens H, Chudasama KK et al (2012) Exome sequencing and genetic testing for MODY. PLoS One 7(5), e38050. doi:10.1371/journal.pone.0038050

Kaye J, Hurles M, Griffin H et al (2014) Managing clinically significant findings in research: the UK10K example. Eur J Hum Genet 22(9):1100–1104. doi:10.1038/ejhg.2013.290

Kere J, Nagaraja R, Mumm S, Ciccodicola A, D'Urso M, Schlessinger D (1992) Mapping human chromosomes by walking with sequence-tagged sites from end fragments of yeast artificial chromosome inserts. Genomics 14(2):241–248, doi:S0888-7543(05)80212-5 [pii]

Koshimizu E, Miyatake S, Okamoto N et al (2013) Performance comparison of bench-top next generation sequencers using microdroplet PCR-based enrichment for targeted sequencing in patients with autism spectrum disorder. PLoS One 8(9), e74167. doi:10.1371/journal.pone.0074167

Liu L, Li Y, Li S et al (2012) Comparison of next-generation sequencing systems. J Biomed Biotechnol 2012:251364. doi:10.1155/2012/251364

Luo TH, Zhao Y, Li G et al (2001) A genome-wide search for type II diabetes susceptibility genes in Chinese Hans. Diabetologia 44(4):501–506. doi:10.1007/s001250051649

Majithia AR, Flannick J, Shahinian P et al (2014) Rare variants in PPARG with decreased activity in adipocyte differentiation are associated with increased risk of type 2 diabetes. Proc Natl Acad Sci USA 111(36):13127–13132. doi:10.1073/pnas.1410428111

Mardis ER (2013) Next-generation sequencing platforms. Annu Rev Anal Chem (Palo Alto Calif) 6:287–303. doi:10.1146/annurev-anchem-062012-092628

Mori Y, Otabe S, Dina C et al (2002) Genome-wide search for type 2 diabetes in Japanese affected sib-pairs confirms susceptibility genes on 3q, 15q, and 20q and identifies two new candidate Loci on 7p and 11p. Diabetes 51(4):1247–1255

Mullis K, Faloona F, Scharf S, Saiki R, Horn G, Erlich H (1986) Specific enzymatic amplification of DNA in vitro: the polymerase chain reaction. Cold Spring Harb Symp Quant Biol 51(Pt 1): 263–273

Murray AW, Szostak JW (1983) Construction of artificial chromosomes in yeast. Nature 305(5931): 189–193

Nyren P (2007) The history of pyrosequencing. Methods Mol Biol 373:1–14, doi:1-59745-377-3:1 [pii]

O'Connor M, Peifer M, Bender W (1989) Construction of large DNA segments in Escherichia coli. Science 244(4910):1307–1312

O'Driscoll A, Daugelaite J, Sleator RD (2013) 'Big data', Hadoop and cloud computing in genomics. J Biomed Inform 46(5):774–781. doi:10.1016/j.jbi.2013.07.001

Pabinger S, Dander A, Fischer M et al (2014) A survey of tools for variant analysis of next-generation genome sequencing data. Brief Bioinform 15(2):256–278. doi:10.1093/bib/bbs086

Permutt MA, Wasson JC, Suarez BK et al (2001) A genome scan for type 2 diabetes susceptibility loci in a genetically isolated population. Diabetes 50(3):681–685

Quail MA, Smith M, Coupland P et al (2012) A tale of three next generation sequencing platforms: comparison of Ion Torrent, Pacific Biosciences and Illumina MiSeq sequencers. BMC Genomics 13:341. doi:10.1186/1471-2164-13-341

Reynisdottir I, Thorleifsson G, Benediktsson R et al (2003) Localization of a susceptibility gene for type 2 diabetes to chromosome 5q34-q35.2. Am J Hum Genet 73(2):323–335, doi:S0002-9297(07)61921-1 [pii]

Rothberg JM, Hinz W, Rearick TM et al (2011) An integrated semiconductor device enabling non-optical genome sequencing. Nature 475(7356):348–352. doi:10.1038/nature10242

Sanger F, Nicklen S, Coulson AR (1977) DNA sequencing with chain-terminating inhibitors. Proc Natl Acad Sci USA 74(12):5463–5467

Schmutz J, Wheeler J, Grimwood J et al (2004) Quality assessment of the human genome sequence. Nature 429(6990):365–368. doi:10.1038/nature02390

Service SK, Teslovich TM, Fuchsberger C et al (2014) Re-sequencing expands our understanding of the phenotypic impact of variants at GWAS loci. PLoS Genet 10(1), e1004147. doi:10.1371/journal.pgen.1004147

Silander K, Mohlke KL, Scott LJ et al (2004) Genetic variation near the hepatocyte nuclear factor-4 alpha gene predicts susceptibility to type 2 diabetes. Diabetes 53(4):1141–1149

Stumvoll M, Wahl HG, Loblein K et al (2001) Pro12Ala polymorphism in the peroxisome proliferator-activated receptor-gamma2 gene is associated with increased antilipolytic insulin sensitivity. Diabetes 50(4):876–881

Summerer D (2009) Enabling technologies of genomic-scale sequence enrichment for targeted high-throughput sequencing. Genomics 94(6):363–368. doi:10.1016/j.ygeno.2009.08.012

Suzuki T, Tsurusaki Y, Nakashima M et al (2014) Precise detection of chromosomal translocation or inversion breakpoints by whole-genome sequencing. J Hum Genet 59(12):649–654. doi:10.1038/jhg.2014.88

Tan R, Wang Y, Kleinstein SE et al (2014) An evaluation of copy number variation detection tools from whole-exome sequencing data. Hum Mutat 35(7):899–907. doi:10.1002/humu.22537

Tattersall RB (1974) Mild familial diabetes with dominant inheritance. Q J Med 43(170):339–357

Tennessen JA, Bigham AW, O'Connor TD et al (2012) Evolution and functional impact of rare coding variation from deep sequencing of human exomes. Science 337(6090):64–69. doi:10.1126/science.1219240

Teslovich TM, Morrison J, Fontanillas P, et al (2012) Whole-exome sequencing of 10,000 type 2 diabetes cases and controls from five major ancestry groups. Presented at the annual meeting of the American society of human genetics, San Francisco, CA, 6–10 November 2012. http://www.ashg.org/2012meeting/abstracts/fulltext/f120120640.htm

Thomas H, Jaschkowitz K, Bulman M et al (2001) A distant upstream promoter of the HNF-4alpha gene connects the transcription factors involved in maturity-onset diabetes of the young. Hum Mol Genet 10(19):2089–2097

Tsuchiya T, Schwarz PE, Bosque-Plata LD et al (2006) Association of the calpain-10 gene with type 2 diabetes in Europeans: results of pooled and meta-analyses. Mol Genet Metab 89(1–2):174–184, doi:S1096-7192(06)00193-4 [pii]

Wu J, Lee WP, Ward A et al (2014) Tangram: a comprehensive toolbox for mobile element insertion detection. BMC Genomics 15:795. doi:10.1186/1471-2164-15-795

Yen CJ, Beamer BA, Negri C et al (1997) Molecular scanning of the human peroxisome proliferator activated receptor gamma (hPPAR gamma) gene in diabetic Caucasians: identification of a Pro12Ala PPAR gamma 2 missense mutation. Biochem Biophys Res Commun 241(2):270–274, doi: S0006-291X(97)97798-6 [pii]

Zouali H, Hani EH, Philippi A et al (1997) A susceptibility locus for early-onset non-insulin dependent (type 2) diabetes mellitus maps to chromosome 20q, proximal to the phosphoenolpyruvate carboxykinase gene. Hum Mol Genet 6(9):1401–1408, doi:dda196 [pii]

Chapter 6
Fine-Mapping of Type 2 Diabetes Loci

Laura J. Scott and Karen L. Mohlke

Abstract Genetic association analyses have identified a large number of loci that are associated with type 2 diabetes (T2D); however, the underlying functional T2D variants at most loci remain unknown. Identification of functional T2D variants can be critical to identifying the target gene(s) and direction of effect by which a locus influences T2D risk. A useful approach to narrow a search for these variants is to perform fine-mapping. In this approach, large sample sizes and defined regions are analyzed using genotypes for as many variants as possible through imputation, arrays of densely spaced markers, or sequencing. Meta-analyses are performed in genetically homogeneous and/or in ancestrally diverse samples, and evidence of multiple association signals in a region is dissected. Credible sets of variants for each association signal are then annotated for potential effects on gene function or regulation that can be examined in laboratory studies. Use of the identified variants may improve power to detect interactions with other variants or the environment and may improve genetic predictions of T2D risk.

6.1 Introduction

As described in earlier chapters, the number of identified type 2 diabetes (T2D)-associated loci continues to increase as the number of T2D cases and controls analyzed and the proportion of the genome assayed increase (Grant et al. 2006; Diabetes Genetics Initiative of Broad Institute of et al. 2007; Scott et al. 2007; Sladek et al. 2007; Zeggini et al. 2007; Wellcome Trust Case Control Consortium 2007; Voight et al. 2010; Morris 2011; Morris et al. 2012). Currently, >80 associated regions have been identified often with multiple apparently independent variants per region (DIAbetes Genetics Replication Meta-analysis Consortium

L.J. Scott (✉)
Department of Biostatistics and Center for Statistical Genetics, University of Michigan, Ann Arbor, MI, USA
e-mail: ljst@umich.edu

K.L. Mohlke
Department of Genetics, University of North Carolina, Chapel Hill, NC, USA

J.C. Florez (ed.), *The Genetics of Type 2 Diabetes and Related Traits*,
DOI 10.1007/978-3-319-01574-3_6

et al. 2014). These lead associated variants (index variants), however, represent a small fraction of the total genetic variation of T2D (estimate of 5.7 % for 63 loci) (Morris et al. 2012), and likely hundreds to thousands of additional variants remain to be identified (Morris et al. 2012; DIAbetes Genetics Replication Meta-analysis Consortium et al. 2014). The underlying goals of T2D genetic association studies are to identify the functional T2D variants, for which one allele increases risk of T2D, and the genes and mechanisms through which they work. Fine-mapping is the process of starting with an index variant association signal and proceeding through a series of steps to localize, and ultimately identify, the functional T2D variants that are responsible for the association signal.

Many large- and small-scale efforts have been undertaken to identify functional T2D variants, but currently, relatively few likely functional T2D variants have been identified (Ng and Gloyn 2013). The ease or difficulty in translating initial association signals into functional T2D variants is influenced by three broad classes of factors: the architecture of functional T2D variants (Sect. 6.2), the resources and methods for narrowing the number of variants of interest (Sects. 6.3.1–6.3.6), and the availability of annotation and biological assays for relevant variant functions (Sect. 6.3.7). The investigator's goal is to use available resources to capture functional T2D variants while reducing the number of nonfunctional variants considered.

6.2 Locus Architecture

The planning and approaches to current fine-mapping studies are based on our understanding of the allelic architecture of functional T2D variants, assuming that we are starting with a common (>5 % minor allele frequency [MAF]) T2D-associated variant and the surrounding genomic neighborhood. These architectural features include the allele frequency(s), number, and effect sizes of the functional T2D variants underlying the original association signal. Other features include the presence of nearby functional T2D variants that do not contribute to the index variant association signal and the extent of linkage disequilibrium (LD), as measured by D' and r^2, between functional T2D variants and surrounding variants.

6.2.1 *Evidence for Rare, Low-Frequency and Common Functional T2D Variants*

The spectrum of views about the number, allele frequency, and types of functional T2D variants underlying an index variant stems from our current knowledge of Mendelian and complex diseases (Dickson et al. 2010; Wang et al. 2010; Visscher et al. 2012).

Much of what we know about rare functional T2D variants in non-autoimmune forms of diabetes comes from MODY and other Mendelian forms of diabetes (Schwitzgebel 2014; Yamagata 2014). In MODY, variants are inherited in an autosomal dominant pattern (Fajans and Bell 2011), and most identified functional T2D variants have been rare and located in the coding or promoter regions of genes (Yamagata 2014). Allelic heterogeneity is common, meaning that many different genetic variants can affect a single gene's function or expression to cause disease. Some genetic heterogeneity exists, meaning that variants in different genes can cause the same or similar disease in different families, but for most Mendelian diseases, a modest number of genes have been implicated.

In contrast, from genome-wide association studies of T2D, it is clear that extensive genetic heterogeneity exists. Hundreds to thousands of functional variants (and their target genes) may be scattered throughout the genome (Morris et al. 2012; DIAbetes Genetics Replication Meta-analysis Consortium et al. 2014; Stahl et al. 2012). In addition, allelic heterogeneity in single genes is likely the cause of the clustering of many independently associated variants within close proximity (Rees et al. 2012b; Bonnefond et al. 2012; DIAbetes Genetics Replication Meta-analysis Consortium et al. 2014).

Most currently known T2D-associated variants are common and noncoding, but more rare and low-frequency T2D-associated variants are beginning to be identified by sequencing and large-scale genotyping. A large Icelandic sequencing and imputation study detected a 1.4 % variant in an intron of *CCND2* and low-frequency and rare coding variants in *PAM-PPIP5K2* and *PDX1* (Steinthorsdottir et al. 2014). Twelve rare protein-truncating variants in *SLC30A8* were associated with lower risk of T2D, suggesting loss-of-function decreases risk (Flannick et al. 2014). The *CCND2* and *SLC30A8* variants are independent from the common T2D index SNP associations at these loci. Although there has been great interest in exploring if rarer variants underlie common index variant signals (Dickson et al. 2010; Wang et al. 2010), with some disappointment, using targeted, exome, or genome sequencing, there has been little evidence of rare variant (s) underlying common T2D and other common disease variant association signals (Visscher et al. 2012; Steinthorsdottir et al. 2014; Shea et al. 2011). This result is consistent with the findings that rarer underlying variation with strong enough effects to have caused the T2D index variant association signal would have resulted in much stronger linkage signals than had been observed previously for T2D (Anderson et al. 2011; Guan et al. 2012), or more broadly for common diseases (Wray et al. 2011). The lack of underlying rare variants is also consistent with observation that common T2D index variants have a strongly consistent direction of effect across ancestries with different LD patterns (DIAbetes Genetics Replication Meta-analysis Consortium et al. 2014).

More generally, genome-wide, a simulation-based study of T2D risk found that models for which T2D risk was highly concentrated in a relatively small number of rare variants were not consistent with the overall pattern of linkage or association observed in published studies, although models with substantial rare variant contribution to T2D risk could not be ruled out (Agarwala et al. 2013). Likewise, models where the bulk of the variance in T2D risk was explained by coding variants in small number of genes were not consistent with the lack of significant associations in an exome sequencing study of 1000 cases/1000 controls (Lohmueller et al. 2013).

6.2.2 Possible Models for Functional T2D Variants Underlying Common T2D Index Variants

Current data suggest that most functional variants underlying common index variants will be common, although it remains possible that rare or low-frequency variant(s) might underlie a small number of signals. Figure 6.1 shows some possible haplotypic relationships between a common index variant and the underlying functional T2D variant(s). In the simplest scenarios, a single common (Fig. 6.1a) or a single low-frequency or rare variant (Fig. 6.1c) underlies the index variant signal. In more complex models, multiple common (Fig. 6.1b) or low-frequency (Fig. 6.1d) functional T2D variants are present on the same haplotype, or multiple rarer functional T2D variants are present on different haplotypes (Fig. 6.1e). Of these scenarios, functional variants can only be strongly suspected from statistical analysis alone when a single variant in a very large thoroughly queried region has a much stronger association signal than other variants, i.e., the functional variant is not in strong r^2 with surrounding variants (a scenario with increased probability of sequencing artifacts for the variant of interest).

6.2.3 Visualization and Features of Index SNP Signals

A beginning assessment of the properties of a locus and a set of potentially functional variants is done by visualization of the T2D-variant association signals plotted by physical distance, often spanning hundreds of kilobases, with annotation of gene transcripts and recombination rates (Pruim et al. 2010; Scott et al. 2007). Loci have very different structures; some may have few T2D-associated variants in well-defined recombination intervals (Fig. 6.2a), others may have hundreds of variants with similar evidence of T2D association that stretches over multiple genes and hundreds of kilobases (Fig. 6.2b), and others may have evidence of multiple independently associated variants in the same region (Fig. 6.2c). This high-level view gives an initial assessment of the LD surrounding the index variant (s) and the strength of the signal(s) but on its own yields few clues as to the allele frequency, effect size, and number of underlying functional T2D variants.

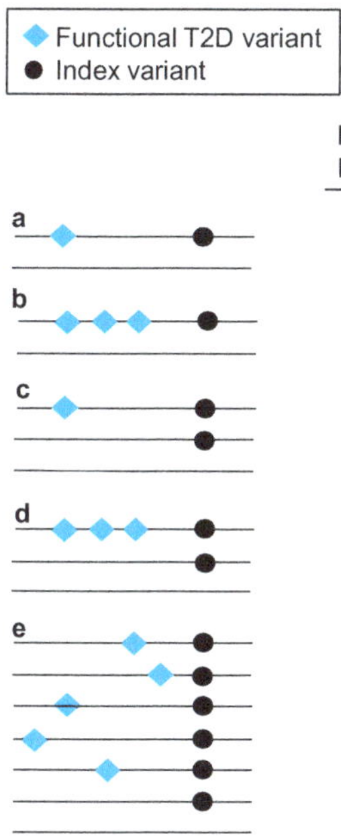

	Haplotype Frequency	Individual Functional T2D Variant RR	Sample Size to Detect Functional T2D Variant
a	.30	1.1	32,956
	.70	1	
b	.30	1.03	32,956
	.70	1	
c	.02	2.5	2,488
	.28	1	
	.70	1	
d	.02	1.4	2,488
	.28	1	
	.70	1	
e	.004	2.5	11,990
	.004	2.5	
	.004	2.5	
	.004	2.5	
	.004	2.5	
	.28	1	
	.70	1	

Fig. 6.1 Possible configurations of a common index SNP and underlying functional T2D variants. Low-frequency and rare functional T2D variants that are responsible for an index variant association signal can be detected in smaller sample sizes than the sample required for the index variant. In all scenarios, the index variant has an effect allele frequency of 0.3 and a relative risk (RR) of 1.1. The sample size to detect a functional T2D variant is the sample size required for 80 % power to detect the variant using a significant threshold of 5×10^{-8}. For each scenario (**a–e**), the sample size required to detect a functional T2D variant with 80 % power using a significance threshold of 5×10^{-8} is shown. (**a**) One common functional T2D variant ($r^2 = 1$ with the index variant); (**b**) three common functional T2D variants on the same haplotype ($r^2 = 1$ with the index variant). The sample size is equivalent to that of (**a**) because we assume that the effects of the three variants are multiplicative ($1.035^3 = 1.1$), and these three variants can't be tested independently of each other; (**c**) one low-frequency variant ($r^2 = 0.048$ with the index variant); (**d**) three low-frequency variants on the same haplotype ($r^2 = 0.048$ with the index variant). The sample size is equivalent to that of **c** because we assume that the effects of the three variants are multiplicative ($1.4^3 = 2.5$), and these three variants can't be tested independently of each other; (**e**) Five rare variants on different haplotypes ($r^2 = 0.0014$ with index variant). In (**e**), for clarity, the sample size is shown for just one of the five functional T2D variants; the same sample size is required for each of the variants

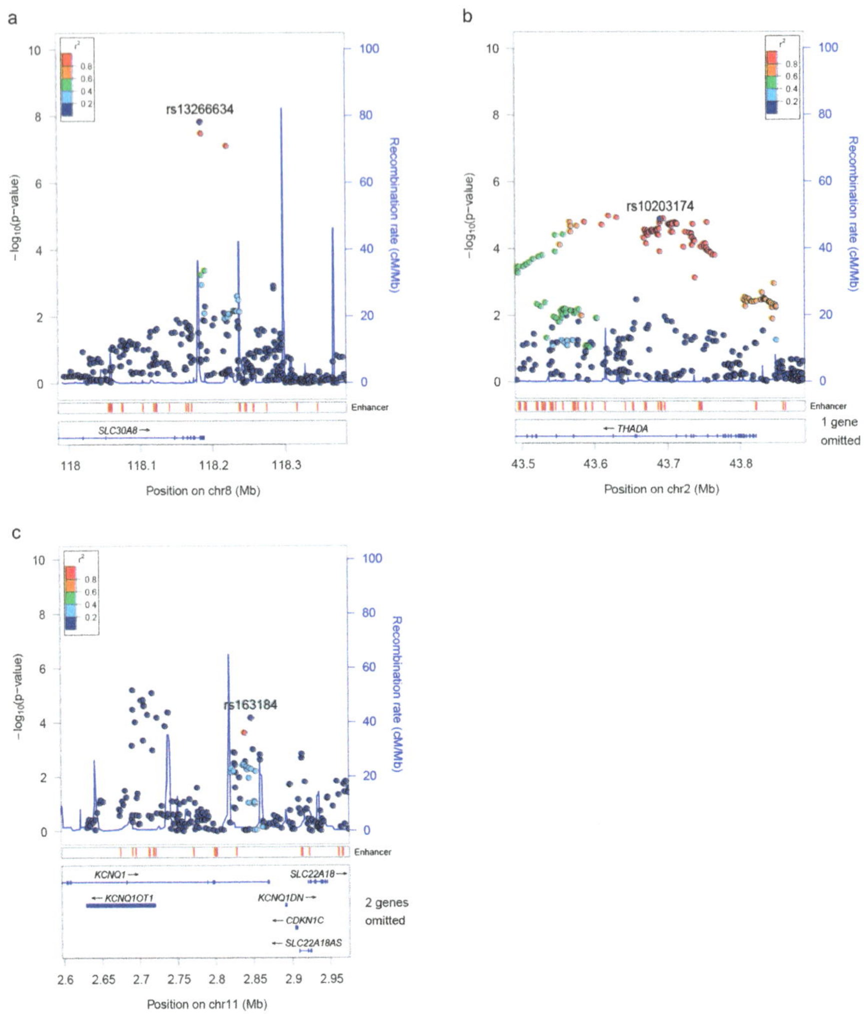

Fig. 6.2 Regional plots of three established T2D loci with different patterns of associated variants. Genotyped and HapMap-imputed variants from an analysis of 8130 cases and 38,987controls (total effective sample size =22,044; Voight et al. 2010) are plotted with their meta-analysis -$\log_{10}$ p-values as a function of genomic position (NCBI Build 37). In each panel, the index association SNP is represented by a diamond. Estimated recombination rates (taken from HapMap) are plotted to reflect the local LD structure around the associated SNPs and their correlated proxies (according to a scale from $r^2 = 0$ to 1, based on pairwise r^2 values from 1000 Genomes Europeans). Gene annotations were extracted from the University of California-Santa Cruz genome browser. ChromHMM enhancer tracks were called from human islet ChIP-seq data (Parker et al. 2013). (**a**) The *SLC30A8* locus has a small number of common strongly associated variants (rare variants not shown); (**b**) the *THADA* locus has a very broad association signal; (**c**) the *KCNQ1* locus includes multiple independent signals, two of which are evident in this plot

6.3 Approaches to Fine-Mapping

Fine-mapping a GWAS locus involves a series of steps that take into account the expected and observed locus architecture to increase the chance that a functional variant(s) will be analyzed and can be distinguished from other variants.

6.3.1 Identify Large Numbers of Samples of Similar and/or Diverse Ancestry

The sample size required depends on several factors, including the belief about functional T2D variant frequency(s). For common functional T2D variants, ideally the sample size would be large enough to have a substantial association signal for each index variant. Fine-mapping sample sizes that are smaller or different from those used for locus discovery may be too weak to distinguish the set of potentially functional T2D variants from surrounding variants in relatively high LD (Wellcome Trust Case Control Consortium et al. 2012; DIAbetes Genetics Replication Meta-analysis Consortium et al. 2014). In contrast, if the assumption is that a single low-frequency or even several rare functional T2D variants underlie a common index variant, then considerably smaller samples would have comparable power to detect a genome-wide significant association of the rare and low-frequency variants (Fig. 6.1c, e).

Combination of samples of diverse ancestry in fine-mapping can help identify new loci and reduce the size of the locus region. Analyses across ethnicities have been successful as most established loci have genetic effect sizes that are remarkably similar across different populations (Ng et al. 2013; DIAbetes Genetics Replication Meta-analysis Consortium et al. 2014; Sim et al. 2011; Waters et al. 2010). Patterns of LD differ across populations due to differing evolutionary and migratory histories, and variants that are in high LD with each other in one population may be in lower LD in another population, which allows narrowing of the set of likely functional T2D variants (DIAbetes Genetics Replication Meta-analysis Consortium et al. 2014). Likewise differences in allele frequencies between ancestries may allow detection of signals due to different functional T2D variants (Unoki et al. 2008; Yasuda et al. 2008). Concerns about inflation of signals through population stratification can be allied through meta-analysis of samples with relatively homogeneous ancestry.

6.3.2 Define Genomic Region to Be Fine-Mapped

Defining a region to analyze is an important step toward identifying a functional T2D variant. The regional boundaries again depend on the researchers' beliefs about the expected allele frequency of the underlying variant(s) and the available resources. In designing the genotyping part of the study, researchers typically choose either to focus on variants within a defined region around the index variant or on a whole genome or whole exome sequencing approach. In the analytical phase, more or less restrictive definitions can be employed depending on the question and data available.

Now that genome sequence-based reference panels contain almost all common variants in designated populations, chances are large that a common functional T2D variant would be both on the panel and well-imputed. In this scenario, the functional T2D variant should be in high r^2 with the index (strongest) variant and located within the same recombination interval. Using these assumptions, regions of interest are often defined based on combination of criteria including LD with index variants (say $r^2 > 0.1$) and/or genetic distance (Wellcome Trust Case Control Consortium et al. 2012; DIAbetes Genetics Replication Meta-analysis Consortium et al. 2014). In regions with extensive LD, the locus size can be large, 500 kb–1 Mb. When there is no cost to searching a larger region, all variants in up to 5–10 Mb surrounding the index variant, can be tested for completeness. Choosing a larger region can impact downstream analysis, particularly the size of credible variant sets (Sect. 6.3.6).

When the goal is to be able to detect underlying functional T2D variants that are rare, a much wider interval needs to be considered because LD between a common variant and rare variants may extend beyond strong recombination hot spots (Huyghe et al. 2013).

6.3.3 *Increase Coverage of Variants at Identified T2D Locus/Loci*

Completeness of variant coverage is also critical to identifying functional variants. The three approaches below, imputation, genotype arrays, and sequencing, require increasing financial resources. The completeness of coverage of the rarest variants increases dramatically from imputation to higher pass sequencing. Each of these approaches is limited by the complexity of the DNA surrounding the functional T2D variant(s). Variants in repetitive or other regions that are difficult to assay will be much less likely to be identified without specific and substantial investment.

6.3.3.1 Imputation

The goal of imputation is to infer nucleotides at un-genotyped variants. Imputation of a study sample requires a scaffold of genotyped variants in that sample and a set of reference haplotypes from more densely spaced variants. Imputation allows association results from disparate genotyped variant sets to be combined and greatly extends the number of variants that can be assayed from even the largest genotyping arrays. Imputation was initially used in T2D to identify association signals (Scott et al. 2007; Zeggini et al. 2008; Voight et al. 2010) but is equally useful in fine-mapping loci (Shea et al. 2011; Morris et al. 2012; Kulzer et al. 2014).

Multiple methods have been developed and have evolved for genotype imputation. Three of the most commonly used methods/software are MaCH, IMPUTE2, and BEAGLE (Browning and Yu 2009; Howie et al. 2009; Li et al. 2010; Marchini and Howie 2010). For large-scale datasets and reference panels, these methods use a two-stage process (Howie et al. 2012). First, GWAS variants from the study samples are phased to produce a "best guess" set of haplotypes. Second, the more densely genotyped reference panel is then used to impute (predict) genotypes in the study sample. Imputation's greatest strength is that it makes an existing dataset much more powerful solely through investment in computational resources. Imputation currently works best for common variants. Because both alleles of common variants are present on multiple reference haplotypes, the chances of correctly imputing a variant are high. Still, direct genotyping can improve an association signal compared to imputation. Imputation quality measures usually accurately reflect information lost compared to direct genotyping for common variants (Li et al. 2010). As the reference panel sizes increase, our ability to accurately impute low-frequency variants has grown and will continue to increase. With current reference panel sizes of <2700 individuals, variants with frequency <0.5 % are poorly imputed, and the imputation quality statistics are not reliable. To improve the quality of low-frequency and particularly of rarer variants, reference panels of >25,000 individuals are currently being constructed; these should allow imputation into allele frequency ranges currently not analyzed in most association studies. Regardless of the reference panel size, imputation will likely not provide a way to recover the very rarest and potentially most deleterious variants because they will not be present in sufficient numbers in the reference panels to provide high-quality imputation. In particular, singleton variants are currently excluded from reference panels, as they cannot be phased statistically.

6.3.3.2 Specialized Arrays

A cost-effective approach to increasing variant coverage in very large numbers of samples is to create a custom genotyping array. Such an array can contain very densely spaced variants, or possibly all known variants, in a specified genome region. Genotyping arrays are most cost-effective when the cost of design can be split among the cost of tens or hundreds of thousands of arrays.

Two arrays used recently to identify or characterize T2D loci are the "Metabochip" (Voight et al. 2012; Morris et al. 2012) and the ITMAT-Broad-CARe (IBC) array (Saxena et al. 2012). The Metabochip included variants to fine-map 257 loci previously associated with one or more of 23 metabolic or cardiovascular traits. The variants were chosen from the HapMap project (International HapMap Consortium et al. 2007) and an early version of the 1000 Genomes Project

(1000 Genomes Project Consortium et al. 2010). Across T2D loci included, the array contains ~90 % of common variants (MAF > 0.05) and ~60 % of low-frequency variants (MAF 0.01–0.05) (Voight et al. 2012). Using this array, two of 36 loci analyzed, *PROX1* and *KLF14*, contained a variant with notably lower allele frequency and stronger odds ratio than the original GWAS variant (Morris et al. 2012). The IBC array contains ~50,000 variants and was designed to capture genetic diversity across 2000 loci previously associated with a cardiovascular, inflammatory, or metabolic trait. Using this array and conditional analysis to fine-map T2D loci, multiple established and new independent signals were identified within five loci in European ancestry samples and within *HMGA2* in African Americans (Saxena et al. 2012).

Practical issues in the use of specialized arrays include the array design and genotype calling for low-frequency variants. Ideally, an array design would include all known variants in all regions to be studied. If resources are limited, a dense but incomplete framework of variants can be selected based on LD between variants, with the expectation that variants not included on the array can be subsequently imputed. Not all variants can be successfully designed into genotyping assays, and not all designed assays succeed to generate high-quality genotypes. To accommodate those genotyping failures, array designs may include redundant or partially redundant variants. Genotyping calling for low-frequency variants can be inaccurate if genotypes are called on small numbers of samples and if rare homozygote genotypes are inaccurately clustered with heterozygotes or inaccurately removed as perceived outliers. Genotype calling can be improved by clustering genotypes across thousands of samples.

6.3.3.3 Sequencing of Candidate Regions, Exomes, and Genomes

Sequencing of candidate regions (Shea et al. 2011), whole exomes (Consortium et al. 2014; Albrechtsen et al. 2013), and whole genomes (Steinthorsdottir et al. 2014) allows a more complete enumeration and analysis of rare and low-frequency variation and can slightly improve common variant genotyping. Whole genome sequencing combined with imputation has led to identification of T2D associations with an associated low-frequency variant in a *CCND2* intron as well as low-frequency and rare coding variants in *PAM-PPIP5K2* and *PDX1* (Steinthorsdottir et al. 2014). Exome sequencing has identified protein-truncating variants in *SLC30A8* (Flannick et al. 2014) and a low-frequency missense variant in *HNF1A* (MODY3) (SIGMA Type 2 Diabetes Consortium et al. 2014).

Other large sequencing efforts are underway, including whole genome sequencing of T2D cases and controls in GoT2D (2657 low-pass genomes (Gaulton et al. 2013)) and in T2D-GENES (590 high-pass genomes (Jun et al. 2013)) and exome sequencing in T2D-GENES and GoT2D (12,940 exomes (Mahajan et al. 2014)). The initial results suggest that a small number of low-frequency variants with strong effects can be detected in samples of this size.

One of the largest challenges in sequencing is the presence of batch effects due to sample source, sample preparation, or technical issues. As with any genotyping/sequencing method that aims to compare cases and controls and particularly given the large potential technical variability in sequencing, balancing cases and controls side-by-side throughout the experiment is strongly advised to help avoid many artifact associations in the analysis stage.

6.3.4 *Analyze and Perform Meta-analysis from Multiple GWAS of Similar and Diverse Ancestry*

Although fine-mapping efforts can be performed within a single study, most studies are not large enough to provide sufficient power to detect and discriminate between variants based on evidence of association. Meta-analysis is used to combine the association results from multiple studies.

6.3.4.1 Meta-analysis of Similar Ancestry Studies

For common variants, meta-analysis is an efficient way to combine results, retaining almost all the power of a combined analysis of all samples (Lin and Zeng 2010). Variants are typically tested using logistic regression and an additive model, with the effect of each allele and significance evaluated using the Wald, likelihood ratio, or score test. Of these, the Wald and likelihood ratio tests also provide estimates of the effect size. Each study's results are then combined using fixed effects meta-analysis, which has the advantage of producing an effect size estimate or by sample size weighting of the z-scores. Multiple quality checks help ensure the results are reliable and include comparing allele frequencies across studies, ensuring standard errors are consistent with the corresponding sample size and allele frequency, and testing for effect size heterogeneity (Winkler et al. 2014).

Study-specific analysis and meta-analysis are less straightforward for low-frequency variants, or more exactly, for variants with lower minor allele count (MAC) in a given study. When stringent significance thresholds are used, the asymptotic assumptions of these tests begin to break down at total MAC $< \sim 400$ and can result in conservative or anti-conservative results dependent on the test and sample (Ma et al. 2013). For studies with a balanced number of case and control samples, the Firth and score tests preserve the most power and are the best calibrated; conversely, the Wald test should not be used because it is highly conservative. For studies with case-to-control sample ratios modestly to highly different from 1:1, all tests can be anti-conservative, and the score test is highly anti-conservative and should not be used (Ma et al. 2013).

6.3.4.2 Meta-analysis of Diverse Ancestry Sample

Trans-ancestry meta-analysis methods assume that the same functional T2D variant exists across populations and use fixed effects, random effects, or modifications of these models. Prior to identifying a functional T2D variant, its effect size across populations is unknown. If the effects are truly similar, then fixed effect models will have higher power. If effects are dissimilar, then random effects' models theoretically perform better. One recently described method improves on a random effects model by relaxing a conservative assumption that effect sizes differ even under the null hypothesis (Han and Eskin 2011). Another method applies a Bayesian approach and assumes that more genetically similar populations will have more similar effect sizes, allowing the effect sizes in less similar populations to vary (Morris 2011). Application of these standard and newer meta-analysis methods to T2D loci showed that the new methods can identify loci not detected by standard fixed or random effects models (Wang et al. 2013).

6.3.5 Identify Likely Independent Signals at Loci Through Direct or Indirect Conditional Analysis

A main goal of conditional analysis is to determine if one or more apparently independently associated signals exist in a region. A second, reciprocal goal is to ask if one or more non-index variants have more evidence for association than the index variant and can explain the index variant signal, that is, if there is a better index variant for the signal. We focus on the first goal, but many of the same principles hold true for the second. Conditional analysis is performed in two main ways, directly using genotype data (exact) (Scott et al. 2007; Zeggini et al. 2008) or indirectly using association results and LD information (Genome-Wide Complex Trait Analysis (GCTA)-based) (Yang et al. 2013).

Conditional analysis has been used to show that an index variant can explain the association signals from variants surrounding it (Scott et al. 2007; Zeggini et al. 2008) and to identify multiple independently associated variants at a number of loci including *KCNQ1*, *CDKN2A/B*, *DGKB*, and *MC4R* (Voight et al. 2010; Morris et al. 2012). GCTA has been used to perform conditional analysis of height (Yang et al. 2012) and in ongoing work in T2D.

6.3.5.1 Exact Conditional Analysis

In exact conditional analysis, each study performs conditional analyses for each variant by including the index variant(s) as a covariate(s) in the logistic regression; the results for each variant are combined using meta-analysis (Sect. 6.3.4). Conditional analysis can be performed with index variants from a specific region (Scott et al. 2007; Zeggini et al. 2008) or with all index variants in the genome (Voight et al. 2010). As the number of associated variants increases, the inclusion of all index SNPs may cause instability in the analysis. In regions with very strong association p-values, some residual (likely false) association signal may remain following conditioning when the conditioning variants are poorly imputed or serve as a poor proxy for a more strongly associated functional T2D variant. Thus, interpretation of secondary signals surrounding a very strong primary signal requires caution. Unless genotype data is centrally analyzed, exact conditional analysis can be logistically challenging and can become impractical with large numbers of studies.

6.3.5.2 GCTA-Based Conditional Analysis

A faster, potentially more feasible alternative is to perform GCTA-based conditional analysis. This approach uses meta-analysis association results and LD data from a representative sample (Yang et al. 2013). Essentially, GCTA removes the proportion of the association signal that could be explained by the index variant given the LD between the variants. The standard errors from the meta-analysis and the GCTA-based (or exact) conditional analysis should be very similar; differences indicate that analytical issues need to be resolved.

6.3.6 Identify Credible Sets of Variants That Are Likely to Contain a Functional T2D Variant

The objective of credible set analysis is to identify the smallest set of analyzed variants that are likely to contain the functional T2D variant. The analysis takes a Bayesian approach, assuming a single functional variant. The Bayes factor for each variant in the region is divided by the sum of the Bayes factors for all variants in the region to obtain a posterior probability for each variant. The credible set is comprised of variants that contain 95 % or 99 % of the posterior probability (Wellcome Trust Case Control Consortium et al. 2012; DIAbetes Genetics Replication Meta-analysis Consortium et al. 2014).

The size of the credible set typically mirrors what is observed in regional association plots (Wellcome Trust Case Control Consortium et al. 2012; DIAbetes Genetics Replication Meta-analysis Consortium et al. 2014). Strong signals in regions with little LD have the smallest credible sets; weak signals in regions with extensive LD have the largest credible sets. In the trans-ethnic meta-analysis, which used HapMap-based imputation, the *JAZF1* and *SLC30A8* loci were substantially narrowed (DIAbetes Genetics Replication Meta-analysis Consortium et al. 2014).

Credible set analysis assumes that the functional T2D variant is present within the set of genotyped variants considered for analysis. Use of sequence-based data or more complete reference panels for imputation typically increases the number of variants in the credible set for a given study (Wellcome Trust Case Control Consortium et al. 2012) but means the functional variant is more likely to be present. To reduce the possibility that variants not responsible for the index signal can decrease the size of the credible set, variants outside of an r^2-defined region may be excluded (Sect. 6.3.2). An alternative approach would be to remove the effects of other known independent signals in the region through conditional analysis.

A second assumption of credible set analysis is that the functional T2D variants association signals are among the strongest of those considered for analysis. This assumption should be true when a sizable index variant association signal exists in the tested sample and when the number of functional variants underlying the signal is not large (see Fig. 6.1). Regions in which the most strongly associated variant has a p-value greater than 5×10^{-8} are less likely to have small credible sets (Wellcome Trust Case Control Consortium et al. 2012; DIAbetes Genetics Replication Meta-analysis Consortium et al. 2014). If a functional T2D variant has a higher standard error than expected for its allele frequency, due to either to smaller sample size or poorer imputation quality than other variants in the region, the variant may not have sufficient association evidence to be included in the credible set. Credible set analysis may also miss functional T2D variants if one variant in a haplotype shows association in single variant analysis, while the second variant is only detected after adjusting for the first variant, such as at the *CDKN2A/B* locus (Wellcome Trust Case Control Consortium et al. 2012).

6.3.7 Incorporate Genome Annotations and Examine Experimentally

Once a credible set of variants has been identified, annotations can be used to predict which of them are more likely to have a functional effect. The annotations can be split broadly into sources that predict the effect of changing protein structure via amino acid substitution and those that predict regulatory activity. Numerous resources exist to annotate both types of variants (Cooper and Shendure 2011). These annotations can be used to prioritize variants for laboratory experimental analysis and to understand more generally which regions of the genome influence T2D risk.

Annotations of protein-coding variants make use of evolutionary conservation, biochemical principles, and consideration of protein structure. Some approaches make predictions based on similarity to a defined property, while other approaches train a classifier based on known functional T2D variants. T2D loci that harbor multiple protein-coding variants include *GCKR* and *MTNR1B* (Bonnefond et al. 2012; Rees et al. 2012a). For these loci, multiple protein-coding variants were tested experimentally and found to have an effect on function. In both examples, the results were only partly predicted by annotations, confirming the need for functional studies (Bonnefond et al. 2012; Rees et al. 2012a).

Regulatory elements can be defined using known transcription factor binding motifs, genomic regions at which the chromatin is open and accessible to transcription factor binding or histone modification, and conservation across species. Annotations based on chromatin accessibility can be cell-type-specific. Many regulatory annotations are available from the ENCODE (ENCODE Project Consortium et al. 2012), Roadmap Epigenomics (Bernstein et al. 2010) and BluePrint projects (Abbott 2011) that provide experimental evidence of regulatory elements in hundreds of cell types. In addition, individual laboratories are generating data to define regulatory elements specifically in tissue types relevant to T2D, such as pancreatic islets (Gaulton et al. 2010; Stitzel et al. 2010; Parker et al. 2013; Pasquali et al. 2014). Regulatory annotations have been used to prioritize among T2D variants for experimental assays to identify effects of alleles on gene regulation (Gaulton et al. 2010; Stitzel et al. 2010; Fogarty et al. 2013; Travers et al. 2013; Kulzer et al. 2014). On a broader scale, identification of other annotated functional classes that show genome-wide enrichment of T2D signals will help guide the choice of regulatory elements and variants to study.

6.4 Summary and Future Directions

Known association signals, our current lampposts (Collins 2006), serve as markers for genomic regions likely to harbor functional T2D variants; fine-mapping is a way of strengthening and focusing this light on the underlying T2D variants and the genes they influence. Successful fine-mapping requires sample sizes large enough to have sufficient association signals to distinguish between variants and requires dense genotyping/imputation/sequencing to assure that the functional T2D variants are tested for association. The width of the interval queried depends both on the LD structure of the locus and on assumptions about the number and allele frequency of those variants. Based on fine-mapping efforts to date, it appears that few, if any, common T2D signals are caused by rare or low-frequency functional variants, although associated rare and low-frequency variants are present in and near loci identified by common T2D index variants. Credible sets of variants can be formed and further annotated with functional elements to help guide selection of variants for functional follow up.

Credible sets with small numbers of plausible variants with predicted functions and practical biological assays are most attractive. As the architecture of the genome is characterized, functional elements and the boundaries of regulatory domains will be identified. These findings should enable easier mapping of regulatory variants to genes, though a direct one-to-one mapping of each regulatory noncoding variant to its target gene(s) is, as of yet, not possible. As the number of identified functional T2D variants grows, we will better understand the classes of variants and genes that influence T2D risk. These discoveries may, in turn, lead to the development of new drugs and a better ability to prevent and treat T2D.

References

1000 Genomes Project Consortium, Abecasis GR, Altshuler D, Auton A, Brooks LD, Durbin RM, Gibbs RA, Hurles ME, McVean GA (2010) A map of human genome variation from population-scale sequencing. Nature 467(7319):1061–1073. doi:10.1038/nature09534

Abbott A (2011) Europe to map the human epigenome. Nature 477(7366):518. doi:10.1038/477518a

Agarwala V, Flannick J, Sunyaev S, GoT2D Consortium, Altshuler D (2013) Evaluating empirical bounds on complex disease genetic architecture. Nat Genet 45(12):1418–1427. doi:10.1038/ng.2804

Anderson CA, Soranzo N, Zeggini E, Barrett JC (2011) Synthetic associations are unlikely to account for many common disease genome-wide association signals. PLoS Biol 9(1), e1000580. doi:10.1371/journal.pbio.1000580

Albrechtsen A, Grarup N, Li Y, Sparso T, Tian G, Cao H, Jiang T, Kim SY, Korneliussen T, Li Q, Nie C, Wu R, Skotte L, Morris AP, Ladenvall C, Cauchi S, Stancakova A, Andersen G, Astrup A, Banasik K, Bennett AJ, Bolund L, Charpentier G, Chen Y, Dekker JM, Doney AS, Dorkhan M, Forsen T, Frayling TM, Groves CJ, Gui Y, Hallmans G, Hattersley AT, He K, Hitman GA, Holmkvist J, Huang S, Jiang H, Jin X, Justesen JM, Kristiansen K, Kuusisto J, Lajer M, Lantieri O, Li W, Liang H, Liao Q, Liu X, Ma T, Ma X, Manijak MP, Marre M, Mokrosinski J, Morris AD, Mu B, Nielsen AA, Nijpels G, Nilsson P, Palmer CN, Rayner NW, Renstrom F, Ribel-Madsen R, Robertson N, Rolandsson O, Rossing P, Schwartz TW, D.E.S.I.R. Study Group, Slagboom PE, Sterner M, the DIAGRAM Consortium, Tang M, Tarnow L, Tuomi T, Van't Riet E, van Leeuwen N, Varga TV, Vestmar MA, Walker M, Wang B, Wang Y, Wu H, Xi F, Yengo L, Yu C, Zhang X, Zhang J, Zhang Q, Zhang W, Zheng H, Zhou Y, Altshuler D, t Hart LM, Franks PW, Balkau B, Froguel P, McCarthy MI, Laakso M, Groop L, Christensen C, Brandslund I, Lauritzen T, Witte DR, Linneberg A, Jorgensen T, Hansen T, Wang J, Nielsen R, Pedersen O (2013) Exome sequencing-driven discovery of coding polymorphisms associated with common metabolic phenotypes. Diabetologia 56(2):298–310. doi:10.1007/s00125-012-2756-1

Bernstein BE, Stamatoyannopoulos JA, Costello JF, Ren B, Milosavljevic A, Meissner A, Kellis M, Marra MA, Beaudet AL, Ecker JR, Farnham PJ, Hirst M, Lander ES, Mikkelsen TS, Thomson JA (2010) The NIH roadmap epigenomics mapping consortium. Nat Biotechnol 28(10):1045–1048. doi:10.1038/nbt1010-1045

Bonnefond A, Clement N, Fawcett K, Yengo L, Vaillant E, Guillaume JL, Dechaume A, Payne F, Roussel R, Czernichow S, Hercberg S, Hadjadj S, Balkau B, Marre M, Lantieri O, Langenberg C, Bouatia-Naji N, Meta-Analysis of G, Insulin-Related Traits C, Charpentier G, Vaxillaire M, Rocheleau G, Wareham NJ, Sladek R, McCarthy MI, Dina C, Barroso I, Jockers R, Froguel P (2012) Rare MTNR1B variants impairing melatonin receptor 1B function contribute to type 2 diabetes. Nat Genet 44(3):297–301. doi:10.1038/ng.1053

Browning BL, Yu Z (2009) Simultaneous genotype calling and haplotype phasing improves genotype accuracy and reduces false-positive associations for genome-wide association studies. Am J Hum Genet 85(6):847–861. doi:10.1016/j.ajhg.2009.11.004

Collins FS (2006) 2005 William Allan Award address. No longer just looking under the lamppost. Am J Hum Genet 79(3):6

Cooper GM, Shendure J (2011) Needles in stacks of needles: finding disease-causal variants in a wealth of genomic data. Nat Rev Genet 12(9):628–640. doi:10.1038/nrg3046

Diabetes Genetics Initiative of Broad Institute of, Harvard and MIT, Lund University, and Novartis Institutes of BioMedical Research, Saxena R, Voight BF, Lyssenko V, Burtt NP, de Bakker PI, Chen H, Roix JJ, Kathiresan S, Hirschhorn JN, Daly MJ, Hughes TE, Groop L, Altshuler D, Almgren P, Florez JC, Meyer J, Ardlie K, Bengtsson Bostrom K, Isomaa B, Lettre G, Lindblad U, Lyon HN, Melander O, Newton-Cheh C, Nilsson P, Orho-Melander M, Rastam L, Speliotes EK, Taskinen MR, Tuomi T, Guiducci C, Berglund A, Carlson J, Gianniny L, Hackett R, Hall L, Holmkvist J, Laurila E, Sjogren M, Sterner M, Surti A, Svensson M, Svensson M, Tewhey R, Blumenstiel B, Parkin M, Defelice M, Barry R, Brodeur W, Camarata J, Chia N, Fava M, Gibbons J, Handsaker B, Healy C, Nguyen K, Gates C, Sougnez C, Gage D, Nizzari M, Gabriel SB, Chirn GW, Ma Q, Parikh H, Richardson D, Ricke D, Purcell S (2007) Genome-wide association analysis identifies loci for type 2 diabetes and triglyceride levels. Science 316(5829):1331–1336. doi:10.1126/science.1142358

DIAbetes Genetics Replication Meta-analysis (DIAGRAM) Consortium, Asian Genetic Epidemiology Network Type 2 Diabetes (AGEN-T2D) Consortium, South Asian Type 2 Diabetes (SAT2D) Consortium, Mexican American Type 2 Diabetes (MAT2D) Consortium, Type 2 Diabetes Genetic Exploration by Next-generation sequencing in multi-Ethnic Samples (T2D-GENES) Consortium, Mahajan A, Go MJ, Zhang W, Below JE, Gaulton KJ, Ferreira T, Horikoshi M, Johnson AD, Ng MC, Prokopenko I, Saleheen D, Wang X, Zeggini E, Abecasis GR, Adair LS, Almgren P, Atalay M, Aung T, Baldassarre D, Balkau B, Bao Y, Barnett AH, Barroso I, Basit A, Been LF, Beilby J, Bell GI, Benediktsson R, Bergman RN, Boehm BO, Boerwinkle E, Bonnycastle LL, Burtt N, Cai Q, Campbell H, Carey J, Cauchi S, Caulfield M, Chan JC, Chang LC, Chang TJ, Chang YC, Charpentier G, Chen CH, Chen H, Chen YT, Chia KS, Chidambaram M, Chines PS, Cho NH, Cho YM, Chuang LM, Collins FS, Cornelis MC, Couper DJ, Crenshaw AT, van Dam RM, Danesh J, Das D, de Faire U, Dedoussis G, Deloukas P, Dimas AS, Dina C, Doney AS, Donnelly PJ, Dorkhan M, van Duijn C, Dupuis J, Edkins S, Elliott P, Emilsson V, Erbel R, Eriksson JG, Escobedo J, Esko T, Eury E, Florez JC, Fontanillas P, Forouhi NG, Forsen T, Fox C, Fraser RM, Frayling TM, Froguel P, Frossard P, Gao Y, Gertow K, Gieger C, Gigante B, Grallert H, Grant GB, Grrop LC, Groves CJ, Grundberg E, Guiducci C, Hamsten A, Han BG, Hara K, Hassanali N, Hattersley AT, Hayward C, Hedman AK, Herder C, Hofman A, Holmen OL, Hovingh K, Hreidarsson AB, Hu C, Hu FB, Hui J, Humphries SE, Hunt SE, Hunter DJ, Hveem K, Hydrie ZI, Ikegami H, Illig T, Ingelsson E, Islam M, Isomaa B, Jackson AU, Jafar T, James A, Jia W, Jockel KH, Jonsson A, Jowett JB, Kadowaki T, Kang HM, Kanoni S, Kao WH, Kathiresan S, Kato N, Katulanda P, Keinanen-Kiukaanniemi KM, Kelly AM, Khan H, Khaw KT, Khor CC, Kim HL, Kim S, Kim YJ, Kinnunen L, Klopp N, Kong A, Korpi-Hyovalti E, Kowlessur S, Kraft P, Kravic J, Kristensen MM, Krithika S, Kumar A, Kumate J, Kuusisto J, Kwak SH, Laakso M, Lagou V, Lakka TA, Langenberg C, Langford C, Lawrence R, Leander K, Lee JM, Lee NR, Li M, Li X, Li Y, Liang J, Liju S, Lim WY, Lind L, Lindgren CM, Lindholm E, Liu CT, Liu JJ, Lobbens S, Long J, Loos RJ, Lu W, Luan J, Lyssenko V, Ma RC, Maeda S, Magi R, Mannisto S, Matthews DR, Meigs JB, Melander O, Metspalu A, Meyer J, Mirza G, Mihailov E, Moebus S, Mohan V, Mohlke KL, Morris AD, Muhleisen TW, Muller-Nurasyid M, Musk B, Nakamura J, Nakashima E, Navarro P, Ng PK, Nica AC, Nilsson PM, Njolstad I, Nothen MM, Ohnaka K, Ong TH, Owen KR, Palmer CN, Pankow JS, Park KS, Parkin M, Pechlivanis S, Pedersen NL, Peltonen L, Perry JR, Peters A, Pinidiyapathirage JM, Platou CG, Potter S, Price JF, Qi L, Radha V, Rallidis L, Rasheed A, Rathman W, Rauramaa R, Raychaudhuri S, Rayner NW, Rees SD, Rehnberg E, Ripatti S, Robertson N, Roden M, Rossin EJ, Rudan I, Rybin D, Saaristo TE, Salomaa V, Saltevo J, Samuel M, Sanghera DK, Saramies J, Scott J, Scott LJ, Scott RA, Segre AV, Sehmi J, Sennblad B, Shah N, Shah S, Shera AS, Shu XO, Shuldiner AR, Sigurdsson G, Sijbrands E, Silveira A, Sim X, Sivapalaratnam S, Small KS, So WY, Stancakova A, Stefansson K, Steinbach G, Steinthorsdottir V, Stirrups K, Strawbridge RJ, Stringham HM, Sun Q, Suo C, Syvanen AC, Takayanagi R, Takeuchi F, Tay WT, Teslovich TM, Thorand B,

Thorleifsson G, Thorsteinsdottir U, Tikkanen E, Trakalo J, Tremoli E, Trip MD, Tsai FJ, Tuomi T, Tuomilehto J, Uitterlinden AG, Valladares-Salgado A, Vedantam S, Veglia F, Voight BF, Wang C, Wareham NJ, Wennauer R, Wickremasinghe AR, Wilsgaard T, Wilson JF, Wiltshire S, Winckler W, Wong TY, Wood AR, Wu JY, Wu Y, Yamamoto K, Yamauchi T, Yang M, Yengo L, Yokota M, Young R, Zabaneh D, Zhang F, Zhang R, Zheng W, Zimmet PZ, Altshuler D, Bowden DW, Cho YS, Cox NJ, Cruz M, Hanis CL, Kooner J, Lee JY, Seielstad M, Teo YY, Boehnke M, Parra EJ, Chambers JC, Tai ES, McCarthy MI, Morris AP (2014) Genome-wide trans-ancestry meta-analysis provides insight into the genetic architecture of type 2 diabetes susceptibility. Nat Genet 46(3):234–244. doi:10.1038/ng.2897

Dickson SP, Wang K, Krantz I, Hakonarson H, Goldstein DB (2010) Rare variants create synthetic genome-wide associations. PLoS Biol 8(1), e1000294. doi:10.1371/journal.pbio.1000294

ENCODE Project Consortium, Bernstein BE, Birney E, Dunham I, Green ED, Gunter C, Snyder M (2012) An integrated encyclopedia of DNA elements in the human genome. Nature 489 (7414):57–74. doi:10.1038/nature11247

Fajans SS, Bell GI (2011) MODY: history, genetics, pathophysiology, and clinical decision making. Diabetes Care 34(8):1878–1884. doi:10.2337/dc11-0035

Flannick J, Thorleifsson G, Beer NL, Jacobs SB, Grarup N, Burtt NP, Mahajan A, Fuchsberger C, Atzmon G, Benediktsson R, Blangero J, Bowden DW, Brandslund I, Brosnan J, Burslem F, Chambers J, Cho YS, Christensen C, Douglas DA, Duggirala R, Dymek Z, Farjoun Y, Fennell T, Fontanillas P, Forsen T, Gabriel S, Glaser B, Gudbjartsson DF, Hanis C, Hansen T, Hreidarsson AB, Hveem K, Ingelsson E, Isomaa B, Johansson S, Jorgensen T, Jorgensen ME, Kathiresan S, Kong A, Kooner J, Kravic J, Laakso M, Lee JY, Lind L, Lindgren CM, Linneberg A, Masson G, Meitinger T, Mohlke KL, Molven A, Morris AP, Potluri S, Rauramaa R, Ribel-Madsen R, Richard AM, Rolph T, Salomaa V, Segre AV, Skarstrand H, Steinthorsdottir V, Stringham HM, Sulem P, Tai ES, Teo YY, Teslovich T, Thorsteinsdottir U, Trimmer JK, Tuomi T, Tuomilehto J, Vaziri-Sani F, Voight BF, Wilson JG, Boehnke M, McCarthy MI, Njolstad PR, Pedersen O, Go-T2D Consortium, T2D-Genes Consortium, Groop L, Cox DR, Stefansson K, Altshuler D (2014) Loss-of-function mutations in SLC30A8 protect against type 2 diabetes. Nat Genet 46(4):357–363. doi:10.1038/ng.2915

Fogarty MP, Panhuis TM, Vadlamudi S, Buchkovich ML, Mohlke KL (2013) Allele-specific transcriptional activity at type 2 diabetes-associated single nucleotide polymorphisms in regions of pancreatic islet open chromatin at the JAZF1 locus. Diabetes 62(5):1756–1762. doi:10.2337/db12-0972

Gaulton KJ, Nammo T, Pasquali L, Simon JM, Giresi PG, Fogarty MP, Panhuis TM, Mieczkowski P, Secchi A, Bosco D, Berney T, Montanya E, Mohlke KL, Lieb JD, Ferrer J (2010) A map of open chromatin in human pancreatic islets. Nat Genet 42(3):255–259. doi:10.1038/ng.530

Gaulton K, Flannick J, Fuchsberger C, Kang HM, Burtt N, Ferrer J, Stitzel M, Kellis M, McCarthy M, Altshuler D, Boehnke M, the GoT2D Consortium (2013) Whole genome sequencing of 2,850 Central-Northern European type 2 diabetes cases and controls reveals insights into functional mechanisms underlying disease pathogenesis. Abstract #175 presented at the 63rd annual meeting of The American Society of Human Genetics, Boston, MA, 24 Oct 2013

Grant SF, Thorleifsson G, Reynisdottir I, Benediktsson R, Manolescu A, Sainz J, Helgason A, Stefansson H, Emilsson V, Helgadottir A, Styrkarsdottir U, Magnusson KP, Walters GB, Palsdottir E, Jonsdottir T, Gudmundsdottir T, Gylfason A, Saemundsdottir J, Wilensky RL, Reilly MP, Rader DJ, Bagger Y, Christiansen C, Gudnason V, Sigurdsson G, Thorsteinsdottir U, Gulcher JR, Kong A, Stefansson K (2006) Variant of transcription factor 7-like 2 (TCF7L2) gene confers risk of type 2 diabetes. Nat Genet 38(3):320–323

Guan W, Boehnke M, Pluzhnikov A, Cox NJ, Scott LJ (2012) Identifying plausible genetic models based on association and linkage results: application to type 2 diabetes. Genet Epidemiol 36 (8):820–828. doi:10.1002/gepi.21668

Han B, Eskin E (2011) Random-effects model aimed at discovering associations in meta-analysis of genome-wide association studies. Am J Hum Genet 88(5):586–598. doi:10.1016/j.ajhg.2011.04.014

Howie BN, Donnelly P, Marchini J (2009) A flexible and accurate genotype imputation method for the next generation of genome-wide association studies. PLoS Genet 5(6), e1000529. doi:10.1371/journal.pgen.1000529

Howie B, Fuchsberger C, Stephens M, Marchini J, Abecasis GR (2012) Fast and accurate genotype imputation in genome-wide association studies through pre-phasing. Nat Genet 44 (8):955–959. doi:10.1038/ng.2354

Huyghe JR, Jackson AU, Fogarty MP, Buchkovich ML, Stancakova A, Stringham HM, Sim X, Yang L, Fuchsberger C, Cederberg H, Chines PS, Teslovich TM, Romm JM, Ling H, McMullen I, Ingersoll R, Pugh EW, Doheny KF, Neale BM, Daly MJ, Kuusisto J, Scott LJ, Kang HM, Collins FS, Abecasis GR, Watanabe RM, Boehnke M, Laakso M, Mohlke KL (2013) Exome array analysis identifies new loci and low-frequency variants influencing insulin processing and secretion. Nat Genet 45(2):197–201. doi:10.1038/ng.2507

International HapMap C, Frazer KA, Ballinger DG, Cox DR, Hinds DA, Stuve LL, Gibbs RA, Belmont JW, Boudreau A, Hardenbol P, Leal SM, Pasternak S, Wheeler DA, Willis TD, Yu F, Yang H, Zeng C, Gao Y, Hu H, Hu W, Li C, Lin W, Liu S, Pan H, Tang X, Wang J, Wang W, Yu J, Zhang B, Zhang Q, Zhao H, Zhao H, Zhou J, Gabriel SB, Barry R, Blumenstiel B, Camargo A, Defelice M, Faggart M, Goyette M, Gupta S, Moore J, Nguyen H, Onofrio RC, Parkin M, Roy J, Stahl E, Winchester E, Ziaugra L, Altshuler D, Shen Y, Yao Z, Huang W, Chu X, He Y, Jin L, Liu Y, Shen Y, Sun W, Wang H, Wang Y, Wang Y, Xiong X, Xu L, Waye MM, Tsui SK, Xue H, Wong JT, Galver LM, Fan JB, Gunderson K, Murray SS, Oliphant AR, Chee MS, Montpetit A, Chagnon F, Ferretti V, Leboeuf M, Olivier JF, Phillips MS, Roumy S, Sallee C, Verner A, Hudson TJ, Kwok PY, Cai D, Koboldt DC, Miller RD, Pawlikowska L, Taillon-Miller P, Xiao M, Tsui LC, Mak W, Song YQ, Tam PK, Nakamura Y, Kawaguchi T, Kitamoto T, Morizono T, Nagashima A, Ohnishi Y, Sekine A, Tanaka T, Tsunoda T, Deloukas P, Bird CP, Delgado M, Dermitzakis ET, Gwilliam R, Hunt S, Morrison J, Powell D, Stranger BE, Whittaker P, Bentley DR, Daly MJ, de Bakker PI, Barrett J, Chretien YR, Maller J, McCarroll S, Patterson N, Pe'er I, Price A, Purcell S, Richter DJ, Sabeti P, Saxena R, Schaffner SF, Sham PC, Varilly P, Altshuler D, Stein LD, Krishnan L, Smith AV, Tello-Ruiz MK, Thorisson GA, Chakravarti A, Chen PE, Cutler DJ, Kashuk CS, Lin S, Abecasis GR, Guan W, Li Y, Munro HM, Qin ZS, Thomas DJ, McVean G, Auton A, Bottolo L, Cardin N, Eyheramendy S, Freeman C, Marchini J, Myers S, Spencer C, Stephens M, Donnelly P, Cardon LR, Clarke G, Evans DM, Morris AP, Weir BS, Tsunoda T, Mullikin JC, Sherry ST, Feolo M, Skol A, Zhang H, Zeng C, Zhao H, Matsuda I, Fukushima Y, Macer DR, Suda E, Rotimi CN, Adebamowo CA, Ajayi I, Aniagwu T, Marshall PA, Nkwodimmah C, Royal CD, Leppert MF, Dixon M, Peiffer A, Qiu R, Kent A, Kato K, Niikawa N, Adewole IF, Knoppers BM, Foster MW, Clayton EW, Watkin J, Gibbs RA, Belmont JW, Muzny D, Nazareth L, Sodergren E, Weinstock GM, Wheeler DA, Yakub I, Gabriel SB, Onofrio RC, Richter DJ, Ziaugra L, Birren BW, Daly MJ, Altshuler D, Wilson RK, Fulton LL, Rogers J, Burton J, Carter NP, Clee CM, Griffiths M, Jones MC, McLay K, Plumb RW, Ross MT, Sims SK, Willey DL, Chen Z, Han H, Kang L, Godbout M, Wallenburg JC, L'Archeveque P, Bellemare G, Saeki K, Wang H, An D, Fu H, Li Q, Wang Z, Wang R, Holden AL, Brooks LD, McEwen JE, Guyer MS, Wang VO, Peterson JL, Shi M, Spiegel J, Sung LM, Zacharia LF, Collins FS, Kennedy K, Jamieson R, Stewart J (2007) A second generation human haplotype map of over 3.1 million SNPs. Nature 449 (7164):851–861. doi:10.1038/nature06258

Jun G, Almeida M, Manning A, Teslovich T, Wood A, Zawistowski M, Won S, Fuchsberger C, Feng S, Gaulton K, Cingolani P, Frayling T, Abecasis G, Blangero J, T2D-GENES Consortium (2013) Deep whole-genome sequencing in pedigrees to quantify the contribution of private variants to type 2 diabetes and related metabolic traits. Abstract 36# presented at the 63rd annual meeting of The American Society of Human Genetics, Boston, MA, 24 Oct 2013

Kulzer JR, Stitzel ML, Morken MA, Huyghe JR, Fuchsberger C, Kuusisto J, Laakso M, Boehnke M, Collins FS, Mohlke KL (2014) A common functional regulatory variant at a type 2 diabetes locus upregulates ARAP1 expression in the pancreatic beta cell. Am J Hum Genet 94(2):186–197. doi:10.1016/j.ajhg.2013.12.011

Li Y, Willer CJ, Ding J, Scheet P, Abecasis GR (2010) MaCH: using sequence and genotype data to estimate haplotypes and unobserved genotypes. Genet Epidemiol 34(8):816–834. doi:10.1002/gepi.20533

Lin DY, Zeng D (2010) Meta-analysis of genome-wide association studies: no efficiency gain in using individual participant data. Genet Epidemiol 34(1):60–66. doi:10.1002/gepi.20435

Lohmueller KE, Sparso T, Li Q, Andersson E, Korneliussen T, Albrechtsen A, Banasik K, Grarup N, Hallgrimsdottir I, Kiil K, Kilpelainen TO, Krarup NT, Pers TH, Sanchez G, Hu Y, Degiorgio M, Jorgensen T, Sandbaek A, Lauritzen T, Brunak S, Kristiansen K, Li Y, Hansen T, Wang J, Nielsen R, Pedersen O (2013) Whole-exome sequencing of 2,000 Danish individuals and the role of rare coding variants in type 2 diabetes. Am J Hum Genet 93 (6):1072–1086. doi:10.1016/j.ajhg.2013.11.005

Ma C, Blackwell T, Boehnke M, Scott LJ, GoT2D investigators (2013) Recommended joint and meta-analysis strategies for case–control association testing of single low-count variants. Genet Epidemiol 37(6):539–550. doi:10.1002/gepi.21742

Mahajan A, Ng HJ, Locke A, Sim X, Manning A, Grarup N, Highland HM, Im HK, Rivas M, Morris AP, Lindgren CM, Gloyn AL, Meigs JB, on behalf of T2D-GENES and GOT2D Consortia (2014) Exome array analysis in >30,000 Europeans identifies novel coding variants associated with glycemic traits. Abstract #1609-P presented at the 74th annual meeting of The American Diabetes Association, Boston, MA, June 2014

Marchini J, Howie B (2010) Genotype imputation for genome-wide association studies. Nat Rev Genet 11(7):499–511. doi:10.1038/nrg2796

Morris AP (2011) Transethnic meta-analysis of genomewide association studies. Genet Epidemiol 35(8):809–822. doi:10.1002/gepi.20630

Morris AP, Voight BF, Teslovich TM, Ferreira T, Segre AV, Steinthorsdottir V, Strawbridge RJ, Khan H, Grallert H, Mahajan A, Prokopenko I, Kang HM, Dina C, Esko T, Fraser RM, Kanoni S, Kumar A, Lagou V, Langenberg C, Luan J, Lindgren CM, Muller-Nurasyid M, Pechlivanis S, Rayner NW, Scott LJ, Wiltshire S, Yengo L, Kinnunen L, Rossin EJ, Raychaudhuri S, Johnson AD, Dimas AS, Loos RJ, Vedantam S, Chen H, Florez JC, Fox C, Liu CT, Rybin D, Couper DJ, Kao WH, Li M, Cornelis MC, Kraft P, Sun Q, van Dam RM, Stringham HM, Chines PS, Fischer K, Fontanillas P, Holmen OL, Hunt SE, Jackson AU, Kong A, Lawrence R, Meyer J, Perry JR, Platou CG, Potter S, Rehnberg E, Robertson N, Sivapalaratnam S, Stancakova A, Stirrups K, Thorleifsson G, Tikkanen E, Wood AR, Almgren P, Atalay M, Benediktsson R, Bonnycastle LL, Burtt N, Carey J, Charpentier G, Crenshaw AT, Doney AS, Dorkhan M, Edkins S, Emilsson V, Eury E, Forsen T, Gertow K, Gigante B, Grant GB, Groves CJ, Guiducci C, Herder C, Hreidarsson AB, Hui J, James A, Jonsson A, Rathmann W, Klopp N, Kravic J, Krjutskov K, Langford C, Leander K, Lindholm E, Lobbens S, Mannisto S, Mirza G, Muhleisen TW, Musk B, Parkin M, Rallidis L, Saramies J, Sennblad B, Shah S, Sigurethsson G, Silveira A, Steinbach G, Thorand B, Trakalo J, Veglia F, Wennauer R, Winckler W, Zabaneh D, Campbell H, van Duijn C, Uitterlinden AG, Hofman A, Sijbrands E, Abecasis GR, Owen KR, Zeggini E, Trip MD, Forouhi NG, Syvanen AC, Eriksson JG, Peltonen L, Nothen MM, Balkau B, Palmer CN, Lyssenko V, Tuomi T, Isomaa B, Hunter DJ, Qi L, Wellcome Trust Case Control Consortium, Meta-Analyses of Glucose and Insulin-related traits Consortium (MAGIC) Investigators, Genetic Investigation of ANthropometric Traits (GIANT) Consortium, Asian Genetic Epidemiology Network–Type 2 Diabetes (AGEN-T2D) Consortium, South Asian Type 2 Diabetes (SAT2D), Shuldiner AR, Roden M, Barroso I, Wilsgaard T, Beilby J, Hovingh K, Price JF, Wilson JF, Rauramaa R, Lakka TA, Lind L, Dedoussis G, Njolstad I, Pedersen NL, Khaw KT, Wareham NJ, Keinanen-Kiukaanniemi SM, Saaristo TE, Korpi-Hyovalti E, Saltevo J, Laakso M, Kuusisto J, Metspalu A, Collins FS, Mohlke KL, Bergman RN, Tuomilehto J, Boehm BO, Gieger C, Hveem K, Cauchi S, Froguel P, Baldassarre D, Tremoli E, Humphries SE, Saleheen D, Danesh J, Ingelsson E, Ripatti S, Salomaa V, Erbel R, Jockel KH, Moebus S,

Peters A, Illig T, de Faire U, Hamsten A, Morris AD, Donnelly PJ, Frayling TM, Hattersley AT, Boerwinkle E, Melander O, Kathiresan S, Nilsson PM, Deloukas P, Thorsteinsdottir U, Groop LC, Stefansson K, Hu F, Pankow JS, Dupuis J, Meigs JB, Altshuler D, Boehnke M, McCarthy MI, DIAbetes Genetics Replication And Meta-analysis (DIAGRAM) Consortium (2012) Large-scale association analysis provides insights into the genetic architecture and pathophysiology of type 2 diabetes. Nat Genet 44(9):981–990

Ng HJ, Gloyn AL (2013) Bridging the gap between genetic associations and molecular mechanisms for type 2 diabetes. Curr Diab Rep 13(6):778–785. doi:10.1007/s11892-013-0429-1

Ng MC, Saxena R, Li J, Palmer ND, Dimitrov L, Xu J, Rasmussen-Torvik LJ, Zmuda JM, Siscovick DS, Patel SR, Crook ED, Sims M, Chen YD, Bertoni AG, Li M, Grant SF, Dupuis J, Meigs JB, Psaty BM, Pankow JS, Langefeld CD, Freedman BI, Rotter JI, Wilson JG, Bowden DW (2013) Transferability and fine mapping of type 2 diabetes loci in African Americans: the Candidate Gene Association Resource Plus Study. Diabetes 62(3):965–976. doi:10.2337/db12-0266

Parker SC, Stitzel ML, Taylor DL, Orozco JM, Erdos MR, Akiyama JA, van Bueren KL, Chines PS, Narisu N, Comparative Sequencing Program NISC, Black BL, Visel A, Pennacchio LA, Collins FS (2013) Chromatin stretch enhancer states drive cell-specific gene regulation and harbor human disease risk variants. Proc Natl Acad Sci U S A 110(44):17921–17926. doi:10.1073/pnas.1317023110

Pasquali L, Gaulton KJ, Rodriguez-Segui SA, Mularoni L, Miguel-Escalada I, Akerman I, Tena JJ, Moran I, Gomez-Marin C, van de Bunt M, Ponsa-Cobas J, Castro N, Nammo T, Cebola I, Garcia-Hurtado J, Maestro MA, Pattou F, Piemonti L, Berney T, Gloyn AL, Ravassard P, Skarmeta JL, Muller F, McCarthy MI, Ferrer J (2014) Pancreatic islet enhancer clusters enriched in type 2 diabetes risk-associated variants. Nat Genet 46(2):136–143. doi:10.1038/ng.2870

Pruim RJ, Welch RP, Sanna S, Teslovich TM, Chines PS, Gliedt TP, Boehnke M, Abecasis GR, Willer CJ (2010) LocusZoom: regional visualization of genome-wide association scan results. Bioinformatics 26(18):2336–2337. doi:10.1093/bioinformatics/btq419

Rees MG, Ng D, Ruppert S, Turner C, Beer NL, Swift AJ, Morken MA, Below JE, Blech I, Program NCS, Mullikin JC, McCarthy MI, Biesecker LG, Gloyn AL, Collins FS (2012a) Correlation of rare coding variants in the gene encoding human glucokinase regulatory protein with phenotypic, cellular, and kinetic outcomes. J Clin Invest 122(1):205–217. doi:10.1172/JCI46425

Rees MG, Wincovitch S, Schultz J, Waterstradt R, Beer NL, Baltrusch S, Collins FS, Gloyn AL (2012b) Cellular characterisation of the GCKR P446L variant associated with type 2 diabetes risk. Diabetologia 55(1):114–122. doi:10.1007/s00125-011-2348-5

Saxena R, Elbers CC, Guo Y, Peter I, Gaunt TR, Mega JL, Lanktree MB, Tare A, Castillo BA, Li YR, Johnson T, Bruinenberg M, Gilbert-Diamond D, Rajagopalan R, Voight BF, Balasubramanyam A, Barnard J, Bauer F, Baumert J, Bhangale T, Bohm BO, Braund PS, Burton PR, Chandrupatla HR, Clarke R, Cooper-DeHoff RM, Crook ED, Davey-Smith G, Day IN, de Boer A, de Groot MC, Drenos F, Ferguson J, Fox CS, Furlong CE, Gibson Q, Gieger C, Gilhuijs-Pederson LA, Glessner JT, Goel A, Gong Y, Grant SF, Grobbee DE, Hastie C, Humphries SE, Kim CE, Kivimaki M, Kleber M, Meisinger C, Kumari M, Langaee TY, Lawlor DA, Li M, Lobmeyer MT, der Zee AH M-v, Meijs MF, Molony CM, Morrow DA, Murugesan G, Musani SK, Nelson CP, Newhouse SJ, O'Connell JR, Padmanabhan S, Palmen J, Patel SR, Pepine CJ, Pettinger M, Price TS, Rafelt S, Ranchalis J, Rasheed A, Rosenthal E, Ruczinski I, Shah S, Shen H, Silbernagel G, Smith EN, Spijkerman AW, Stanton A, Steffes MW, Thorand B, Trip M, van der Harst P, van der A D, van Iperen EP, van Setten J, van Vliet-Ostaptchouk JV, Verweij N, Wolffenbuttel BH, Young T, Zafarmand MH, Zmuda JM, the Look AHEAD Research Group, DIAGRAM consortium, Boehnke M, Altshuler D, McCarthy M, Kao WH, Pankow JS, Cappola TP, Sever P, Poulter N, Caulfield M,

Dominiczak A, Shields DC, Bhatt DL, Zhang L, Curtis SP, Danesh J, Casas JP, van der Schouw YT, Onland-Moret NC, Doevendans PA, Dorn GW, Farrall M 2nd, FitzGerald GA, Hamsten A, Hegele R, Hingorani AD, Hofker MH, Huggins GS, Illig T, Jarvik GP, Johnson JA, Klungel OH, Knowler WC, Koenig W, Marz W, Meigs JB, Melander O, Munroe PB, Mitchell BD, Bielinski SJ, Rader DJ, Reilly MP, Rich SS, Rotter JI, Saleheen D, Samani NJ, Schadt EE, Shuldiner AR, Silverstein R, Kottke-Marchant K, Talmud PJ, Watkins H, Asselbergs FW, de Bakker PI, McCaffery J, Wijmenga C, Sabatine MS, Wilson JG, Reiner A, Bowden DW, Hakonarson H, Siscovick DS, Keating BJ (2012) Large-scale gene-centric meta-analysis across 39 studies identifies type 2 diabetes loci. Am J Hum Genet 90 (3):410–425. doi:10.1016/j.ajhg.2011.12.022

Schwitzgebel VM (2014) Many faces of monogenic diabetes. J Diabetes Investig 5(2):121–133. doi:10.1111/jdi.12197

Scott LJ, Mohlke KL, Bonnycastle LL, Willer CJ, Li Y, Duren WL, Erdos MR, Stringham HM, Chines PS, Jackson AU, Prokunina-Olsson L, Ding CJ, Swift AJ, Narisu N, Hu T, Pruim R, Xiao R, Li XY, Conneely KN, Riebow NL, Sprau AG, Tong M, White PP, Hetrick KN, Barnhart MW, Bark CW, Goldstein JL, Watkins L, Xiang F, Saramies J, Buchanan TA, Watanabe RM, Valle TT, Kinnunen L, Abecasis GR, Pugh EW, Doheny KF, Bergman RN, Tuomilehto J, Collins FS, Boehnke M (2007) A genome-wide association study of type 2 diabetes in Finns detects multiple susceptibility variants. Science 316(5829):1341–1345. doi:10.1126/science.1142382

Shea J, Agarwala V, Philippakis AA, Maguire J, Banks E, Depristo M, Thomson B, Guiducci C, Onofrio RC, Kathiresan S, Gabriel S, Burtt NP, Daly MJ, Groop L, Altshuler D, Myocardial Infarction Genetics C (2011) Comparing strategies to fine-map the association of common SNPs at chromosome 9p21 with type 2 diabetes and myocardial infarction. Nat Genet 43 (8):801–805. doi:10.1038/ng.871

SIGMA Type 2 Diabetes Consortium, Estrada K, Aukrust I, Bjorkhaug L, Burtt NP, Mercader JM, Garcia-Ortiz H, Huerta-Chagoya A, Moreno-Macias H, Walford G, Flannick J, Williams AL, Gomez-Vazquez MJ, Fernandez-Lopez JC, Martinez-Hernandez A, Centeno-Cruz F, Mendoza-Caamal E, Revilla-Monsalve C, Islas-Andrade S, Cordova EJ, Soberon X, Gonzalez-Villalpando ME, Henderson E, Wilkens LR, Le Marchand L, Arellano-Campos O, Ordonez-Sanchez ML, Rodriguez-Torres M, Rodriguez-Guillen R, Riba L, Najmi LA, Jacobs SB, Fennell T, Gabriel S, Fontanillas P, Hanis CL, Lehman DM, Jenkinson CP, Abboud HE, Bell GI, Cortes ML, Boehnke M, Gonzalez-Villalpando C, Orozco L, Haiman CA, Tusie-Luna T, Aguilar-Salinas CA, Altshuler D, Njolstad PR, Florez JC, MacArthur DG (2014) Association of a low-frequency variant in HNF1A with type 2 diabetes in a Latino population. JAMA 311(22):2305–2314. doi:10.1001/jama.2014.6511

Sim X, Ong RT, Suo C, Tay WT, Liu J, Ng DP, Boehnke M, Chia KS, Wong TY, Seielstad M, Teo YY, Tai ES (2011) Transferability of type 2 diabetes implicated loci in multi-ethnic cohorts from Southeast Asia. PLoS Genet 7(4), e1001363. doi:10.1371/journal.pgen.1001363

Sladek R, Rocheleau G, Rung J, Dina C, Shen L, Serre D, Boutin P, Vincent D, Belisle A, Hadjadj S, Balkau B, Heude B, Charpentier G, Hudson TJ, Montpetit A, Pshezhetsky AV, Prentki M, Posner BI, Balding DJ, Meyre D, Polychronakos C, Froguel P (2007) A genome-wide association study identifies novel risk loci for type 2 diabetes. Nature 445 (7130):881–885. doi:10.1038/nature05616

Stahl EA, Wegmann D, Trynka G, Gutierrez-Achury J, Do R, Voight BF, Kraft P, Chen R, Kallberg HJ, Kurreeman FA, Diabetes Genetics Replication and Meta-analysis Consortium; Myocardial Infarction Genetics Consortium, Kathiresan S, Wijmenga C, Gregersen PK, Alfredsson L, Siminovitch KA, Worthington J, de Bakker PI, Raychaudhuri S, Plenge RM (2012) Bayesian inference analyses of the polygenic architecture of rheumatoid arthritis. Nat Genet 44(5):483–489. doi:10.1038/ng.2232

Steinthorsdottir V, Thorleifsson G, Sulem P, Helgason H, Grarup N, Sigurdsson A, Helgadottir HT, Johannsdottir H, Magnusson OT, Gudjonsson SA, Justesen JM, Harder MN, Jorgensen ME, Christensen C, Brandslund I, Sandbaek A, Lauritzen T, Vestergaard H, Linneberg A, Jorgensen T, Hansen T, Daneshpour MS, Fallah MS, Hreidarsson AB, Sigurdsson G, Azizi F, Benediktsson R, Masson G, Helgason A, Kong A, Gudbjartsson DF, Pedersen O, Thorsteinsdottir U, Stefansson K (2014) Identification of low-frequency and rare sequence variants associated with elevated or reduced risk of type 2 diabetes. Nat Genet 46(3):294–298. doi:10.1038/ng.2882

Stitzel ML, Sethupathy P, Pearson DS, Chines PS, Song L, Erdos MR, Welch R, Parker SC, Boyle AP, Scott LJ, Program NCS, Margulies EH, Boehnke M, Furey TS, Crawford GE, Collins FS (2010) Global epigenomic analysis of primary human pancreatic islets provides insights into type 2 diabetes susceptibility loci. Cell Metab 12(5):443–455. doi:10.1016/j.cmet.2010.09.012

Travers ME, Mackay DJ, Dekker Nitert M, Morris AP, Lindgren CM, Berry A, Johnson PR, Hanley N, Groop LC, McCarthy MI, Gloyn AL (2013) Insights into the molecular mechanism for type 2 diabetes susceptibility at the KCNQ1 locus from temporal changes in imprinting status in human islets. Diabetes 62(3):987–992. doi:10.2337/db12-0819

Unoki H, Takahashi A, Kawaguchi T, Hara K, Horikoshi M, Andersen G, Ng DP, Holmkvist J, Borch-Johnsen K, Jorgensen T, Sandbaek A, Lauritzen T, Hansen T, Nurbaya S, Tsunoda T, Kubo M, Babazono T, Hirose H, Hayashi M, Iwamoto Y, Kashiwagi A, Kaku K, Kawamori R, Tai ES, Pedersen O, Kamatani N, Kadowaki T, Kikkawa R, Nakamura Y, Maeda S (2008) SNPs in KCNQ1 are associated with susceptibility to type 2 diabetes in East Asian and European populations. Nat Genet 40(9):1098–1102. doi:10.1038/ng.208

Visscher PM, Brown MA, McCarthy MI, Yang J (2012) Five years of GWAS discovery. Am J Hum Genet 90(1):7–24. doi:10.1016/j.ajhg.2011.11.029

Voight BF, Scott LJ, Steinthorsdottir V, Morris AP, Dina C, Welch RP, Zeggini E, Huth C, Aulchenko YS, Thorleifsson G, McCulloch LJ, Ferreira T, Grallert H, Amin N, Wu G, Willer CJ, Raychaudhuri S, McCarroll SA, Langenberg C, Hofmann OM, Dupuis J, Qi L, Segre AV, van Hoek M, Navarro P, Ardlie K, Balkau B, Benediktsson R, Bennett AJ, Blagieva R, Boerwinkle E, Bonnycastle LL, Bengtsson Bostrom K, Bravenboer B, Bumpstead S, Burtt NP, Charpentier G, Chines PS, Cornelis M, Couper DJ, Crawford G, Doney AS, Elliott KS, Elliott AL, Erdos MR, Fox CS, Franklin CS, Ganser M, Gieger C, Grarup N, Green T, Griffin S, Groves CJ, Guiducci C, Hadjadj S, Hassanali N, Herder C, Isomaa B, Jackson AU, Johnson PR, Jorgensen T, Kao WH, Klopp N, Kong A, Kraft P, Kuusisto J, Lauritzen T, Li M, Lieverse A, Lindgren CM, Lyssenko V, Marre M, Meitinger T, Midthjell K, Morken MA, Narisu N, Nilsson P, Owen KR, Payne F, Perry JR, Petersen AK, Platou C, Proenca C, Prokopenko I, Rathmann W, Rayner NW, Robertson NR, Rocheleau G, Roden M, Sampson MJ, Saxena R, Shields BM, Shrader P, Sigurdsson G, Sparso T, Strassburger K, Stringham HM, Sun Q, Swift AJ, Thorand B, Tichet J, Tuomi T, van Dam RM, van Haeften TW, van Herpt T, van Vliet-Ostaptchouk JV, Walters GB, Weedon MN, Wijmenga C, Witteman J, The Magic investigators, The Giant Consortium, Bergman RN, Cauchi S, Collins FS, Gloyn AL, Gyllensten U, Hansen T, Hide WA, Hitman GA, Hofman A, Hunter DJ, Hveem K, Laakso M, Mohlke KL, Morris AD, Palmer CN, Pramstaller PP, Rudan I, Sijbrands E, Stein LD, Tuomilehto J, Uitterlinden A, Walker M, Wareham NJ, Watanabe RM, Abecasis GR, Boehm BO, Campbell H, Daly MJ, Hattersley AT, Hu FB, Meigs JB, Pankow JS, Pedersen O, Wichmann HE, Barroso I, Florez JC, Frayling TM, Groop L, Sladek R, Thorsteinsdottir U, Wilson JF, Illig T, Froguel P, van Duijn CM, Stefansson K, Altshuler D, Boehnke M, McCarthy MI (2010) Twelve type 2 diabetes susceptibility loci identified through large-scale association analysis. Nat Genet 42(7):579–589. doi:10.1038/ng.609

Voight BF, Kang HM, Ding J, Palmer CD, Sidore C, Chines PS, Burtt NP, Fuchsberger C, Li Y, Erdmann J, Frayling TM, Heid IM, Jackson AU, Johnson T, Kilpelainen TO, Lindgren CM, Morris AP, Prokopenko I, Randall JC, Saxena R, Soranzo N, Speliotes EK, Teslovich TM, Wheeler E, Maguire J, Parkin M, Potter S, Rayner NW, Robertson N, Stirrups K, Winckler W, Sanna S, Mulas A, Nagaraja R, Cucca F, Barroso I, Deloukas P, Loos RJ, Kathiresan S, Munroe

PB, Newton-Cheh C, Pfeufer A, Samani NJ, Schunkert H, Hirschhorn JN, Altshuler D, McCarthy MI, Abecasis GR, Boehnke M (2012) The metabochip, a custom genotyping array for genetic studies of metabolic, cardiovascular, and anthropometric traits. PLoS Genet 8(8), e1002793. doi:10.1371/journal.pgen.1002793

Wang K, Dickson SP, Stolle CA, Krantz ID, Goldstein DB, Hakonarson H (2010) Interpretation of association signals and identification of causal variants from genome-wide association studies. Am J Hum Genet 86(5):730–742. doi:10.1016/j.ajhg.2010.04.003

Wang X, Chua HX, Chen P, Ong RT, Sim X, Zhang W, Takeuchi F, Liu X, Khor CC, Tay WT, Cheng CY, Suo C, Liu J, Aung T, Chia KS, Kooner JS, Chambers JC, Wong TY, Tai ES, Kato N, Teo YY (2013) Comparing methods for performing trans-ethnic meta-analysis of genome-wide association studies. Hum Mol Genet 22(11):2303–2311. doi:10.1093/hmg/ddt064

Waters KM, Stram DO, Hassanein MT, Le Marchand L, Wilkens LR, Maskarinec G, Monroe KR, Kolonel LN, Altshuler D, Henderson BE, Haiman CA (2010) Consistent association of type 2 diabetes risk variants found in Europeans in diverse racial and ethnic groups. PLoS Genet 6 (8):e1001078. doi:10.1371/journal.pgen.1001078

Wellcome Trust Case Control Consortium (2007) Genome-wide association study of 14,000 cases of seven common diseases and 3,000 shared controls. Nature 447(7145):661–678. doi:10.1038/nature05911

Wellcome Trust Case Control Consortium, Maller JB, McVean G, Byrnes J, Vukcevic D, Palin K, Su Z, Howson JM, Auton A, Myers S, Morris A, Pirinen M, Brown MA, Burton PR, Caulfield MJ, Compston A, Farrall M, Hall AS, Hattersley AT, Hill AV, Mathew CG, Pembrey M, Satsangi J, Stratton MR, Worthington J, Craddock N, Hurles M, Ouwehand W, Parkes M, Rahman N, Duncanson A, Todd JA, Kwiatkowski DP, Samani NJ, Gough SC, McCarthy MI, Deloukas P, Donnelly P (2012) Bayesian refinement of association signals for 14 loci in 3 common diseases. Nat Genet 44(12):1294–1301. doi:10.1038/ng.2435

Winkler TW, Day FR, Croteau-Chonka DC, Wood AR, Locke AE, Magi R, Ferreira T, Fall T, Graff M, Justice AE, Luan J, Gustafsson S, Randall JC, Vedantam S, Workalemahu T, Kilpelainen TO, Scherag A, Esko T, Kutalik Z, Heid IM, Loos RJ, Genetic Investigation of Anthropometric Traits (GIANT) Consortium (2014) Quality control and conduct of genome-wide association meta-analyses. Nat Protoc 9(5):1192–1212. doi:10.1038/nprot.2014.071

Wray NR, Purcell SM, Visscher PM (2011) Synthetic associations created by rare variants do not explain most GWAS results. PLoS Biol 9(1), e1000579. doi:10.1371/journal.pbio.1000579

Yamagata K (2014) Roles of HNF1alpha and HNF4alpha in pancreatic beta-cells: lessons from a monogenic form of diabetes (MODY). Vitam Horm 95:407–423. doi:10.1016/B978-0-12-800174-5.00016-8

Yang J, Ferreira T, Morris AP, Medland SE, Genetic Investigation of, ANthropometric Traits (GIANT) Consortium, DIAbetes Genetics Replication Meta-analysis (DIAGRAM) Consortium, Madden PA, Heath AC, Martin NG, Montgomery GW, Weedon MN, Loos RJ, Frayling TM, McCarthy MI, Hirschhorn JN, Goddard ME, Visscher PM (2012) Conditional and joint multiple-SNP analysis of GWAS summary statistics identifies additional variants influencing complex traits. Nat Genet 44(4):369–375. doi:10.1038/ng.2213

Yang J, Lee SH, Goddard ME, Visscher PM (2013) Genome-wide complex trait analysis (GCTA): methods, data analyses, and interpretations. Methods Mol Biol 1019:215–236. doi:10.1007/978-1-62703-447-0_9

Yasuda K, Miyake K, Horikawa Y, Hara K, Osawa H, Furuta H, Hirota Y, Mori H, Jonsson A, Sato Y, Yamagata K, Hinokio Y, Wang HY, Tanahashi T, Nakamura N, Oka Y, Iwasaki N, Iwamoto Y, Yamada Y, Seino Y, Maegawa H, Kashiwagi A, Takeda J, Maeda E, Shin HD, Cho YM, Park KS, Lee HK, Ng MC, Ma RC, So WY, Chan JC, Lyssenko V, Tuomi T, Nilsson P, Groop L, Kamatani N, Sekine A, Nakamura Y, Yamamoto K, Yoshida T, Tokunaga K, Itakura M, Makino H, Nanjo K, Kadowaki T, Kasuga M (2008) Variants in KCNQ1 are associated with susceptibility to type 2 diabetes mellitus. Nat Genet 40 (9):1092–1097. doi:10.1038/ng.207

Zeggini E, Weedon MN, Lindgren CM, Frayling TM, Elliott KS, Lango H, Timpson NJ, Perry JR, Rayner NW, Freathy RM, Barrett JC, Shields B, Morris AP, Ellard S, Groves CJ, Harries LW, Marchini JL, Owen KR, Knight B, Cardon LR, Walker M, Hitman GA, Morris AD, Doney AS, Wellcome Trust Case Control Consortium, McCarthy MI, Hattersley AT (2007) Replication of genome-wide association signals in UK samples reveals risk loci for type 2 diabetes. Science 316(5829):1336–1341. doi:10.1126/science.1142364

Zeggini E, Scott LJ, Saxena R, Voight BF, Marchini JL, Hu T, de Bakker PI, Abecasis GR, Almgren P, Andersen G, Ardlie K, Bostrom KB, Bergman RN, Bonnycastle LL, Borch-Johnsen K, Burtt NP, Chen H, Chines PS, Daly MJ, Deodhar P, Ding CJ, Doney AS, Duren WL, Elliott KS, Erdos MR, Frayling TM, Freathy RM, Gianniny L, Grallert H, Grarup N, Groves CJ, Guiducci C, Hansen T, Herder C, Hitman GA, Hughes TE, Isomaa B, Jackson AU, Jorgensen T, Kong A, Kubalanza K, Kuruvilla FG, Kuusisto J, Langenberg C, Lango H, Lauritzen T, Li Y, Lindgren CM, Lyssenko V, Marvelle AF, Meisinger C, Midthjell K, Mohlke KL, Morken MA, Morris AD, Narisu N, Nilsson P, Owen KR, Palmer CN, Payne F, Perry JR, Pettersen E, Platou C, Prokopenko I, Qi L, Qin L, Rayner NW, Rees M, Roix JJ, Sandbaek A, Shields B, Sjogren M, Steinthorsdottir V, Stringham HM, Swift AJ, Thorleifsson G, Thorsteinsdottir U, Timpson NJ, Tuomi T, Tuomilehto J, Walker M, Watanabe RM, Weedon MN, Willer CJ, Wellcome Trust Case Control C, Illig T, Hveem K, Hu FB, Laakso M, Stefansson K, Pedersen O, Wareham NJ, Barroso I, Hattersley AT, Collins FS, Groop L, McCarthy MI, Boehnke M, Altshuler D (2008) Meta-analysis of genome-wide association data and large-scale replication identifies additional susceptibility loci for type 2 diabetes. Nat Genet 40(5):638–645. doi:10.1038/ng.120

Chapter 7
Genetics of Diabetic Micro- and Macrovascular Complications

Niina Sandholm, Per-Henrik Groop, and Alessandro Doria

Abstract The micro- and macrovascular long-term complications of diabetes account for the majority of mortality and morbidity in diabetes. The diagnosis of diabetic nephropathy is highly correlated with the presence of other complications and associated with an increased risk of mortality, whereas cardiovascular disease is often the final cause of death in diabetes. While environmental factors, especially the blood glucose level, play an important role in the development of diabetic complications, the familial clustering of diabetic complications suggests that genetic factors affect their risk as well. As for most of the common and complex diseases, linkage studies and candidate gene studies have resulted in only a few robust genetic risk factors for diabetic complications. Whereas genome-wide association studies have identified multiple susceptibility loci for chronic kidney disease and cardiovascular disease in the nondiabetic population, the first results are now emerging from genome-wide association studies on micro-and macrovascular complications in persons with diabetes as well.

List of Abbreviations

ACE	Angiotensin I-converting enzyme
AER	Albumin excretion rate

N. Sandholm
Folkhälsan Institute of Genetics, Folkhälsan Research Centre, Helsinki, Finland

Abdominal Center, Nephrology, University of Helsinki and Helsinki University Hospital, Helsinki, Finland

P.-H. Groop
Folkhälsan Institute of Genetics, Folkhälsan Research Centre, Helsinki, Finland

Abdominal Center, Nephrology, University of Helsinki and Helsinki University Hospital, Helsinki, Finland

Baker IDI Heart and Diabetes Institute, Melbourne, VIC, Australia

A. Doria (✉)
Research Division, Joslin Diabetes Center, Boston, MA, USA

Department of Medicine, Harvard Medical School, Boston, MA, USA
e-mail: Alessandro.Doria@joslin.harvard.edu

J.C. Florez (ed.), *The Genetics of Type 2 Diabetes and Related Traits*,
DOI 10.1007/978-3-319-01574-3_7

AGE	Advanced glycation end products
BMI	Body mass index
CAC	Calcium content
CAD	Coronary artery disease
CAN	Cardiovascular autonomic neuropathy
CARe	Candidate gene association resource
CI	Confidence interval
cM	Centimorgan
CVD	Cardiovascular disease
DCCT	Diabetes control and complications trial
DN	Diabetic nephropathy
DR	Diabetic retinopathy
eGFR	Estimated glomerular filtration rate
ERα	Estrogen receptor α
ESRD	End-stage renal disease
GENIE	Genetics of nephropathy and international effort
GoKinD	Genetics of kidneys in diabetes
GWAS	Genome-wide association study
h^2	Heritability
JHS	Joslin heart study
HDL	High-density lipoprotein
LD	Linkage disequilibrium
LDL	Low-density lipoprotein
LOD	Logarithm of odds
MI	Myocardial infarction
OR	Odds ratio
SNP	Single nucleotide polymorphism
T1D	Type 1 diabetes
T2D	Type 2 diabetes

7.1 Introduction

Chronic micro-and macrovascular complications are common in diabetes. The microvascular complications include the sight-threatening diabetic retinopathy (DR), diabetic nephropathy (DN), and peripheral and autonomous neuropathy, whereas the macrovascular complications refer to the cardiovascular complications and strokes. Both environmental and genetic factors affect the risk of the complications, but this chapter overview concentrates on genetic findings behind the diabetic complications.

Identifying the genetic variants' modulating susceptibility to diabetic complications would have important translational implications. First, these variants may point to as yet unidentified molecular pathways linking diabetes to increased atherogenesis and microvascular damage, which may in turn suggest new targets for prevention and/or treatment. Second, they may be used to build algorithms for early identification of diabetic individuals at high risk of complications who may benefit from especially aggressive prevention programs. Finally, such knowledge may allow stratification of diabetic subjects based on their response to antiatherogenic therapies, enabling personalized treatment programs.

7.2 Genetic Factors of Diabetic Microvascular Complications

7.2.1 Diabetic Nephropathy

Diabetic nephropathy is the most devastating microvascular complication of diabetes. In the most severe cases, it can lead to renal failure and end-stage renal disease (ESRD) when regular dialysis treatment or kidney transplantation is required for the patient's survival. In type 1 diabetes (T1D), patients with ESRD carry an 18-fold risk of premature mortality compared with the general population (Groop et al. 2009), whereas the patients with T1D but without DN have no excess of premature mortality compared with the nondiabetic age-matched subjects (Groop et al. 2009; Orchard et al. 2010). Excess mortality has been attributed to DN also in patients with type 2 diabetes (T2D) (Afkarian et al. 2013).

7.2.1.1 Heritability

Persistent hyperglycemia is a major risk factor for DN, and intensive glucose lowering treatment was shown to reduce the occurrence of DN by 54 % in the Diabetes Control and Complications Trial (DCCT) (Reichard et al. 1993). Other important risk factors for DN include high blood pressure, dyslipidemia, male gender, and long duration of diabetes (Parving and Smidt 1986; Tarnow et al. 2008; Tolonen et al. 2009). In addition, familial clustering has been reported for DN (Seaquist et al. 1989; Borch-Johnsen et al. 1992; Harjutsalo et al. 2004) as well as for the mildest form of DN, microalbuminuria (The DCCT Research Group 1997). The sibling recurrence risk for DN after 25 years of T1D duration was estimated to be 2.3-fold, suggesting that genetic factors do affect the risk of DN (Fig. 7.1) (Harjutsalo et al. 2004).

7.2.1.2 Candidate Genes

Candidate gene studies for DN have involved hundreds of candidate genes, with only a few being replicated. In order to critically assess the cumulative evidence for putative candidate genes, Mooyaart et al. performed a literature-based

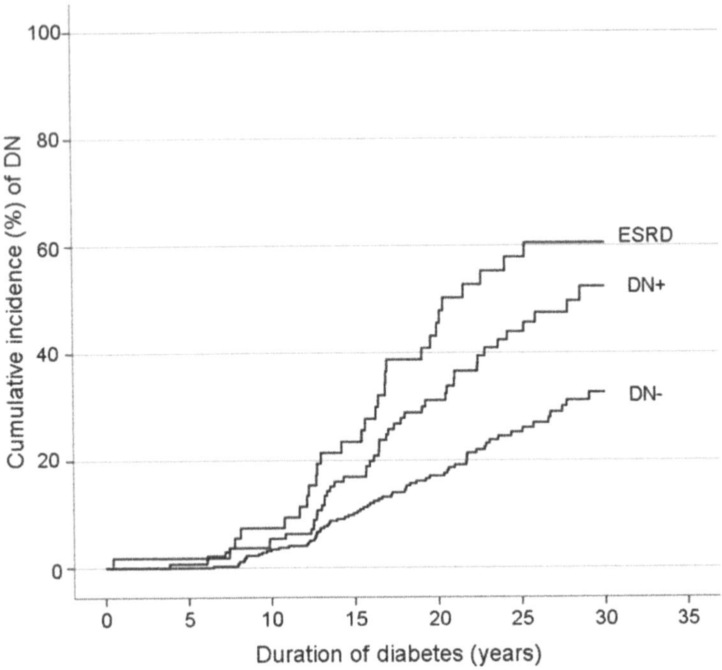

Fig. 7.1 Cumulative incidence of DN in diabetic siblings according to the DN status of probands. Both probands and siblings had T1D. DN-: proband did not have DN. DN+: proband had DN, ESRD excluded. ESRD: proband had ESRD [Modified from Harjutsalo et al. (2004)]

meta-analysis of all the variants that were associated with DN in the initial study and subsequently reproduced in at least one additional study (Mooyaart et al. 2011). Starting with 671 genetic association studies on DN, they identified only 21 variants that remained significant after meta-analysis. The most studied candidate was the rs1799752 insertion/deletion polymorphism in the *ACE* gene encoding the angiotensin I-converting enzyme (ACE), with some of the 42 studies supporting and some refuting the finding. Nevertheless, the meta-analysis indicated that the deletion significantly predisposes to higher risk of DN (Table 7.1). Carriers of the deletion have higher ACE expression (Rigat et al. 1990), leading to elevated plasma angiotensin II levels which promotes podocyte injury (Campbell et al. 2011). Other replicated variants in the meta-analysis implicated genes involved in the polyol pathway, lipid metabolism, inflammatory cytokines, angiogenesis, and oxidative stress. However, a literature-based meta-analysis may be affected by publication bias, potentially leading to overestimation of the effects.

The strongest evidence of association for a candidate gene in terms of a *p*-value is obtained for rs1617640 located on the promoter region of the erythropoietin (*EPO*) gene. The original publication consisted of three studies with participants with T1D and European ancestry and compared the cases with ESRD and proliferative diabetic retinopathy with controls without DN or retinopathy, resulting in a

Table 7.1 Most plausible genes involved in DN

Gene	Variant	Source	Supporting evidence
ACE	rs1799752	Candidate gene	The most studied candidate gene, significantly associated with DN in a literature-based meta-analysis with a pooled odds ratio of 1.24 (95 % CI 1.12–1.37) (Mooyaart et al. 2011)
EPO	rs1617640	Candidate gene	*p*-value 2×10^{-9} in a meta-analysis of five studies (Tong et al. 2008; Williams et al. 2012)
NCK1	rs1866813	Fine-mapping	Strongest association with DN at the 3q21–35 region linked to DN in T1D (Rogus et al. 2008; Moczulski et al. 1998; Imperatore et al. 1998; Bowden et al. 2004; Osterholm et al. 2007; Wessman et al. 2011). *p*-value 7.1×10^{-7}, OR 1.33 (He et al. 2009)
FRMD3	rs10868025	GWAS on DN in T1D	Strongest signal from the GWAS on DN in T1D in GoKinD US, p-value 5.0×10^{-7}, with nominal replication in the DCCT/EDIC study (p-value 0.02) (Pezzolesi et al. 2009a). Supported in a GWAS of African American participants with T2D after adjusting for MYH9 variants (Freedman et al. 2011) and in Japanese participants with T2D (p-value < 0.05) (Maeda et al. 2010a), but not supported in European participants with T1D (Williams et al. 2012)
AFF3	rs7583877	GWAS on ESRD in T1D	Genome-wide significantly associated with ESRD (*p*-value 1.2×10^{-8}) in a meta-analysis of 10,500 participants with T1D (Sandholm et al. 2012)
15q26 *(RGMA—MCTP2)*	rs12437854	GWAS on ESRD in T1D	Genome-wide significantly associated with ESRD (*p*-value 2.0×10^{-9}) in a meta-analysis of 10,500 participants with T1D (Sandholm et al. 2012)
2q31.1 *(CDCA7—SP3)*	rs4972593	GWAS on ESRD in women with T1D	Genome-wide significantly associated with ESRD in women with T1D (*p*-value 3.0×10^{-8}) and replicated in independent participants (*p*-value 0.02) (Sandholm et al. 2013)
GLRA3	rs1564939	GWAS on albuminuria	Genome-wide significantly associated with albuminuria in T1D (5 SNPs with *p*-value $<5 \times 10^{-8}$). Nominally associated with albuminuria (*p*-value $= 0.04$) in seven replication studies, but effect in opposite direction (Sandholm et al. 2014)

p-value of 2.8×10^{-11} (Tong et al. 2008). Evaluation in two additional studies did not replicate the finding, but the association remained significant (*p*-value 2×10^{-9}) after meta-analysis of all five studies (Williams et al. 2012) (Table 7.1).

Candidate genes for DN have been explored also with family-based linkage studies. One of the strongest linkage peaks for DN with the logarithm of odds (LOD) score of 3.1 was obtained in a candidate gene study of the *AGTR1* gene, located at the genetic position 157 cM on chromosome 3q (Moczulski et al. 1998). Interestingly, all the performed genome-wide linkage scans on DN in T1D (Osterholm et al. 2007; Rogus et al. 2008; Wessman et al. 2011) and in T2D (Imperatore et al. 1998; Bowden et al. 2004) suggestively support a linkage peak on the same chromosome 3q region (chromosomal position 134 cM–181 cM). Whereas the region is rather large and the obtained signals were mainly suggestive rather than confirming, this chromosomal region has been a target for intensive fine-mapping efforts. Vionnet et al. examined 69 polymorphisms in 14 genes on the 3q23–q29 region and found the strongest association (p-value 0.01) with DN at variants in the adiponectin (*ADIPOQ*) gene (Vionnet et al. 2006). Adiponectin is a cytokine exclusively produced by adipose tissue that has insulin-sensitizing effects, thus affecting glucose and lipid metabolism; the serum level of adiponectin has been associated with renal function in T1D (Saraheimo et al. 2005). The chromosome 3q21–25 region (134 to 157 cM) which was implicated in the T1D studies was further fine-mapped in nearly 3700 Icelandic, Finnish, and British participants with T1D. The study identified rs1866813 at 3q22 as the strongest factor associated with DN, with an overall p-value of 7.1×10^{-7} (He et al. 2009). The variant is located close to the *NCK1* gene which is expressed in the kidney podocyte where Nck1 links the slit diaphragm protein nephrin to the actin cytoskeleton (Jones et al. 2006) (Table 7.1).

7.2.1.3 Genome-Wide Association Studies (GWASs)

Multiple GWASs on DN have emerged in the last few years, both in participants with T1D and T2D. The first GWAS on DN was performed on Japanese subjects with T2D, reporting a potential association with DN at rs741301 and eight other single nucleotide polymorphisms (SNPs) in the *ELMO1* gene (Shimazaki et al. 2005). The reported variants have not been replicated in other studies (Pezzolesi et al. 2009b; Leak et al. 2009; Williams et al. 2012), but investigations in participants with T1D of European origin and in African American participants with T2D identified other variants in the *ELMO1* gene suggestively associated with DN (Pezzolesi et al. 2009b; Leak et al. 2009).

The first GWAS on DN in T1D, performed in the US Genetics of Kidneys in Diabetes (US GoKinD) study including 1500 participants with T1D, reported multiple suggestive associations ($P < 10^{-5}$) with DN in or near the *CHN2*, *FRMD3*, *CARS*, and *IRS2* genes, although none of the loci reached genome-wide statistical significance (i.e., p-value $<5 \times 10^{-8}$) (Pezzolesi et al. 2009a). Their subsequent analysis using imputed GWAS data resulted in four additional suggestively associated loci in the *SORBS1* gene, near the *TRPS1* gene and between the *CDCA2* and *EBF2* genes, and near the *BUB3* and *GPR26* genes (Pezzolesi et al. 2010). Among these loci, the variants near the *CARS* gene and on the *FRMD3* gene were suggestively replicated in further studies in patients with T1D (Pezzolesi et al. 2009a). The association at the *FRMD3* gene was subsequently supported by a GWAS in African American participants with T2D after adjusting for the major genetic risk factors for nondiabetic kidney disease (Freedman et al. 2011), whereas the association analysis was inconclusive in Japanese participants with T2D (p-value < 0.05 but not withstanding correction for multiple testing) (Maeda et al. 2010a) and not supported in European participants with T1D (Williams et al. 2012). Another GWAS on ESRD using pooled DNA from the same US GoKinD T1D participants suggested associations in the *ZMIZ1* and *MSC* genes and supported the association on chromosome 13q identified by Pezzolesi et al. (Craig et al. 2009).

A large meta-analysis of three GWASs on DN and ESRD was performed in the Genetics of Nephropathy, an International Effort (GENIE) Consortium including 6691 participants with T1D and with European origin. After a combined meta-analysis with nine additional studies on DN in T1D, variants in the *AFF3* gene and between the *RGMA* and *MCTP2* genes were associated with ESRD with genome-wide statistical significance (p-value $<5 \times 10^{-8}$) (Sandholm et al. 2012). *AFF3* encodes a nuclear transcriptional activator that can bind to DNA and RNA (Melko et al. 2011). Variants upstream and in the 5′ end of the *AFF3* gene have been suggestively associated with autoimmune diseases, including juvenile idiopathic arthritis (Hinks et al. 2010), rheumatoid arthritis (Barton et al. 2009), Graves' disease (Todd et al. 2007), and T1D (Todd et al. 2007). In vitro analyses of the *AFF3* expression levels on renal epithelial cells suggested that *AFF*3 may influence the TGF-β1-induced fibrotic responses (Sandholm et al. 2012). In addition to the ESRD findings, a suggestive association with DN (p-value 2.1×10^{-7}) was reported for variants in the *ERBB4* gene. Whereas ErbB4 has been previously implicated in the development of cardiac, mammary gland, and neural tissues (Gassmann et al. 1995; Tidcombe et al. 2003) and associated with cancer (Prickett et al. 2009), recent research on conditional *ERBB4* overexpression and knockout mice models suggests that *ERBB4* is important for the development of the kidneys as well (Veikkolainen et al. 2012). Interestingly, ErbB4 has been suggested as a therapeutic target molecule for cancer and psychiatric and cardiovascular disorders, and ErbB4-binding ligands have been patented for enhancement of the ErbB4 signaling (Paatero and Elenius 2008).

A gender-specific GWAS on ESRD in Finnish participants with T1D identified rs4972593 between the *SP3* and *CDCA7* genes associated with over twofold risk of ESRD in women (p-value 3×10^{-8}). The association was confirmed in independent studies in women with T1D, whereas no association was seen in men or in women with T2D (Sandholm et al. 2013). The RegulomeDB database indicated potential regulatory activity for rs530673 (in strong linkage with rs4972593, $r^2 = 1$) (Boyle et al. 2012), but no direct link has been established between rs4972593 and any of the nearby genes. Nevertheless, the *SP3* gene expression was found higher in the glomeruli of women compared with men (p-value 0.004, fold change -1.45) (Woroniecka et al. 2011), and the Sp3 transcription factor is known to directly interact with the estrogen receptor α (ERα), forming a receptor complex for estradiol (Stoner et al. 2000, 2004). One of the many target genes of Sp3 is the

CD2AP gene encoding a protein that is essential for the glomerular filtration barrier and interacts with nephrin and podocin—two important proteins in the glomerular slit diaphragm (Shih et al. 2001; Schwarz et al. 2001). Therefore, the *SP3* gene seems like a plausible culprit gene to explain the association between rs4972593 and ESRD in women with T1D.

Further, GWASs on DN in Japanese and African American participants with T2D have suggested associations in or near the *ACACB*, *RPS12*, *LIMK2*, *SFI1*, and other genes, but none of the loci have reached genome-wide statistical significance (McDonough et al. 2010; Maeda et al. 2010b). A multimarker data mining approach supported the association between ESRD and the loci between the *RGMA* and *MCTP2* genes and suggested variants between the *WNT4* and *ZBTB40* and between the *SEMA6D* and *SLC24A5* genes as novel risk loci for ESRD (Sambo et al. 2014). However, these loci still warrant further confirmation.

The majority of the published association studies on DN has used a case–control study setting and defined cases as participants with macroalbuminuria or ESRD, or ESRD only, to obtain a more extreme phenotype. Indeed, most of the reported loci with genome-wide statistical significance were obtained with the ESRD case definition rather than a milder case definition, possibly reflecting i) survival bias, i.e., association with the high mortality observed in DN (Forsblom et al. 2011; Groop et al. 2009) instead of ESRD, ii) higher heritability of ESRD than DN (Harjutsalo et al. 2004), iii) clustering of more genetic risk factors for the participants with the most extreme phenotype following the liability model (Gibson 2012), or iv) specific genetic factors predisposing for the transition from macroalbuminuria to ESRD. Only one GWAS has been published thus far using quantitative traits to assess the severity of DN. The study examined genetic risk factors for elevated albumin excretion rate (AER) in 1925 Finnish participants and identified five variants in the *GLRA3* gene associated with elevated AER with p-value $< 5 \times 10^{-8}$. Some evidence of association (p-value 0.028, not significant after correction for multiple testing) was also seen in non-Finnish participants with T1D but with the opposite allele associated with elevated AER. The authors hypothesize that population-specific rare variants may explain the association and the difference in the effect direction but that larger sequencing efforts are required to confirm the finding (Sandholm et al. 2014).

GWASs have been performed on both albuminuria and estimated glomerular filtration rate (eGFR), the two main quantitative traits for evaluation of kidney disease, in the general (mainly nondiabetic) population. A missense mutation rs1801239 in the *CUBN* gene was identified as a risk locus for albuminuria in nondiabetic participants, and the same variant was associated with microalbuminuria in participants with diabetes. *CUBN* encodes cubilin, which is essential for the reuptake of albumin and other low-molecular-weight proteins in the proximal tubules (Böger et al. 2011). Furthermore, multiple loci have been identified for reduced kidney function in nondiabetic subjects, evaluated with eGFR. These include variants in or near the *UMOD*, *SHROOM3*, *GATM-SPATA5L1*, *CST*, and *STC1* genes (Köttgen et al. 2009); *LASS2*, *GCKR*, *ALMS1*, *TFDP2*, *DAB2*, *SLC34A1*, *VEGFA*, *PRKAG2*, *PIP5K1B*, *ATXN2*, *DACH1*, *UBE2Q2*, and *SLC7A9* genes (Köttgen et al. 2010); and *MPPED2*, *DDX1*,

SLC47A1, *CDK12*, *CASP9*, and *INO80* genes (Pattaro et al. 2012). Variants in the *GCKR*, *SHROOM3*, and *UMOD* loci were suggestively associated with eGFR also in participants with T2D (Deshmukh et al. 2013). In addition, variants in the *MYH9* and *APOL1* genes that predispose to chronic kidney disease in the nondiabetic population have been associated with kidney disease in African American participants with T2D as well. In contrast, none of the variants affecting eGFR or risk of chronic kidney disease in the general population were associated with DN or ESRD in the GWAS meta-analysis of participants with T1D (Sandholm et al. 2012). Thus, the overlap between genes predisposing to diabetic and nondiabetic kidney disease seems limited to individuals with T2D.

7.2.2 Diabetic Retinopathy

Diabetic retinopathy belongs to the severe microvascular complications of diabetes (Klein 1987) and is one of the leading causes of blindness worldwide (Resnikoff et al. 2004). Family studies have shown that DR clusters in families in both T1D and T2D, with the heritability estimates ranging from 25 % in T2D to 52 % in T1D (Hietala et al. 2008; Hallman et al. 2005; Rema et al. 2002; The DCCT Research Group 1997; Arar et al. 2008). As for the other diabetic complications, many candidate gene studies have been conducted, but convincing replication of the reported associations has been rare. Abhary et al. published in 2009 a literature-based meta-analysis of the association analyses on DR. They identified 160 publications containing 196 genetic polymorphisms in 65 genes and finally evaluated associations at 34 repeatedly assessed variants from 20 genes. The most consistent association was obtained for variants in the aldose reductase encoding *AKR1B1* gene, whereas some association with specific subtypes was also seen for variants in the *NOS3*, *VEGF*, *ITGA2*, and *ICAM1* genes (Abhary et al. 2009).

Paving the way for GWAS, the Candidate Gene Association Resource (CARe) performed a large-scale candidate gene analysis containing 49,320 SNPs from roughly 2000 genes suspected to affect the risk of cardiovascular, metabolic, and inflammatory diseases. The study included 2691 participants with T2D and fundus photographs available for the classification of DR, and replication of the main findings was attempted in more than 5000 participants with T2D or T1D and with diverse ethnicity. None of the evaluated variants were consistently associated with DR, but among the 39 genes that had been previously associated with DN, DR, or T2D, the strongest evidence of association was obtained for variants in the P-selectin encoding *SELP* gene (p-value 3.1×10^{-6}). Among all the studied variants, the strongest association was obtained for a variant near the *IDUA* gene (p-value 3.1×10^{-6}) (Sobrin et al. 2011).

Four GWASs have been performed on DR. An early GWAS on DR in 290 participants with T2D suggested variants in the *TINAG* and *C6orf170* genes, but the p-values were only moderate (Fu et al. 2010). The largest GWAS on DR to date included 2829 participants with T1D evaluated at 2.5 million SNPs after genotype imputation. None of the loci reached genome-wide statistical significance, but the strongest signal with p-value 1.2×10^{-7} was obtained for rs476141 near the *AKT3* gene. However, the results were not assessed in any independent replication cohorts (Grassi et al. 2011). A GWAS including 174 DR cases and 575 non-DR controls from the Taiwanese T2D population identified two loci with genome-wide statistical significance using a recessive association model: rs17376456 on chromosome 5q with p-value 3.0×10^{-15} and rs2038823 in the *HS6ST3* gene. However, these findings were not replicated in independent participants (Huang et al. 2011). Finally, a GWAS on DR in 1007 Chinese participants with T2D suggested associations on the *TBC1D4-COMMD6-UCHL3*, *LRP2-BBS5*, and *ARL4C-SH3BP4* gene regions but were unable to replicate the findings in 585 Hispanic participants with T2D (Sheu et al. 2013). All in all, the candidate gene studies and GWASs have suggested multiple potential risk factors for DR, but none of them have yet been confirmed in independent studies. Thus, role of these variants remains uncertain.

7.2.3 Diabetic Neuropathy

Diabetic neuropathy is one of the most common but least studied diabetic complications, and the pathophysiology remains poorly understood. In addition, the phenotype is not well defined and can be further divided into peripheral and autonomic neuropathy. Although candidate gene studies on diabetic neuropathy in both T1D and T2D exist, no systematic replication or meta-analysis has been performed to summarize the findings. Many of the studied candidate genes were candidates for diabetes or other diabetic complications as well, such as *APOE* associated with the severity of peripheral neuropathy (Monastiriotis et al. 2013), *VEGF* associated with diabetic neuropathy (Tavakkoly-Bazzaz et al. 2010), and *TCF7L2* variants associated with cardiovascular autonomic neuropathy (CAN) (Ciccacci et al. 2013), but the number of participants with neuropathy has been small, typically ranging from tens to few hundreds.

Despite the scarcity of candidate gene studies, a GWAS on chronic pain was recently performed in 3063 UK-based patients with T2D, including 572 cases with a positive monofilament test result in at least one foot. The strongest evidence of association was obtained for a locus on chromosome 8 near the *GFRA2* gene with the lowest p-value of 1.8×10^{-7} for rs17428041. The authors conclude that their next step is to attempt replication of the association to confirm the finding (Meng et al. 2014). In a pilot GWAS on erectile dysfunction in men with T1D diabetes, the strongest association was obtained for two SNPs near the *ALCAM* gene (p-value 7×10^{-7}). Similar to the other GWAS on neuropathy, the final conclusions regarding this locus are pending replication in independent studies (Hotaling et al. 2012).

7.3 Genetic Factors of Macrovascular Complications

A large proportion of the morbidity and mortality associated with diabetes, especially with the type 2 form, is due to the macrovascular complications of the disease. Relative to nondiabetic subjects, patients with diabetes have a two- to fourfold increased risk of cardiovascular death (Krolewski and Warram 2005; Stamler et al. 1993; Warram et al. 1997). The impact of diabetes on coronary artery disease is so profound that diabetic patients without a previous myocardial infarction (MI) carry the same risk for an acute coronary event as nondiabetic patients with a previous MI (Haffner et al. 1998). Diabetes also worsens early and late outcomes in acute coronary syndromes, determining an increased risk of complications after an MI (Beckman et al. 2002). Part of this increase in cardiovascular risk is fostered by conventional cardiovascular risk factors, such as increased levels of small low-density lipoprotein (LDL), hypertension, low high-density lipoprotein (HDL) cholesterol, and central obesity (Warram et al. 1997), which are particularly frequent in T2D and are often associated with hyperinsulinemia in the so-called insulin resistance syndrome or syndrome X (Reaven 1997). These metabolic abnormalities can antedate the onset of T2D by several decades ("prediabetic" exposures). Once diabetes becomes manifest, prediabetic exposures continue but are augmented by the pro-atherogenic effects of hyperglycemia (Laakso 1999) through the buildup of advanced glycation end products (AGE) (Brownlee 1994); activation of protein kinase C (Rask-Madsen and King 2005); increased production of superoxide, polyols, and hexosamine (Nishikawa et al. 2000); and other as yet unidentified cellular pathways.

While all diabetic subjects are at increased cardiovascular risk, their susceptibility to the detrimental effect of diabetes varies, in part under the control of their genetic background. Genetic factors have been known for many years to modulate the development of coronary artery disease in the general population (Shea et al. 1984). In a landmark study from Sweden published in 1992, 50 % of twins of subjects with early cardiovascular disease (CVD) mortality were reported to die before age 70 as compared to 10 % in unselected subjects (Marenberg et al. 1994). Several reports have extended this evidence to individuals with diabetes. In a study of families with T2D, Wagenknecht et al. found that up to 50 % of the variance of coronary calcium content (CAC, an index of atherosclerotic burden) is accounted for by familial factors (Wagenknecht et al. 2001). Such estimate was minimally affected by adjustment for HDL, body mass index (BMI), and hypertension, indicating that this effect was not due to familial aggregation of traditional risk factors. Similar heritability estimates ($h^2 = 0.41$) have been obtained using carotid intima thickness as a marker of subclinical atherosclerosis (Lange et al. 2002).

7.3.1 Candidate Gene Studies

As with other complex disorders, initial efforts were based on the study of candidate genes chosen on the basis of their postulated function in the etiology of coronary artery disease. Among the most interesting results obtained through this approach are those concerning adiponectin (briefly discussed above) (Scherer et al. 1995). Adiponectin also has direct antiatherogenic effects on the arterial wall by antagonizing monocyte adhesion to the endothelium, smooth muscle cell proliferation, and foam cell formation (Kadowaki and Yamauchi 2005). A SNP in an intron of the adiponectin gene (rs1501299) was reported to be associated with a twofold increase in the risk of coronary artery disease (CAD) in a meta-analysis of four different sets of diabetic subjects (Qi et al. 2006). An effect of this SNP, or other variants in linkage disequilibrium (LD) with it, on circulating adiponectin levels appears to be responsible for this association (Menzaghi et al. 2007). Genetic variability in the adiponectin receptors may also play a significant role. Genetic variants in the 3′ half of *ADIPOR1*—one of the adiponectin receptors described to date—have emerged as being associated with CAD among individuals with T2D from the USA and Italy. These variants were also associated with lower *ADIPOR1* mRNA levels in carriers, suggesting a blunted response to the antiatherogenic effects of adiponectin on the vascular wall as a mechanism mediating this genetic effect (Soccio et al. 2006). Interesting findings have also been obtained for three relatively infrequent non-synonymous variants that had been previously shown to affect insulin-signaling—*ENPP1* K121Q, *IRS1* G972R, and *TRIB3* Q84R (Prudente et al. 2009; Sharma et al. 2011; Bacci et al. 2011, 2013; Morini et al. 2009). The first of these variants, which enhances the inhibitory effects of the phosphodiesterase ENPP1 on insulin receptor signaling (Pizzuti et al. 1999), was found to be significantly associated with an increased risk and a younger age at onset of major cardiovascular events among diabetic subjects, especially in the presence of obesity (p-value for gene x obesity interaction = 0.003) (Bacci et al. 2011). Considered jointly, the three variants were associated with an 18 % increase in CAD risk as well as with whole-body and endothelium-specific insulin sensitivity (Bacci et al. 2013). Altogether, these findings confirm the well-known association between insulin resistance and cardiovascular risk and point to possible targets for new interventions to break this link.

7.3.2 Genome-Wide Association Studies (GWAS)

7.3.2.1 Studies in the General Population

Over 40 loci have been identified to date by GWASs in the general population as being associated with CAD or MI with genome-wide significance. Examining these findings, three general themes emerge that are also common to GWAS for other complex disorders (Schunkert et al. 2011; Coronary Artery Disease (C4D) Genetics Consortium 2011; Deloukas et al. 2013). The first one is that most genetic variants are associated with a relative small increase in cardiovascular risk, with allelic odds ratios (OR) that do not exceed 1.30 and are mostly below 1.20. The second feature is that most of the variants associated with CAD are placed in noncoding regions, suggesting that the genetic effects involve regulatory elements and alteration of gene expression rather than the amino acid sequence. The third one is that many of the genes that are placed in the vicinity of these variants have functions that cannot be easily connected to what we currently know about the pathophysiology of atherosclerosis. Among the few exceptions are *LDLR*, which codes for the LDL receptor, *PCSK9*, which codes for a serine protease modulating the expression of the LDL receptor and mutated in Mendelian forms of hypercholesterolemia (Abifadel et al. 2003), and the *SLC22A3-LPAL2-LPA* cluster, which include the coding sequences for the atherogenic lipoprotein Lp(a).

In all the GWAS for CAD conducted to date, the strongest association signal has been observed at a locus on chromosome 9p21, with odds ratios in the order of 1.30. The increased cardiovascular risk associated with this locus is unaffected by adjustment for other cardiovascular risk factors, implying that this effect is independent from known risk pathways. The SNPs most strongly associated with CAD at this location are placed in a 60 Kb LD block. Given the strong LD, it has been difficult to pinpoint the variant(s) responsible for the association with CAD based solely on the association data. Extensive resequencing of this region has failed to identify coding variants on the CAD-associated haplotypes, suggesting that this association results from allelic variability in gene expression and/or splicing (McPherson et al. 2007). The LD block where the association signal is placed includes the most 3′ exons of the noncoding gene *CDKN2B-AS* (a.k.a. *ANRIL*), which is expressed in many cell types relevant to the atherosclerotic process (Broadbent et al. 2008; Jarinova et al. 2009). The *CDKN2B-AS* gene is transcribed as two alternatively spliced transcripts—a long one including the 3′ exons placed in the CAD-associated block and a short one missing those exons. It has been proposed that the balance of these splice variants is altered in carriers of the 9p21 allele and that this may in turn affect the expression of other genes through mechanisms such as RNA interference or chromatin remodeling. Special attention as potential targets has been given to *CDKN2A* and *CDKN2B*, which are adjacent to and partially overlap with *CDKN2B-AS L*. These two genes code for inhibitors of cyclin-dependent kinases that are involved in the control of cell proliferation, cell aging, and apoptosis and are expressed at high levels in endothelial and inflammatory cells (Kamb et al. 1994; Pomerantz et al. 1998; Hannon and Beach 1994). In support of a role of these genes, a positive correlation, albeit marginally significant, was found between transcript levels of the long variant of *CDKN2B-AS L* and *CDKN2B* mRNA (Jarinova et al. 2009). Thus, the risk allele at 9p21 might predispose to CAD by decreasing the expression of the long form of *CDKN2B-AS*, which in turn decreases *CDKN2B* expression, thereby promoting a proliferative phenotype in cell types relevant to atherogenesis such as smooth muscle cells.

The 9p21 locus influences cardiovascular risk also in the presence of diabetes. In fact, data from the Joslin Heart Study (JHS) suggest that this locus may have a larger effect on CAD risk among individuals with T2D than in the general population (Doria et al. 2008). In this study, individuals with T2D and angiographic evidence of significant CAD were compared to subjects who had a negative cardiovascular history of CAD and a normal exercise treadmill test despite a long-term (>5 years) exposure to T2D. The odds ratios associated with a representative 9p21 variant (rs2383206) were 1.45 (95 % confidence interval (CI) 0.94–2.22) for heterozygotes and 2.37 (1.52–3.70) for homozygotes. The larger odds ratio as compared to the general population suggests the possibility of an interaction with some aspect of diabetes such as hyperglycemia. Consistent with this hypothesis, in the JHS, the odds of CAD for individuals who were homozygotes for the risk allele and were also in poor glycemic control (upper tertile of HbA1c at examination) were increased almost four times as compared to subjects who were not homozygous for the risk allele and were in good glycemic control (Fig. 7.2). By contrast, individuals who had either the homozygote genotype or poor glycemic control had only a small increase in the odds of CAD as compared to subjects who had neither risk factor. Such interaction between homozygous genotype and glycemic control was significant and was also observed with respect to cardiovascular mortality in a separate prospective study of 475 type 2 diabetic subjects from the Joslin Clinic. After 10 years of follow-up, 36 % of risk allele homozygotes with a history of poor glycemic control had died because of a cardiovascular cause as compared to only 15–20 % of all other subjects (Doria et al. 2008).

These findings have several implications. From an epidemiological perspective, they may explain past difficulties in demonstrating an association between glycemic control and cardiovascular outcomes (Beckman et al. 2002; Rask-Madsen and King 2005; Libby and Plutzky 2002; UK Prospective Diabetes Study (UKPDS) Group 1998; Haffner 1999; Wild et al. 1999). If poor glycemic control has a major impact on cardiovascular risk in only the 30 % of individuals with T2D who are homozygous for the 9p21 risk allele, one would expect the association between CAD and poor glycemic control to be quite modest among unselected individuals. From a clinical perspective, they indicate that the 9p21 genotype could be potentially used to identify candidates for intensive glycemic control interventions, and they call for clinical trials to test this hypothesis. From a biological perspective, they suggest that the atherogenic pathways on which the 9p21 variant and hyperglycemia act may intersect at some level(s). Thus, identification of the gene(s) whose function is affected by the risk allele may also provide clues to the molecular mechanisms linking excess glucose to increased atherosclerosis, which are the object of intense debate still (Beckman et al. 2002; Rask-Madsen and King 2005; Libby and Plutzky 2002). As discussed above, the candidate genes at 9p21 include *CDKN2A* and *CDKN2B*, which code for three inhibitors of cyclin-dependent kinases ($p16^{INK4a}$, ARF, and $p15^{INK4b}$) controlling cell proliferation, cell aging, and apoptosis (Kamb et al. 1994; Pomerantz et al. 1998; Hannon and Beach 1994). These functions, which are all potentially relevant to the atherosclerotic process, may also be influenced by high glucose (Zheng et al. 2007) or other conditions associated with it, such as insulin resistance.

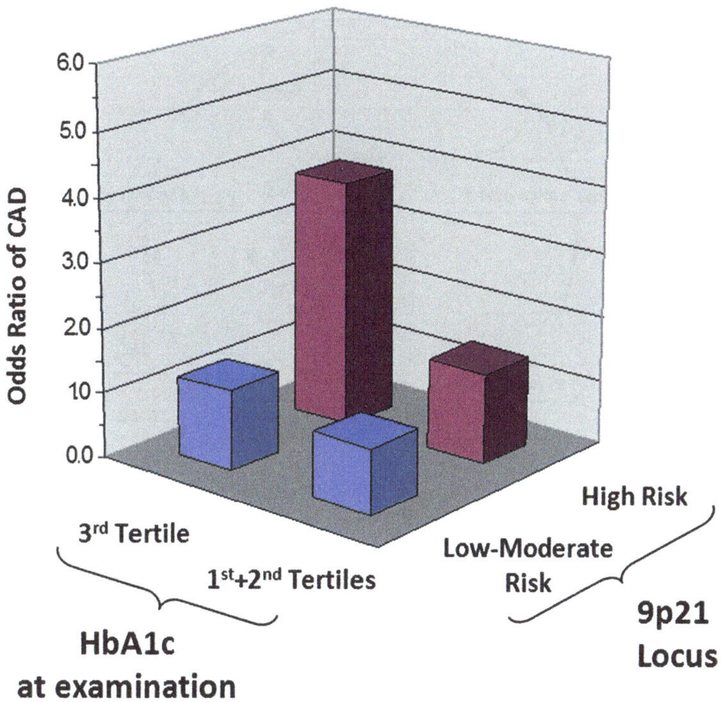

Fig. 7.2 Synergism between poor glycemic control and 9p21 locus on the odds of CAD in T2D. Data are from the study in Doria et al. (2008) and refer to 322 CAD-positive cases and 412 -CAD-negative controls from the Joslin Heart Study. Adjusted odds ratios of CAD are shown according to HbA1c value at examination (top tertile [HbA1c > 7.6] vs. lower two tertiles) and genotypes at rs2383206 (G/G homozygosis vs. other genotypes)

By dissecting the interaction between hyperglycemia and 9p21 locus, it may be possible to identify critical nodes in these pathways that can be used to develop new preventive interventions or treatments.

7.3.2.2 Studies in the Diabetic Population

The modifying effect of glycemic levels on the 9p21 genetic effect raises the possibility that genetic effects may exist that interact so strongly with the diabetic milieu to be observed only among individuals with diabetes. To investigate this hypothesis, a GWAS for CAD was recently conducted specifically among subjects with T2D. The study, consisting of three stages, included a total of 1517 CAD cases and 2671 CAD-negative controls, all with T2D, from the USA and Italy. One SNP (rs10911021) was nominally associated with CAD at each stage and reached genome-wide significance in the three stages combined ($p = 2 \times 10^{-8}$) (Qi et al. 2013). Two features make this finding especially remarkable. First, the

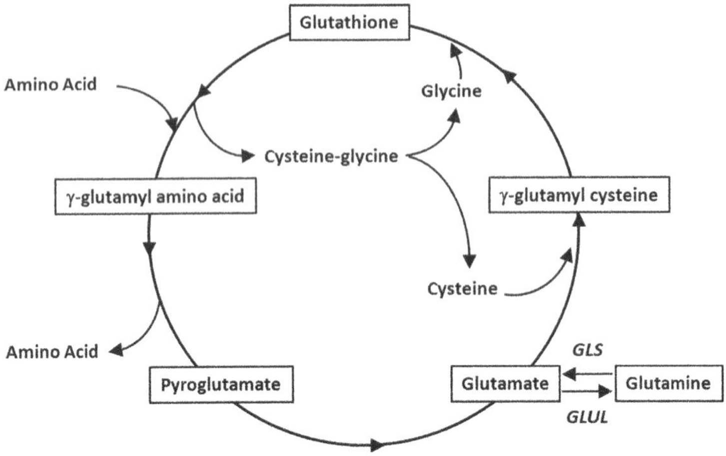

Fig. 7.3 Schematic representation of the γ-glutamyl cycle. *GLUL* glutamate–ammonia ligase (also known as glutamine synthase), *GLS* glutaminase

strength of this association between rs10911021 and CAD (summary OR = 1.36) is in the same order of magnitude as the strongest genetic effect on CAD identified in the general population. Second, this genetic effect appears to be specific for diabetes as no significant association could be identified among nondiabetic individuals (OR = 0.99, 95 % CI 0.87–1.13, *p*-value for SNP × diabetes interaction = 2.6×10^{-4}). In further support of the specificity for diabetes, rs10911021 was found to be associated with CAD in the general population (CARDIoGRAM) at the level that one would expect (OR = 1.04, $p = 0.01$) if the genetic effect was present only in the diabetic subset of the population with the same strength as that observed in our GWAS (Qi et al. 2013).

SNP rs10911021 is located between two genes, *ZNF648* in centromeric direction and *GLUL* in telomeric direction. No missense variants in LD with this variant are present in the HapMap or the 1000 Genome Projects databases, suggesting that the association with CAD is mediated by an effect on gene regulation. Indeed, the rs10911021 risk allele was found to be associated with a significant reduction in the expression of the neighboring *GLUL* gene in endothelial cells. The other flanking gene *(ZNF648)* was not expressed in endothelial cells, and none of the other neighboring genes were associated with rs10911021. The *GLUL* gene encodes the enzyme glutamate–ammonia ligase (also known as glutamine synthase), which catalyzes the conversion of glutamic acid and ammonia into glutamine (Fig. 7.3) (Krebs 1935). No significant association was detected between rs10911021 and plasma glutamic acid or glutamine concentrations in a sample of 100 Joslin Heart Study participants. However, the ratio between plasma pyroglutamic acid (the immediate precursor of glutamic acid in the γ-glutamyl cycle [Fig. 7.3]) and glutamic acid was significantly lower in risk allele homozygotes than in protective allele homozygotes ($p = 0.029$) (Qi et al. 2013). The pyroglutamic-to-glutamic ratio was also significantly lower in CAD cases than in CAD-negative control ($p = 0.02$). The OR for CAD of risk allele homozygotes decreased from 1.83 to 1.39 (a ~50 % reduction in the log scale) after adjustment for the pyroglutamic-to-glutamic acid ratio, suggesting that the effect of this locus on CAD was at least in part mediated by its effect on this parameter.

These findings implicate the γ-glutamyl cycle, of which pyroglutamic acid is an intermediate, in the etiology of CAD in diabetes. One can hypothesize that alterations of this pathway may limit the synthesis of the natural antioxidant glutathione, compounding the known negative effect of diabetes on this metabolite (Yoshida et al. 1995) and decreasing glutathione levels below a critical threshold, under which subjects become more vulnerable to oxidative stress and, consequently, more susceptible to the development of atherosclerosis. In vitro and in vivo studies are in progress to test this hypothesis as well as others based on the role of glutamate and glutamine in several other pathways that are potentially relevant to vascular biology and atherosclerosis.

7.4 Conclusions and Future Aspects

While many of the genetic risk factors for cardiovascular disease identified in the general population seem to play an important role in diabetic participants as well, little overlap has been found between kidney, eye, and neuronal complications in the general and the diabetic populations. The recent advances in the genetic studies on the diabetic complications have revealed novel susceptibility loci especially for diabetic kidney complications. The good news are that large studies are ongoing to further dissect the genetic background of DN, and large GWASs on DR are emerging. Importantly, international consortia, required for increased sample sizes, have already been established. However, larger studies are still needed to identify common genetic risk factors with modest effect or rarer causal variation. Apart from increasing the sample size, improving the phenotype may be another way to find further susceptibility loci especially for the microvascular complications where the disease manifestation is more difficult to diagnose, especially at the early stages of the disease.

As for any complex disease, exome and whole-genome sequencing are the likely next steps in order to identify rare and causal variation in both existing and novel susceptibility loci. Epigenetic factors have also been suggested to affect the risk of diabetic complications, especially due to the phenomenon known as "metabolic memory," i.e., the persistence of adverse effects of high blood glucose despite subsequent intensive diabetes treatment (Keating and El-Osta 2013). Currently, the field of genetics of diabetic complications is in the stage of finding associated variants. Once the susceptibility loci are established, functional studies are needed to define the molecular mechanisms behind the associations.

References

Abhary S, Hewitt AW, Burdon KP, Craig JE (2009) A systematic meta-analysis of genetic association studies for diabetic retinopathy. Diabetes 58(9):2137–2147

Abifadel M, Varret M, Rabes JP, Allard D, Ouguerram K, Devillers M, Cruaud C, Benjannet S, Wickham L, Erlich D, Derre A, Villeger L, Farnier M, Beucler I, Bruckert E, Chambaz J, Chanu B, Lecerf JM, Luc G, Moulin P, Weissenbach J, Prat A, Krempf M, Junien C, Seidah NG, Boileau C (2003) Mutations in PCSK9 cause autosomal dominant hypercholesterolemia. Nat Genet 34(2):154–156

Afkarian M, Sachs MC, Kestenbaum B, Hirsch IB, Tuttle KR, Himmelfarb J, de Boer IH (2013) Kidney disease and increased mortality risk in type 2 diabetes. J Am Soc Nephrol 24 (2):302–308

Arar NH, Freedman BI, Adler SG, Iyengar SK, Chew EY, Davis MD, Satko SG, Bowden DW, Duggirala R, Elston RC, Guo X, Hanson RL, Igo RP, Jr Ipp E, Kimmel PL, Knowler WC, Molineros J, Nelson RG, Pahl MV, Quade SR, Rasooly RS, Rotter JI, Saad MF, Scavini M, Schelling JR, Sedor JR, Shah VO, Zager PG, Abboud HE, Family Investigation of Nephropathy and Diabetes Research Group (2008) Heritability of the severity of diabetic retinopathy: the FIND-Eye study. Invest Ophthalmol Vis Sci 49(9):3839–3845

Bacci S, Rizza S, Prudente S, Spoto B, Powers C, Facciorusso A, Pacilli A, Lauro D, Testa A, Zhang YY, Di SG, Mallamaci F, Tripepi G, Xu R, Mangiacotti D, Aucella F, Lauro R, Gervino EV, Hauser TH, Copetti M, De CS, Pellegrini F, Zoccali C, Federici M, Doria A, Trischitta V (2011) The ENPP1 Q121 variant predicts major cardiovascular events in high-risk individuals: evidence for interaction with obesity in diabetic patients. Diabetes 60(3):1000–1007

Bacci S, Prudente S, Copetti M, Spoto B, Rizza S, Baratta R, di Pietro N, Morini E, Di PR, Testa A, Mallamaci F, Tripepi G, Zhang YY, Mercuri L, Di SS, Lauro R, Malatino L, Consoli A, Pellegrini F, Pandolfi A, Frittitta L, Zoccali C, Federici M, Doria A, Trischitta V (2013) Joint effect of insulin signaling genes on cardiovascular events and on whole body and endothelial insulin resistance. Atherosclerosis 226(1):140–145

Barton A, Eyre S, Ke X, Hinks A, Bowes J, Flynn E, Martin P, Consortium YEAR, Consortium BIRAC, Wilson AG, Morgan AW, Emery P, Steer S, Hocking LJ, Reid DM, Harrison P, Wordsworth P, Thomson W, Worthington J (2009) Identification of AF4/FMR2 family, member 3 (AFF3) as a novel rheumatoid arthritis susceptibility locus and confirmation of two further pan-autoimmune susceptibility genes. Hum Mol Genet 18(13):2518–2522

Beckman JA, Creager MA, Libby P (2002) Diabetes and atherosclerosis: epidemiology, pathophysiology, and management. JAMA 287(19):2570–2581

Böger CA, Chen MH, Tin A, Olden M, Kottgen A, de Boer IH, Fuchsberger C, O'Seaghdha CM, Pattaro C, Teumer A, Liu CT, Glazer NL, Li M, O'Connell JR, Tanaka T, Peralta CA, Kutalik Z, Luan J, Zhao JH, Hwang SJ, Akylbekova E, Kramer H, van der Harst P, Smith AV, Lohman K, de Andrade M, Hayward C, Kollerits B, Tonjes A, Aspelund T, Ingelsson E, Eiriksdottir G, Launer LJ, Harris TB, Shuldiner AR, Mitchell BD, Arking DE, Franceschini N, Boerwinkle E, Egan J, Hernandez D, Reilly M, Townsend RR, Lumley T, Siscovick DS, Psaty BM, Kestenbaum B, Haritunians T, Bergmann S, Vollenweider P, Waeber G, Mooser V, Waterworth D, Johnson AD, Florez JC, Meigs JB, Lu X, Turner ST, Atkinson EJ, Leak TS, Aasarod K, Skorpen F, Syvanen AC, Illig T, Baumert J, Koenig W, Kramer BK, Devuyst O, Mychaleckyj JC, Minelli C, Bakker SJ, Kedenko L, Paulweber B, Coassin S, Endlich K, Kroemer HK, Biffar R, Stracke S, Volzke H, Stumvoll M, Magi R, Campbell H, Vitart V, Hastie ND, Gudnason V, Kardia SL, Liu Y, Polasek O, Curhan G, Kronenberg F, Prokopenko I, Rudan I, Arnlov J, Hallan S, Navis G, CKDGen C, Parsa A, Ferrucci L, Coresh J, Shlipak MG, Bull SB, Paterson NJ, Wichmann HE, Wareham NJ, Loos RJ, Rotter JI, Pramstaller PP, Cupples LA, Beckmann JS, Yang Q, Heid IM, Rettig R, Dreisbach AW, Bochud M, Fox CS, Kao WH (2011) CUBN is a gene locus for albuminuria. J Am Soc Nephrol 22(3):555–570

Borch-Johnsen K, Norgaard K, Hommel E, Mathiesen ER, Jensen JS, Deckert T, Parving HH (1992) Is diabetic nephropathy an inherited complication. Kidney Int 41(4):719–722

Bowden DW, Colicigno CJ, Langefeld CD, Sale MM, Williams A, Anderson PJ, Rich SS, Freedman BI (2004) A genome scan for diabetic nephropathy in African Americans. Kidney Int 66(4):1517–1526

Boyle AP, Hong EL, Hariharan M, Cheng Y, Schaub MA, Kasowski M, Karczewski KJ, Park J, Hitz BC, Weng S, Cherry JM, Snyder M (2012) Annotation of functional variation in personal genomes using RegulomeDB. Genome Res 22(9):1790–1797

Broadbent HM, Peden JF, Lorkowski S, Goel A, Ongen H, Green F, Clarke R, Collins R, Franzosi MG, Tognoni G, Seedorf U, Rust S, Eriksson P, Hamsten A, Farrall M, Watkins H (2008) Susceptibility to coronary artery disease and diabetes is encoded by distinct, tightly linked SNPs in the ANRIL locus on chromosome 9p. Hum Mol Genet 17(6):806–814

Brownlee M (1994) Lilly lecture 1993. Glycation and diabetic complications. Diabetes 43 (6):836–841

Campbell KN, Raij L, Mundel P (2011) Role of angiotensin II in the development of nephropathy and podocytopathy of diabetes. Curr Diabetes Rev 7(1):3–7

Ciccacci C, Di Fusco D, Cacciotti L, Morganti R, D'Amato C, Novelli G, Sangiuolo F, Spallone V, Borgiani P (2013) TCF7L2 gene polymorphisms and type 2 diabetes: association with diabetic retinopathy and cardiovascular autonomic neuropathy. Acta Diabetol 50(5):789–799

Coronary Artery Disease (C4D) Genetics Consortium 2011 (2011) A genome-wide association study in Europeans and South Asians identifies five new loci for coronary artery disease. Nat Genet 43(4):339–344

Craig DW, Millis MP, DiStefano JK (2009) Genome-wide SNP genotyping study using pooled DNA to identify candidate markers mediating susceptibility to end-stage renal disease attributed to Type 1 diabetes. Diabet Med 26(11):1090–1098

Deloukas P, Kanoni S, Willenborg C, Farrall M, Assimes TL, Thompson JR, Ingelsson E, Saleheen D, Erdmann J, Goldstein BA, Stirrups K, Konig IR, Cazier JB, Johansson A, Hall AS, Lee JY, Willer CJ, Chambers JC, Esko T, Folkersen L, Goel A, Grundberg E, Havulinna AS, Ho WK, Hopewell JC, Eriksson N, Kleber ME, Kristiansson K, Lundmark P, Lyytikainen LP, Rafelt S, Shungin D, Strawbridge RJ, Thorleifsson G, Tikkanen E, Van Zuydam N, Voight BF, Waite LL, Zhang W, Ziegler A, Absher D, Altshuler D, Balmforth AJ, Barroso I, Braund PS, Burgdorf C, Claudi-Boehm S, Cox D, Dimitriou M, Do R, Doney AS, El MN, Eriksson P, Fischer K, Fontanillas P, Franco-Cereceda A, Gigante B, Groop L, Gustafsson S, Hager J, Hallmans G, Han BG, Hunt SE, Kang HM, Illig T, Kessler T, Knowles JW, Kolovou G, Kuusisto J, Langenberg C, Langford C, Leander K, Lokki ML, Lundmark A, McCarthy MI, Meisinger C, Melander O, Mihailov E, Maouche S, Morris AD, Muller-Nurasyid M, Nikus K, Peden JF, Rayner NW, Rasheed A, Rosinger S, Rubin D, Rumpf MP, Schafer A, Sivananthan M, Song C, Stewart AF, Tan ST, Thorgeirsson G, van der Schoot CE, Wagner PJ, Wells GA, Wild PS, Yang TP, Amouyel P, Arveiler D, Basart H, Boehnke M, Boerwinkle E, Brambilla P, Cambien F, Cupples AL, de Faire U, Dehghan A, Diemert P, Epstein SE, Evans A, Ferrario MM, Ferrieres J, Gauguier D, Go AS, Goodall AH, Gudnason V, Hazen SL, Holm H, Iribarren C, Jang Y, Kahonen M, Kee F, Kim HS, Klopp N, Koenig W, Kratzer W, Kuulasmaa K, Laakso M, Laaksonen R, Lee JY, Lind L, Ouwehand WH, Parish S, Park JE, Pedersen NL, Peters A, Quertermous T, Rader DJ, Salomaa V, Schadt E, Shah SH, Sinisalo J, Stark K, Stefansson K, Tregouet DA, Virtamo J, Wallentin L, Wareham N, Zimmermann ME, Nieminen MS, Hengstenberg C, Sandhu MS, Pastinen T, Syvanen AC, Hovingh GK, Dedoussis G, Franks PW, Lehtimaki T, Metspalu A, Zalloua PA, Siegbahn A, Schreiber S, Ripatti S, Blankenberg SS, Perola M, Clarke R, Boehm BO, O'Donnell C, Reilly MP, Marz W, Collins R, Kathiresan S, Hamsten A, Kooner JS, Thorsteinsdottir U, Danesh J, Palmer CN, Roberts R, Watkins H, Schunkert H, Samani NJ (2013) Large-scale association analysis identifies new risk loci for coronary artery disease. Nat Genet 45(1):25–33

Deshmukh HA, Palmer CN, Morris AD, Colhoun HM (2013) Investigation of known estimated glomerular filtration rate loci in patients with type 2 diabetes. Diabet Med 30(10):1230–1235

Doria A, Wojcik J, Xu R, Gervino EV, Hauser TH, Johnstone MT, Nolan D, Hu FB, Warram JH (2008) Interaction between poor glycemic control and 9p21 locus on risk of coronary artery disease in type 2 diabetes. JAMA 300(20):2389–2397

Forsblom C, Harjutsalo V, Thorn LM, Waden J, Tolonen N, Saraheimo M, Gordin D, Moran JL, Thomas MC, Groop PH, FinnDiane Study Group (2011) Competing-risk analysis of ESRD and death among patients with type 1 diabetes and macroalbuminuria. J Am Soc Nephrol 22 (3):537–544

Freedman BI, Langefeld CD, Lu L, Divers J, Comeau ME, Kopp JB, Winkler CA, Nelson GW, Johnson RC, Palmer ND, Hicks PJ, Bostrom MA, Cooke JN, McDonough CW, Bowden DW (2011) Differential effects of MYH9 and APOL1 risk variants on FRMD3 Association with Diabetic ESRD in African Americans. PLoS Genet 7(6):e1002150

Fu YP, Hallman DM, Gonzalez VH, Klein BE, Klein R, Hayes MG, Cox NJ, Bell GI, Hanis CL (2010) Identification of diabetic retinopathy genes through a Genome-Wide Association Study among Mexican-Americans from Starr County, Texas. J Ophthalmol 2010:pii: 861291. doi:10. 1155/2010/861291, Epub 2010 Sep 2

Gassmann M, Casagranda F, Orioli D, Simon H, Lai C, Klein R, Lemke G (1995) Aberrant neural and cardiac development in mice lacking the ErbB4 neuregulin receptor. Nature 378 (6555):390–394

Gibson G (2012) Rare and common variants: twenty arguments. Nat Rev Genet 13(2):135–145

Grassi MA, Tikhomirov A, Ramalingam S, Below JE, Cox NJ, Nicolae DL (2011) Genome-wide meta-analysis for severe diabetic retinopathy. Hum Mol Genet 20(12):2472–2481

Groop PH, Thomas MC, Moran JL, Wadèn J, Thorn LM, Mäkinen VP, Rosengård-Bärlund M, Saraheimo M, Hietala K, Heikkilä O, Forsblom C, FinnDiane Study Group (2009) The presence and severity of chronic kidney disease predicts all-cause mortality in type 1 diabetes. Diabetes 58(7):1651

Haffner SM (1999) Epidemiological studies on the effects of hyperglycemia and improvement of glycemic control on macrovascular events in type 2 diabetes. Diabetes Care 22(Suppl 3):C54–C56

Haffner SM, Lehto S, Ronnemaa T, Pyorala K, Laakso M (1998) Mortality from coronary heart disease in subjects with type 2 diabetes and in nondiabetic subjects with and without prior myocardial infarction. N Engl J Med 339(4):229–234

Hallman DM, Huber JC Jr, Gonzalez VH, Klein BE, Klein R, Hanis CL (2005) Familial aggregation of severity of diabetic retinopathy in Mexican Americans from Starr County, Texas. Diabetes Care 28(5):1163–1168

Hannon GJ, Beach D (1994) p15INK4B is a potential effector of TGF-beta-induced cell cycle arrest. Nature 371(6494):257–261

Harjutsalo V, Katoh S, Sarti C, Tajima N, Tuomilehto J (2004) Population-based assessment of familial clustering of diabetic nephropathy in type 1 diabetes. Diabetes 53(9):2449–2454

He B, Österholm AM, Hoverfält A, Forsblom C, Hjörleifsdóttir EE, Nilsson AS, Parkkonen M, Pitkäniemi J, Hreidarsson A, Sarti C, McKnight AJ, Maxwell AP, Tuomilehto J, Groop PH, Tryggvason K (2009) Association of genetic variants at 3q22 with nephropathy in patients with type 1 diabetes mellitus. Am J Hum Genet 84(1):5–13

Hietala K, Forsblom C, Summanen P, Groop PH, FinnDiane Study Group (2008) Heritability of proliferative diabetic retinopathy. Diabetes 57(8):2176–2180

Hinks A, Eyre S, Ke X, Barton A, Martin P, Flynn E, Packham J, Worthington J, Childhood Arthritis Prospective Study, UKRAG Consortium, BSPAR Study Group, Thomson W (2010) Association of the AFF3 gene and IL2/IL21 gene region with juvenile idiopathic arthritis. Genes Immun 11(2):194–198

Hotaling JM, Waggott DR, Goldberg J, Jarvik G, Paterson AD, Cleary PA, Lachin J, Sarma A, Wessells H, DCCT/EDIC Research Group (2012) Pilot genome-wide association search identifies potential loci for risk of erectile dysfunction in type 1 diabetes using the DCCT/ EDIC study cohort. J Urol 188(2):514–520

Huang YC, Lin JM, Lin HJ, Chen CC, Chen SY, Tsai CH, Tsai FJ (2011) Genome-wide association study of diabetic retinopathy in a Taiwanese population. Ophthalmology 118 (4):642–648

Imperatore G, Hanson RL, Pettitt DJ, Kobes S, Bennett PH, Knowler WC (1998) Sib-pair linkage analysis for susceptibility genes for microvascular complications among Pima Indians with type 2 diabetes. Pima Diabetes Genes Group. Diabetes 47(5):821–830

Jarinova O, Stewart AF, Roberts R, Wells G, Lau P, Naing T, Buerki C, McLean BW, Cook RC, Parker JS, McPherson R (2009) Functional analysis of the chromosome 9p21.3 coronary artery disease risk locus. Arterioscler Thromb Vasc Biol 29(10):1671–1677

Jones N, Blasutig IM, Eremina V, Ruston JM, Bladt F, Li H, Huang H, Larose L, Li SS, Takano T, Quaggin SE, Pawson T (2006) Nck adaptor proteins link nephrin to the actin cytoskeleton of kidney podocytes. Nature 440(7085):818–823

Kadowaki T, Yamauchi T (2005) Adiponectin and adiponectin receptors. Endocr Rev 26 (3):439–451

Kamb A, Gruis NA, Weaver-Feldhaus J, Liu Q, Harshman K, Tavtigian SV, Stockert E, Day RS III, Johnson BE, Skolnick MH (1994) A cell cycle regulator potentially involved in genesis of many tumor types. Science 264(5157):436–440

Keating ST, El-Osta A (2013) Glycemic memories and the epigenetic component of diabetic nephropathy. Curr Diabetes Rep 13(4):574–581

Klein R (1987) The epidemiology of diabetic retinopathy: findings from the Wisconsin Epidemiologic Study of Diabetic Retinopathy. Int Ophthalmol Clin 27(4):230–238

Köttgen A, Glazer NL, Dehghan A, Hwang SJ, Katz R, Li M, Yang Q, Gudnason V, Launer LJ, Harris TB, Smith AV, Arking DE, Astor BC, Boerwinkle E, Ehret GB, Ruczinski I, Scharpf RB, Chen YD, de Boer IH, Haritunians T, Lumley T, Sarnak M, Siscovick D, Benjamin EJ, Levy D, Upadhyay A, Aulchenko YS, Hofman A, Rivadeneira F, Uitterlinden AG, van Duijn CM, Chasman DI, Pare G, Ridker PM, Kao WH, Witteman JC, Coresh J, Shlipak MG, Fox CS (2009) Multiple loci associated with indices of renal function and chronic kidney disease. Nat Genet 41(6):712–717

Köttgen A, Pattaro C, Boger CA, Fuchsberger C, Olden M, Glazer NL, Parsa A, Gao X, Yang Q, Smith AV, O'Connell JR, Li M, Schmidt H, Tanaka T, Isaacs A, Ketkar S, Hwang SJ, Johnson AD, Dehghan A, Teumer A, Pare G, Atkinson EJ, Zeller T, Lohman K, Cornelis MC, Probst-Hensch NM, Kronenberg F, Tonjes A, Hayward C, Aspelund T, Eiriksdottir G, Launer LJ, Harris TB, Rampersaud E, Mitchell BD, Arking DE, Boerwinkle E, Struchalin M, Cavalieri M, Singleton A, Giallauria F, Metter J, de Boer IH, Haritunians T, Lumley T, Siscovick D, Psaty BM, Zillikens MC, Oostra BA, Feitosa M, Province M, de Andrade M, Turner ST, Schillert A, Ziegler A, Wild PS, Schnabel RB, Wilde S, Munzel TF, Leak TS, Illig T, Klopp N, Meisinger C, Wichmann HE, Koenig W, Zgaga L, Zemunik T, Kolcic I, Minelli C, Hu FB, Johansson A, Igl W, Zaboli G, Wild SH, Wright AF, Campbell H, Ellinghaus D, Schreiber S, Aulchenko YS, Felix JF, Rivadeneira F, Uitterlinden AG, Hofman A, Imboden M, Nitsch D, Brandstatter A, Kollerits B, Kedenko L, Magi R, Stumvoll M, Kovacs P, Boban M, Campbell S, Endlich K, Volzke H, Kroemer HK, Nauck M, Volker U, Polasek O, Vitart V, Badola S, Parker AN, Ridker PM, Kardia SL, Blankenberg S, Liu Y, Curhan GC, Franke A, Rochat T, Paulweber B, Prokopenko I, Wang W, Gudnason V, Shuldiner AR, Coresh J, Schmidt R, Ferrucci L, Shlipak MG, van Duijn CM, Borecki I, Kramer BK, Rudan I, Gyllensten U, Wilson JF, Witteman JC, Pramstaller PP, Rettig R, Hastie N, Chasman DI, Kao WH, Heid IM, Fox CS (2010) New loci associated with kidney function and chronic kidney disease. Nat Genet 42(5):376–384

Krebs HA (1935) Metabolism of amino-acids: the synthesis of glutamine from glutamic acid and ammonia, and the enzymic hydrolysis of glutamine in animal tissues. Biochem J 29 (8):1951–1969

Krolewski AS, Warram JH (2005) Epidemiology of late complications of diabetes: a basis for the development and evaluation of preventive program. In: Kahn CR, Weir GC, King GL, Jacobson AM, Moses AC, Smith RJ (eds) Joslin's Diabetes Mellitus. Lippincott, Williams & Wilkins, New York

Laakso M (1999) Hyperglycemia and cardiovascular disease in type 2 diabetes. Diabetes 48 (5):937–942
Lange LA, Bowden DW, Langefeld CD, Wagenknecht LE, Carr JJ, Rich SS, Riley WA, Freedman BI (2002) Heritability of carotid artery intima-medial thickness in type 2 diabetes. Stroke 33 (7):1876–1881
Leak TS, Perlegas PS, Smith SG, Keene KL, Hicks PJ, Langefeld CD, Mychaleckyj JC, Rich SS, Kirk JK, Freedman BI, Bowden DW, Sale MM (2009) Variants in intron 13 of the ELMO1 gene are associated with diabetic nephropathy in African Americans. Ann Hum Genet 73 (2):152–159
Libby P, Plutzky J (2002) Diabetic macrovascular disease: the glucose paradox? Circulation 106 (22):2760–2763
Maeda S, Araki S, Babazono T, Toyoda M, Umezono T, Kawai K, Imanishi M, Uzu T, Watada H, Suzuki D, Kashiwagi A, Iwamoto Y, Kaku K, Kawamori R, Nakamura Y (2010a) Replication study for the association between four Loci identified by a genome-wide association study on European American subjects with type 1 diabetes and susceptibility to diabetic nephropathy in Japanese subjects with type 2 diabetes. Diabetes 59(8):2075–2079
Maeda S, Kobayashi MA, Araki S, Babazono T, Freedman BI, Bostrom MA, Cooke JN, Toyoda M, Umezono T, Tarnow L, Hansen T, Gaede P, Jorsal A, Ng DP, Ikeda M, Yanagimoto T, Tsunoda T, Unoki H, Kawai K, Imanishi M, Suzuki D, Shin HD, Park KS, Kashiwagi A, Iwamoto Y, Kaku K, Kawamori R, Parving HH, Bowden DW, Pedersen O, Nakamura Y (2010b) A single nucleotide polymorphism within the acetyl-coenzyme A carboxylase beta gene is associated with proteinuria in patients with type 2 diabetes. PLoS Genet 6(2):e1000842
Marenberg ME, Risch N, Berkman LF, Floderus B, De Faire U (1994) Genetic susceptibility to death from coronary heart disease in a study of twins. N Engl J Med 330(15):1041–1046
McDonough CW, Palmer ND, Hicks PJ, Roh BH, An SS, Cooke JN, Hester JM, Wing MR, Bostrom MA, Rudock ME, Lewis JP, Talbert ME, Blevins RA, Lu L, Ng MC, Sale MM, Divers J, Langefeld CD, Freedman BI, Bowden DW (2010) A genome-wide association study for diabetic nephropathy genes in African Americans. Kidney Int 79(5):563–572
McPherson R, Pertsemlidis A, Kavaslar N, Stewart A, Roberts R, Cox DR, Hinds DA, Pennacchio LA, Tybjaerg-Hansen A, Folsom AR, Boerwinkle E, Hobbs HH, Cohen JC (2007) A common allele on chromosome 9 associated with coronary heart disease. Science 316(5830):1488–1491
Melko M, Douguet D, Bensaid M, Zongaro S, Verheggen C, Gecz J, Bardoni B (2011) Functional characterization of the AFF (AF4/FMR2) family of RNA-binding proteins: insights into the molecular pathology of FRAXE intellectual disability. Hum Mol Genet 20(10):1873–1885
Meng W, Deshmukh HA, van Zuydam NR, Liu Y, Donnelly LA, Zhou K, Wellcome Trust Case Control Consortium 2 (WTCCC2), Surrogate Markers for Micro- and Macro-Vascular Hard Endpoints for Innovative Diabetes Tools (SUMMIT) Study Group, Morris AD, Colhoun HM, Palmer CN, Smith BH (2014) A genome-wide association study suggests an association of Chr8p21.3 (GFRA2) with diabetic neuropathic pain. Eur J Pain 19(3):392–399
Menzaghi C, Trischitta V, Doria A (2007) Genetic influences of adiponectin on insulin resistance, type 2 diabetes, and cardiovascular disease. Diabetes 56(5):1198–1209
Moczulski DK, Rogus JJ, Antonellis A, Warram JH, Krolewski AS (1998) Major susceptibility locus for nephropathy in type 1 diabetes on chromosome 3q: results of novel discordant sib-pair analysis. Diabetes 47(7):1164
Monastiriotis C, Papanas N, Trypsianis G, Karanikola K, Veletza S, Maltezos E (2013) The epsilon4 allele of the APOE gene is associated with more severe peripheral neuropathy in type 2 diabetic patients. Angiology 64(6):451–455
Mooyaart A, Valk EJJ, van Es L, Bruijn J, de Heer E, Freedman B, Dekkers O, Baelde H (2011) Genetic associations in diabetic nephropathy: a meta-analysis. Diabetologia 1–10
Morini E, Prudente S, Succurro E, Chandalia M, Zhang YY, Mammarella S, Pellegrini F, Powers C, Proto V, Dallapiccola B, Cama A, Sesti G, Abate N, Doria A, Trischitta V (2009) IRS1 G972R polymorphism and type 2 diabetes: a paradigm for the difficult ascertainment of

the contribution to disease susceptibility of 'low-frequency-low-risk' variants. Diabetologia 52 (9):1852–1857

Nishikawa T, Edelstein D, Du XL, Yamagishi S, Matsumura T, Kaneda Y, Yorek MA, Beebe D, Oates PJ, Hammes HP, Giardino I, Brownlee M (2000) Normalizing mitochondrial superoxide production blocks three pathways of hyperglycaemic damage. Nature 404(6779):787–790

Orchard TJ, Secrest AM, Miller RG, Costacou T (2010) In the absence of renal disease, 20 year mortality risk in type 1 diabetes is comparable to that of the general population: a report from the Pittsburgh Epidemiology of Diabetes Complications Study. Diabetologia 53 (11):2312–2319

Osterholm AM, He B, Pitkaniemi J, Albinsson L, Berg T, Sarti C, Tuomilehto J, Tryggvason K (2007) Genome-wide scan for type 1 diabetic nephropathy in the Finnish population reveals suggestive linkage to a single locus on chromosome 3q. Kidney Int 71(2):140–145

Paatero I, Elenius K (2008) ErbB4 and its isoforms: patentable drug targets? Recent Pat DNA Gene Seq 2(1):27–33

Parving HH, Smidt UM (1986) Hypotensive therapy reduces microvascular albumin leakage in insulin-dependent diabetic patients with nephropathy. Diabet Med 3(4):312–315

Pattaro C, Kottgen A, Teumer A, Garnaas M, Boger CA, Fuchsberger C, Olden M, Chen MH, Tin A, Taliun D, Li M, Gao X, Gorski M, Yang Q, Hundertmark C, Foster MC, O'Seaghdha CM, Glazer N, Isaacs A, Liu CT, Smith AV, O'Connell JR, Struchalin M, Tanaka T, Li G, Johnson AD, Gierman HJ, Feitosa M, Hwang SJ, Atkinson EJ, Lohman K, Cornelis MC, Johansson A, Tonjes A, Dehghan A, Chouraki V, Holliday EG, Sorice R, Kutalik Z, Lehtimaki T, Esko T, Deshmukh H, Ulivi S, Chu AY, Murgia F, Trompet S, Imboden M, Kollerits B, Pistis G, CARDIoGRAM Consortium, ICBP Consortium, CARe Consortium, Wellcome Trust Case Control Consortium 2 (WTCCC2), Harris TB, Launer LJ, Aspelund T, Eiriksdottir G, Mitchell BD, Boerwinkle E, Schmidt H, Cavalieri M, Rao M, Hu FB, Demirkan A, Oostra BA, de Andrade M, Turner ST, Ding J, Andrews JS, Freedman BI, Koenig W, Illig T, Doring A, Wichmann HE, Kolcic I, Zemunik T, Boban M, Minelli C, Wheeler HE, Igl W, Zaboli G, Wild SH, Wright AF, Campbell H, Ellinghaus D, Nothlings U, Jacobs G, Biffar R, Endlich K, Ernst F, Homuth G, Kroemer HK, Nauck M, Stracke S, Volker U, Volzke H, Kovacs P, Stumvoll M, Magi R, Hofman A, Uitterlinden AG, Rivadeneira F, Aulchenko YS, Polasek O, Hastie N, Vitart V, Helmer C, Wang JJ, Ruggiero D, Bergmann S, Kahonen M, Viikari J, Nikopensius T, Province M, Ketkar S, Colhoun H, Doney A, Robino A, Giulianini F, Kramer BK, Portas L, Ford I, Buckley BM, Adam M, Thun GA, Paulweber B, Haun M, Sala C, Metzger M, Mitchell P, Ciullo M, Kim SK, Vollenweider P, Raitakari O, Metspalu A, Palmer C, Gasparini P, Pirastu M, Jukema JW, Probst-Hensch NM, Kronenberg F, Toniolo D, Gudnason V, Shuldiner AR, Coresh J, Schmidt R, Ferrucci L, Siscovick DS, van Duijn CM, Borecki I, Kardia SL, Liu Y, Curhan GC, Rudan I, Gyllensten U, Wilson JF, Franke A, Pramstaller PP, Rettig R, Prokopenko I, Witteman JC, Hayward C, Ridker P, Parsa A, Bochud M, Heid IM, Goessling W, Chasman DI, Kao WH, Fox CS (2012) Genome-wide association and functional follow-up reveals new loci for kidney function. PLoS Genet 8(3):e1002584

Pezzolesi MG, Poznik GD, Mychaleckyj JC, Paterson AD, Barati MT, Klein JB, Ng DPK, Placha G, Canani LH, Bochenski J, Waggott D, Merchant ML, Krolewski B, Mirea L, Wanic K, Katavetin P, Kure M, Wolkow P, Dunn JS, Smiles A, Walker WH, Boright AP, Bull SB; DCCT/EDIC Research Group, Doria A, Rogus JJ, Rich SS, Warram JH, Krolewski AS (2009a) Genome-wide association scan for diabetic nephropathy susceptibility genes in type 1 diabetes. Diabetes 58(6):1403–1410

Pezzolesi MG, Katavetin P, Kure M, Poznik GD, Skupien J, Mychaleckyj JC, Rich SS, Warram JH, Krolewski AS (2009b) Confirmation of genetic associations at ELMO1 in the GoKinD collection supports its role as a susceptibility gene in diabetic nephropathy. Diabetes 58 (11):2698–2702

Pezzolesi MG, Skupien J, Mychaleckyj JC, Warram JH, Krolewski AS (2010) Insights to the genetics of diabetic nephropathy through a genome-wide association study of the GoKinD collection. Semin Nephrol 30(2):126–140

Pizzuti A, Frittitta L, Argiolas A, Baratta R, Goldfine ID, Bozzali M, Ercolino T, Scarlato G, Iacoviello L, Vigneri R, Tassi V, Trischitta V (1999) A polymorphism (K121Q) of the human glycoprotein PC-1 gene coding region is strongly associated with insulin resistance. Diabetes 48(9):1881–1884

Pomerantz J, Schreiber-Agus N, Liegeois NJ, Silverman A, Alland L, Chin L, Potes J, Chen K, Orlow I, Lee HW, Cordon-Cardo C, DePinho RA (1998) The Ink4a tumor suppressor gene product, p19Arf, interacts with MDM2 and neutralizes MDM2's inhibition of p53. Cell 92 (6):713–723

Prickett TD, Agrawal NS, Wei X, Yates KE, Lin JC, Wunderlich JR, Cronin JC, Cruz P, Rosenberg SA, Samuels Y (2009) Analysis of the tyrosine kinome in melanoma reveals recurrent mutations in ERBB4. Nat Genet 41(10):1127–1132

Prudente S, Scarpelli D, Chandalia M, Zhang YY, Morini E, Del Guerra S, Perticone F, Li R, Powers C, Andreozzi F, Marchetti P, Dallapiccola B, Abate N, Doria A, Sesti G, Trischitta V (2009) The TRIB3 Q84R polymorphism and risk of early-onset type 2 diabetes. J Clin Endocrinol Metab 94(1):190–196

Qi L, Doria A, Manson JE, Meigs JB, Hunter D, Mantzoros CS, Hu FB (2006) Adiponectin genetic variability, plasma adiponectin, and cardiovascular risk in patients with type 2 diabetes. Diabetes 55(5):1512–1516

Qi L, Qi Q, Prudente S, Mendonca C, Andreozzi F, di Pietro N, Sturma M, Novelli V, Mannino GC, Formoso G, Gervino EV, Hauser TH, Muehlschlegel JD, Niewczas MA, Krolewski AS, Biolo G, Pandolfi A, Rimm E, Sesti G, Trischitta V, Hu F, Doria A (2013) Association between a genetic variant related to glutamic acid metabolism and coronary heart disease in individuals with type 2 diabetes. JAMA 310(8):821–828

Rask-Madsen C, King GL (2005) Proatherosclerotic mechanisms involving protein kinase C in diabetes and insulin resistance. Arterioscler Thromb Vasc Biol 25(3):487–496

Reaven GM (1997) Banting lecture 1988. Role of insulin resistance in human disease. Nutrition 13 (1):65

Reichard P, Nilsson BY, Rosenqvist U (1993) The effect of long-term intensified insulin treatment on the development of microvascular complications of diabetes mellitus. N Engl J Med 329 (5):304–309

Rema M, Saravanan G, Deepa R, Mohan V (2002) Familial clustering of diabetic retinopathy in South Indian Type 2 diabetic patients. Diabet Med 19(11):910–916

Resnikoff S, Pascolini D, Etya'ale D, Kocur I, Pararajasegaram R, Pokharel GP, Mariotti SP (2004) Global data on visual impairment in the year 2002. Bull World Health Organ 82 (11):844–851

Rigat B, Hubert C, Alhenc-Gelas F, Cambien F, Corvol P, Soubrier F (1990) An insertion/deletion polymorphism in the angiotensin I-converting enzyme gene accounting for half the variance of serum enzyme levels. J Clin Invest 86(4):1343–1346

Rogus JJ, Poznik GD, Pezzolesi MG, Smiles AM, Dunn J, Walker W, Wanic K, Moczulski D, Canani L, Araki S, Makita Y, Warram JH, Krolewski AS (2008) High-density single nucleotide polymorphism genome-wide linkage scan for susceptibility genes for diabetic nephropathy in type 1 diabetes. Diabetes 57(9):2519

Sambo F, Malovini A, Sandholm N, Stavarachi M, Forsblom C, Makinen VP, Harjutsalo V, Lithovius R, Gordin D, Parkkonen M, Saraheimo M, Thorn LM, Tolonen N, Waden J, He B, Osterholm AM, Tuomilehto J, Lajer M, Salem RM, McKnight AJ, The GENIE Consortium, Tarnow L, Panduru NM, Barbarini N, Di Camillo B, Toffolo GM, Tryggvason K, Bellazzi R, Cobelli C, The FinnDiane Study Group, Groop PH (2014) Novel genetic susceptibility loci for diabetic end-stage renal disease identified through robust naive Bayes classification. Diabetologia 57(8):1611–1622

Sandholm N, Salem RM, McKnight AJ, Brennan EP, Forsblom C, Isakova T, McKay GJ, Williams WW, Sadlier DM, Makinen VP, Swan EJ, Palmer C, Boright AP, Ahlqvist E, Deshmukh HA, Keller BJ, Huang H, Ahola AJ, Fagerholm E, Gordin D, Harjutsalo V, He B, Heikkila O, Hietala K, Kyto J, Lahermo P, Lehto M, Lithovius R, Osterholm AM, Parkkonen M, Pitkaniemi J, Rosengard-Barlund M, Saraheimo M, Sarti C, Soderlund J, Soro-Paavonen A, Syreeni A, Thorn LM, Tikkanen H, Tolonen N, Tryggvason K, Tuomilehto J, Waden J, Gill GV, Prior S, Guiducci C, Mirel DB, Taylor A, Hosseini SM, DCCT/EDIC Research Group, Parving HH, Rossing P, Tarnow L, Ladenvall C, Alhenc-Gelas F, Lefebvre P, Rigalleau V, Roussel R, Tregouet DA, Maestroni A, Maestroni S, Falhammar H, Gu T, Mollsten A, Cimponeriu D, Ioana M, Mota M, Mota E, Serafinceanu C, Stavarachi M, Hanson RL, Nelson RG, Kretzler M, Colhoun HM, Panduru NM, Gu HF, Brismar K, Zerbini G, Hadjadj S, Marre M, Groop L, Lajer M, Bull SB, Waggott D, Paterson AD, Savage DA, Bain SC, Martin F, Hirschhorn JN, Godson C, Florez JC, Groop PH, Maxwell AP (2012) New susceptibility Loci associated with kidney disease in type 1 diabetes. PLoS Genet 8(9):e1002921

Sandholm N, McKnight AJ, Salem RM, Brennan EP, Forsblom C, Harjutsalo V, Makinen VP, McKay GJ, Sadlier DM, Williams WW, Martin F, Panduru NM, Tarnow L, Tuomilehto J, Tryggvason K, Zerbini G, Comeau ME, Langefeld CD, FIND Consortium, Godson C, Hirschhorn JN, Maxwell AP, Florez JC, Groop PH, FinnDiane Study Group and the GENIE Consortium (2013) Chromosome 2q31.1 associates with ESRD in women with type 1 diabetes. J Am Soc Nephrol 24(10):1537–1543

Sandholm N, Forsblom C, Makinen VP, McKnight AJ, Osterholm AM, He B, Harjutsalo V, Lithovius R, Gordin D, Parkkonen M, Saraheimo M, Thorn LM, Tolonen N, Waden J, Tuomilehto J, Lajer M, Ahlqvist E, Mollsten A, Marcovecchio ML, Cooper J, Dunger D, Paterson AD, Zerbini G, Groop L, Consortium SUMMIT, Tarnow L, Maxwell AP, Tryggvason K, Groop PH, FinnDiane Study Group (2014) Genome-wide association study of urinary albumin excretion rate in patients with type 1 diabetes. Diabetologia 57 (6):1143–1153

Saraheimo M, Forsblom C, Fagerudd J, Teppo AM, Pettersson-Fernholm K, Frystyk J, Flyvbjerg A, Groop PH, FinnDiane study group (2005) Serum adiponectin is increased in type 1 diabetic patients with nephropathy. Diabetes Care 28(6):1410–1414

Scherer PE, Williams S, Fogliano M, Baldini G, Lodish HF (1995) A novel serum protein similar to C1q, produced exclusively in adipocytes. J Biol Chem 270(45):26746–26749

Schunkert H, Konig IR, Kathiresan S, Reilly MP, Assimes TL, Holm H, Preuss M, Stewart AF, Barbalic M, Gieger C, Absher D, Aherrahrou Z, Allayee H, Altshuler D, Anand SS, Andersen K, Anderson JL, Ardissino D, Ball SG, Balmforth AJ, Barnes TA, Becker DM, Becker LC, Berger K, Bis JC, Boekholdt SM, Boerwinkle E, Braund PS, Brown MJ, Burnett MS, Buysschaert I, Carlquist JF, Chen L, Cichon S, Codd V, Davies RW, Dedoussis G, Dehghan A, Demissie S, Devaney JM, Diemert P, Do R, Doering A, Eifert S, Mokhtari NE, Ellis SG, Elosua R, Engert JC, Epstein SE, de Faire U, Fischer M, Folsom AR, Freyer J, Gigante B, Girelli D, Gretarsdottir S, Gudnason V, Gulcher JR, Halperin E, Hammond N, Hazen SL, Hofman A, Horne BD, Illig T, Iribarren C, Jones GT, Jukema JW, Kaiser MA, Kaplan LM, Kastelein JJ, Khaw KT, Knowles JW, Kolovou G, Kong A, Laaksonen R, Lambrechts D, Leander K, Lettre G, Li M, Lieb W, Loley C, Lotery AJ, Mannucci PM, Maouche S, Martinelli N, McKeown PP, Meisinger C, Meitinger T, Melander O, Merlini PA, Mooser V, Morgan T, Muhleisen TW, Muhlestein JB, Munzel T, Musunuru K, Nahrstaedt J, Nelson CP, Nothen MM, Olivieri O, Patel RS, Patterson CC, Peters A, Peyvandi F, Qu L, Quyyumi AA, Rader DJ, Rallidis LS, Rice C, Rosendaal FR, Rubin D, Salomaa V, Sampietro ML, Sandhu MS, Schadt E, Schafer A, Schillert A, Schreiber S, Schrezenmeir J, Schwartz SM, Siscovick DS, Sivananthan M, Sivapalaratnam S, Smith A, Smith TB, Snoep JD, Soranzo N, Spertus JA, Stark K, Stirrups K, Stoll M, Tang WH, Tennstedt S, Thorgeirsson G, Thorleifsson G, Tomaszewski M, Uitterlinden AG, van Rij AM, Voight BF, Wareham NJ, Wells GA, Wichmann HE, Wild PS, Willenborg C, Witteman JC, Wright BJ, Ye S, Zeller T,

Ziegler A, Cambien F, Goodall AH, Cupples LA, Quertermous T, Marz W, Hengstenberg C, Blankenberg S, Ouwehand WH, Hall AS, Deloukas P, Thompson JR, Stefansson K, Roberts R, Thorsteinsdottir U, O'Donnell CJ, McPherson R, Erdmann J, Samani NJ (2011) Large-scale association analysis identifies 13 new susceptibility loci for coronary artery disease. Nat Genet 43(4):333–338

Schwarz K, Simons M, Reiser J, Saleem MA, Faul C, Kriz W, Shaw AS, Holzman LB, Mundel P (2001) Podocin, a raft-associated component of the glomerular slit diaphragm, interacts with CD2AP and nephrin. J Clin Invest 108(11):1621–1629

Seaquist ER, Goetz FC, Rich S, Barbosa J (1989) Familial clustering of diabetic kidney disease. N Engl J Med 320(18):1161–1165

Sharma R, Prudente S, Andreozzi F, Powers C, Mannino G, Bacci S, Gervino EV, Hauser TH, Succurro E, Mercuri L, Goheen EH, Shah H, Trischitta V, Sesti G, Doria A (2011) The type 2 diabetes and insulin-resistance locus near IRS1 is a determinant of HDL cholesterol and triglycerides levels among diabetic subjects. Atherosclerosis 216(1):157–160

Shea S, Ottman R, Gabrieli C, Stein Z, Nichols A (1984) Family history as an independent risk factor for coronary artery disease. J Am Coll Cardiol 4(4):793–801

Sheu WH, Kuo JZ, Lee IT, Hung YJ, Lee WJ, Tsai HY, Wang JS, Goodarzi MO, Klein R, Klein BE, Ipp E, Lin SY, Guo X, Hsieh CH, Taylor KD, Fu CP, Rotter JI, Chen YD (2013) Genome-wide association study in a Chinese population with diabetic retinopathy. Hum Mol Genet 22 (15):3165–3173

Shih NY, Li J, Cotran R, Mundel P, Miner JH, Shaw AS (2001) CD2AP localizes to the slit diaphragm and binds to nephrin via a novel C-terminal domain. Am J Pathol 159 (6):2303–2308

Shimazaki A, Kawamura Y, Kanazawa A, Sekine A, Saito S, Tsunoda T, Koya D, Babazono T, Tanaka Y, Matsuda M, Kawai K, Iizumi T, Imanishi M, Shinosaki T, Yanagimoto T, Ikeda M, Omachi S, Kashiwagi A, Kaku K, Iwamoto Y, Kawamori R, Kikkawa R, Nakajima M, Nakamura Y, Maeda S (2005) Genetic variations in the gene encoding ELMO1 are associated with susceptibility to diabetic nephropathy. Diabetes 54(4):1171–1178

Sobrin L, Green T, Sim X, Jensen RA, Tai ES, Tay WT, Wang JJ, Mitchell P, Sandholm N, Liu Y, Hietala K, Iyengar SK, Family Investigation of Nephropathy and Diabetes-Eye Research Group, Brooks M, Buraczynska M, Van Zuydam N, Smith AV, Gudnason V, Doney AS, Morris AD, Leese GP, Palmer CN, Wellcome Trust Case Control Consortium 2, Swaroop A, Taylor HA, Jr Wilson JG, Penman A, Chen CJ, Groop PH, Saw SM, Aung T, Klein BE, Rotter JI, Siscovick DS, Cotch MF, Klein R, Daly MJ, Wong TY (2011) Candidate gene association study for diabetic retinopathy in persons with type 2 diabetes: the Candidate gene Association Resource (CARe). Invest Ophthalmol Vis Sci 52(10):7593–7602

Soccio T, Zhang YY, Bacci S, Mlynarski W, Placha G, Raggio G, Di Paola R, Marucci A, Johnstone MT, Gervino EV, Abumrad NA, Klein S, Trischitta V, Doria A (2006) Common haplotypes at the adiponectin receptor 1 (ADIPOR1) locus are associated with increased risk of coronary artery disease in type 2 diabetes. Diabetes 55(10):2763–2770

Stamler J, Vaccaro O, Neaton JD, Wentworth D (1993) Diabetes, other risk factors, and 12-yr cardiovascular mortality for men screened in the Multiple Risk Factor Intervention Trial. Diabetes Care 16(2):434–444

Stoner M, Wang F, Wormke M, Nguyen T, Samudio I, Vyhlidal C, Marme D, Finkenzeller G, Safe S (2000) Inhibition of vascular endothelial growth factor expression in HEC1A endometrial cancer cells through interactions of estrogen receptor alpha and Sp3 proteins. J Biol Chem 275 (30):22769–22779

Stoner M, Wormke M, Saville B, Samudio I, Qin C, Abdelrahim M, Safe S (2004) Estrogen regulation of vascular endothelial growth factor gene expression in ZR-75 breast cancer cells through interaction of estrogen receptor alpha and SP proteins. Oncogene 23(5):1052–1063

Tarnow L, Groop PH, Hadjadj S, Kazeem G, Cambien F, Marre M, Forsblom C, Parving HH, Tregouet D, Thevard A, Farrall M, Gut I, Gauguier D, Cox R, Matsuda F, Lathrop M, Vionnet N, EURAGEDIC Consortium (2008) European rational approach for the genetics of diabetic complications--EURAGEDIC: patient populations and strategy. Nephrol Dial Transplant 23(1):161–168

Tavakkoly-Bazzaz J, Amoli MM, Pravica V, Chandrasecaran R, Boulton AJ, Larijani B, Hutchinson IV (2010) VEGF gene polymorphism association with diabetic neuropathy. Mol Biol Rep 37(7):3625–3630

The DCCT Research Group (1997) Clustering of long-term complications in families with diabetes in the diabetes control and complications trial. Diabetes 46(11):1829–1839

Tidcombe H, Jackson-Fishert A, Mathers K, Stern DF, Gassmann M, Golding JP (2003) Neural and mammary gland defects in Erbb4 knockout mice genetically rescued from embryonic lethality. Proc Natl Acad Sci U S A 100(14):8281–8286

Todd JA, Walker NM, Cooper JD, Smyth DJ, Downes K, Plagnol V, Bailey R, Nejentsev S, Field SF, Payne F, Lowe CE, Szeszko JS, Hafler JP, Zeitels L, Yang JH, Vella A, Nutland S, Stevens HE, Schuilenburg H, Coleman G, Maisuria M, Meadows W, Smink LJ, Healy B, Burren OS, Lam AA, Ovington NR, Allen J, Adlem E, Leung HT, Wallace C, Howson JM, Guja C, Ionescu-Tirgoviste C, Genetics of Type 1 Diabetes in Finland, Simmonds MJ, Heward JM, Gough SC, Wellcome Trust Case Control C, Dunger DB, Wicker LS, Clayton DG (2007) Robust associations of four new chromosome regions from genome-wide analyses of type 1 diabetes. Nat Genet 39(7):857–864

Tolonen N, Forsblom C, Thorn L, Waden J, Rosengard-Barlund M, Saraheimo M, Feodoroff M, Makinen VP, Gordin D, Taskinen MR, Groop PH, FinnDiane Study Group (2009) Lipid abnormalities predict progression of renal disease in patients with type 1 diabetes. Diabetologia 52(12):2522–2530

Tong Z, Yang Z, Patel S, Chen H, Gibbs D, Yang X, Hau VS, Kaminoh Y, Harmon J, Pearson E, Buehler J, Chen Y, Yu B, Tinkham NH, Zabriskie NA, Zeng J, Luo L, Sun JK, Prakash M, Hamam RN, Tonna S, Constantine R, Ronquillo CC, Sadda S, Avery RL, Brand JM, London N, Anduze AL, King GL, Bernstein PS, Watkins S; Genetics of Diabetes and Diabetic Complication Study Group, Jorde LB, Li DY, Aiello LP, Pollak MR, Zhang K (2008) Promoter polymorphism of the erythropoietin gene in severe diabetic eye and kidney complications. Proc Natl Acad Sci 105(19):6998

UK Prospective Diabetes Study (UKPDS) Group (1998) Intensive blood-glucose control with sulphonylureas or insulin compared with conventional treatment and risk of complications in patients with type 2 diabetes (UKPDS 33). UK Prospective Diabetes Study (UKPDS) Group. Lancet 352(9131):837–853

Veikkolainen V, Naillat F, Railo A, Chi L, Manninen A, Hohenstein P, Hastie N, Vainio S, Elenius K (2012) ErbB4 modulates tubular cell polarity and lumen diameter during kidney development. J Am Soc Nephrol 23(1):112–122

Vionnet N, Tregouet D, Kazeem G, Gut I, Groop PH, Tarnow L, Parving HH, Hadjadj S, Forsblom C, Farrall M, Gauguier D, Cox R, Matsuda F, Heath S, Thevard A, Rousseau R, Cambien F, Marre M, Lathrop M (2006) Analysis of 14 candidate genes for diabetic nephropathy on chromosome 3q in European populations: strongest evidence for association with a variant in the promoter region of the adiponectin gene. Diabetes 55(11):3166–3174

Wagenknecht LE, Bowden DW, Carr JJ, Langefeld CD, Freedman BI, Rich SS (2001) Familial aggregation of coronary artery calcium in families with type 2 diabetes. Diabetes 50 (4):861–866

Warram JH, Kopczynski J, Janka HU, Krolewski AS (1997) Epidemiology of non-insulin-dependent diabetes mellitus and its macrovascular complications. A basis for the development of cost-effective programs. Endocrinol Metab Clin North Am 26(1):165–188

Wessman M, Forsblom C, Kaunisto MA, Soderlund J, Ilonen J, Sallinen R, Hiekkalinna T, Parkkonen M, Maxwell AP, Tarnow L, Parving HH, Hadjadj S, Marre M, Peltonen L, Groop PH, FinnDiane Study Group (2011) Novel susceptibility locus at 22q11 for diabetic nephropathy in type 1 diabetes. PLoS ONE 6(9):e24053

Wild SH, Dunn CJ, McKeigue PM, Comte S (1999) Glycemic control and cardiovascular disease in type 2 diabetes: a review. Diabetes Metab Res Rev 15(3):197–204

Williams WW, Salem RM, McKnight AJ, Sandholm N, Forsblom C, Taylor A, Guiducci C, McAteer JB, McKay GJ, Isakova T, Brennan EP, Sadlier DM, Palmer C, Soderlund J, Fagerholm E, Harjutsalo V, Lithovius R, Gordin D, Hietala K, Kyto J, Parkkonen M, Rosengard-Barlund M, Thorn L, Syreeni A, Tolonen N, Saraheimo M, Waden J, Pitkaniemi J, Sarti C, Tuomilehto J, Tryggvason K, Osterholm AM, He B, Bain S, Martin F, Godson C, Hirschhorn JN, Maxwell AP, Groop PH, Florez JC, GENIE Consortium (2012) Association testing of previously reported variants in a large case–control meta-analysis of diabetic nephropathy. Diabetes 61(8):2187–2194

Woroniecka KI, Park AS, Mohtat D, Thomas DB, Pullman JM, Susztak K (2011) Transcriptome analysis of human diabetic kidney disease. Diabetes 60(9):2354–2369

Yoshida K, Hirokawa J, Tagami S, Kawakami Y, Urata Y, Kondo T (1995) Weakened cellular scavenging activity against oxidative stress in diabetes mellitus: regulation of glutathione synthesis and efflux. Diabetologia 38(2):201–210

Zheng XL, Yuan SG, Peng DQ (2007) Phenotype-specific inhibition of the vascular smooth muscle cell cycle by high glucose treatment. Diabetologia 50(4):881–890

Part II
Interpretation and Data Integration

Chapter 8
Transferability Across Ethnic Groups

Richa Saxena and Nicholette D. Palmer

Abstract Type 2 diabetes (T2D) is a common complex disease increasing in prevalence worldwide. While studies in populations of European ancestry predominated early gene discovery efforts, extension of genome-wide association studies (GWAS) to multiple ethnic groups has identified additional novel loci and insights based on differences in allele frequency, physiologic effect, and evolutionary history between populations. These studies highlight the value of global genetic studies in diverse populations. Transferability studies of T2D, and to a limited extent related traits, are now under way across diverse ethnic groups and have revealed (a) consistent effects for several T2D loci across most ethnic groups implying shared underlying causal variants; (b) allelic heterogeneity at several loci, validating a critical role for these loci in disease risk; (c) the utility of studies in different ethnic groups, especially of those with recent African ancestry, in fine mapping of genetic associations based on regional differences in linkage disequilibrium (LD); (d) heterogeneity of effect between ethnic groups at some T2D loci; and (e) population-specific effects. Discovery and transferability studies across ethnic groups promise to be integral in advancing our understanding of the genetic basis of T2D and providing insights into differences in prevalence and physiology of disease between ethnic groups. Multiethnic genetic studies for T2D are critical

R. Saxena (✉)
Center for Human Genetic Research, Massachusetts General Hospital, Boston, MA, USA

Department of Anesthesia, Critical Care and Pain Medicine, Massachusetts General Hospital, Boston, MA, USA

Department of Anesthesia, Harvard Medical School, Boston, MA, USA

Program in Medical and Population Genetics, Broad Institute, Cambridge, MA, USA
e-mail: rsaxena@broadinstitute.org

N.D. Palmer
Department of Biochemistry, Wake Forest School of Medicine, Winston-Salem, NC, USA

Center for Genomics and Personalized Medicine Research, Wake Forest School of Medicine, Winston-Salem, NC, USA

Center for Diabetes Research, Wake Forest School of Medicine, Winston-Salem, NC, USA

Center for Public Health Genomics, Wake Forest School of Medicine, Winston-Salem, NC, USA

J.C. Florez (ed.), *The Genetics of Type 2 Diabetes and Related Traits*,
DOI 10.1007/978-3-319-01574-3_8

for addressing health disparities and developing preventive and therapeutic strategies to reduce disease burden.

8.1 Introduction

The prevalence of diabetes is estimated at 6.4 % worldwide, with 3.8 % in Africa, 4.7 % in the West Pacific region including East Asia, 6.6 % in South and Central America, 6.9 % in Europe, 7.6 % in South Asia, 9.3 % in the Middle East, and 10.2 % in North America (Shaw et al. 2010). Between 2010 and 2030, there is expected to be a 69 % increase in the number of adults with diabetes in developing countries and a 20 % increase in developed countries, underscoring the need for a global understanding of the etiology of diabetes.

In the USA, compared to a prevalence of 7.1 % in non-Hispanic whites, Asian Americans (8.4 %), Hispanics (11.8 %), and African-Americans (12.6 %) are all at a higher risk for developing T2D. Yet despite a stronger demonstrated genetic component to disease in Hispanics ($h^2 = 53$ %; Duggirala et al. 1999) and African-Americans ($h^2 = 60$ %; Duggirala et al. 1999) as compared to Europeans ($h^2 = 25$ %; Almgren et al. 2011), studies of these populations have been limited. Increased prevalence of disease coupled with a stronger genetic basis emphasizes the importance of genetic studies in minority populations. Studies of T2D in multiple ethnic groups have the potential to refine association signals at established T2D susceptibility loci and identify novel pathways and variants that contribute to disease.

8.2 Transferability of Established T2D Loci

8.2.1 East Asians

T2D in East Asians (including Japanese, Chinese, Korean, and Southeast Asian populations) is often characterized by earlier onset and lower body mass index (BMI) than in Europeans. Whether genetic differences between East Asians and Europeans underlie this differential phenotypic manifestation is still unknown, but analyses of cross ethnic transferability of loci using recent large-scale GWAS are beginning to shed light on genetic similarities and differences.

Approximately 30 % of the novel T2D genetic loci discovered to date are from GWAS in East Asians. In 2010, two independent GWAS in individuals of Japanese ancestry identified novel variants in the potassium voltage-gated channel gene *KCNQ1* (Yasuda et al. 2008; Unoki et al. 2008). Subsequent GWAS identified two novel loci (*UBE2E2* and *C2CD4A-C2CD4B*) in Japanese (Yamauchi et al. 2010) and four loci (*PTPRD*, *SRR*, *SPRY2*, and *CDC123/CAMK1D)* in Han Chinese (Tsai et al. 2010; Shu et al. 2010). The largest East Asian study to date, the

AGEN-T2D consortium meta-analysis of GWAS (6952 cases/11,865 controls), identified eight additional novel signals (*PSMD6*, *MAEA*, *ZFAND3*, *KCNK16*, *GCC1/PAX4*, *GLIS3*, *PEPD*, and *FITM2/R3HDML/HNF4A*; Cho et al. 2012a). Most recently, two Chinese GWAS reported three novel loci (*GRK5*, *RASGRP*, and *PAX4;* Ma et al. 2013; Li et al. 2013), and a Japanese GWAS reported three additional novel loci (*MIR129-LEP*, *GPSM1*, and *SLC16A13;* Hara et al. 2014).

East Asian GWAS have augmented gene discovery in Europeans by revealing novel associations at variants with increased frequency in Asians as compared to Europeans, thus improving power to detect true associations that escaped notice in European populations. Interestingly, East Asian T2D GWAS have also discovered new signals at previously established European T2D loci (e.g. *CDC123/CAMK1D and CDKAL1)* revealing allelic heterogeneity. A novel T2D signal overlapping with a previously established European fasting glucose locus (*GLIS3*) was also found, confirming the role of this fasting glucose locus in the pathogenesis of T2D.

Numerous studies have examined the transferability of index T2D SNPs, i.e., the first reported variant, across European and East Asian populations, with a recent comprehensive survey evaluating 47 SNPs (Cho et al. 2012b; Voight et al. 2010). Transferability was observed for 24 associations (20 loci originally discovered in Europeans and 4 in East Asians), of which most surpassed the genome-wide significance threshold ($P < 5 \times 10^{-8}$) in both populations. Sixteen additional variants were associated in Europeans only and not in East Asians. Of these, the lack of transferability of seven signals could be explained by differences in allele frequency between the two ethnicities; these alleles were rare or monomorphic in East Asians. Conversely, seven variants were associated in East Asians only and did not show consistent effects in Europeans despite excellent power and high allele frequency. Whether this apparent nonreplication between populations is because of LD differences between index SNP and causal variant(s) between the ethnic groups, insufficient power, or because of true nontransferability remains to be determined. Notably, regional variants at three of the East Asian signals that do not show index SNP transferability (*UBE2E2*, *GLIS3*, and *CAMK1D/CDC23)* show nominal associations in large-scale European T2D meta-analysis; therefore, differences in LD at these loci should be explored. A recent large trans-ethnic meta-analysis reports ancestry-specific effects at two loci: *PEPD1*, with an association only in samples of East Asian ancestry, and *KLF14*, with an association only in samples of European ancestry (Diabetes Genetics Replication Meta-analysis Consortium 2014).

The transferability of most European loci to East Asians also extends to diabetes-related quantitative traits, including fasting, post-load glucose and insulin measures, and derived indices of insulin secretion and sensitivity. A recent trans-ethnic fine-mapping study in a Chinese population evaluated SNPs at 50 primarily European loci for T2D and related glycemic traits using the Metabochip, an SNP array specifically designed for fine mapping of metabolic and cardiovascular loci. Remarkable consistency across ethnicities (replication at 38/50 loci) and multiple novel, independent signals were found, highlighting allelic heterogeneity across these populations both in T2D and related traits (Kuo et al. 2013). Functional follow-up of allelic variants should reveal common biological insights as well as those relevant to disease in each population. We expect that imminent fine-mapping efforts using targeted arrays such as the Metabochip will enable refinement of trans-ethnic associations and assist in defining causal variants underlying T2D associations.

8.2.2 South Asians

Like East Asians, South Asians have an earlier onset of T2D at lower BMI than Europeans and, in addition, a faster progression to cardiovascular and renal complications. T2D GWAS in globally assembled individuals of South Asian ancestry have identified six loci (*GRB14*, *ST6GAL1*, *VPS26A*, *HMG20A*, *AP3S2*, and *HNF4A)* (Kooner et al. 2011) for T2D. In addition, GWAS in northern and southern Indian populations have found another two novel loci (*SGCG*, *TMEM163*; Tabassum et al. 2013; Saxena et al. 2013) that merit additional replication.

Transferability of European loci to South Asians has also been addressed in multiple GWAS, with a large proportion of European association signals (>70 %) found to be consistent in direction and magnitude of effect in all three South Asian GWAS (Kooner et al. 2011; Saxena et al. 2013; Tabassum et al. 2013), but genome-wide significance observed only at the transcription factor 7-like 2 (*TCF7L2*) locus. Index SNPs from six South Asian loci are associated in both European and East Asian populations (Cho et al. 2012b; Kooner et al. 2011), but a northern Indian variant in the gene encoding sarcoglycan (*SGCG*) does not replicate in other South or East Asian populations and is absent from Europeans, raising the possibility of an untyped mutation contributing a population-specific effect (Saxena et al. 2013).

8.2.3 Arabs

GWAS for T2D or related traits have not yet been reported in Arabs, but initial studies have examined the transferability of individual SNPs at European GWAS loci and suggest considerable genetic overlap (Almawi et al. 2013; Cauchi et al. 2012). In the largest study to date evaluating 44 SNPs from 37 European loci in a Moroccan Arab population (1193 cases/1055 controls) from North Africa, 15 loci were nominally associated with a consistent direction of effect and, of these, 13 were further confirmed in Tunisians (1446 cases/942 controls). Importantly, replicated associations included those with the strongest effects (*TCF7L2*, *CDKN2A*/*CDKN2B*, and *IGF2BP2*) with similar allele frequencies and effect sizes as in Europeans (Cauchi et al. 2012). A concordant direction of effect, with differences in risk allele frequency and/or regional LD structure between the Moroccan and European populations, was reported for most remaining variants. Nominal replication of index SNPs from eight loci has also been reported in Lebanese Arabs (Almawi et al. 2013), but larger studies with greater coverage at T2D loci in Arab populations are warranted to more completely evaluate transferability. The high levels of consanguinity in many Arab populations may also aid novel gene discovery for T2D and related traits by GWAS and familial sequencing.

8.2.4 African-Americans

African-Americans represent one of the most understudied populations in the GWAS era with only one published evaluation (Palmer et al. 2012). However, parallel with ongoing discovery efforts, several groups have evaluated the transferability of established T2D loci. Among these reports, a recent study utilizing electronic medical records evaluated GWAS-identified T2D loci in a large African-American cohort (1554 cases/2734 controls) from the southeastern United States (Long et al. 2012). Among the SNPs evaluated, >75 % were directionally consistent, with replication of multiple loci especially among larger effect variants in genes such as insulin-like growth factor 2 (*IGF2BP2*) and Wolfram syndrome 1 (*WFS1*) where this study was well powered to observe association. *TCF7L2*, which has one of the largest effects observed across studies, replicated and attained genome-wide significance.

To further explore the fine-scale allelic architecture of this population, more sophisticated analyses have examined locus transferability. These methods seek to overcome limitations to replication arising from differential LD patterns, effect sizes, and allelic heterogeneity while concurrently using these features to fine-map loci to identify causal variants. Using a large meta-analysis from six African-American cohorts (2806 cases/4265 controls), seven of 41 T2D susceptibility index variants replicated with the strongest association observed at *TCF7L2* (Ng et al. 2013). Examination of locus transferability identified significant associations and differential LD with reported susceptibility variants observed at *KLF14*, *HMGA2*, and *KCNQ1*. Interestingly at the Kruppel-like factor 14 (*KLF14*) gene, the strongest signal was observed 38 kb proximal to the index variant, and while these variants were in complete LD in Caucasian populations, only weak correlation was observed in African-Americans with the newly identified SNP accounting for the residual association at the index variant.

In addition to transferability of T2D loci, quantitative intermediate phenotypes of disease that more directly assess glucose homeostasis have been evaluated for effect in the African-American population (5984 individuals) (Liu et al. 2012). All 18 index variants evaluated displayed a consistent direction of effect with prior reports in Europeans with four loci significantly associated with glycemic traits. With the exception of *MTNR1B*, all other variants displayed differential LD between the index and "best" SNP suggesting multiple signals or, more likely, shared haplotype structure with the causal variant. This is exemplified with the gene encoding glucose-6-phosphatase (*G6PC2*). The index SNP and the "best" SNP from this study are correlated with a known functional variant in their respective populations, European and African-American, but not with one another potentially giving rise to the novel signal observed.

These studies take advantage of the ancestry-rich accumulation of genetic variation in this minority population, fine-scale allelic architecture, and power afforded from quantitative trait analysis while not suffering from the limitations of disease classification. These efforts have aided attempts to refine association signals from European populations and provided evidence of shared genetic influences in the African-American population.

8.2.5 Latinos

Latinos are a US minority population with variable ancestral genetic contributions from Europeans, Native Americans, and Africans. This population experiences a disproportionate increase in prevalence with T2D incidence at younger ages (Pleis and Lucas 2009). Moreover, prevalence rates differ based on country of birth with individuals from Mexico having higher rates than those from Central America.

Published studies of T2D, to date, have focused on Latinos from the southern United States and Mexico (Hayes et al. 2007; Parra et al. 2011). Exploration of the genetic ancestry of a Mexico City population identified strong contributions from American Indians (63 %), which varied between cases and controls, and Europeans (34 %) with nominal influence from West Africans (3 %). Signals of association were observed in previously identified loci including *HNF1A* and *KCNQ1*, derived from European and East Asian populations, respectively, and more nominal associations in *CDKN2A/B* and *IGF2BP2*. Notably at the *KCNQ1* locus, studies in Japanese and Europeans have identified two distinct, uncorrelated signals, of which the variant identified in Japanese was associated despite the reduced frequency observed in the American Indian population.

Extension of these findings to quantitative intermediate phenotypes has been facilitated through population-based studies limited to T2D index variants. Among the early evaluations of European-derived susceptibility variants, 17 have been evaluated in a cohort of Hispanic Americans with comprehensive metabolic testing of diabetes-related quantitative traits (Palmer et al. 2008). The most striking association was observed at the *CDKAL1* locus with reduced insulin secretion. Additional nominal associations with consistent direction of effect were observed between *SLC30A8* and *IGF2BP2* and insulin signaling and response as measured by the disposition index. Subsequent evaluation of *TCF7L2* in this (Palmer et al. 2008) and other studies (Watanabe et al. 2007) has demonstrated association with reduced insulin response.

With a renewed focus on the utility of genetic mapping in minority populations, new efforts are under way to extend these findings. More recently, a GWAS of Mexicans was published by the Slim Initiative in Genomic Medicine for the Americas (SIGMA) T2D Consortium (SIGMA Type 2 Diabetes Consortium et al. 2014) that demonstrates progress toward amassing larger numbers of samples to evaluate T2D. The key finding from that effort is discussed in detail in Sect. 8.3.5. Additional studies such as the Genetics UndeRlying DIAbetes in

HispaNics (GUARDIAN) consortium (Goodarzi et al. 2014) will evaluate the physiological metabolic phenotypes of disease. As has been demonstrated (SIGMA Type 2 Diabetes Consortium et al. 2014), these studies have the potential to uncover novel variants involved in T2D and related traits.

8.2.6 American Indians

American Indians display the highest T2D prevalence rates among minority groups in the United States with an estimated 14.2 % of the population diagnosed with disease and increasing to as much as 33.5 % among those living in southern Arizona (Centers for Disease Control and Prevention 2011). The high prevalence rate of disease and strong familial aggregation (Knowler et al. 1990) coupled with limited genetic and environmental variability makes the Pima Indians from southern Arizona an ideal population for the identification of the genetic determinants of T2D.

Early GWAS of T2D in the Pima Indians were inconclusive, with no association of established T2D loci identified primarily from European populations (Hanson et al. 2007). *TCF7L2*, one of the most widely replicated loci for T2D across multiple ethnicities, failed to show association with T2D even following a more comprehensive evaluation (Guo et al. 2007). Moreover, the protective allele in Europeans was associated with increased adiposity potentially mitigating its protective effects. Apparent nontransferability of additional established T2D loci, including *CDKAL1*, *SLC30A8*, and *IGFBP2*, has been subsequently evaluated with additional, focused genotyping (Rong et al. 2009) yielding estimated effects that were consistent with previous publications. Notably, nominal association of *CDKAL1* and *HHEX* was observed with reduced insulin response, a key predictor of T2D. Taken together, the results argue for additional genetic studies in a population disproportionately affected by disease with limited genetic and environmental variability. This will aid in the conclusive evaluation of previously identified loci and identify novel loci that contribute to the highest rates of T2D observed in a US minority population.

8.3 Studies in Diverse Ethnic Groups: Understanding the Genetic Architecture of T2D

8.3.1 Allele Frequency Differences Facilitate Gene Mapping and Fine Mapping: KCNQ1, CDKAL1, CDC/CAMKD1, *and* TCF7L2

Population differences in disease allele frequencies influence the power to detect genetic effects, as illustrated for *KCNQ1*, the first East Asian T2D locus identified independently by two Japanese GWAS in 2008 (Unoki et al. 2008; Yasuda et al. 2008). Both groups used East Asian samples of Chinese and Korean or Singaporean ancestry and European replication datasets to establish that this association was trans-ethnic and not previously detected in large European-ancestry studies due to a much lower risk allele frequency (5 vs. 40 %). GWAS meta-analysis subsequently confirmed association of *KCNQ1* variants in Europeans but at nominal significance levels and identified an independent, common variant in Europeans that reached genome-wide significance (Voight et al. 2010). This second *KCNQ1* signal is in turn less frequent in East Asians (15 vs. 48 %) and has only been nominally replicated in recent East Asian meta-analyses. More recent GWAS in East Asians have identified additional signals at *CDKAL1* (Kuo et al. 2013) and *CDC/CAMKD1* (Shu et al. 2010), pointing to allelic heterogeneity at these loci across populations and confirming a central role for these loci in T2D pathophysiology.

Population differences in LD help to fine-map genetic associations as demonstrated for SNPs at *TCF7L2* with the strongest effect on T2D across most ethnic groups. To refine the association between three highly correlated markers within a 64 kb region of strong LD in Europeans, researchers genotyped these markers in a West African T2D case–control population and found robust replication only of the rs7903146 variant, while association to the other two markers was absent or much weaker (Helgason et al. 2007). More recently, resequencing and fine mapping in a 4.3 kb region of strong LD encompassing this variant in African-Americans (Palmer et al. 2011) points to this variant as the most likely causal variant.

8.3.2 Nontransferability of T2D Association in East Asians Despite Associations with T2D-Related Quantitative Traits: MTNR1B *and* GCK

Two examples of European T2D loci that appear to have diminished T2D effects in East Asians, despite higher allele frequencies, are the lead SNPs at the *MTNR1B* (Wang et al. 2013; Cho et al. 2012a, b) and *GCK* (Wang et al. 2013) loci. Melatonin receptor 1B (*MTNR1B*) rs10830963 is consistently associated with fasting glucose and multiple measures of insulin secretion in studies of nondiabetic individuals from China (Hu et al. 2010; Song et al. 2011; Tam et al. 2010) demonstrating transferability across diabetes-related quantitative traits to East Asians. In locus-transferability analyses, this SNP also shows the most robust replication for fasting glucose in African-Americans (Liu et al. 2012). These results suggest a role for this SNP as the causal variant or as highly correlated with the causal variant at this locus. Despite these trans-ethnic associations to fasting glucose, no association has consistently been observed for T2D in East Asian populations in either SNP-based or locus-based analyses (Xia et al. 2012; Cho et al. 2012b; Wang et al. 2013), raising the possibility that the physiology, genetic background, or environmental conditions in this ethnic group may diminish or abrogate the genetic effect. Interestingly, a similar profile is observed for variants in the glucokinase (*GCK*) gene, with demonstrated association to fasting glucose but not T2D in East Asian populations (Wang et al. 2013). Recently, physiologic analyses of T2D loci demonstrated that *MTNR1B* and *GCK* variants form a cluster that acts through reduced insulin secretion and fasting hyperglycemia, likely through different mechanisms (Dimas et al. 2013) but currently, the basis or significance of this differential impact on T2D is unknown.

8.3.3 BMI-Independent Effect of FTO *on T2D*

In 2007, the fat mass- and obesity-associated (*FTO*) gene was found to contribute to the risk of T2D through BMI in a GWAS of European populations (Frayling et al. 2007). The strong association between *FTO* SNPs and T2D was completely abolished by adjusting for BMI. Since then, association of *FTO* with BMI and waist circumference has been observed in GWAS of multiple ethnic groups, including populations of European, South Asian, and East Asian ancestry.

In recent meta-analyses examining transferability of this locus in East and South Asian populations, the association between *FTO* and T2D was only partly accounted for by association with BMI (Li et al. 2012; Rees et al. 2011; Vasan et al. 2013), suggesting an impact on the risk of T2D independent of BMI in both populations. More recent examination of prevalent and incident T2D in a European cohort confirmed that the association of *FTO* SNP rs9939609 with T2D was partly independent of its effect on BMI in Europeans as well, demonstrating the value of transferability studies in elucidating the differential impact of this variant (Hertel et al. 2011).

8.3.4 A Population with Unique Genetic Effects for T2D: HNF1A *in the American Indian Oji-Cree*

Targeting population isolates with an unusual prevalence of disease for gene discovery can identify founder alleles that rise to detectable frequency, thereby informing the underlying disease biology relevant to all populations and relevant to physiology, epidemiology, and risk of T2D in the specific population. An early example in the field of T2D genetics was the discovery of a novel missense variant (G319S) in the HNF1 homeobox A (*HNF1A*) gene in a Canadian Native American population with an unusually high prevalence of impaired glucose tolerance and T2D (40 %). Rare, often familial mutations in *HNF1A* are associated with maturity-onset diabetes of the young (MODY). Using candidate gene resequencing, investigators identified *HNF1A* G319S in the Sandy Lake Oji-Cree population with a frequency of 20.9 % in adult subjects with T2D and 8.7 % in nondiabetic subjects. G319S was associated with a distinct form of T2D characterized by earlier age of onset, lower body mass, and a higher post-challenge plasma glucose than typical of T2D. More recently, common variants in *HNF1A* have also been associated with T2D in GWAS from Mexican and European populations (Parra et al. 2011; Voight et al. 2010), demonstrating that this gene contributes to T2D across the entire allelic spectrum harboring rare variants with strong deleterious effects and population-specific or trans-ethnic common variants.

8.3.5 *Novel Insights into T2D from Latino Populations:* SLC16A11

GWAS of T2D in ethnic minority populations, particularly in Latino populations, were previously limited to two published reports (Below et al. 2011; Hayes et al. 2007), despite an increased prevalence of disease that is twice that of their non-Hispanic White counterparts (Centers for Disease Control and Prevention 2011). More recently, the SIGMA T2D Consortium has begun to characterize the genetic basis of disease in a large Mexican and Latin American population (3848 cases/4366 controls) (SIGMA Type 2 Diabetes Consortium et al. 2014). While previously reported susceptibility variants in *TCF7L2* and *KCNQ1* were among the most associated variants, a novel association on 17p13.1 spanning the two solute carrier genes (*SLC16A13* and *SLC16A11*) was observed. The strongest signals of association were within the *SLC16A11* locus from one silent and four missense mutations and associated with a 20 % increased risk for T2D. These individuals develop T2D 2.1 years earlier at a decreased BMI. This finding is estimated to explain 20 % (9.2–29 %) of the increased disease burden observed in Mexican and Latin American populations relative to populations of European ancestry.

This discovery is a novel finding despite large-scale GWAS that have been conducted for T2D in European (Voight et al. 2010) and Asian (Cho et al. 2012a; Kooner et al. 2011; Hara et al. 2014) ancestry samples. This lack of association is attributed to mutations that are rare or absent in representative European populations (1 %), of intermediate frequency in East Asian populations (~16 %), and common (~50 %) in the Americas (1000 Genomes Project Consortium et al. 2012). Trends of association in the East Asian population, with additional signals observed (Hara et al. 2014), argued against population stratification which was further supported by high sequence divergence which precedes the "out of Africa" population bottleneck (Li and Durbin 2011). The latter observation is consistent with admixture attributable to Neanderthals and collaborated by high sequence homology with an unpublished Neanderthal genome from Denisova Cave.

Although poorly characterized, *SLC16A11* is a member of the monocarboxylic acid transporter family of solute carriers. Analysis of T2D-related quantitative traits offered little insight in the physiological role of this protein in vivo. However, in vitro metabolomic profiling suggested an effect on T2D risk mediated through effects on lipid metabolism consistent with expression in the liver. This finding demonstrates the utility of disease mapping in diverse populations in an effort to illuminate disease etiology and dissect its genetic architecture.

8.4 Novel Approaches to Disease Identification Using Diverse Populations

8.4.1 Admixture Mapping to Identify Genetic Variants Contributing to T2D

Admixture mapping offers a powerful approach to identifying genetic variants contributing to T2D with considerable cost savings using significantly smaller marker panels (2000–3000 SNPs) than traditional GWAS approaches. Admixture mapping is a genetic association strategy that takes advantage of long haplotype blocks created by admixture among populations with disparate disease prevalence (Stephens et al. 1994). As a consequence of differing prevalence rates, this approach identifies disease-associated alleles derived from one of the ancestral populations, which differ in frequency. As a complex disease with a demonstrated genetic component, T2D represents a disease with a high probability of success for admixture mapping. This is particularly seen in admixed populations such as African-Americans and Latinos where a higher relative risk is observed, e.g., the population relative risk in Africans vs. Europeans is 1.99 (Songer and Zimmet 1995).

African-Americans represent an admixed population derived, on average, with 80 % African and 20 % European genetic ancestry and displaying a twofold increase in T2D prevalence (Centers for Disease Control and Prevention 2011). With knowledge of the ancestral populations, African-Americans have previously represented the most intensively studied population for admixture mapping. The largest admixture mapping study for T2D published to date evaluated greater than 7000 African-Americans (Cheng et al. 2012). As expected, participants displayed increased African ancestry attributable to the higher prevalence of disease in this population despite adjustment for adiposity and socioeconomic status. Moreover, African ancestry was correlated with surrogate measures of insulin resistance consistent with the literature (Karter et al. 1996). As with most common complex diseases, signals of association were observed; however, the study failed to identify a locus of large effect. Given the unexplained (Manolio et al. 2009) yet established genetic component to T2D, this latter observation could suggest that while individual effects are minimal, these may function in aggregate to explain the missing heritability and ethnic disparities of disease. This work is now being extended to Latino populations which represent populations of varied contributions from Native Americans and Europeans and with and without contributions from West Africans (Hanis et al. 1991).

As with other approaches, admixture mapping is not without its limitations. Aside from computational nuances, admixture mapping has the inherent potential for false-negative results when a causal variant is of similar frequency among the ancestral populations. However, it has been demonstrated that power for admixture mapping is relatively consistent over a range of admixture proportions, i.e., 10–90 % (Patterson et al. 2004). Moreover, admixture mapping suffers from the same limitation as linkage analysis in that variant localization is problematic requiring additional genetic makers and alternative analytical approaches. Despite these limitations, admixture mapping has facilitated profound discoveries in complex disease genetics (Genovese et al. 2010).

8.4.2 *Ethnic Transferability and Meta-analyses*

A transferability study in the Multiethnic Cohort (MEC) in 2010 showing consistent association of European T2D loci in African-Americans, Latinos, Japanese Americans, and Native Hawaiians provided strong support for the idea that shared genetic variation contributes to T2D risk in multiple populations (Waters et al. 2010). The authors found consistent association with 14 of 19 loci across populations and consistent direction of effect for the five remaining loci. Concurrently, a genotype risk score based on 16 European fasting glucose-associated variants showed consistent association with fasting glucose in non-Hispanic Whites, non-Hispanic Blacks, and Mexican Americans, despite variations in allele frequency by race/ethnicity, suggesting shared genetic influence on fasting glucose levels across populations (Yang et al. 2010). More recently, a large-scale transferability analysis of ten genetic variants associated with fasting glucose and insulin across an ethnically diverse population of Europeans, African-Americans, Hispanics, American Indians, and Asians revealed that several glucose SNPs were associated across multiple racial and ethnic groups, confirming generalizability and importance of these loci (Fesinmeyer et al. 2013). The transferability of associations across populations also implies that rare variants are unlikely to underlie the common variant associations since rare variants would be expected to differ between ethnic groups. Transferability of signals across ethnic groups are illustrated for signals discovered in subjects of European ancestry (Fig. 8.1), subjects of East Asian ancestry (Fig. 8.2) and subjects of South Asian ancestry (Fig. 8.3).

Larger multi-ethnic GWAS are needed to discover new robust T2D and related-trait loci with trans-ethnic effects. At loci with lack of heterogeneity in genetic effects across multiple populations, trans-ethnic GWAS also offers opportunities for simultaneous fine mapping by taking advantage of differences in the distribution of LD. A recent genome-wide trans-ethnic meta-analysis of European, East Asian, South Asian, Mexican and Mexican-American ancestry demonstrates the promise of this approach for gene discovery, transferability assessment, and fine mapping (Diabetes Genetics Replication Meta-analysis Consortium 2014). This study identified seven new T2D susceptibility loci, found concordant direction of effect across

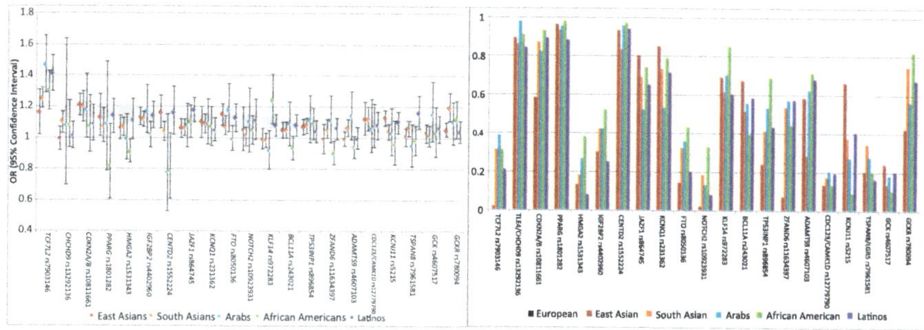

Fig. 8.1 Comparison of effect sizes (Panel A) and allele frequencies (Panel B) for T2D susceptibility variants discovered in populations of European descent across ethnically diverse populations. Panel A compares the odds ratio (OR) and 95 % confidence interval observed in Europeans (*background gray*; Voight et al. 2010) to those evaluated for transferability in populations of East Asians (*red*; Cho et al. 2012a), South Asians (*orange*; Kooner et al. 2011), Arabs (*blue*; Cauchi et al. 2012), African-Americans (*green*; Ng et al. 2013), and Latinos (*purple*; SIGMA Type 2 Diabetes Consortium et al. 2014). Variants that achieved nominal levels of significance ($P < 0.05$) are denoted with a diamond. Panel B compares the allele frequency observed in a representative European population (CEU; *black*) to those evaluated for transferability in populations of East Asians (*red*; Cho et al. 2012a), South Asians (*orange*; Kooner et al. 2011), Arabs (*blue*; Cauchi et al. 2012), African-Americans (*green*; Ng et al. 2013), and Latinos (*purple*; SIGMA Type 2 Diabetes Consortium et al. 2014). Variants for both panels are ordered by descending effect size observed in Europeans

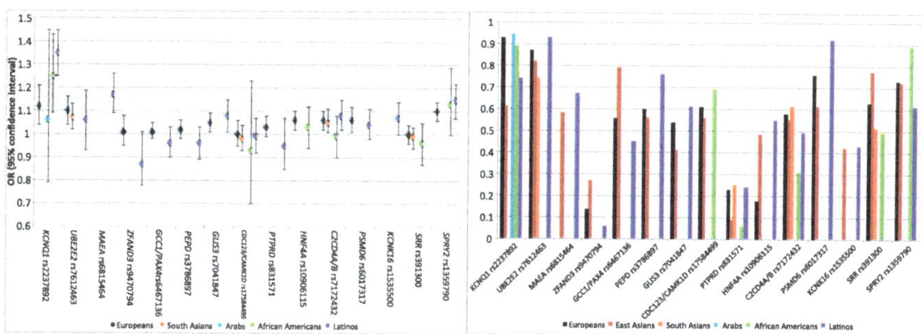

Fig. 8.2 Comparison of effect sizes (Panel A) and allele frequencies (Panel B) for T2D susceptibility variants discovered in populations of East Asian descent across ethnically diverse populations. Panel A compares the odds ratio (OR) and 95 % confidence interval observed in East Asians (*background gray*) to those evaluated for transferability in populations of European (*black*; Voight et al. 2010), South Asians (*orange*; Kooner et al. 2011), Arabs (*blue*; Cauchi et al. 2012), African-Americans (*green*; Ng et al. 2013), and Latinos (*purple*; SIGMA Type 2 Diabetes Consortium et al. 2014). Variants that achieved nominal levels of significance ($P < 0.05$) are denoted with a diamond. Panel B compares the allele frequency observed in the East Asian GWAS (*red*; Cho et al. 2012a) to those evaluated in a representative European population (CEU; black) and in populations of South Asians (*orange*; Kooner et al. 2011), Arabs (blue; Cauchi et al. 2012), African-Americans (green; Ng et al. 2013), and Latinos (*purple*; SIGMA Type 2 Diabetes Consortium et al. 2014). Variants for both panels are ordered by descending effect size observed in East Asians

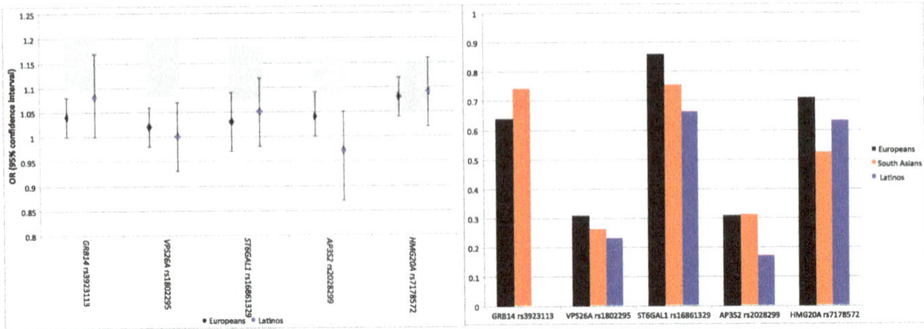

Fig. 8.3 Comparison of effect sizes (Panel A) and allele frequencies (Panel B) for T2D susceptibility variants discovered in populations of South Asian descent across ethnically diverse populations. Panel A compares the odds ratio (OR) and 95 % confidence interval observed in South Asian GWAS (*background gray*; Kooner et al. 2011) to those evaluated for transferability in populations of European ancestry (*black*; Voight et al. 2010) and in Latinos (*purple*; SIGMA Type 2 Diabetes Consortium et al. 2014). Variants that achieved nominal levels of significance ($P < 0.05$) are denoted with a diamond. Panel B compares the allele frequency observed in the South Asian GWAS (*orange*; Kooner et al. 2011) to those evaluated in a representative European population (CEU; *black*) and in Latinos (*purple*; SIGMA Type 2 Diabetes Consortium et al. 2014). Variants for both panels are ordered by descending effect size observed in South Asians

34/52 (65 %) index SNPs across all ethnic groups, and led to reduction in genomic intervals and number of candidate causal variants by fine mapping at 8/10 association signals.

8.4.3 *T2D Genetic Risk Assessment*

Cumulative risk assessment for the effects of T2D susceptibility variants derived predominantly from European populations has been evaluated for their predictive value in diverse ethnic groups which display differential disease prevalence rates. In a multi-ethnic approach, Waters et al. (Waters et al. 2010) evaluated the distribution of risk alleles among 19 variants reproducibly associated with T2D in European populations with transferability to populations of African-Americans, Latinos, Japanese Americans, and Native Hawaiians. Investigators observed a similar incremental increase in risk per allele with all populations except Japanese Americans. Risk per allele was approximately double in this population, which also demonstrated a 3.1-fold increased risk of disease for upper quartile of disease allele carriers. However, compared to Europeans, a greater disease burden was not associated with the genetic risk assessment. A more focused analysis in African-Americans has extended these findings; however, it suggests that the effects observed with *TCF7L2* may be driving the association of a risk score with T2D (Cooke et al. 2012). Taken together, these results hint at potential utility; however, the variants identified to date likely do not represent causal alleles suggesting further mapping studies may yield more fruitful risk assessment as predictive indices of T2D risk.

8.4.4 Transferability of Loci Across Ethnic Groups: The Role of Rare Variants

As demonstrated for MODY genes that harbor an allelic spectrum of common and rare T2D-associated variants, recent examples show that new or known T2D or related-trait loci harbor rare variants of high functional impact in Europeans (e.g., *MTNR1B*, *CCND2*, *PAM*, and *PDX1* for T2D (Bonnefond et al. 2012; Steinthorsdottir et al. 2014) and *SGSM2*, *MADD*, *TBC1D30*, *KANK1*, and *PAM* for insulin processing and secretion traits; Huyghe et al. 2013), but studies in other populations have not yet been reported. Rare variants are expected to be recent and therefore population specific. Discovery of rare high-impact functional variants contributing to one or more populations would be important for the field of diabetes genetics because it would (1) highlight that a gene is causal for diabetes pathophysiology, bypassing difficulties of identifying the causal gene within a region of common variant association; (2) be amenable for functional studies to elucidate disease mechanisms; and (3) add to the growing understanding of the role of rare variants in the genetic architecture and heritability of the disease, perhaps contributing to the differences in prevalence and disease presentation across ethnicities. Future screening of multiple population groups for rare variation is thus a powerful and necessary approach to understand the genetic basis of T2D (McCarthy 2011).

8.5 What We Have Learned

Genetic studies of T2D in ethnically diverse populations, stemming from studies in Europeans, have revealed transferability of effect, while differences in allele frequency resulting in reduced statistical power may have impacted the ability to observe significant associations. A number of established T2D susceptibility loci exhibit allelic heterogeneity with differential signals of association observed across populations. The analysis of common variants in the GWAS era has culminated in the identification of novel variants in diverse ethnicities, which could contribute to the observed ethnic disparity of T2D across populations. The next generation of genetic studies with a focus on rare variants to explain the missing heritability will rely upon studies of T2D in multiethnic cohorts and may reveal novel insights into the genetic architecture of T2D (Table 8.1).

Table 8.1 Advantages of genetic studies in diverse populations (Rosenberg et al. 2010)

Feature varying among ethnicities	Examples in T2D genetics	Advantage
Population-specific alleles	*HNF1A* in American Indian Oji-Cree	**Novel gene discovery**: Risk or protective variants occur in a specific population, allowing for novel insights into etiology of disease relevant globally, but relevant for risk prediction only in the specific population. Founder mutations in population isolates with high inbreeding may lead to increase in allele frequency and lead to a substantial contribution to disease prevalence in the specific population. Rare variants are also expected to be population-specific
Evolutionary history	*SLC16A11* introgressed into modern humans via admixture with Neanderthals and high frequency in Latinos	**Novel gene discovery**: Different ethnic groups have varying prevalence and heritability of disease, as well as unique evolutionary histories, enabling novel genetic discovery
Allele-frequency	*KCNQ1* in East Asians with a signal transferable to other ethnicities; *SLC16A11* in Latinos	**Improved detection**: Variants are more common and thus more easily detected in some populations based on population genetic history; this is also manifest as allelic heterogeneity across populations, which confirms the role of critical loci in susceptibility to disease
Linkage-disequilibrium (LD)	*TCF7L2* fine-mapping in West Africans and African-Americans; trans-ethnic fine-mapping at *JAZF1* and *SLC30A8*	**Fine-Mapping**: Populations with differential LD facilitate fine-mapping of association signal to localize a causal variant
Allelic effect	tentative: *MTNR1B, GCK, PEPD1* in East Asian ancestry; *KLF14* in European ancestry	**Insights on ethnicity-dependent disease presentation**: Risk variants have heterogeneity of effect across populations, based on underlying differences in genetic or environmental background leading to possible differences in disease presentation

References

1000 Genomes Project Consortium, Abecasis GR, Auton A, Brooks LD, DePristo MA, Durbin RM, Handsaker RE, Kang HM, Marth GT, McVean GA (2012) An integrated map of genetic variation from 1,092 human genomes. Nature 491(7422):56–65. doi:10.1038/nature11632

Almawi WY, Nemr R, Keleshian SH, Echtay A, Saldanha FL, AlDoseri FA, Racoubian E (2013) A replication study of 19 GWAS-validated type 2 diabetes at-risk variants in the Lebanese population. Diabetes Res Clin Pract 102(2):117–122. doi:10.1016/j.diabres.2013.09.001

Almgren P, Lehtovirta M, Isomaa B, Sarelin L, Taskinen MR, Lyssenko V, Tuomi T, Groop L, Botnia Study Group (2011) Heritability and familiarity of type 2 diabetes and related quantitative traits in the Botnia Study. Diabetologia 54(11):2811–2819. doi:10.1007/s00125-011-2267-5

Below JE, Gamazon ER, Morrison JV, Konkashbaev A, Pluzhnikov A, McKeigue PM, Parra EJ, Elbein SC, Hallman DM, Nicolae DL, Bell GI, Cruz M, Cox NJ, Hanis CL (2011) Genome-wide association and meta-analysis in populations from Starr County, Texas, and Mexico City identify type 2 diabetes susceptibility loci and enrichment for expression quantitative trait loci in top signals. Diabetologia 54(8):2047–2055. doi:10.1007/s00125-011-2188-3

Bonnefond A, Clement N, Fawcett K, Yengo L, Vaillant E, Guillaume JL, Dechaume A, Payne F, Roussel R, Czernichow S, Hercberg S, Hadjadj S, Balkau B, Marre M, Lantieri O, Langenberg C, Bouatia-Naji N, Meta-Analysis of Glucose and Insulin-Related Traits Consortium (MAGIC), Charpentier G, Vaxillaire M, Rocheleau G, Wareham NJ, Sladek R, McCarthy MI, Dina C, Barroso I, Jockers R, Froguel P (2012) Rare MTNR1B variants impairing melatonin receptor 1B function contribute to type 2 diabetes. Nat Genet 44(3):297–301. doi:10.1038/ng.1053

Cauchi S, Ezzidi I, El Achhab Y, Mtiraoui N, Chaieb L, Salah D, Nejjari C, Labrune Y, Yengo L, Beury D, Vaxillaire M, Mahjoub T, Chikri M, Froguel P (2012) European genetic variants associated with type 2 diabetes in North African Arabs. Diabetes Metab 38(4):316–323. doi:10.1016/j.diabet.2012.02.003

Centers for Disease Control and Prevention (2011) National diabetes fact sheet: national estimates and general information on diabetes and prediabetes in the United States, 2011

Cheng CY, Reich D, Haiman CA, Tandon A, Patterson N, Selvin E, Akylbekova EL, Brancati FL, Coresh J, Boerwinkle E, Altshuler D, Taylor HA, Henderson BE, Wilson JG, Kao WH (2012) African ancestry and its correlation to type 2 diabetes in African Americans: a genetic admixture analysis in three U.S. population cohorts. PLoS ONE 7(3), e32840. doi:10.1371/journal.pone.0032840

Cho YS, Chen CH, Hu C, Long J, Ong RT, Sim X, Takeuchi F, Wu Y, Go MJ, Yamauchi T, Chang YC, Kwak SH, Ma RC, Yamamoto K, Adair LS, Aung T, Cai Q, Chang LC, Chen YT, Gao Y, Hu FB, Kim HL, Kim S, Kim YJ, Lee JJ, Lee NR, Li Y, Liu JJ, Lu W, Nakamura J, Nakashima E, Ng DP, Tay WT, Tsai FJ, Wong TY, Yokota M, Zheng W, Zhang R, Wang C, So WY, Ohnaka K, Ikegami H, Hara K, Cho YM, Cho NH, Chang TJ, Bao Y, Hedman AK, Morris AP, McCarthy MI, DIAGRAM Consortium; MuTHER Consortium, Takayanagi R, Park KS, Jia W, Chuang LM, Chan JC, Maeda S, Kadowaki T, Lee JY, Wu JY, Teo YY, Tai ES, Shu XO, Mohlke KL, Kato N, Han BG, Seielstad M (2012a) Meta-analysis of genome-wide association studies identifies eight new loci for type 2 diabetes in east Asians. Nat Genet 44(1):67–72. doi:10.1038/ng.1019

Cho YS, Lee JY, Park KS, Nho CW (2012b) Genetics of type 2 diabetes in East Asian populations. Curr Diab Rep 12(6):686–696. doi:10.1007/s11892-012-0326-z

Cooke JN, Ng MC, Palmer ND, An SS, Hester JM, Freedman BI, Langefeld CD, Bowden DW (2012) Genetic risk assessment of type 2 diabetes-associated polymorphisms in African Americans. Diabetes Care 35(2):287–292. doi:10.2337/dc11-0957

Diabetes Genetics Replication Meta-analysis Consortium (2014) Genome-wide trans-ancestry meta-analysis provides insight into the genetic architecture of type 2 diabetes susceptibility. Nat Genet. doi:10.1038/ng.2897

Dimas AS, Lagou V, Barker A, Knowles JW, Magi R, Hivert MF, Benazzo A, Rybin D, Jackson AU, Stringham HM, Song C, Fischer-Rosinsky A, Boesgaard TW, Grarup N, Abbasi FA, Assimes TL, Hao K, Yang X, Lecoeur C, Barroso I, Bonnycastle LL, Bottcher Y, Bumpstead S, Chines PS, Erdos MR, Graessler J, Kovacs P, Morken MA, Narisu N, Payne F, Stancakova A, Swift AJ, Tonjes A, Bornstein SR, Cauchi S, Froguel P, Meyre D, Schwarz PE, Haring HU, Smith U, Boehnke M, Bergman RN, Collins FS, Mohlke KL, Tuomilehto J, Quertemous T, Lind L, Hansen T, Pedersen O, Walker M, Pfeiffer AF, Spranger J, Stumvoll M, Meigs JB, Wareham NJ, Kuusisto J, Laakso M, Langenberg C, Dupuis J, Watanabe RM, Florez JC, Ingelsson E, McCarthy MI, Prokopenko I, on behalf of the Mi (2013) Impact of type 2 diabetes susceptibility variants on quantitative glycemic traits reveals mechanistic heterogeneity. Diabetes 63:2158–2171. doi:10.2337/db13-0949

Duggirala R, Blangero J, Almasy L, Dyer TD, Williams KL, Leach RJ, O'Connell P, Stern MP (1999) Linkage of type 2 diabetes mellitus and of age at onset to a genetic location on chromosome 10q in Mexican Americans. Am J Hum Genet 64(4):1127–1140

Fesinmeyer MD, Meigs JB, North KE, Schumacher FR, Buzkova P, Franceschini N, Haessler J, Goodloe R, Spencer KL, Voruganti VS, Howard BV, Jackson R, Kolonel LN, Liu S, Manson JE, Monroe KR, Mukamal K, Dilks HH, Pendergrass SA, Nato A, Wan P, Wilkens LR, Le Marchand L, Ambite JL, Buyske S, Florez JC, Crawford DC, Hindorff LA, Haiman CA, Peters U, Pankow JS (2013) Genetic variants associated with fasting glucose and insulin concentrations in an ethnically diverse population: results from the Population Architecture using Genomics and Epidemiology (PAGE) study. BMC Med Genet 14:98. doi:10.1186/1471-2350-14-98

Frayling TM, Timpson NJ, Weedon MN, Zeggini E, Freathy RM, Lindgren CM, Perry JR, Elliott KS, Lango H, Rayner NW, Shields B, Harries LW, Barrett JC, Ellard S, Groves CJ, Knight B, Patch AM, Ness AR, Ebrahim S, Lawlor DA, Ring SM, Ben-Shlomo Y, Jarvelin MR, Sovio U, Bennett AJ, Melzer D, Ferrucci L, Loos RJ, Barroso I, Wareham NJ, Karpe F, Owen KR, Cardon LR, Walker M, Hitman GA, Palmer CN, Doney AS, Morris AD, Smith GD, Hattersley AT, McCarthy MI (2007) A common variant in the FTO gene is associated with body mass index and predisposes to childhood and adult obesity. Science 316(5826):889–894. doi:10.1126/science.1141634

Genovese G, Friedman DJ, Ross MD, Lecordier L, Uzureau P, Freedman BI, Bowden DW, Langefeld CD, Oleksyk TK, Uscinski Knob AL, Bernhardy AJ, Hicks PJ, Nelson GW, Vanhollebeke B, Winkler CA, Kopp JB, Pays E, Pollak MR (2010) Association of trypanolytic ApoL1 variants with kidney disease in African Americans. Science 329(5993):841–845. doi:10.1126/science.1193032

Goodarzi MO, Langefeld CD, Xiang AH, Chen YD, Guo X, Hanley AJ, Raffel LJ, Kandeel F, Nadler JL, Buchanan TA, Norris JM, Fingerlin TE, Lorenzo C, Rewers MJ, Haffner SM, Bowden DW, Rich SS, Bergman RN, Rotter JI, Watanabe RM, Wagenknecht LE (2014) Insulin sensitivity and insulin clearance are heritable and have strong genetic correlation in mexican americans. Obesity (Silver Spring) 22(4):1157–1164. doi:10.1002/oby.20639

Guo T, Hanson RL, Traurig M, Muller YL, Ma L, Mack J, Kobes S, Knowler WC, Bogardus C, Baier LJ (2007) TCF7L2 is not a major susceptibility gene for type 2 diabetes in Pima Indians: analysis of 3,501 individuals. Diabetes 56(12):3082–3088. doi:10.2337/db07-0621

Hanis CL, Hewett-Emmett D, Bertin TK, Schull WJ (1991) Origins of U.S. Hispanics. Implications for diabetes. Diabetes Care 14(7):618–627

Hanson RL, Bogardus C, Duggan D, Kobes S, Knowlton M, Infante AM, Marovich L, Benitez D, Baier LJ, Knowler WC (2007) A search for variants associated with young-onset type 2 diabetes in American Indians in a 100K genotyping array. Diabetes 56(12):3045–3052. doi:10.2337/db07-0462

Hara K, Fujita H, Johnson TA, Yamauchi T, Yasuda K, Horikoshi M, Peng C, Hu C, Ma RC, Imamura M, Iwata M, Tsunoda T, Morizono T, Shojima N, So WY, Leung TF, Kwan P, Zhang R, Wang J, Yu W, Maegawa H, Hirose H, DIAGRAM consortium, Kaku K, Ito C, Watada H, Tanaka Y, Tobe K, Kashiwagi A, Kawamori R, Jia W, Chan JC, Teo YY, Shyong

TE, Kamatani N, Kubo M, Maeda S, Kadowaki T (2014) Genome-wide association study identifies three novel loci for type 2 diabetes. Hum Mol Genet 23(1):239–246

Hayes MG, Pluzhnikov A, Miyake K, Sun Y, Ng MC, Roe CA, Below JE, Nicolae RI, Konkashbaev A, Bell GI, Cox NJ, Hanis CL (2007) Identification of type 2 diabetes genes in Mexican Americans through genome-wide association studies. Diabetes 56(12):3033–3044. doi:10.2337/db07-0482

Helgason A, Palsson S, Thorleifsson G, Grant SF, Emilsson V, Gunnarsdottir S, Adeyemo A, Chen Y, Chen G, Reynisdottir I, Benediktsson R, Hinney A, Hansen T, Andersen G, Borch-Johnsen K, Jorgensen T, Schafer H, Faruque M, Doumatey A, Zhou J, Wilensky RL, Reilly MP, Rader DJ, Bagger Y, Christiansen C, Sigurdsson G, Hebebrand J, Pedersen O, Thorsteinsdottir U, Gulcher JR, Kong A, Rotimi C, Stefansson K (2007) Refining the impact of TCF7L2 gene variants on type 2 diabetes and adaptive evolution. Nat Genet 39(2):218–225. doi:10.1038/ng1960

Hertel JK, Johansson S, Sonestedt E, Jonsson A, Lie RT, Platou CG, Nilsson PM, Rukh G, Midthjell K, Hveem K, Melander O, Groop L, Lyssenko V, Molven A, Orho-Melander M, Njolstad PR (2011) FTO, type 2 diabetes, and weight gain throughout adult life: a meta-analysis of 41,504 subjects from the Scandinavian HUNT, MDC, and MPP studies. Diabetes 60 (5):1637–1644. doi:10.2337/db10-1340

Hu C, Zhang R, Wang C, Yu W, Lu J, Ma X, Wang J, Jiang F, Tang S, Bao Y, Xiang K, Jia W (2010) Effects of GCK, GCKR, G6PC2 and MTNR1B variants on glucose metabolism and insulin secretion. PLoS ONE 5(7), e11761. doi:10.1371/journal.pone.0011761

Huyghe JR, Jackson AU, Fogarty MP, Buchkovich ML, Stancakova A, Stringham HM, Sim X, Yang L, Fuchsberger C, Cederberg H, Chines PS, Teslovich TM, Romm JM, Ling H, McMullen I, Ingersoll R, Pugh EW, Doheny KF, Neale BM, Daly MJ, Kuusisto J, Scott LJ, Kang HM, Collins FS, Abecasis GR, Watanabe RM, Boehnke M, Laakso M, Mohlke KL (2013) Exome array analysis identifies new loci and low-frequency variants influencing insulin processing and secretion. Nat Genet 45(2):197–201. doi:10.1038/ng.2507

Karter AJ, Mayer-Davis EJ, Selby JV, D'Agostino RB Jr, Haffner SM, Sholinsky P, Bergman R, Saad MF, Hamman RF (1996) Insulin sensitivity and abdominal obesity in African-American, Hispanic, and non-Hispanic white men and women. The Insulin Resistance and Atherosclerosis Study. Diabetes 45(11):1547–1555

Knowler WC, Pettitt DJ, Saad MF, Bennett PH (1990) Diabetes mellitus in the Pima Indians: incidence, risk factors and pathogenesis. Diabetes Metab Rev 6(1):1–27

Kooner JS, Saleheen D, Sim X, Sehmi J, Zhang W, Frossard P, Been LF, Chia KS, Dimas AS, Hassanali N, Jafar T, Jowett JB, Li X, Radha V, Rees SD, Takeuchi F, Young R, Aung T, Basit A, Chidambaram M, Das D, Grundberg E, Hedman AK, Hydrie ZI, Islam M, Khor CC, Kowlessur S, Kristensen MM, Liju S, Lim WY, Matthews DR, Liu J, Morris AP, Nica AC, Pinidiyapathirage JM, Prokopenko I, Rasheed A, Samuel M, Shah N, Shera AS, Small KS, Suo C, Wickremasinghe AR, Wong TY, Yang M, Zhang F, Diagram MT, Abecasis GR, Barnett AH, Caulfield M, Deloukas P, Frayling TM, Froguel P, Kato N, Katulanda P, Kelly MA, Liang J, Mohan V, Sanghera DK, Scott J, Seielstad M, Zimmet PZ, Elliott P, Teo YY, McCarthy MI, Danesh J, Tai ES, Chambers JC (2011) Genome-wide association study in individuals of South Asian ancestry identifies six new type 2 diabetes susceptibility loci. Nat Genet 43(10):984–989. doi:10.1038/ng.921

Kuo JZ, Sheu WH, Assimes TL, Hung YJ, Absher D, Chiu YF, Mak J, Wang JS, Kwon S, Hsu CC, Goodarzi MO, Lee IT, Knowles JW, Miller BE, Lee WJ, Juang JM, Wang TD, Guo X, Taylor KD, Chuang LM, Hsiung CA, Quertermous T, Rotter JI, Chen YD (2013) Trans-ethnic fine mapping identifies a novel independent locus at the 3′ end of CDKAL1 and novel variants of several susceptibility loci for type 2 diabetes in a Han Chinese population. Diabetologia 56 (12):2619–2628. doi:10.1007/s00125-013-3047-1

Li H, Durbin R (2011) Inference of human population history from individual whole-genome sequences. Nature 475(7357):493–496. doi:10.1038/nature10231

Li H, Kilpelainen TO, Liu C, Zhu J, Liu Y, Hu C, Yang Z, Zhang W, Bao W, Cha S, Wu Y, Yang T, Sekine A, Choi BY, Yajnik CS, Zhou D, Takeuchi F, Yamamoto K, Chan JC, Mani KR, Been LF, Imamura M, Nakashima E, Lee N, Fujisawa T, Karasawa S, Wen W, Joglekar CV, Lu W, Chang Y, Xiang Y, Gao Y, Liu S, Song Y, Kwak SH, Shin HD, Park KS, Fall CH, Kim JY, Sham PC, Lam KS, Zheng W, Shu X, Deng H, Ikegami H, Krishnaveni GV, Sanghera DK, Chuang L, Liu L, Hu R, Kim Y, Daimon M, Hotta K, Jia W, Kooner JS, Chambers JC, Chandak GR, Ma RC, Maeda S, Dorajoo R, Yokota M, Takayanagi R, Kato N, Lin X, Loos RJ (2012) Association of genetic variation in FTO with risk of obesity and type 2 diabetes with data from 96,551 East and South Asians. Diabetologia 55(4):981–995. doi:10.1007/s00125-011-2370-7

Li H, Gan W, Lu L, Dong X, Han X, Hu C, Yang Z, Sun L, Bao W, Li P, He M, Sun L, Wang Y, Zhu J, Ning Q, Tang Y, Zhang R, Wen J, Wang D, Zhu X, Guo K, Zuo X, Guo X, Yang H, Zhou X, DIAGRAM Consortium; AGEN-T2D Consortium, Zhang X, Qi L, Loos RJ, Hu FB, Wu T, Liu Y, Liu L, Yang Z, Hu R, Jia W, Ji L, Li Y, Lin X (2013) A genome-wide association study identifies GRK5 and RASGRP1 as type 2 diabetes loci in Chinese Hans. Diabetes 62 (1):291–298. doi:10.2337/db12-0454

Liu CT, Ng MC, Rybin D, Adeyemo A, Bielinski SJ, Boerwinkle E, Borecki I, Cade B, Chen YD, Djousse L, Fornage M, Goodarzi MO, Grant SF, Guo X, Harris T, Kabagambe E, Kizer JR, Liu Y, Lunetta KL, Mukamal K, Nettleton JA, Pankow JS, Patel SR, Ramos E, Rasmussen-Torvik L, Rich SS, Rotimi CN, Sarpong D, Shriner D, Sims M, Zmuda JM, Redline S, Kao WH, Siscovick D, Florez JC, Rotter JI, Dupuis J, Wilson JG, Bowden DW, Meigs JB (2012) Transferability and fine-mapping of glucose and insulin quantitative trait loci across populations: CARe, the Candidate Gene Association Resource. Diabetologia 55 (11):2970–2984. doi:10.1007/s00125-012-2656-4

Long J, Edwards T, Signorello LB, Cai Q, Zheng W, Shu XO, Blot WJ (2012) Evaluation of genome-wide association study-identified type 2 diabetes loci in African Americans. Am J Epidemiol 176(11):995–1001. doi:10.1093/aje/kws176

Ma RC, Hu C, Tam CH, Zhang R, Kwan P, Leung TF, Thomas GN, Go MJ, Hara K, Sim X, Ho JS, Wang C, Li H, Lu L, Wang Y, Li JW, Wang Y, Lam VK, Wang J, Yu W, Kim YJ, Ng DP, Fujita H, Panoutsopoulou K, Day-Williams AG, Lee HM, Ng AC, Fang YJ, Kong AP, Jiang F, Ma X, Hou X, Tang S, Lu J, Yamauchi T, Tsui SK, Woo J, Leung PC, Zhang X, Tang NL, Sy HY, Liu J, Wong TY, Lee JY, Maeda S, Xu G, Cherny SS, Chan TF, Ng MC, Xiang K, Morris AP, DIAGRAM Consortium, Keildson S, MuTHER Consortium, Hu R, Ji L, Lin X, Cho YS, Kadowaki T, Tai ES, Zeggini E, McCarthy MI, Hon KL, Baum L, Tomlinson B, So WY, Bao Y, Chan JC, Jia W (2013) Genome-wide association study in a Chinese population identifies a susceptibility locus for type 2 diabetes at 7q32 near PAX4. Diabetologia 56 (6):1291–1305. doi:10.1007/s00125-013-2874-4

Manolio TA, Collins FS, Cox NJ, Goldstein DB, Hindorff LA, Hunter DJ, McCarthy MI, Ramos EM, Cardon LR, Chakravarti A, Cho JH, Guttmacher AE, Kong A, Kruglyak L, Mardis E, Rotimi CN, Slatkin M, Valle D, Whittemore AS, Boehnke M, Clark AG, Eichler EE, Gibson G, Haines JL, Mackay TF, McCarroll SA, Visscher PM (2009) Finding the missing heritability of complex diseases. Nature 461(7265):747–753. doi:10.1038/nature08494

McCarthy MI (2011) The importance of global studies of the genetics of type 2 diabetes. Diabetes Metab J 35(2):91–100. doi:10.4093/dmj.2011.35.2.91

Ng MC, Saxena R, Li J, Palmer ND, Dimitrov L, Xu J, Rasmussen-Torvik LJ, Zmuda JM, Siscovick DS, Patel SR, Crook ED, Sims M, Chen YD, Bertoni AG, Li M, Grant SF, Dupuis J, Meigs JB, Psaty BM, Pankow JS, Langefeld CD, Freedman BI, Rotter JI, Wilson JG, Bowden DW (2013) Transferability and fine mapping of type 2 diabetes loci in African Americans: the Candidate Gene Association Resource Plus Study. Diabetes 62(3):965–976. doi:10.2337/db12-0266

Palmer ND, Goodarzi MO, Langefeld CD, Ziegler J, Norris JM, Haffner SM, Bryer-Ash M, Bergman RN, Wagenknecht LE, Taylor KD, Rotter JI, Bowden DW (2008) Quantitative trait analysis of type 2 diabetes susceptibility loci identified from whole genome association studies in the Insulin Resistance Atherosclerosis Family Study. Diabetes 57(4):1093–1100. doi:10.2337/db07-1169

Palmer ND, Hester JM, An SS, Adeyemo A, Rotimi C, Langefeld CD, Freedman BI, Ng MC, Bowden DW (2011) Resequencing and analysis of variation in the TCF7L2 gene in African Americans suggests that SNP rs7903146 is the causal diabetes susceptibility variant. Diabetes 60(2):662–668. doi:10.2337/db10-0134

Palmer ND, McDonough CW, Hicks PJ, Roh BH, Wing MR, An SS, Hester JM, Cooke JN, Bostrom MA, Rudock ME, Talbert ME, Lewis JP, DIAGRAM Consortium; MAGIC Investigators, Ferrara A, Lu L, Ziegler JT, Sale MM, Divers J, Shriner D, Adeyemo A, Rotimi CN, Ng MC, Langefeld CD, Freedman BI, Bowden DW, Voight BF, Scott LJ, Steinthorsdottir V, Morris AP, Dina C, Welch RP, Zeggini E, Huth C, Aulchenko YS, Thorleifsson G, McCulloch LJ, Ferreira T, Grallert H, Amin N, Wu G, Willer CJ, Raychaudhuri S, McCarroll SA, Langenberg C, Hofmann OM, Dupuis J, Qi L, Segre AV, van Hoek M, Navarro P, Ardlie K, Balkau B, Benediktsson R, Bennett AJ, Blagieva R, Boerwinkle E, Bonnycastle LL, Bostrom KB, Bravenboer B, Bumpstead S, Burtt NP, Charpentier G, Chines PS, Cornelis M, Couper DJ, Crawford G, Doney AS, Elliott KS, Elliott AL, Erdos MR, Fox CS, Franklin CS, Ganser M, Gieger C, Grarup N, Green T, Griffin S, Groves CJ, Guiducci C, Hadjadj S, Hassanali N, Herder C, Isomaa B, Jackson AU, Johnson PR, Jorgensen T, Kao WH, Klopp N, Kong A, Kraft P, Kuusisto J, Lauritzen T, Li M, Lieverse A, Lindgren CM, Lyssenko V, Marre M, Meitinger T, Midthjell K, Morken MA, Narisu N, Nilsson P, Owen KR, Payne F, Perry JR, Petersen AK, Platou C, Proenca C, Prokopenko I, Rathmann W, Rayner NW, Robertson NR, Rocheleau G, Roden M, Sampson MJ, Saxena R, Shields BM, Shrader P, Sigurdsson G, Sparso T, Strassburger K, Stringham HM, Sun Q, Swift AJ, Thorand B, Tichet J, Tuomi T, van Dam RM, van Haeften TW, van Herpt T, van Vliet-Ostaptchouk JV, Walters GB, Weedon MN, Wijmenga C, Witteman J, Bergman RN, Cauchi S, Collins FS, Gloyn AL, Gyllensten U, Hansen T, Hide WA, Hitman GA, Hofman A, Hunter DJ, Hveem K, Laakso M, Mohlke KL, Morris AD, Palmer CN, Pramstaller PP, Rudan I, Sijbrands E, Stein LD, Tuomilehto J, Uitterlinden A, Walker M, Wareham NJ, Watanabe RM, Abecasis GR, Boehm BO, Campbell H, Daly MJ, Hattersley AT, Hu FB, Meigs JB, Pankow JS, Pedersen O, Wichmann HE, Barroso I, Florez JC, Frayling TM, Groop L, Sladek R, Thorsteinsdottir U, Wilson JF, Illig T, Froguel P, van Duijn CM, Stefansson K, Altshuler D, Boehnke M, McCarthy MI, Soranzo N, Wheeler E, Glazer NL, Bouatia-Naji N, Magi R, Randall J, Johnson T, Elliott P, Rybin D, Henneman P, Dehghan A, Hottenga JJ, Song K, Goel A, Egan JM, Lajunen T, Doney A, Kanoni S, Cavalcanti-Proenca C, Kumari M, Timpson NJ, Zabena C, Ingelsson E, An P, O'Connell J, Luan J, Elliott A, McCarroll SA, Roccasecca RM, Pattou F, Sethupathy P, Ariyurek Y, Barter P, Beilby JP, Ben-Shlomo Y, Bergmann S, Bochud M, Bonnefond A, Borch-Johnsen K, Bottcher Y, Brunner E, Bumpstead SJ, Chen YD, Chines P, Clarke R, Coin LJ, Cooper MN, Crisponi L, Day IN, de Geus EJ, Delplanque J, Fedson AC, Fischer-Rosinsky-A, Forouhi NG, Frants R, Franzosi MG, Galan P, Goodarzi MO, Graessler J, Grundy S, Gwilliam R, Hallmans G, Hammond N, Han X, Hartikainen AL, Hayward C, Heath SC, Hercberg S, Hicks AA, Hillman DR, Hingorani AD, Hui J, Hung J, Jula A, Kaakinen M, Kaprio J, Kesaniemi YA, Kivimaki M, Knight B, Koskinen S, Kovacs P, Kyvik KO, Lathrop GM, Lawlor DA, Le Bacquer O, Lecoeur C, Li Y, Mahley R, Mangino M, Manning AK, Martinez-Larrad MT, McAteer JB, McPherson R, Meisinger C, Melzer D, Meyre D, Mitchell BD, Mukherjee S, Naitza S, Neville MJ, Oostra BA, Orru M, Pakyz R, Paolisso G, Pattaro C, Pearson D, Peden JF, Pedersen NL, Perola M, Pfeiffer AF, Pichler I, Polasek O, Posthuma D, Potter SC, Pouta A, Province MA, Psaty BM, Rayner NW, Rice K, Ripatti S, Rivadeneira F, Rolandsson O, Sandbaek A, Sandhu M, Sanna S, Sayer AA, Scheet P, Seedorf U, Sharp SJ, Shields B, Sijbrands EJ, Silveira A, Simpson L, Singleton A, Smith NL, Sovio U, Swift A, Syddall H, Syvanen AC, Tanaka T, Tonjes A, Uitterlinden AG, van Dijk KW, Varma D, Visvikis-Siest S, Vitart V, Vogelzangs N, Waeber G, Wagner PJ, Walley A, Ward KL, Watkins H, Wild SH, Willemsen G, Witteman JC, Yarnell JW, Zelenika D, Zethelius B, Zhai G, Zhao JH, Zillikens MC, Borecki IB, Loos RJ, Meneton P, Magnusson PK, Nathan DM,

Williams GH, Silander K, Salomaa V, Smith GD, Bornstein SR, Schwarz P, Spranger J, Karpe F, Shuldiner AR, Cooper C, Dedoussis GV, Serrano-Rios M, Lind L, Palmer LJ, Franks PW, Ebrahim S, Marmot M, Kao WH, Pramstaller PP, Wright AF, Stumvoll M, Hamsten A, Buchanan TA, Valle TT, Rotter JI, Siscovick DS, Penninx BW, Boomsma DI, Deloukas P, Spector TD, Ferrucci L, Cao A, Scuteri A, Schlessinger D, Uda M, Ruokonen A, Jarvelin MR, Waterworth DM, Vollenweider P, Peltonen L, Mooser V, Sladek R (2012) A genome-wide association search for type 2 diabetes genes in African Americans. PLoS ONE 7(1):e29202. doi:10.1371/journal.pone.0029202

Parra EJ, Below JE, Krithika S, Valladares A, Barta JL, Cox NJ, Hanis CL, Wacher N, Garcia-Mena J, Hu P, Shriver MD, Diabetes Genetics Replication and Meta-analysis (DIAGRAM) Consortium, Kumate J, McKeigue PM, Escobedo J, Cruz M (2011) Genome-wide association study of type 2 diabetes in a sample from Mexico City and a meta-analysis of a Mexican-American sample from Starr County, Texas. Diabetologia 54(8):2038–2046. doi:10.1007/s00125-011-2172-y

Patterson N, Hattangadi N, Lane B, Lohmueller KE, Hafler DA, Oksenberg JR, Hauser SL, Smith MW, O'Brien SJ, Altshuler D, Daly MJ, Reich D (2004) Methods for high-density admixture mapping of disease genes. Am J Hum Genet 74(5):979–1000. doi:10.1086/420871

Pleis JR, Lucas JW (2009) Summary health statistics for U.S. adults: national health interview survey, 2007. Vital Health Stat 10(240):1–159

Rees SD, Islam M, Hydrie MZ, Chaudhary B, Bellary S, Hashmi S, O'Hare JP, Kumar S, Sanghera DK, Chaturvedi N, Barnett AH, Shera AS, Weedon MN, Basit A, Frayling TM, Kelly MA, Jafar TH (2011) An FTO variant is associated with Type 2 diabetes in South Asian populations after accounting for body mass index and waist circumference. Diabet Med 28(6):673–680. doi:10.1111/j.1464-5491.2011.03257.x

Rong R, Hanson RL, Ortiz D, Wiedrich C, Kobes S, Knowler WC, Bogardus C, Baier LJ (2009) Association analysis of variation in/near FTO, CDKAL1, SLC30A8, HHEX, EXT2, IGF2BP2, LOC387761, and CDKN2B with type 2 diabetes and related quantitative traits in Pima Indians. Diabetes 58(2):478–488. doi:10.2337/db08-0877

Rosenberg NA, Huang L, Jewett EM, Szpiech ZA, Jankovic I, Boehnke M (2010) Genome-wide association studies in diverse populations. Nat Rev Genet 11(5):356–366. doi:10.1038/nrg2760

Saxena R, Saleheen D, Been LF, Garavito ML, Braun T, Bjonnes A, Young R, Ho WK, Rasheed A, Frossard P, Sim X, Hassanali N, Radha V, Chidambaram M, Liju S, Rees SD, Ng DP, Wong TY, Yamauchi T, Hara K, Tanaka Y, Hirose H, McCarthy MI, Morris AP, Diagram MT, Agen BA, Barnett AH, Katulanda P, Matthews D, Mohan V, Wander GS, Singh JR, Mehra NK, Ralhan S, Kamboh MI, Mulvihill JJ, Maegawa H, Tobe K, Maeda S, Cho YS, Tai ES, Kelly MA, Chambers JC, Kooner JS, Kadowaki T, Deloukas P, Rader DJ, Danesh J, Sanghera DK (2013) Genome-wide association study identifies a novel locus contributing to type 2 diabetes susceptibility in Sikhs of Punjabi origin from India. Diabetes 62(5):1746–1755. doi:10.2337/db12-1077

Shaw JE, Sicree RA, Zimmet PZ (2010) Global estimates of the prevalence of diabetes for 2010 and 2030. Diabetes Res Clin Pract 87(1):4–14. doi:10.1016/j.diabres.2009.10.007

Shu XO, Long J, Cai Q, Qi L, Xiang YB, Cho YS, Tai ES, Li X, Lin X, Chow WH, Go MJ, Seielstad M, Bao W, Li H, Cornelis MC, Yu K, Wen W, Shi J, Han BG, Sim XL, Liu L, Qi Q, Kim HL, Ng DP, Lee JY, Kim YJ, Li C, Gao YT, Zheng W, Hu FB (2010) Identification of new genetic risk variants for type 2 diabetes. PLoS Genet 6(9), e1001127. doi:10.1371/journal.pgen.1001127

SIGMA Type 2 Diabetes Consortium, Williams AL, Jacobs SB, Moreno-Macias H, Huerta-Chagoya A, Churchhouse C, Marquez-Luna C, Garcia-Ortiz H, Gomez-Vazquez MJ, Burtt NP, Aguilar-Salinas CA, Gonzalez-Villalpando C, Florez JC, Orozco L, Haiman CA, Tusie-Luna T, Altshuler D (2014) Sequence variants in SLC16A11 are a common risk factor for type 2 diabetes in Mexico. Nature 506(7486):97–101. doi:10.1038/nature12828

Song JY, Wang HJ, Ma J, Xu ZY, Hinney A, Hebebrand J, Wang Y (2011) Association of the rs10830963 polymorphism in MTNR1B with fasting glucose levels in Chinese children and adolescents. Obes Facts 4(3):197–203. doi:10.1159/000329306

Songer TJ, Zimmet PZ (1995) Epidemiology of type II diabetes: an international perspective. Pharmacoeconomics 8(Suppl 1):1–11

Steinthorsdottir V, Thorleifsson G, Sulem P, Helgason H, Grarup N, Sigurdsson A, Helgadottir HT, Johannsdottir H, Magnusson OT, Gudjonsson SA, Justesen JM, Harder MN, Jorgensen ME, Christensen C, Brandslund I, Sandbaek A, Lauritzen T, Vestergaard H, Linneberg A, Jorgensen T, Hansen T, Daneshpour MS, Fallah MS, Hreidarsson AB, Sigurdsson G, Azizi F, Benediktsson R, Masson G, Helgason A, Kong A, Gudbjartsson DF, Pedersen O, Thorsteinsdottir U, Stefansson K (2014) Identification of low-frequency and rare sequence variants associated with elevated or reduced risk of type 2 diabetes. Nat Genet. doi:10.1038/ng.2882

Stephens JC, Briscoe D, O'Brien SJ (1994) Mapping by admixture linkage disequilibrium in human populations: limits and guidelines. Am J Hum Genet 55(4):809–824

Tabassum R, Chauhan G, Dwivedi OP, Mahajan A, Jaiswal A, Kaur I, Bandesh K, Singh T, Mathai BJ, Pandey Y, Chidambaram M, Sharma A, Chavali S, Sengupta S, Ramakrishnan L, Venkatesh P, Aggarwal SK, Ghosh S, Prabhakaran D, Srinath RK, Saxena M, Banerjee M, Mathur S, Bhansali A, Shah VN, Madhu SV, Marwaha RK, Basu A, Scaria V, McCarthy MI, Diagram I, Venkatesan R, Mohan V, Tandon N, Bharadwaj D (2013) Genome-wide association study for type 2 diabetes in Indians identifies a new susceptibility locus at 2q21. Diabetes 62(3):977–986. doi:10.2337/db12-0406

Tam CH, Ho JS, Wang Y, Lee HM, Lam VK, Germer S, Martin M, So WY, Ma RC, Chan JC, Ng MC (2010) Common polymorphisms in MTNR1B, G6PC2 and GCK are associated with increased fasting plasma glucose and impaired beta-cell function in Chinese subjects. PLoS ONE 5(7), e11428. doi:10.1371/journal.pone.0011428

Tsai FJ, Yang CF, Chen CC, Chuang LM, Lu CH, Chang CT, Wang TY, Chen RH, Shiu CF, Liu YM, Chang CC, Chen P, Chen CH, Fann CS, Chen YT, Wu JY (2010) A genome-wide association study identifies susceptibility variants for type 2 diabetes in Han Chinese. PLoS Genet 6(2), e1000847. doi:10.1371/journal.pgen.1000847

Unoki H, Takahashi A, Kawaguchi T, Hara K, Horikoshi M, Andersen G, Ng DP, Holmkvist J, Borch-Johnsen K, Jorgensen T, Sandbaek A, Lauritzen T, Hansen T, Nurbaya S, Tsunoda T, Kubo M, Babazono T, Hirose H, Hayashi M, Iwamoto Y, Kashiwagi A, Kaku K, Kawamori R, Tai ES, Pedersen O, Kamatani N, Kadowaki T, Kikkawa R, Nakamura Y, Maeda S (2008) SNPs in KCNQ1 are associated with susceptibility to type 2 diabetes in East Asian and European populations. Nat Genet 40(9):1098–1102. doi:10.1038/ng.208

Vasan SK, Karpe F, Gu HF, Brismar K, Fall CH, Ingelsson E, Fall T (2013) FTO genetic variants and risk of obesity and type 2 diabetes: a meta-analysis of 28,394 Indians. Obesity (Silver Spring). doi:10.1002/oby.20606

Voight BF, Scott LJ, Steinthorsdottir V, Morris AP, Dina C, Welch RP, Zeggini E, Huth C, Aulchenko YS, Thorleifsson G, McCulloch LJ, Ferreira T, Grallert H, Amin N, Wu G, Willer CJ, Raychaudhuri S, McCarroll SA, Langenberg C, Hofmann OM, Dupuis J, Qi L, Segre AV, van Hoek M, Navarro P, Ardlie K, Balkau B, Benediktsson R, Bennett AJ, Blagieva R, Boerwinkle E, Bonnycastle LL, Bengtsson Bostrom K, Bravenboer B, Bumpstead S, Burtt NP, Charpentier G, Chines PS, Cornelis M, Couper DJ, Crawford G, Doney AS, Elliott KS, Elliott AL, Erdos MR, Fox CS, Franklin CS, Ganser M, Gieger C, Grarup N, Green T, Griffin S, Groves CJ, Guiducci C, Hadjadj S, Hassanali N, Herder C, Isomaa B, Jackson AU, Johnson PR, Jorgensen T, Kao WH, Klopp N, Kong A, Kraft P, Kuusisto J, Lauritzen T, Li M, Lieverse A, Lindgren CM, Lyssenko V, Marre M, Meitinger T, Midthjell K, Morken MA, Narisu N, Nilsson P, Owen KR, Payne F, Perry JR, Petersen AK, Platou C, Proenca C, Prokopenko I, Rathmann W, Rayner NW, Robertson NR, Rocheleau G, Roden M, Sampson MJ, Saxena R, Shields BM, Shrader P, Sigurdsson G, Sparso T, Strassburger K, Stringham HM, Sun Q, Swift AJ, Thorand B, Tichet J, Tuomi T, van Dam RM, van Haeften TW, van Herpt T, van Vliet-Ostaptchouk JV, Walters GB, Weedon MN, Wijmenga C, Witteman J, Bergman RN, Cauchi S, Collins FS, Gloyn AL, Gyllensten U, Hansen T, Hide WA, Hitman GA, Hofman A, Hunter DJ, Hveem K, Laakso M, Mohlke KL, Morris AD, Palmer CN, Pramstaller PP, Rudan I, Sijbrands E, Stein LD, Tuomilehto J, Uitterlinden A, Walker M, Wareham NJ, Watanabe RM, Abecasis GR, Boehm BO, Campbell H, Daly MJ, Hattersley AT, Hu FB, Meigs JB, Pankow JS, Pedersen O, Wichmann HE, Barroso I, Florez JC, Frayling TM, Groop L, Sladek R, Thorsteinsdottir U, Wilson JF, Illig T, Froguel P, van Duijn CM, Stefansson K, Altshuler D, Boehnke M, McCarthy MI, Investigators M, GIANT Consortium (2010) Twelve type 2 diabetes susceptibility loci identified through large-scale association analysis. Nat Genet 42(7):579–589. doi:10.1038/ng.609

Wang H, Liu L, Zhao J, Cui G, Chen C, Ding H, Wang DW (2013) Large scale meta-analyses of fasting plasma glucose raising variants in GCK, GCKR, MTNR1B and G6PC2 and their impacts on type 2 diabetes mellitus risk. PLoS ONE 8(6), e67665. doi:10.1371/journal.pone.0067665

Watanabe RM, Allayee H, Xiang AH, Trigo E, Hartiala J, Lawrence JM, Buchanan TA (2007) Transcription factor 7-like 2 (TCF7L2) is associated with gestational diabetes mellitus and interacts with adiposity to alter insulin secretion in Mexican Americans. Diabetes 56 (5):1481–1485. doi:10.2337/db06-1682

Waters KM, Stram DO, Hassanein MT, Le Marchand L, Wilkens LR, Maskarinec G, Monroe KR, Kolonel LN, Altshuler D, Henderson BE, Haiman CA (2010) Consistent association of type 2 diabetes risk variants found in europeans in diverse racial and ethnic groups. PLoS Genet 6 (8):e1001078. doi:10.1371/journal.pgen.1001078

Xia Q, Chen ZX, Wang YC, Ma YS, Zhang F, Che W, Fu D, Wang XF (2012) Association between the melatonin receptor 1B gene polymorphism on the risk of type 2 diabetes, impaired glucose regulation: a meta-analysis. PLoS ONE 7(11), e50107. doi:10.1371/journal.pone.0050107

Yamauchi T, Hara K, Maeda S, Yasuda K, Takahashi A, Horikoshi M, Nakamura M, Fujita H, Grarup N, Cauchi S, Ng DP, Ma RC, Tsunoda T, Kubo M, Watada H, Maegawa H, Okada-Iwabu M, Iwabu M, Shojima N, Shin HD, Andersen G, Witte DR, Jorgensen T, Lauritzen T, Sandbaek A, Hansen T, Ohshige T, Omori S, Saito I, Kaku K, Hirose H, So WY, Beury D, Chan JC, Park KS, Tai ES, Ito C, Tanaka Y, Kashiwagi A, Kawamori R, Kasuga M, Froguel P, Pedersen O, Kamatani N, Nakamura Y, Kadowaki T (2010) A genome-wide association study in the Japanese population identifies susceptibility loci for type 2 diabetes at UBE2E2 and C2CD4A-C2CD4B. Nat Genet 42(10):864–868. doi:10.1038/ng.660

Yang Q, Liu T, Shrader P, Yesupriya A, Chang MH, Dowling NF, Ned RM, Dupuis J, Florez JC, Khoury MJ, Meigs JB, Investigators M (2010) Racial/ethnic differences in association of fasting glucose-associated genomic loci with fasting glucose, HOMA-B, and impaired fasting glucose in the U.S. adult population. Diabetes Care 33(11):2370–2377. doi:10.2337/dc10-0898

Yasuda K, Miyake K, Horikawa Y, Hara K, Osawa H, Furuta H, Hirota Y, Mori H, Jonsson A, Sato Y, Yamagata K, Hinokio Y, Wang HY, Tanahashi T, Nakamura N, Oka Y, Iwasaki N, Iwamoto Y, Yamada Y, Seino Y, Maegawa H, Kashiwagi A, Takeda J, Maeda E, Shin HD, Cho YM, Park KS, Lee HK, Ng MC, Ma RC, So WY, Chan JC, Lyssenko V, Tuomi T, Nilsson P, Groop L, Kamatani N, Sekine A, Nakamura Y, Yamamoto K, Yoshida T, Tokunaga K, Itakura M, Makino H, Nanjo K, Kadowaki T, Kasuga M (2008) Variants in KCNQ1 are associated with susceptibility to type 2 diabetes mellitus. Nat Genet 40 (9):1092–1097. doi:10.1038/ng.207

Chapter 9
Physiology Insights

Richard M. Watanabe and Torben Hansen

Abstract In recent years, the search for genetic determinants of type 2 diabetes has resulted in identification of numerous type 2 diabetes-associated loci as well as a number of loci associating with related prediabetic traits. These findings have illuminated new biological pathways contributing to the pathogenesis of type 2 diabetes, but have also demonstrated that type 2 diabetes is an extremely heterogeneous disease with limited overlap between genetic loci associating with type 2 diabetes and loci associating with diabetes-related traits, such as body mass index, fasting glucose levels, and fasting insulin levels. Combined, these loci only account for a fraction of the observed familial clustering of type 2 diabetes and only up to about 10 % of the variation in prediabetic quantitative traits. Improved methods are needed to dig deeper into a biological understanding of the pathophysiology of type 2 diabetes.

9.1 Introduction

"There may be 2 types of diabetes, but there's more than 2 types of patients with diabetes." This quote comes from an advertising campaign started in 2009 by Novo Nordisk. This was a simple message highlighting the challenge faced by clinicians across the globe struggling to treat and prevent diabetes. This simple message also highlights the challenges faced by geneticists as we gain more insights into the genetic basis for diabetes and attempt to translate genetic findings into clinically relevant tools or interventions. Diabetes, as a group of related conditions, involves a

R.M. Watanabe (✉)
Department of Preventive Medicine, Keck School of Medicine of USC, 2250 Alcazar Street, CSC-204, Los Angeles, CA 90089-9073, USA
e-mail: rwatanab@usc.edu

T. Hansen
The Novo Nordisk Foundation Center for Basic Metabolic Research, University of Copenhagen, Copenhagen, Denmark
e-mail: torben.hansen@sund.ku.dk

J.C. Florez (ed.), *The Genetics of Type 2 Diabetes and Related Traits*,
DOI 10.1007/978-3-319-01574-3_9

complex interplay of genetic risk, lifestyle, and socioeconomic factors. Although broadly categorized into two major forms, type 1 and type 2 diabetes, monogenic, syndromic, and other rare forms of the disease exists. It is easy to forget that the first description of diabetes as a medical condition was as early as 1500 BC and research into the underlying physiology and causes of diabetes has stretched across centuries. The complexity of the disease led the late James Neel (1915–2000) to declare diabetes as the "geneticists' nightmare" (Neel 1976).

However, recent advances in genetics have resulted in the identification of loci underlying risk for both type 1 and type 2 diabetes mainly through genome-wide association (GWA) studies. These findings have not just illuminated new biological pathways contributing to the pathogenesis of diabetes but have reinforced and provided additional insights into previously known pathways and revealed potential new pharmacologic targets. These new genetic insights have slowly allowed us to awaken from the "geneticists' nightmare."

In this chapter, we will explore the relationship between genetic variation and the pathophysiology of type 2 diabetes. We will examine how physiology can be used to inform genetics and how genetics can inform physiology. The complex relationship between genetic variation and physiology, so-called genotype-phenotype relationships, will be critical to understanding how genetic variation may be leveraged to derive improved interventional strategies, both pharmacologic and lifestyle, that may reduce diabetes-attributable morbidity and mortality.

9.2 Hallmarks of Type 2 Diabetes

Insulin resistance, pancreatic beta-cell dysfunction, and obesity are traditional hallmarks of type 2 diabetes. Parsing the disease to these three components helps to distinguish it from other forms of diabetes, but each component in and of itself is a complex phenotype we have yet to fully understand. Furthermore, each component is part of a larger complex feedback regulatory system designed to maintain glycemia in the normal range. The robust nature of this system is evident when examining the progression toward type 2 diabetes. Buchanan et al. in Mexican Americans (Xiang et al. 2006) and Mason et al. in Pima Indians (Mason et al. 2007) showed that glycemia creeps upward as patients progress toward type 2 diabetes. This progressive, but restrained, increase in glycemia reflects the physiologic system attempting to maintain glucose levels in the normal range. There is then a rapid increase in glycemic levels as beta-cell function deteriorates to a point where normoglycemia cannot be maintained. This simple glycemic pattern in the progression to diabetes provides two important lessons. First, the regulatory feedback system is very robust and can restrain glucose levels for a significant proportion of the temporal trajectory toward hyperglycemia. Second, it reveals that beta-cell function is the primary mechanism used to regulate glucose levels, and despite other pathways being available to control glucose, those pathways cannot replace the critical role of beta-cell function.

9.2.1 Obesity

There is clear evidence that type 2 diabetes and obesity are connected. Overweight and obesity have been estimated to account for 65–80 % of new cases of type 2 diabetes, and it has been demonstrated that lifestyle or pharmacologic intervention can significantly reduce risk for type 2 diabetes (Diabetes Prevention Program Research Group 2002; The Diabetes Prevention Program Research Group 2005; Buchanan et al. 2002). For years, body fat was considered only as an insulin-sensitive energy depot. However, this view has radically changed over the past couple of decades. It is now recognized that adipose tissue acts as an endocrine organ that secretes a wide variety of adipokines signaling a variety of tissues. There is additional evidence that low-grade inflammation in adipose tissue may play a critical role in altering the milieu of adipokines, which changes feedback signaling to the brain to affect feeding behavior.

One might expect significant overlap in loci contributing to risk for obesity, obesity-related traits, and type 2 diabetes, given that obesity is a known risk factor and a hallmark of type 2 diabetes. However, when one compares results from GWA studies of obesity and obesity-related traits with those identified from genetic studies of type 2 diabetes, there is very little overlap in signals (Fig. 9.1). As a simple demonstration, we examined genetic loci firmly established as type 2 diabetes risk loci for overlap with established unique "obesity" loci (adapted from Grarup et al., Diabetologia, accepted). There were only three loci that appeared in both lists: *FTO*, *MC4R*, and *QPCTL/GIPR* (Fig. 9.1). It is of interest to note that *FTO* was first identified as a type 2 diabetes locus, but its relevance to adiposity was quickly identified as the diabetes association signal was significantly attenuated when body mass index (BMI) was included as a covariate in the analysis (Diabetes Genetics Initiative of Broad Institute of Harvard and MIT Lund University and Novartis Institute for Biomedical Research et al. 2007; Scott et al. 2007; Zeggini et al. 2007).

So what is the implication of this lack of overlap? This may, in part, reflect the heterogeneity underlying both obesity and type 2 diabetes. The diagnosis of "obesity" is typically based on BMI. BMI is strongly correlated with measures of body fat, but in reality, BMI is more a measure of body density than it is of adiposity. Too often individuals with the same BMI will have widely different levels of body fat. Also, in terms of genetic studies, it is equivocal whether BMI captures the same genetic variation as body fat. This question has not been adequately addressed by the field and is not restricted to obesity, but extends to other diseases and disease-related phenotypes. In fact, even within different obesity-related anthropometrics, e.g., waist circumference, waist-to-hip ratio, etc., the overlap among GWA signals is lower than one might expect. This lack of overlap suggests that while GWA studies based on direct measures of body fat may reflect the effect of genes on adiposity, signals based on other obesity-related phenotypes may be capturing variation due to other adiposity-related biology. Additional research is needed to

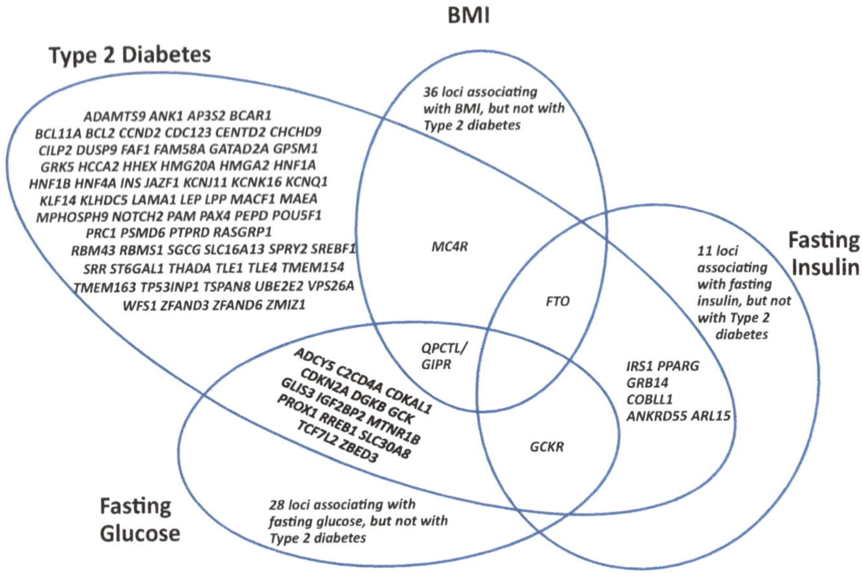

Fig. 9.1 Venn diagram of the intersection between loci associated at genome-wide significance with type 2 diabetes, BMI, fasting glucose, and fasting insulin levels. Shown are genome-wide significant associations for type 2 diabetes and three type 2 diabetes-related metabolic traits. Gene names shown in the plot are as convention the closest gene and not necessarily the functional gene

quantify the degree to which genetic variation underlying one trait is captured by related phenotypes.

Another possibility is etiologic heterogeneity introduced by the clinical definition of obesity. The current clinical standard using BMI cutoffs to classify individuals was established in 1997 by the World Health Organization (2000) and generally groups individuals as being "underweight," "normal," "overweight," "obese," or "morbidly obese." The fact that actual body fat can vary widely across individuals with the same BMI might introduce unwanted heterogeneity in a genetic analysis. Also, it is important to remember that the relationship between body fat and BMI differs by sex, so using a single scale for both sexes further blurs the boundaries between obesity categories. There is also clear evidence that the traditional BMI scale does not apply to Asian populations where individuals who might be classified as "overweight" are actually "obese" (Kanazawa et al. 2002). Thus, clinical classification can be fraught with etiologic heterogeneity that may reduce power to detect genetic loci by conventional association methods. A similar situation exists for the classification of diabetes, whose definition has changed over the last several decades coupled with the fact that the glycemic cutoffs for diagnosis are selected based on risk for retinopathy, which is also likely to result in etiologic heterogeneity.

9.2.2 Insulin Resistance

It is interesting to note that among the loci contributing to risk for type 2 diabetes to date, only a handful can be considered "insulin resistance" loci; among these are *PPARG*, *IRS1*, *GCKR*, and *IGF1* (Fig. 9.1). Why are there so few insulin resistance loci? This is an important question given the central role played by insulin resistance in the pathogenesis of type 2 diabetes coupled with the fact that a significant proportion of investigators consider understanding the biology of insulin resistance the key to "curing" type 2 diabetes. This latter viewpoint is not without merit, especially given pharmacologic intervention trials that have targeted amelioration of insulin resistance show clear reductions in risk for type 2 diabetes. For example, the Diabetes Prevention Program (DPP) showed that metformin monotherapy reduced incidence for type 2 diabetes by 31 % (Diabetes Prevention Program Research Group 2002). Although this reduction was not as dramatic as the intensive lifestyle arm of the study, it clearly demonstrated that treating hepatic insulin resistance using metformin could have an impact on risk for disease. The troglitazone arm of the DPP was prematurely terminated due to the removal of the drug from the market. Troglitazone, a member of the thiazolidinedione (TZD) class of insulin-sensitizing agents, showed a larger 75 % reduction in risk for type 2 diabetes compared to intensive lifestyle (58 % reduction) and metformin (44 % reduction) in the relevant subset of the overall DPP study (The Diabetes Prevention Program Research Group 2005). In fact, Buchanan et al. were the first to show that treating at-risk individuals with a TZD could reduce risk for future diabetes by 55 % (Buchanan et al. 2002), a result that was replicated by the DPP and in larger trials using other members of the TZD class of medications. While one might take these results to suggest that amelioration of insulin resistance was the direct cause of the reduction in risk, Buchanan et al. nicely show that, in reality, the reduction in insulin resistance was accompanied by a concomitant reduction in insulin secretion, an unloading of the pancreatic beta cells, and it is the ability to reduce the demand on the pancreatic beta cell that may be responsible for the reduction in risk and not the reduction in insulin resistance per se (Buchanan et al. 2002). This highlights the importance of carefully considering the closed-loop feedback nature of glucoregulation when considering the effect of a single change in one part of the system.

It is also important to remember that insulin resistance is not a single entity. Insulin resistance can occur in any of the insulin-sensitive tissues. In the case of type 2 diabetes, we need to carefully consider hepatic vs. peripheral insulin resistance, with the latter further partitioned into insulin resistance in primarily fat vs. muscle. Sadly, methods to quantify insulin resistance are time consuming, invasive, and costly and have rarely been applied to large sample studies. Additionally, the ability to distinguish hepatic vs. peripheral resistance requires additional use of stable label isotopes or other specialized techniques, which add an additional barrier for use in large sample studies. The lack of precise measures of insulin resistance has led many studies to rely upon indirect measures of insulin resistance, such as the HOMA-IR (Matthews et al. 1985) or Stumvoll et al. (2000) indices among others. Unfortunately, similar to the discussion regarding measures of adiposity, it is not clear these indirect measures adequately capture genetic variation underlying insulin resistance or whether these various indices capture the same genetic variation underlying insulin resistance. Evidence would suggest such indices are poor reflections of actual insulin resistance. For example, HOMA-IR has a high overall correlation with the insulin sensitivity index (S_I) from the minimal model (Bergman et al. 1981), but a low genetic correlation, suggesting that while the two indices capture the same overall variation, they are capturing different genetic information (Bergman et al. 2003). Similar results have been reported with respect to HOMA-IR and the euglycemic glucose clamp (Rasmussen-Torvik et al. 2007). Additionally, computer simulation studies suggest changes in insulin secretion can confound many of the indirect measures of insulin resistance (Hücking et al. 2008), which in turn can lead to misleading genetic associations (Watanabe 2010). Dimas et al. examined 37 SNPs showing evidence for association with type 2 diabetes from GWA studies in samples of northern European ancestry and tested them for association with a variety of type 2 diabetes-related traits (Dimas et al. 2013). These 37 SNPs were tested for association with five different indirect measures of insulin resistance/sensitivity: Belfiore et al. (1998), Stumvoll et al. (2000), Matsuda & DeFronzo (1999), Gutt et al. (2000), and HOMA-IR (Matthews et al. 1985). Each individual index showed nominal association ($p \leq 0.05$) with one or more SNPs (Dimas et al. 2013): 7 for Belfiore, 5 for Stumvoll, 8 for Matsuda, 8 for Gutt, and 10 for HOMA-IR. The association between indices and SNPs was consistent across all five indices only for *IRS1* and *PPARG*, illustrating the challenges faced when using indirect measures of insulin resistance for genetic association. It should be noted that at the time of this writing, independent GWA studies examining insulin resistance assessed by more direct measures were underway in the **M**eta-**A**nalyses of **G**lucose- and **I**nsulin-related Traits **C**onsortium (MAGIC), **GENE**tic**S** of **I**nsulin **S**ensitivity (GENESIS) consortium, and **G**enetics **U**nde**R**lying **DI**abetes in Hisp**AN**ics (GUaRDIAN) consortium. The limited sample size in these consortia will limit power to detect association, but results from those GWA meta-analyses could potentially illuminate additional loci underlying variation in insulin resistance and type 2 diabetes.

9.2.3 Beta-Cell Dysfunction

A large proportion of the signals from GWA studies of type 2 diabetes are linked to the pancreatic beta cell. This may not be surprising given the studies by Buchanan (Xiang et al. 2006) and Mason et al. (2007) showing that glycemia does not rapidly increase until beta cells fail in their ability to compensate for insulin resistance. Also, numerous studies have shown changes in insulin secretion or beta-cell function to be predictive of type 2 diabetes (Lorenzo et al. 2010; Abdul-Ghani et al. 2007; Weyer et al. 1999; Saad et al. 2005; Lyssenko et al. 2008). Signals from

GWA studies of type 2 diabetes have highlighted almost every known component of insulin secretion, e.g., glycolysis, ATP-sensitive potassium channel, components of insulin processing, transcriptional regulation of genes regulating insulin secretion, and components related to beta-cell turnover (Diabetes Genetics Initiative of Broad Institute of Harvard and MIT Lund University and Novartis Institute for Biomedical Research et al. 2007; Scott et al. 2007; Zeggini et al. 2007, 2008; Sladek et al. 2007; Steinthorsdottir et al. 2007; Florez et al. 2007; Hayes et al. 2007; Hanson et al. 2007; Yasuda et al. 2008; Unoki et al. 2008; Voight et al. 2010; Rung et al. 2009). Despite the number of type 2 diabetes signals mapping to the pancreatic beta cells, when testing 37 SNPs associated with type 2 diabetes for association with fasting insulin levels in nondiabetic individuals, only 10 showed evidence for association at $P \leq 0.05$ (Dimas et al. 2013). Interestingly, loci that are clearly related to the pancreatic beta cell, such as the zinc transporter *SLC30A8* and ATP-sensitive potassium channel *KCNJ11*, were not associated with fasting insulin, suggesting fasting insulin may not capture the effects of these gene variants. This may not be too surprising given the known physiology of insulin secretion. Insulin is essentially directly secreted into the liver where approximately 50 % is extracted before entering the systemic circulation. There is good evidence demonstrating that the level of hepatic extraction is dependent upon both insulin concentration and glucose tolerance status. Thus, the effect of genetic variants at the level of the pancreatic beta cell may not be reflected in fasting levels of insulin, which reflect the net integrated effect of secretion, hepatic extraction, and peripheral clearance. If the threshold for association were set to the genome-wide level, only *GCKR* and *IRS1* would show evidence for association with fasting insulin (Dimas et al. 2013). In fact, GWA studies of fasting insulin have identified only two loci showing convincing evidence for association: *GCKR* and *IGF1* (Dupuis et al. 2010).

The lack of association with fasting insulin suggests other beta-cell-related phenotypes need to be examined to better understand how these loci contribute to insulin secretion and beta-cell function. Indeed, when the insulinogenic index (Phillips et al. 1994) or HOMA-B (Matthews et al. 1985) is examined, a larger number of type 2 diabetes loci show evidence for association, 15 for insulinogenic index and 21 for HOMA-B, but concordance between the two indices is 78 % (Dimas et al. 2013) suggesting they may be capturing different aspects of beta-cell biology. Like indices for insulin resistance, the ability to capture loci contributing to insulin secretion or beta-cell function is highly dependent upon the phenotype examined. There is clear evidence that measures of stimulated insulin from oral glucose tolerance test (OGTT) or meal tolerance test will significantly differ from measures derived from intravenous glucose administration, e.g., hyperglycemic glucose clamp, FSIGT, or amino acid stimulation. Oral glucose engenders the release of gut hormones, primarily glucagon-like peptide-1 (GLP1) and gastric inhibitory polypeptide (GIP), which act as insulin secretagogues and enhance insulin secretion (Nauck et al. 1986). Thus, one must carefully consider whether associations with oral glucose-stimulated measures of insulin reflect biologic effects at the level of the pancreatic beta cell or effects within the incretin signaling pathway. This argues for comparing oral vs. intravenous glucose-stimulated measures of insulin secretion to tease out incretin effects from non-incretin effects and/or to examine stimulated incretin release and incretin effects on beta-cell function.

Amino acid stimulation during a hyperglycemic clamp is typically used to assess beta-cell mass as opposed to insulin secretion. Thus far, there have not been GWA studies examining amino acid-stimulated insulin secretion or studies of more limited sample size. Such studies could differentiate loci involved in regulation of beta-cell mass from those involved in insulin processing or secretion.

Most investigators make no distinction between insulin secretion and pancreatic beta-cell function. However, these represent different aspects of insulin response to stimuli. Insulin secretion represents the amount of insulin released, in total or per unit time, in response to a stimulus. Beta-cell function is a quantitative assessment of the ability of beta cells to respond to a stimulus. Differentiating these concepts is important to the understanding of the role of the pancreatic beta cell in the pathogenesis of type 2 diabetes and in the interpretation of genetic associations. For many years, studies measured and compared insulin secretion across groups to draw conclusions regarding the role of insulin secretion in the pathogenesis of type 2 diabetes. However, this approach ignored the underlying feedback regulatory nature of insulin resistance and secretion. Differences in underlying insulin resistance confounded direct comparison of insulin secretion among groups. Bergman and colleagues introduced the disposition index (DI), which generally states there is a hyperbolic relationship between insulin secretion and insulin sensitivity (Bergman et al. 1981; Bergman 2007), i.e., the pancreatic beta cells adjust their output in a nonlinear relationship to the level of insulin sensitivity. This relationship could be captured by DI, which was derived as the multiplicative relationship between insulin secretion, measured as the acute insulin response (AIR) to glucose, and S_I ($DI = S_I \times AIR$) (Bergman et al. 1981; Bergman 2007). It is important to emphasize that DI is the quantitative representation of the hyperbolic relationship and by definition is unitless. This fact may seem trivial, but is an important distinction of DI as a quantitative measure of beta-cell function. Many have taken to multiplying any measure of insulin secretion with any measure of insulin resistance and calling that quantity a "disposition index." However, indices constructed in this fashion tend not to be unitless and therefore are not hyperbolic in nature. They form what can be considered "disposition index-like" measures of beta-cell function. The fact that these indices are not hyperbolic means they will over- or underestimate true beta-cell function at the extremes of insulin resistance.

The general distinction in conclusions derived from examining insulin secretion vs. beta-cell function can be summarized by examining Fig. 9.2. Results from the Insulin Resistance Atherosclerosis Study (IRAS) are summarized showing the average S_I, AIR, and DI for individuals with normal glucose tolerance (NGT), impaired glucose tolerance (IGT), and type 2 diabetes (Wagenknecht et al. 1995). One might conclude focusing on AIR alone that individuals with IGT have a modest 32 % beta-cell "defect" compared to individuals with NGT, while individuals with type 2 diabetes have a more extreme 87 % "defect" compared to

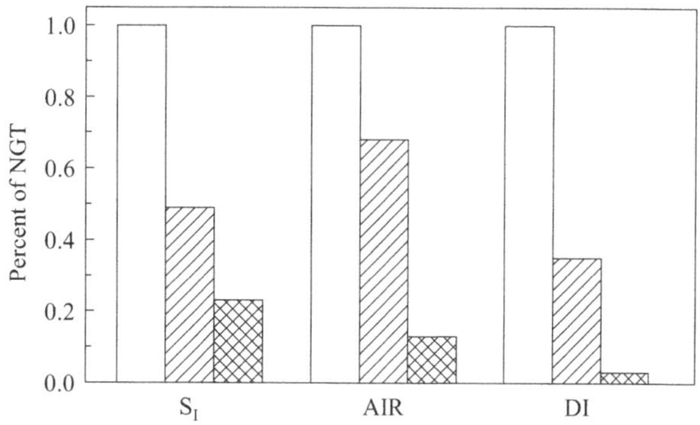

Fig. 9.2 Comparison of S_I, AIR, and DI among individuals with NGT, IGT, or type 2 diabetes from the IRAS. Average data are represented relative to the NGT group for each measurement. NGT ($n = 674$) are represented by clear bars, IGT ($n = 336$) are represented by hatched bars, and type 2 diabetes ($n = 485$) are represented by cross-hatched bars

individuals with NGT. However, note these groups have substantial differences in underlying insulin resistance. This means that beta-cell function, as represented by DI, reveals a substantial beta-cell "defect" in individuals with IGT compared to NGT (65 % lower) and almost nonexistent beta-cell function in individuals with type 2 diabetes compared to NGT (97 % lower) (Fig. 9.2). In other words, both the IGT and type 2 diabetes groups need to be secreting significantly more insulin than the NGT group to compensate for the underlying insulin resistance and maintain normal glucose tolerance. Thus, focusing on AIR alone would result in a substantial underestimation of the existing beta-cell defect. Insulin secretion is diminished, but given the underlying differences in insulin resistance, the pancreatic beta cells in individuals with IGT and type 2 diabetes are functioning significantly below levels necessary to maintain normal glucose tolerance. The distinction between secretion and function has important implications for the physiologic interpretation of the effect of genetic variation on the pancreatic beta cell and the different aspects of beta-cell biology they represent.

9.3 Discovery in Individuals Without Diabetes

Discovery of loci underlying type 2 diabetes need not be restricted to solely examining the disease as an outcome phenotype. Identifying loci underlying variation in type 2 diabetes-related traits has the potential to illuminate other loci that may contribute to risk for the disease. The MAGIC was established with this approach in mind and has performed a series of GWA studies in individuals with fasting glucose <7 mM and has identified numerous loci underlying variation in fasting glucose (Dupuis et al. 2010; Prokopenko et al. 2009), OGTT 2-h glucose (Saxena et al. 2010), and fasting insulin (Dupuis et al. 2010). Although the majority of these loci appear to only contribute to the variation in these phenotypes, several, like *MTNR1B* and *ADCY5*, were found to also be associated with type 2 diabetes (Fig. 9.1), supporting the strategy of examining disease-related traits to further identify loci underlying disease risk. Unlike other traits, when the known type

2 diabetes risk loci were tested for association with fasting glucose in the study by Dimas et al. (2013), a larger proportion, 25 of the 37 SNPs, showed evidence for association with fasting glucose, which is also reflected in Fig. 9.1.

One might ask the question, "why does association with fasting glucose reveal so few additional type 2 diabetes susceptibility loci?" One potential answer is the sample used for such association studies. The MAGIC analyses, which has identified the majority of variants underlying glycemic traits, were restricted to individuals with fasting glucose <7 mM and not taking anti-diabetes medications (Dupuis et al. 2010; Prokopenko et al. 2009; Saxena et al. 2010). This restriction avoided the confounding effects of hyperglycemia per se and effects due to treatment for diabetes. Therefore, the analysis was based on individuals who were normoglycemic or modestly hyperglycemic, the latter comprising individuals with IGT or impaired fasting glucose (IFG) (The Expert Committee on the Diagnosis and Classification of Diabetes Mellitus 1997). The IGT/IFG grouping consists of individuals who are at risk for future development of diabetes and from a clinical perspective forms a "prediabetes" category. However, the proportion of individuals with IGT converting to type 2 diabetes widely varies by population (18–50 %), suggesting a relatively large proportion never covert to type 2 diabetes. Thus, only type 2 diabetes risk variants with relatively strong effect are likely to be identified from the analysis of glycemic traits in a sample that is a mixture of NGT/IGT/IFG individuals. This may explain why many of the variants underlying variation in glycemic traits do not show subsequent association with type 2 diabetes. However, understanding the general genetic architecture underlying day-to-day regulation of fasting glucose can reveal important insights into the pathogenesis of type 2 diabetes. We encourage the reader to examine the study by Ingelsson et al. (2010), who tested the known variants underlying fasting glucose for association with other type 2 diabetes-related traits in order to gain insights into the physiology of glucoregulation.

9.4 What About All the Known Variants?

The study by Dimas et al. (2013) represents one of the early attempts to better understand the underlying physiology of the genetics of type 2 diabetes. However, this study was restricted by the available phenotypes, mainly fasting values and variables constructed from the fasting data. Classic physiologic studies have shown that more detailed phenotype can provide unique insights into the underlying biology of type 2 diabetes that many times cannot be gained through the study of readily available phenotypes. Also, based on our physiologic knowledge of type 2 diabetes, we know that gene–gene and gene-environment interactions are likely to be a critical component of the larger genetic architecture of the disease.

As part of the ongoing attempts to understand the role of gene variants and to push forward discovery, many groups have taken to using the genotype score or genetic score as a single representation of the net effect of genetic variation and test this score for association with phenotypes of interest. There are several assumptions underlying the use of the genotype score, but fundamentally, such a score represents the overall genetic burden for a given phenotype. However, the score also assumes that the effect of each locus is additive and actually contributing information regarding the phenotypic variation. The results of an analysis in which a genotype score based on type 2 diabetes risk loci is created and tested for association with diabetes-related phenotypes to gain physiologic insights could be misleading. As an example, we examined 38 type 2 diabetes risk variants for association with type 2 diabetes-related traits in the BetaGene study (Watanabe et al. 2007). We performed three different analyses: univariate association with each trait and correcting the significance level for the number of SNPs tested, multiple linear regression analysis where each SNP is tested for inclusion in the overall model conditional on the other SNPs followed by an omnibus test of association for SNPs remaining in the model, and genotype score-based association. We restricted our analyses to only samples with complete data across all SNPs to ensure fair comparison among approaches and adjusted for the effects of age and sex.

There were only three SNPs showing evidence for association with three traits in the univariate analysis. This reflects the relatively low power given the modest sample size of approximately 700 individuals and the need to correct for multiple testing. *KCNQ1* rs2237892 was associated with both disposition index (corrected $P = 0.002$) and insulin clearance rate ($P = 0.032$). *GCKR* rs780094 was associated with triglyceride levels ($P = 0.002$) replicating a previously known association, and *MTNR1B* rs10830963 was associated with disposition index ($P = 0.002$), replicating results we previously reported in the complete BetaGene sample (Ren et al. 2014). We can look through the pattern of associations across phenotypes for each individual SNP to make an inference on its physiologic effect, but this is fraught with potential bias, since the inference would be based on our perceived knowledge of the underlying biology. The results of the multiple regression analysis are summarized in Table 9.1. The number of SNPs comprising the final association model ranged from a minimum of 4 to a maximum of 10, depending upon the trait. On average, these models accounted for 3.8 % of the variation in phenotype, with a low of 0.79 % for the 3 SNPs associated with HDL cholesterol ($P = 0.137$) to a high of 7.2 % for the 8 SNPs associated with AIR ($P = 8.0 \times 10^{-7}$). The SNPs associated with AIR represented *ADCY5*, *CDKAL1*, *DGKB/TMEM*, *KCNQ1*, *MTNR1B*, and *PRC1*, all loci that have relevant biology for the pancreatic beta cell and subsequent effects on insulin secretion. Scanning across the results, different subsets of SNPs show statistically significant association with different traits. This approach provides an opportunity to isolate loci that may have the greatest relevance for a given phenotype conditional on the effects of the other SNPs and provide a "best" subset of loci that helps to refine the physiologic interpretation.

In contrast, when the genotype score is tested for association with the phenotypes, on average only 0.44 % of the variation in phenotype is accounted for the by score (Table 9.1). The genotype score accounted for essentially none of the variation in fasting glucose, diastolic blood pressure, and BMI, but accounted for

Table 9.1 Comparison of multiple regression and genotype score association analysis for 38 known type 2 diabetes SNPs

		Multiple regression analysis			Genotype score	
Trait	n	No. of SNPs[a]	% of variation explained[b]	*P*-value*	% of variation explained[b]	*P*-value*
BMI	719	4	3.03	0.002	0	0.92
Percentage body fat	719	8	1.38	9.9×10^{-5}	0.08	0.56
Fasting glucose	727	5	2.27	0.03	0	0.78
2-h glucose	723	5	2.37	0.03	0.24	0.11
Fasting insulin	726	10	4.81	6.3×10^{-5}	0	0.54
2-h insulin	722	10	4.12	3.9×10^{-4}	0.14	0.09
30-min Δinsulin	717	6	3.99	4.0×10^{-4}	0.25	0.22
Glucose effectiveness	718	9	4.69	5.0×10^{-4}	0.68	0.02
Insulin sensitivity	718	7	3.27	0.004	0.01	0.94
AIR	718	8	7.19	8.0×10^{-7}	2.69	4.6×10^{-5}
DI	718	5	7.19	5.5×10^{-9}	2.48	1.6×10^{-5}
Insulin clearance	717	9	5.49	4.9×10^{-5}	0.02	0.82
Cholesterol	718	4	3.39	6.7×10^{-4}	0.40	0.23
HDL cholesterol	718	3	0.79	0.14	0.21	0.17
Triglycerides	718	7	4.44	3.8×10^{-5}	0.04	0.71
Systolic BP	721	3	2.93	2.9×10^{-4}	0.28	0.22
Diastolic BP	721	5	3.36	5.9×10^{-4}	0	0.87

**P*-value for the omnibus test of the effect of SNPs or genotype score
[a]Number of SNPs that entered into the multiple regression analysis
[b]The percentage of total phenotype variation explained by either SNPs or genotype score

2.69 % of the variation in AIR ($P = 4.6 \times 10^{-5}$) and 2.48 % of the variation in DI ($P = 1.6 \times 10^{-5}$). There was also evidence for association with glucose effectiveness ($P = 0.02$), a measure of the ability of glucose per se to enhance glucose uptake and suppress hepatic glucose output at fasting insulin (Bergman et al. 1979, 1981; Ader & Bergman 1985). These associations may not be too surprising given that the genotype score represents the overall genetic load for type 2 diabetes and the key phenotype for the transition to type 2 diabetes based on physiologic studies is insulin secretion or beta-cell function. The result for glucose effectiveness may also not be too surprising, given that this parameter has been shown to be a significant contributor to the regulation of glucose tolerance. However, this example illustrates the need to carefully consider the use of the genetic score.

The indiscriminate use of the genotype score can lead to a misguided physiologic interpretation of the role of genetic variation in the pathophysiology of disease.

9.5 Final Thoughts

The search for genetic risk factors in type 2 diabetes and related traits is now targeting low-frequency and rare variations as well as copy number variations (see Chap. 5). It was expected that combining major data sets will enable discovery of rare variants and structural variations with relatively high effect sizes within the coming few years. However, initial results suggest relatively low effect sizes even for rare variants, which add an additional piece of the puzzle that is the genetics of type 2 diabetes. However, gene variants, both common and rare, have a multitude of effects, and their contributions to underlying disease mechanisms are at present unknown. Systemic integration of complex data obtained from other "omics" techniques such as transcriptomics, proteomics, metabolomics, and microbiomics and modeling of the combined composite impact of common metabolic phenotypes are projected to illuminate further breakthroughs in understanding the genetic determinants of type 2 diabetes and metabolic dysfunctions. Possibly, genetic identification of pathophysiological specific subgroups of patients will pave the road for stratified treatment to a highly heterogeneous patient group.

Acknowledgments RMW gratefully acknowledges the contributions made by colleagues in the BetaGene study and MAGIC that are highlighted in this chapter. RMW also acknowledges the efforts of Nan Wang, who performed the univariate, multivariate, and genotype score-based analyses presented in this chapter.

References

Abdul-Ghani MA, Williams K, DeFronzo RA et al (2007) What is the best predictor of future type 2 diabetes? Diabetes Care 30:1544–1548

Ader M, Bergman RN (1985) Importance of glucose per se to intravenous glucose tolerance. Comparison of the Minimal-Model prediction with direct measurements. Diabetes 34:1092–1103

Belfiore F, Iannello S, Volpicelli G (1998) Insulin sensitivity indices calculated from basal and OGTT-induced insulin, glucose, and FFA levels. Mol Genet Metab 63:134–141

Bergman RN (2007) Orchestration of glucose homeostasis: from a small acorn to the California oak. Diabetes 56:1489–1501

Bergman RN, Ider YZ, Bowden CR et al (1979) Quantitative estimation of insulin sensitivity. Am J Physiol 236:E667–E677

Bergman RN, Phillips LS, Cobelli C (1981) Physiologic evaluation of factors controlling glucose tolerance in man. Measurement of insulin sensitivity and β-cell glucose sensitivity from the response to intravenous glucose. J Clin Invest 68:1456–1467

Bergman RN, Zaccaro DJ, Watanabe RM et al (2003) Minimal model-based insulin sensitivity has greater heritability and a different genetic basis than homeostasis model assessment or fasting insulin. Diabetes 52:2168–2174
Buchanan TA, Xiang AH, Peters RK et al (2002) Preservation of pancreatic β-cell function and prevention of type 2 diabetes by pharmacological treatment of insulin resistance in high-risk hispanic women. Diabetes 51:2796–2803
Diabetes Genetics Initiative of Broad Institute of Harvard and MIT Lund University and Novartis Institute for Biomedical Research, Saxena R, Voight BF et al (2007) Genome-wide association analysis identifies loci for type 2 diabetes and triglyceride levels. Science 316:1331–1336
Diabetes Prevention Program Research Group (2002) Reduction in the incidence of type 2 diabetes with lifestyle intervention or metformin. N Engl J Med 346:393–403
Dimas AS, Lagou V, Barker A et al (2013) Impact of type 2 diabetes susceptibility variants on quantitative glycemic traits reveals mechanistic heterogeneity. Diabetes 63:2158–2171. doi:10.2337/db13-0949
Dupuis J, Langenberg C, Prokopenko I et al (2010) New genetic loci implicated in fasting glucose homeostasis and their impact on type 2 diabetes risk. Nat Genet 42:105–116
Florez JC, Manning AK, Dupuis J et al (2007) A 100K genome-wide association scan for diabetes and related traits in the Framinham Heart Study. Diabetes 56:3063–3074
Gutt M, Davis CL, Spitzer SB et al (2000) Validation of the insulin sensitivity index (ISI(0,120)): comparison with other measures. Diabetes Res Clin Pract 47:177–184
Hanson RL, Bogardus C, Duggan D et al (2007) A search for variants associated with young-onset type 2 diabetes in American Indians in a 100K genotyping array. Diabetes 56:3045–3052
Hayes MG, Pluzhnikov A, Miyake K et al (2007) Identification of type 2 diabetes genes in Mexican Americans through genome-wide association studies. Diabetes 56:3033–3044
Hücking K, Watanabe RM, Stefanovski D et al (2008) OGTT-derived measures of insulin sensitivity are confounded by factors other than insulin sensitivity itself. Obesity 16:1938–1945
Ingelsson E, Langenberg C, Hivert MF et al (2010) Detailed physiologic characterization reveals diverse mechanisms for novel genetic Loci regulating glucose and insulin metabolism in humans. Diabetes 59:1266–1275
Kanazawa M, Yoshiike N, Osaka T et al (2002) Criteria and classification of obesity in Japan and Asia-Oceania. Asia Pac J Clin Nutr 11(Suppl 8):S732–S737, S732–S737
Lorenzo C, Wagenknecht LE, Rewers MJ et al (2010) Disposition index, glucose effectiveness, and conversion to type 2 diabetes: the Insulin Resistance Atherosclerosis Study (IRAS). Diabetes Care 33:2098–2103
Lyssenko V, Jonsson A, Almgren P et al (2008) Clinical risk factors, DNA variants, and the development of type 2 diabetes. N Engl J Med 359:2220–2232
Mason CC, Hanson RL, Knowler WC (2007) Progression to type 2 diabetes characterized by moderate then rapid glucose increases. Diabetes 56:2054–2061
Matsuda M, DeFronzo RA (1999) Insulin sensitivity indices obtained from oral glucose tolerance testing: comparison with the euglycemic insulin clamp. Diabetes Care 22:1462–1470
Matthews DR, Hosker JP, Rudenski AS et al (1985) Homeostasis model assessment: insulin resistance and β-cell function from fasting plasma glucose and insulin concentrations in man. Diabetologia 28:412–419
Nauck MA, Homberger E, Siegel EG et al (1986) Incretin effects of increasing glucose loads in man calculated from venous insulin and C-peptide responses. J Clin Endocrinol Metab 63:492–498
Neel JV (1976) Diabetes mellitus - a geneticist's nightmare. In: Creutzfeldt W, Kobberling J, Neel JV (eds) The genetics of diabetes. Springer, New York, pp 1–11
Phillips DI, Clark PM, Hales CN et al (1994) Understanding oral glucose tolerance: comparison of glucose or insulin measurements during the oral glucose tolerance test with specific measurements of insulin resistance and insulin secretion. Diabet Med 11:286–292

Prokopenko I, Langenberg C, Florez JC et al (2009) Variants in MTNR1B influence fasting glucose levels. Nat Genet 41:77–81

Rasmussen-Torvik LJ, Pankow JS, Jacobs DR et al (2007) Heritability and genetic correlations of insulin sensitivity measured by the euglycaemic clamp. Diabet Med 24:1286–1289

Ren J, Xiang AH, Trigo E et al (2014) Genetic variation in MTNR1B is associated with gestational diabetes mellitus and contributes only to the absolute level of beta cell compensation in Mexican Americans. Diabetologia. doi:10.1007/s00125-014-3239-3:

Rung J, Cauchi S, Albrechtsen A et al (2009) Genetic variant near IRS1 is associated with type 2 diabetes, insulin resistance and hyperinsulinemia. Nat Genet 41:1110–1115

Saad R, Gungor N, Arslanian S (2005) Progression from normal glucose tolerance to type 2 diabetes in a young girl: longitudinal changes in insulin sensitivity and secretion assessed by the clamp technique and surrogate estimates. Pediatr Diabetes 6:95–99

Saxena R, Hivert MF, Langenberg C et al (2010) Genetic variation in GIPR influences the glucose and insulin responses to an oral glucose challenge. Nat Genet 42:142–148

Scott LJ, Mohlke KL, Bonnycastle LL et al (2007) A genome-wide association study of type 2 diabetes in Finns detects multiple susceptibility variants. Science 316:1341–1345

Sladek R, Rocheleau G, Rung J et al (2007) A genome-wide association study identified novel risk loci for type 2 diabetes. Nature 445:881–885

Steinthorsdottir V, Thorleifsson G, Reynisdottir I et al (2007) A variant in CDKAL1 influences insulin response and risk of type 2 diabetes. Nat Genet 39:770–775

Stumvoll M, Mitrakou A, Pimenta W et al (2000) Use of the oral glucose tolerance test to assess insulin release and insulin sensitivity. Diabetes Care 23:295–301

The Diabetes Prevention Program Research Group (2005) Prevention of type 2 diabetes with troglitazone in the Diabetes Prevention Program. Diabetes 54:1150–1156

The Expert Committee on the Diagnosis and Classification of Diabetes Mellitus (1997) Report of the expert committee on the diagnosis and classification of diabetes mellitus. Diabetes Care 20:1183–1197

Unoki H, Takahashi A, Kawaguchi T et al (2008) SNPs in KCNQ1 are associated with susceptibility to type 2 diabetes in East Asian and European populations. Nat Genet 40:1098–1102

Voight BF, Scott LJ, Steinthorsdottir V et al (2010) Twelve type 2 diabetes susceptibility loci identified through large-scale association analysis. Nat Genet 42:579–589

Wagenknecht LE, Mayer EJ, Rewers M et al (1995) The Insulin Resistance Atherosclerosis Study (IRAS) objectives, design, and recruitment results. Ann Epidemiol 5:464–472

Watanabe RM (2010) The genetics of insulin resistance: where's Waldo? Curr Diab Rep 10:476–484

Watanabe RM, Allayee H, Xiang AH et al (2007) Transcription factor 7-like 2 (*TCF7L2*) is associated with gestational diabetes mellitus and interacts with adiposity to alter insulin secretion in Mexican Americans. Diabetes 56:1481–1485

Weyer C, Bogardus C, Mott DM et al (1999) The natural history of insulin secretory dysfunction and insulin resistance in the pathogenesis of type 2 diabetes mellitus. J Clin Invest 104:787–794

World Health Organization (2000) Obesity: preventing and managing the global epidemic. Report of a WHO consultation. World Health Organ Tech Rep Ser 894:i–xii, 1–253; i–253

Xiang AH, Wang C, Peters RK et al (2006) Coordinate changes in plasma glucose and pancreatic β-cell function in Latino women at high risk for type 2 diabetes. Diabetes 55:1074–1079

Yasuda K, Miyake K, Horikawa Y et al (2008) Variants in KCNQ1 are associated with susceptibility to type 2 diabetes mellitus. Nat Genet 40:1092–1097

Zeggini E, Weedon MN, Lindgren CM et al (2007) Replication of genome-wide association signals in U.K. samples reveals risk loci for type 2 diabetes. Science 316:1336–1341

Zeggini E, Scott LJ, Saxena R et al (2008) Meta-analysis of genome-wide association data and large-scale replication identifies additional susceptibility loci for type 2 diabetes. Nat Genet 40:638–645

Chapter 10
Insights from Monogenic Diabetes

Tormund Salvesen Njølstad, Sian Ellard, and Pål Rasmus Njølstad

Abstract Monogenic diabetes arises from single-gene mutations, mostly affecting β-cell physiology and endocrine function. More than 30 different genes in which mutations cause monogenic diabetes have been described. The identification of these genes has shed new light on their biological function as well as etiology and mechanisms of common forms of diabetes. Monogenic diabetes is traditionally subclassified as neonatal diabetes, maturity-onset diabetes of the young (MODY), or syndromic diabetes. The most common forms have few features other than endocrine dysfunction. However, in some families, additional clinical features including neurological abnormalities, urogenital malformations, or exocrine pancreas dysfunction may be present and can aid the diagnostic classification. Correct diagnosis according to the genetic etiology has important implications for both prognosis and treatment. While mutations in *GCK* cause a mild fasting hyperglycemia which rarely needs insulin treatment and has a low risk for complications, *HNF1A* mutations lead to diabetes in that, in severity, treatment and complication risk resembles type 1 diabetes (T1D). Furthermore, the majority of patients with mutations in the MODY genes *HNF1A* and *HNF4A*, as well as the neonatal diabetes genes *ABCC8* and *KCNJ11*, experience that sulfonylurea treatment is superior to insulin. Genome-wide approaches using next-generation DNA sequencing technology have revealed new genetic etiologies, and targeted next-generation sequencing assays now allow simultaneous testing of all monogenic diabetes genes in a single test. We expect that this technology will facilitate further important breakthroughs in unraveling the causes of monogenic diabetes during the next few years.

T.S. Njølstad • P.R. Njølstad, M.D. Ph.D. (✉)
Department of Clinical Science, KG Jebsen Center for Diabetes Research, University of Bergen, Bergen, Norway

Department of Pediatrics, Haukeland University Hospital, Bergen, Norway
e-mail: pal.njolstad@uib.no

S. Ellard
Institute of Biomedical and Clinical Science, University of Exeter Medical School, Exeter, UK

J.C. Florez (ed.), *The Genetics of Type 2 Diabetes and Related Traits*,
DOI 10.1007/978-3-319-01574-3_10

10.1 Introduction

Insight into monogenic causes of non-immunological diabetes has greatly increased over the last decades. An explosion of inquisitive work fueled by vast advances in genetic research methodology and decreasing costs of genetic analysis has provided important understanding into the entire chain of steps controlling glucose homeostasis. This research has proved beneficial for patients affected with specific forms of monogenic diabetes, when in some cases tailoring therapy to an individual's genetic constitution can improve glycemic control and quality of life. Additionally, lessons from monogenic diabetes into disease mechanism by the distinct correlation between genotype and phenotype can be applicable to type 2 diabetes (T2D), suggesting that similar approaches to individualizing treatment could be beneficial. A great number of excellent and well-written reviews describing, reflecting, and elaborating on the topic of monogenic diabetes have been published the recent decade (see Aguilar-Bryan and Bryan (2008), Ashcroft and Rorsman (2012), Molven and Njølstad (2011), Murphy et al. (2008), and Rubio-Cabezas and Ellard (2013) among others). The interested reader is encouraged to consult these for further information.

10.2 Definitions

Monogenic diabetes encompasses relatively rare forms of diabetes with exclusively genetic origin, caused by single gene-gene defects or chromosomal abnormalities affecting normal β-cell physiology, development, and differentiation or the insulin gene itself. The different monogenic forms may be dominantly or recessively inherited or caused spontaneously by a de novo mutation. Traditionally, monogenic diabetes has been phenotypically classified as neonatal diabetes mellitus (NDM), maturity-onset diabetes of the young (MODY), or syndromic diabetes. Neonatal diabetes is characterized by diabetes that develops within the first 6 months of life, while MODY is commonly used to describe patients with non-insulin-dependent, nonketotic diabetes diagnosed at a young age (typically before 25 years old), with autosomal dominant inheritance and lack of autoantibodies (Tattersall et al. 1975). Some argue, however, that the denomination today is misleading, as *maturity-onset* was introduced as a term to distinguish it from *young-onset* diabetes now known as type 1 diabetes (T1D), implying a resemblance to T2D. Although MODY in its classical form is distinct from T1D and early-onset multifactorial T2D, several patients might have features of two or more diabetes types, possibly due to the increasing prevalence of obesity leading to patients that can be difficult to classify. In the 1998 revision of diabetes classification toward a more etiological basis, the term MODY is now included in the group of other forms of diabetes.

10.3 Clinical Clues of Monogenic Diabetes

Although each individual form of monogenic diabetes is much rarer than the multifactorial common subtypes of T1D or T2D, they collectively share an important proportion of total diabetes disease burden, affecting millions of people worldwide.

The majority of patients with genetically proven monogenic diabetes are initially incorrectly diagnosed as having T1D or T2D. Noticeably, a precise diagnosis is of great importance, as it can predict both clinical course and associated clinical manifestations. Most importantly, it can guide improved treatment and disease follow-up based on risk for complications, as this varies with the underlying genetic defect. Several clinical features can indicate that the initial diagnosis of T1D or T2D is incorrect and that further genetic evaluation should be considered. For example, a monogenic origin should be suspected when the diagnosis is made before 6 months of life (suggesting neonatal diabetes), if diabetes is familiar with an affected parent, if there are associated extra-pancreatic features (suggesting syndromic diabetes), and when a child has mild fasting hyperglycemia (5.5–8.5 mmol/l) (Hattersley et al. 2009). Specific examples of "atypical" T1D suggesting the possibility of a monogenic cause include evidence of endogenous insulin production outside the honeymoon period and absence of pancreatic islet antibodies, in addition to the aforementioned early diagnosis and presence of an affected parent. Features suggesting monogenic etiology in children initially thought to have T2D can be the lack of marked obesity, presence of diabetic normal-weight family members, absence of acanthosis nigricans, or lack of insulin resistance with normal fasting C-peptide. Furthermore, in addition to manifesting clinical features uncharacteristic of T1D or T2D, the patient can also have features of a specific genetic subtype of monogenic diabetes, especially in the case of concomitant extra-pancreatic, multisystemic manifestations, pointing toward a particular gene to aid in the diagnostic approach.

10.4 Monogenic Forms of Diabetes

An increasing number of single-gene abnormalities leading to monogenic diabetes have been identified, each leading to a specific disease phenotype with variations in age of onset, severity of hyperglycemia, risk of complications, associated clinical manifestations, and treatment possibilities. These subtypes are hence defined by their specific description of the known genetic defect. Table 10.1 systematically presents different monogenic diabetes genes with high penetrance, corresponding diabetes types, pattern of inheritance, and relevant clinical characteristics. Attempts of subcategorizing monogenic diabetes are multifarious: genetically splitting between impaired insulin secretion and abnormal insulin response, clinically

Table 10.1 Monogenic diabetes genes of high penetrance

Gene, chromosome, or protein affected	Type of diabetes	Inheritance	Mechanism	Prevalence	Relevant clinical characteristics
ABCC8 (SUR1)	NDM, rarely MODY	AD, often de novo, or AR	Ion channel disorder, abnormal β-cell function	~50 % of PNDM[a] ~25 % of TNDM[a]	NDM: IUGR. Most are sensitive to sulfonylurea. Some mutations cause DEND MODY: A few case reports
BLK	MODY	AD	Gene expression disorder, abnormal β-cell function	<1 % of MODY	Only five cases reported
CEL	MODY	AD	Exocrine pancreas disease	<1 % of MODY	Diabetes and pancreatic exocrine dysfunction
CFTR	Syndromic	AR	Exocrine pancreas disease	Rare	Variable age at diagnosis. Pulmonary disease, exocrine pancreas dysfunction, diabetes due to decreased insulin production secondary to exocrine dysfunction, insulin resistance
EIF2AK3	Syndromic (NDM)	AR	ER stress/cell death	~3 % of PNDM	Wolcott-Rallison syndrome. PNDM, epiphyseal dysplasia, recurrent liver dysfunction
FOXP3	Syndromic (NDM)	X-linked, recessive	Autoimmune disorder, increased β-cell destruction	~1 % of PNDM	IPEX syndrome: PNDM, immunodysregulation, polyendocrinopathy, enteropathy
GATA6	NDM	AD, often de novo	Gene expression disorder, abnormal pancreatic development	Rare	PNDM, pancreatic agenesis, congenital heart defects, and biliary abnormalities
GCK	NDM, MODY	AR (PNDM) or AD (MODY)	Abnormal β-cell function affecting set point for glucose metabolism	~4 % of PNDM ~10–80 % of MODY	PNDM: due to homozygous or compound heterozygous mutations. Markedly IUGR, insulin dependent MODY: stable elevated fasting glucose levels. Rarely microvascular complications
GLIS3	Syndromic (NDM)	AR	Gene expression disorder, abnormal pancreatic development	Rare	PNDM with congenital hypothyroidism, glaucoma, hepatic fibrosis, and renal cysts

HFE	Syndromic	AR	Iron deposition within β-cells	Rare	Variable age at diagnosis. Diabetes, hepatic disease, skin pigmentation, arthropathy, cardiomegaly
HNF1A	MODY	AD	Gene expression disorder	~20–50 % of MODY	MODY: progressive insulin secretion defect, sensitive to sulfonylureas
HNF1B	MODY, rarely NDM	AD (50 % de novo)	Gene expression disorder, abnormal pancreatic development	TNDM: rare MODY5: ~5 % of MODY	MODY/RCAD: diabetes, renal cysts, genito-urinary abnormalities, pancreatic atrophy, hyperuricemia TNDM: with pancreatic hypoplasia and renal cysts
HNF4A	MODY	AD	Gene expression disorder	~5 % of MODY	MODY: progressive insulin secretion defect, sulfonylurea sensitive. Macrosomia in ~50 % and neonatal hypoglycemia in ~10 %
IER3IP1	Syndromic (NDM)	AR	ER stress/cell death	Rare	PNDM with microcephaly, lissencephaly, and infantile epileptic encephalopathy
IL2RA	Syndromic (NDM)	AR	Systemic autoimmunity	Rare	IPEX syndrome. PNDM, immunodysre-gulation, polyendocrinopathy, enteropathy
INS	NDM, MODY	AD (often de novo), AR	Insulin synthesis disorder, increased β-cell destruction	~22 % of PNDM MODY10: <1 % of MODY	PNDM, IUGR INS-MODY or MODY10
KCNJ11 (KIR6.2)	NDM	AD (often de novo)	Ion channel disorder, abnormal β-cell function	~50 % of PNDM[a] ~25 % of TNDM[a]	NMD: IUGR; some have developmental delay and seizures (DEND). Most sensitive to sulfonylureas
KLF11	MODY	AD	Gene expression disorder	<1 % of MODY	Only two families reported
LMNA	Syndromic	AD		Rare	Partial lipodystrophy, insulin resistance
MNX1	NDM	AR	Gene expression disorder, abnormal pancreatic development	Rare	PNDM and pancreatic hypoplasia

(continued)

Table 10.1 (continued)

Gene, chromosome, or protein affected	Type of diabetes	Inheritance	Mechanism	Prevalence	Relevant clinical characteristics
MT-TL1, *MT-TE*, *MT-TK* (mtRNA genes)	Mitochondrial diabetes	Maternally inherited	Mitochondrial disorder	~1 % of monogenic diabetes	MIDD: maternally inherited diabetes and deafness syndrome
NEUROD1	NDM, MODY	AR (PNDM) or AD (MODY)	Gene expression disorder	Rare	NEUROMODY: rare PNDM: cerebellar hypoplasia, visual impairment, and deafness
NEUROG3	NDM	AR	Gene expression disorder, abnormal pancreatic development	Rare	PNDM and enteric anendocrinosis (severe malabsorptive diarrhea)
NKX2.2	NDM	AR	Gene expression disorder	Rare	
PAX4	MODY	AD	Gene expression disorder	<1 % of MODY	Only three families reported
PAX6	MODY	AR	Gene expression disorder, abnormal pancreatic development	Rare	PNDM, microphthalmia and brain malformations
PDX1	PNDM, MODY	AR (PNDM) or AD (MODY)	Gene expression disorder, abnormal pancreatic development	<1 % of MODY	MODY: only one family reported PNDM: SGA, diarrhea, malnutrition, pancreatic agenesis. Parent with MODY4
PPARG	Syndromic	AD		Rare	Partial lipodystrophy, diabetes, insulin resistance, cardiovascular complications frequent
PTF1A	Syndromic (NDM)	AR	Gene expression disorder, abnormal pancreatic development	Rare	PNDM, pancreatic and cerebellar hypoplasia/aplasia, and central respiratory dysfunction (coding mutations) or isolated pancreatic agenesis (enhancer mutations)
RFX6	Syndromic (NDM)	AR	Gene expression disorder, abnormal pancreatic development	Rare	PNDM, pancreatic and hepatobiliary abnormalities and intestinal atresia

SLC2A2 (GLUT2)	PNDM	AR	Glucose transport disorder, abnormal β-cell function	Rare	Fanconi-Bickel syndrome; PNDM, hypergalactosemia, and liver dysfunction
SLC19A2	PNDM	AR	Insulin secretion disorder, abnormal β-cell function	Rare	Roger syndrome; PNDM, thiamine-responsive megaloblastic anemia, sensori-neural deafness
TRMT10A	MODY	AR	ER stress/cell death	Rare	MODY, short stature, and microcephaly
WFS1	PNDM	AR	ER stress/cell death	Rare	Wolfram syndrome; PNDM, diabetes insipidus, optic atrophy, deafness
ZFP57	TNDM	AR	Epigenetic disorder, abnormal pancreatic development	Rare	TNDM, multiple hypomethylation syndrome, developmental delay, congenital heart disease
Chromosome 6q24 (*PLAGL1*, *HYMA1*)	NDM	AD for paternal duplication	Epigenetic disorder, abnormal pancreatic development	~70 % of TNDM	TNDM, IUGR, umbilical hernia; some have macroglossia

Adapted from reviews by Ashcroft and Rorsman (2012), Molven and Njølstad (2011), Murphy et al. (2008), Haldorsen et al. (2012)

Abbreviations: *ABCC8* ATP-binding cassette transporter subfamily C member 8, *AD* autosomal dominant, *AR* autosomal recessive, *BLK* B-lymphoid tyrosine kinase, *CEL* carboxylester lipase, *CFTR* cystic fibrosis transmembrane conductance regulator, *DEND* developmental delay, epilepsy, and neonatal diabetes, *EIF2AK3* eukaryotic translation initiation factor 2-alpha kinase 3, *ER* endoplasmic reticulum, *FOXP3* forkhead box P3, *GATA6* GATA-binding protein 6, *GCK* glucokinase, *GLIS3* GLIS family zinc finger protein 3, *GLUT2* glucose transporter 2, *HFE* hemochromatosis, *HNF1A* hepatocyte nuclear factor 1α, *HNF1B* hepatocyte nuclear factor 1β, *HNF4A* hepatocyte nuclear factor 4α, *IER3IP1* intermediate-early response 3-interacting protein 1, *IL2RA* interleukin-2 receptor alpha, *INS* insulin, *IPEX* immunodysregulation polyendocrinopathy enteropathy X-linked, *IUGR* intrauterine growth restriction, *KIR6.2* potassium channel, inwardly rectifying, *KLF11* Kruppel-like factor 11, *MNX1* motor neuron and pancreas homeobox 1, *MODY* maturity-onset diabetes of the young, *mtRNA* mitochondrial transfer ribonucleic acid, *NEUROD1* neurogenic differentiation factor 1, *NEUROG3* neurogenin 3, *NDM* neonatal diabetes mellitus, *NKX2.2* NK2 homeobox 2, *PAX4* paired box 4 gene, *PAX6* paired box 6 gene, *PDX1* pancreatic and duodenal homeobox 1, also known as insulin promoter factor 1, *PNDM* permanent neonatal diabetes mellitus, *PPARG* peroxisome proliferator-activated receptor gamma, *PTF1A* pancreas transcription factor 1A, *RFX6* regulatory factor X6, *SGA* small for gestational age, *SLC19A2* solute carrier family 19, member 2, also known as thiamine transporter 1, *SUR1* sulfonylurea receptor 1, *TNDM* transient neonatal diabetes mellitus, *WFS1* Wolfram syndrome 1, *ZFP57* zinc finger protein 57

[a]Prevalence of *ABCC8* and *KCNJ11* mutations combined

dividing between ages of symptom onset, or by the presence of extra-pancreatic manifestations (syndromic, monogenic diabetes).

In such an attempt of subcategorizing monogenic diabetes, a convenient division is to consider neonatal diabetes-only, young-onset diabetes-only without extra-pancreatic features, and syndromic diabetes independently, although increasing evidence point to some same-gene mutations manifesting effects at different ages.

10.4.1 Neonatal Diabetes Mellitus

Neonatal diabetes, frequently defined as diabetes diagnosed within 6 months of life, is a monogenic disorder caused by mutations affecting β-cell function. Once considered a rare variant of T1D, it is contrasted by the absence of islet antibodies, serving as an important diagnostic clue (one exception, however, is IPEX syndrome). It is now widely accepted that diabetes presenting before 6 months of age is unlikely to be autoimmune T1D, and an assessment for monogenic origin should be pursued (Iafusco et al. 2002). The majority of these presenting patients have an identifiable monogenic disorder responsible for their diabetes and, in some cases, also for other associated extra-pancreatic features. It should be noted that although patients presenting with diabetes at an age between 6 and 12 months have a higher baseline probability of autoimmune T1D, there are reports of neonatal diabetes debuts extending beyond 6 months of life (Mohamadi et al. 2010; Rubio-Cabezas et al. 2012).

Neonatal diabetes is a genetically heterogeneous disease, and multiple genetic abnormalities have been identified as causes. Mutated genes have the joint feature of playing key roles in β-cell function or development, often involving glucokinase, the β-cell ATP-sensitive potassium (K_{ATP}) channel, or insulin itself. Consequently, the majority of cases exhibit impaired insulin secretion rather than impaired insulin sensitivity. As insulin is an important growth factor, children with neonatal diabetes often have intrauterine growth restriction and are born small for gestational age. The diabetic manifestation may portray a permanent (PNDM), transient (TNDM), or relapsing-remitting course and is often classified according to this presentation. Transient neonatal diabetes usually remits within 3 months after presentation on average and within 18 months of age. Interestingly, it frequently relapses later in life in adolescent or young adult age (Temple et al. 2000). Permanent neonatal diabetes, on the other hand, requires lifelong medical therapy.

10.4.1.1 Transient Neonatal Diabetes Mellitus

Children with transient neonatal diabetes are typically of low birth weight, and diabetes is usually diagnosed within the first weeks of life due to development of severe nonketotic hyperglycemia. One-third also have macroglossia, and some have umbilical hernia. The majority of cases remit within 12 weeks of diagnosis with relapse commonly seen in adolescence or early adulthood. During the neonatal phase, patients are conventionally treated with insulin. However, if relapsing after a period of remission, treatment may include dietary modification, oral hypoglycemic agents, and/or insulin (Temple et al. 2000). The most prevalent genetic cause is the overexpression of paternal chromosome 6q24 genes *PLAGL1* and *HYMA1*, accounting for about 70 % of cases. The disease is linked to different abnormalities in an imprinted region on chromosome 6q24 resulting from paternal duplication, paternal uniparental disomy, or abnormal methylation of the maternal allele (Temple et al. 2000). This methylation defect can arise secondary to biallelic mutations in zinc finger protein 57 (ZFP57), resulting in a generalized hypomethylation syndrome. Ultimately, it results in an overexpression of genes *PLAGL1* (pleomorphic adenoma gene-like 1, also often referred to as tumor repressor ZAC) and *HYMA1* (hydatidiform mole-associated and imprinted gene). It should be noted that the transient form also can result from several other mutations in the same genes that are associated with permanent neonatal diabetes, such as mutations in *KCNJ11* and *ABCC8*. This constitutes the majority of the remaining patients with TNDM.

10.4.1.2 Permanent Neonatal Diabetes Mellitus

By definition, permanent neonatal diabetes does not relapse and requires lifelong treatment. Apart from this, the clinical debut differs a little from the transient type. Mutations in more than 20 different genes have been described as causing rare forms of permanent NDM, but sporadic mutations in a few genes account for the majority of cases. The most common causes are activating mutations in the genes *KCNJ11* and *ABCC8*. Encoding subunits KIR6.2 and SUR1 of the K_{ATP} channel (inwardly rectifying potassium channel and sulfonylurea receptor 1, respectively), *KCNJ11* and *ABCC8* play an essential role in insulin secretion. K_{ATP} channel ATP sensitivity determines the tendency to close in response to increased metabolism (specifically, to a high ATP-ADP ratio in the β-cell from increased plasma glucose levels). K_{ATP} closing promotes membrane depolarization, calcium influx, and consequently insulin secretion. Activating mutations reduce ATP sensitivity and thus hinder closing of the K_{ATP} channel despite increased β-cell metabolism, in turn reducing insulin secretion and ultimately leading to diabetes (Proks et al. 2006; Gloyn et al. 2004). Meanwhile, loss-of-function mutations tend to increase ATP sensitivity and facilitate insulin secretion resulting in congenital hyperinsulinism. There is a direct correlation between magnitude of the K_{ATP} channel current and disease severity; mutations that greatly reduce K_{ATP} channel ATP sensitivity can lead to severe forms of neonatal diabetes and associated neurological manifestations such as developmental delay and early-onset epilepsy (DEND syndrome) (Gloyn et al. 2004). This is consistent with observations of genes encoding K_{ATP} channels also being expressed in the central nervous system. Mutations causing smaller alterations in ATP sensitivity cause neonatal diabetes alone, without extra-pancreatic manifestations. Hence, it has been hypothesized that "more active"

PNDM mutations completely suppress insulin release, whereas "less severe" TNDM mutations permit some insulin release, sufficient for diabetes remission.

The most important consequence of diagnosing an activating K_{ATP} channel mutation is one of vital therapeutic impact: the responsiveness to oral sulfonylurea medication, which closes the K_{ATP} channel in an ATP-independent manner. More than 90 % of patients can safely discontinue their insulin regimen and start high-dose oral sulfonylurea without compromising glycemic control and with reduced risk of hypoglycemia (Pearson et al. 2006). Patients with DEND also respond positively on sulfonylurea treatment with neurological improvement (Aguilar-Bryan and Bryan 2008; Shah et al. 2012).

The second most common cause of permanent NDM is a mutation affecting the insulin (INS) gene. Heterozygous missense mutations are thought to affect correct folding and processing of the insulin molecule, leading to accumulation of misfolded proinsulin in the endoplasmic reticulum (ER) causing ER stress and pancreatic β-cell destruction. As β-cell destruction continues gradually, their diabetes is permanent and requires lifelong insulin treatment, but is without extra-pancreatic features other than low birth weight (Støy et al. 2007).

Heterozygous mutations in *KCNJ11*, *ABCC8*, and *INS* most often occur de novo and can hence be passed on to descendants who will each be at 50 % risk of neonatal diabetes.

10.4.2 Young-Onset Diabetes

The term MODY traditionally described a subgroup of diabetes patients with nonketotic, dominantly inherited diabetes with early-onset and (usually) lacking positive islet antibody tests. The characteristics of obesity and insulin resistance, often seen in T2D, are an uncommon feature. However, there is often significant clinical overlap between MODY and both T1D and T2D, leading to frequent misdiagnosis. Today several single-gene mutations have been included to the growing list of MODY causes, generally characterized by being transcription factors important for β-cell function and regulation or encoding the glucokinase enzyme responsible for β-cell response to changes in glycemia. Currently, there are at least 11 MODY genes, and Table 10.2 displays their details and clinical characteristics, restricted to MODY forms listed by the National Center for Biotechnology Information (NCBI), USA, as of March 2014. Some MODY forms are extremely rare, but four subtypes account for 80–90 % of MODY cases with a known genetic origin (Tables 10.1 and 10.2).

10.4.2.1 GCK-MODY, Mild Fasting Hyperglycemia

The first MODY gene identified encodes for the enzyme glucokinase (GCK) and is commonly classified as GCK-MODY (or MODY2) (Froguel et al. 1992; Njølstad

Table 10.2 MODY genes, restricted to MODY forms listed by the National Center for Biotechnology Information (NCBI), USA, as of March 2014

Gene	Location	OMIM number	Prevalence (%)	Clinical features other than diabetes	References
HNF4A	Chromosome 20q13.12	125850	~5	Transient hyperinsulinism of infancy, hypoglycemia, macrosomia, increased lipoproteins	Yamagata et al. (1996a)
GCK	Chromosome 7p13	125851	10–80	None	Vionnet et al. (1992)
HNF1A	Chromosome 12q24.2	600496	20–50	Glycosuria	Yamagata et al. (1996b)
PDX1	Chromosome 13q21.1	606392	<1	None	Stoffers et al. (1997)
HNF1B	Chromosome 17q12	137920	~5	Urogenital malformations, renal cysts, renal dysfunction, pancreatic hypoplasia, exocrine dysfunction	Horikawa et al. (1997)
NEUROD1	Chromosome 2q32	606394	<1		Malecki et al. (1999)
KLF11	Chromosome 2q25	610508	<1	None	Neve et al. (2005)
CEL	Chromosome 9q34	609812	<1	Pancreatic cysts, exocrine pancreatic dysfunction	Ræder et al. (2006)
PAX4	Chromosome 7q32	612225	<1	None	Plengvidhya et al. (2007)
INS	Chromosome 11p15.5	613370	<1	None	Molven et al. (2008), Edghill et al. (2008)
BLK	Chromosome 8p23	613375	<1	Obesity	Borowiec et al. (2009)

Abbreviations: *BLK* B-lymphoid tyrosine kinase, *CEL* carboxyl ester lipase, *GCK* glucokinase, *HNF* hepatocyte nuclear factor, *INS* insulin, *IPF1* insulin promoter factor 1, *KLF11* Kruppel-like factor 11, *MODY* maturity-onset diabetes of the young, *NEUROD1* neurogenic differentiation factor 1, *PAX4* paired box 4 gene, *PDX1* pancreatic and duodenal homeobox 1

et al. 2001). More than 600 different inactivating GCK mutations have been identified, and it is one of the most common monogenic causes of young-onset diabetes. Patients with heterozygous loss-of-function GCK mutations are characterized by having mild fasting hyperglycemia (5.5–8.0 mmol/l) from birth, but are often undiagnosed until incidental testing reveals hyperglycemia (Feigerlova et al. 2006). Disease misclassification is usually determined by the age at diagnosis: as T1D in early childhood, as well-controlled T2D if discovered as an adult, or as gestational diabetes if detected routinely during pregnancy. Consequently, if the diagnosis of T1D is made, it may lead to unnecessary insulin treatment—and thus the correct genetic diagnosis is essential to avoid this. An important and distinct feature of this disease phenotype is the relatively low risk of developing characteristic late complications of diabetes, differing it from the other MODY forms (Steele et al. 2014).

The glucokinase enzyme normally catalyzes glucose phosphorylation from glucose to glucose-6-phosphate, enabling appropriate β-cell response to glycemia. GCK can thus be regarded as the body's glucose sensor, increasing insulin secretion in response to increased glucose levels. In the case of defective glucokinase, glucose-dependent ATP production is impaired, which inhibits the closing of K_{ATP} channels and thus affects insulin secretion (Negahdar et al. 2012, 2014). Despite this, it is only the "set point" of fasting glucose that is increased in GCK mutations; glucose metabolism remains well regulated at the increased set point level, reflected by only small increments in postprandial plasma glucose levels and adequate insulin response. This can explain why microvascular complications are rare and why hemoglobin A1c seldom exceeds 7.5 percent (Steele et al. 2013).

As glucose homeostasis is not severely impaired, pharmacological treatment is rarely needed; the majority of patients can be managed by diet alone (Stride et al. 2014). Despite low risk of microvascular complications, a *GCK* mutation does not protect against concomitant development of multifactorial T2D.

Insulin treatment is sometimes used in cases of pregnancy with a *GCK* mutation if the fetus has excess growth; if the fetus does not inherit the mutation (50 % chance), the fetal response to maternal fasting hyperglycemia is through insulin secretion and hence increased intrauterine growth. If the fetus inherits the GCK mutation, the fetus experiences the same increased "set point" for sensing maternal hyperglycemia, and consequently, it will produce normal amounts of insulin and grow normally despite maternal fasting hyperglycemia.

10.4.2.2 Transcription Factor MODY

Hepatocyte nuclear factors (HNFs) are transcription factors interacting in a complex network regulating gene expression. The HNF proteins are expressed in various organs and tissues; as the name implies, correct development and function of the liver is an important attribute. Mutations in some HNF subtypes can cause young-onset diabetes, with phenotypes differing from T1D and T2D suggesting further genetic evaluation.

Heterozygous mutations in genes encoding *HNF1A* (Yamagata et al. 1996b) and *HNF4A* (Yamagata et al. 1996a) cause what is commonly referred to as HNF1A-MODY (MODY3) and HNF4A-MODY (MODY1), respectively. They collectively portray a similar clinical picture, with progressive β-cell dysfunction leading to increasing hyperglycemia and eventually diabetes. Despite HNF gene expression being present in a variety of tissues, it is primarily the pancreatic activity that is affected. As a consequence of poor glucose control, these patients are at risk of diabetic late complications, such as cardiovascular disease, kidney failure, retinopathy, and neuropathy, and should be followed up accordingly (Isomaa et al. 1998).

HNF1A mutations are the most common HNF subgroup with more than 400 different mutations reported (Colclough et al. 2014). About two-thirds present before the age of 25 (thus fitting traditional MODY criteria), the remainder present later in life. Age of diabetes diagnosis is partly determined by mutation location/isoform affected (Harries et al. 2006). Risk of microvascular complications equals that of T1D or T2D and relates directly to poor glycemic control. An important attribute to patients with *HNF1A* mutations is their high sensitivity to sulfonylurea treatment, facilitating the transfer from insulin to low-dose oral treatment without worsening glycemic control. Low-dose sulfonylurea treatment is therefore recommended as first-line treatment (Pearson et al. 2003). *HNF4A* mutation carriers present in a very similar clinical picture, but tend to differ in terms of their lipid profile—a finding of which clinical significance is not yet determined. Management with oral sulfonylurea is equally effective. Highly sensitive C-reactive protein and altered patterns of plasma protein fucosylation seem promising as biological markers to discriminate those at highest risk of having HNF1A-MODY from common forms of diabetes (Thanabalasingham et al. 2011, 2013).

Mutations in *PDX1* (pancreatic and duodenal homeobox 1 gene; also known as insulin promoter factor 1) or *NEUROD1* (neurogenic differentiation factor 1) can cause MODY but are extremely rare. A small number of families with MODY have been shown to have mutations in *PAX4* (paired box 4 gene) and *CEL* (carboxyl ester lipase).

10.4.3 Syndromic Diabetes

The term *syndromic diabetes* is characterized by additional presence of nondiabetic features dominating the clinical presentation and can by this definition consist of both neonatal and young-onset diabetes.

The most prevalent cause of syndromic diabetes is a mutation in the transcription factor *HNF1B*. As noted, the transcription factor family of HNFs is expressed in both pancreatic and extra-pancreatic tissue. Although *HNF1A* and *HNF4A* mutations usually only display endocrine affection, *HNF1B* mutation is usually systemically manifested. HNF1B-MODY (or MODY5 as it is sometimes known) is characterized by diabetes in addition to renal cysts and other malformations, renal dysfunction, pancreatic hypoplasia, urogenital abnormalities, hypomagnesaemia, and liver and exocrine pancreas dysfunction. Hyperuricemia and gout may also occur. Birth weight is usually reduced due to reduced intrauterine insulin secretion. Coexistence of pancreatic atrophy and insulin resistance makes *HNF1B* mutation carriers insensitive to sulfonylurea treatment and dependent on insulin therapy.

Concurrent exocrine and endocrine dysfunction in monogenic diabetes patients is not uncommon, and identifying contributing genetic factors of this phenotype has been thoroughly searched for. Mutations in the gene coding for carboxyl ester lipase (*CEL*) have been found to be a cause of exocrine pancreas dysfunction, commonly denoted CEL-MODY (or MODY8), more specifically characterized by a single-base deletion in the variable number of tandem repeats containing exon 11 of the CEL gene (Ræder et al. 2006). Other mutations in the *CEL* gene have since been discovered, with similar symptoms of exocrine dysfunction, predominantly from the gastrointestinal tract such as steatorrhea, malabsorption, and episodic abdominal pain. Findings indicating that *CEL* is not expressed in pancreatic β-cells support a theory of diabetogenic effect secondary to this exocrine dysfunction.

Wolcott-Rallison syndrome (WRS) is an autosomal recessive disorder caused by mutations in the gene encoding a regulator of translation initiation factor 2 alpha (*EIF2AK3*), important during the unfolded protein response. It is the most common form of permanent neonatal diabetes in consanguineous families, caused by homozygous EIF2AK3 mutations (Delépine et al. 2000). The syndrome is characterized by neonatal or early infancy insulin-requiring diabetes, epiphyseal dysplasia, osteoporosis, and growth retardation. The clinical picture is also frequently contemplated by other multisystemic manifestations, including acute hepatic failure and renal dysfunction, exocrine pancreas insufficiency, as well as intellectual disability, hypothyroidism, neutropenia with recurrent infections, and cardiovascular abnormalities.

Wolfram syndrome (WS) is a recessively inherited, multisystemic neurodegenerative disorder, also known as diabetes insipidus, insulin-deficient diabetes mellitus, optic atrophy, and deafness (DIDMOAD). Mutations in the gene *WFS1* encoding the protein wolframin cause this disease phenotype (Inoue et al. 1998). *WFS1* is shown to have a protective function on endoplasmic reticulum stress, and observations of *WSF1*-deficient β-cells in mice and WS patients' lymphocytes exhibiting dysregulated ER stress signaling suggest chronic, unresolvable ER stress leading to pancreatic β-cell death in diabetes (Fonseca et al. 2010).

10.4.4 Mitochondrial Diabetes

Mutations in mitochondrial DNA can also be a cause of diabetes. As glucose homeostasis is greatly dependent on mitochondrial function, the task of ATP production is disrupted when mutations causing mitochondrial dysfunction affect the highly metabolic active pancreatic β-cells, in turn reducing β-cell mass and resulting in insulin deficiency. Mutations are characteristic of strict maternal inheritance and often cause sensorineural hearing loss. It is therefore often denoted as maternally inherited diabetes and deafness (MIDD) syndrome, reviewed in Gerbitz et al. (1995). Heteroplasmy of mitochondrial DNA (mtDNA) within an individual, or within a cell, complicates diagnostics as the ratio of normal-to-mutated mitochondria may vary between family members and between tissues within the same patient. Different point mutations in mtDNA are also differently manifested; in severe cases, the patient may present with mitochondrial myopathy, encephalopathy, lactic acidosis, and stroke-like episodes syndrome. The majority of patients resemble the phenotype of T2D and can initially be managed with dietary modification or oral antidiabetic drugs, but a switch to insulin is usually required.

10.5 Insights from Monogenic Diabetes

Advances in the understanding of monogenic causes of diabetes and the discovery of single-gene mutations responsible for different phenotypes have greatly increased our knowledge of β-cell physiology. Insight into the chain of steps controlling glucose homeostasis has thus markedly improved understanding of diabetes pathophysiology. First and foremost, such advances have had implications for the individual patient diagnosed with the specific monogenic cause of diabetes. Findings of sulfonylurea responsiveness in patients with mutated MODY genes *HNF1A* and *HNF4A* and neonatal diabetes genes *ABCC8* and *KCNJ11* are remarkable examples of pharmacogenomic medicine allowing for the discontinuation of insulin among a subgroup of patients with diabetes who were previously insulin-dependent. Such breakthroughs also deliver hope for future antidiabetic drugs, tailored to the patient's diabetes genetic etiology.

Furthermore, combining research into identifying rare, highly penetrant single-gene mutations involved in monogenic forms of diabetes with findings from genome-wide association studies (GWAS) of multifactorial T2D, a significant overlap has been observed (Fig. 10.1). Evidence points toward common

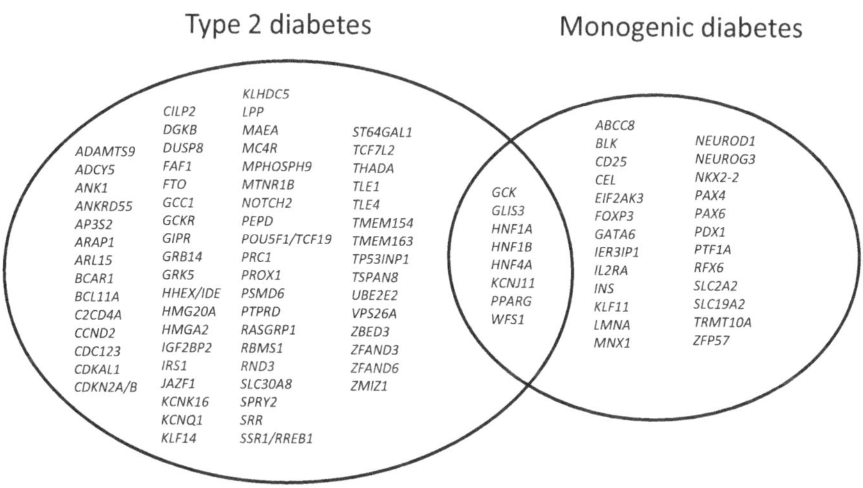

Fig. 10.1 Diagram showing the overlap between genes in which mutations are known to cause monogenic forms of diabetes and those genes where variants increase the risk of T2D. Note that for the T2D genes, the gene listed may be that located nearest to the risk variant and may not necessarily be the causal gene

polymorphism in or near some monogenic diabetes genes may also affect susceptibility to adult T2D. For example, single-nucleotide polymorphisms in and upstream of the *HNF1A* gene have been identified increasing the risk for T2D and accelerating disease onset (Voight et al. 2010). A relationship between *KCNJ11* and *ABCC8* genotypes and responsiveness to sulfonylurea in T2D has also been discovered. Enhanced hypoglycemic effects of sulfonylurea have been observed among subjects carrying specific *KCNJ11/ABCC8* haplotypes, explained by increased sensitivity to gliclazide for the specific K_{ATP} channel subtype (Hamming et al. 2009). Moreover, subjects carrying Ala at position 1369 in the SUR1 protein have lower fasting glucose than those carrying Ser at position 1369 when treated with gliclazide (Feng et al. 2008). Such findings suggest that pharmacogenomic approaches to individualized therapy also can be applicable to T2D.

With these advances in mind, early-onset monogenic diabetes will remain of high value in the search to identify new targets of β-cell dysfunction. Fuelled by advances in genetic research methodology, such as whole-exome/genome sequencing, and gradually decreasing costs, novel genetic causes of diabetes are very likely to emerge, with important translational implications for treatment and disease follow-up.

References

Aguilar-Bryan L, Bryan J (2008) Neonatal diabetes mellitus. Endocr Rev 29(3):265–291

Ashcroft FM, Rorsman P (2012) Diabetes mellitus and the β cell: the last ten years. Cell 148 (6):1160–1171

Colclough K, Saint-Martin C, Timsit J, Ellard S, Bellanné-Chantelot C (2014) Clinical utility gene card for: maturity-onset diabetes of the young. Eur J Hum Genet 22(9). doi: 10.1038/ ejhg.2014.14

Delépine M, Nicolino M, Barrett T, Golamaully M, Lathrop GM, Julier C (2000) EIF2AK3, encoding translation initiation factor 2-α kinase 3, is mutated in patients with Wolcott-Rallison syndrome. Nat Genet 25(4):406–409

Feigerlova E, Pruhová Š, Dittertova L, Lebl J, Pinterova D, Kološtová K, Černá M, Pedersen O, Hansen T (2006) Aetiological heterogeneity of asymptomatic hyperglycaemia in children and adolescents. Eur J Pediatr 165(7):446–452

Feng Y, Mao G, Ren X, Xing H, Tang G, Li Q, Li X, Sun L, Yang J, Ma W (2008) Ser1369Ala variant in sulfonylurea receptor gene ABCC8 is associated with antidiabetic efficacy of gliclazide in Chinese type 2 diabetic patients. Diabetes Care 31(10):1939–1944

Fonseca SG, Ishigaki S, Oslowski CM, Lu S, Lipson KL, Ghosh R, Hayashi E, Ishihara H, Oka Y, Permutt MA (2010) Wolfram syndrome 1 gene negatively regulates ER stress signaling in rodent and human cells. J Clin Invest 120(3):744

Froguel P, Vaxillaire M, Sun F, Velho G, Zouali H, Butel M, Lesage S, Vionnet N, Clement K, Fougerousse F (1992) Close linkage of glucokinase locus on chromosome 7p to early-onset non-insulin-dependent diabetes mellitus. Nature 356(6365):162–164

Gerbitz K-D, van den Ouweland JM, Maassen J, Jaksch M (1995) Mitochondrial diabetes mellitus: a review. Biochim Biophys Acta 1271(1):253–260

Gloyn AL, Pearson ER, Antcliff JF, Proks P, Bruining GJ, Slingerland AS, Howard N, Srinivasan S, Silva JM, Molnes J (2004) Activating mutations in the gene encoding the

ATP-sensitive potassium-channel subunit Kir6. 2 and permanent neonatal diabetes. N Engl J Med 350(18):1838–1849

Haldorsen IS, Ræder H, Vesterhus M, Molven A, Njølstad PR (2012) The role of pancreatic imaging in monogenic diabetes mellitus. Nat Rev Endocrinol 8(3):148–159

Hamming KS, Soliman D, Matemisz LC, Niazi O, Lang Y, Gloyn AL, Light PE (2009) Coexpression of the type 2 diabetes susceptibility gene variants KCNJ11 E23K and ABCC8 S1369A alter the ATP and sulfonylurea sensitivities of the ATP-sensitive K+ channel. Diabetes 58(10):2419–2424

Harries LW, Ellard S, Stride A, Morgan NG, Hattersley AT (2006) Isomers of the TCF1 gene encoding hepatocyte nuclear factor-1 alpha show differential expression in the pancreas and define the relationship between mutation position and clinical phenotype in monogenic diabetes. Hum Mol Genet 15(14):2216–2224

Hattersley A, Bruining J, Shield J, Njolstad P, Donaghue KC (2009) The diagnosis and management of monogenic diabetes in children and adolescents. Pediatr Diabetes 10(s12):33–42

Iafusco D, Stazi M, Cotichini R, Cotellessa M, Martinucci M, Mazzella M, Cherubini V, Barbetti F, Martinetti M, Cerutti F (2002) Permanent diabetes mellitus in the first year of life. Diabetologia 45(6):798–804

Inoue H, Tanizawa Y, Wasson J, Behn P, Kalidas K, Bernal-Mizrachi E, Mueckler M, Marshall H, Donis-Keller H, Crock P (1998) A gene encoding a transmembrane protein is mutated in patients with diabetes mellitus and optic atrophy (Wolfram syndrome). Nat Genet 20 (2):143–148

Isomaa B, Henricsson M, Lehto M, Forsblom C, Karanko S, Sarelin L, Häggblom M, Groop L (1998) Chronic diabetic complications in patients with MODY3 diabetes. Diabetologia 41 (4):467–473

Mohamadi A, Clark LM, Lipkin PH, Mahone EM, Wodka EL, Plotnick LP (2010) Medical and developmental impact of transition from subcutaneous insulin to oral glyburide in a 15-yr-old boy with neonatal diabetes mellitus and intermediate DEND syndrome: extending the age of KCNJ11 mutation testing in neonatal DM. Pediatr Diabetes 11(3):203–207

Molven A, Njølstad PR (2011) Role of molecular genetics in transforming diagnosis of diabetes mellitus. Expert Rev Mol Diagn 11(3):313–320

Murphy R, Ellard S, Hattersley AT (2008) Clinical implications of a molecular genetic classification of monogenic β-cell diabetes. Nat Clin Pract Endocrinol Metab 4(4):200–213

Negahdar M, Aukrust I, Johansson BB, Molnes J, Molven A, Matschinsky FM, Søvik O, Kulkarni RN, Flatmark T, Njølstad PR (2012) GCK-MODY diabetes associated with protein misfolding, cellular self-association and degradation. Biochim Biophys Acta 1822(11):1705–1715

Negahdar M, Aukrust I, Molnes J, Solheim MH, Johansson BB, Sagen JV, Dahl-Jørgensen K, Kulkarni RN, Søvik O, Flatmark T (2014) GCK-MODY diabetes as a protein misfolding disease: the mutation R275C promotes protein misfolding, self-association and cellular degradation. Mol Cell Endocrinol 382(1):55–65

Njølstad PR, Søvik O, Cuesta-Muñoz A, Bjørkhaug L, Massa O, Barbetti F, Undlien DE, Shiota C, Magnuson MA, Molven A (2001) Neonatal diabetes mellitus due to complete glucokinase deficiency. N Engl J Med 344(21):1588–1592

Pearson ER, Starkey BJ, Powell RJ, Gribble FM, Clark PM, Hattersley AT (2003) Genetic cause of hyperglycaemia and response to treatment in diabetes. Lancet 362(9392):1275–1281

Pearson ER, Flechtner I, Njølstad PR, Malecki MT, Flanagan SE, Larkin B, Ashcroft FM, Klimes I, Codner E, Iotova V (2006) Switching from insulin to oral sulfonylureas in patients with diabetes due to Kir6. 2 mutations. N Engl J Med 355(5):467–477

Proks P, Arnold AL, Bruining J, Girard C, Flanagan SE, Larkin B, Colclough K, Hattersley AT, Ashcroft FM, Ellard S (2006) A heterozygous activating mutation in the sulphonylurea receptor SUR1 (ABCC8) causes neonatal diabetes. Hum Mol Genet 15(11):1793–1800

Ræder H, Johansson S, Holm PI, Haldorsen IS, Mas E, Sbarra V, Nermoen I, Eide SÅ, Grevle L, Bjørkhaug L (2006) Mutations in the CEL VNTR cause a syndrome of diabetes and pancreatic exocrine dysfunction. Nat Genet 38(1):54–62

Rubio-Cabezas O, Ellard S (2013) Diabetes mellitus in neonates and infants: genetic heterogeneity, clinical approach to diagnosis, and therapeutic options. Horm Res Paediatr 80(3):137–146

Rubio-Cabezas O, Flanagan SE, Damhuis A, Hattersley AT, Ellard S (2012) KATP channel mutations in infants with permanent diabetes diagnosed after 6 months of life. Pediatr Diabetes 13(4):322–325

Shah RP, Spruyt K, Kragie BC, Greeley SAW, Msall ME (2012) Visuomotor performance in KCNJ11-related neonatal diabetes is impaired in children with DEND-associated mutations and may be improved by early treatment with sulfonylureas. Diabetes Care 35(10):2086–2088

Steele AM, Wensley KJ, Ellard S, Murphy R, Shepherd M, Colclough K, Hattersley AT, Shields BM (2013) Use of HbA1c in the identification of patients with hyperglycaemia caused by a glucokinase mutation: observational case control studies. PLoS ONE 8(6), e65326

Steele AM, Shields BM, Wensley KJ, Colclough K, Ellard S, Hattersley AT (2014) Prevalence of vascular complications among patients with glucokinase mutations and prolonged, mild hyperglycemia. JAMA 311(3):279–286

Støy J, Edghill EL, Flanagan SE, Ye H, Paz VP, Pluzhnikov A, Below JE, Hayes MG, Cox NJ, Lipkind GM (2007) Insulin gene mutations as a cause of permanent neonatal diabetes. Proc Natl Acad Sci USA 104(38):15040–15044

Stride A, Shields B, Gill-Carey O, Chakera AJ, Colclough K, Ellard S, Hattersley AT (2014) Cross-sectional and longitudinal studies suggest pharmacological treatment used in patients with glucokinase mutations does not alter glycaemia. Diabetologia 57(1):54–56

Tattersall RB, Fajans SS, Arbor A (1975) A difference between the inheritance of classical juvenile-onset and maturity-onset type diabetes of young people. Diabetes 24(1):44–53

Temple IK, Gardner RJ, Mackay D, Barber J, Robinson DO, Shield J (2000) Transient neonatal diabetes: widening the understanding of the etiopathogenesis of diabetes. Diabetes 49 (8):1359–1366

Thanabalasingham G, Shah N, Vaxillaire M, Hansen T, Tuomi T, Gašperíková D, Szopa M, Tjora E, James T, Kokko P (2011) A large multi-centre European study validates high-sensitivity C-reactive protein (hsCRP) as a clinical biomarker for the diagnosis of diabetes subtypes. Diabetologia 54(11):2801–2810

Thanabalasingham G, Huffman JE, Kattla JJ, Novokmet M, Rudan I, Gloyn AL, Hayward C, Adamczyk B, Reynolds RM, Muzinic A (2013) Mutations in HNF1A result in marked alterations of plasma glycan profile. Diabetes 62(4):1329–1337

Voight BF, Scott LJ, Steinthorsdottir V, Morris AP, Dina C, Welch RP, Zeggini E, Huth C, Aulchenko YS, Thorleifsson G (2010) Twelve type 2 diabetes susceptibility loci identified through large-scale association analysis. Nat Genet 42(7):579–589

Yamagata K, Furuta H, Oda N, Kaisaki PJ, Menzel S, Cox NJ, Fajans SS, Signorini S, Stoffel M, Bell GI (1996a) Mutations in the hepatocyte nuclear factor-4alpha gene in maturity-onset diabetes of the young (MODY1). Nature 384(6608):458–460. doi:10.1038/384458a0

Yamagata K, Oda N, Kaisaki PJ, Menzel S, Furuta H, Vaxillaire M, Southam L, Cox RD, Lathrop GM, Boriraj VV, Chen X, Cox NJ, Oda Y, Yano H, Le Beau MM, Yamada S, Nishigori H, Takeda J, Fajans SS, Hattersley AT, Iwasaki N, Hansen T, Pedersen O, Polonsky KS, Bell GI et al (1996) Mutations in the hepatocyte nuclear factor-1alpha gene in maturity-onset diabetes of the young (MODY3). Nature 384(6608):455–458, PMID: 8945470

Chapter 11
Epigenetics in Type 2 Diabetes

Charlotte Ling and Lorenzo Pasquali

Abstract Combinations of genetic and environmental factors contribute to the pathogenesis of type 2 diabetes (T2D); however, our knowledge of the molecular mechanisms by which these factors trigger diabetes is still limited. While genome-wide association studies have identified and characterized more than 60 genomic loci associated with T2D, recent methylome charts and reference regulatory maps obtained from tissues central to T2D can help to pinpoint the causative genetic variants. Yet, the proportion of overall trait variance explained by these genetic variants is still modest. Aging, diet, obesity, and physical inactivity represent nongenetic risk factors that may be reflected in epigenetic processes promoting T2D. Recent studies have characterized epigenetic modifications in pancreatic islets, skeletal muscle, and adipose tissue from T2D patients suggesting a central role for epigenetic mechanisms in the pathogenesis of the disease. Altered epigenetic patterns have also been found in first-degree relatives of patients with T2D and in healthy subjects born with a low birth weight suggesting that epigenetic modifications may predispose to diabetes. Lifestyle interventions including exercise and diet have also been shown to alter the epigenome in target tissues for T2D. Overall, these data propose a model where combinations of genetic, epigenetic, and nongenetic factors contribute to the risk of T2D. In this book chapter, we will explore the potential role of epigenetic mechanisms in T2D and discuss how genetics, epigenetics, and environment may interact to define the risk of developing the disease.

C. Ling (✉)
Department of Clinical Sciences, Lund University Diabetes Centre, Lund University, CRC, Scania University Hospital, Malmö, Sweden
e-mail: charlotte.ling@med.lu.se

L. Pasquali
Genomic Programming of Beta-cells Laboratory, Institut d'Investigacions Biomèdiques August Pi i Sunyer (IDIBAPS), Barcelona, Spain

CIBER de Diabetes y Enfermedades Metabólicas Asociadas (CIBERDEM), Barcelona, Spain
e-mail: pasquali@clinic.ub.es

J.C. Florez (ed.), *The Genetics of Type 2 Diabetes and Related Traits*,
DOI 10.1007/978-3-319-01574-3_11

11.1 Introduction

Type 2 diabetes (T2D) is a polygenic, multifactorial disease characterized by chronic hyperglycemia due to impaired insulin secretion in combination with insulin resistance in target tissues such as skeletal muscle, adipose tissue, and the liver (Alberti and Zimmet 1998). While it is well established that a sedentary lifestyle and high-calorie food intake increase the risk of T2D, family-based studies have shown that genetic factors also contribute to disease susceptibility (Kobberling and Tillil 1990). Recently, many studies have focused on the genetic contribution to common diseases such as diabetes. T2D genome-wide association studies (GWAS) uncovered many loci each containing genetic variants associated to the diabetic phenotype (McCarthy 2010) (see Chap. 2). Although important improvements have been made into defining loci and variants contributing to individual risk of T2D, even when combined, these established loci account for only a modest proportion of the observed familial aggregation (Willems et al. 2011; Voight et al. 2010; Manolio et al. 2009; Billings and Florez 2010).

Additional factors therefore remain to be found to elucidate the remaining genetic contributions to the disease susceptibility. These may include rare variants, copy number variation, or epigenetic modifications. Epigenetics was first introduced by Conrad Waddington who in the 1940s (Holliday 2006) focused his research on combining embryology and genetics. Later Griffith and Mahler suggested that DNA methylation may have an important biological role by contributing to long-term memory in the brain. In 1975, it was further proposed that DNA methylation could regulate gene expression, explaining the changes in expression taking place during development. At the same time, it was suggested that the DNA methylation pattern could be inherited. At a more recent Cold Spring Harbor meeting, epigenetics was defined as "a stably heritable phenotype resulting from changes in a chromosome without alterations in the DNA sequence" (Berger et al. 2009). The epigenome includes DNA methylation, hydroxymethylation, histone modifications, and noncoding RNA. These factors contribute to cell differentiation during development, parental imprinting, X-chromosome inactivation, and cell-specific gene expression. This book chapter will mainly focus on DNA methylation and histone modifications.

In differentiated mammalian cells, DNA methylation mainly occurs on cytosines in CG dinucleotides (Lister et al. 2009). These are called CpG sites. Enzymes responsible for adding methyl groups to the mammalian DNA include DNMT1, which is responsible for copying the methylation pattern during replication, and DNMT3a and 3b, which are responsible for de novo methylation. Demethylation may occur if the DNMT1 activity is low during replication. Recently, TET enzymes were suggested to contribute to demethylation by oxidizing methyl groups (Wu and Zhang 2014).

DNA methylation of promoter regions has been found to decrease the transcriptional activity of the corresponding genes (Bird 2007; Suzuki and Bird 2008). Today, it is known that DNA methylation of different genomic regions also affect alternative splicing events, the recombination rate, and it may increase the transcriptional elongation process (Jones 2012). The chromatin is built of nucleosomes, which is made up of approximately 147 bp DNA and an octamer of histones. The N-terminal tails of these histones can be chemically modified by numerous enzymes which are responsible for adding methyl, acetyl, and phosphor groups to histones. These histone modifications affect the chromatin structure and can subsequently control the chromatin accessibility at certain genomic locations. While some histone modifications such as H3K9me3 contribute to a dense, closed chromatin structure, others are enriched at active genes (e.g., H3K9ac and H3K4me3) or at distal regulatory elements (e.g., H3K27ac and H3K4me1) (Fig. 11.1). As aging and environmental factors can alter the DNA methylation pattern and introduce histone modifications (Fraga et al. 2005), epigenetics may affect the development of a complex, multifactorial disease such as T2D. Indeed, during the last years

Histone modification	*Functional association*
H3K4me1	Enhancers
H3K4me2	Enhancers, Promoters
H3K4me3	Promoters
H3K9me2	Inactive chromatin
H3K9me3	Inactive chromatin
H3K27me3	Inactive chromatin
H3K27ac	Enhancers, Promoters
H3K36me3	Transcription
H3K79me2	Transcription

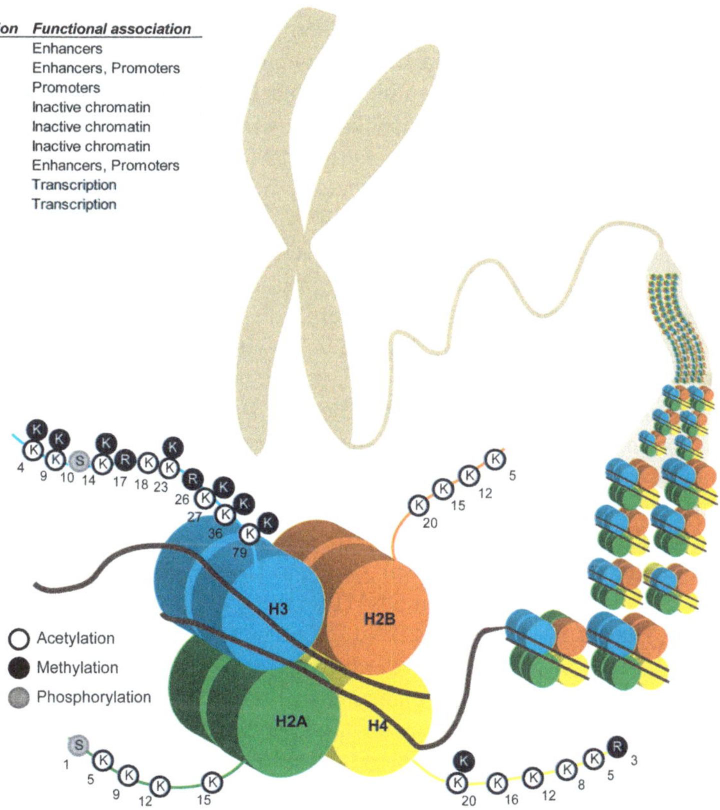

Fig. 11.1 Schematic representation of the main posttranslational modifications of the four core histones. Different combinations of posttranslational histone modifications result in a "histone code" that contributes to establishing the global and local chromatin states that eventually determine gene expression. *Highlighted* in the table are the major histone modifications and their functional association

numerous studies have tried to identify epigenetic modifications that play a role in the pathogenesis of T2D (Ling and Groop 2009). This book chapter aims at summarizing current knowledge in the field of epigenetics in T2D.

11.2 The Epigenetic Basis of T2D

In 2008, the first epigenetic modifications were identified in pancreatic islets from patients with T2D (Ling et al. 2008). Here, the promoter of the transcriptional co-activator *PPARGC1A* was found to have increased DNA methylation in human islets from subjects with T2D. Importantly, the diabetic islets also had decreased *PPARGC1A* expression in parallel with lower glucose-stimulated insulin secretion in vitro compared with nondiabetic controls. Additionally, silencing of *PPARGC1A* in human islets using siRNA resulted in decreased insulin secretion. Later, Barres et al. found increased DNA methylation of *PPARGC1A* also in skeletal muscle from patients with T2D compared with nondiabetic subjects (Barres et al. 2009). More recent studies have identified altered DNA methylation patterns in skeletal muscle from diabetic versus nondiabetic twins, using twin studies with monozygotic twin pairs discordant for T2D (Ribel-Madsen et al. 2012; Nitert et al. 2012). Additional studies have found epigenetic modifications in pancreatic islets from human donors with T2D (Yang et al. 2011, 2012; Volkmar et al. 2012; Dayeh et al. 2014). Using a candidate gene approach, *INS* and *PDX1* were found to have increased DNA methylation and decreased mRNA expression in islets from human donors with T2D (Yang et al. 2011, 2012). Additionally, the degree of DNA methylation correlated negatively with expression of these two genes, suggesting a potential regulatory role of methylation on expression. Indeed, reporter assays, where promoter regions of *PDX1* were inserted into a luciferase vector and then methylated with methyltransferases, showed that increased DNA methylation of the *PDX1* promoter and its enhancer region reduced the transcriptional activity (Yang et al. 2012). Hyperglycemia, measured by HbA1c, was associated with increased DNA methylation and decreased expression of *INS* and *PDX1*. Importantly, glucose was shown to have a direct effect on methylation of *INS* and *PDX1* as well as on DNMT1 expression in clonal beta cells cultured in vitro (Yang et al. 2011, 2012). Two additional studies have analyzed DNA methylation of ~27,000 and ~480,000 in human pancreatic islets from donors with T2D and controls, respectively (Volkmar et al. 2012; Dayeh et al. 2014). While Volkmar et al. found differential DNA methylation of 276 CpG sites at $P < 0.01$ (Volkmar et al. 2012), Dayeh et al. found altered methylation of 3116 CpG sites based on a false discovery rate analysis (Dayeh et al. 2014). Importantly, none of these two studies found any differences in beta-cell content between diabetic and nondiabetic human islets, supporting that the epigenetic differences seen in diabetic islets are not due to an altered cell composition. Interestingly, both studies found that the majority of the differentially methylated CpG sites show decreased methylation in diabetic versus nondiabetic islets. Moreover, Dayeh et al. found altered methylation in sites connected to genes that previously have been shown to play a role in pancreatic islets, in the exocytosis process and in apoptosis. They also found decreased methylation of genes associated with T2D, e.g., *TCF7L2*, *FTO*, and *KCNQ1*, in islets from donors with T2D compared with controls. Noncoding RNAs do also contribute to epigenetic regulation (Moran et al. 2012), and a recent study found altered expression of microRNAs in parallel with differential DNA methylation in human diabetic islets (Kameswaran et al. 2014). The epigenome is also altered in adipose tissue from subjects with T2D (Nilsson et al. 2014). There was a significant enrichment of genes involved in inflammation and glycan degradation as well as pathways in cancer, Wnt signaling, and MAPK signaling. Additionally, numerous genes previously associated with T2D including *KCNQ1*, *IRS1*, *PPARG*, and *TCF7L2* showed differential DNA methylation in adipose tissue from diabetic compared with nondiabetic subjects. While these studies demonstrate that epigenetic modifications exist in diabetic patients, the data does not show whether epigenetic modifications are the cause or consequence of the disease. However, altered DNA methylation patterns have also been found in healthy first-degree relatives of patients with T2D (Nitert et al. 2012). Additionally, healthy young men with an increased risk for diabetes due to their low birth weight demonstrate epigenetic modification in skeletal muscle (Brons et al. 2010; Jacobsen et al. 2014). Moreover, aging, obesity, and hyperglycemia are associated with differential DNA methylation in human pancreatic islets from nondiabetic subjects of CpG sites/genes that also show altered methylation in islets from subjects with T2D (Dayeh et al. 2014). Together these data support that epigenetic modifications contribute to the pathogenesis of T2D.

11.3 Epigenetics, Diabetes, and Insulin Secretion

Impaired insulin secretion from pancreatic islets is a hallmark of diabetes. As epigenetic modifications have been identified in pancreatic islets from patients with T2D (Ling et al. 2008; Yang et al. 2011, 2012; Volkmar et al. 2012; Dayeh et al. 2014; Kameswaran et al. 2014), which show impaired insulin secretion, it is likely that epigenetic mechanisms also affect the secretion of insulin. PDX1 is a transcription factor that plays a key role during development of the pancreas. It does also regulate beta-cell expression of insulin in postnatal life. Animals lacking islet Pdx1 expression develop diabetes, and mutations in *PDX1* can give a monogenic form of diabetes (Ahlgren et al. 1998; Stoffers et al. 1997). Animal studies have also shown that an impaired intrauterine environment causes epigenetic modifications of *Pdx1*, which result in decreased islet Pdx1 expression, perturbed insulin secretion, and eventually diabetes later in life (Park et al. 2008). In a different animal study, Sandovici et al. showed that a low-protein intake in pregnant mothers causes epigenetic modifications of *Hnf4a* in islets from the offspring and consequently decreased Hnf4a expression, perturbed insulin secretion, and eventually hyperglycemia in adult life (Sandovici et al. 2011). These studies suggest that epigenetic modifications that take place in pancreatic islets during development can contribute to impaired insulin secretion and diabetes in postnatal life. A recent genome-wide epigenetic study followed up identified target genes, which showed differential DNA methylation and gene expression in pancreatic islets from patients with T2D, in clonal beta cells (Dayeh et al. 2014). This study showed that genes with both altered DNA methylation and expression in human diabetic islets directly affect insulin secretion and exocytosis in beta cells. Interestingly, while over-expression of Cdkn1a and Pde7b resulted in decreased glucose-stimulated insulin secretion, Exoc3l deficiency resulted in decreased exocytosis. Additionally, increased age was associated with decreased DNA methylation of *CDKN1A* and increased DNA methylation of *EXOC3L2*, which is in line with what is seen in T2D islets. Obesity increases the risk of T2D, and diabetic subjects often have elevated circulating lipid levels. To mimic the situation in people with diabetes, clonal beta cells were exposed to palmitate, and insulin secretion, metabolomics, gene expression, and histone modifications were analyzed (Malmgren et al. 2013). Lipotoxicity impaired glucose-stimulated insulin secretion and induced histone modifications of genes showing differential gene expression, e.g., *Insig1*. Importantly, palmitate also regulated the histone acetyl transferase activity and the H3K27me3 methyltransferase activity in the clonal beta cells. Together, these data show that epigenetic modifications of numerous genes affect islets' function and insulin secretion and potentially are linked to the development of diabetes.

11.4 Epigenetics, Diabetes, and Insulin Resistance

Insulin resistance and decreased insulin action in target tissues such as skeletal muscle, adipose tissue, and the liver contribute to the development of T2D. Multiple studies have therefore tried to dissect the impact of epigenetic modifications on insulin sensitivity in these tissues. While regular exercise is known to have beneficial effects, a high-fat diet, physical inactivity, and aging may reduce insulin sensitivity. Interestingly, skeletal muscle from elderly subjects was found to exhibit increased DNA methylation and decreased expression of genes involved in oxidative phosphorylation, and these subjects had decreased glucose uptake during clamp compared to young subjects (Ling et al. 2007; Ronn et al. 2008). Also a high-fat diet for 5 days induces epigenetic changes in human skeletal muscle and increased insulin resistance in the young men exposed to the high-fat diet (Brons et al. 2012; Jacobsen et al. 2012). Additionally, a 9-day bed rest in young men was found to decrease expression and increase DNA methylation of *PPARGC1A* in skeletal muscle and to deteriorate whole-body insulin action (Alibegovic et al. 2009, 2010). In contrast, both acute and regular exercise decreased DNA methylation of *PPARGC1A* in human muscle (Nitert et al. 2012; Barres et al. 2012). Moreover, an exercise intervention induced epigenetic modifications of thousands of genes, including genes with known function in diabetes and obesity, in both skeletal muscle and adipose tissue from middle aged sedentary men (Nitert et al.

2012; Ronn et al. 2013). This took place in parallel with altered gene expression and improved metabolism. Functional in vitro studies of genes that exhibit decreased expression and increased DNA methylation in adipose tissue in response to exercise further showed that knockdown of identified genes, e.g., Ncor2 and Hdac4 in an adipocyte cell line, increased the insulin-stimulated lipogenesis (Ronn et al. 2013). It subsequently seems that epigenetic modifications in muscle and fat contribute to whole-body insulin sensitivity.

11.5 Genetics, Epigenetics, and Environment Interplay

As a prototype of a multifactorial complex disease, T2D arises from an intricate interaction of environment factors and inherited predisposition. The attempt to partition individual propensity to develop T2D among genetic and environmental components may be often frustrated by the intimate connections between them. Subsequently, our knowledge about the molecular mechanisms linking genetic variation and environmental factors with T2D remains limited. In this scenario, epigenetics may play an important role in interfacing the molecular response of an organism to an external influence by orchestrating and modulating tissue-specific gene expression (Fig. 11.2). In this view, understanding the epigenetic processes in

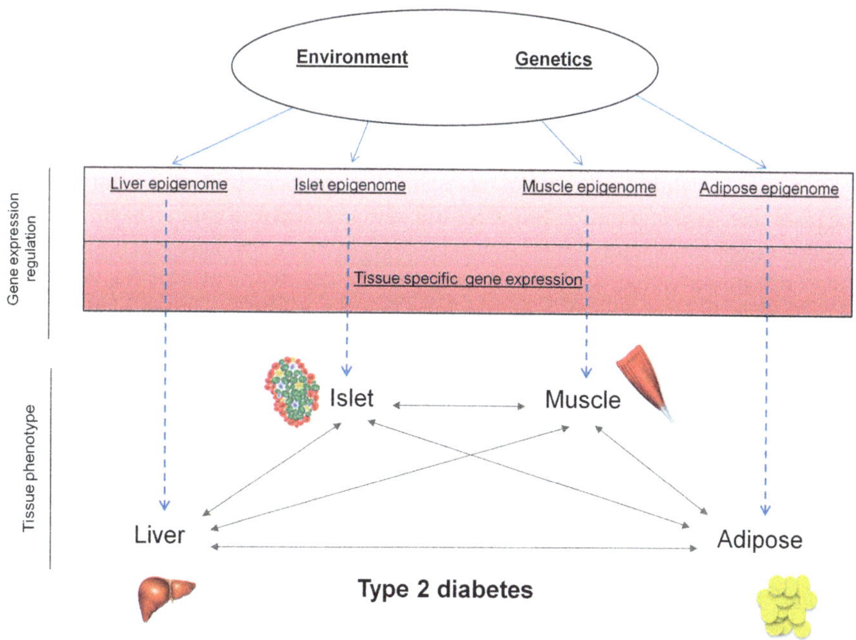

Fig. 11.2 Model proposing that epigenetic mechanisms may play an important role in the pathogenesis of T2D by mediating the impact of genetic and environmental factors on the phenotype of organs central to the disease

the context of T2D will likely shed light on the molecular mechanisms underlying the development of the disease.

11.6 Linking GWAS and Epigenetics

So far GWAS revealed broad genomic regions, and only in very few cases it was possible to pinpoint the causal sequence variant and identify a possible mechanism linking the associated polymorphism with the development of T2D (Pasquali et al. 2014; Gaulton et al. 2010; Stitzel et al. 2010; Dayeh et al. 2013; Kulzer et al. 2014). In summary, several important observations emerged from the GWAS:

1. For a significant fraction of loci, genetic variation impacts insulin secretion indicating the central role of pancreatic islet as a relevant tissue to study in order to understand the genetic mechanisms underlying this disease.
2. Most of the loci do not harbor variations residing in coding regions suggesting that the risk variant may influence transcript regulation rather than altering the sequence of the protein itself.
3. For most loci, it is unclear which of the genes in the region is mechanistically responsible for the association effect observed.
4. The discovered common variant signals have at most a modest effect accounting for no more than 5–10 % of overall trait variance even when considered in combination, fractions that make them unsuitable for use in clinic for disease prediction (Willems et al. 2011). Nevertheless whole-chip heritability estimates indicate that as much as 50 % of the heritability of T2D may be explained by common variants, accounting for an estimate of 3000 variants (Stahl et al. 2012; Morris et al. 2012).

The fact that almost all the variants identified are located in noncoding regions suggests that gene regulation has a central role in the development of the disease and that epigenetic factors such as protein-DNA interactions, DNA methylation, or chromatin modification might be involved in the development of T2D. A logical path to shed light on the disease mechanisms that modify gene regulation goes thereby through the analysis of chromatin structure and posttranslational modifications of nucleosomal histones.

In the past, human geneticists took advantage of relevant genetic maps such as the exon annotation to identify coding mutations causing monogenic forms of diabetes (Fajans et al. 2001). For T2D as well as for other complex diseases where the associated signals do not reside in the coding part of the genome, we instead need maps of the active regulatory elements in disease-relevant tissues. Such tissue-specific regulatory maps could in turn help us to identify the associated variants that may have a causative role by disrupting the tissue-specific transcriptional regulatory activity.

Large consortia such as ENCODE (Dunham et al. 2012) and the Epigenome Roadmap (Bernstein et al. 2010) provided extensive epigenetic maps allowing annotation of the noncoding regions of the human genome for a large amount of tissues including several relevant to T2D such as adipose tissue and skeletal muscle, while other less accessible primary tissues such as the endocrine pancreas were not prioritized in these studies. For their central role in diabetes pathogenesis, different laboratories embarked in profiling the epigenetic landscape of human pancreatic islet cells (Dayeh et al. 2014; Pasquali et al. 2014; Gaulton et al. 2010; Stitzel et al. 2010; Bhandare et al. 2010; Parker et al. 2013) in an ongoing effort to dissect the molecular mechanisms of human T2D (Ashcroft and Rorsman 2012).

Early this year, the first assembly of a complete human pancreatic islet *cis*-regulome map unmasked an enrichment of T2D-associated variants in islet-specific distal regulatory elements, implicating genetic variation at islet enhancers in the susceptibility for T2D (Pasquali et al. 2014). These findings confirm the primary role of the islet-cell gene regulation in the pathogenesis of T2D and demonstrate the potential of integrating different levels of experimental data such as genetics, epigenetics, and transcripts maps in shedding light on the molecular mechanisms underlying complex polygenic diseases such as diabetes. While these advances unmasked the tissue-specific gene expression regulatory framework in nondiabetic islet cells and enable the identification of T2D-associated variants with a potential regulatory function, a major question still remains open: "how does genetic variation affect the pancreatic islet tissue-specific epigenetic landscape in the context of T2D?" Proof-of-concept examples showed that at least in some cases, genetic variants do impact the human islet epigenetic landscape (Gaulton et al. 2010; Dayeh et al. 2013) or disrupt the regulatory potential of a genetic site in vitro, implicating a functional role for those variants (Pasquali et al. 2014; Gaulton et al. 2010; Stitzel et al. 2010; Dayeh et al. 2013).

In particular, along these lines in a recent publication, Dayeh et al. demonstrated that about 50 % of SNPs associated with T2D in GWAS introduce or remove CpG sites. Importantly, all these diabetes-associated CpG-SNPs were associated with differential DNA methylation in the SNP site as well as altered expression, alternative splicing events, and hormone secretion in human pancreatic islets (Dayeh et al. 2013).

Future progress in understanding the impact of genetic variants on the tissue-specific epigenome in the context of T2D will necessarily need to go through:

(a) Whole-genome sequencing of patients with T2D resulting in the identification of low-frequency variants associated with the disease.
(b) Charting epigenetic maps in T2D-relevant tissues including early and late stages of development, as well as pertinent metabolic states.

These advances will enable us to dissect the contribution of genetic variation to the disease development and unmask mechanistic links with the tissue-specific gene regulation.

The expansion of association studies to rare or personal sequence variants will certainly improve the estimates of variance explained, but will unlikely achieve a complete explanation of the predisposition. An open road in an attempt to unmask the fraction of disease variance still unexplained by the GWAS may pass through the epigenetic characterization of humans at risk of T2D. Few studies mostly focusing on the methylation status of selected CpG sites have so far attempted this challenge using pancreatic islets from T2D-affected individuals (Ling et al. 2008; Yang et al. 2011, 2012). The first epigenome-wide association studies (EWAS) are also now starting to be performed with tissues obtained from people with T2D (Toperoff et al. 2012; Rakyan et al. 2011; Bell et al. 2010), but only in few cases it was possible to use relevant tissues (Barres et al. 2009; Ribel-Madsen et al. 2012; Volkmar et al. 2012; Dayeh et al. 2014). These first studies, as it had been initially for the GWAS, are limited by low statistical power and rare follow-up replication. EWAS will almost certainly rely on centralized community efforts due to the high costs of the experiments and the difficulty of accessing to sufficient numbers of samples from disease-relevant tissues and/or cell types. These studies will allow understanding the contribution of the epigenome rather than the sequence composition to the disease development. Finally integration of genetics and epigenetic data will allow a clearer picture of the molecular mechanisms behind the development of T2D.

11.7 Linking Environment and Epigenetics

It is well established that the environment plays a central role in the development of complex polygenic diseases such as T2D with the onset arising after a long latency period from the environmental trigger. In particular, early life events such as the diet during pregnancy and lactation may have an important effect not only on the mother's health but also on the health of her offspring. The correct development of an organism and its continuous interaction with the environment requires a plastic adaptation of the cells that continuously modulates and fine-tunes their gene expression. This process is at least in part mediated by epigenetic mechanisms such as methylation and histone modification placing epigenetics as an interface between the environment stimuli and the molecular adaptation of an organism (Fig. 11.1). In this view, epigenome and genome interactively influence the mature phenotype and may determine sensitivity to the development of a disease later in life.

Studies of individuals who were in utero during famine periods have shown a correlation between maternal nutrient deficiency and increased risk of T2D in the offspring (Ravelli et al. 1998; Li et al. 2010). These observations are supported by animal models showing that maternal diet during pregnancy influence the long-term risk of developing T2D (Ozanne and Hales 2002; Gluckman et al. 2007).

Rodent maternal nutrition is able to change epigenetic marks during fetal growth, but yet a small number of studies have examined DNA methylation in the context of diabetes. An example of how environmental exposure can influence diabetic risk through epigenetic mechanisms was shown by Sandovici et al. In this study, carried out in rats, maternal low-protein diet silenced, through an epigenetic mechanism involving the lifelong inhibition of a distal regulatory element, pancreatic islets' expression of the transcription factor HNF4A which in turn was associated with an increased risk of T2D (Sandovici et al. 2011). In another example on the same line, Ozanne et al. showed a reduced expression of the insulin signal-transduction pathway components in skeletal muscle such as GLUT4 in rats' offspring of mothers on a protein-restricted diet. These abnormalities could contribute to insulin resistance later in life (Ozanne et al. 2005).

Environment may also influence the disease metabolic risk through epigenetic mechanisms not only during the development of the organisms but also in adult individuals. For example, as described in a previous section, several studies showed that physical activity has an effect on the epigenome of organs central to T2D and does influence the disease pathogenesis (Nitert et al. 2012; Barres et al. 2012; Ronn et al. 2013).

Finally, the impact of the environment may also be observed over multiple generations in humans and animal models (Jimenez-Chillaron et al. 2009; Carone et al. 2010). Along this line, paternal high-fat diet consumption was shown to induce transcriptome alterations and islets dysfunction in female rat offspring. Importantly, epigenetic marks such as the promoter methylation of IL13 receptor were found to be decreased in the offspring with concomitant increase in its expression (Ng et al. 2010). While it is now clear that epigenetics can confer stable functional genomic configurations within somatic lineages, this and other epidemiological observations (Carone et al. 2010; Kaati et al. 2007) place epigenetic mechanisms as possible mediators of environmental effects that may impact and influence disease related phenotypes over multiple generations. From these observations, it could be speculated the possibility of a transgenerational transmission of metabolic risk (Pembrey et al. 2006). However, a clear molecular mechanism for such inheritance is still missing. The relative contribution of genetic and epigenetic variation to transgenerational epigenetic inheritance will be better understood in the future through the use of isogenic strains in animal studies.

Epigenetic mechanisms may mediate the gene-environment dialogue in early life and adult individuals influencing the metabolic risk eventually resulting in T2D. Understanding how early and adult life experiences can give rise to lasting epigenetic marks conferring an increased risk for T2D is a challenge that is now in an important focus for the diabetes research field. An attractive working hypothesis is to consider epigenetic modifications in a final common pathway through which both genetic and environmental effects can impact T2D risk.

The downstream consequences of alterations in gene expression influencing glucose homeostasis could be the result of genetic variants modifying the regulatory potential of a sequence (e.g., disruption of a transcription factor binding site or introducing a CpG site) but could as well be caused by environmental cues able to alter the methylation or histone modifications' profiles at the same genetic loci. To which extent genetic variants rather than environmental factors are responsible of the altered gene regulation that reflects the diabetic phenotype is an open question that the diabetic field will address in the upcoming years.

References

Ahlgren U, Jonsson J, Jonsson L, Simu K, Edlund H (1998) beta-cell-specific inactivation of the mouse Ipf1/Pdx1 gene results in loss of the beta-cell phenotype and maturity onset diabetes. Genes Dev 12(12):1763–1768

Alberti KG, Zimmet PZ (1998) Definition, diagnosis and classification of diabetes mellitus and its complications. Part 1: diagnosis and classification of diabetes mellitus provisional report of a WHO consultation. Diabet Med 15(7):539–553

Alibegovic AC, Hojbjerre L, Sonne MP, van Hall G, Stallknecht B, Dela F, Vaag A (2009) Impact of 9 days of bed rest on hepatic and peripheral insulin action, insulin secretion, and whole-body lipolysis in healthy young male offspring of patients with type 2 diabetes. Diabetes 58(12): 2749–2756

Alibegovic AC, Sonne MP, Hojbjerre L, Bork-Jensen J, Jacobsen S, Nilsson E, Faerch K, Hiscock N, Mortensen B, Friedrichsen M, Stallknecht B, Dela F, Vaag A (2010) Insulin resistance induced by physical inactivity is associated with multiple transcriptional changes in skeletal muscle in young men. Am J Physiol Endocrinol Metab 299(5):E752–E763. doi:10.1152/ajpendo.00590.2009

Ashcroft FM, Rorsman P (2012) Diabetes mellitus and the beta cell: the last ten years. Cell 148(6): 1160–1171. doi:10.1016/j.cell.2012.02.010

Barres R, Osler ME, Yan J, Rune A, Fritz T, Caidahl K, Krook A, Zierath JR (2009) Non-CpG methylation of the PGC-1alpha promoter through DNMT3B controls mitochondrial density. Cell Metab 10(3):189–198

Barres R, Yan J, Egan B, Treebak JT, Rasmussen M, Fritz T, Caidahl K, Krook A, O'Gorman DJ, Zierath JR (2012) Acute exercise remodels promoter methylation in human skeletal muscle. Cell Metab 15(3):405–411. doi:10.1016/j.cmet.2012.01.001

Bell CG, Finer S, Lindgren CM, Wilson GA, Rakyan VK, Teschendorff AE, Akan P, Stupka E, Down TA, Prokopenko I, Morison IM, Mill J, Pidsley R, Deloukas P, Frayling TM, Hattersley AT, McCarthy MI, Beck S, Hitman GA (2010) Integrated genetic and epigenetic analysis identifies haplotype-specific methylation in the FTO type 2 diabetes and obesity susceptibility locus. PLoS ONE 5(11), e14040

Berger SL, Kouzarides T, Shiekhattar R, Shilatifard A (2009) An operational definition of epigenetics. Genes Dev 23(7):781–783. doi:10.1101/gad.1787609

Bernstein BE, Stamatoyannopoulos JA, Costello JF, Ren B, Milosavljevic A, Meissner A, Kellis M, Marra MA, Beaudet AL, Ecker JR, Farnham PJ, Hirst M, Lander ES, Mikkelsen TS, Thomson JA (2010) The NIH roadmap epigenomics mapping consortium. Nat Biotechnol 28(10):1045–1048. doi:10.1038/nbt1010-1045

Bhandare R, Schug J, Le Lay J, Fox A, Smirnova O, Liu C, Naji A, Kaestner KH (2010) Genome-wide analysis of histone modifications in human pancreatic islets. Genome Res 20(4):428–433

Billings LK, Florez JC (2010) The genetics of type 2 diabetes: what have we learned from GWAS? Ann N Y Acad Sci 1212:59–77. doi:10.1111/j.1749-6632.2010.05838.x

Bird A (2007) Perceptions of epigenetics. Nature 447(7143):396–398

Brons C, Jacobsen S, Nilsson E, Ronn T, Jensen CB, Storgaard H, Poulsen P, Groop L, Ling C, Astrup A, Vaag A (2010) Deoxyribonucleic acid methylation and gene expression of PPARGC1A in human muscle is influenced by high-fat overfeeding in a birth-weight-dependent manner. J Clin Endocrinol Metab 95(6):3048–3056. doi:10.1210/jc.2009-2413

Brons C, Jacobsen S, Hiscock N, White A, Nilsson E, Dunger D, Astrup A, Quistorff B, Vaag A (2012) Effects of high-fat overfeeding on mitochondrial function, glucose and fat metabolism, and adipokine levels in low-birth-weight subjects. Am J Physiol Endocrinol Metab 302(1): E43–E51. doi:10.1152/ajpendo.00095.2011

Carone BR, Fauquier L, Habib N, Shea JM, Hart CE, Li R, Bock C, Li C, Gu H, Zamore PD, Meissner A, Weng Z, Hofmann HA, Friedman N, Rando OJ (2010) Paternally induced transgenerational environmental reprogramming of metabolic gene expression in mammals. Cell 143(7):1084–1096. doi:10.1016/j.cell.2010.12.008

Dayeh TA, Olsson AH, Volkov P, Almgren P, Ronn T, Ling C (2013) Identification of CpG-SNPs associated with type 2 diabetes and differential DNA methylation in human pancreatic islets. Diabetologia 56(5):1036–1046. doi:10.1007/s00125-012-2815-7

Dayeh TVP, Salö S, Hall E, Nilsson E, Olsson AH, Kirkpatrick CL, Wollheim C, Eliasson L, Rönn T, Bacos K, Ling C (2014) Genome-wide DNA methylation analysis of human pancreatic islets from type 2 diabetic and non-diabetic donors identifies candidate genes that influence insulin secretion. PLoS Genet 10(3), e1004160

Dunham I, Kundaje A, Aldred SF, Collins PJ, Davis CA, Doyle F, Epstein CB, Frietze S, Harrow J, Kaul R, Khatun J, Lajoie BR, Landt SG, Lee BK, Pauli F, Rosenbloom KR, Sabo P, Safi A, Sanyal A, Shoresh N, Simon JM, Song L, Trinklein ND, Altshuler RC, Birney E, Brown JB, Cheng C, Djebali S, Dong X, Ernst J, Furey TS, Gerstein M, Giardine B, Greven M, Hardison RC, Harris RS, Herrero J, Hoffman MM, Iyer S, Kelllis M, Kheradpour P, Lassman T, Li Q, Lin X, Marinov GK, Merkel A, Mortazavi A, Parker SC, Reddy TE, Rozowsky J, Schlesinger F, Thurman RE, Wang J, Ward LD, Whitfield TW, Wilder SP, Wu W, Xi HS, Yip KY, Zhuang J, Bernstein BE, Green ED, Gunter C, Snyder M, Pazin MJ, Lowdon RF, Dillon LA, Adams LB, Kelly CJ, Zhang J, Wexler JR, Good PJ, Feingold EA, Crawford GE, Dekker J, Elinitski L, Farnham PJ, Giddings MC, Gingeras TR, Guigo R, Hubbard TJ, Kellis M, Kent WJ, Lieb JD, Margulies EH, Myers RM, Starnatoyannopoulos JA, Tennebaum SA, Weng Z, White KP, Wold B, Yu Y, Wrobel J, Risk BA, Gunawardena HP, Kuiper HC, Maier CW, Xie L, Chen X, Mikkelsen TS, Gillespie S, Goren A, Ram O, Zhang X, Wang L, Issner R, Coyne MJ, Durham T, Ku M, Truong T, Eaton ML, Dobin A, Lassmann T, Tanzer A, Lagarde J, Lin W, Xue C, Williams BA, Zaleski C, Roder M, Kokocinski F, Abdelhamid RF, Alioto T, Antoshechkin I, Baer MT, Batut P, Bell I, Bell K, Chakrabortty S, Chrast J, Curado J, Derrien T, Drenkow J, Dumais E, Dumais J, Duttagupta R, Fastuca M, Fejes-Toth K, Ferreira P, Foissac S, Fullwood MJ, Gao H, Gonzalez D, Gordon A, Howald C, Jha S, Johnson R, Kapranov P, King B, Kingswood C, Li G, Luo OJ, Park E, Preall JB, Presaud K, Ribeca P, Robyr D, Ruan X, Sammeth M, Sandu KS, Schaeffer L, See LH, Shahab A, Skancke J, Suzuki AM, Takahashi H, Tilgner H, Trout D, Walters N, Wang H, Hayashizaki Y, Reymond A, Antonarakis SE, Hannon GJ, Ruan Y, Carninci P, Sloan CA, Learned K, Malladi VS, Wong MC, Barber GP, Cline MS, Dreszer TR, Heitner SG, Karolchik D, Kirkup VM, Meyer LR, Long JC, Maddren M, Raney BJ, Grasfeder LL, Giresi PG, Battenhouse A, Sheffield NC, Showers KA, London D, Bhinge AA, Shestak C, Schaner MR, Kim SK, Zhang ZZ, Mieczkowski PA, Mieczkowska JO, Liu Z, McDaniell RM, Ni Y, Rashid NU, Kim MJ, Adar S, Zhang Z, Wang T, Winter D, Keefe D, Iyer VR, Sandhu KS, Zheng M, Wang P, Gertz J, Vielmetter J, Partridge EC, Varley KE, Gasper C, Bansal A, Pepke S, Jain P, Amrhein H, Bowling KM, Anaya M, Cross MK, Muratet MA, Newberry KM, McCue K, Nesmith AS, Fisher-Aylor KI, Pusey B, DeSalvo G, Parker SL, Balasubramanian S, Davis NS, Meadows SK, Eggleston T, Newberry JS, Levy SE, Absher DM, Wong WH, Blow MJ, Visel A, Pennachio LA, Elnitski L, Petrykowska HM, Abyzov A, Aken B, Barrell D, Barson G, Berry A, Bignell A, Boychenko V, Bussotti G, Davidson C, Despacio-Reyes G, Diekhans M, Ezkurdia I, Frankish A, Gilbert J, Gonzalez JM, Griffiths E, Harte R, Hendrix DA, Hunt T, Jungreis I, Kay M, Khurana E, Leng J, Lin MF, Loveland J, Lu Z, Manthravadi D, Mariotti M, Mudge J, Mukherjee G, Notredame C, Pei B, Rodriguez JM, Saunders G, Sboner A, Searle S, Sisu C, Snow C, Steward C, Tapanan E, Tress ML, van Baren MJ, Washieti S, Wilming L, Zadissa A, Zhengdong Z, Brent M, Haussler D, Valencia A, Raymond A, Addleman N, Alexander RP, Auerbach RK, Bettinger K, Bhardwaj N, Boyle AP, Cao AR, Cayting P, Charos A, Cheng Y, Eastman C, Euskirchen G, Fleming JD, Grubert F, Habegger L, Hariharan M, Harmanci A, Iyenger S, Jin VX, Karczewski KJ, Kasowski M, Lacroute P, Lam H, Larnarre-Vincent N, Lian J, Lindahl-Allen M, Min R, Miotto B, Monahan H, Moqtaderi Z, Mu XJ, O'Geen H, Ouyang Z, Patacsil D, Raha D, Ramirez L, Reed B, Shi M, Slifer T, Witt H, Wu L, Xu X, Yan KK, Yang X, Struhl K, Weissman SM, Tenebaum SA, Penalva LO, Karmakar S, Bhanvadia RR, Choudhury A, Domanus M, Ma L, Moran J, Victorsen A, Auer T,

Centarin L, Eichenlaub M, Gruhl F, Heerman S, Hoeckendorf B, Inoue D, Kellner T, Kirchmaier S, Mueller C, Reinhardt R, Schertel L, Schneider S, Sinn R, Wittbrodt B, Wittbrodt J, Jain G, Balasundaram G, Bates DL, Byron R, Canfield TK, Diegel MJ, Dunn D, Ebersol AK, Frum T, Garg K, Gist E, Hansen RS, Boatman L, Haugen E, Humbert R, Johnson AK, Johnson EM, Kutyavin TM, Lee K, Lotakis D, Maurano MT, Neph SJ, Neri FV, Nguyen ED, Qu H, Reynolds AP, Roach V, Rynes E, Sanchez ME, Sandstrom RS, Shafer AO, Stergachis AB, Thomas S, Vernot B, Vierstra J, Vong S, Weaver MA, Yan Y, Zhang M, Akey JA, Bender M, Dorschner MO, Groudine M, MacCoss MJ, Navas P, Stamatoyannopoulos G, Stamatoyannopoulos JA, Beal K, Brazma A, Flicek P, Johnson N, Lukk M, Luscombe NM, Sobral D, Vaquerizas JM, Batzoglou S, Sidow A, Hussami N, Kyriazopoulou-Panagiotopoulou S, Libbrecht MW, Schaub MA, Miller W, Bickel PJ, Banfai B, Boley NP, Huang H, Li JJ, Noble WS, Bilmes JA, Buske OJ, Sahu AO, Kharchenko PV, Park PJ, Baker D, Taylor J, Lochovsky L (2012) An integrated encyclopedia of DNA elements in the human genome. Nature 489(7414):57–74. doi:10.1038/nature11247

Fajans SS, Bell GI, Polonsky KS (2001) Molecular mechanisms and clinical pathophysiology of maturity-onset diabetes of the young. N Engl J Med 345(13):971–980. doi:10.1056/NEJMra002168

Fraga MF, Ballestar E, Paz MF, Ropero S, Setien F, Ballestar ML, Heine-Suner D, Cigudosa JC, Urioste M, Benitez J, Boix-Chornet M, Sanchez-Aguilera A, Ling C, Carlsson E, Poulsen P, Vaag A, Stephan Z, Spector TD, Wu YZ, Plass C, Esteller M (2005) Epigenetic differences arise during the lifetime of monozygotic twins. Proc Natl Acad Sci USA 102(30):10604–10609

Gaulton KJ, Nammo T, Pasquali L, Simon JM, Giresi PG, Fogarty MP, Panhuis TM, Mieczkowski P, Secchi A, Bosco D, Berney T, Montanya E, Mohlke KL, Lieb JD, Ferrer J (2010) A map of open chromatin in human pancreatic islets. Nat Genet 42(3):255–259

Gluckman PD, Lillycrop KA, Vickers MH, Pleasants AB, Phillips ES, Beedle AS, Burdge GC, Hanson MA (2007) Metabolic plasticity during mammalian development is directionally dependent on early nutritional status. Proc Natl Acad Sci U S A 104(31):12796–12800. doi:10.1073/pnas.0705667104

Holliday R (2006) Epigenetics: a historical overview. Epigenetics 1(2):76–80

Jacobsen SC, Brons C, Bork-Jensen J, Ribel-Madsen R, Yang B, Lara E, Hall E, Calvanese V, Nilsson E, Jorgensen SW, Mandrup S, Ling C, Fernandez AF, Fraga MF, Poulsen P, Vaag A (2012) Effects of short-term high-fat overfeeding on genome-wide DNA methylation in the skeletal muscle of healthy young men. Diabetologia. doi:10.1007/s00125-012-2717-8

Jacobsen SC, Gillberg L, Bork-Jensen J, Ribel-Madsen R, Lara E, Calvanese V, Ling C, Fernandez AF, Fraga MF, Poulsen P, Brons C, Vaag A (2014) Young men with low birthweight exhibit decreased plasticity of genome-wide muscle DNA methylation by high-fat overfeeding. Diabetologia. doi:10.1007/s00125-014-3198-8

Jimenez-Chillaron JC, Isganaitis E, Charalambous M, Gesta S, Pentinat-Pelegrin T, Faucette RR, Otis JP, Chow A, Diaz R, Ferguson-Smith A, Patti ME (2009) Intergenerational transmission of glucose intolerance and obesity by in utero undernutrition in mice. Diabetes 58(2):460–468. doi:10.2337/db08-0490

Jones PA (2012) Functions of DNA methylation: islands, start sites, gene bodies and beyond. Nat Rev 13(7):484–492. doi:10.1038/nrg3230

Kaati G, Bygren LO, Pembrey M, Sjostrom M (2007) Transgenerational response to nutrition, early life circumstances and longevity. Eur J Hum Genet 15(7):784–790

Kameswaran V, Bramswig NC, McKenna LB, Penn M, Schug J, Hand NJ, Chen Y, Choi I, Vourekas A, Won KJ, Liu C, Vivek K, Naji A, Friedman JR, Kaestner KH (2014) Epigenetic regulation of the DLK1-MEG3 microRNA cluster in human type 2 diabetic islets. Cell Metab 19(1):135–145. doi:10.1016/j.cmet.2013.11.016

Kobberling J, Tillil H (1990) Genetic and nutritional factors in the etiology and pathogenesis of diabetes mellitus. World Rev Nutr Diet 63:102–115

Kulzer JR, Stitzel ML, Morken MA, Huyghe JR, Fuchsberger C, Kuusisto J, Laakso M, Boehnke M, Collins FS, Mohlke KL (2014) A common functional regulatory variant at a type 2 diabetes locus upregulates ARAP1 expression in the pancreatic beta cell. Am J Hum Genet 94(2): 186–197. doi:10.1016/j.ajhg.2013.12.011

Li Y, He Y, Qi L, Jaddoe VW, Feskens EJ, Yang X, Ma G, Hu FB (2010) Exposure to the Chinese famine in early life and the risk of hyperglycemia and type 2 diabetes in adulthood. Diabetes 59(10):2400–2406. doi:10.2337/db10-0385

Ling C, Groop L (2009) Epigenetics: a molecular link between environmental factors and type 2 diabetes. Diabetes 58(12):2718–2725

Ling C, Poulsen P, Simonsson S, Ronn T, Holmkvist J, Almgren P, Hagert P, Nilsson E, Mabey AG, Nilsson P, Vaag A, Groop L (2007) Genetic and epigenetic factors are associated with expression of respiratory chain component NDUFB6 in human skeletal muscle. J Clin Invest 117(11):3427–3435. doi:10.1172/JCI30938

Ling C, Del Guerra S, Lupi R, Ronn T, Granhall C, Luthman H, Masiello P, Marchetti P, Groop L, Del Prato S (2008) Epigenetic regulation of PPARGC1A in human type 2 diabetic islets and effect on insulin secretion. Diabetologia 51(4):615–622

Lister R, Pelizzola M, Dowen RH, Hawkins RD, Hon G, Tonti-Filippini J, Nery JR, Lee L, Ye Z, Ngo QM, Edsall L, Antosiewicz-Bourget J, Stewart R, Ruotti V, Millar AH, Thomson JA, Ren B, Ecker JR (2009) Human DNA methylomes at base resolution show widespread epigenomic differences. Nature 462(7271):315–322. doi:10.1038/nature08514

Malmgren S, Spegel P, Danielsson AP, Nagorny CL, Andersson LE, Dekker Nitert M, Ridderstrale M, Mulder H, Ling C (2013) Coordinate changes in histone modifications, mRNA levels and metabolite profiles in clonal INS-1 832/13 beta-cells accompany functional adaptations to lipotoxicity. J Biol Chem. doi:10.1074/jbc.M112.422527

Manolio TA, Collins FS, Cox NJ, Goldstein DB, Hindorff LA, Hunter DJ, McCarthy MI, Ramos EM, Cardon LR, Chakravarti A, Cho JH, Guttmacher AE, Kong A, Kruglyak L, Mardis E, Rotimi CN, Slatkin M, Valle D, Whittemore AS, Boehnke M, Clark AG, Eichler EE, Gibson G, Haines JL, Mackay TF, McCarroll SA, Visscher PM (2009) Finding the missing heritability of complex diseases. Nature 461(7265):747–753. doi:10.1038/nature08494

McCarthy M (2010) Genomics, type 2 diabetes, and obesity. N Engl J Med 363(24):2339–2350

Moran I, Akerman I, van de Bunt M, Xie R, Benazra M, Nammo T, Arnes L, Nakic N, Garcia-Hurtado J, Rodriguez-Segui S, Pasquali L, Sauty-Colace C, Beucher A, Scharfmann R, van Arensbergen J, Johnson PR, Berry A, Lee C, Harkins T, Gmyr V, Pattou F, Kerr-Conte J, Piemonti L, Berney T, Hanley N, Gloyn AL, Sussel L, Langman L, Brayman KL, Sander M, McCarthy MI, Ravassard P, Ferrer J (2012) Human beta cell transcriptome analysis uncovers lncRNAs that are tissue-specific, dynamically regulated, and abnormally expressed in type 2 diabetes. Cell Metab 16(4):435–448. doi:10.1016/j.cmet.2012.08.010

Morris AP, Voight BF, Teslovich TM, Ferreira T, Segre AV, Steinthorsdottir V, Strawbridge RJ, Khan H, Grallert H, Mahajan A, Prokopenko I, Kang HM, Dina C, Esko T, Fraser RM, Kanoni S, Kumar A, Lagou V, Langenberg C, Luan J, Lindgren CM, Muller-Nurasyid M, Pechlivanis S, Rayner NW, Scott LJ, Wiltshire S, Yengo L, Kinnunen L, Rossin EJ, Raychaudhuri S, Johnson AD, Dimas AS, Loos RJ, Vedantam S, Chen H, Florez JC, Fox C, Liu CT, Rybin D, Couper DJ, Kao WH, Li M, Cornelis MC, Kraft P, Sun Q, van Dam RM, Stringham HM, Chines PS, Fischer K, Fontanillas P, Holmen OL, Hunt SE, Jackson AU, Kong A, Lawrence R, Meyer J, Perry JR, Platou CG, Potter S, Rehnberg E, Robertson N, Sivapalaratnam S, Stancakova A, Stirrups K, Thorleifsson G, Tikkanen E, Wood AR, Almgren P, Atalay M, Benediktsson R, Bonnycastle LL, Burtt N, Carey J, Charpentier G, Crenshaw AT, Doney AS, Dorkhan M, Edkins S, Emilsson V, Eury E, Forsen T, Gertow K, Gigante B, Grant GB, Groves CJ, Guiducci C, Herder C, Hreidarsson AB, Hui J, James A, Jonsson A, Rathmann W, Klopp N, Kravic J, Krjutskov K, Langford C, Leander K, Lindholm E, Lobbens S, Mannisto S, Mirza G, Muhleisen TW, Musk B, Parkin M, Rallidis L, Saramies J, Sennblad B, Shah S, Sigurethsson G, Silveira A, Steinbach G, Thorand B, Trakalo J, Veglia F, Wennauer R, Winckler W, Zabaneh D, Campbell H, van Duijn C, Uitterlinden AG, Hofman A, Sijbrands E, Abecasis GR, Owen KR, Zeggini E, Trip MD, Forouhi NG, Syvanen AC, Eriksson JG, Peltonen L, Nothen MM, Balkau B,

Palmer CN, Lyssenko V, Tuomi T, Isomaa B, Hunter DJ, Qi L, Shuldiner AR, Roden M, Barroso I, Wilsgaard T, Beilby J, Hovingh K, Price JF, Wilson JF, Rauramaa R, Lakka TA, Lind L, Dedoussis G, Njolstad I, Pedersen NL, Khaw KT, Wareham NJ, Keinanen-Kiukaanniemi SM, Saaristo TE, Korpi-Hyovalti E, Saltevo J, Laakso M, Kuusisto J, Metspalu A, Collins FS, Mohlke KL, Bergman RN, Tuomilehto J, Boehm BO, Gieger C, Hveem K, Cauchi S, Froguel P, Baldassarre D, Tremoli E, Humphries SE, Saleheen D, Danesh J, Ingelsson E, Ripatti S, Salomaa V, Erbel R, Jockel KH, Moebus S, Peters A, Illig T, de Faire U, Hamsten A, Morris AD, Donnelly PJ, Frayling TM, Hattersley AT, Boerwinkle E, Melander O, Kathiresan S, Nilsson PM, Deloukas P, Thorsteinsdottir U, Groop LC, Stefansson K, Hu F, Pankow JS, Dupuis J, Meigs JB, Altshuler D, Boehnke M, McCarthy MI (2012) Large-scale association analysis provides insights into the genetic architecture and pathophysiology of type 2 diabetes. Nat Genet 44(9):981–990. doi:10.1038/ng.2383

Ng SF, Lin RC, Laybutt DR, Barres R, Owens JA, Morris MJ (2010) Chronic high-fat diet in fathers programs beta-cell dysfunction in female rat offspring. Nature 467(7318):963–966

Nilsson E, Jansson P, Perfilyev A, Volkov P, Pedersen M, Svensson M, Poulsen P, Ribel-Madsen R, Almgren P, Fadista J, Rönn T, Klarlund-Pedersen B, Scheele C, Vaag A, Ling C (2014) Altered DNA methylation and differential expression of genes influencing metabolism and inflammation in adipose tissue from subjects with type 2 diabetes. Diabetes 63(9):2962–2976

Nitert MD, Dayeh T, Volkov P, Elgzyri T, Hall E, Nilsson E, Yang BT, Lang S, Parikh H, Wessman Y, Weishaupt H, Attema J, Abels M, Wierup N, Almgren P, Jansson PA, Ronn T, Hansson O, Eriksson KF, Groop L, Ling C (2012) Impact of an exercise intervention on DNA methylation in skeletal muscle from first-degree relatives of patients with type 2 diabetes. Diabetes 61(12):3322–3332. doi:10.2337/db11-1653

Ozanne SE, Hales CN (2002) Pre- and early postnatal nongenetic determinants of type 2 diabetes. Expert Rev Mol Med 4(24):1–14. doi:10.1017/S1462399402005240

Ozanne SE, Jensen CB, Tingey KJ, Storgaard H, Madsbad S, Vaag AA (2005) Low birthweight is associated with specific changes in muscle insulin-signalling protein expression. Diabetologia 48(3):547–552

Park JH, Stoffers DA, Nicholls RD, Simmons RA (2008) Development of type 2 diabetes following intrauterine growth retardation in rats is associated with progressive epigenetic silencing of Pdx1. J Clin Invest 118(6):2316–2324

Parker SC, Stitzel ML, Taylor DL, Orozco JM, Erdos MR, Akiyama JA, van Bueren KL, Chines PS, Narisu N, Black BL, Visel A, Pennacchio LA, Collins FS (2013) Chromatin stretch enhancer states drive cell-specific gene regulation and harbor human disease risk variants. Proc Natl Acad Sci U S A 110(44):17921–17926. doi:10.1073/pnas.1317023110

Pasquali L, Gaulton KJ, Rodriguez-Segui S, Mularoni L, Miguel-Escalada I, Akerman I, Tena JJ, Moran I, Gómez-Marín C, van de Bunt M, Ponsa-Cobas J, Castro N, Nammo T, Cebola I, Garcia-Hurtado J, Maestro MA, Pattou F, Piemonti L, Berney T, Gloyn AL, Ravassard P, Muller F, McCarthy MI, Ferrer J (2014) Pancreatic islet enhancer clusters enriched in type 2 diabetes risk–associated variants. Nat Genet. doi:10.1038/ng.2870

Pembrey ME, Bygren LO, Kaati G, Edvinsson S, Northstone K, Sjostrom M, Golding J (2006) Sex-specific, male-line transgenerational responses in humans. Eur J Hum Genet 14(2): 159–166. doi:10.1038/sj.ejhg.5201538

Rakyan VK, Down TA, Balding DJ, Beck S (2011) Epigenome-wide association studies for common human diseases. Nat Rev 12(8):529–541. doi:10.1038/nrg3000

Ravelli AC, van der Meulen JH, Michels RP, Osmond C, Barker DJ, Hales CN, Bleker OP (1998) Glucose tolerance in adults after prenatal exposure to famine. Lancet 351(9097):173–177

Ribel-Madsen R, Fraga MF, Jacobsen S, Bork-Jensen J, Lara E, Calvanese V, Fernandez AF, Friedrichsen M, Vind BF, Hojlund K, Beck-Nielsen H, Esteller M, Vaag A, Poulsen P (2012) Genome-wide analysis of DNA methylation differences in muscle and fat from monozygotic twins discordant for type 2 diabetes. PLoS ONE 7(12), e51302. doi:10.1371/journal.pone.0051302

Ronn T, Poulsen P, Hansson O, Holmkvist J, Almgren P, Nilsson P, Tuomi T, Isomaa B, Groop L, Vaag A, Ling C (2008) Age influences DNA methylation and gene expression of COX7A1 in human skeletal muscle. Diabetologia 51(7):1159–1168

Ronn T, Volkov P, Davegardh C, Dayeh T, Hall E, Olsson AH, Nilsson E, Tornberg A, Dekker Nitert M, Eriksson KF, Jones HA, Groop L, Ling C (2013) A six months exercise intervention influences the genome-wide DNA methylation pattern in human adipose tissue. PLoS Genet 9(6), e1003572. doi:10.1371/journal.pgen.1003572

Sandovici I, Smith NH, Nitert MD, Ackers-Johnson M, Uribe-Lewis S, Ito Y, Jones RH, Marquez VE, Cairns W, Tadayyon M, O'Neill LP, Murrell A, Ling C, Constância M, Ozanne SE (2011) Maternal diet and aging alter the epigenetic control of a promoter-enhancer interaction at the Hnf4a gene in rat pancreatic islets. Proc Natl Acad Sci USA 108(13):5449–5454. doi:10.1073/pnas.1019007108, Epub 2011 Mar 8

Stahl EA, Wegmann D, Trynka G, Gutierrez-Achury J, Do R, Voight BF, Kraft P, Chen R, Kallberg HJ, Kurreeman FA, Kathiresan S, Wijmenga C, Gregersen PK, Alfredsson L, Siminovitch KA, Worthington J, de Bakker PI, Raychaudhuri S, Plenge RM (2012) Bayesian inference analyses of the polygenic architecture of rheumatoid arthritis. Nat Genet 44(5):483–489. doi:10.1038/ng.2232

Stitzel ML, Sethupathy P, Pearson DS, Chines PS, Song L, Erdos MR, Welch R, Parker SC, Boyle AP, Scott LJ, Margulies EH, Boehnke M, Furey TS, Crawford GE, Collins FS (2010) Global epigenomic analysis of primary human pancreatic islets provides insights into type 2 diabetes susceptibility loci. Cell Metab 12(5):443–455

Stoffers DA, Ferrer J, Clarke WL, Habener JF (1997) Early-onset type-II diabetes mellitus (MODY4) linked to IPF1. Nat Genet 17(2):138–139

Suzuki MM, Bird A (2008) DNA methylation landscapes: provocative insights from epigenomics. Nat Rev 9(6):465–476

Toperoff G, Aran D, Kark JD, Rosenberg M, Dubnikov T, Nissan B, Wainstein J, Friedlander Y, Levy-Lahad E, Glaser B, Hellman A (2012) Genome-wide survey reveals predisposing diabetes type 2-related DNA methylation variations in human peripheral blood. Hum Mol Genet 21(2):371–383. doi:10.1093/hmg/ddr472

Voight BF, Scott LJ, Steinthorsdottir V, Morris AP, Dina C, Welch RP, Zeggini E, Huth C, Aulchenko YS, Thorleifsson G, McCulloch LJ, Ferreira T, Grallert H, Amin N, Wu G, Willer CJ, Raychaudhuri S, McCarroll SA, Langenberg C, Hofmann OM, Dupuis J, Qi L, Segre AV, van Hoek M, Navarro P, Ardlie K, Balkau B, Benediktsson R, Bennett AJ, Blagieva R, Boerwinkle E, Bonnycastle LL, Bengtsson Bostrom K, Bravenboer B, Bumpstead S, Burtt NP, Charpentier G, Chines PS, Cornelis M, Couper DJ, Crawford G, Doney AS, Elliott KS, Elliott AL, Erdos MR, Fox CS, Franklin CS, Ganser M, Gieger C, Grarup N, Green T, Griffin S, Groves CJ, Guiducci C, Hadjadj S, Hassanali N, Herder C, Isomaa B, Jackson AU, Johnson PR, Jorgensen T, Kao WH, Klopp N, Kong A, Kraft P, Kuusisto J, Lauritzen T, Li M, Lieverse A, Lindgren CM, Lyssenko V, Marre M, Meitinger T, Midthjell K, Morken MA, Narisu N, Nilsson P, Owen KR, Payne F, Perry JR, Petersen AK, Platou C, Proenca C, Prokopenko I, Rathmann W, Rayner NW, Robertson NR, Rocheleau G, Roden M, Sampson MJ, Saxena R, Shields BM, Shrader P, Sigurdsson G, Sparso T, Strassburger K, Stringham HM, Sun Q, Swift AJ, Thorand B, Tichet J, Tuomi T, van Dam RM, van Haeften TW, van Herpt T, van Vliet-Ostaptchouk JV, Walters GB, Weedon MN, Wijmenga C, Witteman J, Bergman RN, Cauchi S, Collins FS, Gloyn AL, Gyllensten U, Hansen T, Hide WA, Hitman GA, Hofman A, Hunter DJ, Hveem K, Laakso M, Mohlke KL, Morris AD, Palmer CN, Pramstaller PP, Rudan I, Sijbrands E, Stein LD, Tuomilehto J, Uitterlinden A, Walker M, Wareham NJ, Watanabe RM, Abecasis GR, Boehm BO, Campbell H, Daly MJ, Hattersley AT, Hu FB, Meigs JB, Pankow JS, Pedersen O, Wichmann HE, Barroso I, Florez JC, Frayling TM, Groop L, Sladek R, Thorsteinsdottir U, Wilson JF, Illig T, Froguel P, van Duijn CM, Stefansson K, Altshuler D, Boehnke M, McCarthy MI (2010) Twelve type 2 diabetes susceptibility loci identified through large-scale association analysis. Nat Genet 42(7):579–589

Volkmar M, Dedeurwaerder S, Cunha DA, Ndlovu MN, Defrance M, Deplus R, Calonne E, Volkmar U, Igoillo-Esteve M, Naamane N, Del Guerra S, Masini M, Bugliani M, Marchetti P, Cnop M, Eizirik DL, Fuks F (2012) DNA methylation profiling identifies epigenetic dysregulation in pancreatic islets from type 2 diabetic patients. EMBO J 31(6):1405–1426. doi:10.1038/emboj.2011.503

Willems SM, Mihaescu R, Sijbrands EJ, van Duijn CM, Janssens AC (2011) A methodological perspective on genetic risk prediction studies in type 2 diabetes: recommendations for future research. Curr Diab Rep 11(6):511–518. doi:10.1007/s11892-011-0235-6

Wu H, Zhang Y (2014) Reversing DNA methylation: mechanisms, genomics, and biological functions. Cell 156(1–2):45–68. doi:10.1016/j.cell.2013.12.019

Yang BT, Dayeh TA, Kirkpatrick CL, Taneera J, Kumar R, Groop L, Wollheim CB, Nitert MD, Ling C (2011) Insulin promoter DNA methylation correlates negatively with insulin gene expression and positively with HbA(1c) levels in human pancreatic islets. Diabetologia 54(2): 360–367. doi:10.1007/s00125-010-1967-6

Yang BT, Dayeh TA, Volkov PA, Kirkpatrick CL, Malmgren S, Jing X, Renstrom E, Wollheim CB, Nitert MD, Ling C (2012) Increased DNA methylation and decreased expression of PDX-1 in pancreatic islets from patients with type 2 diabetes. Mol Endocrinol 26(7): 1203–1212. doi:10.1210/me.2012-1004

Chapter 12
Gene-Environment Interaction: Methods and Examples in Type 2 Diabetes and Obesity

Alisa K. Manning

Abstract Although both genetics and environmental factors play important roles in the etiology of type 2 diabetes (T2D), the extent to which genetics influence the environment (or vice versa) is still an open question. In this chapter, we first motivate the study of gene-environment ($G \times E$) interaction with T2D, present statistical models of interaction, and then share illustrating examples of $G \times E$ interaction in the literature: *PPGARG* with T2D, *COBLL1/GRB14* with fasting insulin levels and *FTO* with T2D, obesity, and physical activity.

12.1 Introduction

Debates on the causes of type 2 diabetes (T2D) often center on the question of "nature vs nurture"—to what extent is the disease caused by genetics or exposure to environmental risk factors? In this chapter, we explore the idea of incorporating these external, non-genetic variables in genetic association studies of diabetes and diabetes-related traits with the goals of identifying novel genetic associations which might be modulated by environmental factors (or vice versa) and thus better understand disease mechanisms. We will present the methodology and current best practices of gene-environment interaction studies.

It is important to distinguish between biological and statistical interaction. Biological interaction is a term used to describe dependent biological systems. The simplest example relevant to diabetes physiology concerns the canonical relationship between insulin secretion and insulin action, captured by the molecular interaction between circulating insulin and its receptor.

The statistical study of gene-environment interaction is concerned with finding genetic variants for which the association effect between an outcome and the

A.K. Manning (✉)
Department of Medicine, Harvard Medical School, Boston, MA 02138, USA

Program in Medical and Population Genetics, Broad Institute, Cambridge, MA 02141, USA

Center for Human Genetics Research, Massachusetts General Hospital, Boston, MA 02114, USA
e-mail: amanning@broadinstitute.org

J.C. Florez (ed.), *The Genetics of Type 2 Diabetes and Related Traits*,
DOI 10.1007/978-3-319-01574-3_12

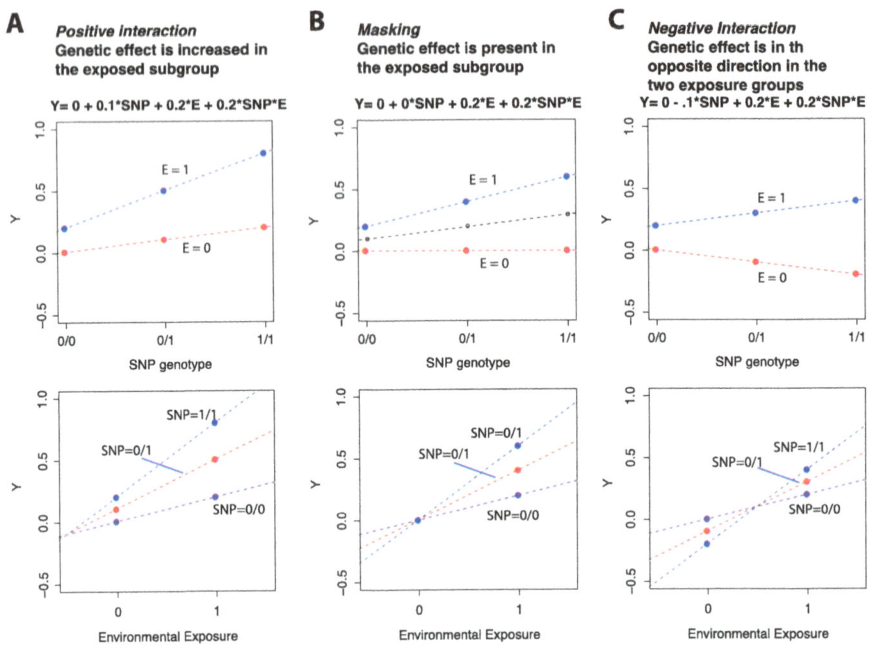

Fig. 12.1 Interpreting gene-environment interaction: For each panel, two plots are presented. In the *top plot*, the outcome (Y) value is displayed for the three genotypes under an additive genetic model. The association in the exposed group (*blue*) is compared to the genetic association in the unexposed subgroup (*red*). In the *bottom plot*, Y values are displayed for the two exposure groups with the *three lines* showing the values in the three genotype groups (0/0, *purple*; 0/1, *red*; and 1/1, *blue*). Three different interaction models are presented: (**a**) positive interaction occurs when the genetic effect and the interaction effect are in the same direction. Here, there is a genetic association in the absence of the exposure variable (when $E=0$, $\beta_{SNP}=0.1$). In the exposed group, the association is strengthened (when $E=1$, $\beta_{SNP}=0.3$). (**b**) Masking can occur when the genetic association is not present in one exposure group. Here, there is no genetic association in the non-exposed group (when $E=0$, $\beta_{SNP}=0$). In a test of only the genetic main effect (*black line*), the association may still be detectable ($\beta_{SNP}=0.09$) but it is much stronger in the exposed group (when $E=1$, $\beta_{SNP}=0.2$). (**c**) The third example illustrates a situation where the genetic effect in the exposed group ($\beta_{SNP}=0.1$) is in the opposite direction as the genetic effect in the unexposed group ($\beta_{SNP}=-0.1$)

variant is modified by an additional covariate, the environmental variable. In the presence of statistical interaction, the association between the genetic variant and the outcome is different for the various levels of the environmental variable (see Fig. 12.1). Note that the relationship can be viewed in a reciprocal manner, that is, the effect of the environmental variable on the outcome differs by genotype (or degree of genetic exposure).

Questions related to gene-environment interaction include:

- Are there individuals with particular genetic profiles who are more likely to develop T2D when exposed to sedentary lifestyles and/or poor nutrition? (See Chaps. 27 and 28.)

- Can we personalize treatment knowing that certain medications work best in individuals with particular genetic profiles? (See Chaps. 24 and 25.)

12.1.1 Type 2 Diabetes, Glycemic Traits, and Gene-Environment Interactions

Type 2 diabetes is a disease of deteriorating beta-cell function and increasing insulin resistance with lifestyle risk factors such as lack of physical activity and obesity (McCarthy 2010). Candidate gene studies of genes in known diabetes pathways or previously implicated in neonatal diabetes or maturity-onset diabetes of the young (MODY) (Altshuler et al. 2000; Gloyn et al. 2004; Winckler et al. 2007; Sandhu et al. 2007) and genome-wide association studies (GWAS) of common genetic variants (Voight et al. 2010; Saxena et al. 2012; Morris et al. 2012) point to two mechanistic pathways of T2D disease progression: (1) variants near genes such as *CDKAL1*, *CDKN2A*, and *CDKN2B* reduce beta-cell mass and variants near genes such as *MTNR1B*, *TCF7L2*, and *KCNJ11* influence beta-cell dysfunction, both of which result in reduced insulin secretion to lower glucose levels, and (2) insulin resistance, where cells and tissues become resistant to the effects of insulin, with association in or near genes such as *FTO* (related to obesity), *IRS1*, and *PPARG* (McCarthy 2010). Genetic variants associated with fasting glucose levels (related to beta-cell dysfunction) and fasting insulin levels (related to insulin resistance) have also been published (Prokopenko et al. 2008; Dupuis et al. 2010; Manning et al. 2012; Scott et al. 2012) (see Chaps. 2 and 3).

Studies show that both dietary fats and free fatty acids impact insulin resistance, possibly through mediating genetic factors such as *PPARG* variation (Roden et al. 1996; Kubota et al. 1999; Haag and Dippenaar 2005). External environmental variables (lifestyle factors such as diet or exercise) that impact T2D disease progression, beta-cell deterioration, and/or insulin sensitivity have been proposed as environmental exposures that may interact with genetics in the etiology of T2D. Furthermore, obesity contributes to insulin resistance by creating an "obesogenic environment," making continuous body mass index (BMI) or categorical obesity (as defined by BMI $\geq$ 30 kg/m^2) attractive candidate variables for interaction studies.

The measurement of lifestyle variables (physical activity, diet, and smoking status) can differ between studies, making meta-analysis and replication of genetic associations more difficult. For example, physical activity is a measure of an individual's energy expenditure that can be summarized through questionnaires or more direct means such as continuous heart rate monitoring. Crude categories (sedentary, active, and/or very active) are often used in order to reach concordance between the various measures of physical activity used across the study designs, resulting in a loss of information in the subset of studies using sophisticated measurement tools. Other exposures such as smoking status and diet may be inconsistently measured across studies.

The InterAct project was designed to investigate lifestyle interactions in the development of T2D (The InterAct Consortium 2011) using a case-cohort sample of 10,901 incident diabetes cases from eight EPIC countries and a control cohort of 15,352 participants including 736 cases of incident T2D. Findings include an increased incidence of T2D with high total protein intake (HR = 1.13, 95 % confidence interval [CI]: 1.08–1.19) and high animal protein intake (HR = 1.12 95 % CI: 1.07–1.17), where effect modification was observed by sex ($P < 0.001$) and BMI among women ($P < 0.001$).

12.2 Statistical Model, Study Designs, and Interpretation

Numerous reviews of the design, implementation, and interpretation of GWAS of gene-environment interaction are available (Ottman 1996; Thomas 2010; Ober and Vercelli 2011; Aschard et al. 2012; Gauderman et al. 2013). Here, we first present the basic methodology of gene-environment interaction analyses and then describe several popular extensions.

12.2.1 Type 2 Diabetes as the Outcome

Although other models might be appropriate for the scientific question at hand, and complex diseases can be studied with a variety of models (Clayton 2012), the association between T2D and genetic factors is often assessed using logistic regression models in appropriate samples.

The term "main-effects model" refers to a test of the marginal association between a genetic variant and the outcome (without interaction). Here, disease status is dichotomous and coded as T2D = 1 for individuals with T2D and T2D = 0 otherwise. Along with the independent genetic variable G, coded for an appropriate genetic model, additional covariates such as age and sex are usually included in the main-effects model:

$$\log(\text{odds of T2D}) = \beta_0 + \beta_1 \text{SEX} + \beta_2 \text{AGE} + \beta_3 G \tag{12.1}$$

Using the regression estimates, an estimate of the odds ratio for the association between G and T2D can be obtained: $OR_G = e^{\hat{\beta}_3}$.

Statistical interaction by an independent variable E is defined as a departure from the multiplicative odds ratio model for the joint effect of G and E. Using the odds ratio for the association between T2D and E, OR_E, a relationship between OR_G and OR_E can be defined: if there is no interaction, and the association between G and T2D is the same for all levels of E, then the two variables are independent and

$OR_{G,E} = OR_G \times OR_E$. When $OR_{G,E} \neq OR_G \times OR_E$, or $\frac{OR_{G,E}}{OR_G OR_E} \neq 1$, statistical interaction is present.

One statistical test for interaction can be performed by including term for G, E, and the product of the two variables in the regression model:

$$\log(\text{odds of T2D}) = \beta_0 + \beta_1\text{SEX} + \beta_2\text{AGE} + \beta_3 G + \beta_4 E + \beta_5 G \times E \quad (12.2)$$

In this model, the odds ratio estimate for the increased or decreased risk of T2D must be derived using both $\hat{\beta}_3$ and $\hat{\beta}_5$ (see Fig. 12.1).

12.2.2 Quantitative Outcomes

With quantitative outcomes such as glucose or insulin levels, linear regression models can be used to assess G × E interaction effects. There are several assumptions of linear regression that should be considered and are discussed in many statistical texts.

As with T2D, covariates such as age and BMI are commonly included in the genetic association tests of diabetes-related quantitative traits.

The main-effects regression model describes the relationship between Y and G:

$$Y = \beta_0 + \beta_1\text{SEX} + \beta_2\text{AGE} + \beta_3 G + \varepsilon \quad (12.3)$$

A test for gene-environment interaction can be performed by adding an interaction term to the main-effects regression model:

$$Y = \beta_0 + \beta_1\text{SEX} + \beta_2\text{AGE} + \beta_3 G + \beta_4 E + \beta_5 G \times E + \varepsilon \quad (12.4)$$

When fit to data, there is evidence for statistical interaction if the estimate of the regression estimate $\hat{\beta}_5$ is significantly different from zero.

12.2.3 Dichotomous Environmental Variable

Alternatively, if the E is dichotomous, the sample can be split into two strata, one with $E = 1$ and another with $E = 0$. The main-effects regression models, in which only the association between G and the outcome is considered, are assessed within strata of the environmental variable.

For logistic regression:

$$\begin{aligned}\log(\text{odds of T2D}) &= \beta_0^{E_1} + \beta_1\text{SEX} + \beta_2\text{AGE} + \beta_3^{E=1}G \\ \log(\text{odds of T2D}) &= \beta_0^{E_2} + \beta_1\text{SEX} + \beta_2\text{AGE} + \beta_3^{E=0}G\end{aligned} \tag{12.5}$$

For linear regression:

$$\begin{aligned}Y &= \beta_0^{E_1} + \beta_1\text{SEX} + \beta_2\text{AGE} + \beta_3^{E=1}G + \varepsilon \\ Y &= \beta_0^{E_2} + \beta_1\text{SEX} + \beta_2\text{AGE} + \beta_3^{E=0}G + \varepsilon\end{aligned} \tag{12.6}$$

Here, statistical interaction is present if the estimate $\hat{\beta}_3^{E=1}$ differs significantly from the estimate $\hat{\beta}_3^{E=0}$ (Aschard et al. 2010).

The interpretation of interaction for linear regression is in terms of the different slopes of the relationship between G and Y for different values of E, when E is a continuous measure, or different strata of E, when it is dichotomous (see Fig. 12.1).

12.3 Genome-Wide Interaction Tests

12.3.1 Screening for Interaction and "Case-Only" Tests

Often genome-wide main-effects analyses are performed prior to interaction studies being undertaken. This approach runs the risk of type II error (the failure to reject a false null hypothesis), as genetic variants with strong environmental interactions may have weak overall main effects and greater heterogeneity in main-effects testing, hindering their detection by main-effects screens. One proposed method studying gene-environment interaction while reducing the number of interaction tests performed was to choose a threshold P_0 (e.g., 0.05 or 0.01) and only investigate the interaction model (testing only the interaction term) on those genome-wide variants with a main-effects P-value less than P_0 (Kooperberg and LeBlanc 2008).

Another statistical test for interaction is the case-only test for gene-environment interaction (Piegorsch et al. 1994), which tests for an association between the environmental exposure (E) and the genetic variant (G) in a sample of "cases," people with T2D, for example. Under the assumption that the genetic variable is not associated with the environmental variable, the estimated odds ratio, OR_E, from this model is mathematically identical to the interaction odds ratio $OR_{G=1,E=1}$ from formula (12.2). Although a large increase in statistical power is observed with the case-only test (Yang et al. 1997), there is a strong assumption that G is independent of E in the overall population. The case-only method and other methods that leverage the G and E independence assumption can exhibit both an increase in type I error (the false-positive rate) and a decrease in statistical power (Wu et al. 2013; Gauderman et al. 2013). Several methods that use the case-only test, but retain statistical power when the G-E independence assumption is violated, have been proposed. These include screening methods (Murcray et al. 2009, 2011) and "cocktail" methods (Mukherjee et al. 2012; Hsu et al. 2012) that combine the case–control test for interaction with case-only methods by using screening or model averaging (Mukherjee et al. 2008; Li and Conti 2009) approaches. Another method proposed for testing for interactions include a screening method that jointly assess the significance of the environment-gene association (as in the case-only test) and the disease-gene association in the screening step (Gauderman et al. 2013).

12.3.2 Joint Test of Main and Interaction Effects

The joint test of marginal G and interaction effects was introduced as a flexible test for genome-wide discovery of genetic associations when the underlying interaction model is suspected but unknown (Kraft et al. 2007). A statistical test is constructed to test if either or both of the genetic terms in the interaction model are significantly different from zero ($H_0 : \beta_3 = \beta_5 = 0$). The statistic can be constructed using a likelihood ratio test or Wald test; it follows a 2 degree-of-freedom chi-square distribution and remains valid when the gene-environment independence assumption is violated. Over a range of models, the joint test has comparable or better power than the interaction or case-only test, making it an attractive approach for genome-wide analysis, as only one statistical model needs to be applied to the genetic data.

12.3.3 Meta-Analysis Methods

Meta-analysis has become the de facto standard for performing genetic discovery analyses when the genetic effects are too small for detection with individual cohorts. Most common genetic discoveries were possible only when consortia were formed to conduct these meta-analysis (Prokopenko et al. 2008; Dupuis et al. 2010; Voight et al. 2010). In order to detect genetic interactions, much larger samples are required than that needed to detect comparable main effects (Aschard et al. 2010). One recent efficient and powerful meta-analysis method for testing the interaction effect across multiple studies has been proposed (Li et al. 2014). This method uses summary statistics from the individual studies (as in other meta-analysis methods) and a meta-regression approach to adaptively estimate the gene-environment interaction effect.

The joint test has been extended to a meta-analysis framework (Aschard et al. 2010; Manning et al. 2011). The joint meta-analysis, or JMA, is a meta-analysis method that allows individual cohorts to submit regression statistics from the interaction model: $Y = \beta_0 + \beta_1\text{SEX} + \beta_2\text{AGE} + \beta_3 G + \beta_4 E + \beta_5 G \times E + \varepsilon$. The statistics that need to be submitted for meta-analysis are the estimates of $\hat{\beta}_4$ and $\hat{\beta}_5$, the robust standard error and robust covariance of these estimates. The method is implemented in a modified version of the METAL software (Willer et al. 2010), available from the corresponding authors of the JMA paper (Manning et al. 2011), which produces summarized regression estimates of β_3 and β_5 and a 2 degree-of-freedom chi-square test of significance.

If the environmental variable is dichotomous, a simplified version of the joint test can be applied using a score test approach (Aschard et al. 2010). Regressions are performed in the stratum-specific main-effects models and the regression estimates of $\beta_3^{E=1}$ and $\beta_3^{E=0}$ are meta-analyzed using a standard inverse variance approach (de Bakker et al. 2008; Zeggini and Ioannidis 2009). The joint test, constructed with a sum of the fixed-effects tests for $\beta_3^{E=1}=0$ and $\beta_3^{E=0}=0$, follows a chi-square distribution with 2 degrees-of-freedom.

12.3.4 Statistical Issues and Best Practices

Statistical tests for interaction can be dependent on the trait scale—the interactions can be present in one scale (after a log-transformation, for example) and undetectable if modeled on another scale. Issues of environmental exposure and departures from the gene-environment independence assumption have been recently discussed (Lindström et al. 2009; Cornelis et al. 2012). Generally, the joint test performs well in the presence of environmental misspecification, a problem that can be somewhat controlled for through the use of robust standard error estimates (Cornelis et al. 2012). Furthermore, the use of robust standard error estimates corrects an issue of apparent QQ-plot inflation, from violations of assumptions such as linearity and homoscedasticity between Y and E, observed when comparing the P-values for a test of β_5, to those from the expected P-value distribution (Voorman et al. 2011).

Finally, a recent paper discusses the implication of confounding on the interaction term in the interaction model (12.2) (Keller 2014). The $G \times E$ term will be biased if either (a) a covariate (C) is associated with the SNP and the relationship between E and Y differs according to C ($\beta_{C \times E} \neq 0$) or (b) C is associated with E and the relationship between G and Y differs according to C ($\beta_{C \times G} \neq 0$). This implies that if either of these relationships holds, then the interaction terms $\beta_{C \times E}$ or $\beta_{C \times \mathrm{SNP}}$ should be included in the model for each covariate considered. These models should be considered on a case-by-case basis, depending on the outcome, environmental variable, and whether or not the additional covariates could be independently associated with the SNP or be candidates for $G \times E$ interaction.

12.4 Illustrating Examples

12.4.1 PPARG

Of the candidate gene associations previously described, one of the early confirmed genetic associations with T2D was the Pro12Ala polymorphism in the *PPARG* gene (Altshuler et al. 2000; McCarthy 2010). Replication of this association was not universal, with several studies confirming the association and other studies failing to replicate it (see Ludovico et al. for a comprehensive citation list of the *PPARG* studies and Gouda et al. for a comprehensive review). Among the first attempts to explain this heterogeneity of effects, some groups found that along with increasing the risk for diabetes, the Pro121 allele also decreased insulin sensitivity, possibly lowered BMI, and was associated with increased adipose tissue formation (Roden et al. 1996; Kubota et al. 1999; Haag and Dippenaar 2005; Cecil et al. 2006).

In an analysis of "time to onset of diabetes" in the Diabetes Prevention Program, a significant gene-environment interaction was found between the Pro12Ala variant and obesity traits (Florez et al. 2007). Self-reported ethnicity was considered as an additional variable in a test for potential interaction but was not found to be significant. The Pro121 carriers progressed more quickly to diabetes (HR, 1.24; 95 % CI, 0.99–1.57; $P = 0.07$), and in models with P121Q-adiposity interactions with BMI (interaction $P = 0.03$) and waist circumference (interaction $P = 0.002$), the incidence of diabetes increased for higher mean BMI levels, showing that the protective effects of the alanine allele were attenuated at higher BMI levels.

A large meta-analysis ($N = 42{,}910$) was conducted based on 41 published studies and 2 unpublished studies to determine possible sources of the effect heterogeneity in the association of *PPARG* Pro121Ala with T2D (Ludovico et al. 2007). The association was confirmed (Ala12 OR = 0.81, $P = 0.005$), and population-specific differences in the reduced risk of T2D due to the Ala12 variant were also reported: the odds ratio was 0.65 in the Asian subgroup, 0.82 in the North American subgroup, and 0.85 in the European subgroup. Although the authors describe that the difference in the Asian subgroup could be due to BMI (48 % of the heterogeneity was explained by the BMI in the control groups), different population-specific genetic backgrounds were stated as a more likely cause for the heterogeneity observed in the European and North American studies.

In a subsequent meta-analysis, of 60 studies with up to 32,849 type 2 diabetes cases and 47,456 controls, the estimated odds ratio for the 121Ala allele was 0.86 (95 % CI: 0.91–0.90) and 0.85 (CI: 0.82–0.88) for random-effects and fixed-effects meta-analyses, respectively (Gouda et al. 2010). The authors report a moderate degree of inconsistency among the studies contributing to this meta-analysis ($I^2 = 37$ %, 95 % CI 9–54; $P = 0.003$). Ethnicity accounted for some of the heterogeneity (14 % of the between-study variance), but mean BMI levels among the T2D cases in the studies varied widely: although not significant, a trend was observed such that the protective effect of the variant was strongest (the odds ratio was lowest) for studies with mean case BMI < 25 kg/m^2, and the protective effect was attenuated (the odds ratio increased toward the null) as the mean BMI in cases increased.

12.4.2 BMI Interactions with Fasting Insulin

Initial publications from the Meta-Analysis of Glucose and Insulin-related traits Consortium (MAGIC) described 16 loci associated with fasting glucose levels compared to two loci associated with fasting insulin levels (Prokopenko et al. 2008; Dupuis et al. 2010), indicating differences in the genetic architectures for beta-cell dysfunction and insulin resistance. The marginal models investigated by MAGIC included minimal adjustments for age and sex. Subsequently, two efforts were undertaken to investigate the role of obesity in the variation of quantitative glycemic traits: (1) interaction models on a subset of MAGIC cohorts for which obesity measures were available (Manning et al. 2012) and (2) meta-analyses of main-effects models adjusting for obesity measures including larger sample sizes with the inclusion of Metabochip genotype data (Scott et al. 2012) (see Chap. 3).

For the first analysis, two terms were added to models for fasting glucose and log-transformed fasting insulin: the adjustment for body mass index (BMI) and the interaction between a genetic variant and BMI.

$$\text{Fasting Glucose} = \beta_0 + \beta_1\text{SEX} + \beta_2\text{AGE} + \beta_3 G + \beta_4\text{BMI} + \beta_5 G \times \text{BMI} + \varepsilon$$
$$\log(\text{Fasting Insulin}) = \beta_0 + \beta_1\text{SEX} + \beta_2\text{AGE} + \beta_3 G + \beta_4\text{BMI} + \beta_5 G \times \text{BMI} + \varepsilon$$

The joint meta-analysis was applied in a genome-wide analysis of 2.4 million single nucleotide polymorphisms (SNPs) and six and seven additional loci were found to be associated with fasting insulin and fasting glucose, respectively (Manning et al. 2012), with one locus, *PPP1R3B*, showing association with both fasting insulin and fasting glucose levels. Of these loci, one fasting insulin association (rs7607980 in the *COBLL1/GRB14* locus, joint $P = 4.3 \times 10^{-20}$) and three fasting glucose loci displayed a greater degree of significance in the joint test compared to a model that only included an adjustment for BMI. All loci were reported as significant in the second analysis, demonstrating that either larger sample sizes or adjustment for BMI was necessary for their discovery (Scott et al. 2012). Although a number of the loci reported in Manning et al. showed differential evidence for significance and effect sizes in the high BMI group compared to the low BMI group (as defined by a BMI cutoff of 28 kg/m^2), only rs7607980 showed evidence for an interaction effect when a meta-analysis of the interaction term was performed ($P = 0.0002$). The additive main effect of rs7607980 on log-transformed fasting insulin levels was 0.02 (with standard error 0.0033), similar to the BMI-adjusted main effect of 0.028 (0.0033). In the subset with high BMI, the effect was 0.041 (0.0064) with $P = 3.0 \times 10^{-10}$, while in the lower BMI stratum, the effect was weaker and less significant at 0.0175

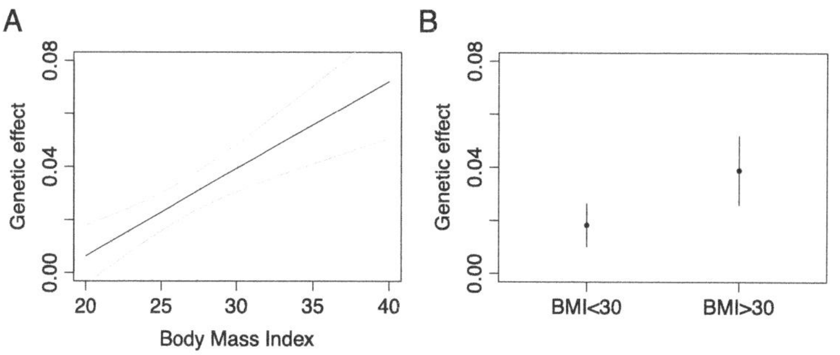

Fig. 12.2 The genetic effect estimate of rs7607980 from the *COBLL1/GRB14* locus accounting for the interaction with BMI. The additive genetic effect of rs7607980 changes for different BMI levels. (**a**) For the joint meta-analysis, where BMI is a continuous exposure variable, the estimate (*solid black line*) and 95 % confidence interval (*gray curves*), the estimate is $\hat{\beta}_{\mathrm{SNP}} = -0.06 + 0.003 \times \mathrm{BMI}$. (**b**) The studies were dichotomized into high- and low-BMI groups and the estimate of the genetic effect was obtained within each subgroup. The additive genetic effect is displayed with the circles, with the 95 % confidence interval of the estimate shown by the *vertical lines*. In the subset with high BMI, $\hat{\beta}_{\mathrm{SNP}} = 0.04$, and in the subset with the lower BMI, $\hat{\beta}_{\mathrm{SNP}} = 0.02$

(0.0041) with $P = 1.8 \times 10^{-5}$. The stratum-specific effect was consistent with the jointly estimated effect from the interaction model (Fig. 12.2). These findings support the sensible assumption that taking adiposity into account can augment discovery of genetic variants that underlie insulin resistance.

12.4.3 FTO: *Type 2 Diabetes, Body Mass Index, and Interaction with Physical Activity*

In 2007, the Wellcome Trust Case Control Consortium (WCCC) performed a genome-wide association study for type 2 diabetes and described a strong increase of risk for T2D associated with SNPs in the first intron of the *FTO* gene (rs9939609, OR = 1.27; $P = 5 \times 10^{-8}$ in 1,924 T2D cases and 2,938 controls) which was replicated in an independent sample (OR = 1.15; $P = 9 \times 10^{-6}$ in 3,757 T2D cases and 5,346 controls) (Frayling et al. 2007). A strong association with BMI was also observed ($P = 3 \times 10^{-35}$ in 30,081 individuals with BMI values). As a classic demonstration of confounding, the T2D association was abolished in subsequent analyses that adjusted for BMI as a covariate in the regression procedure (OR = 1.03; $P = 0.44$). *FTO* is now recognized as a locus harboring strong associations with obesity (Frayling et al. 2007; Scuteri et al. 2007) with associations with T2D appearing because the typical T2D cases are more obese than typical nondiabetic controls.

An analysis was performed in the Danish Inter99 cohort exploring SNP by physical activity interactions at the *FTO* locus (Andreasen et al. 2008). First the association between the *FTO* SNP rs9939609 and BMI was established: the AA genotype group had 1.1 kg/m^2 higher BMI levels on average compared to the TT genotype group, and those in the AT genotype had BMI levels 0.3 kg/m2 higher than the TT genotype group (additive effect $P = 1 \times 10^{-9}$ with $N = 5{,}722$). Physical activity status was assessed by questionnaire and individuals were classified into three groups: physically inactive ($N = 1{,}914$), lightly or moderately physically active ($N = 3{,}224$), and very physically active ($N = 416$). A statistically significant interaction was observed ($P = 0.007$). The genetic effect between rs9939609 and BMI was weaker in the individuals with the highest physical activity: the average BMI in the AA genotype group was 0.47 kg/m^2 (*not significant*) higher compared to the TT group. This association was stronger in the physically inactive group: here, the average BMI in the AA genotype group was 1.95 kg/m^2 higher than the TT group, a much larger increase than 0.38 kg/m^2 difference observed between the AT group and the TT group.

A careful exploration of this interaction effect was reported in a meta-analysis of 218,166 adults (Kilpeläinen et al. 2011). The physical activity interaction was replicated ($P = 0.005$), although the effect was not as strong as originally reported, an example of a possible *winner's curse*. In all individuals, the additive effect of the BMI-increasing allele (A) of rs9939609 was 0.36 kg/m^2 ($P = 1.8 \times 10^{-75}$). In the two physical activity strata applied across the study samples, the additive effect of the BMI-increasing allele was 0.46 kg/m^2 in the inactive group ($P = 3.7 \times 10^{-23}$, $N = 54{,}611$) and 0.32 kg/m^2 in the active group ($P = 4.5 \times 10^{-69}$, $N = 163{,}555$). Interestingly, heterogeneity was observed ($I^2 = 36\ \%$), mainly from cohorts of European origin. When the North American cohorts were analyzed on their own, the interaction was much stronger ($P = 1.6 \times 10^{-9}$): the additive effects were 0.82 kg/m^2 ($P = 2.7 \times 10^{-21}$, $N = 9{,}438$) and 0.34 kg/m^2 ($P = 6.1 \times 10^{-12}$, $N = 38{,}500$) in the inactive and active groups, respectively, with no measurable heterogeneity ($I^2 = 0\ \%$). The authors of this study carefully consider sources of bias and confounding in this association, and although they note that this result has importance for public health (being physically active can further alleviate a genetic predisposition toward obesity beyond the obvious health benefits), they further note that the changes in the genetic association due to physical activity could be confounded by correlated lifestyle and environmental factors. The observed interaction does not imply causation—as in other studies of genetic effects, the appropriate epidemiological interpretations apply.

12.5 Summary

In this chapter, we have introduced the concept of statistical interaction by exposure variables in the study of the genetic determinants of T2D and related traits. The basic methodology of gene-environment interaction studies was presented along with several extensions that have been recently proposed. Finally, three relevant examples of gene-environment interaction in the literature were described.

Of the greatest importance for future studies of gene-lifestyle interaction are the following. First, we suggest a careful consideration of the epidemiological design and hypotheses to be tested in the study—if the harmonization of exposure variables and outcomes increases noise and heterogeneity in your study data, then the potential gain in power from larger sample sizes might be obliterated. Second, appropriate statistical tests must be applied—if there is a reasonable expectation that there is a genetic basis for the exposure variable, then methods that depend upon gene-exposure independence may not be ideal.

Studies that account for differences in genetic effects due to environmental exposures will continue to be important as genetic association studies query low-frequency and rare genetic variants. Testing for interaction, accounting for the variability in the outcome due to the exposure (by using it as a covariate) or looking for genetic associations in distinct subgroups (revealing masking effects), may reveal additional genetic susceptibility loci that could illuminate biological pathways in the pathophysiology of type 2 diabetes.

References

Altshuler D, Hirschhorn JN, Klannemark M (2000) The common PPAR[gamma] Pro12Ala polymorphism is associated with decreased risk of type 2 diabetes. Nat Genet 26(1):76–80

Andreasen CH, Stender-Petersen KL, Mogensen MS et al (2008) Low physical activity accentuates the effect of the FTO rs9939609 polymorphism on body fat accumulation. Diabetes 57:95–101. doi:10.2337/db07-0910

Aschard H, Hancock DB, London SJ, Kraft P (2010) Genome-wide meta-analysis of joint tests for genetic and gene-environment interaction effects. Hum Hered 70:292–300. doi:10.1159/000323318

Aschard H, Lutz S, Maus B et al (2012) Challenges and opportunities in genome-wide environmental interaction (GWEI) studies. Hum Genet 131:1591–1613. doi:10.1007/s00439-012-1192-0

Cecil JE, Watt P, Palmer CN, Hetherington M (2006) Energy balance and food intake: the role of PPARγ gene polymorphisms. Physiol Behav 88(3):227–233

Clayton D (2012) Link functions in multi-locus genetic models: implications for testing, prediction, and interpretation. Genet Epidemiol 36:409–418. doi:10.1002/gepi.21635

Cornelis MC, Tchetgen EJT, Liang L et al (2012) Gene-environment interactions in genome-wide association studies: a comparative study of tests applied to empirical studies of type 2 diabetes. Am J Epidemiol 175:191–202. doi:10.1093/aje/kwr368

de Bakker PIW, Ferreira MAR, Jia X et al (2008) Practical aspects of imputation-driven meta-analysis of genome-wide association studies. Hum Mol Genet 17:R122–R128. doi:10.1093/hmg/ddn288

Dupuis J, Langenberg C, Prokopenko I et al (2010) New genetic loci implicated in fasting glucose homeostasis and their impact on type 2 diabetes risk. Nat Genet 42:105–116. doi:10.1038/ng.520

Florez JC, Jablonski KA, Sun MW et al (2007) Effects of the type 2 diabetes-associated PPARG P12A polymorphism on progression to diabetes and response to troglitazone. J Clin Endocrinol Metab 92:1502–1509. doi:10.1210/jc.2006-2275

Frayling TM, Timpson NJ, Weedon MN et al (2007) A common variant in the FTO gene is associated with body mass index and predisposes to childhood and adult obesity. Science 316:889–894. doi:10.1126/science.1141634

Gauderman WJ, Zhang P, Morrison JL, Lewinger JP (2013) Finding novel genes by testing G × E interactions in a genome-wide association study. Genet Epidemiol 37:603–613. doi:10.1002/gepi.21748

Gloyn AL, Pearson ER, Antcliff JF et al (2004) Activating mutations in the gene encoding the ATP-sensitive potassium-channel subunit Kir6.2 and permanent neonatal diabetes. N Engl J Med 350:1838–1849. doi:10.1056/NEJMoa032922

Gouda HN, Sagoo GS, Harding AH et al (2010) The association between the peroxisome proliferator-activated receptor- 2 (PPARG2) Pro12Ala gene variant and type 2 diabetes mellitus: a HuGE review and meta-analysis. Am J Epidemiol 171:645–655. doi:10.1093/aje/kwp450

Haag M, Dippenaar NG (2005) Dietary fats, fatty acids and insulin resistance: short review of a multifaceted connection. Med Sci Monit 11:RA359–RA367

Hsu L, Jiao S, Dai JY et al (2012) Powerful cocktail methods for detecting genome-wide gene-environment interaction. Genet Epidemiol 36:183–194. doi:10.1002/gepi.21610

Keller MC (2014) Gene × environment interaction studies have not properly controlled for potential confounders: the problem and the (simple) solution. Biol Psychiatry 75:18–24. doi:10.1016/j.biopsych.2013.09.006

Kilpeläinen TO, Qi L, Brage S et al (2011) Physical activity attenuates the influence of FTO variants on obesity risk: a meta-analysis of 218,166 adults and 19,268 children. PLoS Med 8: e1001116–e1001116. doi:10.1371/journal.pmed.1001116

Kooperberg C, LeBlanc M (2008) Increasing the power of identifying gene × gene interactions in genome-wide association studies. Genet Epidemiol 32(3):255–263

Kraft P, Yen Y-C, Stram DO et al (2007) Exploiting gene-environment interaction to detect genetic associations. Hum Hered 63:111–119. doi:10.1159/000099183

Kubota N, Terauchi Y, Miki H et al (1999) PPARγ mediates high-fat diet–induced adipocyte hypertrophy and insulin resistance. Mol Cell 4:597–609. doi:10.1016/S1097-2765(00)80210-5

Li D, Conti DV (2009) Detecting gene-environment interactions using a combined case-only and case–control approach. Am J Epidemiol 169:497–504. doi:10.1093/aje/kwn339

Li S, Mukherjee B, Taylor JMG et al (2014) The role of environmental heterogeneity in meta-analysis of gene-environment interactions with quantitative traits. Genet Epidemiol 38:416–429. doi:10.1002/gepi.21810

Lindström S, Yen Y-C, Spiegelman D, Kraft P (2009) The impact of gene-environment dependence and misclassification in genetic association studies incorporating gene-environment interactions. Hum Hered 68:171–181. doi:10.1159/000224637

Ludovico O, Pellegrini F, Paola R et al (2007) Heterogeneous effect of peroxisome proliferator-activated receptor γ2 Ala12 variant on type 2 diabetes risk. Obesity 15:1076–1081. doi:10.1038/oby.2007.617

Manning AK, LaValley M, Liu C-T et al (2010) Gene–environment-wide association studies: emerging approaches. Nat Publ Group 11:259–272. doi:10.1038/nrg2764

Manning AK, LaValley M, Liu C-T et al (2011) Meta-analysis of gene-environment interaction: joint estimation of SNP and SNP × environment regression coefficients. Genet Epidemiol 35:11–18. doi:10.1002/gepi.20546

Manning AK, Hivert M-F, Scott RA et al (2012) A genome-wide approach accounting for body mass index identifies genetic variants influencing fasting glycemic traits and insulin resistance. Nat Genet 44:659–669. doi:10.1038/ng.2274

McCarthy MI (2010) Genomics, type 2 diabetes, and obesity. N Engl J Med 363:2339–2350. doi:10.1056/NEJMra0906948

Morris AP, Voight BF, Teslovich TM et al (2012) Large-scale association analysis provides insights into the genetic architecture and pathophysiology of type 2 diabetes. Nat Genet 44:981–990. doi:10.1038/ng.2383

Mukherjee B, Ahn J, Gruber SB et al (2008) Tests for gene-environment interaction from case–control data: a novel study of type I error, power and designs. Genet Epidemiol 32:615–626. doi:10.1002/gepi.20337

Mukherjee B, Ahn J, Gruber SB, Chatterjee N (2012) Testing gene-environment interaction in large-scale case–control association studies: possible choices and comparisons. Am J Epidemiol 175:177–190. doi:10.1093/aje/kwr367

Murcray CE, Lewinger JP, Gauderman WJ (2009) Gene-environment interaction in genome-wide association studies. Am J Epidemiol 169:219–226. doi:10.1093/aje/kwn353

Murcray CE, Lewinger JP, Conti DV et al (2011) Sample size requirements to detect gene-environment interactions in genome-wide association studies. Genet Epidemiol 35:201–210. doi:10.1002/gepi.20569

Ober C, Vercelli D (2011) Gene–environment interactions in human disease: nuisance or opportunity? Trends Genet 27:107–115. doi:10.1016/j.tig.2010.12.004

Ottman R (1996) Gene–environment interaction: definitions and study designs. Prev Med 25:764

Piegorsch WW, Weinberg CR, Taylor JA (1994) Non-hierarchical logistic models and case-only designs for assessing susceptibility in population-based case–control studies. Stat Med 13:153–162

Prokopenko I, Langenberg C, Florez JC et al (2008) Variants in MTNR1B influence fasting glucose levels. Nat Genet 41:77–81. doi:10.1038/ng.290

Roden M, Price TB, Perseghin G et al (1996) Mechanism of free fatty acid-induced insulin resistance in humans. J Clin Invest 97:2859. doi:10.1172/JCI118742

Sandhu MS, Weedon MN, Fawcett KA et al (2007) Common variants in WFS1 confer risk of type 2 diabetes. Nat Genet 39:951–953. doi:10.1038/ng2067

Saxena R, Elbers CC, Guo Y et al (2012) Large-scale gene-centric meta-analysis across 39 studies identifies type 2 diabetes loci. Am J Hum Genet 90:410–425. doi:10.1016/j.ajhg.2011.12.022

Scott RA, Lagou V, Welch RP et al (2012) Large-scale association analyses identify new loci influencing glycemic traits and provide insight into the underlying biological pathways. Nat Genet 44:991–1005. doi:10.1038/ng.2385

Scuteri A, Sanna S, Chen W-M et al (2007) Genome-wide association scan shows genetic variants in the FTO gene are associated with obesity-related traits. PLoS Genet 3, e115. doi:10.1371/journal.pgen.0030115

The InterAct Consortium (2011) Design and cohort description of the InterAct Project: an examination of the interaction of genetic and lifestyle factors on the incidence of type 2 diabetes in the EPIC Study. Diabetologia 54:2272–2282. doi:10.1007/s00125-011-2182-9

Thomas D (2010) Gene–environment-wide association studies: emerging approaches. Nat Publ Group 11:259–272. doi:10.1038/nrg2764

Voight BF, Scott LJ, Steinthorsdottir V et al (2010) Twelve type 2 diabetes susceptibility loci identified through large-scale association analysis. Nat Genet 42:579–589. doi:10.1038/ng.609

Voorman A, Lumley T, McKnight B, Rice K (2011) Behavior of QQ-plots and genomic control in studies of gene-environment interaction. PLoS ONE 6, e19416. doi:10.1371/journal.pone.0019416

Willer CJ, Li Y, Abecasis GR (2010) METAL: fast and efficient meta-analysis of genomewide association scans. Bioinformatics 26:2190–2191. doi:10.1093/bioinformatics/btq340

Winckler W, Weedon MN, Graham RR et al (2007) Evaluation of common variants in the six known maturity-onset diabetes of the young (MODY) genes for association with type 2 diabetes. Diabetes 56:685–693. doi:10.2337/db06-0202

Wu C, Chang J, Ma B et al (2013) The case-only test for gene-environment interaction is not uniformly powerful: an empirical example. Genet Epidemiol 37:402–407. doi:10.1002/gepi.21713

Yang Q, Khoury MJ, Flanders WD (1997) Sample size requirements in case-only designs to detect gene-environment interaction. Am J Epidemiol 146:713–720

Zeggini E, Ioannidis JPA (2009) Meta-analysis in genome-wide association studies. Pharmacogenomics 10:191–201. doi:10.2217/14622416.10.2.191

Chapter 13
The Gut Microbiota in Type 2 Diabetes

Trine Nielsen, Kristine H. Allin, and Oluf Pedersen

Abstract The exploration of the gut microbiota has intensified within the past decade with the introduction of cultivation-independent methods. By investigation of the gut bacterial genes, our understanding of the compositional and functional capability of the gut microbiome has increased. It is now widely recognized that the gut microbiota has profound effect on host metabolism and recently changes in the gut microbiota have been associated with type 2 diabetes. Animal models and human studies have linked changes in the gut microbiota to the induction of low-grade inflammation, altered immune response, and changes in lipid and glucose metabolism. Several factors have been identified that might affect the healthy microbiota, potentially inducing a dysbiotic microbiota associated with a disease state. This increased understanding of the gut microbiota might potentially contribute to targeted intervention strategies to prevent or treat type 2 diabetes.

13.1 Introduction to the Gut Microbiota

The study of microbes has intrigued scientists for centuries. With the introduction of cultivation-independent methods, the focus has expanded from investigations of pathogenic organisms to a deeper interest in the symbiotic, nonpathogenic, i.e., commensal, microbes and their interaction with their human host. By far, the largest concentrations of microbes are located in the human distal gut, where the vast majority is of bacterial origin. The gut *microbiota*, which is the collective microbial community in the gut, exhibits profound effects on human health. The gut microbiota can be regarded as a microbial organ within the human body, and while the host provides for an appropriate environment, the gut microbes execute pertinent functions that the host is unable to perform by itself. These functions include production of short chain fatty acids (SCFA) by fermentation of otherwise indigestible polysaccharides from the diet and synthesis and absorption of vitamins

T. Nielsen • K.H. Allin • O. Pedersen (✉)
Novo Nordisk Foundation Center for Basic Metabolic Research, Faculty of Health and Medical Sciences, University of Copenhagen, Copenhagen, Denmark
e-mail: oluf@sund.ku.dk

J.C. Florez (ed.), *The Genetics of Type 2 Diabetes and Related Traits*,
DOI 10.1007/978-3-319-01574-3_13

such as B and K vitamins. Importantly, the gut microbiota also plays a significant role in the maturation and adaptation of the immune system and in the defense against pathogens, in regulation of intestinal hormone secretion, and in gastrointestinal nerve activity.

The gut microbiota can be described in terms of its composition (which bacteria are present?) and in terms of its functional capability (what functions do the bacteria exert?). To describe the composition of the gut microbiota, microbial DNA is usually extracted from fecal samples, as these generally are considered to be representative of the distal colon microbiota and sequenced either in a targeted or untargeted manner. Describing the functional capability of the gut microbiota is typically done by applying whole-genome shotgun sequencing which allows characterization of the collective genome of the gut microbiota—the *microbiome*. The human gut microbiota is estimated to consist of approximately 1000 prevalent bacterial species, and each individual harbors at least 160 such species (Qin et al. 2010). It has been suggested that a core microbiome exists as gut bacterial species seem to be shared among individuals (Qin et al. 2010). Yet, based on current evidence, this "core microbiome" probably represents a core of common functions necessary for correct functioning of the gut (Tap et al. 2009), rather than a core of common organisms. The concept of grouping individuals based on their gut bacterial composition was elaborated with the introduction of enterotypes (Arumugam et al. 2011). The bacterial composition of the three identified enterotypes is driven by the genera *Bacteroides*, *Prevotella*, or *Ruminococcus* and they seem to be independent of age, body mass index, and nationality. Not all studies have been able to reproduce these three clusters but suggest gradients of genera (Zupancic et al. 2012; Claesson et al. 2011), and future research will reveal which functions clusters or gradients of bacteria might possess.

13.2 Assessment of the Gut Microbiota

Historically, the microbial community was studied by culture-based methods where the presence of certain microorganisms was identified by their growth on specialized media. However, it has proven difficult to culture the majority of the commensal microorganisms and to overcome this limitation, culture-independent methods were developed. Whereas culture-dependent methods are based on the culturing of single microorganisms, culture-independent methods are based on studying DNA extracted collectively from the microbial community present in a sample. Today, DNA-based studies of the gut microbial community are dominated by two methods: targeted 16S rRNA gene sequencing focusing on the composition and diversity of the microbiota and whole-genome shotgun sequencing allowing in addition for a deeper evaluation of the functional capability of the gut microbiota (Fig. 13.1).

16S rRNA gene sequencing includes targeted amplification and subsequent next-generation sequencing of phylogenetically informative marker sequences of

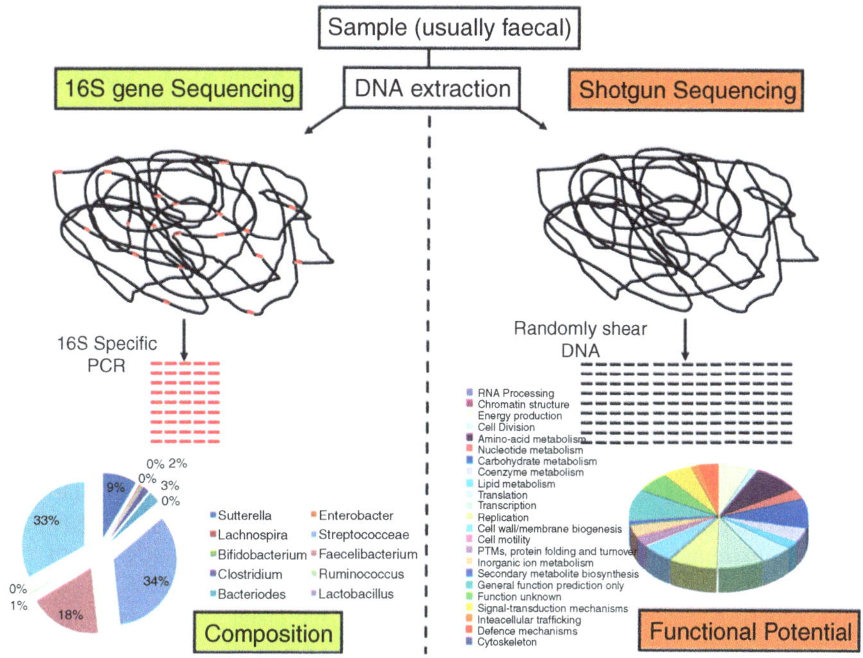

Fig. 13.1 16S rRNA gene sequencing and whole-genome shotgun sequencing. *Left panel*: targeted 16S rRNA gene amplification using specific PCR primers followed by next-generation sequencing to reveal the microbial composition. *Right panel*: untargeted next-generation sequencing of community DNA to reveal the functional potential of the microbiota
Reprinted with permission from Paul D. Cotter rights holder of: Clarke SF et al.: The gut microbiota and its relationship to diet and obesity. Gut microbes 3(3), 2012:186–202

microbial DNA (Kuczynski et al. 2012). Prokaryotic ribosomes contain three types of rRNA: 23S, 16S, and 5S. The 16S rRNA gene sequence is about 1550 base pair long and is composed of eight highly conserved regions and nine hypervariable regions that differ across bacterial taxa. Universal primers complementary to the conserved regions are used to amplify one or more of the variable regions, and the sequences of the variable regions are used to classify bacteria, typically at the family or genus levels (Kuczynski et al. 2012). From 16S data the number of different taxa present in a sample and their relative abundance can be estimated by comparisons with 16S rRNA sequence databases or by using software packages to cluster highly similar sequences into operational taxonomic units (OTUs) (Weinstock 2012). Thus, targeted 16S rRNA gene sequencing can describe the bacterial community in terms of which taxa/OTUs are present, their relative abundance, and their phylogenetic relationships, but when it comes to genetic and functional information, the 16S method as of today falls short. Another approach which has been used to describe the bacterial community based on 16S rRNA is phylogenetic oligonucleotide microarrays (phylochips) by which a large number of pre-specified microbes can be identified at high throughput (Rajilic-Stojanovic et al. 2009).

To describe the genetic content and the functional potential of a community, whole-genome shotgun sequencing must be applied. In contrast to targeted 16S rRNA gene sequencing, whole-genome shotgun sequencing or metagenomic sequencing involves directly next-generation sequencing of community DNA without amplification of certain regions of the DNA. This method is far more expensive than targeted sequencing but may provide a less biased and a high-resolution view of the bacterial composition compared to targeted sequencing of 16S marker genes, which depends on the choice of primers (Kuczynski et al. 2012). Moreover, given its high level of resolution, it enables a complete description of the genes present and consequently the functional capabilities of the community as well as identification of single-nucleotide polymorphisms and other variant sequences (Morgan and Huttenhower 2012). In order to identify which organisms and genes are present and their relative abundances, shotgun reads are compared to bacterial reference genomes and gene catalogues whereas comparisons with functional databases enable prediction of pathways (Weinstock 2012). Although metagenomic sequencing is an extremely powerful tool to describe in detail the composition and the functional capabilities of the gut microbiota, it creates an overwhelming amount of data to be analyzed. In 2010 the first catalogue of the distal gut microbiome was released and 3.3 million genes were identified (Qin et al. 2010) based upon discoveries in Danish and Spanish individuals. This is more than 100 times the number of genes in the human genome. The genes were estimated to belong to ~1000 different abundant bacterial species, where each individual was believed to harbor at least 160 of these abundant species, many of which were believed to be shared among individuals. As more samples are sequenced at high sequencing depth, the number of identified genes increases (The Human Microbiome Project 2012; Qin et al. 2012), and current gut microbiome reference catalogues have identified about 10 million microbial genes and are believed to cover the majority of abundant microbial genes shared among individuals (Li et al. 2014; Karlsson et al. 2014). As a considerable number of the genes can't be mapped to taxonomically known bacteria, methods have been developed to enable grouping of genes that vary in abundance in a similar manner across a dataset into gene clusters. The identified groups or clusters of certain sizes, also termed metagenomic clusters or units, are believed to represent yet taxonomically unknown bacterial species, whereas other clusters might represent bacteriophages (Karlsson et al. 2013; Qin et al. 2012; Nielsen et al. 2014).

Before undertaking expensive sequencing and complex quantitative bioinformatics of the gut microbiome, it is very important to consider how the biological samples are best obtained, how they are best stored, and which method is the optimal one for microbial DNA extraction. All three factors may have major impact on whether the extracted basis material DNA accurately reflects the microbial diversity in the samples (Yuan et al. 2012). Hopefully, future exhaustive studies will relate to this issue and consensus standards will be agreed upon and generally applied internationally within the research field. Additionally, in the near future, more microbiota studies will likely include data on microbial transcriptomics and proteomics as well as information on the metabolic activity of the microbial community. Lastly, more emphasis will also be put on the study of human viruses, bacteriophages, and fungi which are also crucial members of the gut microbial community.

13.3 Factors Influencing the Gut Microbiota

The human fetus is believed to be colonized from birth by microbes from the mother and the surroundings. Thus, the human gut is presumably sterile at the time of birth, and colonization depends on the mode of delivery and feeding and is narrow and highly changeable until weaning. Along with the introduction of solid food, the child's microbiota changes over the first 4 years to achieve the characteristics of the adult microbiota, which remains relatively stable and rich throughout adulthood until immune function-related changes of gut microbiota occur at advanced age (Spor et al. 2011). Thus, despite high interindividual variation, the composition of the gut microbiota of an individual remains relatively stable over time in the absence of perturbation (Dethlefsen and Relman 2011). Host genotype, dietary factors, smoking, antibiotics and other medications, age, and probably several additional yet unidentified factors all seem to influence the composition of the gut microbiota. Of these, diet and antibiotics are to date the most investigated factors that seem to exert profound effects on host metabolism via changes in gut microbiota.

Mice studies have clearly demonstrated the dynamics of the gut microbiota in response to changes in diet (Turnbaugh et al. 2009b; Hildebrandt et al. 2009; de Wit et al. 2012), yet there are still only few reports on the effect of long-term dietary habits on the gut microbiota in humans. The composition and functional capabilities of the human gut microbiota seem to adapt to the diet consumed. Dietary habits enriched in plants products are reflected by an increased abundance of gut bacteria known to degrade plant polysaccharides (De Filippo et al. 2010), whereas, a diet rich in protein and fat is reflected by increased activity of gut bacterial pathways needed for digestion of these macronutrients (Yatsunenko et al. 2012). Recently, it was demonstrated that the human gut microbiota can adapt to changes in dietary patterns within 2–4 days, as illustrated by the consumption of a diet entirely composed of either plant or animal products for 4 days (David et al. 2014). On the other hand, the enterotypes seemed to be resistant to a 10-day controlled dietary intervention with either high fat or high fiber, but was found to be associated with differences in long-term dietary patterns (Wu et al. 2011). Finally, there is evidence that a diet rich in fibers and low in fat is associated with a more diverse gut microbiota compared to a diet rich in fat and low in fiber and that a permanent change in diet could lead to long-term effects on the composition of the gut microbiota (Claesson et al. 2012).

Current speculations on whether the introduction of antibiotics in the early twentieth century may contribute to the obesity epidemic are supported by a few, but large epidemiological studies showing that exposure to antibiotics early in life increases the risk of being overweight at 7 years old (Ajslev et al. 2011; Trasande et al. 2012; Bailey et al. 2014; Murphy et al. 2014). Other (nonrandomized) studies have also observed increased adiposity in humans after antibiotic treatment (Francois et al. 2011; Thuny et al. 2010). In industrial farming subtherapeutic doses of antibiotics have been given to farm animals to spur growth for decades and experiments in mice given low doses of antibiotics increased total fat mass and bone mineral density and shifted the composition of the gut microbiota (Cho et al. 2012). An increased cecal content of SCFA and lower caloric output in feces in the mice treated with antibiotics, despite similar caloric intake as the non-treated mice, support previous suggestions of an increased energy harvest by microbiota leading to obesity (Turnbaugh et al. 2006; Cho et al. 2012). In contrast to these observations, other mice studies using germ-free or conventionalized mice have indicated improved glucose tolerance independent of weight changes, lower levels of circulating lipopolysaccharide (LPS), as well as a lower bacterial count, following antibiotic treatment as indications of an improved metabolic state (Carvalho et al. 2012; Membrez et al. 2008). While the possible role of antibiotics in development of metabolic diseases needs further elucidation, present studies indicate that antibiotics have not only short- but also long-term effect on the diversity and/or composition of the gut microbiota (Jernberg et al. 2007) and consequently a potential association between antibiotics and development of metabolic diseases may partly be mediated through perturbations of the gut microbiota.

13.4 Genotype-Microbiota Interaction

It is likely that variation in host genes influences the composition of the gut microbiota as evidenced in mice knockout models and twin studies in humans. On the other hand, the microbiota is expected to regulate host gene transcription through epigenetic modifications. So far the evidence to support a contribution of host genetics to the diversity of the microbial community has been limited, mainly due to low numbers of investigated samples in available datasets, but data are slowly emerging. Monozygotic twins have a higher degree of similarity in their gut microbiota than dizygotic twins, and the microbiota of family members is more similar than unrelated subjects (Stewart et al. 2005; Palmer et al. 2007; Zoetendal 2001; Turnbaugh et al. 2009a; Tims et al. 2013). Just recently, the first study attempting to estimate the heritability of the human gut microbiota was published, supporting a role of host genetics controlling the gut microbiota. Heritability estimates of 0.40 of certain taxa were reported, whereas other taxa seemed to be more environmentally determined (Goodrich et al. 2014). Interestingly, based on their finding in twins, the authors demonstrated that the most heritable family of bacteria had an impact on host metabolism and was associated with a lean phenotype. Few earlier studies have taken a more traditional candidate gene approach to investigate the genotype-microbiota interaction. Most evidence comes from knockout mice models, but few studies in humans have found the investigated variant to be associated with a shift in the bacterial composition (Spor et al. 2011). All of these reports support a role of host genetics in determining the gut microbiota.

A quantitative trait locus (QTL) detection approach has been attempted to elucidate the effect of host genotype on the features of the gut microbiota in mice (Benson et al. 2010). The authors found a common measurable core of 64 conserved taxonomic groups identified in 16S RNA gene sequences from 645 mice. By using QTL analysis, they searched for taxa that co-segregated with 530 genomic markers and revealed 13 loci that showed significant linkage with the relative abundance of the specific taxa. Another study using a similar approach on a different mouse model could also detect linkage between loci and variation across taxa (McKnite et al. 2012). Interestingly, loci identified in both studies included genes with immune function or the potential to affect the gut community structure. Lessons from the candidate genes studies and genome-wide linkage scans in the search for susceptibility genes in human metabolism should guide studies of genotype-microbiota interactions, implying that investigators should aim for statistically well-powered studies and replication of findings in separate cohorts. Likely, genome-wide association approaches will also be undertaken to relate variation in host genotype to variation within the gut microbiota. Considering the gut microbiota as a phenotypic trait, this approach might be feasible, once the gut microbiome profiles are available in a large number of individuals probably needed for these studies.

13.5 The Gut Microbiota in Obesity, Insulin Resistance, and Type 2 Diabetes

Within the past decade, the focus has been on defining the *dysbiotic* microbiota; that is the altered microbial composition or the altered functional capacity of the microbiota associated with disease, and less focus has been on defining the healthy microbiota. However, with the current evidence, it is not possible to tell whether a dysbiotic state is a cause or consequence—or both—of disease.

Obesity, insulin resistance, and type 2 diabetes have been correlated with an altered gut microbiota composition. The first studies in this area reported that obesity was associated with changes in the ratio of the two main phyla in the gut, Bacteroidetes and Firmicutes, with increased levels of Firmicutes associated with the obese state. However, some but not all studies have since been able to reproduce these compositional changes at the phylum level (Furet et al. 2010; Duncan et al. 2008; Schwiertz et al. 2010; Zhang et al. 2009). Whereas one study has reported a higher proportion of Actinobacteria in obese individuals (Turnbaugh et al. 2009a), other reported an inverse relationship of the Bacteroidetes/Firmicutes ratio (Schwiertz et al. 2010).

While inconsistent findings at the phylum level have been reported, reduced bacterial diversity has been associated with obesity (Turnbaugh et al. 2009a). Reduced diversity has also been associated with inflammatory bowel diseases and recently also with markers of low-grade inflammation and insulin resistance. In 292 nondiabetic Danish individuals investigated using shotgun sequencing and quantitative metagenomics, 23 % of the population had low gene richness (a marker of diversity) and had on average a 40 % lower bacterial gene count than individuals with sustained bacterial gene count (Le Chatelier et al. 2013). Individuals with low richness were characterized by low-grade inflammation, insulin resistance, dyslipidemia, and overall adiposity, which put them at a high metabolic risk of progression to obesity-related comorbidities.

It has been reported that individuals with type 2 diabetes have an altered taxonomic composition and functional potential of their gut microbiota compared to nondiabetic individuals (Larsen et al. 2010; Qin et al. 2012; Graessler et al. 2012; Furet et al. 2010; Karlsson et al. 2013; Zhang et al. 2013). The largest metagenomic-based study to date included 368 Chinese individuals and identified a moderate degree of gut bacterial dysbiosis, with a decline in butyrate-producing bacteria and an increase in opportunistic bacteria among type 2 diabetes patients (Qin et al. 2012). However, no difference in diversity between type 2 diabetes patients and healthy controls was observed. Butyrate-producing bacteria have been associated with increased insulin sensitivity (Vrieze et al. 2012) and in other studies with lower levels of markers of low-grade inflammation (Furet et al. 2010) and are now generally believed to be associated with a healthy gut. A study of 145 Swedish women with normal, impaired, or diabetic glucose regulation also applying a metagenomics approach found similar, but not identical, compositional changes associated with type 2 diabetes suggesting ethnic or regional differences in microbiota characteristics (Karlsson et al. 2013). In this study, the type 2 diabetes-associated changes in the gut microbiota were used to subgroup 49 women with impaired glucose regulation into a type 2 diabetes or normal glucose tolerance like metabolism. This stratification could potentially help to identify the ~40 % of individuals with prediabetes who are at high risk of progression to type 2 diabetes and hence need careful attention. In both the Swedish and the Chinese study (Karlsson et al. 2013; Qin et al. 2012), it was possible with few metagenomic markers (gene clusters) to discriminate between type 2 diabetes patients and nondiabetic controls with high accuracy corresponding to an area under the curve of ~0.8 in receiver operating characteristic analyses. Such observations may pave the way for future biomarker developments where few gut microbial signatures with high accuracy predict risk of diabetes development.

Interestingly, the Swedish study found a few known species to be associated with metformin treatment, yet the significance of this is unsolved. No information on the intake of metformin was reported in any of the other studies investigating type 2 diabetes patients. However, metformin, which is known to have gastrointestinal side effects, may have a substantial impact on the composition of the gut microbiota and therefore could seriously confound the reported findings of type 2 diabetes. In support of this hypothesis, a study of high-fat diet-fed mice showed that metformin treatment in parallel with its beneficial effects on glucose metabolism also resulted in an altered composition of the gut microbiota, including a higher abundance of *Akkermansia*. Moreover, oral administration of *Akkermansia muciniphila* to high-fat diet-fed mice that were not treated with metformin improved glucose tolerance suggesting a novel mechanism for the glucose-lowering effect of metformin (Shin et al. 2013). *A.muciniphila* have in other studies been reported to correlate with a healthy metabolic phenotype (Everard et al. 2013), underlining the beneficial properties of this bacteria.

While investigations of the gut microbiota in humans have indicated that a dysbiotic state is associated with development of metabolic diseases, animal studies have generated hypotheses of the role of the microbiota in the pathogeneses of obesity, insulin resistance, and type 2 diabetes. Early studies in mice suggested that the gut microbiota might function as an environmental factor that could regulate energy metabolism and affect development of obesity (Backhed et al. 2004). Most of the evidence came from a series of elegant studies of germ-free and conventionalized mice, showing that germ-free mice were leaner than the conventional mice and were resistant to diet-induced obesity when introduced to a high-fat western style diet (Backhed et al. 2004; Rabot et al. 2010). By colonizing germ-free mice with microbiota from normal mice, the amount of body fat increased and the insulin sensitivity decreased in the germ-free mice (Backhed et al. 2004).

It was furthermore discovered that obese mice had an altered composition of their microbiota compared to lean mice, characterized by a higher content of Firmicutes and fewer Bacteroidetes (Ley et al. 2005). These changes in the microbiota were associated with an increased capacity for energy harvest from otherwise indigestible polysaccharides in the diet (Turnbaugh et al. 2006). The obesity phenotype was shown to be transmissible when lean germ-free mice were colonized with gut microbiota from the obese mice, suggesting a causal relation for the microbiota in the pathogenesis of obesity (Turnbaugh et al. 2006). Moreover, it was discovered that the gut microbiota suppress the secretion of the lipoprotein lipase inhibitor angiopoietin-like protein 4 (Angptl4), thereby inducing deposition of triglycerides in adipocytes (Backhed et al. 2004). In addition, inhibition of Angptl4 in conventionalized mice was associated with de novo hepatic lipogenesis (Backhed et al. 2007).

The induction of endotoxemia is another widely discussed hypothesis on how the gut microbiota can promote obesity and obesity-associated low-grade inflammation. LPS is produced in the intestine from Gram-negative bacteria and triggers secretion of pro-inflammatory cytokines (Wright et al. 1990). Animal models have shown that a high-fat diet increased the absorption of LPS either through an increased uptake in chylomicrons or through increased gut epithelium permeability, and infusion of LPS in mice can induce weight gain, insulin resistance, and deposition of triglycerides in the liver (Cani et al. 2007). Hence, based on mice studies, it has been hypothesized that a high-fat diet induces changes in the gut microbiota in favor of the Gram-negative bacteria, which increases circulating levels of LPS through an increased intestinal permeability contributing to low-grade inflammation, obesity, and type 2 diabetes (Cani et al. 2007, 2008). The microbiota could modulate the intestinal permeability through a GLP-2 dependent mechanism, as decreased permeability and reduced LPS levels were observed after an endogenous GLP-2 increase (Cani et al. 2009). However, the exact mechanism by which an altered microbiota triggers inflammation is unknown, but there is evidence of an interaction with the innate immune system. LPS binds to the Toll-like receptor 4 (TLR4)-CD14 complex, which activates pro-inflammatory pathways (McGettrick and O'Neill 2010). Toll-like receptors are part of pattern recognition receptors (PRRs) that activate the innate immune system through recognition of microbial molecules and are suggested to have a substantial role in the cross talk between the immune system, inflammation, and metabolism (Burcelin 2012). Other TLRs like the flagellin receptor TLR5 and the TLR2 have both been reported to be involved in development of the metabolic syndrome or components of the syndrome (Caricilli et al. 2011; Vijay-Kumar et al. 2010). *Tlr5*-deficient mice have been shown to exhibit signs of the metabolic syndrome, a feature that was transmissible with transplantation of gut microbiota to wild-type mice (Vijay-Kumar et al. 2010), again suggesting that the microbiota itself may mediate disease. Also other PRRs, the NOD-like receptors, recognize peptidoglycans from Gram-negative bacteria, activate pro-inflammatory pathways, and are believed to mediate immune responses (Burcelin 2012). Finally, a study of gut inflammasomes, which are multi-protein complexes that recognize microbiota-associated molecular patterns and are involved in the pro-inflammatory IL-1β and IL-18 pathways, revealed a connection between altered gut microbiota and development of nonalcoholic fatty liver disease (NAFLD), which is often seen in obese individuals, probably through altered gut permeability (Henao-Mejia et al. 2012). The gut microbiota has also been reported to influence the circulating levels of the incretins GLP-1 and GIP—important hormones regulating the postprandial insulin response (Wichmann et al. 2013; Lin et al. 2012), most likely through the production of SCFA. SFCA, especially butyrate, are recognized for their beneficial effects on host metabolism and have just recently been shown to play a role in the activation of intestinal gluconeogenesis (De Vadder et al. 2014). Hence, SCFA along with their G-protein-coupled receptors GPR41 and GPR43 are currently being intensively investigated, as they offer novel targets not only for modulation of the gut microbiota toward a healthy state but also for treatment or prevention of obesity and type 2 diabetes.

Bile acids have also been described as signaling molecules in the interplay between gut microbiota and glucose metabolism in both mouse models and human intervention studies (Vrieze et al. 2014; Ryan et al. 2014). The gut bacteria mediate the conversion of primary bile acids to secondary bile acids, and when the balance of the gut microbiota is disrupted, changes in bile acids have been observed. This was reported after one week of vancomycin treatment, where an increase in primary bile acids was reported along with a decrease in insulin sensitivity (Vrieze et al. 2014). In a mouse model of gastric bypass, here using a vertical sleeve model, the improvement in glucose metabolism co-occurred with changes in bile acid concentrations. Furthermore, it was shown that the ability to lose weight and improve glucose homeostasis was dependent on the bile acid receptor FXR (Ryan et al. 2014).

While convincing pieces of evidence for a causal role of gut microbiota in disease development are present in animal studies, only few human studies have addressed this issue. Based on the concept of stool transplantation, a Dutch study investigated whether fecal transplantation could be used to correct the disturbances observed in metabolic disease. This study investigated 18 insulin-resistant men that were randomized to receive either autologous or allogeneic gut microbiota infusion from healthy, lean donors (Vrieze et al. 2012). Insulin sensitivity as estimated from hyperinsulinemic, euglycemic clamp studies was improved 6 weeks after transfusion from lean donors, along with increased diversity of the microbiota. Despite the lack of reported long-term effects on insulin sensitivity, the concept proposes the most interesting target for the future and fecal matter transplantation has with great success been implemented in the treatment for recurrent *Clostridium difficile* infection (van Nood et al. 2013).

From the current evidence, it may be *hypothesized* that an adverse combination of host genetic susceptibility, lifestyle, and environment in a broad sense may trigger a change in the gut microbiota from a healthy microbiota to an unhealthy perturbed microbiota which may result in increased energy extraction from dietary fibers, a leaky mucosa, increased gut permeability with absorption of macromolecules from intestinal content triggering systemic immune alterations, altered lipid metabolism, and low-grade inflammation leading to insulin resistance and eventually contributing to type 2 diabetes pathogenesis (Fig. 13.2) (de Vos and Nieuwdorp 2013).

To fully understand whether the gut microbiota is a cause or consequence—or both—of type 2 diabetes, well-designed prospective studies of humans are needed. A possibility could be to follow carefully phenotyped prediabetic individuals in order to identify the dysbiotic microbiota associated with the progression to type 2 diabetes. Prospective studies could optimally be followed up by mechanistic studies in suitable mice models. By inoculating mice with human gut microbiota from prediabetic individuals who rapidly have progressed to overt type 2 diabetes and with gut microbiota from prediabetics who have regressed to normal glucose tolerance during follow-up, we could possible solve some of the current discussion of whether altered gut microbiota by itself can mediate dysglycemia in genetically susceptible individuals.

13.6 Perspectives

The first landmark studies of quantitative metagenomics published within the area of metabolic research have contributed significant knowledge both in the context of biological understanding of the role of the gut microbiota in metabolic disease and in the context of data handling and analyses. From the latter point of view,

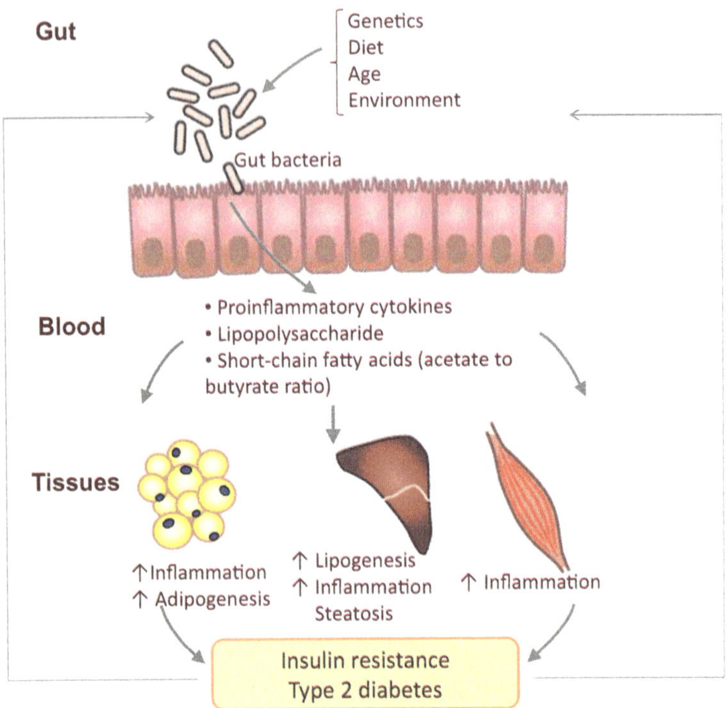

Fig. 13.2 Potential mechanisms underlying an association between the gut microbiota and type 2 diabetes. The composition of the gut microbiota is influenced by host genetics (single-nucleotide polymorphisms, copy-number variations, and mutations), diet (intake of fat, carbohydrates, fruit, and vegetables), age, and environmental factors such as mode of birth (vaginal or caesarean section) and exposure to antibiotics, especially in early life. An adverse gut microbiota may result in elevated circulating levels of pro-inflammatory cytokines and lipopolysaccharide originating from Gram-negative bacteria and decreased butyrate levels. This may induce inflammation in adipose, liver, and muscle tissue as well as increased adipogenesis in adipose tissue and lipogenesis and steatosis in the liver which may lead to insulin resistance and type 2 diabetes. Finally, type 2 diabetes may also potentially adversely influence the composition of the gut microbial community creating a vicious circle

metagenomic studies have revealed that the majority of sequenced microbial genes cannot be mapped to known taxonomic or functional references, and current efforts to generate new bacterial genome sequences will most certainly facilitate subsequent taxonomic and functional annotation for future studies. Concurrent development and improvement of bioinformatic tools are needed to reduce and comprehend the impressive amount of data produced with deep shotgun sequencing. Finally, international efforts to implement standardized approaches for sample processing and extraction of DNA and RNA are highly needed to enable reliable comparisons across studies.

While convincing evidence suggests that a dysbiotic state of the gut microbiota composition is associated with metabolic disease, we do not yet understand the underlying mechanisms. Although we cannot claim causality, several studies have documented that metabolic health or disease is transmissible in both animal and humans studies when stools are transplanted, but also that modulation of a dysbiotic microbiota toward a healthier microbiota is possible. An elegant proof of principle of this was reported by the Gordon laboratory (Ridaura et al. 2013). In this study, it was firstly shown that gut microbiota and its associated phenotype—in this case obesity—are transmissible between mice. Secondly, the authors used the fact that mice are coprophagic and they went on to show in cohousing experiments that development of obesity could be prevented by lean microbiota. Finally, it was shown that this phenotypic rescue was diet dependent and occurred only in mice eating a "humanized diet" low in fat and high in fruits and vegetables (Ridaura et al. 2013). Thus, it is clear that disease-promoting communities of gut microbes can be transmitted from a donor to a recipient and that the recipient in a reversible way and dependent on environmental conditions adapts to the phenotype of the donor. Obviously, in the context of obesity, findings as discussed above inspire to hunt for discovery of anorexigenic microbial gut communities to be tested for their potential as bacterio-therapeutical tools to eliminate or diminish obesity phenotypes. Another example of an upcoming task is characterization and culture of health-promoting complex bacterial communities of multiple specific and interdependent healthy and in most cases anaerobically living gut bacterial species. These may in randomized clinical trials be administered daily as slow-release encapsulated microbial cultures along with a diet high in prebiotics (nondigestible but fermentable food ingredients that selectively stimulate the growth or activity of beneficial gut microbes) to test their preventive and therapeutic potentials.

The quantitative metagenomics studies also suggest that we need much focus at the microbial community functions instead of only focusing on the specific composition of the gut microbiota. Future systems biology studies will integrate microbial metabolic pathways identified and featured through metatranscriptomics, metaproteomics, and metabolomics to enable discovery of the huge reservoir of bioactive compounds produced by gut microbes including bacterial metabolites, neurotransmitters, immunological factors, and noncoding RNAs which impact the host biology in multiple ways. Some of these bacterial compounds that due to microbe-human coevolution are familiar to the host may as well be tested for their therapeutic or disease-preventive potentials.

References

Ajslev TA, Andersen CS, Gamborg M, Sorensen TI, Jess T (2011) Childhood overweight after establishment of the gut microbiota: the role of delivery mode, pre-pregnancy weight and early administration of antibiotics. Int J Obes (Lond) 35(4):522–529. doi:10.1038/ijo.2011.27

Arumugam M, Raes J, Pelletier E, Le Paslier D, Yamada T, Mende DR, Fernandes GR, Tap J, Bruls T, Batto JM, Bertalan M, Borruel N, Casellas F, Fernandez L, Gautier L, Hansen T, Hattori M, Hayashi T, Kleerebezem M, Kurokawa K, Leclerc M, Levenez F, Manichanh C, Nielsen HB, Nielsen T, Pons N, Poulain J, Qin J, Sicheritz-Ponten T, Tims S, Torrents D,

Ugarte E, Zoetendal EG, Wang J, Guarner F, Pedersen O, de Vos WM, Brunak S, Dore J, Antolin M, Artiguenave F, Blottiere HM, Almeida M, Brechot C, Cara C, Chervaux C, Cultrone A, Delorme C, Denariaz G, Dervyn R, Foerstner KU, Friss C, van de Guchte M, Guedon E, Haimet F, Huber W, van Hylckama-Vlieg J, Jamet A, Juste C, Kaci G, Knol J, Lakhdari O, Layec S, Le Roux K, Maguin E, Merieux A, Melo Minardi R, M'Rini C, Muller J, Oozeer R, Parkhill J, Renault P, Rescigno M, Sanchez N, Sunagawa S, Torrejon A, Turner K, Vandemeulebrouck G, Varela E, Winogradsky Y, Zeller G, Weissenbach J, Ehrlich SD, Bork P (2011) Enterotypes of the human gut microbiome. Nature 473(7346):174–180. doi:10.1038/nature09944

Backhed F, Ding H, Wang T, Hooper LV, Koh GY, Nagy A, Semenkovich CF, Gordon JI (2004) The gut microbiota as an environmental factor that regulates fat storage. Proc Natl Acad Sci USA 101(44):15718–15723. doi:10.1073/pnas.0407076101

Backhed F, Manchester JK, Semenkovich CF, Gordon JI (2007) Mechanisms underlying the resistance to diet-induced obesity in germ-free mice. Proc Natl Acad Sci USA 104 (3):979–984. doi:10.1073/pnas.0605374104

Bailey LC, Forrest CB, Zhang P, Richards TM, Livshits A, DeRusso PA (2014) Association of antibiotics in infancy with early childhood obesity. JAMA Pediatr 168(11):1063–1069. doi:10.1001/jamapediatrics.2014.1539

Benson AK, Kelly SA, Legge R, Ma F, Low SJ, Kim J, Zhang M, Oh PL, Nehrenberg D, Hua K, Kachman SD, Moriyama EN, Walter J, Peterson DA, Pomp D (2010) Individuality in gut microbiota composition is a complex polygenic trait shaped by multiple environmental and host genetic factors. Proc Natl Acad Sci USA 107(44):18933–18938. doi:10.1073/pnas.1007028107

Burcelin R (2012) Regulation of metabolism: a cross talk between gut microbiota and its human host. Physiology 27(5):300–307. doi:10.1152/physiol.00023.2012

Cani PD, Amar J, Iglesias MA, Poggi M, Knauf C, Bastelica D, Neyrinck AM, Fava F, Tuohy KM, Chabo C, Waget A, Delmee E, Cousin B, Sulpice T, Chamontin B, Ferrieres J, Tanti JF, Gibson GR, Casteilla L, Delzenne NM, Alessi MC, Burcelin R (2007) Metabolic endotoxemia initiates obesity and insulin resistance. Diabetes 56(7):1761–1772. doi:10.2337/db06-1491

Cani PD, Bibiloni R, Knauf C, Waget A, Neyrinck AM, Delzenne NM, Burcelin R (2008) Changes in gut microbiota control metabolic endotoxemia-induced inflammation in high-fat diet-induced obesity and diabetes in mice. Diabetes 57(6):1470–1481. doi:10.2337/db07-1403

Cani PD, Possemiers S, Van de Wiele T, Guiot Y, Everard A, Rottier O, Geurts L, Naslain D, Neyrinck A, Lambert DM, Muccioli GG, Delzenne NM (2009) Changes in gut microbiota control inflammation in obese mice through a mechanism involving GLP-2-driven improvement of gut permeability. Gut 58(8):1091–1103. doi:10.1136/gut.2008

Caricilli AM, Picardi PK, de Abreu LL, Ueno M, Prada PO, Ropelle ER, Hirabara SM, Castoldi A, Vieira P, Camara NO, Curi R, Carvalheira JB, Saad MJ (2011) Gut microbiota is a key modulator of insulin resistance in TLR 2 knockout mice. PLoS Biol 9(12), e1001212. doi:10.1371/journal.pbio.1001212

Carvalho BM, Guadagnini D, Tsukumo DM, Schenka AA, Latuf-Filho P, Vassallo J, Dias JC, Kubota LT, Carvalheira JB, Saad MJ (2012) Modulation of gut microbiota by antibiotics improves insulin signalling in high-fat fed mice. Diabetologia 55(10):2823–2834. doi:10.1007/s00125-012-2648-4

Cho I, Yamanishi S, Cox L, Methe BA, Zavadil J, Li K, Gao Z, Mahana D, Raju K, Teitler I, Li H, Alekseyenko AV, Blaser MJ (2012) Antibiotics in early life alter the murine colonic microbiome and adiposity. Nature 488(7413):621–626. doi:10.1038/nature11400

Claesson MJ, Cusack S, O'Sullivan O, Greene-Diniz R, de Weerd H, Flannery E, Marchesi JR, Falush D, Dinan T, Fitzgerald G, Stanton C, van Sinderen D, O'Connor M, Harnedy N, O'Connor K, Henry C, O'Mahony D, Fitzgerald AP, Shanahan F, Twomey C, Hill C, Ross RP, O'Toole PW (2011) Composition, variability, and temporal stability of the intestinal microbiota of the elderly. Proc Natl Acad Sci USA 108(Suppl 1):4586–4591. doi:10.1073/pnas.1000097107

Claesson MJ, Jeffery IB, Conde S, Power SE, O'Connor EM, Cusack S, Harris HM, Coakley M, Lakshminarayanan B, O'Sullivan O, Fitzgerald GF, Deane J, O'Connor M, Harnedy N, O'Connor K, O'Mahony D, van Sinderen D, Wallace M, Brennan L, Stanton C, Marchesi JR, Fitzgerald AP, Shanahan F, Hill C, Ross RP, O'Toole PW (2012) Gut microbiota composition correlates with diet and health in the elderly. Nature 488(7410):178–184. doi:10.1038/nature11319

David LA, Maurice CF, Carmody RN, Gootenberg DB, Button JE, Wolfe BE, Ling AV, Devlin AS, Varma Y, Fischbach MA, Biddinger SB, Dutton RJ, Turnbaugh PJ (2014) Diet rapidly and reproducibly alters the human gut microbiome. Nature 505(7484):559–563. doi:10.1038/nature12820

De Filippo C, Cavalieri D, Di Paola M, Ramazzotti M, Poullet JB, Massart S, Collini S, Pieraccini G, Lionetti P (2010) Impact of diet in shaping gut microbiota revealed by a comparative study in children from Europe and rural Africa. Proc Natl Acad Sci USA 107 (33):14691–14696. doi:10.1073/pnas.1005963107

De Vadder F, Kovatcheva-Datchary P, Goncalves D, Vinera J, Zitoun C, Duchampt A, Backhed F, Mithieux G (2014) Microbiota-generated metabolites promote metabolic benefits via gut-brain neural circuits. Cell 156(1–2):84–96. doi:10.1016/j.cell.2013.12.016

de Vos WM, Nieuwdorp M (2013) Genomics: a gut prediction. Nature 498(7452):48–49. doi:10.1038/nature12251

de Wit N, Derrien M, Bosch-Vermeulen H, Oosterink E, Keshtkar S, Duval C, de Vogel-van den Bosch J, Kleerebezem M, Muller M, van der Meer R (2012) Saturated fat stimulates obesity and hepatic steatosis and affects gut microbiota composition by an enhanced overflow of dietary fat to the distal intestine. Am J Physiol Gastrointest Liver Physiol 303(5):G589–599. doi:10.1152/ajpgi.00488.2011

Dethlefsen L, Relman DA (2011) Incomplete recovery and individualized responses of the human distal gut microbiota to repeated antibiotic perturbation. Proc Natl Acad Sci USA 108(Suppl 1):4554–4561. doi:10.1073/pnas.1000087107

Duncan SH, Lobley GE, Holtrop G, Ince J, Johnstone AM, Louis P, Flint HJ (2008) Human colonic microbiota associated with diet, obesity and weight loss. Int J Obes (Lond) 32 (11):1720–1724. doi:10.1038/ijo.2008.155

Everard A, Belzer C, Geurts L, Ouwerkerk JP, Druart C, Bindels LB, Guiot Y, Derrien M, Muccioli GG, Delzenne NM, de Vos WM, Cani PD (2013) Cross-talk between Akkermansia muciniphila and intestinal epithelium controls diet-induced obesity. Proc Natl Acad Sci USA 110(22):9066–9071. doi:10.1073/pnas.1219451110

Francois F, Roper J, Joseph N, Pei Z, Chhada A, Shak JR, de Perez AZ, Perez-Perez GI, Blaser MJ (2011) The effect of H. pylori eradication on meal-associated changes in plasma ghrelin and leptin. BMC Gastroenterol 11:37. doi:10.1186/1471-230X-11-37

Furet JP, Kong LC, Tap J, Poitou C, Basdevant A, Bouillot JL, Mariat D, Corthier G, Dore J, Henegar C, Rizkalla S, Clement K (2010) Differential adaptation of human gut microbiota to bariatric surgery-induced weight loss: links with metabolic and low-grade inflammation markers. Diabetes 59(12):3049–3057. doi:10.2337/db10-0253

Goodrich JK, Waters JL, Poole AC, Sutter JL, Koren O, Blekhman R, Beaumont M, Van Treuren W, Knight R, Bell JT, Spector TD, Clark AG, Ley RE (2014) Human genetics shape the gut microbiome. Cell 159(4):789–799. doi:10.1016/j.cell.2014.09.053

Graessler J, Qin Y, Zhong H, Zhang J, Licinio J, Wong ML, Xu A, Chavakis T, Bornstein AB, Ehrhart-Bornstein M, Lamounier-Zepter V, Lohmann T, Wolf T, Bornstein SR (2012) Metagenomic sequencing of the human gut microbiome before and after bariatric surgery in obese patients with type 2 diabetes: correlation with inflammatory and metabolic parameters. Pharmacogenomics J. doi:10.1038/tpj.2012.43

Henao-Mejia J, Elinav E, Jin C, Hao L, Mehal WZ, Strowig T, Thaiss CA, Kau AL, Eisenbarth SC, Jurczak MJ, Camporez JP, Shulman GI, Gordon JI, Hoffman HM, Flavell RA (2012) Inflammasome-mediated dysbiosis regulates progression of NAFLD and obesity. Nature 482 (7384):179–185. doi:10.1038/nature10809

Hildebrandt MA, Hoffmann C, Sherrill-Mix SA, Keilbaugh SA, Hamady M, Chen YY, Knight R, Ahima RS, Bushman F, Wu GD (2009) High-fat diet determines the composition of the murine gut microbiome independently of obesity. Gastroenterology 137(5):1716–1724. doi:10.1053/j.gastro.2009.08.042

The Human Microbiome Project Consortium (2012) A framework for human microbiome research. Nature 486(7402):215–221. doi:10.1038/nature11209

Jernberg C, Lofmark S, Edlund C, Jansson JK (2007) Long-term ecological impacts of antibiotic administration on the human intestinal microbiota. ISME J 1(1):56–66. doi:10.1038/ismej.2007.3

Karlsson FH, Tremaroli V, Nookaew I, Bergstrom G, Behre CJ, Fagerberg B, Nielsen J, Backhed F (2013) Gut metagenome in European women with normal, impaired and diabetic glucose control. Nature 498(7452):99–103. doi:10.1038/nature12198

Karlsson FH, Nookaew I, Nielsen J (2014) Metagenomic data utilization and analysis (MEDUSA) and construction of a global gut microbial gene catalogue. PLoS Comput Biol 10(7), e1003706. doi:10.1371/journal.pcbi.1003706

Kuczynski J, Lauber CL, Walters WA, Parfrey LW, Clemente JC, Gevers D, Knight R (2012) Experimental and analytical tools for studying the human microbiome. Nat Rev Genet 13 (1):47–58. doi:10.1038/nrg3129

Larsen N, Vogensen FK, van den Berg FW, Nielsen DS, Andreasen AS, Pedersen BK, Al-Soud WA, Sorensen SJ, Hansen LH, Jakobsen M (2010) Gut microbiota in human adults with type 2 diabetes differs from non-diabetic adults. PLoS One 5(2), e9085. doi:10.1371/journal.pone.0009085

Le Chatelier E, Nielsen T, Qin J, Prifti E, Hildebrand F, Falony G, Almeida M, Arumugam M, Batto JM, Kennedy S, Leonard P, Li J, Burgdorf K, Grarup N, Jorgensen T, Brandslund I, Nielsen HB, Juncker AS, Bertalan M, Levenez F, Pons N, Rasmussen S, Sunagawa S, Tap J, Tims S, Zoetendal EG, Brunak S, Clement K, Dore J, Kleerebezem M, Kristiansen K, Renault P, Sicheritz-Ponten T, de Vos WM, Zucker JD, Raes J, Hansen T, MetaHIT Consortium, Bork P, Wang J, Ehrlich SD, Pedersen O (2013) Richness of human gut microbiome correlates with metabolic markers. Nature 500(7464):541–546. doi:10.1038/nature12506

Ley RE, Backhed F, Turnbaugh P, Lozupone CA, Knight RD, Gordon JI (2005) Obesity alters gut microbial ecology. Proc Natl Acad Sci USA 102(31):11070–11075. doi:10.1073/pnas.0504978102

Li J, Jia H, Cai X, Zhong H, Feng Q, Sunagawa S, Arumugam M, Kultima JR, Prifti E, Nielsen T, Juncker AS, Manichanh C, Chen B, Zhang W, Levenez F, Wang J, Xu X, Xiao L, Liang S, Zhang D, Zhang Z, Chen W, Zhao H, Al-Aama JY, Edris S, Yang H, Wang J, Hansen T, Nielsen HB, Brunak S, Kristiansen K, Guarner F, Pedersen O, Dore J, Ehrlich SD, MetaHIT Consortium, Bork P, Wang J (2014) An integrated catalog of reference genes in the human gut microbiome. Nat Biotechnol 32(8):834–841. doi:10.1038/nbt.2942

Lin HV, Frassetto A, Kowalik EJ Jr, Nawrocki AR, Lu MM, Kosinski JR, Hubert JA, Szeto D, Yao X, Forrest G, Marsh DJ (2012) Butyrate and propionate protect against diet-induced obesity and regulate gut hormones via free fatty acid receptor 3-independent mechanisms. PLoS One 7(4), e35240. doi:10.1371/journal.pone.0035240

McGettrick AF, O'Neill LA (2010) Regulators of TLR4 signaling by endotoxins. Subcell Biochem 53:153–171. doi:10.1007/978-90-481-9078-2_7

McKnite AM, Perez-Munoz ME, Lu L, Williams EG, Brewer S, Andreux PA, Bastiaansen JW, Wang X, Kachman SD, Auwerx J, Williams RW, Benson AK, Peterson DA, Ciobanu DC (2012) Murine gut microbiota is defined by host genetics and modulates variation of metabolic traits. PLoS One 7(6), e39191. doi:10.1371/journal.pone.0039191

Membrez M, Blancher F, Jaquet M, Bibiloni R, Cani PD, Burcelin RG, Corthesy I, Mace K, Chou CJ (2008) Gut microbiota modulation with norfloxacin and ampicillin enhances glucose tolerance in mice. FASEB J 22(7):2416–2426. doi:10.1096/fj.07-102723

Morgan XC, Huttenhower C (2012) Chapter 12: human microbiome analysis. PLoS Comput Biol 8(12), e1002808. doi:10.1371/journal.pcbi.1002808

Murphy R, Stewart AW, Braithwaite I, Beasley R, Hancox RJ, Mitchell EA, Group IPTS (2014) Antibiotic treatment during infancy and increased body mass index in boys: an international cross-sectional study. Int J Obes (Lond) 38(8):1115–1119. doi:10.1038/ijo.2013.218

Nielsen HB, Almeida M, Juncker AS, Rasmussen S, Li J, Sunagawa S, Plichta DR, Gautier L, Pedersen AG, Le Chatelier E, Pelletier E, Bonde I, Nielsen T, Manichanh C, Arumugam M, Batto JM, Quintanilha Dos Santos MB, Blom N, Borruel N, Burgdorf KS, Boumezbeur F, Casellas F, Dore J, Dworzynski P, Guarner F, Hansen T, Hildebrand F, Kaas RS, Kennedy S, Kristiansen K, Kultima JR, Leonard P, Levenez F, Lund O, Moumen B, Le Paslier D, Pons N, Pedersen O, Prifti E, Qin J, Raes J, Sorensen S, Tap J, Tims S, Ussery DW, Yamada T, Meta HITC, Renault P, Sicheritz-Ponten T, Bork P, Wang J, Brunak S, Ehrlich SD, MetaHIT Consortium (2014) Identification and assembly of genomes and genetic elements in complex metagenomic samples without using reference genomes. Nat Biotechnol 32(8):822–828. doi:10.1038/nbt.2939

Palmer C, Bik EM, DiGiulio DB, Relman DA, Brown PO (2007) Development of the human infant intestinal microbiota. PLoS Biol 5(7), e177. doi:10.1371/journal.pbio.0050177

Qin J, Li R, Raes J, Arumugam M, Burgdorf KS, Manichanh C, Nielsen T, Pons N, Levenez F, Yamada T, Mende DR, Li J, Xu J, Li S, Li D, Cao J, Wang B, Liang H, Zheng H, Xie Y, Tap J, Lepage P, Bertalan M, Batto JM, Hansen T, Le Paslier D, Linneberg A, Nielsen HB, Pelletier E, Renault P, Sicheritz-Ponten T, Turner K, Zhu H, Yu C, Jian M, Zhou Y, Li Y, Zhang X, Qin N, Yang H, Wang J, Brunak S, Dore J, Guarner F, Kristiansen K, Pedersen O, Parkhill J, Weissenbach J, MetaHIT Consortium, Bork P, Ehrlich SD, Wang J (2010) A human gut microbial gene catalogue established by metagenomic sequencing. Nature 464 (7285):59–65. doi:10.1038/nature08821

Qin J, Li Y, Cai Z, Li S, Zhu J, Zhang F, Liang S, Zhang W, Guan Y, Shen D, Peng Y, Zhang D, Jie Z, Wu W, Qin Y, Xue W, Li J, Han L, Lu D, Wu P, Dai Y, Sun X, Li Z, Tang A, Zhong S, Li X, Chen W, Xu R, Wang M, Feng Q, Gong M, Yu J, Zhang Y, Zhang M, Hansen T, Sanchez G, Raes J, Falony G, Okuda S, Almeida M, LeChatelier E, Renault P, Pons N, Batto JM, Zhang Z, Chen H, Yang R, Zheng W, Li S, Yang H, Wang J, Ehrlich SD, Nielsen R, Pedersen O, Kristiansen K, Wang J (2012) A metagenome-wide association study of gut microbiota in type 2 diabetes. Nature 490(7418):55–60. doi:10.1038/nature11450

Rabot S, Membrez M, Bruneau A, Gerard P, Harach T, Moser M, Raymond F, Mansourian R, Chou CJ (2010) Germ-free C57BL/6J mice are resistant to high-fat-diet-induced insulin resistance and have altered cholesterol metabolism. FASEB J 24(12):4948–4959. doi:10.1096/fj.10-164921

Rajilic-Stojanovic M, Heilig HG, Molenaar D, Kajander K, Surakka A, Smidt H, de Vos WM (2009) Development and application of the human intestinal tract chip, a phylogenetic microarray: analysis of universally conserved phylotypes in the abundant microbiota of young and elderly adults. Environ Microbiol 11(7):1736–1751. doi:10.1111/j.1462-2920.2009.01900.x

Ridaura VK, Faith JJ, Rey FE, Cheng J, Duncan AE, Kau AL, Griffin NW, Lombard V, Henrissat B, Bain JR, Muehlbauer MJ, Ilkayeva O, Semenkovich CF, Funai K, Hayashi DK, Lyle BJ, Martini MC, Ursell LK, Clemente JC, Van Treuren W, Walters WA, Knight R, Newgard CB, Heath AC, Gordon JI (2013) Gut microbiota from twins discordant for obesity modulate metabolism in mice. Science 341(6150):1241214. doi:10.1126/science.1241214

Ryan KK, Tremaroli V, Clemmensen C, Kovatcheva-Datchary P, Myronovych A, Karns R, Wilson-Perez HE, Sandoval DA, Kohli R, Backhed F, Seeley RJ (2014) FXR is a molecular target for the effects of vertical sleeve gastrectomy. Nature 509(7499):183–188. doi:10.1038/nature13135

Schwiertz A, Taras D, Schafer K, Beijer S, Bos NA, Donus C, Hardt PD (2010) Microbiota and SCFA in lean and overweight healthy subjects. Obesity 18(1):190–195. doi:10.1038/oby.2009.167

Shin NR, Lee JC, Lee HY, Kim MS, Whon TW, Lee MS, Bae JW (2013) An increase in the Akkermansia spp. population induced by metformin treatment improves glucose homeostasis in diet-induced obese mice. Gut. doi:10.1136/gutjnl-2012-303839

Spor A, Koren O, Ley R (2011) Unravelling the effects of the environment and host genotype on the gut microbiome. Nat Rev Microbiol 9(4):279–290. doi:10.1038/nrmicro2540

Stewart JA, Chadwick VS, Murray A (2005) Investigations into the influence of host genetics on the predominant eubacteria in the faecal microflora of children. J Med Microbiol 54 (Pt 12):1239–1242. doi:10.1099/jmm.0.46189-0

Tap J, Mondot S, Levenez F, Pelletier E, Caron C, Furet JP, Ugarte E, Munoz-Tamayo R, Paslier DL, Nalin R, Dore J, Leclerc M (2009) Towards the human intestinal microbiota phylogenetic core. Environ Microbiol 11(10):2574–2584. doi:10.1111/j.1462-2920.2009.01982.x

Thuny F, Richet H, Casalta JP, Angelakis E, Habib G, Raoult D (2010) Vancomycin treatment of infective endocarditis is linked with recently acquired obesity. PLoS One 5(2), e9074. doi:10.1371/journal.pone.0009074

Tims S, Derom C, Jonkers DM, Vlietinck R, Saris WH, Kleerebezem M, de Vos WM, Zoetendal EG (2013) Microbiota conservation and BMI signatures in adult monozygotic twins. ISME J 7 (4):707–717. doi:10.1038/ismej.2012.146

Trasande L, Blustein J, Liu M, Corwin E, Cox LM, Blaser MJ (2012) Infant antibiotic exposures and early-life body mass. Int J Obes (Lond). doi:10.1038/ijo.2012.132

Turnbaugh PJ, Ley RE, Mahowald MA, Magrini V, Mardis ER, Gordon JI (2006) An obesity-associated gut microbiome with increased capacity for energy harvest. Nature 444 (7122):1027–1031. doi:10.1038/nature05414

Turnbaugh PJ, Hamady M, Yatsunenko T, Cantarel BL, Duncan A, Ley RE, Sogin ML, Jones WJ, Roe BA, Affourtit JP, Egholm M, Henrissat B, Heath AC, Knight R, Gordon JI (2009a) A core gut microbiome in obese and lean twins. Nature 457(7228):480–484. doi:10.1038/nature07540

Turnbaugh PJ, Ridaura VK, Faith JJ, Rey FE, Knight R, Gordon JI (2009b) The effect of diet on the human gut microbiome: a metagenomic analysis in humanized gnotobiotic mice. Sci Transl Med 1(6):6ra14. doi:10.1126/scitranslmed.3000322

van Nood E, Vrieze A, Nieuwdorp M, Fuentes S, Zoetendal EG, de Vos WM, Visser CE, Kuijper EJ, Bartelsman JF, Tijssen JG, Speelman P, Dijkgraaf MG, Keller JJ (2013) Duodenal infusion of donor feces for recurrent Clostridium difficile. N Engl J Med 368(5):407–415. doi:10.1056/NEJMoa1205037

Vijay-Kumar M, Aitken JD, Carvalho FA, Cullender TC, Mwangi S, Srinivasan S, Sitaraman SV, Knight R, Ley RE, Gewirtz AT (2010) Metabolic syndrome and altered gut microbiota in mice lacking Toll-like receptor 5. Science 328(5975):228–231. doi:10.1126/science.1179721

Vrieze A, Van Nood E, Holleman F, Salojarvi J, Kootte RS, Bartelsman JF, Dallinga-Thie GM, Ackermans MT, Serlie MJ, Oozeer R, Derrien M, Druesne A, Van Hylckama Vlieg JE, Bloks VW, Groen AK, Heilig HG, Zoetendal EG, Stroes ES, de Vos WM, Hoekstra JB, Nieuwdorp M (2012) Transfer of intestinal microbiota from lean donors increases insulin sensitivity in individuals with metabolic syndrome. Gastroenterology 143(4):913–916. doi:10.1053/j.gastro.2012.06.031

Vrieze A, Out C, Fuentes S, Jonker L, Reuling I, Kootte RS, van Nood E, Holleman F, Knaapen M, Romijn JA, Soeters MR, Blaak EE, Dallinga-Thie GM, Reijnders D, Ackermans MT, Serlie MJ, Knop FK, Holst JJ, van der Ley C, Kema IP, Zoetendal EG, de Vos WM, Hoekstra JB, Stroes ES, Groen AK, Nieuwdorp M (2014) Impact of oral vancomycin on gut microbiota, bile acid metabolism, and insulin sensitivity. J Hepatol 60(4):824–831. doi:10.1016/j.jhep.2013.11.034

Weinstock GM (2012) Genomic approaches to studying the human microbiota. Nature 489 (7415):250–256. doi:10.1038/Nature11553

Wichmann A, Allahyar A, Greiner TU, Plovier H, Lunden GO, Larsson T, Drucker DJ, Delzenne NM, Cani PD, Backhed F (2013) Microbial modulation of energy availability in the colon regulates intestinal transit. Cell Host Microbe 14(5):582–590. doi:10.1016/j.chom.2013.09.012

Wright SD, Ramos RA, Tobias PS, Ulevitch RJ, Mathison JC (1990) CD14, a receptor for complexes of lipopolysaccharide (LPS) and LPS binding protein. Science 249 (4975):1431–1433

Wu GD, Chen J, Hoffmann C, Bittinger K, Chen YY, Keilbaugh SA, Bewtra M, Knights D, Walters WA, Knight R, Sinha R, Gilroy E, Gupta K, Baldassano R, Nessel L, Li H, Bushman FD, Lewis JD (2011) Linking long-term dietary patterns with gut microbial enterotypes. Science 334(6052):105–108. doi:10.1126/science.1208344

Yatsunenko T, Rey FE, Manary MJ, Trehan I, Dominguez-Bello MG, Contreras M, Magris M, Hidalgo G, Baldassano RN, Anokhin AP, Heath AC, Warner B, Reeder J, Kuczynski J, Caporaso JG, Lozupone CA, Lauber C, Clemente JC, Knights D, Knight R, Gordon JI (2012) Human gut microbiome viewed across age and geography. Nature 486 (7402):222–227. doi:10.1038/nature11053

Yuan S, Cohen DB, Ravel J, Abdo Z, Forney LJ (2012) Evaluation of methods for the extraction and purification of DNA from the human microbiome. PLoS One 7(3), e33865. doi:10.1371/journal.pone.0033865

Zhang H, DiBaise JK, Zuccolo A, Kudrna D, Braidotti M, Yu Y, Parameswaran P, Crowell MD, Wing R, Rittmann BE, Krajmalnik-Brown R (2009) Human gut microbiota in obesity and after gastric bypass. Proc Natl Acad Sci USA 106(7):2365–2370. doi:10.1073/pnas.0812600106

Zhang X, Shen D, Fang Z, Jie Z, Qiu X, Zhang C, Chen Y, Ji L (2013) Human gut microbiota changes reveal the progression of glucose intolerance. PLoS One 8(8), e71108. doi:10.1371/journal.pone.0071108

Zoetendal EG (2001) The host genotype affects the bacterial community in the human gastrointestinal tract. Microb Ecol Health Dis 13(3):129–134

Zupancic ML, Cantarel BL, Liu Z, Drabek EF, Ryan KA, Cirimotich S, Jones C, Knight R, Walters WA, Knights D, Mongodin EF, Horenstein RB, Mitchell BD, Steinle N, Snitker S, Shuldiner AR, Fraser CM (2012) Analysis of the gut microbiota in the old order Amish and its relation to the metabolic syndrome. PLoS One 7(8), e43052. doi:10.1371/journal.pone.0043052

Part III
From Association to Function

Chapter 14
Transcription Factor 7-Like 2 (TCF7L2)

Struan F.A. Grant and Leif Groop

Abstract In 2006 a genetic signal within the gene encoding transcription factor 7-like 2 (*TCF7L2*) was first reported to be associated with type 2 diabetes. Since then multiple genome-wide association studies have revealed this signal to be among the most strongest associations reported with this disease to date. Furthermore, multiple studies around the world have revealed the global relevance of this locus in intron 3 of the gene and have helped pinpoint, with relative confidence, the causal lesion at this locus. However, despite this association being beyond doubt, there is still a lack of consensus with respect to mechanisms of action and which tissue(s) it actually exerts its influence on the pathogenesis of type 2 diabetes.

14.1 *TCF7L2*: The Genetic Hallmark of T2D

Type 2 diabetes (T2D) is a complex trait that clearly results from the interaction between a number of factors. Despite this complexity, there is strong evidence for a genetic component contributing to an individual's risk for presenting with T2D. This evidence largely comes from the observation that the disease is heritable, although with an approximate sibling risk of 3.5-fold (Rich 1990), its heritability is somewhat modest when contrasted with type 1 diabetes (T1D), which exhibits an

---

S.F.A. Grant (✉)
Division of Human Genetics, The Children's Hospital of Philadelphia, Philadelphia, PA 19104, USA

Department of Genetics and Institute of Diabetes, Obesity and Metabolism, Perelman School of Medicine, University of Pennsylvania, Philadelphia, PA 19104, USA

Department of Pediatrics, Perelman School of Medicine, University of Pennsylvania, Philadelphia, PA 19104, USA
e-mail: grants@email.chop.edu

L. Groop
Department of Clinical Sciences, Diabetes and Endocrinology, Lund University Diabetes Centre, Skåne University Hospital Malmö, 205 02 Malmö, Sweden

Finnish Institute for Molecular Medicine (FIMM), Helsinki University, Helsinki, Finland

J.C. Florez (ed.), *The Genetics of Type 2 Diabetes and Related Traits*,
DOI 10.1007/978-3-319-01574-3_14

approximate sibling risk of 10–15 times greater than the general population (Clayton 2009). As such, given the relatively weak genetic contribution to T2D, it has proven historically challenging to identify genes associated with this very common disease, with T2D earning the nickname of the "geneticist's nightmare" prior to the advent of genome-wide association studies (GWAS). Indeed, before the GWAS era, only a handful of loci had been reported to be associated with T2D, primarily as a consequence of candidate gene and family-based linkage studies. The genetic basis of T2D has therefore been largely elusive until relatively recently, when the utility of high-throughput single-nucleotide polymorphism (SNP) genotyping arrays became a reality. This technology has allowed investigators to carry out non-hypothesis-driven GWAS analyses to uncover key genetic components to the pathogenesis of most complex traits. Indeed, GWAS has now clearly revealed many tens of loci driving T2D risk (Sladek et al. 2007; Wellcome Trust Case Control Consortium 2007; Saxena et al. 2007; Zeggini et al. 2007; Scott et al. 2007; Steinthorsdottir et al. 2007; Salonen et al. 2007; Gudmundsson et al. 2007; Zeggini et al. 2008a; Unoki et al. 2008; Yasuda et al. 2008; Rung et al. 2009; Voight et al. 2010; Dupuis et al. 2010; Cho et al. 2011; Kooner et al. 2011; Morris et al. 2012; DIAbetes Genetics Replication And Meta-analysis (DIAGRAM) Consortium 2014).

Just before the dawn of GWAS, the first author of this chapter uncovered the strong association of variants in the transcription factor 7-like 2 (*TCF7L2*) gene with T2D (Grant et al. 2006). Other investigative groups rapidly and independently replicated this finding in different ethnicities, and, interestingly, from the first GWAS reports of T2D, undertaken in European ancestry populations, the strongest association was indeed with *TCF7L2* (Sladek et al. 2007; Wellcome Trust Case Control Consortium 2007; Saxena et al. 2007; Zeggini et al. 2007; Scott et al. 2007); as such this is now considered one of the most significant genetic findings in T2D reported to date (Zeggini and McCarthy 2007; Weedon 2007; Hattersley 2007).

Although variation (primarily represented by rs7903146) in the third intron of the *TCF7L2* gene confers the strongest inherited risk of T2D among common variants worldwide, many questions remain regarding the mechanisms by which it increases susceptibility to this disease. First, by which molecular mechanisms does an intronic variant exert its effect? Second, what are the target tissues for its detrimental effects on glucose metabolism? While pancreatic islets represent the logical explanation for a defect leading to diabetes, discrepant results from cell lines, animal models, and humans do not provide a unifying explanation whether it is due to gain or reduced function of TCF7L2. Finally can genetic variants in *TCF7L2* help elucidate the origin of the diabetic subgroup referred to as "latent autoimmune diabetes in adults" (LADA)?

This chapter reviews what has been learned about the *TCF7L2* locus thus far and what it can teach us about how to address the increasing number of loci emerging from ever larger genome-wide surveys of T2D.

14.2 Discovery in Iceland

The strong association between *TCF7L2* and T2D was first detected in a 2006 report (Grant et al. 2006) as a result of a fine-mapping exercise, leveraging highly polymorphic microsatellite markers, of a linkage region on chromosome 10 (Reynisdottir et al. 2003). This region of linkage was observed as a consequence of extensive recruitment and genotyping of large extended Icelandic pedigrees consisting of multiple cases of T2D.

The variation found at this locus turned out to be common in the population, occurring in a noncoding region within the *TCF7L2* gene and resulting in an approximate 40 % increase in T2D risk (Grant et al. 2006) when considered in isolation. Within the same study, this signal was robustly replicated in European ancestry cohorts from Denmark and the USA, revealing a very similar frequency and relative risk, leading to the calculation of an overall population attributable risk of 21 %. The risk alleles of the variants in linkage disequilibrium with each other within the gene that captured the association were approximately 1.5 times more common in patients than in controls; this corresponded to an approximately 50 % increase in risk of T2D per copy carried. As such, more than one third of individuals in the European ancestry populations studied carried one copy of the at-risk variant and were at an approximately 45 % increased risk of the disease compared to controls, while 7 % carried two copies and were at a 141 % greater risk.

This genetic finding was all the more remarkable as the genetic component to T2D risk has always been considered so mild for this phenotype (Rich 1990) (see above). As a consequence of this unexpected discovery and the great thirst at that time to uncover a robust genetic component to T2D, the finding received extensive media attention, including being featured on the front page of the *New York Times* (http://www.nytimes.com/2006/01/16/science/16gene.html).

As early as 2007, a meta-analysis of published follow-up studies of this genetic relationship estimated a pooled odds ratio, using a Mantel–Haenszel procedure, of 1.46 worldwide (with a highly notable $P = 5.4 \times 10^{-140}$) (Cauchi et al. 2007a), making it one of the most statistically significant genetic findings in T2D to date. Indeed, it has now been confirmed via several T2D GWAS (Sladek et al. 2007; Wellcome Trust Case Control Consortium 2007; Saxena et al. 2007; Zeggini et al. 2007; Scott et al. 2007) (see below).

14.3 Observations from GWAS

T2D has been the focus of more GWAS efforts than any other complex trait studied to date, primarily due to the sizeable proportion of the many populations it blights during an individual's lifetime (see Chap. 2). The first batch of such studies, published in *Nature* (Sladek et al. 2007; Wellcome Trust Case Control Consortium 2007) and *Science* (Saxena et al. 2007; Zeggini et al. 2007; Scott et al. 2007) and performed in European ancestry cohorts, revealed multiple loci, with the strongest association by far residing within *TCF7L2*. Interestingly, when GWAS analyses were subsequently carried out in East Asian populations, *TCF7L2* was by no means the strongest associated locus, rather a signal coinciding with the gene encoding potassium voltage-gated channel, KQT-like subfamily, member 1 (*KCNQ1*), which yielded the lowest *P*-value (Unoki et al. 2008; Yasuda et al. 2008; Cho et al. 2011). Although the *TCF7L2* locus was detected, these observations suggest that it exerts its effect differently depending on the population being considered, potentially as a consequence of differences in diet (Helgason et al. 2007a).

Despite the many breakthroughs, the proportion of the predicted genetic contribution to T2D pathogenesis uncovered is still very much in the minority. Although variation at the *TCF7L2* locus confers a highly statistically significant contribution to the trait, when combined with all the other GWAS-implicated loci subsequently reported, together they only explain approximately 10 % of the predicted genetic component to the disease. Therefore, much more work is going to be required to uncover the remaining genetic etiology of T2D, primarily by leveraging new advances in whole-genome sequencing technologies.

14.4 Leveraging Global Patterns to Infer the Causal Variant

Although GWAS has uncovered genetic signals robustly associated with various complex traits over recent years, the approach was never intended to interrogate the actual underlying causal variant in the first instance, but rather simply "tag" the approximate location of a disease-associated variant (McCarthy et al. 2008; Manolio et al. 2008; Altshuler et al. 2008), typically down to a few 100 kb. As such it has turned out to be very challenging to distill out the underlying causal variant driving a particular signal. In this sense, the T2D association with *TCF7L2* may be at a more mature stage, as there is strong consensus that the causal variant has actually been identified through a process of statistical elimination and functional work, reflecting the longer time that this associated locus has been known in comparison with the typical T2D GWAS signal.

This association was first refined and narrowed down utilizing a West African patient cohort (Helgason et al. 2007b), where this population harbors greater haplotype diversity. As such, the associated region was contained within a smaller region of linkage disequilibrium and thus aided the pinpointing of the putative functional variant. In brief, when this association was reported in European ancestry populations, it was noted that rs12255372 and rs7903146 both captured the association well (Grant et al. 2006), but the African ancestry study showed that rs12255372 was a less optimal tag SNP and revealed rs7903146 to be the better proxy for the signal detected (Helgason et al. 2007a; Palmer et al. 2011).

The T allele of the rs7903146 SNP within *TCF7L2* is now widely considered to be the causal variant for this disease through such studies in multiple ethnicities (Helgason et al. 2007a; Palmer et al. 2011; Tong et al. 2009; Cauchi et al. 2007b) as well as a Bayesian modeling approach (Maller et al. 2012). Therefore it is viewed as the best variant to test globally when searching for the association with T2D. Interestingly, reflecting what was observed in GWAS analyses of East Asians (Chang et al. 2007; Ng et al. 2007, 2008; Ren et al. 2008; Yu et al. 2010; Zheng et al. 2012), the risk T allele of rs7903146 is common in European and African populations but not in the Han Chinese. This latter observation may also explain some of the inconsistent results of replication studies using Chinese populations, where many may have been statistically underpowered to detect the association at the diminished minor allele frequency.

The variant is intronic so clearly does not directly influence the amino acid sequence of the TCF7L2 protein. Although there is strong evidence of an allelic difference in open chromatin across this SNP in pancreatic islets (Gaulton et al. 2010) (more details below) suggesting an influence on transcription control, the overall mechanism of action of this variant still needs to be fully elucidated.

14.4.1 Molecular Mechanisms

TCF7L2 is a high-mobility-group box-containing transcription factor in the canonical Wnt signaling cascade, regulating the expression of downstream target genes (Clevers 2006) and driving key aspects of embryonic development (Nusse and Varmus 1992; Paul and Dey 2008). In the absence of β-catenin, TCF7L2 binds to Wnt-responsive elements to repress target gene transcription, while β-catenin binding to TCF7L2 activates gene expression. *TCF7L2* is expressed in most tissues including islets and adipose and in organs such as the liver, gut, and brain. It is, though, unclear whether the Wnt signaling cascade plays a role in regulating insulin secretion in human pancreatic islets.

Like the vast majority of common variants associated with complex traits uncovered by GWAS, rs7903146 is located within a noncoding region, i.e., intron 3 of *TCF7L2*. A number of hypotheses have been proposed to explain the causal mechanism influenced by this SNP; several studies have shown that rs7903146 resides in an active enhancer region with open chromatin (Gaulton et al. 2010; Savic et al. 2013; Stitzel et al. 2010). Using FAIRE sequencing to identify active regulatory elements in islets, Gaulton et al. (2010) demonstrated that the SNP resides in islet-selective open chromatin. The SNP showed allelic imbalance in islet FAIRE signals and the risk T allele induced much greater enhancer activity than the C allele in MIN6 cells. The simplest explanation implicates the intronic variant as a cis eQTL influencing expression of the *TCF7L2* gene in islets. However, expression of TCF7L2 is not decreased in human islets, if anything it is increased, and expression is not influenced by the SNP (Lyssenko et al. 2007;

Taneera et al. 2012) despite the correlation seen with open chromatin across this key region in human pancreatic islets (Gaulton et al. 2010).

14.4.2 *TCF7L2* Is a Master Regulator of Insulin Production

When it was first implicated in T2D, it was shown that the *TCF7L2* variant is associated with impaired glucose and GLP-1-stimulated insulin secretion (Grant et al. 2006; Yi et al. 2005a; Hansson et al. 2010). Knockout mice for *TCF7L2* are embryonically lethal and among other things characterized by defective proliferation of crypt stem cells of the small intestine (Korinek et al. 1998a). TCF7L2 also binds to the promoter region of the proglucagon gene to control its transcriptional activity in the intestinal GLUTag cell line, where the dominant-negative mutant of TCF7L2 abolishes proglucagon mRNA levels (Yi et al. 2005b). However, this role in humans remains elusive. TCF7L2 though seems to play a key role in regulating pancreatic β-cell function. This was clearly demonstrated in a recent paper showing that disruption of *tcf7l2* in mice, a rat insulinoma cell line (INS1), and in human pancreatic islets resulted in downregulation not only of genes and proteins in the proinsulin pathway (MAFA, ISL-1, etc.) but also of genes and proteins regulating processing of proinsulin (PCSK1 and PCSK2) (Zhou et al. 2014a). These multiple hits on insulin (both synthesis and processing) could explain why variants (with most likely inhibitory effects) confer the greatest risk of T2D. Disruption of *Tcf7l2* in primary mouse islets and MIN6 and INS-1 cells results in impaired glucose-stimulated insulin secretion and promotes a diabetic phenotype (da Silva Xavier et al. 2009; Shu et al. 2008, 2009; Zhou et al. 2012). The *Tcf7l2* mutant mice die within 24 h of birth displaying defects in crypt stem cells of the small intestine (Korinek et al. 1998b). However, they were also characterized by hypoglycemia at birth. Two studies with β-cell-specific knockdown of *Tcf7l2* using *Pdx1* (da Silva Xavier et al. 2012) or *Ins2* (Takamoto et al. 2014) promoters have also described islet dysfunction, impaired insulin secretion, and reduced β-cell mass. In contrast, disrupting or overexpressing the 92 bp genomic interval carrying the rs7903146 variant in mice yielded opposite phenotypes, i.e., impaired glucose tolerance (Savic et al. 2011). A recent paper applying a tamoxifen-inducible β-cell-specific disruption of *Tcf7l2* added further confusion to the field by showing no diabetic phenotype even during high-fat feeding (Boj et al. 2012). However, in a series of follow-up studies, the authors presented results pointing at a key role of the liver. First, they reassessed the total body knockout of *Tcf7l2* showing that newborn mice die of hypoglycemia because they could not turn on genes necessary for glycogen breakdown (e.g., *Gys2*) and gluconeogenesis (e.g., *Pck1*, *G6pc*). During the embryonic period the fetus is dependent upon energy metabolism of the mother but after birth upregulation of these liver genes is dependent upon Tcf7l2 and the Wnt pathway. This was confirmed by liver-specific disruption and overexpression of *Tcf7l2*. Similar results were obtained by Norton et al. (2011) by disrupting *Tcf7l2* in rat hepatoma cell line, H4IIE. These data are also supported by the clinical observation of elevated rates of basal hepatic glucose production in carriers of the risk T allele (Lyssenko et al. 2007). While targeting two tissues (pancreas, liver) could explain the strong diabetogenic effect of the *TCF7L2* variant, but it would require a reduced function in pancreatic islets and gain of function in the liver. Data on expression of *TCF7L2* (including splice isoforms) in the human liver in relation to genotype is urgently needed.

Can we rule out effects in other tissues? In adipose tissue expression of *TCF7L2* is downregulated by insulin (Ahlzen et al. 2008) and expression of a short isoform lacking exons 12,13, and 13a decreased after bariatric surgery in adipose tissue (Kaminska et al. 2012a), but these effects were unrelated to the genotype. TCF7L2 and Wnt signaling has also been ascribed a role in the brain, where a dominant-negative *Tcf7l2* resulted in decreased expression of the proglucagon gene (and thereby Glp1) in the brain and gut and impaired glucose-stimulated insulin secretion and disposal (Savic et al. 2012; Shao et al. 2012). On the other hand, risk genotype has not been associated with an impairment in insulin sensitivity (Lyssenko et al. 2007). Given the link between the brain and islet/liver in regulation of glucose metabolism, more research is required to explore the potential of TCF7L2 to influence these central mechanisms.

14.4.3 The Missing Link Between rs7903146 and Expression in Islets and Liver?

To explain the effects described above in human islets and liver, one would also need to assume opposite effects in the two tissues. Therefore the question has risen: Could there be different splice isoforms with activating or inactivating actions in different tissues? The *TCF7L2* gene is comprised of 17 exons, five of which have been shown to be alternative, at least in colon cancer cells (exons 4 and 5 in the 5′ region and exons 13–16 in the 3′ region) (Duval et al. 2000a).

TCF7L2 transcripts are heavily spliced at the 3′ terminus, with the 5′ end being the most stable region among isoforms which encode the beta-catenin binding domain (Duval et al. 2000b). To make the picture even more complicated, tissue specific isoforms have been widely reported (Mondal et al. 2010; Prokunina-Olsson and Hall 2010; Kaminska et al. 2012b; Prokunina-Olsson et al. 2009a, b).

Despite the strong body of evidence pointing to the extensive alternative splicing of *TCF7L2* transcripts, it is much less clear if such isoforms track with rs7903146 genotype, as large sample populations of the correct tissue types would be required to uncover any putative allele-specific effects on splicing patterns (Osmark et al. 2009). It is, however, possible that the *TCF7L2* locus exerts its effect in more than one setting and perhaps with a different set of isoforms in play in each instance, particularly in pancreatic islets and liver.

Using absolute qPCR, we could show that an isoform including exon 4 dominated in human pancreatic islets and its incorporation increased with increasing glycated hemoglobin of the donors (Osmark et al. 2009). In the liver, loss of exon 4 promotes development of tumors (Tomimaru et al. 2013). Others have suggested, using relative qPCR in a small number of human islets, that a low abundant exon 16 containing isoform could be exclusive to islets but no relation to the SNP has been demonstrated (Mondal et al. 2010; Prokunina-Olsson and Hall 2010). Follow-up studies have been hampered by the paucity of good antibodies for this region of TCF7L2. To find the missing link between the SNP and expression, whether it influences splicing or something else, we need to focus on the variant. This would be possible today using CRISPR followed by measurements of insulin secretion and RNA sequencing at high read length of appropriate human islet and liver cell lines exposed to high glucose.

14.4.4 Epigenomic Studies

The question remains: How does a variant in intron 3 of the *TCF7L2* gene increase susceptibility to T2D? In addition to confirming the open chromatin observations described above (Gaulton et al. 2010; Savic et al. 2013; Stitzel et al. 2010) through exploring DNAse hypersensitive sites as well as sites of activating H3 lysine modifications (H3K4me1, H3K4me3, H3K79me2), Stitzel et al. (2010) also demonstrated a threefold increased transcriptional activity of the T allele than the C allele. In support of these findings, Savic et al. (2013) observed allele-specific properties for trs7903146, which was not restricted to the β-cell line MIN6 cells but also seen in myoblasts. Taken together with previous findings of increased expression of *TCF7L2* in human islets from particularly TT carriers (Lyssenko et al. 2007), the data suggest that excess of the T allele has negative effects on insulin secretion and glucose tolerance.

When carrying out pathway analyses of the nearest gene to a given TCF7L2 occupancy site derived from ChIP-seq, which combines ChIP and massively parallel sequencing, strong enrichment was found for molecular networks involved in metabolism (Zhao et al. 2010). Furthermore, the list of genes bound by TCF7L2 revealed an unexpected enrichment of GWAS-implicated loci, in particular for diseases related to metabolic and cardiovascular trait areas. As such, it has been suggested that TCF7L2 operates as a central node in the genetic basis for T2D through the regulation of a number of genes that contribute to the pathogenesis of this disease.

The question also remains: What binds to the T allele in the enhancer region to turn on transcription? This question was recently addressed by performing pull-down followed by mass spectrometry in a colon cancer cell line, HCT116 (Xia et al. 2014). Strong binding of poly (ADP-ribose) polymerase 1 (PARP1) to the region of open chromatin was observed, and by overexpressing TCF7L2 with β-catenin, there was a marked allele-specific increase in transcriptional activity. The main caveat was that it occurred in the opposite direction than seen in vivo, with stronger activity seen for the C allele. In addition, as PARP1 is a very abundant protein, it cannot be excluded that PARP-1 is simply the dominant partner, from a detection point of view, in a larger protein complex that binds to the region. Also, the experiments need to be replicated in human islets and liver. However, interestingly, it is well established that PARP-1 plays a key role in DNA damage detection and repair, while other studies have shown that PARP-1 plays an important role in chromatin and transcriptional regulation (Tulin et al. 2003; Kraus and Lis 2003). Furthermore, PARP activation has been detected in healthy subjects who are known to be at risk of developing diabetes plus in established T2D cases (Szabo et al. 2002). Combined with the facts that TCF7L2 and PARP-1 have been shown to be known protein partners (Idogawa et al. 2005), PARP-deficient mice are protected from streptozotocin-induced diabetes (Pieper et al. 1999), and PARP-1 inhibitors are well established to delay diabetic complications (Ilnytska et al. 2006; Li et al. 2005; Obrosova et al. 2004), these findings merit further investigation.

Finally, to understand how the transcriptional signal is transformed in an allele-specific manner from the enhancer region to the promoter, where the effect on polymerase should occur, may require studies like chromosomal conformational capture (4C).

14.5 Variant Correlation with Drug Response

Many therapeutic agents currently available for the treatment of T2D are often associated with a number of side effects, including an elevation in body mass index (BMI) and an increased risk of hypoglycemia and gastrointestinal complications. Furthermore, it is well known that these drugs are often not highly efficacious, with their optimal impact occurring in the early stages of T2D, where fasting blood glucose levels are relatively low. Indeed, approximately 40 % of patients with T2D who are prescribed oral antidiabetes agents fail to actually control their blood glucose and end up having to use insulin. Therefore it is abundantly clear that there is an unmet need for improved therapeutics with better response rates; indeed, this is where gene discovery efforts hold promise. In terms of the established T2D association with the *TCF7L2* locus, there have already been a number of exploratory efforts in this context.

Studies in humans have clearly shown that carriers of the risk T allele of the rs7903146 variant are characterized by impaired glucose- and incretin-stimulated insulin secretion, increased hepatic glucose production, and elevated circulating levels of proinsulin and GIP (Lyssenko et al. 2007; Loos et al. 2007; Pilgaard et al. 2009a; Villareal et al. 2010; Gjesing et al. 2010). The variant also predicts progression from normal or impaired glucose tolerance to diabetes (Florez et al. 2006; Lyssenko et al. 2008; Wang et al. 2007). Thereby, islet dysfunction seems to be the culprit of deteriorated glucose tolerance in risk T-allele carriers. This is also supported by findings of reduced glucose-stimulated insulin secretion and exocytosis (Rosengren et al. 2012), islet density (Le Bacquer et al. 2012), and insulin content (Zhou et al. 2014b) in islets of human carriers with the T risk allele.

Exploring how *TCF7L2* genotypic status influences drug response is a reasonable endeavor given the magnitude of its association with T2D. Indeed, it has been noted that subjects that have both impaired glucose tolerance and carry the *TCF7L2* risk variant are more likely to present with T2D, where the effect is more pronounced in untreated individuals than in metformin and lifestyle-intervention groups (Florez et al. 2006). Treatment options should also be considered in the context of the observation that the variant is primarily associated with decreased insulin secretion and not insulin resistance (Florez et al. 2006), suggesting impaired β-cell function. Interesting, clinic-based studies have revealed that *TCF7L2* risk variants are correlated with response to sulfonylureas but not with metformin among patients with T2D; however with this former treatment group, it was clear that the T2D-associated allele was also associated with a higher rate for sulfonylurea treatment failure (Pearson et al. 2007; Holstein et al. 2011) opening an opportunity for personalized therapeutics. These findings may also provide insight into how the *TCF7L2* locus exerts its effect on T2D risk, where sulfonylureas act on insulin production from pancreatic β-cells, while metformin influences glucose via the liver by improving insulin action (Kirpichnikov et al. 2002).

Similarly to insulin, the insulinotropic hormone, glucagon-like peptide 1 (GLP-1), exerts a strong influence on blood glucose homeostasis (Yi et al. 2005a); indeed, GLP-1 analogs and inhibitors of dipeptidyl peptidase IV are of intense clinical interest in the context of T2D. It is has also been speculated that TCF7L2 could influence levels of GLP-1 (Villareal et al. 2010; Vilsboll et al. 2003; Yu et al. 2009) because this transcription factor is known to regulate the expression of proglucagon in enteroendocrine cells. Furthermore, it has been reported that the TCF7L2 protein is detectable in β-cells where it may regulate expression of receptors to glucose-dependent insulinotropic polypeptide (GIP) and GLP-1 (Lyssenko et al. 2007; Shu et al. 2009; Qu and Polychronakos 2007; Pilgaard et al. 2009b). TCF7L2-regulated incretins could also influence the brain to affect appetite regulation (Shao et al. 2013; Nobrega 2013).

As such there is clinical interest in understanding the role of the *TCF7L2* locus in the context of T2D in order to tailor better therapy for individuals, either at risk or already presenting with this disease.

14.6 What Can *TCF7L2* Variants Tell About the Genetic Origin of LADA?

As T1D and T2D both result from the metabolic consequences of suboptimal insulin action, with similar complications, an overlap in the genetic predisposition to these two diseases has been long suggested (Wilkin 2001). However none of the genes identified to date in each of these given disorders have been conclusively shown to be associated with the other disease (Schernthaner et al. 2001; Qu et al. 2008; Raj et al. 2009). And in particular, no evidence has been found for the role of the *TCF7L2* locus in T1D (Qu and Polychronakos 2007; Raj et al. 2009; Field et al. 2007), suggesting distinct biological mechanisms underpinning these two phenotypes.

"Latent autoimmune diabetes in adults" (LADA) was first described as an autoimmune form of diabetes with onset in adult age (over 35 years of age), characterized by presence of GAD autoantibodies and no absolute dependence of insulin therapy during the first 6 months of disease (Tuomi et al. 1993, 1999); as such, LADA patients often exhibit a clinical picture similar to T2D but also exhibit circulating islet autoantibodies similar to those seen in T1D (Alberti and Zimmet 1998), earning it the nickname "type 1.5 diabetes" (Schernthaner et al. 2001). Indeed, on average 8–10 % of patients suspected to be presenting with T2D are thought to be in fact misdiagnosed LADA cases.

A genetic relationship to T1D was demonstrated by showing that LADA patients shared an increased frequency of T1D-associated HLA genotypes with T1D (Tuomi et al. 1999) and LADA was considered by the World Health Organization to be a slowly progressing form of autoimmune type 1 diabetes. At that time it was not possible to discern a potential influence of a T2D genetic background on LADA, but this became possible with the discovery of the association between variation in *TCF7L2* and T2D (Grant et al. 2006).

Since then, several studies have shown also an increased frequency of the T allele of rs7903146 in patients with LADA (Cervin et al. 2008; Bakhtadze et al. 2008; Tuomi et al. 2014) suggesting that LADA is either an admixture of the two forms of diabetes rather than a subtype of T1D (Tuomi et al. 2014) or driven by some other form of mechanism at play (Grant et al. 2010).

Additionally, the *TCF7L2* rs7903146 T allele has been reported to be overrepresented in European LADA cases (Lukacs et al. 2012; Szepietowska et al. 2009). Furthermore, association of *TCF7L2* variation with low GAD autoantibody titer in LADA subjects has been reported (Zampetti et al. 2010). In addition, it was shown that the *TCF7L2* observation could be used to distinguish middle-aged from young antibody-positive patients (Bakhtadze et al. 2008).

Given the increasing prevalence of both T1D and T2D, a large number of patients should share the genetic susceptibility to both diabetic subforms and understanding how this will influence the phenotype may help to disentangle the phenotypic heterogeneity of diabetes. Indeed, diabetes genetics has clearly not been totally resolved, with additional breakthroughs in T1D and T2D becoming increasingly hard to come by. The more recent loci uncovered for T2D through consortia-based analyses (DIAbetes Genetics Replication And Meta-analysis (DIAGRAM) Consortium 2014) have revealed them all to have increasingly small effect sizes; however the fact that such studies do not typically account for LADA cases, that will invariably be present in the study populations, begs the question: Given what is observed with *TCF7L2*, could some of these association signals be due to the genetics of LADA rather than the genetics of T2D?

14.7 Pediatric Studies

The health of a child has implications for disease risk later in life. Indeed a genetic predisposition for T2D in adults may manifest itself in childhood traits, and through studying pediatric phenotypes, one can gain insight in to the early genetic actors for disease risk and therefore identify options for both prevention and early intervention. Epidemiological studies have clearly established that low birth weight confers risk for development of metabolic disease in adulthood (Whincup et al. 2008), possibly reflecting the important combination of interacting genetic, epigenetic, intrauterine, nutritional, developmental, hormonal, and environmental factors that play a role in this complex setting.

A study of over 15,000 individuals plus approximately 8000 mothers showed that each maternal risk-conferring allele of rs7903146 within *TCF7L2* increased birth weight by 31 g; indeed, when combined with the alleles at the *GCK* locus, this birth weight increase went up to 119 g (Freathy et al. 2007). This observation supports the "fetal insulin hypothesis" where it has been proposed that the intrauterine insulin environment has implications for birth weight and subsequent disease risk in later life (Hattersley and Tooke 1999); indeed, a maternal T2D risk genotype might lead to maternal hyperglycemia, causing fetal hyperinsulinemia and macrosomia, whereas a fetal T2D risk genotype might lead to lower insulin levels in the fetus and lower growth but eventual T2D later in life. However, some follow-up studies have not observed an effect of fetal *TCF7L2* genotype on fetal and postnatal growth (Mook-Kanamori et al. 2009) or an interaction between *TCF7L2* genotype and birth weight that ultimately influences disease risk (Pulizzi et al. 2009). Other studies in pediatric populations, albeit with more limited numbers, did suggest an interaction between *TCF7L2* variation (Dabelea et al. 2011), lifestyle modification in obese children (de Kort et al. 2010), and growth hormone therapy in lean children born small for gestational age (Reinehr et al. 2008); however larger studies with greater phenotypic depth and ideally longitudinal data are required to fully validate these potentially interesting findings.

Of particular note is that the same genetic variants of *TCF7L2* that drive T2D risk also increase susceptibility to cystic fibrosis-related diabetes (CFRD) (Blackman et al. 2009, 2013; Furgeri et al. 2012), with the data suggesting that the T allele of rs7903146 in fact exerts a more pronounced effect in this setting. As such, a therapeutic developed in the context of the *TCF7L2* locus may have a particularly beneficial effect for CFRD cases.

Given the reported association in LADA patients, who often express a single islet autoantibody, one study found in children with autoimmune T1D that the *TCF7L2* rs7903146 T allele was more frequent among cases with a single islet autoantibody than in those with $\geq$2 islet autoantibodies (Redondo et al. 2014). These results resonated with a previous study in children which reported that this allele is more frequent in autoantibody-negative than autoantibody-positive T1D children (Yu et al. 2009). However, an investigation in nondiabetic children showed that the *TCF7L2* rs7903146 T allele did not increase the risk of islet autoantibody expression (Winkler et al. 2012).

14.8 *TCF7L2* Also a Cancer Gene: Possible Connections?

Before it was implicated in the pathogenesis of T2D, *TCF7L2* (formerly known as *TCF4*) was already a well-characterized cancer gene. Indeed, missense mutations in *TCF7L2* have been known for some time to be playing a role in conferring colorectal cancer risk (Duval et al. 2000a, b; Yochum et al. 2007). Furthermore, the TCF7L2 protein regulates the expression of several genes, including those that encode cyclin D1 (*CCND1*) and c-myc (*MYC*) which are involved in the control of the G1 to S phase transition in the cell cycle (Baker et al. 2000; Tetsu and McCormick 1999).

Indeed, this connection intensified following reports that the 8q24 locus revealed by GWAS of a number of cancers, including colorectal carcinomas (Duval et al. 1999, 2000b), was due to an extreme upstream TCF7L2-binding element driving the transcription of *MYC* (Pomerantz et al. 2009; Tuupanen et al. 2009; Sur et al. 2012). In addition, it has been shown that when *TCF7L2* recurrently fuses with its neighboring gene, *VTI1A*, colorectal adenocarcinomas result. Furthermore, a recent GWAS of breast cancer found four *TCF7L2* SNPs (including rs7903146) significantly associated with the trait (Connor et al. 2012). Curiously, many of the T2D GWAS-derived risk-conferring alleles have now also been shown to protect against prostate cancer (Frayling et al. 2008), including *THADA*, *JAZF1*, and *HNF1B* (formerly *TCF2*) (Frayling et al. 2008; Gudmundsson et al. 2007; Thomas et al. 2008; Zeggini et al. 2008b). Thus, some of the T2D associated loci, in particular *TCF7L2*, also appear to influence cancer pathogenesis and therefore there may be a specific yin and yang relationship between cancer and T2D.

Although there are apparent genetic links between cancer and T2D, the importance of this relationship is still far from understood. Given that TCF7L2 regulates *MYC* at the key cancer GWAS-implicated locus, it is conceivable that TCF7L2 in fact regulates the *TCF7L2* gene itself. Functional assay work using a reporter system showed that TCF7L2 and β-catenin are both required for canonical Wnt pathway signaling and that Wnt signaling plays a role in the regulation of *TCF7L2* expression itself (Xia et al. 2014), suggesting that rs7903146 may perturb the function of *TCF7L2* in the context of conferring T2D risk.

14.9 Summary/Conclusions

Large-scale GWAS efforts have transformed the way investigators search for loci involved in the pathogenesis of complex traits. Indeed, the past 7 years have revealed multiple loci robustly associated with T2D, with *TCF7L2* having the largest effect size. However, as illustrated by the challenges in elucidating how this key locus affects T2D risk functionally, it is clear that it will take some time to decipher the mechanism of action of all the T2D loci uncovered to date and the optimal route to genetically informed therapeutic intervention (Visscher et al. 2012).

As there is great demand for improved therapeutic strategies for T2D, work on key gene products like TCF7L2 holds great promise for more individualized treatments in the future. Clearly, in addition to careful functional appraisals, more detailed genotype-phenotype studies, such as those seen with *TCF7L2*, are going to be needed to be carried out to elucidate both the mechanisms of action involved and how they might be involved in other disorders. Only once we've conquered these challenges, with perhaps work on the *TCF7L2* locus leading the way, can such work ultimately benefit actual patient care.

References

Ahlzen M et al (2008) Expression of the transcription factor 7-like 2 gene (TCF7L2) in human adipocytes is down regulated by insulin. Biochem Biophys Res Commun 370:49–52

Alberti KG, Zimmet PZ (1998) Definition, diagnosis and classification of diabetes mellitus and its complications. Part 1: diagnosis and classification of diabetes mellitus provisional report of a WHO consultation. Diabet Med 15:539–553

Altshuler D, Daly MJ, Lander ES (2008) Genetic mapping in human disease. Science 322:881–888

Baker N, Morin PJ, Clevers H (2000) The Ying-Yang of TCF/beta-catenin signaling. Adv Cancer Res 77:1–24

Bakhtadze E et al (2008) Common variants in the TCF7L2 gene help to differentiate autoimmune from non-autoimmune diabetes in young (15–34 years) but not in middle-aged (40–59 years) diabetic patients. Diabetologia 51:2224–2232

Blackman SM et al (2009) A susceptibility gene for type 2 diabetes confers substantial risk for diabetes complicating cystic fibrosis. Diabetologia 52:1858–1865

Blackman SM et al (2013) Genetic modifiers of cystic fibrosis-related diabetes. Diabetes 62:3627–3635

Boj SF et al (2012) Diabetes risk gene and Wnt effector Tcf7l2/TCF4 controls hepatic response to perinatal and adult metabolic demand. Cell 151:1595–1607

Cauchi S et al (2007) TCF7L2 is reproducibly associated with type 2 diabetes in various ethnic groups: a global meta-analysis. J Mol Med (Berl) 85:777–782

Cervin C et al (2008) Genetic similarities between latent autoimmune diabetes in adults, type 1 diabetes, and type 2 diabetes. Diabetes 57:1433–1437

Chang YC et al (2007) Association study of the genetic polymorphisms of the transcription factor 7-like 2 (TCF7L2) gene and type 2 diabetes in the Chinese population. Diabetes 56:2631–2637

Cho YS et al (2011) Meta-analysis of genome-wide association studies identifies eight new loci for type 2 diabetes in east Asians. Nat Genet 44:67–72

Clayton DG (2009) Prediction and interaction in complex disease genetics: experience in type 1 diabetes. PLoS Genet 5, e1000540

Clevers H (2006) Wnt/beta-catenin signaling in development and disease. Cell 127:469–480

Connor AE et al (2012) Associations between TCF7L2 polymorphisms and risk of breast cancer among Hispanic and non-Hispanic white women: the Breast Cancer Health Disparities Study. Breast Cancer Res Treat 136:593–602 da Silva Xavier G et al (2009) TCF7L2 regulates late events in insulin secretion from pancreatic islet beta-cells. Diabetes 58:894–905

da Silva Xavier G et al (2012) Abnormal glucose tolerance and insulin secretion in pancreas-specific Tcf7l2-null mice. Diabetologia 55:2667–2676

Dabelea D et al (2011) Association testing of *TCF7L2* polymorphisms with type 2 diabetes in multi-ethnic youth. Diabetologia 54:535–539

de Kort SW, Mook-Kanamori DO, Jaddoe VW, Hokken-Koelega AC (2010) Interactions between TCF7L2 genotype and growth hormone-induced changes in glucose homeostasis in small for gestational age children. Clin Endocrinol (Oxf) 72:47–52

DIAbetes Genetics Replication And Meta-analysis (DIAGRAM) Consortium (2014) Genome-wide trans-ancestry meta-analysis provides insight into the genetic architecture of type 2 diabetes susceptibility. Nat Genet 46:234–244

Dupuis J et al (2010) New genetic loci implicated in fasting glucose homeostasis and their impact on type 2 diabetes risk. Nat Genet 42:105–116

Duval A et al (1999) Frequent frameshift mutations of the TCF-4 gene in colorectal cancers with microsatellite instability. Cancer Res 59:4213–4215

Duval A, Busson-Leconiat M, Berger R, Hamelin R (2000a) Assignment of the TCF-4 gene (TCF7L2) to human chromosome band 10q25.3. Cytogenet Cell Genet 88:264–265

Duval A et al (2000b) The human T-cell transcription factor-4 gene: structure, extensive characterization of alternative splicings, and mutational analysis in colorectal cancer cell lines. Cancer Res 60:3872–3879

Field SF et al (2007) Analysis of the type 2 diabetes gene, TCF7L2, in 13,795 type 1 diabetes cases and control subjects. Diabetologia 50:212–213

Florez JC et al (2006) TCF7L2 polymorphisms and progression to diabetes in the Diabetes Prevention Program. N Engl J Med 355:241–250

Frayling TM, Colhoun H, Florez JC (2008) A genetic link between type 2 diabetes and prostate cancer. Diabetologia 51:1757–1760

Freathy RM et al (2007) Type 2 diabetes TCF7L2 risk genotypes alter birth weight: a study of 24,053 individuals. Am J Hum Genet 80:1150–1161

Furgeri DT, Marson FA, Ribeiro AF, Bertuzzo CS (2012) Association between the IVS4G > T mutation in the TCF7L2 gene and susceptibility to diabetes in cystic fibrosis patients. BMC Res Notes 5:561

Gaulton KJ et al (2010) A map of open chromatin in human pancreatic islets. Nat Genet 42:255–259

Gjesing AP et al (2010) Carriers of the TCF7L2 rs7903146 TT genotype have elevated levels of plasma glucose, serum proinsulin and plasma gastric inhibitory polypeptide (GIP) during a meal test. Diabetologia 54:103–110

Grant SF et al (2006) Variant of transcription factor 7-like 2 (TCF7L2) gene confers risk of type 2 diabetes. Nat Genet 38:320–323

Grant SF, Hakonarson H, Schwartz S (2010) Can the genetics of type 1 and type 2 diabetes shed light on the genetics of latent autoimmune diabetes in adults? Endocr Rev 31:183–193

Gudmundsson J et al (2007) Two variants on chromosome 17 confer prostate cancer risk, and the one in *TCF2* protects against type 2 diabetes. Nat Genet 39:977–983

Hansson O, Zhou Y, Renström E, Osmark P (2010) Molecular function of TCF7L2: consequences of TCF7L2 splicing for molecular function and risk for type 2 diabetes. Curr Diab Rep 10:444–451

Hattersley AT (2007) Prime suspect: the TCF7L2 gene and type 2 diabetes risk. J Clin Invest 117:2077–2079

Hattersley AT, Tooke JE (1999) The fetal insulin hypothesis: an alternative explanation of the association of low birthweight with diabetes and vascular disease. Lancet 353:1789–1792

Helgason A et al (2007a) Refining the impact of *TCF7L2* gene variants on type 2 diabetes and adaptive evolution. Nat Genet 39:218–225

Helgason A et al (2007b) Refining the impact of TCF7L2 gene variants on type 2 diabetes and adaptive evolution. Nat Genet
Holstein A, Hahn M, Korner A, Stumvoll M, Kovacs P (2011) TCF7L2 and therapeutic response to sulfonylureas in patients with type 2 diabetes. BMC Med Genet 12:30
Idogawa M et al (2005) Poly(ADP-ribose) polymerase-1 is a component of the oncogenic T-cell factor-4/beta-catenin complex. Gastroenterology 128:1919–1936
Ilnytska O et al (2006) Poly(ADP-ribose) polymerase inhibition alleviates experimental diabetic sensory neuropathy. Diabetes 55:1686–1694
Kaminska D et al (2012) Adipose tissue TCF7L2 splicing is regulated by weight loss and associates with glucose and fatty acid metabolism. Diabetes 61:2807–2813
Kirpichnikov D, McFarlane SI, Sowers JR (2002) Metformin: an update. Ann Intern Med 137:25–33
Kooner JS et al (2011) Genome-wide association study in individuals of South Asian ancestry identifies six new type 2 diabetes susceptibility loci. Nat Genet 43:984–989
Korinek V et al (1998) Depletion of epithelial stem-cell compartments in the small intestine of mice lacking Tcf-4. Nat Genet 19:379–383
Kraus WL, Lis JT (2003) PARP goes transcription. Cell 113:677–683
Le Bacquer O et al (2012) TCF7L2 rs7903146 impairs islet function and morphology in non-diabetic individuals. Diabetologia 55:2677–2681
Li F, Drel VR, Szabo C, Stevens MJ, Obrosova IG (2005) Low-dose poly(ADP-ribose) polymerase inhibitor-containing combination therapies reverse early peripheral diabetic neuropathy. Diabetes 54:1514–1522
Loos RJ et al (2007) TCF7L2 polymorphisms modulate proinsulin levels and beta-cell function in a British Europid population. Diabetes 56:1943–1947
Lukacs K et al (2012) The type 2 diabetes-associated variant in TCF7L2 is associated with latent autoimmune diabetes in adult Europeans and the gene effect is modified by obesity: a meta-analysis and an individual study. Diabetologia 55:689–693
Lyssenko V et al (2007) Mechanisms by which common variants in the TCF7L2 gene increase risk of type 2 diabetes. J Clin Invest 117:2155–2163
Lyssenko V et al (2008) Clinical risk factors, DNA variants, and the development of type 2 diabetes. N Engl J Med 359:2220–2232
Maller JB et al (2012) Bayesian refinement of association signals for 14 loci in 3 common diseases. Nat Genet 44:1294–1301
Manolio TA, Brooks LD, Collins FS (2008) A HapMap harvest of insights into the genetics of common disease. J Clin Invest 118:1590–1605
McCarthy MI et al (2008) Genome-wide association studies for complex traits: consensus, uncertainty and challenges. Nat Rev Genet 9:356–369
Mondal AK et al (2010) Genotype and tissue-specific effects on alternative splicing of the transcription factor 7-like 2 gene in humans. J Clin Endocrinol Metab 95:1450–1457
Mook-Kanamori DO et al (2009) Type 2 diabetes gene TCF7L2 polymorphism is not associated with fetal and postnatal growth in two birth cohort studies. BMC Med Genet 10:67
Morris AP et al (2012) Large-scale association analysis provides insights into the genetic architecture and pathophysiology of type 2 diabetes. Nat Genet
Ng MC et al (2007) Replication and identification of novel variants at TCF7L2 associated with type 2 diabetes in Hong Kong Chinese. J Clin Endocrinol Metab 92:3733–3737
Ng MC et al (2008) Implication of genetic variants near TCF7L2, SLC30A8, HHEX, CDKAL1, CDKN2A/B, IGF2BP2, and FTO in type 2 diabetes and obesity in 6,719 Asians. Diabetes 57:2226–2233
Nobrega MA (2013) TCF7L2 and glucose metabolism: time to look beyond the pancreas. Diabetes 62:706–708
Norton L et al (2011) Chromatin occupancy of transcription factor 7-like 2 (TCF7L2) and its role in hepatic glucose metabolism. Diabetologia 54:3132–3142
Nusse R, Varmus HE (1992) Wnt genes. Cell 69:1073–1087

Obrosova IG et al (2004) Role of poly(ADP-ribose) polymerase activation in diabetic neuropathy. Diabetes 53:711–720

Osmark P et al (2009) Unique splicing pattern of the TCF7L2 gene in human pancreatic islets. Diabetologia 52:850–854

Palmer ND et al (2011) Resequencing and analysis of variation in the TCF7L2 gene in African Americans suggests that SNP rs7903146 is the causal diabetes susceptibility variant. Diabetes 60:662–668

Paul S, Dey A (2008) Wnt signaling and cancer development: therapeutic implication. Neoplasma 55:165–176

Pearson ER et al (2007) Variation in TCF7L2 influences therapeutic response to sulfonylureas: a GoDARTs study. Diabetes 56:2178–2182

Pieper AA et al (1999) Poly(ADP-ribose) polymerase-deficient mice are protected from streptozotocin-induced diabetes. Proc Natl Acad Sci U S A 96:3059–3064

Pilgaard K et al (2009) The T allele of rs7903146 *TCF7L2* is associated with impaired insulinotropic action of incretin hormones, reduced 24 h profiles of plasma insulin and glucagon, and increased hepatic glucose production in young healthy men. Diabetologia 52:1298–1307

Pomerantz MM et al (2009) The 8q24 cancer risk variant rs6983267 shows long-range interaction with MYC in colorectal cancer. Nat Genet 41:882–884

Prokunina-Olsson L, Hall JL (2010) Evidence for neuroendocrine function of a unique splicing form of TCF7L2 in human brain, islets and gut. Diabetologia 53:712–716

Prokunina-Olsson L, Kaplan LM, Schadt EE, Collins FS (2009a) Alternative splicing of TCF7L2 gene in omental and subcutaneous adipose tissue and risk of type 2 diabetes. PLoS ONE 4, e7231

Prokunina-Olsson L et al (2009b) Tissue-specific alternative splicing of TCF7L2. Hum Mol Genet 18:3795–3804

Pulizzi N et al (2009) Interaction between prenatal growth and high-risk genotypes in the development of type 2 diabetes. Diabetologia 52:825–829

Qu HQ, Polychronakos C (2007) The TCF7L2 locus and type 1 diabetes. BMC Med Genet 8:51

Qu HQ et al (2008) Association analysis of type 2 diabetes Loci in type 1 diabetes. Diabetes 57:1983–1986

Raj SM et al (2009) No association of multiple type 2 diabetes loci with type 1 diabetes. Diabetologia

Redondo MJ et al (2014) Association of TCF7L2 variation with single islet autoantibody expression in children with type 1 diabetes. BMJ Open Diabetes Res Care 2, e000008

Reinehr T et al (2008) Evidence for an influence of TCF7L2 polymorphism rs7903146 on insulin resistance and sensitivity indices in overweight children and adolescents during a lifestyle intervention. Int J Obes (Lond) 32:1521–1524

Ren Q et al (2008) Exon sequencing and association analysis of polymorphisms in TCF7L2 with type 2 diabetes in a Chinese population. Diabetologia 51:1146–1152

Reynisdottir I et al (2003) Localization of a susceptibility gene for type 2 diabetes to chromosome 5q34-q35.2. Am J Hum Genet 73:323–335

Rich SS (1990) Mapping genes in diabetes. Genetic epidemiological perspective. Diabetes 39:1315–1319

Rosengren AH et al (2012) Reduced insulin exocytosis in human pancreatic beta-cells with gene variants linked to type 2 diabetes. Diabetes 61:1726–1733

Rung J et al (2009) Genetic variant near *IRS1* is associated with type 2 diabetes, insulin resistance and hyperinsulinemia. Nat Genet 41:1110–1115

Salonen JT et al (2007) Type 2 diabetes whole-genome association study in four populations: the DiaGen consortium. Am J Hum Genet 81:338–345

Savic D et al (2011) Alterations in TCF7L2 expression define its role as a key regulator of glucose metabolism. Genome Res 21:1417–1425

Savic D, Park SY, Bailey KA, Bell GI, Nobrega MA (2012) In vitro scan for enhancers at the TCF7L2 locus. Diabetologia 56:121–125

Savic D, Park SY, Bailey KA, Bell GI, Nobrega MA (2013) In vitro scan for enhancers at the TCF7L2 locus. Diabetologia 56:121–125

Saxena R et al (2007) Genome-wide association analysis identifies loci for type 2 diabetes and triglyceride levels. Science 316:1331–1336

Schernthaner G et al (2001) Progress in the characterization of slowly progressive autoimmune diabetes in adult patients (LADA or type 1.5 diabetes). Exp Clin Endocrinol Diabetes 109 (Suppl 2):S94–S108

Scott LJ et al (2007) A genome-wide association study of type 2 diabetes in Finns detects multiple susceptibility variants. Science 316:1341–1345

Shao W et al (2012) The Wnt signaling pathway effector TCF7L2 controls gut and brain proglucagon gene expression and glucose homeostasis. Diabetes 62:789–800

Shao W et al (2013) The Wnt signaling pathway effector TCF7L2 controls gut and brain proglucagon gene expression and glucose homeostasis. Diabetes 62:789–800

Shu L et al (2008) Transcription factor 7-like 2 regulates beta-cell survival and function in human pancreatic islets. Diabetes 57:645–653

Shu L et al (2009) Decreased TCF7L2 protein levels in type 2 diabetes mellitus correlate with downregulation of GIP- and GLP-1 receptors and impaired beta-cell function. Hum Mol Genet 18:2388–2399

Sladek R et al (2007) A genome-wide association study identifies novel risk loci for type 2 diabetes. Nature 445:881–885

Steinthorsdottir V et al (2007) A variant in *CDKAL1* influences insulin response and risk of type 2 diabetes. Nat Genet 39:770–775

Stitzel ML et al (2010) Global epigenomic analysis of primary human pancreatic islets provides insights into type 2 diabetes susceptibility loci. Cell Metab 12:443–455

Sur IK et al (2012) Mice lacking a Myc enhancer that includes human SNP rs6983267 are resistant to intestinal tumors. Science 338:1360–1363

Szabo C et al (2002) Poly(ADP-Ribose) polymerase is activated in subjects at risk of developing type 2 diabetes and is associated with impaired vascular reactivity. Circulation 106:2680–2686

Szepietowska B et al (2009) Transcription factor 7-like 2-gene polymorphism is related to fasting C peptide in latent autoimmune diabetes in adults (LADA). Acta Diabetol 47(1):83–86

Takamoto I et al (2014) TCF7L2 in mouse pancreatic beta cells plays a crucial role in glucose homeostasis by regulating beta cell mass. Diabetologia 57:542–553

Taneera J et al (2012) A systems genetics approach identifies genes and pathways for type 2 diabetes in human islets. Cell Metab 16:122–134

Tetsu O, McCormick F (1999) Beta-catenin regulates expression of cyclin D1 in colon carcinoma cells. Nature 389:422–426

Thomas G et al (2008) Multiple loci identified in a genome-wide association study of prostate cancer. Nat Genet 40:310–315

Tomimaru Y et al (2013) Loss of exon 4 in a human T-cell factor-4 isoform promotes hepatic tumourigenicity. Liver Int 33:1536–1548

Tong Y et al (2009) Association between TCF7L2 gene polymorphisms and susceptibility to type 2 diabetes mellitus: a large Human Genome Epidemiology (HuGE) review and meta-analysis. BMC Med Genet 10:15

Tulin A, Chinenov Y, Spradling A (2003) Regulation of chromatin structure and gene activity by poly(ADP-ribose) polymerases. Curr Top Dev Biol 56:55–83

Tuomi T et al (1993) Antibodies to glutamic acid decarboxylase reveal latent autoimmune diabetes mellitus in adults with a non-insulin-dependent onset of disease. Diabetes 42:359–362

Tuomi T et al (1999) Clinical and genetic characteristics of type 2 diabetes with and without GAD antibodies. Diabetes 48:150–157

Tuomi T et al (2014) The many faces of diabetes: a disease with increasing heterogeneity. Lancet 383:1084–1094

Tuupanen S et al (2009) The common colorectal cancer predisposition SNP rs6983267 at chromosome 8q24 confers potential to enhanced Wnt signaling. Nat Genet 41:885–890

Unoki H et al (2008) SNPs in *KCNQ1* are associated with susceptibility to type 2 diabetes in East Asian and European populations. Nat Genet 40:1098–1102

Villareal DT et al (2010) *TCF7L2* variant rs7903146 affects the risk of type 2 diabetes by modulating incretin action. Diabetes 59:479–485

Vilsboll T et al (2003) Incretin secretion in relation to meal size and body weight in healthy subjects and people with type 1 and type 2 diabetes mellitus. J Clin Endocrinol Metab 88:2706–2713

Visscher PM, Brown MA, McCarthy MI, Yang J (2012) Five years of GWAS discovery. Am J Hum Genet 90:7–24

Voight BF et al (2010) Twelve type 2 diabetes susceptibility loci identified through large-scale association analysis. Nat Genet 42:579–589

Wang J et al (2007) Variants of transcription factor 7-like 2 (*TCF7L2*) gene predict conversion to type 2 diabetes in the Finnish Diabetes Prevention Study and are associated with impaired glucose regulation and impaired insulin secretion. Diabetologia 50:1192–1200

Weedon MN (2007) The importance of TCF7L2. Diabet Med 24:1062–1066

Wellcome Trust Case Control Consortium (2007) Genome-wide association study of 14,000 cases of seven common diseases and 3,000 shared controls. Nature 447:661–678

Whincup PH et al (2008) Birth weight and risk of type 2 diabetes: a systematic review. JAMA 300:2886–2897

Wilkin TJ (2001) The accelerator hypothesis: weight gain as the missing link between Type I and Type II diabetes. Diabetologia 44:914–922

Winkler C, Raab J, Grallert H, Ziegler AG (2012) Lack of association of type 2 diabetes susceptibility genotypes and body weight on the development of islet autoimmunity and type 1 diabetes. PLoS ONE 7, e35410

Xia Q, Deliard S, Yuan CX, Johnson ME, Grant SF (2014) Characterization of the transcriptional machinery bound across the widely presumed type 2 diabetes causal variant, rs7903146, within TCF7L2. Eur J Hum Genet

Yasuda K et al (2008) Variants in *KCNQ1* are associated with susceptibility to type 2 diabetes mellitus. Nat Genet 40:1092–1097

Yi F, Brubaker PL, Jin T (2005) TCF-4 mediates cell type-specific regulation of proglucagon gene expression by beta-catenin and glycogen synthase kinase-3beta. J Biol Chem 280:1457–1464

Yochum GS et al (2007) Serial analysis of chromatin occupancy identifies beta-catenin target genes in colorectal carcinoma cells. Proc Natl Acad Sci U S A 104:3324–3329

Yu J et al (2009) Single nucleotide transcription factor 7-like 2 (TCF7L2) gene polymorphisms in antiislet autoantibody-negative patients at onset of diabetes. J Clin Endocrinol Metab 94:504–510

Yu M et al (2010) KCNJ11 Lys23Glu and TCF7L2 rs290487(C/T) polymorphisms affect therapeutic efficacy of repaglinide in Chinese patients with type 2 diabetes. Clin Pharmacol Ther 87:330–335

Zampetti S et al (2010) Association of TCF7L2 gene variants with low GAD autoantibody titre in LADA subjects (NIRAD Study 5). Diabet Med 27:701–704

Zeggini E, McCarthy MI (2007) *TCF7L2*: the biggest story in diabetes genetics since HLA? Diabetologia 50:1–4

Zeggini E et al (2007) Replication of genome-wide association signals in UK samples reveals risk loci for type 2 diabetes. Science 316:1336–1341

Zeggini E et al (2008) Meta-analysis of genome-wide association data and large-scale replication identifies additional susceptibility loci for type 2 diabetes. Nat Genet 40:638–645

Zhao J, Schug J, Li M, Kaestner KH, Grant SF (2010) Disease-associated loci are significantly over-represented among genes bound by transcription factor 7-like 2 (TCF7L2) in vivo. Diabetologia 53:2340–2346

Zheng X et al (2012) Association of type 2 diabetes susceptibility genes (TCF7L2, SLC30A8, PCSK1 and PCSK2) and proinsulin conversion in a Chinese population. Mol Biol Rep 39:17–23

Zhou Y et al (2012) Survival of pancreatic beta cells is partly controlled by a TCF7L2-p53-p53INP1-dependent pathway. Hum Mol Genet 21:196–207

Zhou Y et al (2014a) TCF7L2 is a master regulator of insulin production and processing. Hum Mol Genet

Zhou Y et al (2014b) Transcriptomic analysis reveals key regulators of mammogenesis and the pregnancy-lactation cycle. Sci China Life Sci 57:340–355

Chapter 15
GCKR: How Genetic Variation Across the Allelic Spectrum Influences Protein Function and Metabolic Traits in Humans

Marju Orho-Melander and Anna L. Gloyn

Abstract Genome-wide association studies (GWAS) have generated considerable interest in glucokinase regulatory protein (GKRP; gene name *GCKR*) which is an inhibitor of hepatic glucokinase (GCK), an enzyme that plays a critical role in glucose update and disposal in liver. From the initial discovery of *GCKR* variants associated with triglyceride and glucose levels through the identification of pleiotropic associations with a wide variety of metabolic phenotypes, we have learned a great deal about the importance of GKRP as a critical node in hepatic metabolism. GKRP remains one of the few well-studied GWAS loci where attempts have been made to understand the functional as well as the phenotypic impact of genetic variants across the allelic spectrum. Given the interest in developing liver-specific glucokinase activators and small molecules which disrupt the GKRP:GCK interaction for the treatment of type 2 diabetes, these genetic insights provide a wealth of information regarding efficacy and potential adverse on-target effects in humans.

15.1 Genome-Wide Association Studies Uncover a Locus on Chromosome 2 Associated with an Inverse Relationship Between Fasting Glucose and Triglyceride Levels

In 2007, one of the first genome-wide association studies (GWAS) performed by the Diabetes Genetics Initiative (DGI) for a total of 19 traits including plasma lipids and fasting glucose levels identified the intronic rs780094 single-nucleotide

M. Orho-Melander
Department of Clinical Sciences in Malmö, Lund University Diabetes Center, Lund University, Malmö, Sweden

A.L. Gloyn (✉)
Oxford Centre for Diabetes Endocrinology and Metabolism, University of Oxford, Churchill Hospital, Headington, Oxford OX3 7LE, UK
e-mail: Anna.Gloyn@drl.ox.ac.uk

J.C. Florez (ed.), *The Genetics of Type 2 Diabetes and Related Traits*,
DOI 10.1007/978-3-319-01574-3_15

polymorphism (SNP) in the gene encoding glucokinase regulatory protein (GKRP; gene name *GCKR*) as the first novel quantitative trait locus associated with plasma triglyceride concentrations in humans (Saxena et al. 2007). This finding has since then been consistently replicated in other large population cohorts (Kathiresan et al. 2008; Willer et al. 2008; Sparso et al. 2008; Scott et al. 2007; Vaxillaire et al. 2008). Although the rs780094 variant was in a large block of linkage disequilibrium (LD), spanning >400 kb and covering 16 protein-coding genes, the location of the SNP within the highly plausible biological candidate gene, *GCKR*, suggested that the causal effect could originate from this gene. Indeed, both imputation and genotype fine mapping of the *GCKR* locus highlighted the nonsynonymous coding variant (Pro446Leu; rs1260326) as the strongest association signal, suggesting that this coding variant as the causal variant (Orho-Melander et al. 2008). In fact these efforts at the *GCKR* locus were one of the first fine-mapping approaches for loci identified in GWAS and demonstrated that at least within regions of high LD, genotypes predicted by imputation were highly accurate and provided a good starting point for genotype fine mapping (Orho-Melander et al. 2008). Association studies in African-Americans, where rs780094 and Pro446Leu do not correlate as strongly as in other populations, provided further evidence for Pro446Leu, which showed a stronger association with triglycerides than rs780094 (Deo et al. 2009).

Prior to this discovery, it was already well recognized that GKRP regulates glucose disposal in the liver through its function as a negative regulator of hepatic glucokinase (GCK) activity. An earlier study of adenoviral-mediated overexpression of *Gckr* in mouse hepatocytes resulted in increased GCK activity and reduced fasting blood glucose levels in these animals (Slosberg et al. 2001). In line with this, two earlier studies had observed that overexpression of *Gck* in liver reduced fasting glucose levels but unfortunately this reduction in glucose was accompanied by elevated triglyceride levels (O'Doherty et al. 1999; Ferre et al. 1996). Furthermore, although *Gckr*$^{-/-}$deficient mice had reduced *Gck* expression and activity at basal glucose levels following glucose stimulation, they displayed impaired glucose clearance due to insufficient hepatic nuclear Gck reserves (Grimsby et al. 2000). Encouraged by these earlier findings, associations between rs780094 and glucose homeostasis were investigated in both the DGI and replication samples, and consistent with these earlier studies in mice, the triglyceride-raising (T) allele of rs780094 was associated with lower fasting glucose levels, improved insulin sensitivity, and a reduced risk of type 2 diabetes (T2D) (Saxena et al. 2007). These opposing effects on glucose and triglyceride levels were subsequently robustly replicated in a Danish study (Sparso et al. 2008) and in a number of other population-based cohorts including those from outside Europe (Orho-Melander et al. 2008; Vaxillaire et al. 2008; Teslovich et al. 2010; Johansen et al. 2010).

The association between *GCKR* variants with multiple glucose-related metabolic traits is one of the most robust observations and has been confirmed in large GWAS meta-analyses: the triglyceride-raising allele at this locus is associated with lower plasma glucose concentrations, improved insulin sensitivity, and decreased

T2D risk (Saxena et al. 2010; Dupuis et al. 2010). However, despite strong effects on plasma triglycerides, *GCKR* variants do not associate with an increased risk for CAD (Teslovich et al. 2010; Varbo et al. 2011). This special pattern of associations suggests a pleiotropic role of GKRP in linking carbohydrate and lipoprotein metabolism in a fashion that was previously unrecognized in humans.

15.2 Evidence for Further Pleiotropic Associations

Pleiotropy can be described as the effect of a single genetic variant or genetic region on multiple phenotypic traits. As outlined above, the possibility of pleiotropic associations at the *GCKR* locus was first proposed by the DGI study where the triglyceride-increasing allele had an opposite associated effect on fasting glucose and insulin resistance by homeostasis model assessment (HOMA-IR) (Saxena et al. 2007). It was rapidly recognized that the pleiotropic effects of this locus were not restricted to glucose and lipid metabolism (Table 15.1). In 2008, Ridker et al. published a GWAS for C-reactive protein (CRP) levels, and one of the associated loci was *GCKR*, where the Pro446Leu T allele (Leu) was associated with higher CRP levels (Ridker et al. 2008). Elevated CRP levels are a known hallmark of low-grade systemic inflammation, but the link between GKRP and inflammation had not been previously recognized. This finding has been widely replicated in multiple ancestries and the CRP association has been shown to be independent of triglyceride levels (Orho-Melander et al. 2008; Dehghan et al. 2011; Reiner et al. 2012).

Further evidence for pleiotropy at the *GCKR* locus comes from GWAS for uric acid (urate) levels (Kolz et al. 2009; Yang et al. 2010). Serum uric acid concentrations are mainly determined by the balance between production and renal excretion of urate. Genetic associations with serum urate levels could therefore originate from defects in purine metabolism and renal function or could simply represent a proxy for another highly correlated trait. Insulin resistance has been proposed as one such potential metabolic link between these traits and serum uric acid concentrations. However, since the *GCKR* rs780094 T-allele associated with higher urate levels is also associated with reduced insulin resistance, it is unlikely that insulin resistance is driving this association and therefore mostly likely alternative mechanisms are involved. Indeed, one study which replicated the association between *GCKR* variants and uric acid levels also observed that adjustment for triglycerides levels considerably attenuated the association with uric acid concentrations while the association with triglyceride levels remained highly significant when adjusted for both uric acid concentration and fractional uric acid excretion (van der Harst et al. 2010). Of all the loci associated with uric acid concentrations, *GCKR* is unique in that it is also associated with triglyceride levels. These observations are consistent with the current understanding of the function of GKRP as both a regulator of GCK activity and intracellular localization. Glucose-6-phosphate (G6P) is a known precursor for both hepatic glycogen synthesis and de novo purine synthesis and is produced from glucose by GCK, which in turn is regulated by GKRP. This

Table 15.1 Pleiotropic GWAS significant associations of *GCKR* locus

Phenotype	Variant	Minor allele	Direct effect of minor allele	References
Plasma triglycerides	rs780094 rs1260326/ (Pro446Leu)	T T	+	Saxena et al. (2007), Orho-Melander et al. (2008), Kathiresan et al. (2008), Sabatti et al. (2009)
CRP	rs1260326	T	+	Ridker et al. (2008)
Fasting glucose	rs780094	T	–	Orho-Melander et al. (2008), Dupuis et al. (2010)
Serum uric acid	rs780094	T	+	Kolz et al. (2009), Yang et al. (2010)
Fasting insulin	rs780094	T	–	Dupuis et al. (2010)
Type 2 diabetes	rs780094	T	–	Dupuis et al. (2010)
HOMA	rs780094	T	–	Dupuis et al. (2010)
Serum creatinine	rs1260326	T	+	Kottgen et al. (2010)
Serum albumin	rs1260326	T	+	Kamatani et al. (2010), Franceschini et al. (2012)
Nonalcoholic fatty liver disease	rs780094	T	+	Speliotes et al. (2011)
γ-Glutamyl transpeptidase (GTT)	rs1260326	T	+	Chambers et al. (2011)
Serum calcium	rs780094	T	+	O'Seaghdha et al. (2013)

mechanism is also supported by symptoms of glycogen storage disease type 1 (von Gierke disease), a genetic deficiency of glucose-6-phosphatase, which leads to accumulation of G6P followed by hypoglycemia and excess glycogen storage but also hypertriglyceridemia and hyperuricemia (Yang Chou and Mansfield 1999).

Creatine levels have also been reported to be associated with variants *GCKR* locus. In 2010, the chronic kidney disease consortium CKDGen performed a meta-analysis of genome-wide association data in 67,093 Caucasian individuals from 20 population-based studies and identified several new susceptibility loci, including *GCKR*, for reduced renal function, estimated by serum creatinine and cystatin C levels (Kottgen et al 2010).

In 2011, the *GCKR* locus was identified as a region associated with nonalcoholic fatty liver disease (NAFLD) (Speliotes et al. 2011). Further support for the importance of GKRP in hepatic metabolism came from Chambers et al., who carried out a GWAS in 61,089 study participants to identify genetic loci influencing liver function measured by concentrations of blood alanine transaminase (ALT), alkaline phosphatase (ALP), and γ-glutamyl transferase (GGT) (Chambers et al. 2011). Several loci were found associated with liver enzymes including *GCKR* which was associated with both GGT and ALP levels.

Furthermore, the triglyceride-raising T allele (Leu) of the rs1260326 (Pro446Leu) variant was also associated with HDL, IDL, and LDL cholesterol levels, VLDL particle size, omega-3 and omega-6 fatty acid levels, and concentrations of the metabolic substrates citrate, pyruvate, and branched-chain amino acids. These data generated by NMR confirmed and extended previous mass spectroscopy-based studies, which showed strong association of *GCKR* variants with absolute and relative abundances of polyunsaturated fatty acids and hepatic steatosis (Gieger et al. 2008; Illig et al. 2010; Chambers et al. 2011).

Variants at the *GCKR* locus have also been reported to associate with serum albumin levels and this association is independent of effects on triglyceride levels (Kamatani et al. 2010; Franceschini et al. 2012). The association with albumin levels is interesting as a wide range of conditions including liver and kidney diseases, acute and chronic inflammatory states, and cancer manifest as reduced plasma albumin concentrations.

More recently, O'Seaghdha and coworkers identified *GCKR* as one of six novel loci associated with serum calcium levels (O'Seaghdha et al. 2010). Calcium is vital to many biological processes and its serum concentration is tightly regulated by three major hormones acting on their corresponding receptors in the gut, kidney, and bone: parathyroid hormone, calcitonin, and the active metabolite of vitamin D, 1,25(OH)2-D. Consistent with the association with calcium levels, the calcium-raising allele at rs780094 is also nominally associated with increased bone mineral density (BMD) at the lumbar spine (O'Seaghdha et al. 2013). Since *Gckr* is not actually expressed in any primary calcium handling organs (the duodenum, kidney, and bone), and the association between *GCKR* variants and serum calcium concentration is attenuated after adjustment for albumin levels, it is likely that the association with calcium levels is a consequence of effects on albumin levels (Franceschini et al. 2012).

Based on published GWAS data, *GCKR* represents a very pleiotropic locus with both direct and indirect effects on many human physiological systems (Table 15.1). In the past, pleiotropic genes have often been recognized to either be involved in environmental adaptation or to reside in central node positions in protein–protein interaction networks (Foster et al. 2004; Zou et al. 2008). Recently, a systematic analysis of the entire GWAS catalog indicated 57 loci with a high pleiotropy index of ≥ 5. Interestingly, the top two loci with most evidence for pleiotropic associations were *GCKR* and the obesity locus *FTO* (fat and obesity associated) (Huang et al. 2011). Both loci had a very high pleiotropy index of 18, much higher than, for example, the already earlier well-known pleiotropic *ABO* region which had a pleiotropy index of 9 (Huang et al. 2011). In addition, while the pleiotropic-associated effects of the *FTO* locus may indicate mostly independent associations with different adiposity traits, the *GCKR* locus not only associates with traits related to glucose and lipid metabolism but also with very distinct traits such as CRP, serum uric acid, and albumin levels. GWAS have thus identified the *GCKR* locus as an extreme example of classical pleiotropy (Table 15.1).

15.3 Intervention Studies

The *GCKR* Pro446Leu variant has been investigated in both short- and longer-term intervention studies. The first was a short-term high-fat challenge study, which demonstrated that subjects with the T allele (Leu) had higher absolute plasma postprandial Triglyceride (TG) and incremental TG concentrations, as well as postprandial VLDL-Cholesterol levels following exposure to a high-fat diet (Shen et al. 2009). In the Diabetes Prevention Program (DPP), which is a multicenter trial designed to test whether intensive lifestyle modification or pharmacologic interventions with metformin or troglitazone prevent or delay the onset of diabetes in individuals at high risk (The Diabetes Prevention Program. Design and methods for a clinical trial in the prevention of type 2 diabetes 1999), the intensive lifestyle intervention markedly lowered incidence of diabetes (Knowler et al. 2002) and triglyceride levels (Ratner et al. 2005). In 3346 individuals participating in the DPP, intensive lifestyle change (increased physical activity, dietary changes, and weight loss) reduced the deleterious impact of the *GCKR* Leu446 variant on triglyceride levels without significantly affecting its relationship with insulin sensitivity (Fulgencio et al. 2001). However, carriers of the glucose-raising allele (Pro446) showed greater responsiveness to metformin as indicated by more decreased HOMA-IR index indicating a potentially relevant pharmacogenetic interaction as GCK has earlier been observed to have a stimulatory effect on metformin in hepatocytes (Fulgencio et al. 2001). The Leu446 variant has previously been shown to be associated with greater increases in triglycerides over a 23-years period (Orho-Melander et al. 2008) and the findings in DPP suggest that increased physical activity and weight loss could partially mitigate this effect.

15.4 Interaction with Diet

Although environmental and genetic factors together are known to determine lipid traits and the risk of multifactorial diseases including obesity, T2D, and cardiovascular disease, still very little is known about the environmental influences on genetic susceptibility or how genetic factors may modify the effects of environmental exposures. The pivotal role of GKRP as a regulator of GCK in the liver and hence hepatic glucose uptake and de novo lipogenesis makes the genetic variation in *GCKR* that influences these key metabolic functions, particularly interesting for studying the effects of modification by dietary exposures. In one of the largest "diet × gene" interaction studies performed to date, a meta-analysis of data from 14 cohorts with a total of 48,000 participants of European descent, interactions between whole-grain intake and genetic variants associated with fasting glucose and/or insulin levels were investigated for their interaction (Nettleton et al. 2010). In line with several earlier studies, higher whole-grain intake was associated with lower fasting glucose and insulin levels independent of demographics, other dietary and lifestyle factors, and body mass index (BMI). For the majority of loci included, the inverse association between whole-grain intake and fasting glucose or insulin was not observed to be modified by allelic variation at the analyzed genetic loci. However, effects were observed for carriers of the *GCKR* rs780094 on fasting insulin where carriers of the insulin-raising allele (i.e., the allele that associates with lower triglycerides) displayed a far less reduction in fasting insulin levels for their whole-grain intake (Nettleton et al. 2010). This observed interaction may suggest that genetic variation in *GCKR* could diminish the beneficial effects of whole-grain foods on insulin homeostasis, possibly as a result of the marked effects of the *GCKR* variant on both triglyceride and glucose levels and by extrapolation that the mechanisms by which whole-grain intake improves insulin resistance could potentially involve hepatic GCK. A further, albeit smaller study of 399 subjects reported that the *GCKR* Pro466Leu variant may influence insulin resistance by interacting with plasma omega-3-polyunsaturated fatty acid levels in patients classified as having the metabolic syndrome (Perez-Martinez et al. 2009). In a recent study of 20,986 individuals of the European Prospective Investigation into Cancer (EPIC)-Norfolk study, the association between the *GCKR* Pro446Leu with triglyceride levels was not modified by the Mediterranean Diet Score (Sotos-Prieto et al. 2014).

15.5 GKRP and Its Role in Hepatic Glucose Storage and Disposal

The central role for GCK in hepatic glucose storage and disposal and the importance of GKRP in regulating glucose phosphorylation by GCK have been understood for some time. GKRP forms a cytosolic protein complex with GCK and inhibits activity through nuclear import and storage. This process is initiated at low glucose concentrations and then reversed at high glucose levels (Fig. 15.1a). The cytosolic GCK–GKRP protein complex assembly and nuclear trafficking are further modulated by phosphate esters. Both glucose and fructose 1-phosphate (F1P), a product of fructose and sorbitol metabolism, oppose GCK–GKRP complex formation, nuclear sequestration, and the subsequent inhibition of GCK activity, while fructose 6-phosphate (F6P), an intermediate of glycolysis, glycogenolysis, and gluconeogenesis, enhances these actions in humans. Critically, studies working with recombinant forms of rat (rGKRP) and human GKRP (hGKRP) have highlighted differences in their ability to inhibit GCK (Brocklehurst et al. 2004). While rGKRP does not appreciably inhibit GCK in the absence of F6P, the human variety of the protein is much less dependent on F6P and can strongly inhibit GCK

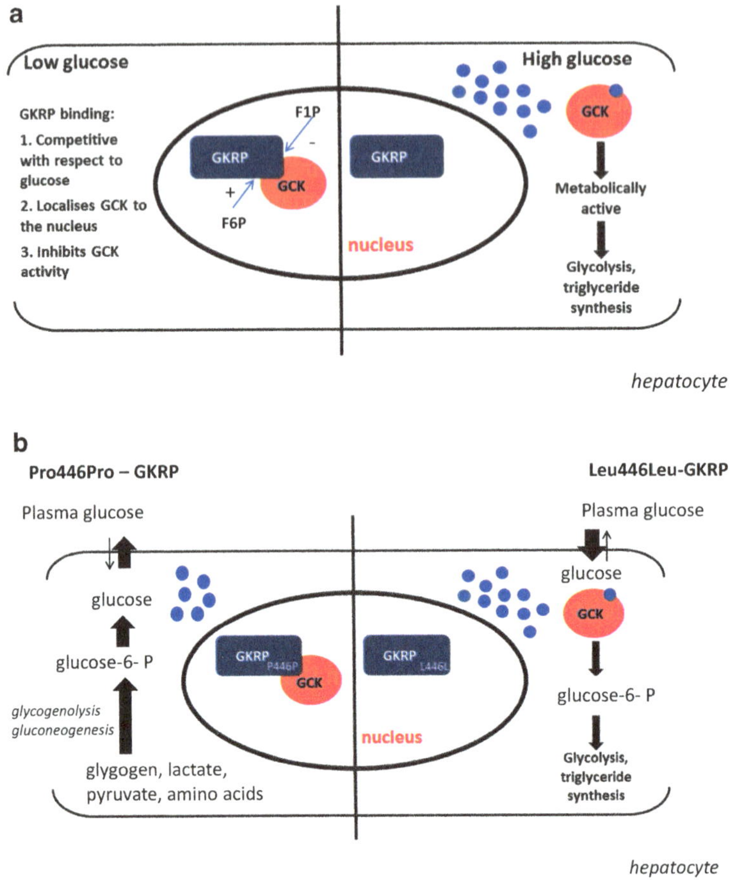

Fig. 15.1 Schematic representation of hepatic GCK regulation by GKRP and the molecular mechanism for the Pro446Leu-GKRP variant. (**a**) At low glucose concentrations GKRP binds to GCK and sequesters it in the nucleus resulting in reduced GCK activity in the cytoplasm. At high glucose concentrations GCK and GKRP disassociate and GCK is released into the cytoplasm where it is metabolically active and available for glycolysis generating the precursors for de novo lipogenesis. The interaction between GCK and GKRP is further regulated by the phosphate esters fructose-6-phosphate (F6P) and fructose-1-phosphate (F1P) which bind at the same motif on GKRP. GKRP inhibition of GCK is enhanced by F6P while F1P opposes the interaction between the two proteins. (**b**) The Leu446Leu-GKRP protein is less able to both sequester GCK in the nucleus and directly inhibit the regulatory protein. GCK is metabolically active leading to increased glycolysis and triglyceride synthesis

even in the absence of this modulator. Furthermore hGKRP has a 5–10-fold higher affinity for F6P than rGKRP (Brocklehurst et al. 2004).

15.6 Functional Characterization of the Pro446Leu-GKRP Protein

The early success with fine-mapping efforts and the fact the lead variant altered the protein sequence of a strong biological candidate greatly facilitated efforts to identify the molecular mechanisms driving the association of the Pro446Leu variant with triglyceride and glucose levels. Kinetic characterization of human recombinant proteins demonstrated that the Leu446-hGKRP protein resulted in a reduced ability to inhibit GCK and that its regulation by the phosphate ester F6P was impaired leading to a diminished ability of Leu446-hGKRP to inhibit GCK and consequently glycolysis (Beer et al. 2009). These findings were entirely consistent with the observed phenotype of the triglyceride-raising Leu446 allele also reducing glucose levels, since increased glycolysis would simultaneously reduce fasting glucose levels and increase synthesis of energy storage molecules including glycogen, triglycerides, and cholesterol. Importantly, these findings were in stark contrast to an earlier study using rGKRP where no differences were observed between Leu446 and Pro446 rGKRP (Veiga-da-Cunha et al. 2003). The discrepancy is most likely due to the differences in the kinetic characteristics between rGKRP and hGKRP which were not appreciated at the time of the study (Brocklehurst et al. 2004).

It is well established that kinetic mechanisms alone are not sufficient to explain GKRP regulation of GCK in vivo. *Gckr*$^{-/-}$ knockout mice display impaired postprandial glucose handling and paradoxical reductions in GCK protein levels and activity in spite of the removal of GKRP inhibition (Farrelly et al. 1999; Grimsby et al. 2000). These observations strongly support a role for the nuclear sequestration in regulating GCK protein concentration and stability and maintenance of glucose homeostasis. In a follow-up study, the impact of the Pro446Leu variant on intracellular localization of both GKRP and GCK was investigated in HeLa cells and mouse primary hepatocytes (Rees et al. 2012b). Using a quantitative method to assess nuclear and cytoplasmic fluorescence of tagged GKRP and GCK proteins, the authors demonstrated reduced nuclear localization of Leu446-hGKRP in HeLa cells and mouse primary hepatocytes compared to Pro446-hGKRP. Critically, the reduction in Leu446-hGKRP levels was accompanied by a reduced nuclear sequestration of GCK compared to Pro446-hGRKP. Furthermore, a decrease in the interaction between Leu4446-hGKRP and GCK compared to Pro446-hGKRP was also demonstrated using fluorescence resonance energy transfer (FRET) in both HeLa and mouse primary hepatocytes (Rees et al. 2012b). Taken together these data suggest that in addition to the kinetic defects, Leu446-hGKRP also leads to reduced glucose dependence in its cellular interaction with GCK (Fig. 15.1b). Increased cytoplasmic levels of both hGRKP and GCK are also consistent with increased glycolysis and glucose flux.

Similar to the species differences that have been observed for the kinetics of GKRP, investigation of the cellular localization of rodent and human GKRP revealed differences. The degree of nuclear sequestration of rGKRP was significantly reduced compared to hGKRP in both HeLa cells and mouse primary hepatocytes and critically the abnormality in nuclear GCK sequestration for Leu446-GKRP was only observed with hGKRP and not rGKRP (Rees et al. 2012b). These studies further emphasize the differences between human and rodent GKRP and highlight the need for appropriate assays to decipher complex molecular mechanisms driving human genetic association signals.

15.7 The Phenotypic and Functional Impact of Rare *GCKR* Variants

The majority of studies performed to date have focused on triglycerides as a continuous trait; however one GWAS compared individuals with extremely high levels of triglycerides (>95th centile) to those with normal triglyceride levels using a case–control design (Johansen et al. 2010). The authors reported an association of the *GCKR* rs1260326 (Pro446Leu) variant with triglyceride levels and identified three further genes (*APOA5*, *APOB*, *LPL*) associated with hypertriglyceridemia. In one of the first studies to assess the role of rare variants in genes implicated in a trait, they performed targeted exome sequencing of all four genes in additional cases of hypertriglyceridemia and controls, identifying an enrichment of rare (defined as minor allele frequency <0.01) nonsynonymous variants in cases. The absolute count of rare *GCKR* variants was increased in cases compared to controls (20 versus 5) suggesting that rare variants in *GCKR* may contribute to hypertriglyceridemia. This study implicated for the first time a role for rare *GCKR* variants in lipid phenotypes.

GCKR has also served as an important paradigm for exploring the challenges of interpreting the impact and significance of rare coding variants on disease risk and clinical phenotypes. Given the importance of both common and rare *GCKR* variants in influencing lipids, T2D risk, and other metabolic phenotypes and the availability of robust analytical pipelines for functional characterization, it became an obvious gene to explore some of these issues. As part of the NIH ClinSeq project, *GCKR* was selected as one of several genes for a targeted sequencing project in subjects at increased risk of developing coronary artery disease (CAD) (Rees et al. 2012a). Sequencing of 800 individuals identified 19 nonsynonymous (16 missense, one nonsense, and two frameshift) *GCKR* variants; all but two (Pro446Leu, Arg540Gln) had minor allele frequencies <0.01 in the cohort. At the time of the study, 10 of these variants were novel. Having established that the Pro446Leu variant was associated with triglyceride levels in this cohort, the potential impact of rare coding variants on lipid phenotypes was explored. The initial comparison of the collective group of individuals harboring a rare variant compared to the wild-type (Pro446Pro) reference group suggested a relationship with triglyceride and cholesterol levels; however this approach included all variants whether they were neutral or had altered function and did not distinguish between those that cause loss (LOF) or gain of function (GOF), which could diminish statistical power. Given the rarity of these variants, it was not practical to find large numbers of subjects with a specific *GCKR* variant for detailed phenotyping, so to evaluate the contribution of each variant to triglyceride levels, it was necessary to explore evolutionary, cell biological, and biochemical effects of each *GCKR* variant.

At the time of the study, no crystal structure for GKRP was available, although this has recently been published (Pautsch et al. 2013). Structural information was extrapolated from homology to bacterial proteins of the sugar isomerase (SIS) family which contains two separate SIS domains that combine to form a single site capable of binding to F1P or F6P (Veiga-da-Cunha and Van Schaftingen 2002; Veiga-da-Cunha et al. 2009). From these analyses it was possible to shed insight on some variants which were predicted to reside within a sugar binding motif (e.g., Val103Met) and hence impact on phosphate ester binding. A comprehensive range of in silico programs were employed to assess the potential impact of the amino acid substitutions on GKRP protein function revealing little agreement between them and poor concordance with empirical data (Rees et al. 2012a). However, using a suite of cellular and kinetic assays, similar to those employed earlier for the Pro446Leu variant, it was possible to group the variants into those which were wild-type like (neutral) and those which led to either LOF or GOF.

As a group, individuals with rare LOF *GCKR* variants showed a significant increase in total cholesterol, LDL cholesterol, and triglyceride levels for this subgroup compared with the wild-type (Pro446) reference group (Rees et al. 2012a). The variants in this group resembled Leu446-hGKPR, showing varying degrees of reduced expression, nuclear localization, GCK sequestration, and interaction with phosphate esters. However, while the functional effects resembled those of Leu446-hGKRP, they displayed a range in magnitude of functional defects with the most severe LOF variants (e.g., Val103Met) forming very little if any functional protein, similar to a null mutation. On the other hand, the potential GOF mutations were found in the conserved C-terminal restudies of GKRP and greatly reduced GKRP nuclear localization and GCK sequestration while maintaining high protein expression and cytoplasmic interaction with GCK. Furthermore, the kinetics of these variants (e.g., Arg612Cys) showed no differences to wild-type hGKRP. These findings highlighted a potential and previously unknown role for the C-terminal residues in the molecular mechanisms for localizing GKRP to the nucleus (Rees et al. 2012a).

In addition to the valuable mechanistic insights gained from this painstaking work, the study served two key points. First it demonstrated the poor performance of currently available and widely used in silico prediction programs in assigning functionality to rare nonsynonymous variants. Second, it showed the value of functional studies for the correct assignment of pathogenicity as evidenced by an improved ability to detect genetic associations (Rees et al. 2012a). Unfortunately, these kinetic and cellular assays are expensive and time consuming, making their wider use impractical. However, recent efforts to establish a series of high-throughput cell-free miniaturized assays for characterizing the GCK:GKRP interaction offer the promise of a platform to simultaneously test multiple rare nonsynonymous *GCKR* variants in the future (Rees et al. 2014).

15.8 Outlook

The genetic discoveries described above have highlighted a previously underappreciated central role for GKRP in a wide range of metabolic processes in the liver. Through the initial serendipitous finding of a coding variant, driving these genetic associations, substantial progress has been made in our understanding of the molecular mechanism driving GKRP dysfunction. However, many questions remain unanswered, particularly with reference to the more unexpected pleiotropic associations (e.g., CRP levels) and how altering a regulator of hepatic glucose disposal influences a marker of inflammation. Given the emerging interest in small molecular disruptors of the GKRP:GCK complex as a potential therapeutic agent to treat T2D, close scrutiny of available human genetic data is likely to be of considerable value when evaluating the potential for adverse on-target side effects (Lloyd et al. 2013). Many of the rare variants studied to date exert their impact on GKRP function in a similar manner to these small molecular disruptors by preventing the GKRP:GCK complex (Lloyd et al. 2013; Rees et al. 2012a, b). Increasing our understanding of how deterministic these functional variants are for influencing triglyceride levels and other relevant phenotypes is likely to be of considerable interest. Furthermore, a more comprehensive understanding of the spectrum of GKRP defects could identify an optimal therapeutic window for manipulation of hepatic glucose homeostasis.

References

Beer NL, Tribble ND, McCulloch LJ, Roos C, Johnson PR, Orho-Melander M, Gloyn AL (2009) The P446L variant in GCKR associated with fasting plasma glucose and triglyceride levels exerts its effect through increased glucokinase activity in liver. Hum Mol Genet 18(21):4081–4088. doi:10.1093/hmg/ddp357

Brocklehurst KJ, Davies RA, Agius L (2004) Differences in regulatory properties between human and rat glucokinase regulatory protein. Biochem J 378(Pt 2):693–697. doi:10.1042/BJ20031414

Chambers JC, Zhang W, Sehmi J, Li X, Wass MN, Van der Harst P, Holm H, Sanna S, Kavousi M, Baumeister SE, Coin LJ, Deng G, Gieger C, Heard-Costa NL, Hottenga JJ, Kuhnel B, Kumar V, Lagou V, Liang L, Luan J, Vidal PM, Mateo Leach I, O'Reilly PF, Peden JF, Rahmioglu N, Soininen P, Speliotes EK, Yuan X, Thorleifsson G, Alizadeh BZ, Atwood LD, Borecki IB, Brown MJ, Charoen P, Cucca F, Das D, de Geus EJ, Dixon AL, Doring A, Ehret G, Eyjolfsson GI, Farrall M, Forouhi NG, Friedrich N, Goessling W, Gudbjartsson DF, Harris TB, Hartikainen AL, Heath S, Hirschfield GM, Hofman A, Homuth G, Hypponen E, Janssen HL, Johnson T, Kangas AJ, Kema IP, Kuhn JP, Lai S, Lathrop M, Lerch MM, Li Y, Liang TJ, Lin JP, Loos RJ, Martin NG, Moffatt MF, Montgomery GW, Munroe PB, Musunuru K, Nakamura

Y, O'Donnell CJ, Olafsson I, Penninx BW, Pouta A, Prins BP, Prokopenko I, Puls R, Ruokonen A, Savolainen MJ, Schlessinger D, Schouten JN, Seedorf U, Sen-Chowdhry S, Siminovitch KA, Smit JH, Spector TD, Tan W, Teslovich TM, Tukiainen T, Uitterlinden AG, Van der Klauw MM, Vasan RS, Wallace C, Wallaschofski H, Wichmann HE, Willemsen G, Wurtz P, Xu C, Yerges-Armstrong LM, Abecasis GR, Ahmadi KR, Boomsma DI, Caulfield M, Cookson WO, van Duijn CM, Froguel P, Matsuda K, McCarthy MI, Meisinger C, Mooser V, Pietilainen KH, Schumann G, Snieder H, Sternberg MJ, Stolk RP, Thomas HC, Thorsteinsdottir U, Uda M, Waeber G, Wareham NJ, Waterworth DM, Watkins H, Whitfield JB, Witteman JC, Wolffenbuttel BH, Fox CS, Ala-Korpela M, Stefansson K, Vollenweider P, Volzke H, Schadt EE, Scott J, Jarvelin MR, Elliott P, Kooner JS (2011) Genome-wide association study identifies loci influencing concentrations of liver enzymes in plasma. Nat Genet 43(11):1131–1138. doi:10.1038/ng.970

Dehghan A, Dupuis J, Barbalic M, Bis JC, Eiriksdottir G, Lu C, Pellikka N, Wallaschofski H, Kettunen J, Henneman P, Baumert J, Strachan DP, Fuchsberger C, Vitart V, Wilson JF, Pare G, Naitza S, Rudock ME, Surakka I, de Geus EJ, Alizadeh BZ, Guralnik J, Shuldiner A, Tanaka T, Zee RY, Schnabel RB, Nambi V, Kavousi M, Ripatti S, Nauck M, Smith NL, Smith AV, Sundvall J, Scheet P, Liu Y, Ruokonen A, Rose LM, Larson MG, Hoogeveen RC, Freimer NB, Teumer A, Tracy RP, Launer LJ, Buring JE, Yamamoto JF, Folsom AR, Sijbrands EJ, Pankow J, Elliott P, Keaney JF, Sun W, Sarin AP, Fontes JD, Badola S, Astor BC, Hofman A, Pouta A, Werdan K, Greiser KH, Kuss O, Meyer zu Schwabedissen HE, Thiery J, Jamshidi Y, Nolte IM, Soranzo N, Spector TD, Volzke H, Parker AN, Aspelund T, Bates D, Young L, Tsui K, Siscovick DS, Guo X, Rotter JI, Uda M, Schlessinger D, Rudan I, Hicks AA, Penninx BW, Thorand B, Gieger C, Coresh J, Willemsen G, Harris TB, Uitterlinden AG, Jarvelin MR, Rice K, Radke D, Salomaa V, Willems van Dijk K, Boerwinkle E, Vasan RS, Ferrucci L, Gibson QD, Bandinelli S, Snieder H, Boomsma DI, Xiao X, Campbell H, Hayward C, Pramstaller PP, van Duijn CM, Peltonen L, Psaty BM, Gudnason V, Ridker PM, Homuth G, Koenig W, Ballantyne CM, Witteman JC, Benjamin EJ, Perola M, Chasman DI (2011) Meta-analysis of genome-wide association studies in >80 000 subjects identifies multiple loci for C-reactive protein levels. Circulation 123(7):731–738. doi:10.1161/circulationaha.110.948570

Deo RC, Reich D, Tandon A, Akylbekova E, Patterson N, Waliszewska A, Kathiresan S, Sarpong D, Taylor HA Jr, Wilson JG (2009) Genetic differences between the determinants of lipid profile phenotypes in African and European Americans: the Jackson Heart Study. PLoS Genet 5(1), e1000342. doi:10.1371/journal.pgen.1000342

Dupuis J, Langenberg C, Prokopenko I, Saxena R, Soranzo N, Jackson AU, Wheeler E, Glazer NL, Bouatia-Naji N, Gloyn AL, Lindgren CM, Magi R, Morris AP, Randall J, Johnson T, Elliott P, Rybin D, Thorleifsson G, Steinthorsdottir V, Henneman P, Grallert H, Dehghan A, Hottenga JJ, Franklin CS, Navarro P, Song K, Goel A, Perry JR, Egan JM, Lajunen T, Grarup N, Sparso T, Doney A, Voight BF, Stringham HM, Li M, Kanoni S, Shrader P, Cavalcanti-Proenca C, Kumari M, Qi L, Timpson NJ, Gieger C, Zabena C, Rocheleau G, Ingelsson E, An P, O'Connell J, Luan J, Elliott A, McCarroll SA, Payne F, Roccasecca RM, Pattou F, Sethupathy P, Ardlie K, Ariyurek Y, Balkau B, Barter P, Beilby JP, Ben-Shlomo Y, Benediktsson R, Bennett AJ, Bergmann S, Bochud M, Boerwinkle E, Bonnefond A, Bonnycastle LL, Borch-Johnsen K, Bottcher Y, Brunner E, Bumpstead SJ, Charpentier G, Chen YD, Chines P, Clarke R, Coin LJ, Cooper MN, Cornelis M, Crawford G, Crisponi L, Day IN, de Geus EJ, Delplanque J, Dina C, Erdos MR, Fedson AC, Fischer-Rosinsky A, Forouhi NG, Fox CS, Frants R, Franzosi MG, Galan P, Goodarzi MO, Graessler J, Groves CJ, Grundy S, Gwilliam R, Gyllensten U, Hadjadj S, Hallmans G, Hammond N, Han X, Hartikainen AL, Hassanali N, Hayward C, Heath SC, Hercberg S, Herder C, Hicks AA, Hillman DR, Hingorani AD, Hofman A, Hui J, Hung J, Isomaa B, Johnson PR, Jorgensen T, Jula A, Kaakinen M, Kaprio J, Kesaniemi YA, Kivimaki M, Knight B, Koskinen S, Kovacs P, Kyvik KO, Lathrop GM, Lawlor DA, Le Bacquer O, Lecoeur C, Li Y, Lyssenko V, Mahley R, Mangino M, Manning AK, Martinez-Larrad MT, McAteer JB, McCulloch LJ, McPherson R, Meisinger C, Melzer D, Meyre D, Mitchell BD, Morken MA, Mukherjee S, Naitza S, Narisu N, Neville MJ, Oostra BA,

Orru M, Pakyz R, Palmer CN, Paolisso G, Pattaro C, Pearson D, Peden JF, Pedersen NL, Perola M, Pfeiffer AF, Pichler I, Polasek O, Posthuma D, Potter SC, Pouta A, Province MA, Psaty BM, Rathmann W, Rayner NW, Rice K, Ripatti S, Rivadeneira F, Roden M, Rolandsson O, Sandbaek A, Sandhu M, Sanna S, Sayer AA, Scheet P, Scott LJ, Seedorf U, Sharp SJ, Shields B, Sigurethsson G, Sijbrands EJ, Silveira A, Simpson L, Singleton A, Smith NL, Sovio U, Swift A, Syddall H, Syvanen AC, Tanaka T, Thorand B, Tichet J, Tonjes A, Tuomi T, Uitterlinden AG, van Dijk KW, van Hoek M, Varma D, Visvikis-Siest S, Vitart V, Vogelzangs N, Waeber G, Wagner PJ, Walley A, Walters GB, Ward KL, Watkins H, Weedon MN, Wild SH, Willemsen G, Witteman JC, Yarnell JW, Zeggini E, Zelenika D, Zethelius B, Zhai G, Zhao JH, Zillikens MC, Borecki IB, Loos RJ, Meneton P, Magnusson PK, Nathan DM, Williams GH, Hattersley AT, Silander K, Salomaa V, Smith GD, Bornstein SR, Schwarz P, Spranger J, Karpe F, Shuldiner AR, Cooper C, Dedoussis GV, Serrano-Rios M, Morris AD, Lind L, Palmer LJ, Hu FB, Franks PW, Ebrahim S, Marmot M, Kao WH, Pankow JS, Sampson MJ, Kuusisto J, Laakso M, Hansen T, Pedersen O, Pramstaller PP, Wichmann HE, Illig T, Rudan I, Wright AF, Stumvoll M, Campbell H, Wilson JF, Bergman RN, Buchanan TA, Collins FS, Mohlke KL, Tuomilehto J, Valle TT, Altshuler D, Rotter JI, Siscovick DS, Penninx BW, Boomsma DI, Deloukas P, Spector TD, Frayling TM, Ferrucci L, Kong A, Thorsteinsdottir U, Stefansson K, van Duijn CM, Aulchenko YS, Cao A, Scuteri A, Schlessinger D, Uda M, Ruokonen A, Jarvelin MR, Waterworth DM, Vollenweider P, Peltonen L, Mooser V, Abecasis GR, Wareham NJ, Sladek R, Froguel P, Watanabe RM, Meigs JB, Groop L, Boehnke M, McCarthy MI, Florez JC, Barroso I (2010) New genetic loci implicated in fasting glucose homeostasis and their impact on type 2 diabetes risk. Nat Genet 42(2):105–116. doi:10.1038/ng.520

Farrelly D, Brown KS, Tieman A, Ren J, Lira SA, Hagan D, Gregg R, Mookhtiar KA, Hariharan N (1999) Mice mutant for glucokinase regulatory protein exhibit decreased liver glucokinase: a sequestration mechanism in metabolic regulation. Proc Natl Acad Sci USA 96(25):14511–14516

Ferre T, Riu E, Bosch F, Valera A (1996) Evidence from transgenic mice that glucokinase is rate limiting for glucose utilization in the liver. FASEB J 10(10):1213–1218

Foster KR, Shaulsky G, Strassmann JE, Queller DC, Thompson CR (2004) Pleiotropy as a mechanism to stabilize cooperation. Nature 431(7009):693–696. doi:10.1038/nature02894

Franceschini N, van Rooij FJ, Prins BP, Feitosa MF, Karakas M, Eckfeldt JH, Folsom AR, Kopp J, Vaez A, Andrews JS, Baumert J, Boraska V, Broer L, Hayward C, Ngwa JS, Okada Y, Polasek O, Westra HJ, Wang YA, Del Greco MF, Glazer NL, Kapur K, Kema IP, Lopez LM, Schillert A, Smith AV, Winkler CA, Zgaga L, Bandinelli S, Bergmann S, Boban M, Bochud M, Chen YD, Davies G, Dehghan A, Ding J, Doering A, Durda JP, Ferrucci L, Franco OH, Franke L, Gunjaca G, Hofman A, Hsu FC, Kolcic I, Kraja A, Kubo M, Lackner KJ, Launer L, Loehr LR, Li G, Meisinger C, Nakamura Y, Schwienbacher C, Starr JM, Takahashi A, Torlak V, Uitterlinden AG, Vitart V, Waldenberger M, Wild PS, Kirin M, Zeller T, Zemunik T, Zhang Q, Ziegler A, Blankenberg S, Boerwinkle E, Borecki IB, Campbell H, Deary IJ, Frayling TM, Gieger C, Harris TB, Hicks AA, Koenig W, O'Donnell CJ, Fox CS, Pramstaller PP, Psaty BM, Reiner AP, Rotter JI, Rudan I, Snieder H, Tanaka T, van Duijn CM, Vollenweider P, Waeber G, Wilson JF, Witteman JC, Wolffenbuttel BH, Wright AF, Wu Q, Liu Y, Jenny NS, North KE, Felix JF, Alizadeh BZ, Cupples LA, Perry JR, Morris AP (2012) Discovery and fine mapping of serum protein loci through transethnic meta-analysis. Am J Hum Genet 91(4):744–753. doi:10.1016/j.ajhg.2012.08.021

Fulgencio JP, Kohl C, Girard J, Pegorier JP (2001) Effect of metformin on fatty acid and glucose metabolism in freshly isolated hepatocytes and on specific gene expression in cultured hepatocytes. Biochem Pharmacol 62(4):439–446

Gieger C, Geistlinger L, Altmaier E, Hrabe de Angelis M, Kronenberg F, Meitinger T, Mewes HW, Wichmann HE, Weinberger KM, Adamski J, Illig T, Suhre K (2008) Genetics meets metabolomics: a genome-wide association study of metabolite profiles in human serum. PLoS Genet 4(11), e1000282. doi:10.1371/journal.pgen.1000282

Grimsby J, Coffey JW, Dvorozniak MT, Magram J, Li G, Matschinsky FM, Shiota C, Kaur S, Magnuson MA, Grippo JF (2000) Characterization of glucokinase regulatory protein-deficient mice. J Biol Chem 275(11):7826–7831

Huang J, Johnson AD, O'Donnell CJ (2011) PRIMe: a method for characterization and evaluation of pleiotropic regions from multiple genome-wide association studies. Bioinformatics 27 (9):1201–1206. doi:10.1093/bioinformatics/btr116

Illig T, Gieger C, Zhai G, Romisch-Margl W, Wang-Sattler R, Prehn C, Altmaier E, Kastenmuller G, Kato BS, Mewes HW, Meitinger T, de Angelis MH, Kronenberg F, Soranzo N, Wichmann HE, Spector TD, Adamski J, Suhre K (2010) A genome-wide perspective of genetic variation in human metabolism. Nat Genet 42(2):137–141. doi:10.1038/ng.507

Johansen CT, Wang J, Lanktree MB, Cao H, McIntyre AD, Ban MR, Martins RA, Kennedy BA, Hassell RG, Visser ME, Schwartz SM, Voight BF, Elosua R, Salomaa V, O'Donnell CJ, Dallinga-Thie GM, Anand SS, Yusuf S, Huff MW, Kathiresan S, Hegele RA (2010) Excess of rare variants in genes identified by genome-wide association study of hypertriglyceridemia. Nat Genet 42(8):684–687. doi:10.1038/ng.628

Kamatani Y, Matsuda K, Okada Y, Kubo M, Hosono N, Daigo Y, Nakamura Y, Kamatani N (2010) Genome-wide association study of hematological and biochemical traits in a Japanese population. Nat Genet 42(3):210–215. doi:10.1038/ng.531

Kathiresan S, Musunuru K, Orho-Melander M (2008) Defining the spectrum of alleles that contribute to blood lipid concentrations in humans. Curr Opin Lipidol 19(2):122–127. doi:10.1097/MOL.0b013e3282f70296

Knowler WC, Barrett-Connor E, Fowler SE, Hamman RF, Lachin JM, Walker EA, Nathan DM (2002) Reduction in the incidence of type 2 diabetes with lifestyle intervention or metformin. N Engl J Med 346(6):393–403. doi:10.1056/NEJMoa012512

Kolz M, Johnson T, Sanna S, Teumer A, Vitart V, Perola M, Mangino M, Albrecht E, Wallace C, Farrall M, Johansson A, Nyholt DR, Aulchenko Y, Beckmann JS, Bergmann S, Bochud M, Brown M, Campbell H, Connell J, Dominiczak A, Homuth G, Lamina C, McCarthy MI, Meitinger T, Mooser V, Munroe P, Nauck M, Peden J, Prokisch H, Salo P, Salomaa V, Samani NJ, Schlessinger D, Uda M, Volker U, Waeber G, Waterworth D, Wang-Sattler R, Wright AF, Adamski J, Whitfield JB, Gyllensten U, Wilson JF, Rudan I, Pramstaller P, Watkins H, Doering A, Wichmann HE, Spector TD, Peltonen L, Volzke H, Nagaraja R, Vollenweider P, Caulfield M, Illig T, Gieger C (2009) Meta-analysis of 28,141 individuals identifies common variants within five new loci that influence uric acid concentrations. PLoS Genet 5(6), e1000504. doi:10.1371/journal.pgen.1000504

Kottgen A, Pattaro C, Boger CA, Fuchsberger C, Olden M, Glazer NL, Parsa A, Gao X, Yang Q, Smith AV, O'Connell JR, Li M, Schmidt H, Tanaka T, Isaacs A, Ketkar S, Hwang SJ, Johnson AD, Dehghan A, Teumer A, Pare G, Atkinson EJ, Zeller T, Lohman K, Cornelis MC, Probst-Hensch NM, Kronenberg F, Tonjes A, Hayward C, Aspelund T, Eiriksdottir G, Launer LJ, Harris TB, Rampersaud E, Mitchell BD, Arking DE, Boerwinkle E, Struchalin M, Cavalieri M, Singleton A, Giallauria F, Metter J, de Boer IH, Haritunians T, Lumley T, Siscovick D, Psaty BM, Zillikens MC, Oostra BA, Feitosa M, Province M, de Andrade M, Turner ST, Schillert A, Ziegler A, Wild PS, Schnabel RB, Wilde S, Munzel TF, Leak TS, Illig T, Klopp N, Meisinger C, Wichmann HE, Koenig W, Zgaga L, Zemunik T, Kolcic I, Minelli C, Hu FB, Johansson A, Igl W, Zaboli G, Wild SH, Wright AF, Campbell H, Ellinghaus D, Schreiber S, Aulchenko YS, Felix JF, Rivadeneira F, Uitterlinden AG, Hofman A, Imboden M, Nitsch D, Brandstatter A, Kollerits B, Kedenko L, Magi R, Stumvoll M, Kovacs P, Boban M, Campbell S, Endlich K, Volzke H, Kroemer HK, Nauck M, Volker U, Polasek O, Vitart V, Badola S, Parker AN, Ridker PM, Kardia SL, Blankenberg S, Liu Y, Curhan GC, Franke A, Rochat T, Paulweber B, Prokopenko I, Wang W, Gudnason V, Shuldiner AR, Coresh J, Schmidt R, Ferrucci L, Shlipak MG, van Duijn CM, Borecki I, Kramer BK, Rudan I, Gyllensten U, Wilson JF, Witteman JC, Pramstaller PP, Rettig R, Hastie N, Chasman DI, Kao WH, Heid IM, Fox CS (2010) New loci associated with kidney function and chronic kidney disease. Nat Genet 42(5):376–384. doi:10.1038/ng.568

Lloyd DJ, St Jean DJ Jr, Kurzeja RJ, Wahl RC, Michelsen K, Cupples R, Chen M, Wu J, Sivits G, Helmering J, Komorowski R, Ashton KS, Pennington LD, Fotsch C, Vazir M, Chen K, Chmait S, Zhang J, Liu L, Norman MH, Andrews KL, Bartberger MD, Van G, Galbreath EJ, Vonderfecht SL, Wang M, Jordan SR, Veniant MM, Hale C (2013) Antidiabetic effects of glucokinase regulatory protein small-molecule disruptors. Nature 504(7480):437–440. doi:10.1038/nature12724

Nettleton JA, McKeown NM, Kanoni S, Lemaitre RN, Hivert MF, Ngwa J, van Rooij FJ, Sonestedt E, Wojczynski MK, Ye Z, Tanaka T, Garcia M, Anderson JS, Follis JL, Djousse L, Mukamal K, Papoutsakis C, Mozaffarian D, Zillikens MC, Bandinelli S, Bennett AJ, Borecki IB, Feitosa MF, Ferrucci L, Forouhi NG, Groves CJ, Hallmans G, Harris T, Hofman A, Houston DK, Hu FB, Johansson I, Kritchevsky SB, Langenberg C, Launer L, Liu Y, Loos RJ, Nalls M, Orho-Melander M, Renstrom F, Rice K, Riserus U, Rolandsson O, Rotter JI, Saylor G, Sijbrands EJ, Sjogren P, Smith A, Steingrimsdottir L, Uitterlinden AG, Wareham NJ, Prokopenko I, Pankow JS, van Duijn CM, Florez JC, Witteman JC, Dupuis J, Dedoussis GV, Ordovas JM, Ingelsson E, Cupples L, Siscovick DS, Franks PW, Meigs JB (2010) Interactions of dietary whole-grain intake with fasting glucose- and insulin-related genetic loci in individuals of European descent: a meta-analysis of 14 cohort studies. Diabetes Care 33(12):2684–2691. doi:10.2337/dc10-1150

O'Doherty RM, Lehman DL, Telemaque-Potts S, Newgard CB (1999) Metabolic impact of glucokinase overexpression in liver: lowering of blood glucose in fed rats is accompanied by hyperlipidemia. Diabetes 48(10):2022–2027

O'Seaghdha CM, Yang Q, Glazer NL, Leak TS, Dehghan A, Smith AV, Kao WH, Lohman K, Hwang SJ, Johnson AD, Hofman A, Uitterlinden AG, Chen YD, GEFOS Consortium, Brown EM, Siscovick DS, Harris TB, Psaty BM, Coresh J, Gudnason V, Witteman JC, Liu YM, Kestenbaum BR, Fox CS, Köttgen A (2010) Common variants in the calcium-sensing receptor gene are associated with total serum calcium levels. Hum Mol Genet 19(21):4296–4303. doi:10.1093/hmg/ddq342

O'Seaghdha CM, Wu H, Yang Q, Kapur K, Guessous I, Zuber AM, Kottgen A, Stoudmann C, Teumer A, Kutalik Z, Mangino M, Dehghan A, Zhang W, Eiriksdottir G, Li G, Tanaka T, Portas L, Lopez LM, Hayward C, Lohman K, Matsuda K, Padmanabhan S, Firsov D, Sorice R, Ulivi S, Brockhaus AC, Kleber ME, Mahajan A, Ernst FD, Gudnason V, Launer LJ, Mace A, Boerwinckle E, Arking DE, Tanikawa C, Nakamura Y, Brown MJ, Gaspoz JM, Theler JM, Siscovick DS, Psaty BM, Bergmann S, Vollenweider P, Vitart V, Wright AF, Zemunik T, Boban M, Kolcic I, Navarro P, Brown EM, Estrada K, Ding J, Harris TB, Bandinelli S, Hernandez D, Singleton AB, Girotto G, Ruggiero D, d'Adamo AP, Robino A, Meitinger T, Meisinger C, Davies G, Starr JM, Chambers JC, Boehm BO, Winkelmann BR, Huang J, Murgia F, Wild SH, Campbell H, Morris AP, Franco OH, Hofman A, Uitterlinden AG, Rivadeneira F, Volker U, Hannemann A, Biffar R, Hoffmann W, Shin SY, Lescuyer P, Henry H, Schurmann C, Munroe PB, Gasparini P, Pirastu N, Ciullo M, Gieger C, Marz W, Lind L, Spector TD, Smith AV, Rudan I, Wilson JF, Polasek O, Deary IJ, Pirastu M, Ferrucci L, Liu Y, Kestenbaum B, Kooner JS, Witteman JC, Nauck M, Kao WH, Wallaschofski H, Bonny O, Fox CS, Bochud M (2013) Meta-analysis of genome-wide association studies identifies six new Loci for serum calcium concentrations. PLoS Genet 9(9), e1003796. doi:10.1371/journal.pgen.1003796

Orho-Melander M, Melander O, Guiducci C, Perez-Martinez P, Corella D, Roos C, Tewhey R, Rieder MJ, Hall J, Abecasis G, Tai ES, Welch C, Arnett DK, Lyssenko V, Lindholm E, Saxena R, de Bakker PI, Burtt N, Voight BF, Hirschhorn JN, Tucker KL, Hedner T, Tuomi T, Isomaa B, Eriksson KF, Taskinen MR, Wahlstrand B, Hughes TE, Parnell LD, Lai CQ, Berglund G, Peltonen L, Vartiainen E, Jousilahti P, Havulinna AS, Salomaa V, Nilsson P, Groop L, Altshuler D, Ordovas JM, Kathiresan S (2008) Common missense variant in the glucokinase regulatory protein gene is associated with increased plasma triglyceride and C-reactive protein but lower fasting glucose concentrations. Diabetes 57(11):3112–3121. doi:10.2337/db08-0516

Pautsch A, Stadler N, Lohle A, Rist W, Berg A, Glocker L, Nar H, Reinert D, Lenter M, Heckel A, Schnapp G, Kauschke SG (2013) Crystal structure of glucokinase regulatory protein. Biochemistry 52(20):3523–3531. doi:10.1021/bi4000782

Perez-Martinez P, Corella D, Shen J, Arnett DK, Yiannakouris N, Tai ES, Orho-Melander M, Tucker KL, Tsai M, Straka RJ, Province M, Kai CS, Perez-Jimenez F, Lai CQ, Lopez-Miranda J, Guillen M, Parnell LD, Borecki I, Kathiresan S, Ordovas JM (2009) Association between glucokinase regulatory protein (GCKR) and apolipoprotein A5 (APOA5) gene polymorphisms and triacylglycerol concentrations in fasting, postprandial, and fenofibrate-treated states. Am J Clin Nutr 89(1):391–399. doi:10.3945/ajcn.2008.26363

Ratner R, Goldberg R, Haffner S, Marcovina S, Orchard T, Fowler S, Temprosa M, Diabetes Prevention Program Research Group (2005) Impact of intensive lifestyle and metformin therapy on cardiovascular disease risk factors in the diabetes prevention program. Diabetes Care 28(4):888–894

Rees MG, Ng D, Ruppert S, Turner C, Beer NL, Swift AJ, Morken MA, Below JE, Blech I, Mullikin JC, McCarthy MI, Biesecker LG, Gloyn AL, Collins FS (2012a) Correlation of rare coding variants in the gene encoding human glucokinase regulatory protein with phenotypic, cellular, and kinetic outcomes. J Clin Invest 122(1):205–217. doi:10.1172/JCI46425

Rees MG, Wincovitch S, Schultz J, Waterstradt R, Beer NL, Baltrusch S, Collins FS, Gloyn AL (2012b) Cellular characterisation of the GCKR P446L variant associated with type 2 diabetes risk. Diabetologia 55(1):114–122. doi:10.1007/s00125-011-2348-5

Rees MG, Davis MI, Shen M, Titus S, Raimondo A, Barrett A, Gloyn AL, Collins FS, Simeonov A (2014) A panel of diverse assays to interrogate the interaction between glucokinase and glucokinase regulatory protein, two vital proteins in human disease. PLoS One 9(2), e89335. doi:10.1371/journal.pone.0089335

Reiner AP, Beleza S, Franceschini N, Auer PL, Robinson JG, Kooperberg C, Peters U, Tang H (2012) Genome-wide association and population genetic analysis of C-reactive protein in African American and Hispanic American women. Am J Hum Genet 91(3):502–512. doi:10.1016/j.ajhg.2012.07.023

Ridker PM, Pare G, Parker A, Zee RY, Danik JS, Buring JE, Kwiatkowski D, Cook NR, Miletich JP, Chasman DI (2008) Loci related to metabolic-syndrome pathways including LEPR, HNF1A, IL6R, and GCKR associate with plasma C-reactive protein: the Women's Genome Health Study. Am J Hum Genet 82(5):1185–1192. doi:10.1016/j.ajhg.2008.03.015

Sabatti C, Service SK, Hartikainen AL, Pouta A, Ripatti S, Brodsky J, Jones CG, Zaitlen NA, Varilo T, Kaakinen M, Sovio U, Ruokonen A, Laitinen J, Jakkula E, Coin L, Hoggart C, Collins A, Turunen H, Gabriel S, Elliot P, McCarthy MI, Daly MJ, Jarvelin MR, Freimer NB, Peltonen L (2009) Genome-wide association analysis of metabolic traits in a birth cohort from a founder population. Nat Genet 41(1):35–46. doi:10.1038/ng.271

Saxena R, Voight BF, Lyssenko V, Burtt NP, de Bakker PI, Chen H, Roix JJ, Kathiresan S, Hirschhorn JN, Daly MJ, Hughes TE, Groop L, Altshuler D, Almgren P, Florez JC, Meyer J, Ardlie K, Bengtsson Bostrom K, Isomaa B, Lettre G, Lindblad U, Lyon HN, Melander O, Newton-Cheh C, Nilsson P, Orho-Melander M, Rastam L, Speliotes EK, Taskinen MR, Tuomi T, Guiducci C, Berglund A, Carlson J, Gianniny L, Hackett R, Hall L, Holmkvist J, Laurila E, Sjogren M, Sterner M, Surti A, Svensson M, Tewhey R, Blumenstiel B, Parkin M, Defelice M, Barry R, Brodeur W, Camarata J, Chia N, Fava M, Gibbons J, Handsaker B, Healy C, Nguyen K, Gates C, Sougnez C, Gage D, Nizzari M, Gabriel SB, Chirn GW, Ma Q, Parikh H, Richardson D, Ricke D, Purcell S (2007) Genome-wide association analysis identifies loci for type 2 diabetes and triglyceride levels. Science 316(5829):1331–1336. doi:10.1126/science.1142358

Saxena R, Hivert MF, Langenberg C, Tanaka T, Pankow JS, Vollenweider P, Lyssenko V, Bouatia-Naji N, Dupuis J, Jackson AU, Kao WH, Li M, Glazer NL, Manning AK, Luan J, Stringham HM, Prokopenko I, Johnson T, Grarup N, Boesgaard TW, Lecoeur C, Shrader P, O'Connell J, Ingelsson E, Couper DJ, Rice K, Song K, Andreasen CH, Dina C, Kottgen A, Le Bacquer O, Pattou F, Taneera J, Steinthorsdottir V, Rybin D, Ardlie K, Sampson M, Qi L, van

Hoek M, Weedon MN, Aulchenko YS, Voight BF, Grallert H, Balkau B, Bergman RN, Bielinski SJ, Bonnefond A, Bonnycastle LL, Borch-Johnsen K, Bottcher Y, Brunner E, Buchanan TA, Bumpstead SJ, Cavalcanti-Proenca C, Charpentier G, Chen YD, Chines PS, Collins FS, Cornelis M, J Crawford G, Delplanque J, Doney A, Egan JM, Erdos MR, Firmann M, Forouhi NG, Fox CS, Goodarzi MO, Graessler J, Hingorani A, Isomaa B, Jorgensen T, Kivimaki M, Kovacs P, Krohn K, Kumari M, Lauritzen T, Levy-Marchal C, Mayor V, McAteer JB, Meyre D, Mitchell BD, Mohlke KL, Morken MA, Narisu N, Palmer CN, Pakyz R, Pascoe L, Payne F, Pearson D, Rathmann W, Sandbaek A, Sayer AA, Scott LJ, Sharp SJ, Sijbrands E, Singleton A, Siscovick DS, Smith NL, Sparso T, Swift AJ, Syddall H, Thorleifsson G, Tonjes A, Tuomi T, Tuomilehto J, Valle TT, Waeber G, Walley A, Waterworth DM, Zeggini E, Zhao JH, Illig T, Wichmann HE, Wilson JF, van Duijn C, Hu FB, Morris AD, Frayling TM, Hattersley AT, Thorsteinsdottir U, Stefansson K, Nilsson P, Syvanen AC, Shuldiner AR, Walker M, Bornstein SR, Schwarz P, Williams GH, Nathan DM, Kuusisto J, Laakso M, Cooper C, Marmot M, Ferrucci L, Mooser V, Stumvoll M, Loos RJ, Altshuler D, Psaty BM, Rotter JI, Boerwinkle E, Hansen T, Pedersen O, Florez JC, McCarthy MI, Boehnke M, Barroso I, Sladek R, Froguel P, Meigs JB, Groop L, Wareham NJ, Watanabe RM (2010) Genetic variation in GIPR influences the glucose and insulin responses to an oral glucose challenge. Nat Genet 42(2):142–148. doi:10.1038/ng.521

Scott LJ, Mohlke KL, Bonnycastle LL, Willer CJ, Li Y, Duren WL, Erdos MR, Stringham HM, Chines PS, Jackson AU, Prokunina-Olsson L, Ding CJ, Swift AJ, Narisu N, Hu T, Pruim R, Xiao R, Li XY, Conneely KN, Riebow NL, Sprau AG, Tong M, White PP, Hetrick KN, Barnhart MW, Bark CW, Goldstein JL, Watkins L, Xiang F, Saramies J, Buchanan TA, Watanabe RM, Valle TT, Kinnunen L, Abecasis GR, Pugh EW, Doheny KF, Bergman RN, Tuomilehto J, Collins FS, Boehnke M (2007) A genome-wide association study of type 2 diabetes in Finns detects multiple susceptibility variants. Science 316(5829):1341–1345. doi:10.1126/science.1142382

Shen H, Pollin TI, Damcott CM, McLenithan JC, Mitchell BD, Shuldiner AR (2009) Glucokinase regulatory protein gene polymorphism affects postprandial lipemic response in a dietary intervention study. Hum Genet 126(4):567–574. doi:10.1007/s00439-009-0700-3

Slosberg ED, Desai UJ, Fanelli B, St Denny I, Connelly S, Kaleko M, Boettcher BR, Caplan SL (2001) Treatment of type 2 diabetes by adenoviral-mediated overexpression of the glucokinase regulatory protein. Diabetes 50(8):1813–1820

Sotos-Prieto M, Luben R, Khaw KT, Wareham NJ, Forouhi NG (2014) The association between mediterranean diet score and glucokinase regulatory protein gene variation on the markers of cardiometabolic risk: an analysis in the European prospective investigation into cancer (EPIC)-Norfolk study. Br J Nutr 112(1):1–10. doi:10.1017/S0007114514000580

Sparso T, Andersen G, Nielsen T, Burgdorf KS, Gjesing AP, Nielsen AL, Albrechtsen A, Rasmussen SS, Jorgensen T, Borch-Johnsen K, Sandbaek A, Lauritzen T, Madsbad S, Hansen T, Pedersen O (2008) The GCKR rs780094 polymorphism is associated with elevated fasting serum triacylglycerol, reduced fasting and OGTT-related insulinaemia, and reduced risk of type 2 diabetes. Diabetologia 51(1):70–75. doi:10.1007/s00125-007-0865-z

Speliotes EK, Yerges-Armstrong LM, Wu J, Hernaez R, Kim LJ, Palmer CD, Gudnason V, Eiriksdottir G, Garcia ME, Launer LJ, Nalls MA, Clark JM, Mitchell BD, Shuldiner AR, Butler JL, Tomas M, Hoffmann U, Hwang SJ, Massaro JM, O'Donnell CJ, Sahani DV, Salomaa V, Schadt EE, Schwartz SM, Siscovick DS, Voight BF, Carr JJ, Feitosa MF, Harris TB, Fox CS, Smith AV, Kao WH, Hirschhorn JN, Borecki IB (2011) Genome-wide association analysis identifies variants associated with nonalcoholic fatty liver disease that have distinct effects on metabolic traits. PLoS Genet 7(3), e1001324. doi:10.1371/journal.pgen.1001324

Teslovich TM, Musunuru K, Smith AV, Edmondson AC, Stylianou IM, Koseki M, Pirruccello JP, Ripatti S, Chasman DI, Willer CJ, Johansen CT, Fouchier SW, Isaacs A, Peloso GM, Barbalic M, Ricketts SL, Bis JC, Aulchenko YS, Thorleifsson G, Feitosa MF, Chambers J, Orho-Melander M, Melander O, Johnson T, Li X, Guo X, Li M, Shin Cho Y, Jin Go M, Jin Kim Y, Lee JY, Park T, Kim K, Sim X, Twee-Hee Ong R, Croteau-Chonka DC, Lange LA, Smith

JD, Song K, Hua Zhao J, Yuan X, Luan J, Lamina C, Ziegler A, Zhang W, Zee RY, Wright AF, Witteman JC, Wilson JF, Willemsen G, Wichmann HE, Whitfield JB, Waterworth DM, Wareham NJ, Waeber G, Vollenweider P, Voight BF, Vitart V, Uitterlinden AG, Uda M, Tuomilehto J, Thompson JR, Tanaka T, Surakka I, Stringham HM, Spector TD, Soranzo N, Smit JH, Sinisalo J, Silander K, Sijbrands EJ, Scuteri A, Scott J, Schlessinger D, Sanna S, Salomaa V, Saharinen J, Sabatti C, Ruokonen A, Rudan I, Rose LM, Roberts R, Rieder M, Psaty BM, Pramstaller PP, Pichler I, Perola M, Penninx BW, Pedersen NL, Pattaro C, Parker AN, Pare G, Oostra BA, O'Donnell CJ, Nieminen MS, Nickerson DA, Montgomery GW, Meitinger T, McPherson R, McCarthy MI, McArdle W, Masson D, Martin NG, Marroni F, Mangino M, Magnusson PK, Lucas G, Luben R, Loos RJ, Lokki ML, Lettre G, Langenberg C, Launer LJ, Lakatta EG, Laaksonen R, Kyvik KO, Kronenberg F, Konig IR, Khaw KT, Kaprio J, Kaplan LM, Johansson A, Jarvelin MR, Janssens AC, Ingelsson E, Igl W, Kees Hovingh G, Hottenga JJ, Hofman A, Hicks AA, Hengstenberg C, Heid IM, Hayward C, Havulinna AS, Hastie ND, Harris TB, Haritunians T, Hall AS, Gyllensten U, Guiducci C, Groop LC, Gonzalez E, Gieger C, Freimer NB, Ferrucci L, Erdmann J, Elliott P, Ejebe KG, Doring A, Dominiczak AF, Demissie S, Deloukas P, de Geus EJ, de Faire U, Crawford G, Collins FS, Chen YD, Caulfield MJ, Campbell H, Burtt NP, Bonnycastle LL, Boomsma DI, Boekholdt SM, Bergman RN, Barroso I, Bandinelli S, Ballantyne CM, Assimes TL, Quertermous T, Altshuler D, Seielstad M, Wong TY, Tai ES, Feranil AB, Kuzawa CW, Adair LS, Taylor HA Jr, Borecki IB, Gabriel SB, Wilson JG, Holm H, Thorsteinsdottir U, Gudnason V, Krauss RM, Mohlke KL, Ordovas JM, Munroe PB, Kooner JS, Tall AR, Hegele RA, Kastelein JJ, Schadt EE, Rotter JI, Boerwinkle E, Strachan DP, Mooser V, Stefansson K, Reilly MP, Samani NJ, Schunkert H, Cupples LA, Sandhu MS, Ridker PM, Rader DJ, van Duijn CM, Peltonen L, Abecasis GR, Boehnke M, Kathiresan S (2010) Biological, clinical and population relevance of 95 loci for blood lipids. Nature 466(7307):707–713. doi:10.1038/nature09270

The Diabetes Prevention Program (1999) Design and methods for a clinical trial in the prevention of type 2 diabetes. Diabetes Care 22(4):623–634

van der Harst P, Bakker SJ, de Boer RA, Wolffenbuttel BH, Johnson T, Caulfield MJ, Navis G (2010) Replication of the five novel loci for uric acid concentrations and potential mediating mechanisms. Hum Mol Genet 19(2):387–395. doi:10.1093/hmg/ddp489

Varbo A, Benn M, Tybjaerg-Hansen A, Grande P, Nordestgaard BG (2011) TRIB1 and GCKR polymorphisms, lipid levels, and risk of ischemic heart disease in the general population. Arterioscler Thromb Vasc Biol 31(2):451–457. doi:10.1161/atvbaha.110.216333

Vaxillaire M, Cavalcanti-Proenca C, Dechaume A, Tichet J, Marre M, Balkau B, Froguel P (2008) The common P446L polymorphism in GCKR inversely modulates fasting glucose and triglyceride levels and reduces type 2 diabetes risk in the DESIR prospective general French population. Diabetes 57(8):2253–2257. doi:10.2337/db07-1807

Veiga-da-Cunha M, Van Schaftingen E (2002) Identification of fructose 6-phosphate- and fructose 1-phosphate-binding residues in the regulatory protein of glucokinase. J Biol Chem 277 (10):8466–8473. doi:10.1074/jbc.M105984200

Veiga-da-Cunha M, Delplanque J, Gillain A, Bonthron DT, Boutin P, Van Schaftingen E, Froguel P (2003) Mutations in the glucokinase regulatory protein gene in 2p23 in obese French caucasians. Diabetologia 46(5):704–711. doi:10.1007/s00125-003-1083-y

Veiga-da-Cunha M, Sokolova T, Opperdoes F, Van Schaftingen E (2009) Evolution of vertebrate glucokinase regulatory protein from a bacterial N-acetylmuramate 6-phosphate etherase. Biochem J 423(3):323–332. doi:10.1042/BJ20090986

Willer CJ, Sanna S, Jackson AU, Scuteri A, Bonnycastle LL, Clarke R, Heath SC, Timpson NJ, Najjar SS, Stringham HM, Strait J, Duren WL, Maschio A, Busonero F, Mulas A, Albai G, Swift AJ, Morken MA, Narisu N, Bennett D, Parish S, Shen H, Galan P, Meneton P, Hercberg S, Zelenika D, Chen WM, Li Y, Scott LJ, Scheet PA, Sundvall J, Watanabe RM, Nagaraja R, Ebrahim S, Lawlor DA, Ben-Shlomo Y, Davey-Smith G, Shuldiner AR, Collins R, Bergman RN, Uda M, Tuomilehto J, Cao A, Collins FS, Lakatta E, Lathrop GM, Boehnke M,

Schlessinger D, Mohlke KL, Abecasis GR (2008) Newly identified loci that influence lipid concentrations and risk of coronary artery disease. Nat Genet 40(2):161–169. doi:10.1038/ng.76

Yang Chou J, Mansfield BC (1999) Molecular genetics of type 1 glycogen storage diseases. Trends Endocrinol Metab 10(3):104–113

Yang Q, Kottgen A, Dehghan A, Smith AV, Glazer NL, Chen MH, Chasman DI, Aspelund T, Eiriksdottir G, Harris TB, Launer L, Nalls M, Hernandez D, Arking DE, Boerwinkle E, Grove ML, Li M, Linda Kao WH, Chonchol M, Haritunians T, Li G, Lumley T, Psaty BM, Shlipak M, Hwang SJ, Larson MG, O'Donnell CJ, Upadhyay A, van Duijn CM, Hofman A, Rivadeneira F, Stricker B, Uitterlinden AG, Pare G, Parker AN, Ridker PM, Siscovick DS, Gudnason V, Witteman JC, Fox CS, Coresh J (2010) Multiple genetic loci influence serum urate levels and their relationship with gout and cardiovascular disease risk factors. Circ Cardiovasc Genet 3 (6):523–530. doi:10.1161/CIRCGENETICS.109.934455

Zou L, Sriswasdi S, Ross B, Missiuro PV, Liu J, Ge H (2008) Systematic analysis of pleiotropy in C. elegans early embryogenesis. PLoS Comput Biol 4(2), e1000003. doi:10.1371/journal.pcbi.1000003

Chapter 16
Genetic and Functional Studies Implicate *G6PC2* in the Regulation of Fasting Blood Glucose

Nabila Bouatia-Naji and Richard M. O'Brien

Abstract Genome-wide association studies (GWAS) have shown that single-nucleotide polymorphisms (SNPs) in *G6PC2* are the strongest common determinants of variations in fasting blood glucose (FBG) levels. Despite significant allelic heterogeneity, studies in diverse genetic backgrounds confirmed the role of this gene as a determinant of FBG in a manner independent from the risk of type 2 diabetes. Molecular studies examining the functional impact of these SNPs on *G6PC2* gene transcription and splicing suggest that they affect FBG by directly modulating *G6PC2* expression. This conclusion is supported by studies on *G6pc2* knockout (KO) mice showing that G6pc2 represents a negative regulator of basal glucose-stimulated insulin secretion that acts by hydrolyzing glucose-6-phosphate, thereby reducing glycolytic flux and opposing the action of glucokinase. Suppression of G6PC2 activity might therefore represent a novel therapy to lower FBG and thereby perhaps influence the risk of cardiovascular-associated mortality. GWAS and *G6pc2* KO mouse studies also suggest that G6PC2 affects other aspects of beta-cell function. The evolutionary benefit conferred by G6PC2 remains unclear, but it is likely to be related to its ability to acutely modulate glycolytic flux rather than establish the long-term set point for FBG.

N. Bouatia-Naji
INSERM UMR970, Paris Cardiovascular Research Center PARCC, 56 rue Leblanc, 75015 Paris, France

University Paris-Descartes, Sorbonne Paris Cité, 12 rue de l'école de medicine, 75006 Paris, France

R.M. O'Brien (✉)
Department of Molecular Physiology and Biophysics, Vanderbilt University School of Medicine, 8415 MRB IV, 2213 Garland Ave, Nashville, TN 37232-0615, USA
e-mail: richard.obrien@vanderbilt.edu

J.C. Florez (ed.), *The Genetics of Type 2 Diabetes and Related Traits*,
DOI 10.1007/978-3-319-01574-3_16

16.1 Introduction

It is well established that type 2 diabetes (T2D) and its related quantitative traits, such as fasting blood glucose (FBG), are highly heritable, with heritability estimates for FBG that are higher than those for T2D (Mills et al. 2004; Poulsen et al. 2001). There is also important epidemiological evidence that supports a linear relationship between glucose levels and cardiovascular-associated mortality, even below the glucose threshold for T2D (Balkau et al. 1999; Coutinho et al. 1999). On the other hand, important determinants of T2D risk did not initially appear to markedly impact the variability in FBG in large general populations (Cauchi et al. 2006), which motivated studies into the genetic basis of FBG using hypothesis-free approaches. Over the past 5 years, through the power of genome-wide association studies (GWAS), there has been an explosion in our knowledge with respect to the identity of the genetic loci that influence quantitative traits such as fasting blood glucose (FBG) levels, glycated hemoglobin A1C (HbA1c) levels (Chap. 3), and the loci associated with increased risk for the development of T2D (Chap. 2) and obesity (Chap. 4).

With respect to the genes linked to variations in FBG, multiple GWAS have shown that the *G6PC2* locus harbors the strongest common genetic determinant of FBG levels in terms of significance and effect size with a common single-nucleotide polymorphism (SNP), rs560887, explaining ~1 % of the total variance in FBG (Chap. 3) (Bouatia-Naji et al. 2008; Chen et al. 2008b; Prokopenko et al. 2008; Reiling et al. 2009; Dupuis et al. 2010; Hu et al. 2009, 2010; Tam et al. 2010). Common variants in the *GCK* gene, which encodes glucokinase, have also been linked to variations in FBG, but the influence of these common *GCK* variants on FBG is less than that of the common variants in *G6PC2* (Bouatia-Naji et al. 2008; Dupuis et al. 2010). This observation highlights a critical point, namely, that the magnitude of the effect of common gene variants identified through GWAS does not necessarily correlate with the importance of the gene in relation to the parameter under investigation. With respect to *G6PC2* and *GCK*, deletion of the *G6pc2* gene in mice has a mild metabolic phenotype (Wang et al. 2007; Pound et al. 2013), and rare mutations in *G6PC2* are not a cause of monogenic forms of diabetes (Bonnefond et al. 2009). In contrast, deletion of the *Gck* gene in mice is lethal (Grupe et al. 1995) and rare heterozygous inactivating mutations in *GCK* are a cause of maturity-onset diabetes of the young, which is characterized by mild fasting hyperglycemia, whereas homozygous inactivating glucokinase mutations result in permanent neonatal diabetes mellitus, which is characterized by severe hyperglycemia (Osbak et al. 2009). In contrast, glucokinase activating mutations result in hyperinsulinemia leading to hypoglycemia (Osbak et al. 2009). These rare *GCK* mutations have provided fascinating molecular insights into the function of glucokinase (Osbak et al. 2009) and, along with mouse models of *Gck* overexpression (Magnuson et al. 2003) and tissue-specific deletion (Grupe et al. 1995; Postic et al. 1999), have contributed greatly to the recognition that glucokinase is the pancreatic islet beta-cell glucose sensor (Matschinsky 2005). Far less is known about the *G6PC2* gene, which is the focus of this chapter.

16.2 *G6PC2* Was the First Genetic Determinant of FBG Identified by GWAS

In 2008, two large collaborative studies in European populations were published describing for the first time a very strong statistical signal of association with FBG on Chr2q31.1. One study, conducted by Chen et al. (2008b), identified rs560887 as the strongest FBG-associated signal based on two GWAS including a total of 5088 nondiabetic individuals from Finland and Sardinia. The SNP followed up for replication was rs563694, which is in very high linkage disequilibrium (LD) with rs560887. This study also highlighted the strong LD that exists between several SNPs that show very low statistical *P*-values for association with FBG. These SNPs are located not only in *G6PC2* but also in the intronic regions of the adjacent *ABCB11* gene, which encodes the ATP-binding cassette subfamily B (MDR/TAP) member 11, making both genes putative candidates to explain the association with FBG observed on Chr2q31.1 (Chen et al. 2008b).

A second study, conducted by Bouatia-Naji et al. (2008), independently reported the association of rs560887 with FBG following a GWAS in 654 normoglycemic participants from the French cohort DESIR and a replication sample in ~10,000 people, including young Finnish adolescents and obese French children. rs560887 is located 21 bp upstream the third intron of the glucose-6-phosphatase catalytic subunit 2 (*G6PC2*) gene (Bouatia-Naji et al. 2008). In addition, this study showed that no SNP genotyped or imputed in the Chr2q31.1 region that associates with FBG at a $P < 0.05$ remained associated after conditional regression analyses on rs560887 (Bouatia-Naji et al. 2008). In a very large meta-analysis of GWAS data that included ~120,000 participants from MAGIC, rs560887 was confirmed as the most significantly associated variant with FBG with a 0.075 mmol/l increase per G allele ($P = 8.5 \times 10^{-122}$) (Dupuis et al. 2010). Both Chen et al. (2008b) and Bouatia-Naji et al. (2008) demonstrated that a common variant in the *GCK* gene, which encodes glucokinase, was also associated with FBG at the genomic level, consistent with previous candidate gene studies (Weedon et al. 2006).

16.3 Confirmation of *G6PC2* as a Genetic Determinant of FBG in Non-Europeans and a Genetic Isolate

Several later studies supported the association of variants in regulatory and coding regions of *G6PC2* with FBG. These included confirmatory studies in European cohorts (Reiling et al. 2009; Rose et al. 2009) and several large-scale multiethnic studies in populations of Asian and African descent. Of note, rs560887 exhibits major differences in allele frequency across ethnic groups. According to data from the 1000 Genomes Project, rs560887 is not polymorphic in the Yoruba population and the G allele is at a frequency of 0.97 in Han Chinese. Despite this complication, multiethnic studies supported a role for *G6PC2* in the regulation of FBG based on the association of other *G6PC2* variants, that are more frequent in these ethnicities, with FBG. In a GWAS conducted in ~5700 nondiabetic Indian Asians, several SNPs were associated with FBG, including rs1402837, which is located 5′ of the *G6PC2* gene, that reached genome-wide significant association ($P = 2 \times 10^{-8}$) (Chambers et al. 2009). This study also described the replication of the effect of rs560887 with FBG in Indian Asians ($P = 9 \times 10^{-5}$) and highlighted the important difference ($P = 1.4 \times 10^{-146}$) in terms of frequency for the effect allele of rs560887 between Europeans (frequency of G allele =0.69) and Indian Asians (frequency = 0.85), which explains the lack of power in this ethnic group for detecting a stronger association for this SNP as achieved in the European cohorts (Chambers et al. 2009). In two Chinese populations, an association with FBG was reported for rs16856187, a genetic variant located 3′ region of the *G6PC2* gene, which is present with a higher frequency in Chinese (minor allele frequency [MAF] = 0.29) compared to Europeans (MAF < 0.01) (Hu et al. 2009; Tam et al. 2010).

In a study including Japanese and Sri Lankan populations, rs3755157, located in an intronic region of *ABCB11*, showed stronger association with FBG than r560887, which is not surprising given the allelic heterogeneity and low frequency of the variant allele at rs560887 in Asian populations. In the only study reported in an African American population, rs560887 was not polymorphic, and the analyses of the effect of 46 additional SNPs spanning 61.2 kb at *G6PC2* locus showed marginal contribution of this locus to FBG (Ramos et al. 2011), although the ability to detect true signals may have been affected by the modest sample size ($n = 927$). Larger GWAS and sequencing studies are required to assess the role of this locus in this ethnic group.

A recent study by Service et al. (2014) conducted in a Finnish population, which is a known genetic isolate, demonstrated the utility of the latter approach. Using targeted resequencing of 78 genes located close to 17 GWAS loci associated with lipids and glycemic traits, the authors identified a variant in *G6CP2* with a sixfold higher frequency in Finnish (His177Tyr; rs138726309; MAF = 0.014 in Finnish vs. MAF = 0.0023 in the European ancestry samples in the Exome Variant Server: http://evs.gs.washington.edu/EVS/, $P < 10^{-16}$). This variant changes a highly conserved amino acid and strongly associates with FBG and is predicted in silico to be functional (Service et al. 2014). This association was independent from rs560887, though the effect is in the same direction with the low-frequency alleles associating with lower FBG (Service et al. 2014). No other variant among the 97 SNPs identified by the resequencing of the five genes at this locus, including those identified in *ABCB11*, associates with FBG (Service et al. 2014). This result for a coding variant in *G6PC2* itself effectively pinpoints *G6PC2* as the causative gene at this associated locus.

Altogether, these genetic data, which are based on multiple cohorts and ethnic groups, all show that a Chr2q31.1 locus is a major genetic determinant of FBG with almost all these studies pointing to variants in the *G6PC2* as the gene explaining this genetic association (Fig. 16.1). The molecular and physiological studies described below suggest that G6PC2 directly modulates FBG.

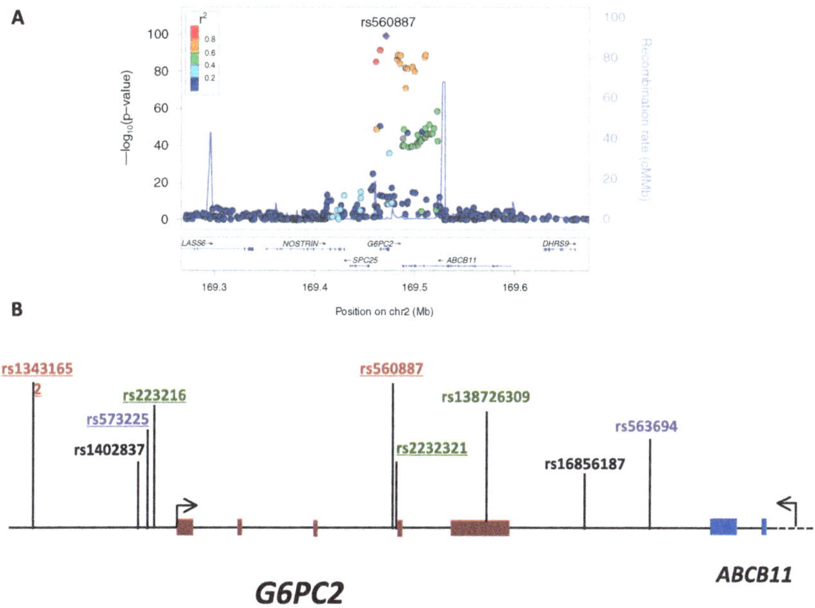

Fig. 16.1 Genomic association of the *G6PC2* locus with FBG.
Panel A shows the genomic context of the SNPs on Chrq31.1 that are associated with FBG and was generated using LocusZoom (Pruim et al. 2010) based on the latest MAGIC meta-analysis of GWAS data (Scott et al. 2012). Panel B shows key *G6PC2* SNPs with those that have been assessed in functional experiments underlined. The SNPs shown in red are the most likely to be causal: rs560887 shows the highest level of statistical association and alters *G6PC2* RNA splicing (Baerenwald et al. 2013); rs13431652, which is in strong LD with rs560887, alters NF-Y binding and *G6PC2*-luciferase fusion gene expression in vitro (Bouatia-Naji et al. 2010) and is located within an islet enhancer (Pasquali et al. 2014). The SNPs shown in green are low-frequency variants that have been demonstrated (rs223216 and rs2232321) or are predicted in silico (rs138726309) to be functional. The relationship between those SNPs shown in purple and FBG is controversial at either a functional (rs573225) or genetic (rs563694) level

16.4 Functional Analyses of SNPs That Modulate *G6PC2* Splicing and Gene Transcription

Several studies have examined the molecular effects of SNPs on *G6PC2* splicing and gene transcription. Two SNPs in the *G6PC2* promoter, rs13431652 and rs2232316, were shown to affect *G6PC2* fusion gene expression by modulating NF-Y and Foxa2 binding, respectively (Bouatia-Naji et al. 2010; Baerenwald et al. 2013). In addition, two SNPs in the third *G6PC2* intron, rs560887 and rs2232321, were shown to affect *G6PC2* RNA splicing (Baerenwald et al. 2013), likely by modulating the strength of a branch point sequence, a key element in RNA splicing (Sharp 1994; Solis et al. 2008). The in vitro and in situ molecular data suggest that all four SNPs are potentially causative since the allele that results in elevated *G6PC2* expression is associated with elevated FBG (Bouatia-Naji et al. 2010; Baerenwald et al. 2013). In contrast, for another *G6PC2* promoter SNP, rs573225, that also affects *G6PC2* fusion gene expression by modulating Foxa2 binding, the allele that results in elevated *G6PC2* expression is associated with reduced FBG (Bouatia-Naji et al. 2010; Baerenwald et al. 2013), suggesting that rs573225 is a functional SNP that opposes the action of causative SNPs on *G6PC2* expression (Bouatia-Naji et al. 2010; Baerenwald et al. 2013), a conclusion that contrasts with an earlier study (Dos Santos et al. 2009). A recent study that mapped and examined the function of human islet *cis*-regulatory networks showed that the rs13431652 promoter variant, though not the rs573225 promoter variant, is located within an enhancer cluster (Pasquali et al. 2014). This finding further supports the functional role of this promoter variant in the regulation of *G6PC2*.

16.5 Challenges in the Identification of Causative *G6PC2* SNPs

There are several key limitations in the analysis of *G6PC2* causative SNPs. First, because these SNPs are in high LD, it is difficult to definitely determine whether one or all of these SNPs are truly causative (Bouatia-Naji et al. 2010; Baerenwald et al. 2013). Second, for *G6PC2*, testing for the association of these SNPs in non-European populations in an attempt to identify a minimal haplotype that could include the functional variant is compromised by the very high allelic heterogeneity of this locus across ethnic groups (Hu et al. 2010; Ramos et al. 2011). Third, because the *G6PC2* gene is only expressed in pancreatic islet beta cells and because the effects of these SNPs are subtle, the lack of sufficient human samples has limited the ability to correlate genotypes with endogenous *G6PC2* expression. Finally, multiple caveats are associated with analyzing *G6PC2* promoter SNPs using fusion genes in islet-derived cell lines (Bouatia-Naji et al. 2010; Baerenwald et al. 2013). Furthermore, most of these cell lines are derived from rodent islets, and recent studies suggest the existence of significant differences between rodent and human islets (Dai et al. 2012).

16.6 Challenges in the Demonstration of a Causative Role for G6PC2 in the Regulation of FBG

As described in Part II of this book, following up on the genes implicated by GWAS in an effort to understand how particular genes contribute to disease risk is a difficult proposition because in many instances, the SNPs that have been linked to disease risk fall in intergenic regions. As such, identifying the disease-related gene(s) associated with these SNPs is a significant challenge. In some cases, these intergenic SNPs may impact the function of transcriptional control structures, such as enhancers and silencers (Cecchini et al. 2009), such that the genes whose expression are affected may be located a considerable distance from the SNP, though in the literature the nearest gene is often assumed to be the likely candidate. Indeed, as noted above, the *G6PC2* rs13431652 promoter variant is located within an enhancer cluster (Pasquali et al. 2014). This implies that this SNP might not only modulate *G6PC2* expression but also that of distant genes. Similarly, even though molecular studies show that rs560887 affects *G6PC2* RNA splicing (Baerenwald et al. 2013), Taneera et al. (2012) have suggested that rs560887 also acts in *trans* to modulate the expression of multiple other genes.

The complexity of these genetic and molecular data demonstrates the importance of physiological and biochemical experiments to investigate the function of G6PC2.

16.7 Correlation Between Genetic Association in Humans and Knockout Mouse Data with Respect to the Regulation of FBG by G6PC2

Data from knockout (KO) mouse studies strongly support the hypothesis that genetic variation within the *G6PC2* gene, rather than surrounding genes, directly contributes to variations in FBG in humans. Thus, a ~15 % decrease in FBG is observed following a global KO of *G6pc2* in mice (Wang et al. 2007; Pound et al. 2013). This decrease in FBG is observed when *G6pc2* KO mice are studied on a mixed (Wang et al. 2007) or pure C57BL/6J (Pound et al. 2013) genetic background. Importantly, the direction of the effect of *G6pc2* deletion on FBG is consistent with the molecular analyses of the impact of SNPs on *G6PC2* expression. For example, the rs560887 allele that is associated with reduced *G6PC2* RNA splicing, required to generate a functional transcript, is associated with reduced FBG (Baerenwald et al. 2013).

16.8 Correlation Between Genetic Association in Humans and Biochemical Data with Respect to the Regulation of FBG by G6PC2

16.8.1 G6PC2 Encodes a Glucose-6-Phosphatase Catalytic Subunit Expressed Exclusively in Pancreatic Islet Beta Cells

Glucose-6-phosphatase catalyzes the hydrolysis of glucose-6-phosphate (G6P) to glucose and inorganic phosphate (Mithieux 1997; Foster et al. 1997; van de Werve et al. 2000; Van Schaftingen and Gerin 2002; Hutton and O'Brien 2009). It exists as a multicomponent system located in the endoplasmic reticulum and is composed of several integral membrane proteins, namely, a catalytic subunit (G6PC), a glucose transporter, and a G6P/inorganic phosphate antiporter (Mithieux 1997; Foster et al. 1997; van de Werve et al. 2000; Van Schaftingen and Gerin 2002; Hutton and O'Brien 2009). Three G6PC isoforms have been identified, designated G6PC, G6PC2, and G6PC3 (Hutton and O'Brien 2009). Each isoform is encoded by a separate gene with a distinct pattern of tissue-specific expression (Hutton and O'Brien 2009). *G6PC2* was originally named *IGRP*, which stands for islet-specific glucose-6-phosphatase catalytic subunit-related protein (Arden et al. 1999; Ebert et al. 1999). The gene is expressed exclusively in pancreatic islet beta cells (Hutton and Eisenbarth 2003). G6PC2 is a major autoantigen in both mouse (Lieberman et al. 2003; Han et al. 2005; Mukherjee et al. 2005) and human (Yang et al. 2006; Jarchum et al. 2008) type 1 diabetes, but interestingly, *G6PC2* SNPs are not associated with type 1 diabetes risk (https://www.t1dbase.org/page/Regions/display/species/human/disease/T1D/type/assoc).

16.8.2 The Mechanism of FBG Regulation by G6PC2

A comparison of glucose-6-phosphatase activity in islets isolated from wild-type and *G6pc2* KO mice indicates that activity is abolished in the latter (Pound et al. 2013). These data led to the simple hypothesis that G6pc2 acts as a negative regulator of basal glucose-stimulated insulin secretion (GSIS) by hydrolyzing G6P and thereby opposing the action of the glucose sensor, glucokinase (Matschinsky 1996; Iynedjian 2009) (Fig. 16.2). This glucokinase/G6pc2 futile substrate cycle is predicted to reduce glycolytic flux and hence insulin secretion (Pound et al. 2013). Consistent with this model and human genetic association data, a reduction in *G6pc2* expression results in a leftward shift in the dose–response curve for GSIS explaining why, under fasting conditions, blood glucose levels are reduced (Pound et al. 2013).

16.9 Do Genetic Association Studies and Knockout Mouse Data Resolve the Historical Controversy Over Islet Glucose-6-Phosphatase?

Since liver is the organ primarily responsible for glycogenolysis and gluconeogenesis during fasting (Mithieux 1997; Foster et al. 1997; van de Werve et al. 2000; Van Schaftingen and Gerin 2002; Hutton and O'Brien 2009) and islets do not express the key gluconeogenic enzyme phosphoenolpyruvate carboxykinase (MacDonald et al. 1992), historically the role of glucose-6-phosphatase activity in

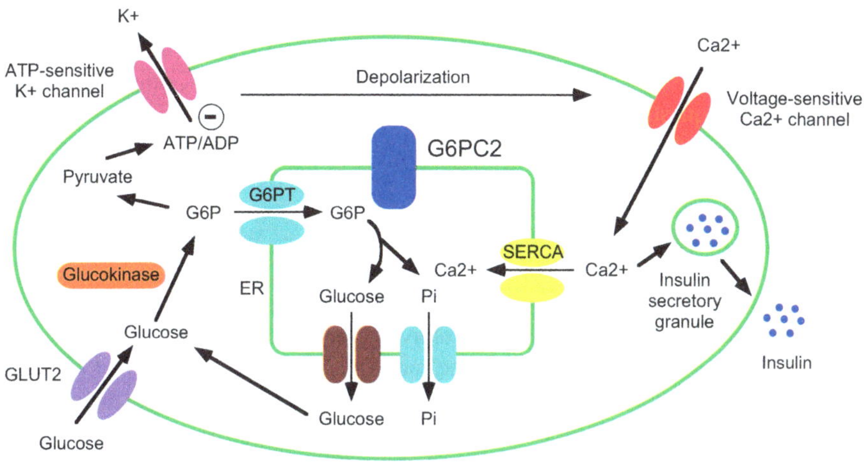

Fig. 16.2 G6PC2 regulates GSIS by opposing the action of glucokinase.
The model shows the best-characterized pathway for GSIS, though studies on potassium channel mutations indicate that other pathways clearly contribute (Jensen et al. 2008). The glucose-6-phosphate (G6P) transporter (G6PT) and inorganic phosphate transporter are shown in the same color to indicate the fact that G6PT is actually a G6P:Pi antiporter (Chen et al. 2008a). The model proposes that G6PC2 regulates GSIS by opposing the action of glucokinase, but it also suggests that G6PC2 might modulate islet calcium metabolism through its ability to promote the generation of inorganic phosphate in the endoplasmic reticulum lumen resulting in the retention of calcium (Wolf et al. 1986) (Modified from Pound LD, Oeser JK, O'Brien TP, Wang Y, Faulman CJ, Dadi PK, Jacobson DA, Hutton JC, McGuinness OP, Shiota M, O'Brien RM: G6PC2: A Negative Regulator of Basal Glucose-Stimulated Insulin Secretion. Diabetes 2013;62:1547–1556) (Pound et al. 2013)

pancreatic islets has been highly controversial. Initially there was a debate over whether such activity even existed in islets though over time the majority of studies found that activity was detectable but at a lower level than that found in the liver (Waddell and Burchell 1988; Foster et al. 1997; Arden et al. 1999; Perales et al. 1991; Trandaburu 1977; Sweet et al. 1997). However, the harder issue to resolve is whether the level of activity is enough to affect glycolytic flux and hence GSIS and therefore be of physiological significance. Several groups have investigated this question through studies on isolated rat and mouse islets.

16.9.1 Glucose Cycling in Isolated Rat Islets

Sweet et al. (1997) concluded that while glucose-6-phosphatase activity is present in rat islets, the level of activity is not enough to result in sufficient G6P hydrolysis so as to affect GSIS. However, the relevance of these rat islet data to the human GWAS data is unclear because, in contrast to all other vertebrate species examined (see http://genome.ucsc.edu/), *G6PC2* is a pseudogene in rats (Martin et al. 2001).

The observations of Sweet et al. (1997) and the absence of G6PC2 in rats raise the question as to what benefit islet glucose-6-phosphatase activity confers to mice and humans that is dispensable to rats. However, three observations suggest that the premise of this question is not well founded. First, rat islets express low levels of the G6PC isoform, whereas microarray data show no expression of *G6pc* or *G6pc3* in mouse islets (unpublished data). *G6PC* is predominantly expressed in the liver and kidney where it catalyzes the final step in the gluconeogenic and glycogenolytic pathways (Mithieux 1997; Foster et al. 1997; van de Werve et al. 2000; Van Schaftingen and Gerin 2002; Hutton and O'Brien 2009), but in various rat models associated with impaired glucose tolerance, *G6PC* expression is induced (Khan et al. 1990b; Laybutt et al. 2003; Tokuyama et al. 1995). Second, Pedersen et al. (2007) have demonstrated that the rat *G6PC* promoter is activated strongly by glucose, much more so than the mouse or human *G6PC* promoters. This implies that the estimates of G6P hydrolysis in rat islets will be very dependent on the glucose concentration in the culture medium and hence the level of induction of *G6PC*. Third, G6PC is ~20-fold more active than G6PC2 (Petrolonis et al. 2004) such that much less G6PC is required in rat islets to catalyze an equivalent rate of G6P hydrolysis as observed in mouse islets. It therefore appears that G6PC may play the same role in rat islets as G6PC2 does in human and mouse islets.

16.9.2 Glucose Cycling in Isolated Mouse Islets

Several groups have also examined G6P hydrolysis in mouse, rather than rat islets. One early study suggested that even though glucose-6-phosphatase activity is present in mouse islets, G6P hydrolysis does not occur (Ashcroft and Randle 1968). While this conclusion appears counterintuitive, it could be explained if G6P entry into the ER lumen was restricted. Nevertheless, this conclusion was challenged by later studies, which showed that the measurement of G6P hydrolysis within islets is critically dependent on experimental conditions (Khan et al. 1989; Chandramouli et al. 1991). The *G6pc2* KO mouse data described above, and especially the demonstration that G6pc2 accounts for the low glucose-6-phosphatase enzyme activity detected in mouse islets (Pound et al. 2013), would seem to resolve the historical controversy over the importance of glucose-6-phosphatase activity in islets. However, two potential caveats remain.

The first potential caveat relates to the fact that estimates of glucose cycling in mouse pancreatic islets are very low (Khan et al. 1990a). However, these estimates of glucose cycling were generated using radioisotopes, and the methodology involved is associated with a number of assumptions (Khan et al. 1989; Chandramouli et al. 1991). To avoid these assumptions, it will be essential to reassess the level of glucose cycling in pancreatic islets using more recently developed stable isotope methodology (Jazmin and Young 2013). If the rates of glucose cycling calculated using this technology are greater than previously estimated using radioisotopes, then this would support the hypothesis that G6PC2 directly influences GSIS through its ability to hydrolyze G6P. The second potential caveat is that because the glucose-6-phosphatase activity of G6PC2 is ~20-fold lower than that of G6PC (Petrolonis et al. 2004; Pound et al. 2013), it heightens the concern that G6PC2 may be influencing GSIS through a mechanism independent of its ability to hydrolyze G6P. Indeed, all three G6PC isoforms possesses a phosphatidic acid phosphatase domain (Martin et al. 2002) raising the possibility that they may also have lipid substrates.

16.10 Human Genetic Association Studies Provide Novel Insights into the Influence of G6PC2 on GSIS

Another potential caveat concerning the function of G6PC2 in pancreatic islets relates to the unexpected effects of altered *G6PC2* expression on insulin secretion during glucose tolerance tests. This issue was initially uncovered through the analysis of GWAS data rather than the analysis of *G6pc2* KO mice. This provides an interesting example of how genetic association data can not only provide insight into genes linked to the initial parameter under investigation but also provide insight into other functions of the genes identified.

16.10.1 G6PC2 Does Not Influence Glucose Tolerance in Mice

As mentioned above, a reduction in *G6pc2* expression results in a leftward shift in the dose–response curve for GSIS explaining why, under fasting conditions, blood glucose levels are reduced (Pound et al. 2013). But under conditions of elevated blood glucose, this same leftward shift arising from a reduction in *G6pc2* expression should result in increased GSIS. Indeed, in perfused pancreas studies comparing pancreata from wild-type and *G6pc2* KO mice, GSIS is increased in the KO pancreata at submaximal glucose concentrations (Pound et al. 2013). Likewise, in isolated islet studies comparing islets from wild-type and *G6pc2* KO mice, GSIS is increased in the KO islets at submaximal glucose concentrations (Pound et al. 2013). This increased insulin secretion at submaximal glucose concentrations might be predicted to result in improved glucose tolerance in *G6pc2* KO mice. However, neither intraperitoneal and oral glucose tolerance tests show major differences in glucose tolerance or insulin secretion between WT and *G6pc2* KO mice over a range of glucose concentrations (Pound et al. 2013). This result likely reflects the fact that glucose tolerance tests, in which blood glucose levels vary over time, are not the optimal assays for detecting the influence of G6PC2 on GSIS. Instead, examining GSIS in wild type and KO mice at submaximal glucose concentrations, as used in the isolated islet and perfused pancrease studies described above, is required.

16.10.2 G6PC2 Does Not Influence Glucose Tolerance in Humans

These observations in *G6pc2* KO mice are consistent with earlier human genetic data showing no association between *G6PC2* and glucose tolerance (Li et al. 2009; Rose et al. 2009; Ingelsson et al. 2010; Heni et al. 2010). Human genetic data also showed that *G6PC2* is not associated with variations in insulin sensitivity in humans (Li et al. 2009; Rose et al. 2009; Ingelsson et al. 2010; Heni et al. 2010), an observation that was confirmed in *G6pc2* KO mice (Pound et al. 2013). To further complicate the situation, human GWAS data show that the SNP within the *G6PC2* gene that is associated with reduced *G6PC2* expression (Baerenwald et al. 2013) and reduced FBG (Bouatia-Naji et al. 2008) is actually associated with a reduction in insulin secretion during glucose tolerance tests rather than the expected increase (Li et al. 2009; Rose et al. 2009; Ingelsson et al. 2010; Heni et al. 2010). So in humans, reduced *G6PC2* expression appears to promote glycolytic flux leading to reduced FBG, but this enhanced flux not only fails to enhance glucose tolerance during a glucose challenge but it is actually associated with a decrease in insulin secretion during that glucose challenge. The data with *G6pc2* KO mice appear slightly different to the human GWAS data in that a similar reduction in insulin secretion was not observed in *G6pc2* KO mice during glucose tolerance tests (Pound et al. 2013). However, this may simply be due to the relatively low number of animals analyzed (Pound et al. 2013) relative to the vast number of humans analyzed in genetic association studies (Li et al. 2009). Thus mouse data are inherently noisy in that significant variations in insulin sensitivity, and hence insulin secretion, are observed even within inbred C57BL/6 J mice (Koza et al. 2006). The key consistent observation is that in both mice and humans, reduced *G6PC2* expression does not lead to an improvement in glucose tolerance.

16.10.3 G6PC2 May Influence the Pulsatility of Insulin Secretion

The unexpected association between reduced *G6PC2* expression and reduced insulin secretion during glucose tolerance tests in humans has been hypothesized to indicate that either G6PC2 affects the pulsatility of insulin secretion (Li et al. 2009) or that it affects hepatic glucose production rather than beta-cell function (Heni et al. 2010). The latter explanation appears highly unlikely since human *G6PC2* (Martin et al. 2001; Bouatia-Naji et al. 2008) and mouse *G6pc2* (Arden et al. 1999; Hutton and O'Brien 2009) are only expressed in islets and not in the liver. Furthermore, while the neighboring *ABCB11* gene is expressed in the liver, its potential role as the molecular mediator of the genetic association between variants at Chr2q.31.2 and FBG has been ruled out as a result of transethnic genetic studies (Chambers et al. 2009; Hu et al. 2010; Takeuchi et al. 2010) and through resequencing of candidate genes at this locus (Service et al. 2014). In contrast a change in the pulsatility of insulin secretion, and hence the efficacy of insulin signaling, would provide an elegant explanation as to how reduced *G6PC2* expression could lead to a reduction in insulin secretion that is not associated with a counterbalancing change in glucose tolerance or insulin sensitivity (Li et al. 2009; Rose et al. 2009; Ingelsson et al. 2010; Heni et al. 2010).

16.10.4 G6PC2 May Influence Islet Calcium Localization

Whether the absence of G6pc2 in mice affects the pulsatility of insulin secretion is a key question that remains to be addressed. However, if this is the explanation for the reduced insulin secretion during glucose tolerance tests in humans, then the question arises as to whether the ability of G6PC2 to influence the pulsatility of insulin secretion is dependent on its ability to hydrolyze G6P or some other function, perhaps connected with the phosphatidic acid phosphatase domain mentioned above (Martin et al. 2002). The generation of a transgenic model in which a mutated form of G6pc2 lacking glucose-6-phosphatase activity is expressed in the *G6pc2* KO mice might provide insight into this question. Merrins et al. (2010) have elegantly shown that pulsatile insulin secretion is driven by metabolic oscillations and that the magnitude of the pulses can be amplified by raising intracellular calcium levels. This then raises the question as to whether G6PC2 might affect metabolic oscillations, intracellular calcium levels, or both. Merrins et al. (2012) have recently shown that metabolic oscillations in islets are initiated at an early stage in glycolysis with the mechanism likely involving the autocatalytic feedback of fructose 1,6-bisphosphate onto phosphofructokinase 1, with phosphofructokinase 1 being activated by its product resulting in the subsequent depletion of its substrate (Bertram et al. 2010). Based on these data, it would appear more likely that G6PC2 affects pulsatile insulin secretion through an action on intracellular islet calcium levels rather than metabolic oscillations. Indeed, basal cytoplasmic calcium levels are enhanced in islets isolated from *G6pc2* KO mice (Pound et al. 2013). This increase was interpreted as a secondary event resulting from the enhanced rate of glycolytic flux (Pound et al. 2013), but, based on the studies of Merrins et al. (2010), if this difference was instead, at least in part, a primary consequence of *G6pc2* deletion, then a difference in intracellular calcium between wild-type and *G6pc2* KO islets would be predicted to be associated with altered pulsatile insulin secretion. Indeed G6PC2 might modulate islet calcium metabolism through its ability to promote the generation of inorganic phosphate in the endoplasmic reticulum lumen resulting in the retention of calcium (Wolf et al. 1986) (Fig. 16.2).

16.11 The Relationship Between G6PC2, FBG, and Type 2 Diabetes Risk

16.11.1 G6PC2 Is Not Associated with T2D risk

Since *G6PC2* is associated with variations in FBG and HbA1c (Bouatia-Naji et al. 2008; Chen et al. 2008b; Prokopenko et al. 2008; Reiling et al. 2009; Dupuis et al. 2010; Hu et al. 2009, 2010; Tam et al. 2010) and because the accepted dogma is that elevated FBG and HbA1c are associated with an increased risk for the development of type 2 diabetes (Droumaguet et al. 2006; Abdul-Ghani and DeFronzo 2009; Edelman et al. 2004; Ye 2013), one would logically have expected that *G6PC2* would also be associated with increased risk for the development of type 2 diabetes. Indeed studies on Chinese individuals have shown such an association (Hu et al. 2009, 2010). However, the sample sizes used in these studies were relatively small, and large meta-analyses of GWAS data in this ethnic group do not support this result with no SNP in or near *G6PC2* associated with T2D risk with a $P < 10^{-4}$ (Cho et al. 2012; Sim et al. 2011). Similarly, in European populations, variants in *G6PC2* have no significant effect on the risk for the development of T2D (Bouatia-Naji et al. 2008; Dupuis et al. 2010; Reiling et al. 2009). Interestingly, while this lack of association between a gene with a confirmed role in controlling glucose variability in healthy populations and T2D was a surprise, it is now apparent that *G6PC2* is only the first example of a large series of genetic determinants of FBG that have no effect on T2D risk. In the latest meta-analysis from MAGIC, only 22 loci (out of 36) that are genetic determinants of FBG are also genetic determinants of T2D risk (Scott et al. 2012). This analysis confirmed that common genetic variations in *GCK* but not in *G6PC2* are associated with T2D risk (Scott et al. 2012).

16.11.2 Glucose Tolerance Versus FBG as a Predictor of T2D Risk

A study by Abdul-Ghani et al. (2010) provides a potentially elegant resolution to this paradox of the dissociation of the genetic architecture of FBG and T2D, two highly correlated and interrelated traits. They have reported that the 1-h glucose level in a glucose tolerance test is a better predictor of type 2 diabetes risk than FBG such that, after correcting for this variable, the association between FBG levels and type 2 diabetes risk is lost. The authors suggest that the apparent correlation between elevated FBG and type 2 diabetes risk is not due to the increase in FBG per se but is due instead to the correlation between FBG and 1-h glycemia (Abdul-Ghani et al. 2010). Based on the observations of Abdul-Ghani et al. (2010), one would predict that variations in *GCK* would affect both 2-h and 1-h glucose levels in a glucose tolerance test, in addition to FBG, whereas variations in *G6PC2* would

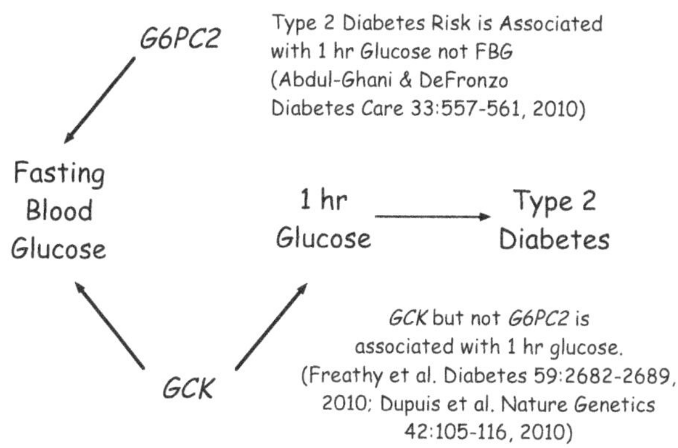

Fig. 16.3 Variations in *GCK* but not *G6PC2* affect type 2 diabetes risk. GWAS data show that SNPs in both *G6PC2* and *GCK* are associated variations in FBG but only SNPs in *GCK* are associated with type 2 diabetes risk. The model proposes that this observation is explained by the fact that SNPs in *GCK* are associated variations in 2-h and 1-h glucose levels during a glucose tolerance test, whereas SNPs in *G6PC2* are not. This concept is based on the studies of Abdul-Ghani et al. (2010) who showed that the 1-h glucose level in a glucose tolerance test is a better predictor of type 2 diabetes risk than FBG such that, after correcting for this variable, the association between FBG levels and type 2 diabetes risk is lost

affect only the latter (Fig. 16.3). Indeed, the rs1799884 *GCK* variant is associated with higher 2-h and 1-h glucose levels (Freathy et al. 2010). In some populations (Freathy et al. 2010), though not others (Li et al. 2009), this variant is also associated with higher 2-h glucose levels. In contrast, the rs560887 *G6PC2* variant is not associated with altered glucose tolerance (Li et al. 2009; Rose et al. 2009; Ingelsson et al. 2010; Heni et al. 2010). In support of this explanation, almost all (eight out of nine) genetic determinants of 2-h glucose levels identified by GWAS are associated with T2D risk (Scott et al. 2012).

The observations of Abdul-Ghani et al. (2010) also lead to the conclusion that the observed decrease in *G6PC2* expression in islets from donors with type 2 diabetes (Taneera et al. 2012) is likely to be a secondary event, specifically a response to the diabetic environment, rather than a causative event that contributes to the development of type 2 diabetes. Thus based on *G6pc2* knockout mouse data (Wang et al. 2007; Pound et al. 2013), a decrease in *G6PC2* expression would lead to enhanced insulin secretion, which would make sense in terms of a compensatory attempt by unhealthy islets to maintain insulin secretion.

Elevated FBG, HbAlc, and body mass have not only been correlated with the risk of developing type 2 diabetes but also cardiovascular-associated mortality (Coutinho et al. 1999; Lawes et al. 2004; Poirier et al. 2006). Just as the studies of Abdul-Ghani et al. (2010) challenge the precise relationships between FBG and type 2 diabetes, the studies of Sarwar et al. (2010) suggest that the precise relationship between FBG and cardiovascular-associated mortality requires further investigation.

As described in Sect. 16.4, the rs560887 *G6PC2* variant linked by GWAS to FBG but not type 2 diabetes risk is predicted to have only a small effect on *G6PC2* expression (Baerenwald et al. 2013). However, even though small changes in *G6PC2* expression are not associated by GWAS with the risk for type 2 diabetes, this would not exclude the possibility that rare variants, such as the His177Tyr rs13872630 variant recently discovered in Finnish individuals (Service et al. 2014), which markedly elevated *G6PC2* expression or increased activity, might be associated with altered risk for type 2 diabetes.

16.12 Evolution and the Function of G6PC2

An intriguing aspect of the human GWAS data discussed above is the conclusion that the absence of G6PC2 would be beneficial to several aspects of human health. In relation to FBG, it would be lowered by the absence of G6PC2, which would be predicted to perhaps reduce the risk of cardiovascular-associated mortality and type 2 diabetes (but see above). Similarly, the human GWAS data imply that if G6PC2 is affecting the pulsatility of insulin secretion and hence the efficacy of insulin signaling, then this pulsatility is actually enhanced by the absence of G6PC2, explaining the reduced insulin secretion during glucose tolerance tests (Li et al. 2009).

Since the presence of G6PC2 has been retained though mammalian evolution, with the exception of rats, this implies that there must be beneficial effects conferred by G6PC2 that are not apparent from the human GWAS data. The health benefits conferred by reduced FBG are unlikely to be relevant to reproductive potential since the diseases associated with elevated FBG typically occur later in life long after an individual has passed on their genetic material to their offspring. Furthermore, such diseases are only prevalent to the modern world and would not even have been a factor during the course of evolution.

The biological benefit(s) conferred by the presence of G6PC2 are currently unknown, but one possibility is that *G6PC2* expression and G6PC2 enzyme activity are activated under specific physiological conditions. This would have the effect of shifting the dose–response curve for GSIS to the right, resulting in reduced insulin secretion (Fig. 16.4). A number of studies have suggested that the activity of hepatic G6PC is altered by insulin signaling (Barzilai and Rossetti 1993), though because the mechanisms involved are unknown, it is unclear whether the same signaling pathway might regulate G6PC2 enzyme activity. Similarly, while multiple transcription factors that contribute to the islet beta cell-specific expression of G6PC2 have been identified (Hutton and O'Brien 2009), there is currently no evidence that *G6PC2* expression is modulated in vivo under different physiological conditions. Nonetheless, there is circumstantial data to support the potential impact of altered *G6PC2* expression. First, in various rat models associated with impaired glucose tolerance, *G6PC* expression is induced such that G6P hydrolysis would be elevated and GSIS blunted (Khan et al. 1990b; Laybutt et al. 2003; Tokuyama et al. 1995).

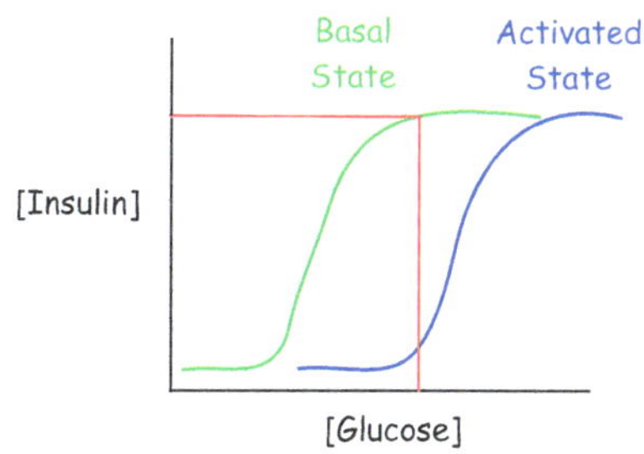

Fig. 16.4 Elevated *G6PC2* expression or enzyme activity would alter the sensitivity of GSIS. Elevated *G6PC2* expression or enzyme activity would alter the sensitivity of GSIS. The evolutionary function of G6PC2 is unknown but is unlikely to be related to the control of FBG. The model speculates that under specific physiological conditions, *G6PC2* expression and enzyme activity are elevated changing the sensitivity of GSIS. This model is consistent with the observations that a reduction in *G6PC2* expression (Pound et al. 2013) or *Gck* overexpression (Wang and Iynedjian 1997b, 1997a) augments glycolytic flux and causes a leftward shift in the dose–response curve for GSIS

This induction may play a protective role against excessive stimulation of the beta cells, which is a concern given their susceptibility to ER (Back and Kaufman 2012) and oxidative (Karunakaran and Park 2013) stress. Second, experiments in which *G6PC* was overexpressed in pancreatic islet beta cell-derived cell lines using adenovirus (Trinh et al. 1997) or stable transfection (Iizuka et al. 2000) have directly demonstrated that altered expression of this single gene is sufficient to inhibit insulin secretion.

16.13 Conclusions and Future Directions

The genetic association and molecular studies described above strongly suggest that G6PC2 modulates FBG by hydrolyzing G6P, thereby opposing the action of the beta-cell glucose sensor, glucokinase. However, these studies suggest that G6PC2 has other unexplained effects on islet beta-cell function that merit further investigation. The evolutionary benefit conferred by G6PC2 remains unclear, but it is likely to be related to its ability to acutely modulate glycolytic flux rather than establish the long-term set point for FBG. The analysis of rare SNPs that markedly affect G6PC2 enzyme activity and the analysis of the biological impact of these SNPs might provide further insight into G6PC2 function, as have similar studies with glucokinase. Finally, because G6PC2 opposes the action of glucokinase, suppression of G6PC2 activity might represent a novel therapy to lower FBG and HbA1c levels and thereby perhaps influence the risk of cardiovascular-associated mortality. It is noteworthy that the alternate strategy, the use of GCK activators, is under investigation in clinical trials (Matschinsky 2009; Iynedjian 2009).

Acknowledgments We would like to thank Les Satin (the University of Michigan) for helpful comments on the mechanism of pulsatile insulin secretion. Research in the laboratory of R. O'B. was supported by NIH grant DK92589.

Recent studies using stable isotopes to examine glucose cycling in pancreatic islets isolated from wild type mice have shown much higher levels of glucose cycling than previously observed using radioisotopes. In addition, these studies have shown that glucose cycling is abolished in islets isolated from G6pc2 KO mice. Both observations support the model presented for the function of G6PC2 in islets. Reference: Wall ML, Pound LD, Trenary I, O'Brien RM, Young JD (2015) Novel stable isotope analyses demonstrate significant rates of glucose cycling in mouse pancreatic islets. Diabetes 64:2129-2137. PMC4439557.

References

Abdul-Ghani MA, DeFronzo RA (2009) Plasma glucose concentration and prediction of future risk of type 2 diabetes. Diabetes Care 32(Suppl 2):S194–S198. doi:10.2337/dc09-S309, 32/suppl_2/S194 [pii]

Abdul-Ghani MA, Stern MP, Lyssenko V, Tuomi T, Groop L, Defronzo RA (2010) Minimal contribution of fasting hyperglycemia to the incidence of type 2 diabetes in subjects with normal 2-h plasma glucose. Diabetes Care 33(3):557–561. doi:10.2337/dc09-1145, dc09-1145 [pii]

Arden SD, Zahn T, Steegers S, Webb S, Bergman B, O'Brien RM, Hutton JC (1999) Molecular cloning of a pancreatic islet-specific glucose-6-phosphatase catalytic subunit-related protein. Diabetes 48(3):531–542

Ashcroft SJ, Randle PJ (1968) Glucose-6-phosphatase activity of mouse pancreatic islets. Nature 219(5156):857–858

Back SH, Kaufman RJ (2012) Endoplasmic reticulum stress and type 2 diabetes. Annu Rev Biochem 81:767–793. doi:10.1146/annurev-biochem-072909-095555

Baerenwald DA, Bonnefond A, Bouatia-Naji N, Flemming BP, Umunakwe OC, Oeser JK, Pound LD, Conley NL, Cauchi S, Lobbens S, Eury E, Balkau B, Lantieri O, Dadi PK, Jacobson DA, Froguel P, O'Brien RM (2013) Multiple functional polymorphisms in the G6PC2 gene contribute to the association with higher fasting plasma glucose levels. Diabetologia 56 (6):1306–1316. doi:10.1007/s00125-013-2875-3

Balkau B, Bertrais S, Ducimetiere P, Eschwege E (1999) Is there a glycemic threshold for mortality risk? Diabetes Care 22(5):696–699

Barzilai N, Rossetti L (1993) Role of glucokinase and glucose-6-phosphatase in the acute and chronic regulation of hepatic glucose fluxes by insulin. J Biol Chem 268(33):25019–25025

Bertram R, Sherman A, Satin LS (2010) Electrical bursting, calcium oscillations, and synchronization of pancreatic islets. Adv Exp Med Biol 654:261–279. doi:10.1007/978-90-481-3271-3_12

Bonnefond A, Bouatia-Naji N, Simon A, Saint-Martin C, Dechaume A, de Lonlay P, Polak M, Bellanne-Chantelot C, Froguel P, Vaxillaire M (2009) Mutations in G6PC2 do not contribute to monogenic forms of early infancy diabetes and beta cell dysfunction. Diabetologia 52 (5):982–985. doi:10.1007/s00125-009-1299-6

Bouatia-Naji N, Rocheleau G, Van Lommel L, Lemaire K, Schuit F, Cavalcanti-Proenca C, Marchand M, Hartikainen AL, Sovio U, De Graeve F, Rung J, Vaxillaire M, Tichet J, Marre M, Balkau B, Weill J, Elliott P, Jarvelin MR, Meyre D, Polychronakos C, Dina C, Sladek R, Froguel P (2008) A polymorphism within the G6PC2 gene is associated with fasting plasma glucose levels. Science 320(5879):1085–1088

Bouatia-Naji N, Bonnefond A, Baerenwald DA, Marchand M, Bugliani M, Marchetti P, Pattou F, Printz RL, Flemming BP, Umunakwe OC, Conley NL, Vaxillaire M, Lantieri O, Balkau B, Marre M, Levy-Marchal C, Elliott P, Jarvelin MR, Meyre D, Dina C, Oeser JK, Froguel P,

O'Brien RM (2010) Genetic and functional assessment of the role of the rs13431652-A and rs573225-A alleles in the G6PC2 promoter that strongly associate with elevated fasting glucose levels. Diabetes 59(10):2662–2671. doi:10.2337/db10-0389, db10-0389 [pii]

Cauchi S, Meyre D, Choquet H, Dina C, Born C, Marre M, Balkau B, Froguel P (2006) TCF7L2 variation predicts hyperglycemia incidence in a French general population: the data from an epidemiological study on the insulin resistance syndrome (DESIR) study. Diabetes 55 (11):3189–3192. doi:10.2337/db06-0692

Cecchini KR, Raja Banerjee A, Kim TH (2009) Towards a genome-wide reconstruction of cis-regulatory networks in the human genome. Semin Cell Dev Biol 20(7):842–848. doi:10.1016/j.semcdb.2009.06.005

Chambers JC, Zhang W, Zabaneh D, Sehmi J, Jain P, McCarthy MI, Froguel P, Ruokonen A, Balding D, Jarvelin MR, Scott J, Elliott P, Kooner JS (2009) Common genetic variation near melatonin receptor MTNR1B contributes to raised plasma glucose and increased risk of type-2 diabetes amongst Indian Asians and European whites. Diabetes 58(11):2703–2708

Chandramouli V, Khan A, Ostenson CG, Berggren PO, Low H, Landau BR, Efendic SE (1991) Quantification of glucose cycling and the extent of equilibration of glucose 6-phosphate with fructose 6-phosphate in islets from ob/ob mice. Biochem J 278(Pt 2):353–359

Chen SY, Pan CJ, Nandigama K, Mansfield BC, Ambudkar SV, Chou JY (2008a) The glucose-6-phosphate transporter is a phosphate-linked antiporter deficient in glycogen storage disease type Ib and Ic. FASEB J 22:2206–2213

Chen WM, Erdos MR, Jackson AU, Saxena R, Sanna S, Silver KD, Timpson NJ, Hansen T, Orru M, Grazia Piras M, Bonnycastle LL, Willer CJ, Lyssenko V, Shen H, Kuusisto J, Ebrahim S, Sestu N, Duren WL, Spada MC, Stringham HM, Scott LJ, Olla N, Swift AJ, Najjar S, Mitchell BD, Lawlor DA, Smith GD, Ben-Shlomo Y, Andersen G, Borch-Johnsen K, Jorgensen T, Saramies J, Valle TT, Buchanan TA, Shuldiner AR, Lakatta E, Bergman RN, Uda M, Tuomilehto J, Pedersen O, Cao A, Groop L, Mohlke KL, Laakso M, Schlessinger D, Collins FS, Altshuler D, Abecasis GR, Boehnke M, Scuteri A, Watanabe RM (2008b) Variations in the G6PC2/ABCB11 genomic region are associated with fasting glucose levels. J Clin Invest 118:2620–2628

Cho YS, Chen CH, Hu C, Long J, Ong RT, Sim X, Takeuchi F, Wu Y, Go MJ, Yamauchi T, Chang YC, Kwak SH, Ma RC, Yamamoto K, Adair LS, Aung T, Cai Q, Chang LC, Chen YT, Gao Y, Hu FB, Kim HL, Kim S, Kim YJ, Lee JJ, Lee NR, Li Y, Liu JJ, Lu W, Nakamura J, Nakashima E, Ng DP, Tay WT, Tsai FJ, Wong TY, Yokota M, Zheng W, Zhang R, Wang C, So WY, Ohnaka K, Ikegami H, Hara K, Cho YM, Cho NH, Chang TJ, Bao Y, Hedman AK, Morris AP, McCarthy MI, Takayanagi R, Park KS, Jia W, Chuang LM, Chan JC, Maeda S, Kadowaki T, Lee JY, Wu JY, Teo YY, Tai ES, Shu XO, Mohlke KL, Kato N, Han BG, Seielstad M (2012) Meta-analysis of genome-wide association studies identifies eight new loci for type 2 diabetes in East Asians. Nat Genet 44(1):67–72. doi:10.1038/ng.1019

Coutinho M, Gerstein HC, Wang Y, Yusuf S (1999) The relationship between glucose and incident cardiovascular events. A metaregression analysis of published data from 20 studies of 95,783 individuals followed for 12.4 years. Diabetes Care 22(2):233–240

Dai C, Brissova M, Hang Y, Thompson C, Poffenberger G, Shostak A, Chen Z, Stein R, Powers AC (2012) Islet-enriched gene expression and glucose-induced insulin secretion in human and mouse islets. Diabetologia 55(3):707–718. doi:10.1007/s00125-011-2369-0

Dos Santos C, Bougneres P, Fradin D (2009) An SNP in a methylatable Foxa2 binding site of the G6PC2 promoter is associated with insulin secretion in vivo and increased promoter activity in vitro. Diabetes 58:489–492

Droumaguet C, Balkau B, Simon D, Caces E, Tichet J, Charles MA, Eschwege E (2006) Use of HbA1c in predicting progression to diabetes in French men and women: data from an epidemiological study on the insulin resistance syndrome (DESIR). Diabetes Care 29 (7):1619–1625. doi:10.2337/dc05-2525

Dupuis J, Langenberg C, Prokopenko I, Saxena R, Soranzo N, Jackson AU, Wheeler E, Glazer NL, Bouatia-Naji N, Gloyn AL, Lindgren CM, Magi R, Morris AP, Randall J, Johnson T, Elliott P,

Rybin D, Thorleifsson G, Steinthorsdottir V, Henneman P, Grallert H, Dehghan A, Hottenga JJ, Franklin CS, Navarro P, Song K, Goel A, Perry JR, Egan JM, Lajunen T, Grarup N, Sparso T, Doney A, Voight BF, Stringham HM, Li M, Kanoni S, Shrader P, Cavalcanti-Proenca C, Kumari M, Qi L, Timpson NJ, Gieger C, Zabena C, Rocheleau G, Ingelsson E, An P, O'Connell J, Luan J, Elliott A, McCarroll SA, Payne F, Roccasecca RM, Pattou F, Sethupathy P, Ardlie K, Ariyurek Y, Balkau B, Barter P, Beilby JP, Ben-Shlomo Y, Benediktsson R, Bennett AJ, Bergmann S, Bochud M, Boerwinkle E, Bonnefond A, Bonnycastle LL, Borch-Johnsen K, Bottcher Y, Brunner E, Bumpstead SJ, Charpentier G, Chen YD, Chines P, Clarke R, Coin LJ, Cooper MN, Cornelis M, Crawford G, Crisponi L, Day IN, de Geus EJ, Delplanque J, Dina C, Erdos MR, Fedson AC, Fischer-Rosinsky A, Forouhi NG, Fox CS, Frants R, Franzosi MG, Galan P, Goodarzi MO, Graessler J, Groves CJ, Grundy S, Gwilliam R, Gyllensten U, Hadjadj S, Hallmans G, Hammond N, Han X, Hartikainen AL, Hassanali N, Hayward C, Heath SC, Hercberg S, Herder C, Hicks AA, Hillman DR, Hingorani AD, Hofman A, Hui J, Hung J, Isomaa B, Johnson PR, Jorgensen T, Jula A, Kaakinen M, Kaprio J, Kesaniemi YA, Kivimaki M, Knight B, Koskinen S, Kovacs P, Kyvik KO, Lathrop GM, Lawlor DA, Le Bacquer O, Lecoeur C, Li Y, Lyssenko V, Mahley R, Mangino M, Manning AK, Martinez-Larrad MT, McAteer JB, McCulloch LJ, McPherson R, Meisinger C, Melzer D, Meyre D, Mitchell BD, Morken MA, Mukherjee S, Naitza S, Narisu N, Neville MJ, Oostra BA, Orru M, Pakyz R, Palmer CN, Paolisso G, Pattaro C, Pearson D, Peden JF, Pedersen NL, Perola M, Pfeiffer AF, Pichler I, Polasek O, Posthuma D, Potter SC, Pouta A, Province MA, Psaty BM, Rathmann W, Rayner NW, Rice K, Ripatti S, Rivadeneira F, Roden M, Rolandsson O, Sandbaek A, Sandhu M, Sanna S, Sayer AA, Scheet P, Scott LJ, Seedorf U, Sharp SJ, Shields B, Sigurethsson G, Sijbrands EJ, Silveira A, Simpson L, Singleton A, Smith NL, Sovio U, Swift A, Syddall H, Syvanen AC, Tanaka T, Thorand B, Tichet J, Tonjes A, Tuomi T, Uitterlinden AG, van Dijk KW, van Hoek M, Varma D, Visvikis-Siest S, Vitart V, Vogelzangs N, Waeber G, Wagner PJ, Walley A, Walters GB, Ward KL, Watkins H, Weedon MN, Wild SH, Willemsen G, Witteman JC, Yarnell JW, Zeggini E, Zelenika D, Zethelius B, Zhai G, Zhao JH, Zillikens MC, Borecki IB, Loos RJ, Meneton P, Magnusson PK, Nathan DM, Williams GH, Hattersley AT, Silander K, Salomaa V, Smith GD, Bornstein SR, Schwarz P, Spranger J, Karpe F, Shuldiner AR, Cooper C, Dedoussis GV, Serrano-Rios M, Morris AD, Lind L, Palmer LJ, Hu FB, Franks PW, Ebrahim S, Marmot M, Kao WH, Pankow JS, Sampson MJ, Kuusisto J, Laakso M, Hansen T, Pedersen O, Pramstaller PP, Wichmann HE, Illig T, Rudan I, Wright AF, Stumvoll M, Campbell H, Wilson JF, Bergman RN, Buchanan TA, Collins FS, Mohlke KL, Tuomilehto J, Valle TT, Altshuler D, Rotter JI, Siscovick DS, Penninx BW, Boomsma DI, Deloukas P, Spector TD, Frayling TM, Ferrucci L, Kong A, Thorsteinsdottir U, Stefansson K, van Duijn CM, Aulchenko YS, Cao A, Scuteri A, Schlessinger D, Uda M, Ruokonen A, Jarvelin MR, Waterworth DM, Vollenweider P, Peltonen L, Mooser V, Abecasis GR, Wareham NJ, Sladek R, Froguel P, Watanabe RM, Meigs JB, Groop L, Boehnke M, McCarthy MI, Florez JC, Barroso I (2010) New genetic loci implicated in fasting glucose homeostasis and their impact on type 2 diabetes risk. Nat Genet 42(2):105–116

Ebert DH, Bischof LJ, Streeper RS, Chapman SC, Svitek CA, Goldman JK, Mathews CE, Leiter EH, Hutton JC, O'Brien RM (1999) Structure and promoter activity of an islet-specific glucose-6- phosphatase catalytic subunit-related gene. Diabetes 48(3):543–551

Edelman D, Olsen MK, Dudley TK, Harris AC, Oddone EZ (2004) Utility of hemoglobin A1c in predicting diabetes risk. J Gen Intern Med 19(12):1175–1180. doi:10.1111/j.1525-1497.2004.40178.x, JGI40178 [pii]

Foster JD, Pederson BA, Nordlie RC (1997) Glucose-6-phosphatase structure, regulation, and function: an update. Proc Soc Exp Biol Med 215(4):314–332

Freathy RM, Hayes MG, Urbanek M, Lowe LP, Lee H, Ackerman C, Frayling TM, Cox NJ, Dunger DB, Dyer AR, Hattersley AT, Metzger BE, Lowe WL Jr (2010) Hyperglycemia and adverse pregnancy outcome (HAPO) study: common genetic variants in GCK and TCF7L2 are associated with fasting and postchallenge glucose levels in pregnancy and with the new consensus definition of gestational diabetes mellitus from the international association of diabetes and pregnancy study groups. Diabetes 59(10):2682–2689. doi:10.2337/db10-0177

Grupe A, Hultgren B, Ryan A, Ma YH, Bauer M, Stewart TA (1995) Transgenic knockouts reveal a critical requirement for pancreatic beta cell glucokinase in maintaining glucose homeostasis. Cell 83(1):69–78

Han B, Serra P, Amrani A, Yamanouchi J, Maree AF, Edelstein-Keshet L, Santamaria P (2005) Prevention of diabetes by manipulation of anti-IGRP autoimmunity: high efficiency of a low-affinity peptide. Nat Med 11(6):645–652

Heni M, Ketterer C, t Hart LM, Ranta F, van Haeften TW, Eekhoff EM, Dekker JM, Boomsma DI, Nijpels G, Kramer MH, Diamant M, Simonis-Bik AM, Heine RJ, de Geus EJ, Schafer SA, Machicao F, Ullrich S, Thamer C, Stefan N, Staiger H, Haring HU, Fritsche A (2010) The impact of genetic variation in the G6PC2 gene on insulin secretion depends on glycemia. J Clin Endocrinol Metab 95:E479–E484. doi:10.1210/jc.2010-0860, jc.2010-0860 [pii]

Hu C, Zhang R, Wang C, Ma X, Wang C, Fang Q, Bao Y, Xiang K, Jia W (2009) A genetic variant of G6PC2 is associated with type 2 diabetes and fasting plasma glucose level in the Chinese population. Diabetologia 52(3):451–456

Hu C, Zhang R, Wang C, Yu W, Lu J, Ma X, Wang J, Jiang F, Tang S, Bao Y, Xiang K, Jia W (2010) Effects of GCK, GCKR, G6PC2 and MTNR1B variants on glucose metabolism and insulin secretion. PLoS One 5(7), e11761. doi:10.1371/journal.pone.0011761

Hutton JC, Eisenbarth GS (2003) A pancreatic beta-cell-specific homolog of glucose-6-phosphatase emerges as a major target of cell-mediated autoimmunity in diabetes. Proc Natl Acad Sci USA 100(15):8626–8628

Hutton JC, O'Brien RM (2009) The glucose-6-phosphatase catalytic subunit gene family. J Biol Chem 284:29241–29245

Iizuka K, Nakajima H, Ono A, Okita K, Miyazaki J, Miyagawa J, Namba M, Hanafusa T, Matsuzawa Y (2000) Stable overexpression of the glucose-6-phosphatase catalytic subunit attenuates glucose sensitivity of insulin secretion from a mouse pancreatic beta-cell line. J Endocrinol 164(3):307–314

Ingelsson E, Langenberg C, Hivert MF, Prokopenko I, Lyssenko V, Dupuis J, Magi R, Sharp S, Jackson AU, Assimes TL, Shrader P, Knowles JW, Zethelius B, Abbasi FA, Bergman RN, Bergmann A, Berne C, Boehnke M, Bonnycastle LL, Bornstein SR, Buchanan TA, Bumpstead SJ, Bottcher Y, Chines P, Collins FS, Cooper CC, Dennison EM, Erdos MR, Ferrannini E, Fox CS, Graessler J, Hao K, Isomaa B, Jameson KA, Kovacs P, Kuusisto J, Laakso M, Ladenvall C, Mohlke KL, Morken MA, Narisu N, Nathan DM, Pascoe L, Payne F, Petrie JR, Sayer AA, Schwarz PE, Scott LJ, Stringham HM, Stumvoll M, Swift AJ, Syvanen AC, Tuomi T, Tuomilehto J, Tonjes A, Valle TT, Williams GH, Lind L, Barroso I, Quertermous T, Walker M, Wareham NJ, Meigs JB, McCarthy MI, Groop L, Watanabe RM, Florez JC (2010) Detailed physiologic characterization reveals diverse mechanisms for novel genetic Loci regulating glucose and insulin metabolism in humans. Diabetes 59(5):1266–1275. doi:10.2337/db09-1568, db09-1568 [pii]

Iynedjian PB (2009) Molecular physiology of mammalian glucokinase. Cell Mol Life Sci 66 (1):27–42

Jarchum I, Nichol L, Trucco M, Santamaria P, DiLorenzo TP (2008) Identification of novel IGRP epitopes targeted in type 1 diabetes patients. Clin Immunol 127(3):359–365

Jazmin LJ, Young JD (2013) Isotopically nonstationary 13C metabolic flux analysis. Methods Mol Biol 985:367–390. doi:10.1007/978-1-62703-299-5_18

Jensen MV, Joseph JW, Ronnebaum SM, Burgess SC, Sherry AD, Newgard CB (2008) Metabolic cycling in control of glucose-stimulated insulin secretion. Am J Physiol Endocrinol Metab 295 (6):E1287–E1297

Karunakaran U, Park KG (2013) A systematic review of oxidative stress and safety of antioxidants in diabetes: focus on islets and their defense. Diabetes Metab J 37(2):106–112. doi:10.4093/dmj.2013.37.2.106

Khan A, Chandramouli V, Ostenson CG, Ahren B, Schumann WC, Low H, Landau BR, Efendic S (1989) Evidence for the presence of glucose cycling in pancreatic islets of the ob/ob mouse. J Biol Chem 264(17):9732–9733

Khan A, Chandramouli V, Ostenson CG, Berggren PO, Low H, Landau BR, Efendic S (1990a) Glucose cycling is markedly enhanced in pancreatic islets of obese hyperglycemic mice. Endocrinology 126(5):2413–2416

Khan A, Chandramouli V, Ostenson CG, Low H, Landau BR, Efendic S (1990b) Glucose cycling in islets from healthy and diabetic rats. Diabetes 39(4):456–459

Koza RA, Nikonova L, Hogan J, Rim JS, Mendoza T, Faulk C, Skaf J, Kozak LP (2006) Changes in gene expression foreshadow diet-induced obesity in genetically identical mice. PLoS Genet 2(5), e81

Lawes CM, Parag V, Bennett DA, Suh I, Lam TH, Whitlock G, Barzi F, Woodward M (2004) Blood glucose and risk of cardiovascular disease in the Asia Pacific region. Diabetes Care 27 (12):2836–2842

Laybutt DR, Glandt M, Xu G, Ahn YB, Trivedi N, Bonner-Weir S, Weir GC (2003) Critical reduction in beta-cell mass results in two distinct outcomes over time. Adaptation with impaired glucose tolerance or decompensated diabetes. J Biol Chem 278(5):2997–3005

Li X, Shu YH, Xiang AH, Trigo E, Kuusisto J, Hartiala J, Swift AJ, Kawakubo M, Stringham HM, Bonnycastle LL, Lawrence JM, Laakso M, Allayee H, Buchanan TA, Watanabe RM (2009) Additive effects of genetic variation in Gck and G6pc2 on insulin secretion and fasting glucose. Diabetes 58:2946–2953

Lieberman SM, Evans AM, Han B, Takaki T, Vinnitskaya Y, Caldwell JA, Serreze DV, Shabanowitz J, Hunt DF, Nathenson SG, Santamaria P, DiLorenzo TP (2003) Identification of the beta cell antigen targeted by a prevalent population of pathogenic CD8+ T cells in autoimmune diabetes. Proc Natl Acad Sci USA 100(14):8384–8388

MacDonald MJ, McKenzie DI, Walker TM, Kaysen JH (1992) Lack of glyconeogenesis in pancreatic islets: expression of gluconeogenic enzyme genes in islets. Horm Metab Res 24 (4):158–160. doi:10.1055/s-2007-1003284

Magnuson MA, She P, Shiota M (2003) Gene-altered mice and metabolic flux control. J Biol Chem 278(35):32485–32488

Martin CC, Bischof LJ, Bergman B, Hornbuckle LA, Hilliker C, Frigeri C, Wahl D, Svitek CA, Wong R, Goldman JK, Oeser JK, Lepretre F, Froguel P, O'Brien RM, Hutton JC (2001) Cloning and characterization of the human and rat islet-specific glucose-6-phosphatase Catalytic subunit-related protein (IGRP) genes. J Biol Chem 276(27):25197–25207

Martin CC, Oeser JK, Svitek CA, Hunter SI, Hutton JC, O'Brien RM (2002) Identification and characterization of a human cDNA and gene encoding a ubiquitously expressed glucose-6-phosphatase catalytic subunit-related protein. J Mol Endocrinol 29:205–222

Matschinsky FM (1996) Banting lecture 1995. A lesson in metabolic regulation inspired by the glucokinase glucose sensor paradigm. Diabetes 45(2):223–241

Matschinsky FM (2005) Glucokinase, glucose homeostasis, and diabetes mellitus. Curr Diab Rep 5(3):171–176

Matschinsky FM (2009) Assessing the potential of glucokinase activators in diabetes therapy. Nat Rev Drug Discov 8(5):399–416

Merrins MJ, Fendler B, Zhang M, Sherman A, Bertram R, Satin LS (2010) Metabolic oscillations in pancreatic islets depend on the intracellular Ca2+ level but not Ca2+ oscillations. Biophys J 99(1):76–84. doi:10.1016/j.bpj.2010.04.012

Merrins MJ, Bertram R, Sherman A, Satin LS (2012) Phosphofructo-2-kinase/fructose-2,6-bisphosphatase modulates oscillations of pancreatic islet metabolism. PLoS One 7(4), e34036. doi:10.1371/journal.pone.0034036

Mills GW, Avery PJ, McCarthy MI, Hattersley AT, Levy JC, Hitman GA, Sampson M, Walker M (2004) Heritability estimates for beta cell function and features of the insulin resistance syndrome in UK families with an increased susceptibility to type 2 diabetes. Diabetologia 47 (4):732–738. doi:10.1007/s00125-004-1338-2

Mithieux G (1997) New knowledge regarding glucose-6 phosphatase gene and protein and their roles in the regulation of glucose metabolism. Eur J Endocrinol 136(2):137–145

Mukherjee R, Wagar D, Stephens TA, Lee-Chan E, Singh B (2005) Identification of CD4+ T cell-specific epitopes of islet-specific glucose-6-phosphatase catalytic subunit-related protein: a novel beta cell autoantigen in type 1 diabetes. J Immunol 174(9):5306–5315

Osbak KK, Colclough K, Saint-Martin C, Beer NL, Bellanne-Chantelot C, Ellard S, Gloyn AL (2009) Update on mutations in glucokinase (GCK), which cause maturity-onset diabetes of the young, permanent neonatal diabetes, and hyperinsulinemic hypoglycemia. Hum Mutat 30 (11):1512–1526. doi:10.1002/humu.21110

Pasquali L, Gaulton KJ, Rodriguez-Segui SA, Mularoni L, Miguel-Escalada I, Akerman I, Tena JJ, Moran I, Gomez-Marin C, van de Bunt M, Ponsa-Cobas J, Castro N, Nammo T, Cebola I, Garcia-Hurtado J, Maestro MA, Pattou F, Piemonti L, Berney T, Gloyn AL, Ravassard P, Gomez-Skarmeta JL, Muller F, McCarthy MI, Ferrer J (2014) Pancreatic islet enhancer clusters enriched in type 2 diabetes risk-associated variants. Nat Genet 46(2):136–143. doi:10.1038/ng.2870

Pedersen KB, Zhang P, Doumen C, Charbonnet M, Lu D, Newgard CB, Haycock JW, Lange AJ, Scott DK (2007) The promoter for the gene encoding the catalytic subunit of rat glucose-6-phosphatase contains two distinct glucose-responsive regions. Am J Physiol Endocrinol Metab 292(3):E788–E801. doi:10.1152/ajpendo.00510.2006, 00510.2006 [pii]

Perales MA, Sener A, Malaisse WJ (1991) Hexose metabolism in pancreatic islets: the glucose-6-phosphatase riddle. Mol Cell Biochem 101(1):67–71

Petrolonis AJ, Yang Q, Tummino PJ, Fish SM, Prack AE, Jain S, Parsons TF, Li P, Dales NA, Ge L, Langston SP, Schuller AG, An WF, Tartaglia LA, Chen H, Hong SB (2004) Enzymatic characterization of the pancreatic islet-specific glucose-6-phosphatase-related protein (IGRP). J Biol Chem 279:13976–13983

Poirier P, Giles TD, Bray GA, Hong Y, Stern JS, Pi-Sunyer FX, Eckel RH (2006) Obesity and cardiovascular disease: pathophysiology, evaluation, and effect of weight loss. Arterioscler Thromb Vasc Biol 26(5):968–976. doi:10.1161/01.ATV.0000216787.85457.f3

Postic C, Shiota M, Niswender KD, Jetton TL, Chen Y, Moates JM, Shelton KD, Lindner J, Cherrington AD, Magnuson MA (1999) Dual roles for glucokinase in glucose homeostasis as determined by liver and pancreatic beta cell-specific gene knock-outs using Cre recombinase. J Biol Chem 274(1):305–315

Poulsen P, Vaag A, Kyvik K, Beck-Nielsen H (2001) Genetic versus environmental aetiology of the metabolic syndrome among male and female twins. Diabetologia 44(5):537–543. doi:10.1007/s001250051659

Pound LD, Oeser JK, O'Brien TP, Wang Y, Faulman CJ, Dadi PK, Jacobson DA, Hutton JC, McGuinness OP, Shiota M, O'Brien RM (2013) G6PC2: a negative regulator of basal glucose-stimulated insulin secretion. Diabetes 62:1547–1556. doi:10.2337/db12-1067

Prokopenko I, Langenberg C, Florez JC, Saxena R, Soranzo N, Thorleifsson G, Loos RJ, Manning AK, Jackson AU, Aulchenko Y, Potter SC, Erdos MR, Sanna S, Hottenga JJ, Wheeler E, Kaakinen M, Lyssenko V, Chen WM, Ahmadi K, Beckmann JS, Bergman RN, Bochud M, Bonnycastle LL, Buchanan TA, Cao A, Cervino A, Coin L, Collins FS, Crisponi L, de Geus EJ, Dehghan A, Deloukas P, Doney AS, Elliott P, Freimer N, Gateva V, Herder C, Hofman A, Hughes TE, Hunt S, Illig T, Inouye M, Isomaa B, Johnson T, Kong A, Krestyaninova M, Kuusisto J, Laakso M, Lim N, Lindblad U, Lindgren CM, McCann OT, Mohlke KL, Morris AD, Naitza S, Orru M, Palmer CN, Pouta A, Randall J, Rathmann W, Saramies J, Scheet P, Scott LJ, Scuteri A, Sharp S, Sijbrands E, Smit JH, Song K, Steinthorsdottir V, Stringham HM, Tuomi T, Tuomilehto J, Uitterlinden AG, Voight BF, Waterworth D, Wichmann HE, Willemsen G, Witteman JC, Yuan X, Zhao JH, Zeggini E, Schlessinger D, Sandhu M, Boomsma DI, Uda M, Spector TD, Penninx BW, Altshuler D, Vollenweider P, Jarvelin MR, Lakatta E, Waeber G, Fox CS, Peltonen L, Groop LC, Mooser V, Cupples LA, Thorsteinsdottir U, Boehnke M, Barroso I, Van Duijn C, Dupuis J, Watanabe RM,

Stefansson K, McCarthy MI, Wareham NJ, Meigs JB, Abecasis GR (2008) Variants in MTNR1B influence fasting glucose levels. Nat Genet 41:77–81
Pruim RJ, Welch RP, Sanna S, Teslovich TM, Chines PS, Gliedt TP, Boehnke M, Abecasis GR, Willer CJ (2010) LocusZoom: regional visualization of genome-wide association scan results. Bioinformatics 26(18):2336–2337. doi:10.1093/bioinformatics/btq419
Ramos E, Chen G, Shriner D, Doumatey A, Gerry NP, Herbert A, Huang H, Zhou J, Christman MF, Adeyemo A, Rotimi C (2011) Replication of genome-wide association studies (GWAS) loci for fasting plasma glucose in African-Americans. Diabetologia 54(4):783–788. doi:10.1007/s00125-010-2002-7
Reiling E, van 't Riet E, Groenewoud MJ, Welschen LM, van Hove EC, Nijpels G, Maassen JA, Dekker JM, t Hart LM (2009) Combined effects of single-nucleotide polymorphisms in GCK, GCKR, G6PC2 and MTNR1B on fasting plasma glucose and type 2 diabetes risk. Diabetologia 52:1866–1870
Rose CS, Grarup N, Krarup NT, Poulsen P, Wegner L, Nielsen T, Banasik K, Faerch K, Andersen G, Albrechtsen A, Borch-Johnsen K, Clausen JO, Jorgensen T, Vaag A, Pedersen O, Hansen T (2009) A variant in the G6PC2/ABCB11 locus is associated with increased fasting plasma glucose, increased basal hepatic glucose production and increased insulin release after oral and intravenous glucose loads. Diabetologia 52(10):2122–2129
Sarwar N, Gao P, Seshasai SR, Gobin R, Kaptoge S, Di Angelantonio E, Ingelsson E, Lawlor DA, Selvin E, Stampfer M, Stehouwer CD, Lewington S, Pennells L, Thompson A, Sattar N, White IR, Ray KK, Danesh J (2010) Diabetes mellitus, fasting blood glucose concentration, and risk of vascular disease: a collaborative meta-analysis of 102 prospective studies. Lancet 375 (9733):2215–2222. doi:10.1016/S0140-6736(10)60484-9, S0140-6736(10)60484-9 [pii]
Scott RA, Lagou V, Welch RP, Wheeler E, Montasser ME, Luan J, Magi R, Strawbridge RJ, Rehnberg E, Gustafsson S, Kanoni S, Rasmussen-Torvik LJ, Yengo L, Lecoeur C, Shungin D, Sanna S, Sidore C, Johnson PC, Jukema JW, Johnson T, Mahajan A, Verweij N, Thorleifsson G, Hottenga JJ, Shah S, Smith AV, Sennblad B, Gieger C, Salo P, Perola M, Timpson NJ, Evans DM, Pourcain BS, Wu Y, Andrews JS, Hui J, Bielak LF, Zhao W, Horikoshi M, Navarro P, Isaacs A, O'Connell JR, Stirrups K, Vitart V, Hayward C, Esko T, Mihailov E, Fraser RM, Fall T, Voight BF, Raychaudhuri S, Chen H, Lindgren CM, Morris AP, Rayner NW, Robertson N, Rybin D, Liu CT, Beckmann JS, Willems SM, Chines PS, Jackson AU, Kang HM, Stringham HM, Song K, Tanaka T, Peden JF, Goel A, Hicks AA, An P, Muller-Nurasyid M, Franco-Cereceda A, Folkersen L, Marullo L, Jansen H, Oldehinkel AJ, Bruinenberg M, Pankow JS, North KE, Forouhi NG, Loos RJ, Edkins S, Varga TV, Hallmans G, Oksa H, Antonella M, Nagaraja R, Trompet S, Ford I, Bakker SJ, Kong A, Kumari M, Gigante B, Herder C, Munroe PB, Caulfield M, Antti J, Mangino M, Small K, Miljkovic I, Liu Y, Atalay M, Kiess W, James AL, Rivadeneira F, Uitterlinden AG, Palmer CN, Doney AS, Willemsen G, Smit JH, Campbell S, Polasek O, Bonnycastle LL, Hercberg S, Dimitriou M, Bolton JL, Fowkes GR, Kovacs P, Lindstrom J, Zemunik T, Bandinelli S, Wild SH, Basart HV, Rathmann W, Grallert H, Maerz W, Kleber ME, Boehm BO, Peters A, Pramstaller PP, Province MA, Borecki IB, Hastie ND, Rudan I, Campbell H, Watkins H, Farrall M, Stumvoll M, Ferrucci L, Waterworth DM, Bergman RN, Collins FS, Tuomilehto J, Watanabe RM, de Geus EJ, Penninx BW, Hofman A, Oostra BA, Psaty BM, Vollenweider P, Wilson JF, Wright AF, Hovingh GK, Metspalu A, Uusitupa M, Magnusson PK, Kyvik KO, Kaprio J, Price JF, Dedoussis GV, Deloukas P, Meneton P, Lind L, Boehnke M, Shuldiner AR, van Duijn CM, Morris AD, Toenjes A, Peyser PA, Beilby JP, Korner A, Kuusisto J, Laakso M, Bornstein SR, Schwarz PE, Lakka TA, Rauramaa R, Adair LS, Smith GD, Spector TD, Illig T, de Faire U, Hamsten A, Gudnason V, Kivimaki M, Hingorani A, Keinanen-Kiukaanniemi SM, Saaristo TE, Boomsma DI, Stefansson K, van der Harst P, Dupuis J, Pedersen NL, Sattar N, Harris TB, Cucca F, Ripatti S, Salomaa V, Mohlke KL, Balkau B, Froguel P, Pouta A, Jarvelin MR, Wareham NJ, Bouatia-Naji N, McCarthy MI, Franks PW, Meigs JB, Teslovich TM, Florez JC, Langenberg C, Ingelsson E, Prokopenko I, Barroso I (2012) Large-scale association

analyses identify new loci influencing glycemic traits and provide insight into the underlying biological pathways. Nat Genet 44(9):991–1005. doi:10.1038/ng.2385

Service SK, Teslovich TM, Fuchsberger C, Ramensky V, Yajnik P, Koboldt DC, Larson DE, Zhang Q, Lin L, Welch R, Ding L, McLellan MD, O'Laughlin M, Fronick C, Fulton LL, Magrini V, Swift A, Elliott P, Jarvelin MR, Kaakinen M, McCarthy MI, Peltonen L, Pouta A, Bonnycastle LL, Collins FS, Narisu N, Stringham HM, Tuomilehto J, Ripatti S, Fulton RS, Sabatti C, Wilson RK, Boehnke M, Freimer NB (2014) Re-sequencing expands our understanding of the phenotypic impact of variants at GWAS loci. PLoS Genet 10(1), e1004147. doi:10.1371/journal.pgen.1004147

Sharp PA (1994) Split genes and RNA splicing. Cell 77(6):805–815

Sim X, Ong RT, Suo C, Tay WT, Liu J, Ng DP, Boehnke M, Chia KS, Wong TY, Seielstad M, Teo YY, Tai ES (2011) Transferability of type 2 diabetes implicated loci in multi-ethnic cohorts from Southeast Asia. PLoS Genet 7(4), e1001363. doi:10.1371/journal.pgen.1001363

Solis AS, Shariat N, Patton JG (2008) Splicing fidelity, enhancers, and disease. Front Biosci 13:1926–1942

Sweet IR, Najafi H, Li G, Grodberg J, Matschinsky FM (1997) Measurement and modeling of glucose-6-phosphatase in pancreatic islets. Am J Physiol 272(4 Pt 1):E696–E711

Takeuchi F, Katsuya T, Chakrewarthy S, Yamamoto K, Fujioka A, Serizawa M, Fujisawa T, Nakashima E, Ohnaka K, Ikegami H, Sugiyama T, Nabika T, Kasturiratne A, Yamaguchi S, Kono S, Takayanagi R, Yamori Y, Kobayashi S, Ogihara T, de Silva A, Wickremasinghe R, Kato N (2010) Common variants at the GCK, GCKR, G6PC2-ABCB11 and MTNR1B loci are associated with fasting glucose in two Asian populations. Diabetologia 53(2):299–308

Tam CH, Ho JS, Wang Y, Lee HM, Lam VK, Germer S, Martin M, So WY, Ma RC, Chan JC, Ng MC (2010) Common polymorphisms in MTNR1B, G6PC2 and GCK are associated with increased fasting plasma glucose and impaired beta-cell function in Chinese subjects. PLoS One 5(7), e11428. doi:10.1371/journal.pone.0011428

Taneera J, Lang S, Sharma A, Fadista J, Zhou Y, Ahlqvist E, Jonsson A, Lyssenko V, Vikman P, Hansson O, Parikh H, Korsgren O, Soni A, Krus U, Zhang E, Jing XJ, Esguerra JL, Wollheim CB, Salehi A, Rosengren A, Renstrom E, Groop L (2012) A systems genetics approach identifies genes and pathways for type 2 diabetes in human islets. Cell Metab 16 (1):122–134. doi:10.1016/j.cmet.2012.06.006

Tokuyama Y, Sturis J, DePaoli AM, Takeda J, Stoffel M, Tang J, Sun X, Polonsky KS, Bell GI (1995) Evolution of beta-cell dysfunction in the male Zucker diabetic fatty rat. Diabetes 44 (12):1447–1457

Trandaburu T (1977) Fine structural localization of glucose-6-phosphatase activity in the pancreatic islets of two amphibian species (Salamandra salamandra L. and Rana esculenta L.). Acta Histochem 59(2):246–253

Trinh K, Minassian C, Lange AJ, O'Doherty RM, Newgard CB (1997) Adenovirus-mediated expression of the catalytic subunit of glucose-6- phosphatase in INS-1 cells. Effects on glucose cycling, glucose usage, and insulin secretion. J Biol Chem 272(40):24837–24842

van de Werve G, Lange A, Newgard C, Mechin MC, Li Y, Berteloot A (2000) New lessons in the regulation of glucose metabolism taught by the glucose 6-phosphatase system. Eur J Biochem 267(6):1533–1549

Van Schaftingen E, Gerin I (2002) The glucose-6-phosphatase system. Biochem J 362 (Pt 3):513–532

Waddell ID, Burchell A (1988) The microsomal glucose-6-phosphatase enzyme of pancreatic islets. Biochem J 255(2):471–476

Wang H, Iynedjian PB (1997a) Acute glucose intolerance in insulinoma cells with unbalanced overexpression of glucokinase. J Biol Chem 272(41):25731–25736

Wang H, Iynedjian PB (1997b) Modulation of glucose responsiveness of insulinoma beta-cells by graded overexpression of glucokinase. Proc Natl Acad Sci USA 94(9):4372–4377

Wang Y, Martin CC, Oeser JK, Sarkar S, McGuinness OP, Hutton JC, O'Brien RM (2007) Deletion of the gene encoding the islet-specific glucose-6-phosphatase catalytic subunit-related protein autoantigen results in a mild metabolic phenotype. Diabetologia 50:774–778

Weedon MN, Clark VJ, Qian Y, Ben-Shlomo Y, Timpson N, Ebrahim S, Lawlor DA, Pembrey ME, Ring S, Wilkin TJ, Voss LD, Jeffery AN, Metcalf B, Ferrucci L, Corsi AM, Murray A, Melzer D, Knight B, Shields B, Smith GD, Hattersley AT, Di Rienzo A, Frayling TM (2006) A common haplotype of the glucokinase gene alters fasting glucose and birth weight: association in six studies and population-genetics analyses. Am J Hum Genet 79(6):991–1001. doi:10.1086/509517

Wolf BA, Colca JR, Comens PG, Turk J, McDaniel ML (1986) Glucose 6-phosphate regulates Ca2+ steady state in endoplasmic reticulum of islets. A possible link in glucose-induced insulin secretion. J Biol Chem 261(35):16284–16287

Yang J, Danke NA, Berger D, Reichstetter S, Reijonen H, Greenbaum C, Pihoker C, James EA, Kwok WW (2006) Islet-specific glucose-6-phosphatase catalytic subunit-related protein-reactive CD4+ T cells in human subjects. J Immunol 176(5):2781–2789

Ye J (2013) Mechanisms of insulin resistance in obesity. Front Med 7(1):14–24. doi:10.1007/s11684-013-0262-6

Chapter 17
From Association to Function: *KCNJ11* and *ABCC8*

Michael N. Weedon and Peter Light

Abstract ATP-sensitive potassium (K_{ATP}) channels regulate the secretion of insulin from pancreatic β-cells in response to glucose. These channels consist of the sulphonylurea receptor (SUR1, encoded by *ABCC8*) and the inwardly rectifying potassium channel (Kir6.2, encoded by *KCNJ11*) channel subunits. Loss-of-function mutations in *ABCC8* or *KCNJ11* cause familial hyperinsulinism, whereas activating mutations of these genes are the commonest cause of neonatal diabetes. Mutations in these genes have also recently been shown to cause maturity-onset diabetes of the young. Given the spectrum of diabetes that can be caused by mutations in *KCNJ11* and *ABCC8*, they are excellent candidate genes for harbouring variants that predispose to later-onset diabetes, and a common haplotype of these genes has been robustly associated with an increased risk of type 2 diabetes. In this review, we discuss the role of *ABCC8* and *KCNJ11* variation in type 2 diabetes.

17.1 ATP-Sensitive Potassium Channels, the ABCC8 and KCNJ11 Genes, and Insulin Secretion

ATP-sensitive potassium (K_{ATP}) channels regulate the secretion of insulin from pancreatic β-cells by coupling metabolism to membrane electrical activity. Two types of subunit make up the β-cell hetero-octameric K_{ATP} channel complex: the sulphonylurea receptor (SUR1, encoded by the gene *ABCC8*) and the inwardly rectifying potassium channel (Kir6.2, encoded by the gene *KCNJ11*; Fig. 17.1) (Inagaki et al. 1995; Aguilar-Bryan and Bryan 1999; Bryan et al 2004). The primary role for β-cell K_{ATP} channels is to transduce changes in the cytosolic ATP/ADP

M.N. Weedon (✉)
Institute of Biomedical and Clinical Science, University of Exeter Medical School, Exeter, UK
e-mail: M.N.Weedon@exeter.ac.uk

P. Light
Department of Pharmacology, Alberta Diabetes Institute, University of Alberta, Edmonton, AB, Canada
e-mail: plight@ualberta.ca

J.C. Florez (ed.), *The Genetics of Type 2 Diabetes and Related Traits*,
DOI 10.1007/978-3-319-01574-3_17

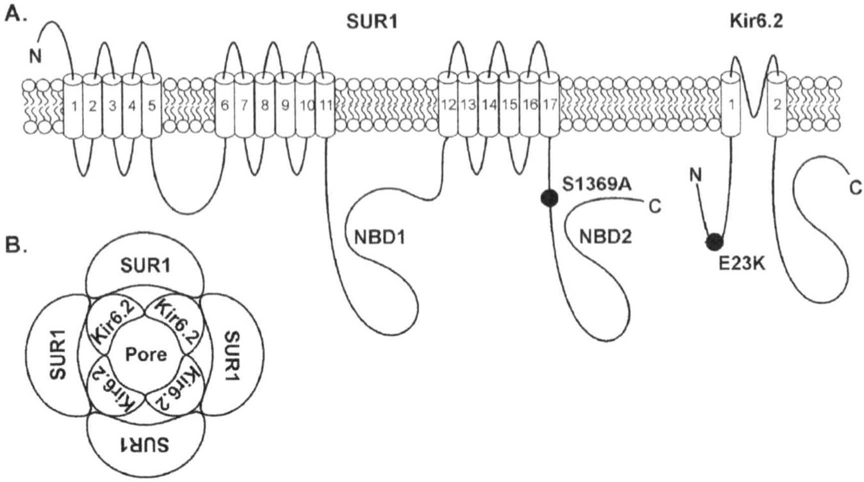

Fig. 17.1 (**a**) Pancreatic K_{ATP} channels are comprised of the 17 transmembrane domain sulphonylurea receptor 1 (SUR1) and the inwardly rectifying potassium channel Kir6.2 subunits. SUR1 is a member of the ATP-binding cassette family of membrane proteins and possesses two nucleotide-binding folds (NBD1 and 2) that dimerize to form the ADP sensing and MgATPase catalytic sites. SUR1 is also the high-affinity binding site for the sulphonylurea class of insulin secretagogues. Note that the S1369A variant (rs757110) is located adjacent to NBD2. Kir6.2 forms the potassium-selective pore region of the K_{ATP} channel complex. The E23K variant (rs5215) is located intracellularly close to the N-terminus of Kir6.2. (**b**) SUR1 and Kir6.2 form a hetero-octamer with 4 Kir6.2 subunits being surrounded by 4 SUR1 subunits

ratio, resulting from glucose metabolism (Ashcroft et al. 1984), into electrical activity and the control of the β-cell membrane potential. As glucose-stimulated insulin secretion is predominately controlled by the β-cell membrane potential, K_{ATP} channels serve to couple glucose metabolism to insulin secretion (Ashcroft and Rorsman 1989; Ashcroft et al. 1994). When plasma glucose levels are low, the cytosolic ATP/ADP ratio is lowered resulting in a basal efflux of potassium ions from the cell via K_{ATP} channel activity that maintains the membrane potential of the β-cell at approximately −70 mV. This polarized membrane potential prevents calcium entry through voltage-gated calcium channels. As elevations in cytosolic calcium are the primary trigger for insulin granule exocytosis, insulin secretion is suppressed when plasma glucose levels are low. However, when plasma glucose levels rise, glucose enters the β-cells via the glucose transporter (GLUT), although the precise GLUT isoform in human beta cells is currently under debate (Van de Brunt and Gloyn 2012). The resultant increase in glucose metabolism leads to an increase in the ratio of cytosolic ATP/ADP ratio, promoting K_{ATP} channel closure. The resulting decrease in potassium ion efflux depolarizes the β-cell membrane potential leading to activation of voltage-gated calcium channels, calcium influx, and calcium-stimulated insulin granule exocytosis (MacDonald and Rorsman 2007). Graded increases in plasma glucose and subsequent metabolism lead to proportional decreases in K_{ATP} channel activity resulting in an appropriate insulin secretory response that is tightly coupled to the plasma glucose concentration. As the electrical resistance of β-cell is high, only small changes in K_{ATP} channel activity are necessary to change β-cell excitability (and hence insulin secretion) via alterations in the β-cell membrane potential. It is of direct clinical relevance to note that a number of clinically used insulin secretagogues, including the sulphonylureas, bind to and inhibit the K_{ATP} channel complex with high affinity, leading to insulin secretion (Bryan et al 2005). Indeed, the sulphonylurea receptor was first named due to its high affinity for sulphonylurea binding and it was then subsequently cloned (Aguilar-Bryan et al. 1995) leading to the elucidation of the complete beta-cell K_{ATP} channel complex by co-expression with Kir6.2 (Inagaki et al. 1995). Interestingly, K_{ATP} channels are also expressed in pancreatic α-cells where they play an important role in regulating the secretion of glucagon (Rorsman et al. 2008; MacDonald et al. 2007) and alterations in α-cell K_{ATP} channel activity, by whatever mechanism, may contribute to elevated blood glucose levels if glucagon secretion is elevated.

17.2 Rare Mutations in ABCC8 and KCNJ11 Cause Familial Hyperinsulinism and Neonatal Diabetes

Mutations within the K_{ATP} channel complex that alter their trafficking, intrinsic activity, or ability to sense changes in either ATP or ADP will result in altered K_{ATP} channel activity (and hence islet hormone secretion) that is correlated to the specific effects of the individual mutation on K_{ATP} channel activity. Positional cloning and candidate genes studies first demonstrated that recessive loss-of-function mutations of these genes cause familial hyperinsulinemic hypoglycaemia of infancy (Thomas et al. 1995; Dunne et al. 1997). Subsequently, candidate gene studies demonstrated that heterozygous activating mutations of *KCNJ11* and *ABCC8*, which result in the K_{ATP} channel being constantly open, are the commonest cause of neonatal diabetes (diabetes diagnosed before the age of 6 months) (Gloyn et al. 2004). More recently, mutations in *KCNJ11* and ABCC8 have been shown to be a cause of maturity-onset diabetes of the young (Bonnefond et al. 2012; Bowman et al. 2012; Liu et al. 2013), an autosomal dominant form of diabetes that is typically diagnosed before the age of 25. While many of these mutations drastically alter K_{ATP} channel activity leading to an overt phenotype (Lang and Light 2010), the potential role of common variants in the development of type 2 diabetes is less clear.

17.3 Candidate Gene Association Studies of KCNJ11 and ABCC8 in Type 2 Diabetes

The known biology of the K_{ATP} channel and the monogenic conditions associated with rare, severe, mutations in *KCNJ11* and *ABCC8* make them excellent candidate genes for harbouring other hypomorphic variants that predispose to late-onset type 2 diabetes. Early candidate gene studies re-sequenced the protein-coding parts of these genes in small samples of patients with type 2 diabetes and then genotyped these variants in a few hundred cases and controls to test for association with the disease. As with most early candidate gene studies in type 2 diabetes (and complex disease more generally), this led to several putative associations being claimed, most of which have not subsequently been replicated (Hirschhorn et al. 2002). This lack of replication has been attributed to a number of factors including multiple hypothesis testing, population stratification, and publication bias (Hirschhorn et al. 2002). In hindsight, one of the biggest reasons for the irreproducibility of genetic association studies was the small effect that individual variants have on type 2 diabetes risk. This meant that the sample sizes used were not sufficiently large to provide robust statistical support for variants (Lohmueller et al. 2003). It became clear that large sample sizes, more robust statistics, and replication are essential to claim a variant as truly associated with type 2 diabetes. This was clearly demonstrated by Altshuler et al. (2000) when they used a meta-analysis and large-scale association study approach to show the Pro12Ala variant of *PPARG* as robustly associated with type 2 diabetes. We took a similar approach for three variants of *ABCC8* and *KCNJ11* that were among the most promising variants from these initial candidate gene screens.

17.4 Large-Scale Association Study and Meta-Analysis Demonstrate the Robust Association of E23K with Type 2 Diabetes

We studied the *ABCC8* exon 16–3t/c, the exon 18T/C, and the *KCNJ11* E23K variants. Each of these variants had been inconsistently reported to be associated with type 2 diabetes (Altshuler et al. 2000; Gloyn et al. 2001; Hani et al. 1997, 1998; Hansen et al. 1997, 1998; Inoue et al. 1996, 1997; Sakura et al. 1996; 't Hart et al. 1999; van Tilburg et al. 2000; Rissanen et al. 2000). Additionally, there was some functional study support for the E23K variant that suggested it increased the threshold ATP concentration required for insulin release (Schwanstecher et al. 2002a). We performed a meta-analysis of all six published studies on the three variants (Gloyn et al. 2003). We found strong evidence for association of the E23K variant with type 2 diabetes, but not the other variants. Meta-analysis of published studies is prone to positive findings due to publication bias. To confirm the association of E23K with type 2 diabetes, we genotyped the three variants in

854 cases and 1182 population-based controls, as well as 150 parent-affected offspring trios. Although dwarfed by the size of current genome-wide association studies, at the time this was a large-scale association study. Consistent with the meta-analyses, we found no evidence for an association of the SUR1 variants, but we did find nominal evidence for association of *KCNJ11* E23K with type 2 diabetes ($P = 0.025$), with an odds ratio consistent with the meta-analysis result (OR = 1.18). This work confirmed E23K as associated with type 2 diabetes. Subsequent studies, including large-scale genome-wide association studies, have replicated this association in European populations and in other ancestry groups (Morris et al. 2012; Replication and Meta-analysis 2014). Other studies have demonstrated that E23K is associated with insulin secretion response to glucose (Nielsen et al. 2003; Florez et al. 2004).

17.5 Is E23K the Causal Variant?

While E23K is robustly associated with type 2 diabetes, it is not clear if it is the causal variant that explains the association signal. This is because E23K is inherited with two other protein-coding variants, V270L in *KCNJ11* (rs5215) and S1369A in *ABCC8* (rs757110), and so it is impossible to separate out, using genetics alone, which of these is the functional variant. This is a common problem in genetic association studies—identifying causal variants is very difficult because of the correlation structure of variants in the genome. Florez et al. (2004) performed a comprehensive haplotype analysis in 3400 individuals and concluded that E23K or S1369A is the most likely functional variants. In fact, the V270L (rs5215) variant is the most strongly associated variant from the largest genome-wide association studies (34,840 cases and 114,981 controls) to date (Morris et al. 2012). But, the results are statistically indistinguishable from the S1369A or E23K variants (see Fig. 17.2), with any difference in association signal between these variants probably due to technical variation (e.g., E23K is not present in the HapMap study and so has not been imputable). Functional studies have therefore been undertaken in order to separate out which of these variants are causal. However, early attempts to establish E23K causality, through mechanistic studies on recombinant channels, did not account for the S1369A variant (Sakura et al. 1996; Schwanstecher et al. 2002b) and these studies were performed using *ABCC8* clones from different species in which either S1369 or A1369 was present and not controlled for. To specifically address the issue of causality, mechanistic studies have recently been performed on recombinant human K_{ATP} channels containing either the E23/S1369 non-risk or E23K/S1369A type 2 diabetes-risk haplotypes. This study determined that the E23K/S1369A type 2 diabetes-risk haplotype channels possess reduced ATP sensitivity and that this effect was bestowed upon the channel complex by the S1369A variant and not the E23K variant (Hamming et al. 2009). Further molecular studies revealed that the S169A variant increases the intrinsic catalytic MgATPase activity of the channel complex resulting in the increased generation of stimulatory ADP

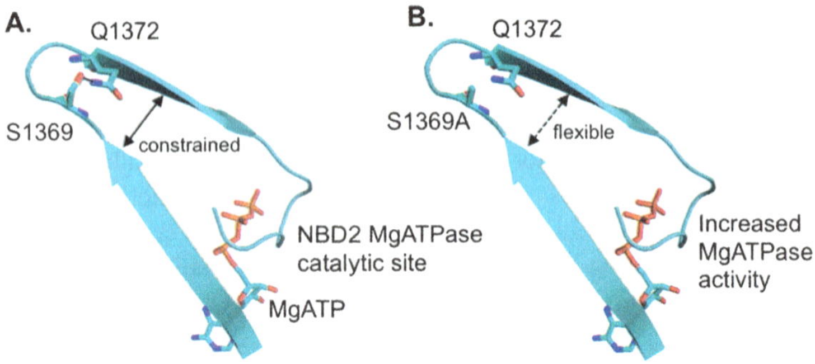

Fig. 17.2 Putative molecular mechanism by which the S1369A variant increases KATP channel activity. (**a**) The S1369 residue is located close to the NBD2 MgATPase catalytic site and predicted to form a hydrogen bond with the adjacent Q1372 residue, constraining movement in this region. (**b**) The S1369A variant removes the hydroxyl group from serine and prevents the formation of the hydrogen bond, leading to increased flexibility in this region and an increase in MgATPase activity

(Fatehi et al. 2012). Based on in silico homology modelling of the nucleotide-binding domains of SUR1 (Masia and Nichols 2008), residue 1369 is located close to the top of a hairpin loop adjacent to the major MgATPase site in nucleotide-binding domain 2 and S1369 can form a hydrogen bond with Q1372 on the opposite side of the loop, thus stabilizing the adjacent catalytic site (Fig. 17.3). In the presence of the A1369 variant, the hydrogen bond is lost and this may result in increased flexibility of the loop resulting in a higher rate of ATP hydrolysis to ADP (Fatehi et al. 2012). From a physiological perspective, the increase in MgATPase activity leads to increased K_{ATP} channel activity that would suppress insulin secretion and perhaps also alter glucagon release (Rorsman et al. 2008; MacDonald et al. 2007). Furthermore, earlier studies have shown altered glucose-induced insulin secretory response in carriers of the E23K/S1369A type 2 diabetes-risk haplotype (Nielsen et al. 2003; Florez et al. 2004). While the A1369 variant may alter the intrinsic nucleotide regulation of the channel, the E23K variant also alters the K_{ATP} channel sensitivity to intracellular long-chain acyl coenzyme As (LC-CoAs). The LC-CoAs are the intracellular esters of fatty acids and are known potent activators of K_{ATP} channels (Riedel et al. 2003). It has previously been shown that K_{ATP} channels containing the E23K variant in the Kir6.2 subunit (KCNJ11 gene) are more sensitive to activation by LC-CoAs in a side-chain length and saturation-dependent manner with longer, more saturated, and trans LC-CoAs being more potent activators (Riedel and Light 2005). These results suggest that there may also be an environmental/dietary component to the cellular mechanisms involved in increased type 2 diabetes risk, with a diet rich in saturated or trans fats leading to increased K_{ATP} channel activity especially in K_{ATP} channels containing E23K in the Kir6.2 subunit (Riedel et al. 2005). In summary, it is difficult to ascribe causality to either variant alone and it is likely that both E23K and S1369A confer alterations in the properties of K_{ATP} channel in response to metabolites and that

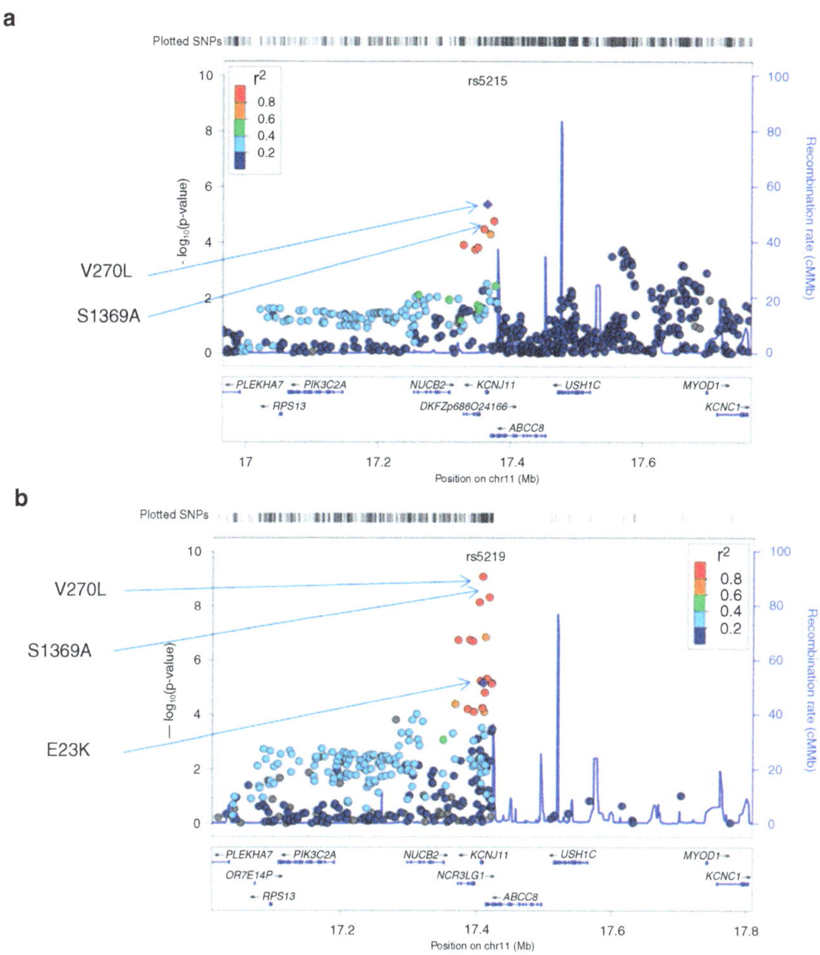

Fig. 17.3 Association results from the latest DIAGRAM consortium genome-wide association study meta-analysis (Morris et al. 2012). Plots generated in LocusZoom (Pruim et al. 2010) from DIAGRAM data available at http://diagram-consortium.org/downloads.html. (**a**) HapMap imputed results from 12,171 cases and 56,862 controls. (**b**) Metabochip results from 22,669 cases and 58,119 controls. E23K is not present in the HapMap so results are not available in (**a**). E23K was only genotyped in 54,597 samples compared to 86,193 for S1369A and V270L which explains its weaker association results. However, the odds ratios for the SNPs were 1.075 (V270L), 1.074 (S1369A), and 1.070 (E23K). These results are not statistically different from each other

both variants contribute to the observed increase in odds ratio. It should be noted that K_{ATP} channels are expressed in many tissue types and can be assembled as hetero-octamers from multiple gene products such as Kir6.1, Kir6.2, SUR1, and SUR2A/B. Therefore, there is much we still don't know regarding the contributions of the E23K and S1369A to systemic metabolic homeostasis in extra-pancreatic tissues such as the heart, skeletal muscle, and central nervous system.

17.6 The Clinical Utility of the E23K/S1369A Variant Haplotype

The current best estimate of the odds ratio for diabetes associated with the K allele of E23K is 1.07 (95 % CI: 1.05, 1.10) from the DIAGRAM consortium. This is significantly lower than the odds ratio of 1.23 that we initially reported (Gloyn et al. 2003), perhaps due to the Winner's curse (Lohmueller et al. 2003). This low odds ratio means that this variant has essentially no predictive power for determining whether an individual will get diabetes, even when in combination with other known type 2 diabetes variants (Weedon et al. 2006; Lango et al. 2008; Lyssenko et al. 2005, 2008; Meigs et al. 2008; Mihaescu et al. 2011). The potential for gene-environment interactions may also further reduce the predictive power as the increase in risk for type 2 diabetes with these variants may involve altered dietary or lifestyle choices as discussed further below. However, patients with neonatal diabetes due to specific activating mutations in *KCNJ11* or *ABCC8* can be effectively treated with sulphonylureas to close mutant K_{ATP} channels, rather than insulin injections (Pearson et al. 2006). So an obvious question is whether the E23K/S1369A variant haplotype is associated with response to sulphonylurea treatment. Several studies have found some association of E23K/S1369A with response to sulphonylureas. Feng et al. (2008) showed an association in two Chinese populations of the S1369A variant affecting glycaemic control on sulphonylureas. A/A homozygotes had 7.7 % higher fasting glucose levels than S/S homozygotes after 8 weeks of treatment on gliclazide. Sesti et al. (2006) studied 525 UK Caucasian patients with diabetes and found nominal evidence for an increase in sulphonylurea failure for K allele carriers (OR = 1.45). Holstein et al. (2009) found that carriers of the K allele had reduced incidence of severe hypoglycaemic events. However, the evidence from other studies is not strong (Holstein et al. 2012; Ragia et al. 2012; Sato et al. 2010). In the Diabetes Prevention Program study, Florez et al. (2007) found that K allele carriers were less likely to develop diabetes but also had a worse response to 1 year of metformin treatment. Whether all of these associations are true associations or due to relatively small sample sizes and publication bias remains to be seen as larger pharmacogenetic studies are performed. Interestingly, recent mechanistic studies on K_{ATP} channels containing E23K and S1369A have revealed differential sulphonylurea sensitivities to gliclazide but not glyburide (Lang et al. 2012; Hamming et al. 2009). These studies also found that the increased inhibition by gliclazide was conferred upon the channel complex by the S1369A variant. Whether this latter observation has implications for the pharmacogenetic management of type 2 diabetes has yet to be determined by clinical studies.

17.6.1 Why Is the E23K/S1369A Haplotype So Common?

It is clear that the E23K and S1369A variants are associated with type 2 diabetes and that they may act in concert to elicit subtle yet important alterations in pancreatic endocrine function that may predispose homozygous individuals to type 2 diabetes. What remains to be determined are the effects of these two variants, either alone or in haplotypic combination, on cellular function in extra-pancreatic tissues as well as their precise interactions with environmental factors such as diet and lifestyle. In this regard, it is worth speculating why ~10 % of individuals are homozygous for the E23K/S1369A haplotype. We have previously suggested that these variants may form part of a haplotype offering some evolutionary advantage that accounts for their stable representation in most ethnic groups tested to date (Riedel et al. 2005). The E23K/S1369A variant may be a "thrifty utility" gene improving athletic endurance in early hunter-gatherer societies by maintaining blood glucose levels via increased glucagon release and decreased insulin release (pancreatic *KCNJ11* and *ABCC8*) and by enhancing skeletal and cardiac muscle function (with E23K in *KCNJ11*). In modern westernized societies that are associated with excess calories and inactivity, these variants may be disadvantageous in homozygous individuals, therefore increasing susceptibility to type 2 diabetes.

17.7 Do Rare Coding Variants of ABCC8 and KCNJ11 Predispose to Diabetes?

Advances in sequencing technologies now allow the rapid and cost-effective sequencing of all protein-coding regions of the genome in large sample sizes. Several whole-exome as well as whole-genome sequence association studies are ongoing to identify new causes of type 2 diabetes. The first published large-scale deep whole-exome sequence study was recently published and analysed 1000 type 2 diabetes cases and 1000 controls (Lohmueller et al. 2013). No novel individual variants were identified from this analysis that had not been identified from previous GWA studies. Sequencing analysis, as opposed to microarray-based association studies, allows the detection of all variants in a patient's exome or genome. This allows a statistical test of rare variants by seeing whether there is an excess of new variants in cases compared to controls. Given the spectrum of effects of different variants in *KCNJ11* and *ABCC8* on diabetes risk—from the most severe V59M mutation which causes syndromic neonatal diabetes to variants that cause transient neonatal diabetes to the type 2 diabetes-risk variants (Fig. 17.4)—it seems likely that additional variants in *KCNJ11* and *ABCC8* may be discovered which also predispose to type 2 diabetes. However, the work by Lohmueller et al. (2013) shows how difficult it will be, as well as showing that coding variants of *KCNJ11* and *ABCC8* will not be a common cause of type 2 diabetes, but rather a contributory component in haplotypic combination with other gene variants.

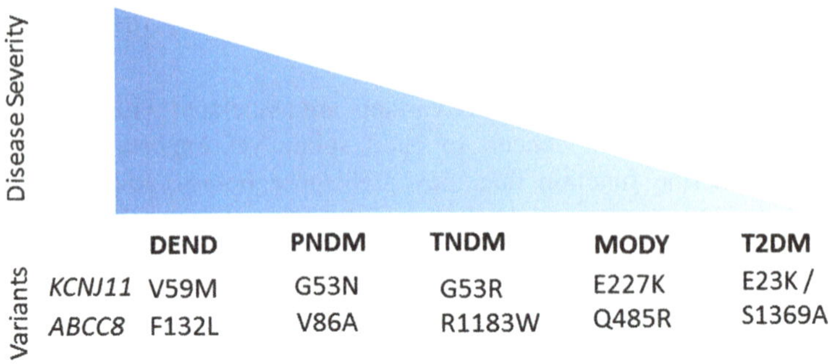

Fig. 17.4 The range of clinical presentations with different mutations of the *KCNJ11* and *ABCC8* genes. A selection of mutations are provided [a more comprehensive list of mutations in these genes causing diabetes or hyperinsulinism is provided in (Flanagan et al. 2009)]. The mutations range from those that cause DEND syndrome, a syndrome that includes neonatal diabetes as well as severe neurological disease, to the E23K/S1369A variant which predispose to type 2 diabetes. *TNDM*, transient neonatal diabetes; *PNDM*, permanent neonatal diabetes. This figure is adapted from Gloyn et al. (2005)

References

Aguilar-Bryan L, Bryan J (1999) Molecular biology of adenosine triphosphate-sensitive potassium channels. Endocr Rev 20(2):101–135

Aguilar-Bryan L, Nichols CG, Wechsler SW, Clement JP 4th, Boyd AE 3rd, González G, Herrera-Sosa H, Nguy K, Bryan J, Nelson DA (1995) Cloning of the beta cell high-affinity sulfonylurea receptor: a regulator of insulin secretion. Science 268(5209):423–426

Altshuler D, Hirschhorn JN, Klannemark M, Lindgren CM, Vohl M-C, Nemesh J, Lane CR, Schaffner F, Bolk S, Brewer C, Tuomi T, Gaudet D, Hudson TJ, Daly M, Groop L, Lander ES (2000) The common PPARg Pro12Ala polymorphism is associated with decreased risk of type 2 diabetes. Nat Genet 26:76–80

Ashcroft FM, Rorsman P (1989) Electrophysiology of the pancreatic beta-cell. Prog Biophys Mol Biol 54(2):87–143

Ashcroft FM, Harrison DE, Ashcroft SJ (1984) Glucose induces closure of single potassium channels in isolated rat pancreatic beta cells. Nature 312(5993):446–448

Ashcroft FM, Proks P, Smith PA, Ammala C, Bokvist K, Rorsman P (1994) Stimulus-secretion coupling in pancreatic beta cells. J Cell Biochem 55(Suppl):54–65

Bonnefond A, Philippe J, Durand E, Dechaume A, Huyvaert M, Montagne L, Marre M, Balkau B, Fajardy I, Vambergue A, Vatin V, Delplanque J, Le Guilcher D, De Graeve F, Lecoeur C, Sand O, Vaxillaire M, Froguel P (2012) Whole-exome sequencing and high throughput genotyping identified KCNJ11 as the thirteenth MODY gene. PLoS One 7(6), e37423

Bowman P, Flanagan SE, Edghill EL, Damhuis A, Shepherd MH, Paisey R, Hattersley AT, Ellard S (2012) Heterozygous ABCC8 mutations are a cause of MODY. Diabetologia 55(1):123–127

Bryan J, Vila-Carriles WH, Zhao G, Babenko AP, Aguilar-Bryan L (2004) Toward linking structure with function in ATP-sensitive K+ channels. Diabetes 53(Suppl 3):S104–S112

Bryan J, Crane A, Vila-Carriles WH, Babenko AP, Aguilar-Bryan L (2005) Insulin secretagogues, sulfonylurea receptors and K(ATP) channels. Curr Pharm Des 11(21):2699–2716

Dunne MJ, Kane C, Shepherd RM, Sanchez JA, James RF, Johnson PR, Aynsley-Green A, Lu S, Clement JP, Lindley KJ, Seino S, Aguilar-Bryan L (1997) Familial persistent hyperinsulinemic hypoglycemia of infancy and mutations in the sulfonylurea receptor. N Engl J Med 336 (10):703–706

Fatehi M, Raja M, Carter C, Soliman D, Holt A, Light PE (2012) The ATP-sensitive K(+) channel ABCC8 S1369A type 2 diabetes risk variant increases MgATPase activity. Diabetes 61 (1):241–249

Feng Y, Mao G, Ren X, Xing H, Tang G, Li Q, Li X, Sun L, Yang J, Ma W, Wang X, Xu X (2008) Ser1369Ala variant in sulfonylurea receptor gene ABCC8 is associated with antidiabetic efficacy of gliclazide in Chinese type 2 diabetic patients. Diabetes Care 31(10):1939–1944

Flanagan SE, Clauin S, Bellanne-Chantelot C, de Lonlay P, Harries LW, Gloyn AL, Ellard S (2009) Update of mutations in the genes encoding the pancreatic beta-cell K(ATP) channel subunits Kir6.2 (KCNJ11) and sulfonylurea receptor 1 (ABCC8) in diabetes mellitus and hyperinsulinism. Hum Mutat 30(2):170–180

Florez JC, Burtt N, de Bakker PI, Almgren P, Tuomi T, Holmkvist J, Gaudet D, Hudson TJ, Schaffner SF, Daly MJ, Hirschhorn JN, Groop L, Altshuler D (2004) Haplotype structure and genotype-phenotype correlations of the sulfonylurea receptor and the islet ATP-sensitive potassium channel gene region. Diabetes 53(5):1360–1368

Florez JC, Jablonski KA, Kahn SE, Franks PW, Dabelea D, Hamman RF, Knowler WC, Nathan DM, Altshuler D (2007) Type 2 diabetes-associated missense polymorphisms KCNJ11 E23K and ABCC8 A1369S influence progression to diabetes and response to interventions in the diabetes prevention program. Diabetes 56(2):531–536

Gloyn AL, Hashim Y, Ashcroft SJH, Ashfield R, Wiltshire S, Turner RC (2001) Association studies of variants in promoter and coding regions of beta-cell ATP-sensitive K-channel genes SUR1 and Kir6.2 with Type 2 diabetes mellitus (UKPDS 53). Diabet Med 18:206–212

Gloyn AL, Weedon MN, Owen KR, Turner MJ, Knight BA, Hitman G, Walker M, Levy JC, Sampson M, Halford S, McCarthy MI, Hattersley AT, Frayling TM (2003) Large-scale association studies of variants in genes encoding the pancreatic beta-cell KATP channel subunits Kir6.2 (KCNJ11) and SUR1 (ABCC8) confirm that the KCNJ11 E23K variant is associated with type 2 diabetes. Diabetes 52(2):568–572

Gloyn AL, Pearson ER, Antcliff JF, Proks P, Bruining GJ, Slingerland AS, Howard N, Srinivasan S, Silva JM, Molnes J, Edghill EL, Frayling TM, Temple IK, Mackay D, Shield JP, Sumnik Z, van Rhijn A, Wales JK, Clark P, Gorman S, Aisenberg J, Ellard S, Njolstad PR, Ashcroft FM, Hattersley AT (2004) Activating mutations in the gene encoding the ATP-sensitive potassium-channel subunit Kir6.2 and permanent neonatal diabetes. N Engl J Med 350(18):1838–1849

Gloyn AL, Reimann F, Girard C, Edghill EL, Proks P, Pearson ER, Temple IK, Mackay DJ, Shield JP, Freedenberg D, Noyes K, Ellard S, Ashcroft FM, Gribble FM, Hattersley AT (2005) Relapsing diabetes can result from moderately activating mutations in KCNJ11. Hum Mol Genet 14(7):925–934

Hamming KS, Soliman D, Matemisz LC, Niazi O, Lang Y, Gloyn AL, Light PE (2009) Coexpression of the type 2 diabetes susceptibility gene variants KCNJ11 E23K and ABCC8 S1369A alter the ATP and sulfonylurea sensitivities of the ATP-sensitive K(+) channel. Diabetes 58(10):2419–2424

Hani EH, Clement K, Velho G, Vionnet N, Hager J, Philippi A, Dina C, Inoue H, Permutt MA, Basdevant A, North M, Demenais F, Guy-Grand B, Froguel P (1997) Genetic studies of the sulphonylurea receptor gene locus in NIDDM and in morbid obesity among French Caucasians. Diabetes 46:688–694

Hani EH, Suaud L, Boutin P, Chevre JC, Durand E, Philippi A, Demenais F, Vionnet N, Furuta H, Velho G, Bell GI, Laine B, Froguel P (1998) A missense mutation in hepatocyte nuclear factor-4 alpha, resulting in a reduced transactivation activity, in human late-onset non-insulin-dependent diabetes mellitus. J Clin Invest 101(3):521–526

Hansen L, Echwald SM, Hansen T, Urhammer SA, Clausen JO, Pedersen O (1997) Amino acid polymorphisms in the ATP-regulatable inward rectifier Kir6.2 and their relationships to glucose- and tolbutamide-induced insulin secretion, the insulin sensitivity index, and NIDDM. Diabetes 46(3):508–512

Hansen T, Echwald SM, Hansen L, Moller AM, Almind K, Clausen JO, Urhammer SA, Inoue H, Ferrer J, Bryan J, Aguilar-Bryan L, Permutt MA, Pedersen O (1998) Decreased tolbutamide-stimulated insulin secretion in healthy subjects with sequence variants in the high-affinity sulfonylurea receptor gene. Diabetes 47(4):598–605

Hirschhorn JN, Lohmueller K, Byrne E, Hirschhorn K (2002) A comprehensive review of genetic association studies. Genet Med 4(2):45–61

Holstein A, Hahn M, Stumvoll M, Kovacs P (2009) The E23K variant of KCNJ11 and the risk for severe sulfonylurea-induced hypoglycemia in patients with type 2 diabetes. Horm Metab Res 41(5):387–390

Holstein JD, Kovacs P, Patzer O, Stumvoll M, Holstein A (2012) The Ser1369Ala variant of ABCC8 and the risk for severe sulfonylurea-induced hypoglycemia in German patients with Type 2 diabetes. Pharmacogenomics 13(1):5–7

Inagaki N, Gonoi T, Clement JP, Namba N, Inazawa J, Gonzalez G, Aguilar-Bryan L, Seino S, Bryan J (1995) Reconstitution of IKATP: an inward rectifier subunit plus the sulfonylurea receptor. Science 270(5239):1166–1170

Inoue H, Ferrer J, Welling CM, Elbein SC, Hoffman M, Mayorga R, Warren-Perry M, Zhang Y, Millns H, Turner R, Province M, Bryan J, Permutt MA, Aguilar-Bryan L (1996) Sequence variants in the sulfonylurea receptor (SUR) gene are associated with NIDDM in Caucasians. Diabetes 45(6):825–831

Inoue H, Ferrer J, Warren-Perry M, Zhang Y, Millns H, Turner RC, Elbein SC, Hampe CL, Suarez BK, Inagaki N, Seino S, Permutt MA (1997) Sequence variants in the pancreatic islet beta-cell inwardly rectifying K+ channel Kir6.2 (Bir) gene: identification and lack of role in Caucasian patients with NIDDM. Diabetes 46(3):502–507

Lang V, Light PE (2010) The molecular mechanisms and pharmacotherapy of ATP-sensitive potassium channel gene mutations underlying neonatal diabetes. Pharmacogenomics Pers Med 3:145–161. doi:10.2147/PGPM.S6969

Lang VY, Fatehi M, Light PE (2012) Pharmacogenomic analysis of ATP-sensitive potassium channels coexpressing the common type 2 diabetes risk variants E23K and S1369A. Pharmacogenet Genomics 22(3):206–214

Lango H, Palmer CN, Morris AD, Zeggini E, Hattersley AT, McCarthy MI, Frayling TM, Weedon MN (2008) Assessing the combined impact of 18 common genetic variants of modest effect sizes on type 2 diabetes risk. Diabetes 57(11):3129–3135

Liu L, Nagashima K, Yasuda T, Liu Y, Hu HR, He G, Feng B, Zhao M, Zhuang L, Zheng T, Friedman TC, Xiang K (2013) Mutations in KCNJ11 are associated with the development of autosomal dominant, early-onset type 2 diabetes. Diabetologia 56(12):2609–2618

Lohmueller KE, Pearce CL, Pike M, Lander ES, Hirschhorn JN (2003) Meta-analysis of genetic association studies supports a contribution of common variants to susceptibility to common disease. Nat Genet 33(2):177–182

Lohmueller KE, Sparso T, Li Q, Andersson E, Korneliussen T, Albrechtsen A, Banasik K, Grarup N, Hallgrimsdottir I, Kiil K, Kilpelainen TO, Krarup NT, Pers TH, Sanchez G, Hu Y, Degiorgio M, Jorgensen T, Sandbaek A, Lauritzen T, Brunak S, Kristiansen K, Li Y, Hansen T, Wang J, Nielsen R, Pedersen O (2013) Whole-exome sequencing of 2,000 Danish individuals and the role of rare coding variants in type 2 diabetes. Am J Hum Genet 93 (6):1072–1086

Lyssenko V, Almgren P, Anevski D, Orho-Melander M, Sjogren M, Saloranta C, Tuomi T, Groop L (2005) Genetic prediction of future type 2 diabetes. PLoS Med 2(12):1299–1308

Lyssenko V, Jonsson A, Almgren P, Pulizzi N, Isomaa B, Tuomi T, Berglund G, Altshuler D, Nilsson P, Groop L (2008) Clinical risk factors, DNA variants, and the development of type 2 diabetes. N Engl J Med 359(21):2220–2232

MacDonald PE, Rorsman P (2007) The ins and outs of secretion from pancreatic beta-cells: control of single-vesicle exo- and endocytosis. Physiology 22:113–121

MacDonald PE, De Marinis YZ, Ramracheya R, Salehi A, Ma X, Johnson PR, Cox R, Eliasson L, Rorsman P (2007) A K ATP channel-dependent pathway within alpha cells regulates glucagon release from both rodent and human islets of Langerhans. PLoS Biol 5(6), e143

Masia R, Nichols CG (2008) Functional clustering of mutations in the dimer interface of the nucleotide binding folds of the sulfonylurea receptor. J Biol Chem 283(44):30322–30329

Meigs JB, Shrader P, Sullivan LM, McAteer JB, Fox CS, Dupuis J, Manning AK, Florez JC, Wilson PW, D'Agostino RB Sr, Cupples LA (2008) Genotype score in addition to common risk factors for prediction of type 2 diabetes. N Engl J Med 359(21):2208–2219

Mihaescu R, Meigs J, Sijbrands E, Janssens AC (2011) Genetic risk profiling for prediction of type 2 diabetes. PLoS Curr 3, RRN1208

Morris AP, Voight BF, Teslovich TM, Ferreira T, Segre AV, Steinthorsdottir V, Strawbridge RJ, Khan H, Grallert H, Mahajan A, Prokopenko I, Kang HM, Dina C, Esko T, Fraser RM, Kanoni S, Kumar A, Lagou V, Langenberg C, Luan J, Lindgren CM, Muller-Nurasyid M, Pechlivanis S, Rayner NW, Scott LJ, Wiltshire S, Yengo L, Kinnunen L, Rossin EJ, Raychaudhuri S, Johnson AD, Dimas AS, Loos RJ, Vedantam S, Chen H, Florez JC, Fox C, Liu CT, Rybin D, Couper DJ, Kao WH, Li M, Cornelis MC, Kraft P, Sun Q, van Dam RM, Stringham HM, Chines PS, Fischer K, Fontanillas P, Holmen OL, Hunt SE, Jackson AU, Kong A, Lawrence R, Meyer J, Perry JR, Platou CG, Potter S, Rehnberg E, Robertson N, Sivapalaratnam S, Stancakova A, Stirrups K, Thorleifsson G, Tikkanen E, Wood AR, Almgren P, Atalay M, Benediktsson R, Bonnycastle LL, Burtt N, Carey J, Charpentier G, Crenshaw AT, Doney AS, Dorkhan M, Edkins S, Emilsson V, Eury E, Forsen T, Gertow K, Gigante B, Grant GB, Groves CJ, Guiducci C, Herder C, Hreidarsson AB, Hui J, James A, Jonsson A, Rathmann W, Klopp N, Kravic J, Krjutskov K, Langford C, Leander K, Lindholm E, Lobbens S, Mannisto S, Mirza G, Muhleisen TW, Musk B, Parkin M, Rallidis L, Saramies J, Sennblad B, Shah S, Sigurethsson G, Silveira A, Steinbach G, Thorand B, Trakalo J, Veglia F, Wennauer R, Winckler W, Zabaneh D, Campbell H, van Duijn C, Uitterlinden AG, Hofman A, Sijbrands E, Abecasis GR, Owen KR, Zeggini E, Trip MD, Forouhi NG, Syvanen AC, Eriksson JG, Peltonen L, Nothen MM, Balkau B, Palmer CN, Lyssenko V, Tuomi T, Isomaa B, Hunter DJ, Qi L, Wellcome Trust Case Control C, Meta-Analyses of G, Insulin-related traits Consortium I, Genetic Investigation of ATC, Asian Genetic Epidemiology Network-Type 2 Diabetes C, South Asian Type 2 Diabetes C, Shuldiner AR, Roden M, Barroso I, Wilsgaard T, Beilby J, Hovingh K, Price JF, Wilson JF, Rauramaa R, Lakka TA, Lind L, Dedoussis G, Njolstad I, Pedersen NL, Khaw KT, Wareham NJ, Keinanen-Kiukaanniemi SM, Saaristo TE, Korpi-Hyovalti E, Saltevo J, Laakso M, Kuusisto J, Metspalu A, Collins FS, Mohlke KL, Bergman RN, Tuomilehto J, Boehm BO, Gieger C, Hveem K, Cauchi S, Froguel P, Baldassarre D, Tremoli E, Humphries SE, Saleheen D, Danesh J, Ingelsson E, Ripatti S, Salomaa V, Erbel R, Jockel KH, Moebus S, Peters A, Illig T, de Faire U, Hamsten A, Morris AD, Donnelly PJ, Frayling TM, Hattersley AT, Boerwinkle E, Melander O, Kathiresan S, Nilsson PM, Deloukas P, Thorsteinsdottir U, Groop LC, Stefansson K, Hu F, Pankow JS, Dupuis J, Meigs JB, Altshuler D, Boehnke M, McCarthy MI, Replication DIG, Meta-analysis C (2012) Large-scale association analysis provides insights into the genetic architecture and pathophysiology of type 2 diabetes. Nat Genet 44(9):981–990. doi:10.1038/ng.2383

Nielsen EM, Hansen L, Carstensen B, Echwald SM, Drivsholm T, Glumer C, Thorsteinsson B, Borch-Johnsen K, Hansen T, Pedersen O (2003) The E23K variant of Kir6.2 associates with impaired post-OGTT serum insulin response and increased risk of type 2 diabetes. Diabetes 52 (2):573–577

Pearson ER, Flechtner I, Njolstad PR, Malecki MT, Flanagan SE, Larkin B, Ashcroft FM, Klimes I, Codner E, Iotova V, Slingerland AS, Shield J, Robert J-J, Holst JJ, Clark PM, Ellard S, Sovik O, Polak M, Hattersley AT, the Neonatal Diabetes International Collaborative G (2006) Switching from insulin to oral sulfonylureas in patients with diabetes due to Kir6.2 mutations. N Engl J Med 355(5):467–477

Pruim RJ, Welch RP, Sanna S, Teslovich TM, Chines PS, Gliedt TP, Boehnke M, Abecasis GR, Willer CJ (2010) LocusZoom: regional visualization of genome-wide association scan results. Bioinformatics 26(18):2336–2337

Ragia G, Tavridou A, Petridis I, Manolopoulos VG (2012) Association of KCNJ11 E23K gene polymorphism with hypoglycemia in sulfonylurea-treated type 2 diabetic patients. Diabetes Res Clin Pract 98(1):119–124

Replication DIG, Meta-analysis C (2014) Genome-wide trans-ancestry meta-analysis provides insight into the genetic architecture of type 2 diabetes susceptibility. Nat Genet 46(3):234–244

Riedel MJ, Light PE (2005) Saturated and cis/trans unsaturated acyl CoA esters differentially regulate wild-type and polymorphic beta-cell ATP-sensitive K+ channels. Diabetes 54 (7):2070–2079

Riedel MJ, Boora P, Steckley D, de Vries G, Light PE (2003) Kir6.2 polymorphisms sensitize beta-cell ATP-sensitive potassium channels to activation by acyl CoAs: a possible cellular mechanism for increased susceptibility to type 2 diabetes? Diabetes 52(10):2630–2635

Riedel MJ, Steckley DC, Light PE (2005) Current status of the E23K Kir6.2 polymorphism: implications for type-2 diabetes. Hum Genet 116(3):133–145

Rissanen J, Markkanen A, Karkkainen P, Pihlajamaki J, Kekalainen P, Mykkanen L, Kuusisto J, Karhapaa P, Niskanen L, Laakso M (2000) Sulfonylurea receptor 1 gene variants are associated with gestational diabetes and type 2 diabetes but not with altered secretion of insulin. Diabetes Care 23(1):70–73

Rorsman P, Salehi SA, Abdulkader F, Braun M, MacDonald PE (2008) K(ATP)-channels and glucose-regulated glucagon secretion. Trends Endocrinol Metab 19(8):277–284

Sakura H, Wat N, Horton V, Millns H, Turner RC, Ashcroft FM (1996) Sequence variations in the human Kir6.2 gene, a subunit of the beta-cell ATP-sensitive K-channel: no association with NIDDM in while Caucasian subjects or evidence of abnormal function when expressed in vitro. Diabetologia 39(10):1233–1236

Sato R, Watanabe H, Genma R, Takeuchi M, Maekawa M, Nakamura H (2010) ABCC8 polymorphism (Ser1369Ala): influence on severe hypoglycemia due to sulfonylureas. Pharmacogenomics 11(12):1743–1750

Schwanstecher C, Meyer U, Schwanstecher M (2002a) K(IR)6.2 polymorphism predisposes to type 2 diabetes by inducing overactivity of pancreatic beta-cell ATP-sensitive K(+) channels. Diabetes 51(3):875–879

Schwanstecher C, Neugebauer B, Schulz M, Schwanstecher M (2002b) The common single nucleotide polymorphism E23K in K(IR)6.2 sensitizes pancreatic beta-cell ATP-sensitive potassium channels toward activation through nucleoside diphosphates. Diabetes 51(Suppl 3):S363–S367

Sesti G, Laratta E, Cardellini M, Andreozzi F, Del Guerra S, Irace C, Gnasso A, Grupillo M, Lauro R, Hribal ML, Perticone F, Marchetti P (2006) The E23K variant of KCNJ11 encoding the pancreatic beta-cell adenosine 5′-triphosphate-sensitive potassium channel subunit Kir6.2 is associated with an increased risk of secondary failure to sulfonylurea in patients with type 2 diabetes. J Clin Endocrinol Metab 91(6):2334–2339

't Hart LM, de Knijff P, Dekker JM, Stolk RP, Nijpels G, can der Does FE, Ruige JB, Grobbee DE, Heine RJ, Maassen JA (1999) Variants in the sulphonylurea receptor gene: association of the exon 16-3t variant with Type II diabetes mellitus in Dutch Caucasians. Diabetologia 42:617–620

Thomas PM, Cote GJ, Wohllk N, Haddad B, Mathew PM, Rabl W, Aguilar-Bryan L, Gagel RF, Bryan J (1995) Mutations in the sulfonylurea receptor gene in familial persistent hyperinsulinemic hypoglycemia of infancy. Science 268(5209):426–429

Van de Brunt M, Gloyn AL (2012) A tale of two glucose transporters: how GLUT2 re-emerged as a contender for glucose transport into the human beta cell. Diabetologia 55(9):2312–2315

van Tilburg JH, Rozeman LB, van Someren H, Rigters-Aris CA, Freriks JP, Pearson PL, Sandkuijl LA, van Haeften TW, Wijmenga C (2000) The exon 16-3t variant of the sulphonylurea receptor gene is not a risk factor for Type II diabetes mellitus in the Dutch Breda cohort. Diabetologia 43(5):681–682

Weedon MN, McCarthy MI, Hitman G, Walker M, Groves CJ, Zeggini E, Rayner NW, Shields B, Owen KR, Hattersley AT, Frayling TM (2006) Combining Information from common Type 2 diabetes risk polymorphisms improves disease prediction. PLoS Med 3(10), e374

Chapter 18
SLC30A8: A Complex Road from Association to Function

Jason Flannick and William L. Lowe

Abstract One of the more compelling common variants associated with type 2 diabetes (T2D) in the initial wave of genome-wide association studies was a missense variant (rs13266634, p.Trp325Arg) in *SLC30A8*, a gene that encodes a zinc transporter (ZnT8) expressed primarily in pancreatic islets. This widely replicated association, together with the importance of ZnT8 in the process of zinc transport into insulin secretory granules, provided a potentially straightforward path toward insight into T2D pathogenesis. However, since this initial observation, the path from association to function for *SLC30A8* has been far from straightforward. In this chapter, we review (a) the initial genetic evidence that identified variation in *SLC30A8* as a risk factor for T2D, (b) the biological function of its protein product (ZnT8), (c) associations between *SLC30A8* variation and other diabetes-related intermediate phenotypes, (d) experiments to discern the mechanism whereby perturbation of ZnT8 influences T2D risk, and (e) recent results from large-scale sequencing and genotyping to identify a broader allelic series within *SLC30A8*. Although both genetic and experimental evidences point to clear link between altered *SLC30A8* function and T2D in human populations, the specific mechanism of pathogenesis is still far from clear.

18.1 Introduction

Genetic mapping and positional cloning together form a paradigm to dissect the genetic basis of disease. For type 2 diabetes (T2D), genome-wide association studies (GWAS) have made significant progress mapping loci that harbor disease-causing

J. Flannick (✉)
Program in Medical and Population Genetics, Broad Institute, Cambridge, MA, USA

Department of Molecular Biology, Massachusetts General Hospital, Boston, MA, USA
e-mail: flannick@broadinstitute.org

W.L. Lowe
Division of Endocrinology, Metabolism, and Molecular Medicine, Department of Medicine, Northwestern University Feinberg School of Medicine, Chicago, IL, USA

J.C. Florez (ed.), *The Genetics of Type 2 Diabetes and Related Traits*,
DOI 10.1007/978-3-319-01574-3_18

variation—identifying over 80 signals and highlighting potential pathways and biology relevant to disease. The next step, to translate an association signal to mechanistic insight and potentially a therapeutic hypothesis, requires the identification of the causal variant and gene underlying the signal, phasing of the directional relationship between gene activity and disease risk, and quantification of the range of phenotypic effects of a broad series of alleles. However, most common disease-associated variation lies in noncoding regions, of which biological understanding is limited, presenting a challenge to translate most association signals to function.

Given our better understanding of the coding genome, disease-associated common variants that alter protein sequence are of particular value toward the goal of causal gene and variant identification. For this reason, one of the more compelling common variants associated with T2D in the initial wave of GWAS was a common variant (rs13266634, p.Trp325Arg) in *SLC30A8*, a gene that encodes a zinc transporter (ZnT8) active in pancreatic islets. Given the high relevance of zinc and islets to T2D pathophysiology, the observation of a T2D-associated protein-altering variant in *SLC30A8* provided one of the potentially more straightforward paths toward insight into T2D pathogenesis and, perhaps, a candidate drug target.

In reality, the path from association to function for *SLC30A8* has been far from straightforward: spanning GWAS of increasing size and diversity, interrogation of variant carriers for a wide range of phenotypes, cellular and animal model systems with at times equivocal results, and resequencing and genotyping in hundreds of thousands of individuals. In this chapter, we will review (a) the initial genetic evidence that identified variation in *SLC30A8* as a risk factor for T2D, (b) the biological function of ZnT8, (c) associations of *SLC30A8* variation with diabetes-related intermediate phenotypes, (d) experiments to discern the mechanism whereby perturbation of ZnT8 influences T2D risk, and (e) recent results from large-scale sequencing and genotyping to identify a broader allelic series and lend further insight into the relationship between *SLC30A8* activity and human physiology.

18.2 Association of rs13266634 with Risk of Type 2 Diabetes

In 2007, the first series of GWAS for T2D were published, ushering in an era that would dramatically expand the number of variants associated with disease risk. The first such study, by Sladek et al. (2007), replicated the strongest previously reported association with T2D (mapped to *TCF7L2*) and also identified four additional signals. Of these newly reported associations, the strongest was a single-nucleotide polymorphism (SNP, rs13266634) in *SLC30A8*, with an estimated frequency of 30 % and an odds ratio of 1.18.

Although most GWAS associations are expected to tag, rather than represent, truly causal SNPs, the *SLC30A8* association was notable given that rs13266634 is a missense SNP, causing a change from arginine to tryptophan at residue 325 (p.Trp3235Arg) in the protein product (ZnT8) of *SLC30A8*. Furthermore, the authors commented on the biological relevance of ZnT8 to T2D, noting previous experiments demonstrating increased glucose-stimulated insulin secretion following overexpression of *SLC30A8* in insulinoma cells (Chimienti et al. 2006). The rs13266634 association thus presented, at an early stage, a candidate causal gene and variant that altered risk of T2D.

Soon after the first publication of the *SLC30A8* association, four additional GWAS independently replicated the *SLC30A8* association at nominal significance, producing similar risk estimates as in the original publication (Saxena et al. 2007; Zeggini et al. 2007; Scott et al. 2007; Steinthorsdottir et al. 2007). Since this first wave of GWAS, rs13266634 has been analyzed in many association studies, spanning multiple cohorts, ancestries, and phenotypes. Association was confirmed in other European (Hertel et al. 2008) and East Asian populations (Tabara et al. 2009), although association in African populations has been more elusive (Xu et al. 2011)—perhaps due to the lower frequency of the minor allele (Myles et al. 2008). Thus, conclusive evidence for association of rs13266634 and T2D was one of the earliest findings to emerge from the era of GWAS for T2D.

Most recently, increasingly large meta-analyses have propagated rs13266634 into hundreds of thousands of samples, progressively increasing the resolution of frequency and effect size estimates for the SNP. The largest study in Europeans has reached ~35,000 cases and ~115,000 controls (Morris et al. 2012) and ranked rs13266634 as the sixth strongest association ($p = 1.3 \times 10^{-21}$), with an estimated risk-allele frequency of 66 %, odds ratio of 1.13 (95 % confidence interval 1.09–1.16), and liability-scale variance explained of 0.185 % (sibling relative risk 1.003). Among the 63 reported T2D-associated SNPs, rs13266634 was estimated to have the fourth highest odds ratio.

In addition, meta-analyses have started to expand to include individuals of increasingly diverse ancestries. The largest trans-ethnic analysis includes ~27,000 cases and ~84,000 controls of European, East Asian, South Asian, and Mexican and Mexican American ancestry (Consortium DGRAM-aD 2014). In it, consistent effect size estimates were observed for rs13266634 across all ethnicities (ranging from 1.11 to 1.16), with independent evidence of association in each ancestry other than Mexican. Furthermore, the increased sample size narrowed the set of candidate causal SNPs potentially responsible for the association to only two, with rs13266634 clearly separated from the rest.

In summary, rs13266634 was one of the earliest, strongest, and most reproducible common variants associated with risk of T2D. Furthermore, there is strong statistical and biological evidence that the causal variant responsible for the signal is a missense variant p.Arg325Trp in *SLC30A8*. Given the clearer functional interpretation of coding variation relative to noncoding variation, rs13266634 offers one of the more promising routes to functional insight into T2D pathogenesis and possibly treatment.

18.3 Function of *SLC30A8*

About 10 years ago, prior to the discovery of the association between rs13266634 and T2D risk, *SLC30A8* was cloned based on homology to members of the ZnT family of zinc transporters (Chimienti et al. 2004) and was subsequently localized to chromosome 8 at 8q24.11. Proteins in the ZnT family transport zinc from the cytoplasm into intracellular vesicles or extracellular fluid and are one of two families of eukaryotic zinc transporters; the other, the ZIP (Zrt- and Irt-like proteins (39A)) family, transport zinc into the cytoplasm from either the extracellular fluid or intracellular vesicles (Cousins et al. 2006; Eide 2006; Kambe 2012; Huang and Tepaamorndech 2013; Mocchegiani et al. 2008). To date, 24 members of these two families have been described, with some being expressed relatively ubiquitously and others having a more tissue-specific pattern of expression (Cousins et al. 2006; Eide 2006; Kambe 2012; Mocchegiani et al. 2008; Chimienti 2013). These transporters play a key role in many biological functions of the human body, as zinc is an essential trace element that binds to a large number of proteins, estimated to be as many as 3000 or 10–20 % of the human proteome, and is critical for the function of over 300 enzymes, serving as a key structural component or cofactor (Chasapis et al. 2012; Jansen et al. 2009). About 1 % of total body zinc is replenished daily through dietary zinc intake and/or vitamins and other supplements (Jansen et al. 2009; Cousins et al. 2006).

SLC30A8 contains eight exons, which span 37 kilobases and encode ZnT8: a 369-amino-acid protein predicted to exist as a homodimer (Chimienti et al. 2004; Kambe 2012; Huang and Tepaamorndech 2013). The three-dimensional structure of the mammalian ZnT family has not been determined, but members of the family are predicted to have N- and C-termini, separated by six transmembrane domains, oriented toward the cytoplasm (Kambe 2012; Weijers 2010; Wijesekara et al. 2009) (Fig. 18.1). The cytoplasmic C-terminus plays a role in the sensing of zinc and contains a zinc binding site, while a cytosolic histidine-rich domain between the fourth and fifth transmembrane domains may serve as an additional zinc binding site or modulate zinc transport activity (Kambe 2012; Weijers 2010; Wijesekara et al. 2009; Huang and Tepaamorndech 2013).

Unique among the ZnT family, *SLC30A8* is expressed almost exclusively in pancreatic islets, representing the most highly expressed zinc transporter in this tissue for both humans and mice (Wijesekara et al. 2009; Chimienti et al. 2004; Nicolson et al. 2009). Furthermore, localization studies in human islets and pancreatic beta-cell lines have demonstrated expression of ZnT8 on secretory granules, where it is important for zinc accumulation into insulin cell secretory granules (Chimienti et al. 2004, 2006; Nicolson et al. 2009; Lemaire et al. 2009) (Fig. 18.2). Based on study of cadaveric pancreata, T2D appears to correlate with decreased expression of *SLC30A8* (Marselli et al. 2010). Low levels of *SLC30A8* expression have also been demonstrated in adipose, thyroid follicular, and adrenal cortical cells (Chimienti 2013; Smidt et al. 2007); however, the role of ZnT8 in these cell types has not been determined.

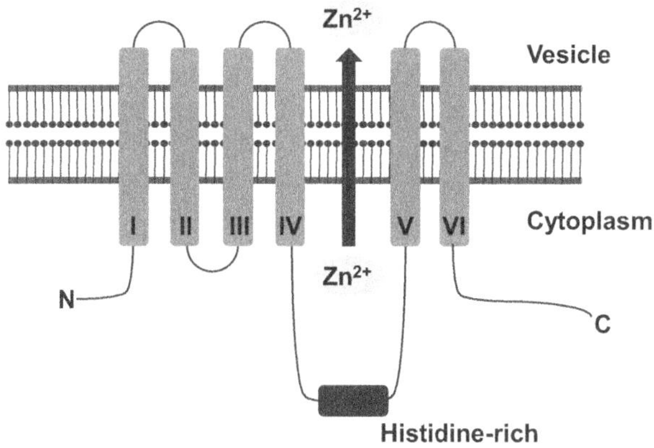

Fig. 18.1 Schematic diagram of ZnT8 monomer

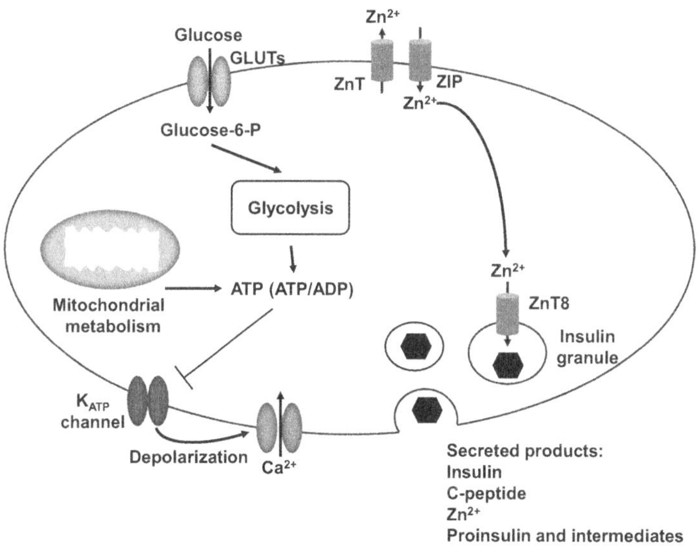

Fig. 18.2 Role of ZnT8 and zinc in pancreatic beta-cell biology. Intracellular zinc levels are modulated by transport of zinc into and out of the cell by members of the ZIP and ZnT family, respectively. One destination for intracellular zinc is transport into insulin secretory granules by ZnT8. In the granules, a hexameric insulin-zinc crystal is formed. Insulin secretion is stimulated by glucose transport into the cell via members of the glucose transporter (GLUT) family, including GLUT1, 2, and 3, and the generation of ATP from glucose via glycolysis and mitochondrial metabolism. An increase in the ATP/ADP ratio closes an ATP-regulated potassium channel, leading to cellular depolarization and entry of calcium via an L-type voltage-gated calcium channel. This, in turn, leads to the secretion of the contents of the secretory granules, which, in addition to insulin and zinc, includes C-peptide and unprocessed proinsulin

Pancreatic islets contain five cell types, with beta-cells—in which *SLC30A8* is most highly expressed—responsible for insulin secretion. Beta-cells contain among the highest concentration of zinc in the body (Wijesekara et al. 2009), in part due to the important role of zinc in insulin processing and storage (Chimienti 2013; Dodson and Steiner 1998; Dunn 2005). Specifically, after cleavage of its signal peptide and transport of proinsulin form the endoplasmic reticulum into the aqueous and zinc-rich Golgi apparatus, soluble zinc-containing hexamers of proinsulin are formed. Proinsulin is then converted to insulin by prohormone convertases, and the hexameric zinc-insulin complex becomes insoluble and crystallizes. These stable hexameric insulin crystals are retained in storage vesicles until they are secreted. Upon secretion, the crystals are released into the intercellular space where the crystal rapidly dissolves—leading to cosecretion of insulin monomers and zinc. The cosecreted zinc may have local paracrine effects in islets independent of insulin: for example, zinc may contribute to the inhibition of glucagon secretion from neighboring alpha cells (Robertson et al. 2011; Zhou et al. 2007).

Consistent with this molecular role of zinc in insulin storage and processing, well-established alterations in zinc metabolism in diabetes have been described, including increased urinary zinc excretion related to hyperglycemia and glucosuria (Jansen et al. 2009). Lower plasma zinc levels in T2D have also been reported, with some evidence suggesting that zinc supplementation improves the metabolic state in T2D (Jansen et al. 2009). A recent meta-analysis demonstrated a significant reduction in fasting glucose and hemoglobin A1c in individuals with T2D taking zinc supplements (Jayawardena et al. 2012), whereas a large prospective study of women demonstrated an inverse association between total zinc intake and T2D risk (Sun et al. 2009).

Given this relevance of zinc metabolism to insulin and diabetes and that rs13266634 results in a missense mutation that alters the protein sequence of *SLC30A8* (p.Arg325Trp) (Chimienti 2013), a potential mechanism for the observed association with T2D might be through a molecular effect on the zinc transport activity of ZnT8. Indeed, early in silico modeling was interpreted to suggest that, although the high-risk Arg325 variant would have minimal impact on protein folding, the variant is present at the interface between ZnT8 monomers and, thus, may impact the zinc transporting function of the protein (Nicolson et al. 2009). However, separate in silico homology modeling of ZnT8's three-dimensional structure predicted that the zinc-sensing capacity of the ZnT8 C-terminal domain—in which the variant lies—would not be affected by the high-risk Arg325 allele (Weijers 2010). Similarly, position 325 was thought to be at too great a distance from the domains of ZnT8 responsible for zinc docking and transport to affect these functions. Thus, the impact of p.Arg325Trp on ZnT8 structure or function is still incompletely understood.

Intriguingly, ZnT8 and p.Arg325Trp also are implicated in the pathophysiology of a different form of diabetes, type 1 diabetes (T1D). Autoantibodies directed against ZnT8 have been identified as an early biomarker of the T1D autoimmune response (Wenzlau et al. 2007, 2008), and the main epitopes of these autoantibodies include the region around amino acid 325 in the C-terminus. In some cases, the specificity of the ZnT8 autoantibodies is in fact dictated by the amino acid residue present at position 325 (Wenzlau et al. 2007, 2008), and patients with onset of T1D before age 5 years have an increased prevalence of the T2D risk-increasing Arg325 allele compared to those with T1D onset after age 5 (Wenzlau et al. 2007, 2008). Whether these observations demonstrate that p.Arg325Trp has an impact on ZnT8 protein structure, or whether they have any relationship to the pathophysiological mechanism of increased T2D risk, is an open question.

In summary, *SLC30A8* is a zinc transporter expressed almost exclusively in pancreatic islets and most highly in beta-cells. Its potential importance as a T2D susceptibility gene is thus augmented by its strong biological plausibility as a candidate gene, being responsible for transport of a key diabetes-relevant metabolite within a key diabetes-relevant cell type. Although the extent to which rs13266634 might impact the function of ZnT8 is unclear, one hypothesis for the observed T2D association is through altered zinc transport activity.

18.4 Association of rs13266634 with Diabetes-Related Traits

As one step toward refining understanding of the physiological impact of rs13266634, its association with a number of additional metabolic and diabetes-related traits has been examined. These studies have investigated traits that intend to measure physiological, cellular, and molecular functions relevant to *SLC30A8* and diabetes, including, among others, blood glucose levels and progression from hyperglycemia to diabetes, insulin processing and secretion from beta-cells, and zinc transport.

One relevant predictor of future diabetes risk is blood glucose levels. In nondiabetic controls, rs13266634 is associated with fasting glucose in adults and children, 2-h glucose during an oral glucose tolerance test (OGTT), and hemoglobin A1c (Dupuis et al. 2010; Kelliny et al. 2009; Nettleton et al. 2013). Interestingly, the high-risk C allele of rs13266634 is also associated with a higher rate of progression from normal fasting glucose to impaired fasting glucose, although not to progression from impaired fasting glucose to T2D, suggesting that the impact of the high-risk variant may be predominantly on prediabetic fasting glucose levels with a lesser effect on the transition to T2D (Walford et al. 2012).

To gain insight into a possible mechanism of action for rs13266634, association with several measures of beta-cell function has been assessed. One consistent finding has been decreased insulin secretion in response to intravenous glucose in nondiabetic individuals of European ancestry (Boesgaard et al. 2008; Ingelsson et al. 2010; Staiger et al. 2007). When this was examined in detail, the deficit appeared to be in first-phase as opposed to second-phase insulin release (Boesgaard et al. 2008), a characteristic of early T2D. Consistent with this finding, the association of the risk allele with reduced disposition index, a measure of beta-cell compensation for increasing insulin resistance, has been demonstrated in European as well as African-American and Hispanic ancestry groups (Palmer et al. 2008; Stancakova et al. 2009). The association of the variant with insulin secretion during an OGTT is less clear: Two studies have reported association of the C allele of rs13266634 with decreased early-phase insulin secretion during an OGTT (Stancakova et al. 2009; Steinthorsdottir et al. 2007), but several studies failed to demonstrate a similar association (Boesgaard et al. 2008; Staiger et al. 2007; Kirchhoff et al. 2008; Pascoe et al. 2007). The reason for this apparent difference in the effect of the risk allele on an intravenous compared to oral glucose load remains to be determined.

In addition to these relatively broad measures of beta-cell function, more specific hypotheses about the rs13266634 mechanism of action have been interrogated. Given the role of ZnT8 in zinc transport into insulin secretory granules, the site of proinsulin to insulin processing, the association of rs13266634 with proinsulin levels or processing has been examined. In general, the high-risk C allele of rs13266634 (or a perfect proxy for it) is associated with a number of related phenotypes that are consistent with impaired proinsulin to insulin conversion and beta-cell dysfunction. These include a higher ratio of proinsulin to insulin release during an OGTT, higher fasting proinsulin levels together with the absence of an increase in the insulinogenic index (a measure of insulin secretion) as fasting proinsulin levels increase, and higher levels of 32,33-split proinsulin, the first cleavage product generated during proinsulin processing to insulin, together with a lower insulinogenic index (Kirchhoff et al. 2008; Stancakova et al. 2009; Ingelsson et al. 2010; Strawbridge et al. 2011). As an association with fasting insulin levels was not demonstrated in the latter case, this finding is consistent with a defect in insulin processing and secretion distal to the first enzymatic step in insulin processing (Strawbridge et al. 2011). In the Diabetes Prevention Program cohort, which included individuals at high risk for developing T2D who underwent interventions to reduce insulin resistance and progression to diabetes, the high-risk C allele was associated with higher fasting proinsulin levels at entry into the study, but after 1 year of intervention, fasting proinsulin levels decreased and were no longer associated with rs13266634 (Majithia et al. 2011). While various possibilities could explain this finding, one possibility is that the intervention decreased the effect of the risk allele. Finally, cluster analyses which grouped diabetes risk loci into different categories based on their relationship to various continuous glycemic phenotypes placed *SLC30A8* into a cluster defined by loci influencing insulin processing and secretion without changes in fasting glucose (Dimas et al. 2013). Together, the above studies in nondiabetic individuals who carry the high-risk C allele of rs13266634 are consistent with the association of the risk variant with beta-cell dysfunction, suggesting that the primary effect of the risk allele is at the level of the beta-cell. Consistent with this, multiple studies have failed to demonstrate an association of rs13266634 with insulin sensitivity (Boesgaard et al. 2008; Staiger et al. 2007; Stancakova et al. 2009; Ingelsson et al. 2010; Dupuis et al. 2010).

Given the relationship of zinc intake to metabolic control in T2D and diabetes risk, there has also been interest in the interaction of zinc intake with rs13266634. A meta-analysis of 14 cohorts of nondiabetic individuals demonstrated an association of total, but not dietary, zinc intake with lower fasting glucose (Kanoni et al. 2011). When an interaction of zinc intake with a proxy for rs13266634 was tested, the glucose-raising effect of the risk allele was attenuated, in part, with higher total zinc intake (Kanoni et al. 2011). In a recent study of Asians, higher plasma zinc levels were associated with a lower risk of T2D or impaired glucose regulation (impaired fasting glucose and/or impaired glucose tolerance); however, in those bearing one or two copies of the diabetes risk allele, the inverse association between plasma zinc concentration and risk of T2D, alone or in combination with impaired glucose regulation, was weaker (Shan et al. 2013). At the same time, the higher risk for T2D in those homozygous for the risk allele was partially attenuated in those with higher plasma zinc concentrations. Together, these studies suggest that the impact of the rs13266634 risk allele on T2D risk or glucose levels is attenuated, in part, by higher zinc levels or intake.

Finally, given early hypotheses about the potential candidacy of *SLC30A8* as a therapeutic target, some studies have investigated the potential interaction of rs13266634 with diabetes treatment. In the Diabetes Prevention Program, the presence or absence of the *SLC30A8* risk allele did not impact the effectiveness of interventions that were tested: lifestyle interventions, metformin, or troglitazone (Majithia et al. 2011). In contrast, in a different study, a cohort of individuals with T2D who carried at least one copy of the protective T allele of rs13266634 exhibited an enhanced response of fasting and postprandial insulin levels after 8 weeks of repaglinide therapy (Huang et al. 2010). Study of a different cohort of newly diagnosed patients with T2D did not replicate this effect of rs13266634 after 48 weeks of repaglinide treatment (Jiang et al. 2012), although a differential genotype-dependent response of insulin secretion and fasting proinsulin levels was demonstrated after treatment with rosiglitazone. Thus, the full range of interactions between rs13266634 and diabetes treatment remains to be determined.

In summary, the bulk of data suggest that the diabetes risk variant of rs13266634 is associated with altered beta-cell function, manifested primarily as impaired proinsulin to insulin processing and insulin secretion in response to an intravenous glucose load. Thus, one initial model for the mechanism of rs13266634 action was through reduced function as a zinc transporter, with a subsequent decrease in insulin processing and secretion.

18.5 Cellular and Animal Models of *SLC30A8*

Given (a) the clear association of rs13266634 with T2D risk, (b) the biological role of ZnT8, and (c) phenotypic measurements suggesting an impact of rs13266634 on beta-cell function, several groups have sought further mechanistic insight through model systems of ZnT8 activity. Both cellular models and *Slc30a8* knockout mice have enabled investigation of the impact of *SLC30A8* overexpression, perturbation, or deletion on multiple aspects of beta-cell- or T2D-relevant biology, including zinc transport, insulin crystallization, insulin secretion, blood insulin and glucose levels, and glucose tolerance.

Both overexpression and knockdown of ZnT8 have been investigated in cellular models. One study overexpressed ZnT8 in rodent beta-cell lines and observed enhanced insulin secretory capacity (Chimienti et al. 2006). Conversely, two studies examined the acute effect of ZnT8 knockdown in rodent beta-cells: Cellular insulin content was decreased in one study, while both demonstrated decreased cellular zinc content and glucose-stimulated insulin secretion (El Muayed et al. 2010; Fu et al. 2009). KCl-stimulated insulin secretion was not affected (Fu et al. 2009). Although limited, these studies seem to suggest that ZnT8 activity is positively correlated with zinc transport and insulin secretion.

The impact of the p.Arg325Trp variant on rodent beta-cell lines has also been investigated. In one study, overexpression of the Arg325 and Trp325 variants did not lead to differences in glucose-stimulated insulin secretion, but the high-risk Arg325 variant did have lower zinc transport activity and zinc accumulation in secretory granules—suggesting reduced function as a zinc transporter (Nicolson et al. 2009). In a second study, insulin and glucagon mRNA levels, as well as basal- and glucose-stimulated insulin secretion, were similar in human islets homozygous or heterozygous for the Arg325 risk variant of ZnT8 compared to those homozygous for the Trp325 variant (Cauchi et al. 2010). These two studies thus suggest that while p.Arg325Trp may affect transport activity of ZnT8, the impact on cellular measures of beta-cell function is less clear.

As cellular models lack physiological context, a complementary line of research has been to investigate the phenotypic effects of *Slc30a8* deletion in mice. The first three knockout mice studies, in 2009, were broadly interpreted as providing evidence that loss of *Slc30a8* led to defects in insulin secretion (Rutter 2010). All three studies were consistent in showing reduced islet zinc content and malformed insulin granules (Lemaire et al. 2009; Nicolson et al. 2009; Pound et al. 2009). Similarly, in all three studies, mice exhibited abnormal insulin secretion or glucose homeostasis under some conditions. In one study by Lemaire et al., animals became glucose intolerant or diabetic on a high-fat diet, proposing a model in which ZnT8 is essential for normal glucose homeostasis under prolonged beta-cell stress (Lemaire et al. 2009). In another study by Nicolson et al., young mice exhibited impaired glucose tolerance, the degree of which was highly dependent on age, sex, and diet (Nicolson et al. 2009). In the third, null mice had lower plasma insulin levels, and isolated islets had lower glucose-stimulated insulin secretion (Pound et al. 2009).

Inspired by these findings, additional studies examined the effects of conditional knockout of *Slc30a8*. To remove confounding effects of ZnT8 activity in tissues other than the pancreas, mice lacking *Slc30a8* solely in either beta-cells or alpha cells were generated (Wijesekara et al. 2010). Consistent with expectations, beta-cell specific knockouts had reduced islet zinc content and first-phase insulin response and became glucose intolerant; alpha-cell-specific knockouts, on the other hand, showed no clear abnormalities in glucagon levels or glucose homeostasis. A second study sought to investigate the interactions between a high-fat diet and ZnT8 deficiency (Hardy et al. 2012). This study suggested that, in response to a high-fat diet, global *Slc30a8* knockouts had markedly higher rates of obesity, hyperinsulinemia, and hyperglycemia than beta-cell-specific knockouts—which exhibited only modest hyperinsulinemia and hyperglycemia—proposing global loss of *Slc30a8* as necessary to fully exacerbate the effects of metabolic stress due to a high-fat diet.

Despite these potentially promising insights into the impact of ZnT8 on mouse physiology, heterogeneity in the details of the observations clouds the picture. Although Lemaire et al. (2009) observed glucose intolerance in null mice on a high-fat diet, mice on a normal diet were phenotypically normal. Similarly, Nicolson et al. (2009) did not observe consistent effects on insulin levels or glucose homeostasis, with only some genders and ages of mice exhibiting elevated fasting glucose levels or glucose intolerance; in some cases, they even observed an increase in insulin secretion from isolated islets. Pound et al. (2009), working independently from the others, observed the least evidence for an effect of ZnT8 knockout on glucose homeostasis, observing unaltered blood glucose levels in spite of lower insulin levels and secretion from isolated islets.

This observed heterogeneity in mouse phenotypes suggested that small sample sizes, different environments, or variable genetic backgrounds might influence conclusions about the effects of *Slc30a8* knockout in mice. To investigate this possibility, Pound and colleagues (Pound et al. 2012) examined the impact of *Slc30a8* deletion on a pure genetic background. While they again observed null mice to have clearly reduced islet zinc content, only females showed lower insulin levels, and all mice had normal glucose levels and glucose-stimulated insulin secretion. Thus, after controlling for environment and genetic background, the authors concluded that loss of ZnT8 had a minimal impact on mouse physiology. The interpretation of these results, in contrast to those obtained on a mixed genetic background, is unclear (da Silva et al. 2013).

To further clarify the physiological role of secreted zinc, most recently Tamaki et al. examined potential effects of *Slc30a8* deletion on tissues other than the pancreas (Tamaki et al. 2013). In this study, beta-cell specific knockouts had low peripheral blood glucose levels, in spite of insulin hypersecretion. Unexpectedly, the explanation was that knockouts had enhanced hepatic insulin clearance, which the authors showed to be due to the removal of inhibitory effects of zinc on hepatic insulin uptake. In the same study, human carriers of the rs13266634 risk allele were also shown to have higher insulin clearance. Thus, despite clear localization of *SLC30A8* to the pancreas, there is compelling evidence that its physiological effects extend to other tissues and behave in complex ways.

In summary, cellular and mouse models have not conclusively determined the physiological role of *SLC30A8* and its relation to T2D. Although there are some signs of altered insulin secretion or glucose intolerance in certain cases, the data are not wholly consistent, and the findings in some genetic backgrounds suggest a limited impact of gene deletion on mouse physiology. Future work will be necessary to resolve these inconsistencies and determine the full effect of *Slc30a8* deletion on the pancreas as well as other tissues.

18.6 Next-Generation Sequencing and *SLC30A8*

Although mouse and cellular models provide essential tools to understand disease mechanisms, they have the same limitations as all models: they simplify complexities and thus cannot represent the full context of human physiology. The value of human genetic studies is that naturally occurring variation can model the effects of gene perturbation in vivo, bridging the gap between molecular or cellular mechanism and ultimate impact on physiology.

These "experiments of nature" have particular value toward the goal of target validation for drug development (Plenge et al. 2013). Perhaps the most famous example involves *PCSK9*, a gene with mutations known to cause hypercholesterolemia (Abifadel et al. 2003). In 2005, it was discovered that loss-of-function mutations in *PCSK9*—resulting in ablation or severe reduction of protein activity—reduce LDL levels in carriers (Cohen et al. 2005). Further identification and study of additional loss-of-function mutations showed carriers to have long-term reduced risk of cardiovascular disease, with no apparent adverse phenotypes (Cohen et al. 2006; Kathiresan and Myocardial Infarction Genetics 2008; Kotowski et al. 2006). The desirable clinical endpoint, as well as the presence of a relevant biomarker, gave confidence that inactivation of *PCSK9* in humans would be an effective therapy—which has since led to the development of a promising candidate drug (Stein et al. 2012). Similar success stories exist for Nav1.7 (pain) (Nassar et al. 2004) and CCR5 (HIV) (De Clercq 2007).

The increasing accessibility of DNA sequencing has enabled studies to identify an increasing number of variants and explore the phenotypic impact of a wider allelic series for many genes. The first such study for *SLC30A8* involved sequencing its exons and surrounding regions in 380 individuals from five ethnicities, with subsequent genotyping of identified variants in 3445 individuals (Billings et al. 2014). Association analysis of 61 variants (44 of which were previously unidentified) identified several common variants as individually associated with oral disposition or insulinogenic index (two measures of beta-cell function). In addition, some measures of the aggregate "load" of rare variation showed modest association with T2D or changes in insulinogenic index over a 1-year period. Although this study had limited power to detect low-frequency or rare-variant associations, it demonstrates that a wider catalog of variation might lend understanding to the effects of a range of perturbations of *SLC30A8* in humans.

A second study used much larger sample sizes to explicitly study a special subclass of mutations in *SLC30A8*: those that cause premature protein truncation and thus a high likelihood of loss of function (Flannick et al. 2014). Through a series of studies which ultimately sequenced 15,282 individuals and genotyped an additional 133,852 individuals, a spectrum of 12 predicted protein-truncating variants was identified. Two of these—a nonsense mutation p.Arg138* and a frameshift mutation p.Lys34Serfs*50—were individually associated with T2D risk and shown to encode unstable ZnT8 proteins.

Surprisingly, however, heterozygosity for any of the 12 protein-truncating variants in *SLC30A8* was associated with *decreased* risk of T2D: an estimated 65 % reduction, at levels of significance sufficient to correct for 20,000 genes in the genome ($p = 1.7 \times 10^{-6}$). Furthermore, the p.Lys34Serfs*50 variant was associated with *lower* glucose levels in nondiabetic controls. These observations were at striking odds with the previous model for the directional relationship between *SLC30A8* activity and disease risk—based on biological hypotheses, functional studies of rs13266634, and mouse models—in which decreased ZnT8 activity would increase disease risk. Although confounding factors can never be completely removed from genetic association studies, the observation of multiple independent protective protein-truncating variants, in multiple ethnicities, provided evidence that the mechanism linking variation in *SLC30A8* to T2D risk is more complex than previously assumed. Furthermore, they suggested that therapeutics directed against *SLC30A8* might be designed to inhibit the protein, rather than activate it.

In summary, the era of next-generation sequencing enables exploration of the full allelic series for many genes, including *SLC30A8*. Initial studies have demonstrated that multiple variants in the gene are associated with risk of T2D or related phenotypes (Table 18.1). Strikingly, the evidence from study of protein-truncating variants in human populations suggests that *SLC30A8* loss of function might reduce risk of T2D, rather than increase it as previously assumed. Although further research is necessary to elucidate the potential mechanism for this protective effect, the observed human genetics data must be reconciled with any molecular or cellular studies in model systems.

18.7 Summary

SLC30A8 represents a promising gene to advance understanding of T2D pathophysiology, as well as a potential therapeutic target. The association of a missense variant with T2D, glucose, and proinsulin—at statistical significance far beyond genome-wide thresholds—clearly implicates it as relevant to T2D. The biological role of its protein product (ZnT8) and more detailed phenotypic interrogation of variant carriers suggest that altered zinc transport within the beta-cell may influence T2D risk. Cellular and animal models have been viewed as supportive of this view, although not as conclusively as might have initially been expected, and have recently suggested a role for *SLC30A8* in tissues other than the pancreas.

However, the mechanism whereby *SLC30A8* perturbation affects T2D risk remains far from clear (Rutter and Chimienti 2015). Recent evidence from human genetics is at odds with the model where reduced activity increases risk and suggests that inhibition, rather than activation, might be the desired therapeutic strategy. These data, together with heterogeneity in mouse phenotypes, the relatively limited functional study of p.Arg325Trp, and recent evidence for an unexpected role of zinc and *SLC30A8* in hepatic insulin clearance, suggest complex disease mechanisms and a fertile path for future research (Table 18.2).

Table 18.1 Associations between variants in *SLC30A8* and human phenotypes

Variant	MAF	Trait	Effect	95 % CI	*P*-value	References
p.Trp325Arg	34 %	**T2D**	**1.13**	**(1.09–1.16)**	$\mathbf{1.3 \times 10^{-21}}$	Consortium DGRAM-aD (2014)
		Fasting glucose	0.03	(0.02–0.04)	5.5×10^{-10}	Dupuis et al. (2010)
		Proinsulin	0.03	(0.02–0.04)	4.2×10^{-13}	Strawbridge et al. (2011)
		2-h glucose	0.09	(0.05–0.13)	6.6×10^{-5}	Saxena et al. (2010)
39 Rare missense	Each <1 %	**T2D**	**1.27**	**(1.00–1.61)**	$\mathbf{5.0 \times 10^{-2}}$	Billings et al. (2014)
		Δ Insulinogenic index	0.03	(−0.03–0.10)	3.0×10^{-2}	Billings et al. (2014)
p.Arg138*	<0.1 %	**T2D**	**0.47**	**(0.27–0.81)**	$\mathbf{6.7 \times 10^{-3}}$	Flannick et al. (2014)
p.Lys34Serfs*50	<0.1 %	**T2D**	**0.17**	**(0.05–0.52)**	$\mathbf{1.9 \times 10^{-3}}$	Flannick et al. (2014)
		Nonfasting glucose	−0.23	(−0.35–−0.10)	4.6×10^{-4}	Flannick et al. (2014)
		1-h glucose	−0.73	(−1.62–−0.17)	5.0×10^{-2}	Flannick et al. (2014)
Ten rare protein truncating	<0.1 %	**T2D**	**0.30**	**(0.14–0.64)**	$\mathbf{2.1 \times 10^{-3}}$	Flannick et al. (2014)

Numerous association studies have tested for association between the common rs13266634 *SLC30A8* variant and T2D or related traits, and more recent resequencing studies have explored the phenotypic effects of a wider allelic series. Shown in the table are published association results for variants in *SLC30A8* and different phenotypes. Columns (from left): a description of the impact on protein sequence of each variant or class of variants (Variant), the minor allele frequency (MAF), the phenotype analyzed (Trait), the estimated effect size of the variant minor allele on the trait analyzed (Effect), the 95 % confidence interval of the estimated effect (95 % CI), the p-value of a test between genotype and phenotype (P-value), and the reference in which the association is reported (Reference). Effect sizes are given in odds ratios for T2D (except for the 39 rare missense variants, in which the value represents a hazard ratio) and in change relative to the major allele for quantitative traits; units are taken from the original reference. For the 39 missense variants, the effect on T2D is based on a genomic risk score, and the effect on Δinsulinogenic index is averaged across carriers of any of the 39 variants

Table 18.2 *SLC30A8:* gain of function or loss of function?

Does loss of *SLC30A8* increase T2D risk?	Or is there a more complex story?
Zinc plays a key role in insulin processing, crystallization, and secretion in the beta-cell (Lemaire et al. 2009)	Recent knockout mouse studies have suggested *SLC30A8* has physiological impact beyond the beta-cell (Tamaki et al. 2013)
Depending on age, gender, diet, and genetic background, knockout mice demonstrate altered insulin secretion or glucose tolerance (Rutter 2010)	Knockout on uniform genetic background fails to have a marked effect on mouse physiology (Pound et al. 2012)
The risk allele of the common variant p. Trp325Arg decreases ZnT8 activity in cellular models (Nicolson et al. 2009)	A series of 12 protein-truncating, and putative loss-of-function, variants, across multiple ancestries, associates with reduced T2D risk in humans at exome-wide significance (Flannick et al. 2014)
A series of rare missense variants is nominally associated with increased T2D risk (Billings et al. 2014)	Effect of perturbation due to missense variant might be markedly different from complete loss of function (Flannick et al. 2014)

Although initial genetic and functional studies of *SLC30A8* and the common rs13266634 variant suggested that loss of function might increase T2D risk and that gain of function might represent a potential therapeutic strategy in the treatment of T2D, recent evidence from human genetic association studies suggests that the picture is much more complicated. The table presents arguments on either side of the debate as to whether *SLC30A8* gain of function or loss of function increases T2D risk in human populations

Nonetheless, *SLC30A8* exemplifies perhaps a canonical strategy to translate common variant associations to mechanistic insight and ultimately treatment. Specific biological hypotheses, and cellular and mouse models, are key tools to gain insight into molecular and cellular mechanisms, which are necessary to identify disease pathways and insight into pathophysiology. Observations from human genetics are powerful data to help understand the impact of gene perturbation on clinical endpoints. Additional study through, and refinement of, all of these techniques will be necessary to determine whether *SLC30A8* represents a promising therapeutic target in the treatment of T2D.

References

Abifadel M, Varret M, Rabes JP, Allard D, Ouguerram K, Devillers M, Cruaud C, Benjannet S, Wickham L, Erlich D, Derre A, Villeger L, Farnier M, Beucler I, Bruckert E, Chambaz J, Chanu B, Lecerf JM, Luc G, Moulin P, Weissenbach J, Prat A, Krempf M, Junien C, Seidah NG, Boileau C (2003) Mutations in PCSK9 cause autosomal dominant hypercholesterolemia. Nat Genet 34(2):154–156. doi:10.1038/ng1161

Billings LK, Jablonski KA, Ackerman RJ, Taylor A, Fanelli RR, McAteer JB, Guiducci C, Delahanty LM, Dabelea D, Kahn SE, Franks PW, Hanson RL, Maruthur NM, Shuldiner A, Mayer-Davis EJ, Knowler WC, Florez JC, for the Diabetes Prevention Program Research Group RMD (2014) The INFLUENCE of rare Genetic variation in SLC30A8 on diabetes incidence and beta-cell function. J Clin Endocrinol Metab 99(5):E926–E930. doi:10.1210/jc.2013-2378

Boesgaard TW, Zilinskaite J, Vanttinen M, Laakso M, Jansson PA, Hammarstedt A, Smith U, Stefan N, Fritsche A, Haring H, Hribal M, Sesti G, Zobel DP, Pedersen O, Hansen T (2008) The common SLC30A8 Arg325Trp variant is associated with reduced first-phase insulin release in 846 non-diabetic offspring of type 2 diabetes patients--the EUGENE2 study. Diabetologia 51(5):816–820. doi:10.1007/s00125-008-0955-6

Cauchi S, Del Guerra S, Choquet H, D'Aleo V, Groves CJ, Lupi R, McCarthy MI, Froguel P, Marchetti P (2010) Meta-analysis and functional effects of the SLC30A8 rs13266634 polymorphism on isolated human pancreatic islets. Mol Genet Metab 100(1):77–82. doi:10.1016/j.ymgme.2010.01.001

Chasapis CT, Loutsidou AC, Spiliopoulou CA, Stefanidou ME (2012) Zinc and human health: an update. Arch Toxicol 86(4):521–534. doi:10.1007/s00204-011-0775-1

Chimienti F (2013) Zinc, pancreatic islet cell function and diabetes: new insights into an old story. Nutr Res Rev 26(1):1–11. doi:10.1017/S0954422412000212

Chimienti F, Devergnas S, Favier A, Seve M (2004) Identification and cloning of a beta-cell-specific zinc transporter, ZnT-8, localized into insulin secretory granules. Diabetes 53 (9):2330–2337

Chimienti F, Devergnas S, Pattou F, Schuit F, Garcia-Cuenca R, Vandewalle B, Kerr-Conte J, Van Lommel L, Grunwald D, Favier A, Seve M (2006) In vivo expression and functional characterization of the zinc transporter ZnT8 in glucose-induced insulin secretion. J Cell Sci 119 (Pt 20):4199–4206. doi:10.1242/jcs.03164

Cohen J, Pertsemlidis A, Kotowski IK, Graham R, Garcia CK, Hobbs HH (2005) Low LDL cholesterol in individuals of African descent resulting from frequent nonsense mutations in PCSK9. Nat Genet 37(2):161–165. doi:10.1038/ng1509

Cohen JC, Boerwinkle E, Mosley TH Jr, Hobbs HH (2006) Sequence variations in PCSK9, low LDL, and protection against coronary heart disease. N Engl J Med 354(12):1264–1272. doi:10.1056/NEJMoa054013

Consortium DGRAM-aD (2014) Genome-wide trans-ancestry meta-analysis provides insight into the genetic architecture of type 2 diabetes susceptibility. Nat Genet 46:234–244

Cousins RJ, Liuzzi JP, Lichten LA (2006) Mammalian zinc transport, trafficking, and signals. J Biol Chem 281(34):24085–24089. doi:10.1074/jbc.R600011200

da Silva XG, Bellomo EA, McGinty JA, French PM, Rutter GA (2013) Animal Models of GWAS-Identified Type 2 Diabetes Genes. J Diabetes Res 2013:906590. doi:10.1155/2013/906590

De Clercq E (2007) The design of drugs for HIV and HCV. Nat Rev Drug Discov 6 (12):1001–1018. doi:10.1038/nrd2424

Dimas AS, Lagou V, Barker A, Knowles JW, Magi R, Hivert MF, Benazzo A, Rybin D, Jackson AU, Stringham HM, Song C, Fischer-Rosinsky A, Boesgaard TW, Grarup N, Abbasi FA, Assimes TL, Hao K, Yang X, Lecoeur C, Barroso I, Bonnycastle LL, Bottcher Y, Bumpstead S, Chines PS, Erdos MR, Graessler J, Kovacs P, Morken MA, Narisu N, Payne F, Stancakova A, Swift AJ, Tonjes A, Bornstein SR, Cauchi S, Froguel P, Meyre D, Schwarz PE, Haring HU, Smith U, Boehnke M, Bergman RN, Collins FS, Mohlke KL, Tuomilehto J, Quertemous T, Lind L, Hansen T, Pedersen O, Walker M, Pfeiffer AF, Spranger J, Stumvoll M, Meigs JB, Wareham NJ, Kuusisto J, Laakso M, Langenberg C, Dupuis J, Watanabe RM, Florez JC, Ingelsson E, McCarthy MI, Prokopenko I, on behalf of the Mi (2013) Impact of type 2 diabetes susceptibility variants on quantitative glycemic traits reveals mechanistic heterogeneity. Diabetes. doi:10.2337/db13-0949

Dodson G, Steiner D (1998) The role of assembly in insulin's biosynthesis. Curr Opin Struct Biol 8 (2):189–194

Dunn MF (2005) Zinc-ligand interactions modulate assembly and stability of the insulin hexamer -- a review. Biometals 18(4):295–303. doi:10.1007/s10534-005-3685-y

Dupuis J, Langenberg C, Prokopenko I, Saxena R, Soranzo N, Jackson AU, Wheeler E, Glazer NL, Bouatia-Naji N, Gloyn AL, Lindgren CM, Magi R, Morris AP, Randall J, Johnson T, Elliott P, Rybin D, Thorleifsson G, Steinthorsdottir V, Henneman P, Grallert H, Dehghan A, Hottenga JJ, Franklin CS, Navarro P, Song K, Goel A, Perry JR, Egan JM, Lajunen T, Grarup N,

Sparso T, Doney A, Voight BF, Stringham HM, Li M, Kanoni S, Shrader P, Cavalcanti-Proenca C, Kumari M, Qi L, Timpson NJ, Gieger C, Zabena C, Rocheleau G, Ingelsson E, An P, O'Connell J, Luan J, Elliott A, McCarroll SA, Payne F, Roccasecca RM, Pattou F, Sethupathy P, Ardlie K, Ariyurek Y, Balkau B, Barter P, Beilby JP, Ben-Shlomo Y, Benediktsson R, Bennett AJ, Bergmann S, Bochud M, Boerwinkle E, Bonnefond A, Bonnycastle LL, Borch-Johnsen K, Bottcher Y, Brunner E, Bumpstead SJ, Charpentier G, Chen YD, Chines P, Clarke R, Coin LJ, Cooper MN, Cornelis M, Crawford G, Crisponi L, Day IN, de Geus EJ, Delplanque J, Dina C, Erdos MR, Fedson AC, Fischer-Rosinsky A, Forouhi NG, Fox CS, Frants R, Franzosi MG, Galan P, Goodarzi MO, Graessler J, Groves CJ, Grundy S, Gwilliam R, Gyllensten U, Hadjadj S, Hallmans G, Hammond N, Han X, Hartikainen AL, Hassanali N, Hayward C, Heath SC, Hercberg S, Herder C, Hicks AA, Hillman DR, Hingorani AD, Hofman A, Hui J, Hung J, Isomaa B, Johnson PR, Jorgensen T, Jula A, Kaakinen M, Kaprio J, Kesaniemi YA, Kivimaki M, Knight B, Koskinen S, Kovacs P, Kyvik KO, Lathrop GM, Lawlor DA, Le Bacquer O, Lecoeur C, Li Y, Lyssenko V, Mahley R, Mangino M, Manning AK, Martinez-Larrad MT, McAteer JB, McCulloch LJ, McPherson R, Meisinger C, Melzer D, Meyre D, Mitchell BD, Morken MA, Mukherjee S, Naitza S, Narisu N, Neville MJ, Oostra BA, Orru M, Pakyz R, Palmer CN, Paolisso G, Pattaro C, Pearson D, Peden JF, Pedersen NL, Perola M, Pfeiffer AF, Pichler I, Polasek O, Posthuma D, Potter SC, Pouta A, Province MA, Psaty BM, Rathmann W, Rayner NW, Rice K, Ripatti S, Rivadeneira F, Roden M, Rolandsson O, Sandbaek A, Sandhu M, Sanna S, Sayer AA, Scheet P, Scott LJ, Seedorf U, Sharp SJ, Shields B, Sigurethsson G, Sijbrands EJ, Silveira A, Simpson L, Singleton A, Smith NL, Sovio U, Swift A, Syddall H, Syvanen AC, Tanaka T, Thorand B, Tichet J, Tonjes A, Tuomi T, Uitterlinden AG, van Dijk KW, van Hoek M, Varma D, Visvikis-Siest S, Vitart V, Vogelzangs N, Waeber G, Wagner PJ, Walley A, Walters GB, Ward KL, Watkins H, Weedon MN, Wild SH, Willemsen G, Witteman JC, Yarnell JW, Zeggini E, Zelenika D, Zethelius B, Zhai G, Zhao JH, Zillikens MC, Borecki IB, Loos RJ, Meneton P, Magnusson PK, Nathan DM, Williams GH, Hattersley AT, Silander K, Salomaa V, Smith GD, Bornstein SR, Schwarz P, Spranger J, Karpe F, Shuldiner AR, Cooper C, Dedoussis GV, Serrano-Rios M, Morris AD, Lind L, Palmer LJ, Hu FB, Franks PW, Ebrahim S, Marmot M, Kao WH, Pankow JS, Sampson MJ, Kuusisto J, Laakso M, Hansen T, Pedersen O, Pramstaller PP, Wichmann HE, Illig T, Rudan I, Wright AF, Stumvoll M, Campbell H, Wilson JF, Bergman RN, Buchanan TA, Collins FS, Mohlke KL, Tuomilehto J, Valle TT, Altshuler D, Rotter JI, Siscovick DS, Penninx BW, Boomsma DI, Deloukas P, Spector TD, Frayling TM, Ferrucci L, Kong A, Thorsteinsdottir U, Stefansson K, van Duijn CM, Aulchenko YS, Cao A, Scuteri A, Schlessinger D, Uda M, Ruokonen A, Jarvelin MR, Waterworth DM, Vollenweider P, Peltonen L, Mooser V, Abecasis GR, Wareham NJ, Sladek R, Froguel P, Watanabe RM, Meigs JB, Groop L, Boehnke M, McCarthy MI, Florez JC, Barroso I (2010) New genetic loci implicated in fasting glucose homeostasis and their impact on type 2 diabetes risk. Nat Genet 42(2):105–116, doi:ng.520 [pii]10.1038/ng.520

Eide DJ (2006) Zinc transporters and the cellular trafficking of zinc. Biochim Biophys Acta 1763 (7):711–722. doi:10.1016/j.bbamcr.2006.03.005

El Muayed M, Billings LK, Raja MR, Zhang X, Park PJ, Newman MV, Kaufman DB, O'Halloran TV, Lowe WL Jr (2010) Acute cytokine-mediated downregulation of the zinc transporter ZnT8 alters pancreatic beta-cell function. J Endocrinol 206(2):159–169. doi:10.1677/JOE-09-0420

Flannick J, Thorleifsson G, Beer NL, Jacobs SBR, Grarup N, Burtt NP, Mahajan A, Fuchsberger C, Atzmon G, Benediktsson R, Blangero J, Bowden DW, Brandslund I, Brosnan J, Burslem F, Chambers J, Shin Cho YS, Christensen C, Douglas DA, Duggirala R, Dymek Z, Farjoun Y, Fennell T, Fontanillas P, Forsén T, Gabriel S, Glaser B, Gudbjartsson DF, Hanis C, Hansen T, Hreidarsson AB, Hveem K, Ingelsson E, Isomaa B, Johansson S, Jørgensen T, Jørgensen ME, Kathiresan S, Kong A, Kooner J, Kravic J, Laakso M, Lee J-Y, Lind L, Lindgren CM, Linneberg A, Masson G, Meitinger T, Mohlke KL, Molven A, Morris AP, Potluri S, Rauramaa R, Ribel-Madsen R, Richard A, Rolph T, Salomaa V, Segrè AV, Skärstrand H, Steinthorsdottir V, Stringham HM, Sulem P, Tai ES, Teo YY, Teslovich T,

Thorsteinsdottir U, Trimmer JK, Tuomi T, Tuomilehto J, Vaziri-Sani F, Voight BF, Wilson JG, Boehnke M, McCarthy MI, Njølstad PR, Pedersen O, Consortium G-TD, Consortium TD-G, Groop L, Cox DR, Stefansson K, Altshuler D (2014) Loss-of-function mutations in SLC30A8 protect against type 2 diabetes. Nat Genet 46:357–363

Fu Y, Tian W, Pratt EB, Dirling LB, Shyng SL, Meshul CK, Cohen DM (2009) Down-regulation of ZnT8 expression in INS-1 rat pancreatic beta cells reduces insulin content and glucose-inducible insulin secretion. PLoS ONE 4(5), e5679. doi:10.1371/journal.pone.0005679

Hardy AB, Wijesekara N, Genkin I, Prentice KJ, Bhattacharjee A, Kong D, Chimienti F, Wheeler MB (2012) Effects of high-fat diet feeding on Znt8-null mice: differences between beta-cell and global knockout of Znt8. Am J Physiol Endocrinol Metab 302(9):E1084–E1096. doi:10.1152/ajpendo.00448.2011

Hertel JK, Johansson S, Raeder H, Midthjell K, Lyssenko V, Groop L, Molven A, Njolstad PR (2008) Genetic analysis of recently identified type 2 diabetes loci in 1,638 unselected patients with type 2 diabetes and 1,858 control participants from a Norwegian population-based cohort (the HUNT study). Diabetologia 51(6):971–977. doi:10.1007/s00125-008-0982-3

Huang L, Tepaamorndech S (2013) The SLC30 family of zinc transporters - a review of current understanding of their biological and pathophysiological roles. Mol Asp Med 34 (2–3):548–560. doi:10.1016/j.mam.2012.05.008

Huang Q, Yin JY, Dai XP, Wu J, Chen X, Deng CS, Yu M, Gong ZC, Zhou HH, Liu ZQ (2010) Association analysis of SLC30A8 rs13266634 and rs16889462 polymorphisms with type 2 diabetes mellitus and repaglinide response in Chinese patients. Eur J Clin Pharmacol 66 (12):1207–1215. doi:10.1007/s00228-010-0882-6

Ingelsson E, Langenberg C, Hivert MF, Prokopenko I, Lyssenko V, Dupuis J, Magi R, Sharp S, Jackson AU, Assimes TL, Shrader P, Knowles JW, Zethelius B, Abbasi FA, Bergman RN, Bergmann A, Berne C, Boehnke M, Bonnycastle LL, Bornstein SR, Buchanan TA, Bumpstead SJ, Bottcher Y, Chines P, Collins FS, Cooper CC, Dennison EM, Erdos MR, Ferrannini E, Fox CS, Graessler J, Hao K, Isomaa B, Jameson KA, Kovacs P, Kuusisto J, Laakso M, Ladenvall C, Mohlke KL, Morken MA, Narisu N, Nathan DM, Pascoe L, Payne F, Petrie JR, Sayer AA, Schwarz PE, Scott LJ, Stringham HM, Stumvoll M, Swift AJ, Syvanen AC, Tuomi T, Tuomilehto J, Tonjes A, Valle TT, Williams GH, Lind L, Barroso I, Quertermous T, Walker M, Wareham NJ, Meigs JB, McCarthy MI, Groop L, Watanabe RM, Florez JC, Investigators M (2010) Detailed physiologic characterization reveals diverse mechanisms for novel genetic Loci regulating glucose and insulin metabolism in humans. Diabetes 59 (5):1266–1275. doi:10.2337/db09-1568

Jansen J, Karges W, Rink L (2009) Zinc and diabetes--clinical links and molecular mechanisms. J Nutr Biochem 20(6):399–417. doi:10.1016/j.jnutbio.2009.01.009

Jayawardena R, Ranasinghe P, Galappatthy P, Malkanthi R, Constantine G, Katulanda P (2012) Effects of zinc supplementation on diabetes mellitus: a systematic review and meta-analysis. Diabetol Metab Syndr 4(1):13. doi:10.1186/1758-5996-4-13

Jiang F, Li Q, Hu C, Zhang R, Wang CR, Yu WH, Lu JY, Tang SS, Bao YQ, Xiang KS, Jia WP (2012) Association of a SLC30A8 genetic variant with monotherapy of repaglinide and rosiglitazone effect in newly diagnosed type 2 diabetes patients in China. Biomed Environ Sci 25(1):23–29. doi:10.3967/0895-3988.2012.01.004

Kambe T (2012) Molecular architecture and function of ZnT transporters. Curr Top Membr 69:199–220. doi:10.1016/B978-0-12-394390-3.00008-2

Kanoni S, Nettleton JA, Hivert MF, Ye Z, van Rooij FJ, Shungin D, Sonestedt E, Ngwa JS, Wojczynski MK, Lemaitre RN, Gustafsson S, Anderson JS, Tanaka T, Hindy G, Saylor G, Renstrom F, Bennett AJ, van Duijn CM, Florez JC, Fox CS, Hofman A, Hoogeveen RC, Houston DK, Hu FB, Jacques PF, Johansson I, Lind L, Liu Y, McKeown N, Ordovas J, Pankow JS, Sijbrands EJ, Syvanen AC, Uitterlinden AG, Yannakoulia M, Zillikens MC, Investigators M, Wareham NJ, Prokopenko I, Bandinelli S, Forouhi NG, Cupples LA, Loos RJ, Hallmans G, Dupuis J, Langenberg C, Ferrucci L, Kritchevsky SB, McCarthy MI, Ingelsson E, Borecki IB, Witteman JC, Orho-Melander M, Siscovick DS, Meigs JB, Franks

PW, Dedoussis GV (2011) Total zinc intake may modify the glucose-raising effect of a zinc transporter (SLC30A8) variant: a 14-cohort meta-analysis. Diabetes 60(9):2407–2416. doi:10.2337/db11-0176

Kathiresan S, Myocardial Infarction Genetics C (2008) A PCSK9 missense variant associated with a reduced risk of early-onset myocardial infarction. N Engl J Med 358(21):2299–2300. doi:10.1056/NEJMc0707445

Kelliny C, Ekelund U, Andersen LB, Brage S, Loos RJ, Wareham NJ, Langenberg C (2009) Common genetic determinants of glucose homeostasis in healthy children: the European Youth Heart Study. Diabetes 58(12):2939–2945. doi:10.2337/db09-0374

Kirchhoff K, Machicao F, Haupt A, Schafer SA, Tschritter O, Staiger H, Stefan N, Haring HU, Fritsche A (2008) Polymorphisms in the TCF7L2, CDKAL1 and SLC30A8 genes are associated with impaired proinsulin conversion. Diabetologia 51(4):597–601. doi:10.1007/s00125-008-0926-y

Kotowski IK, Pertsemlidis A, Luke A, Cooper RS, Vega GL, Cohen JC, Hobbs HH (2006) A spectrum of PCSK9 alleles contributes to plasma levels of low-density lipoprotein cholesterol. Am J Hum Genet 78(3):410–422. doi:10.1086/500615

Lemaire K, Ravier MA, Schraenen A, Creemers JW, Van de Plas R, Granvik M, Van Lommel L, Waelkens E, Chimienti F, Rutter GA, Gilon P, in't Veld PA, Schuit FC (2009) Insulin crystallization depends on zinc transporter ZnT8 expression, but is not required for normal glucose homeostasis in mice. Proc Natl Acad Sci U S A 106(35):14872–14877. doi:10.1073/pnas.0906587106

Majithia AR, Jablonski KA, McAteer JB, Mather KJ, Goldberg RB, Kahn SE, Florez JC, Group DPPR (2011) Association of the SLC30A8 missense polymorphism R325W with proinsulin levels at baseline and after lifestyle, metformin or troglitazone intervention in the Diabetes Prevention Program. Diabetologia 54(10):2570–2574. doi:10.1007/s00125-011-2234-1

Marselli L, Thorne J, Dahiya S, Sgroi DC, Sharma A, Bonner-Weir S, Marchetti P, Weir GC (2010) Gene expression profiles of Beta-cell enriched tissue obtained by laser capture microdissection from subjects with type 2 diabetes. PLoS ONE 5(7), e11499. doi:10.1371/journal.pone.0011499

Mocchegiani E, Giacconi R, Malavolta M (2008) Zinc signalling and subcellular distribution: emerging targets in type 2 diabetes. Trends Mol Med 14(10):419–428. doi:10.1016/j.molmed.2008.08.002

Morris AP, Voight BF, Teslovich TM, Ferreira T, Segre AV, Steinthorsdottir V, Strawbridge RJ, Khan H, Grallert H, Mahajan A, Prokopenko I, Kang HM, Dina C, Esko T, Fraser RM, Kanoni S, Kumar A, Lagou V, Langenberg C, Luan J, Lindgren CM, Muller-Nurasyid M, Pechlivanis S, Rayner NW, Scott LJ, Wiltshire S, Yengo L, Kinnunen L, Rossin EJ, Raychaudhuri S, Johnson AD, Dimas AS, Loos RJ, Vedantam S, Chen H, Florez JC, Fox C, Liu CT, Rybin D, Couper DJ, Kao WH, Li M, Cornelis MC, Kraft P, Sun Q, van Dam RM, Stringham HM, Chines PS, Fischer K, Fontanillas P, Holmen OL, Hunt SE, Jackson AU, Kong A, Lawrence R, Meyer J, Perry JR, Platou CG, Potter S, Rehnberg E, Robertson N, Sivapalaratnam S, Stancakova A, Stirrups K, Thorleifsson G, Tikkanen E, Wood AR, Almgren P, Atalay M, Benediktsson R, Bonnycastle LL, Burtt N, Carey J, Charpentier G, Crenshaw AT, Doney AS, Dorkhan M, Edkins S, Emilsson V, Eury E, Forsen T, Gertow K, Gigante B, Grant GB, Groves CJ, Guiducci C, Herder C, Hreidarsson AB, Hui J, James A, Jonsson A, Rathmann W, Klopp N, Kravic J, Krjutskov K, Langford C, Leander K, Lindholm E, Lobbens S, Mannisto S, Mirza G, Muhleisen TW, Musk B, Parkin M, Rallidis L, Saramies J, Sennblad B, Shah S, Sigurethsson G, Silveira A, Steinbach G, Thorand B, Trakalo J, Veglia F, Wennauer R, Winckler W, Zabaneh D, Campbell H, van Duijn C, Uitterlinden AG, Hofman A, Sijbrands E, Abecasis GR, Owen KR, Zeggini E, Trip MD, Forouhi NG, Syvanen AC, Eriksson JG, Peltonen L, Nothen MM, Balkau B, Palmer CN, Lyssenko V, Tuomi T, Isomaa B, Hunter DJ, Qi L, Wellcome Trust Case Control Consortium; Meta-Analyses of Glucose and Insulin-related traits Consortium (MAGIC) Investigators; Genetic Investigation of ANthropometric Traits (GIANT) Consortium;

Asian GeneticEpidemiology Network–Type 2 Diabetes (AGEN-T2D) Consortium; South Asian Type 2 Diabetes (SAT2D) Consortium, Shuldiner AR, Roden M, Barroso I, Wilsgaard T, Beilby J, Hovingh K, Price JF, Wilson JF, Rauramaa R, Lakka TA, Lind L, Dedoussis G, Njolstad I, Pedersen NL, Khaw KT, Wareham NJ, Keinanen-Kiukaanniemi SM, Saaristo TE, Korpi-Hyovalti E, Saltevo J, Laakso M, Kuusisto J, Metspalu A, Collins FS, Mohlke KL, Bergman RN, Tuomilehto J, Boehm BO, Gieger C, Hveem K, Cauchi S, Froguel P, Baldassarre D, Tremoli E, Humphries SE, Saleheen D, Danesh J, Ingelsson E, Ripatti S, Salomaa V, Erbel R, Jockel KH, Moebus S, Peters A, Illig T, de Faire U, Hamsten A, Morris AD, Donnelly PJ, Frayling TM, Hattersley AT, Boerwinkle E, Melander O, Kathiresan S, Nilsson PM, Deloukas P, Thorsteinsdottir U, Groop LC, Stefansson K, Hu F, Pankow JS, Dupuis J, Meigs JB, Altshuler D, Boehnke M, McCarthy MI, DIAbetes Genetics Replication And Meta-analysis (DIAGRAM) Consortium (2012) Large-scale association analysis provides insights into the genetic architecture and pathophysiology of type 2 diabetes. Nat Genet 44 (9):981–990. doi:10.1038/ng.2383

Myles S, Davison D, Barrett J, Stoneking M, Timpson N (2008) Worldwide population differentiation at disease-associated SNPs. BMC Med Genet 1:22. doi:10.1186/1755-8794-1-22

Nassar MA, Stirling LC, Forlani G, Baker MD, Matthews EA, Dickenson AH, Wood JN (2004) Nociceptor-specific gene deletion reveals a major role for Nav1.7 (PN1) in acute and inflammatory pain. Proc Natl Acad Sci U S A 101(34):12706–12711. doi:10.1073/pnas.0404915101

Nettleton JA, Hivert MF, Lemaitre RN, McKeown NM, Mozaffarian D, Tanaka T, Wojczynski MK, Hruby A, Djousse L, Ngwa JS, Follis JL, Dimitriou M, Ganna A, Houston DK, Kanoni S, Mikkila V, Manichaikul A, Ntalla I, Renstrom F, Sonestedt E, van Rooij FJ, Bandinelli S, de Koning L, Ericson U, Hassanali N, Kiefte-de Jong JC, Lohman KK, Raitakari O, Papoutsakis C, Sjogren P, Stirrups K, Ax E, Deloukas P, Groves CJ, Jacques PF, Johansson I, Liu Y, McCarthy MI, North K, Viikari J, Zillikens MC, Dupuis J, Hofman A, Kolovou G, Mukamal K, Prokopenko I, Rolandsson O, Seppala I, Cupples LA, Hu FB, Kahonen M, Uitterlinden AG, Borecki IB, Ferrucci L, Jacobs DR Jr, Kritchevsky SB, Orho-Melander M, Pankow JS, Lehtimaki T, Witteman JC, Ingelsson E, Siscovick DS, Dedoussis G, Meigs JB, Franks PW (2013) Meta-analysis investigating associations between healthy diet and fasting glucose and insulin levels and modification by loci associated with glucose homeostasis in data from 15 cohorts. Am J Epidemiol 177(2):103–115. doi:10.1093/aje/kws297

Nicolson TJ, Bellomo EA, Wijesekara N, Loder MK, Baldwin JM, Gyulkhandanyan AV, Koshkin V, Tarasov AI, Carzaniga R, Kronenberger K, Taneja TK, da Silva XG, Libert S, Froguel P, Scharfmann R, Stetsyuk V, Ravassard P, Parker H, Gribble FM, Reimann F, Sladek R, Hughes SJ, Johnson PR, Masseboeuf M, Burcelin R, Baldwin SA, Liu M, Lara-Lemus R, Arvan P, Schuit FC, Wheeler MB, Chimienti F, Rutter GA (2009) Insulin storage and glucose homeostasis in mice null for the granule zinc transporter ZnT8 and studies of the type 2 diabetes-associated variants. Diabetes 58(9):2070–2083. doi:10.2337/db09-0551

Palmer ND, Goodarzi MO, Langefeld CD, Ziegler J, Norris JM, Haffner SM, Bryer-Ash M, Bergman RN, Wagenknecht LE, Taylor KD, Rotter JI, Bowden DW (2008) Quantitative trait analysis of type 2 diabetes susceptibility loci identified from whole genome association studies in the Insulin Resistance Atherosclerosis Family Study. Diabetes 57(4):1093–1100

Pascoe L, Tura A, Patel SK, Ibrahim IM, Ferrannini E, Zeggini E, Weedon MN, Mari A, Hattersley AT, McCarthy MI, Frayling TM, Walker M (2007) Common variants of the novel type 2 diabetes genes CDKAL1 and HHEX/IDE are associated with decreased pancreatic beta-cell function. Diabetes 56(12):3101–3104, doi:db07-0634 [pii]10.2337/db07-0634

Plenge RM, Scolnick EM, Altshuler D (2013) Validating therapeutic targets through human genetics. Nat Rev Drug Discov 12(8):581–594. doi:10.1038/nrd4051

Pound LD, Sarkar SA, Benninger RK, Wang Y, Suwanichkul A, Shadoan MK, Printz RL, Oeser JK, Lee CE, Piston DW, McGuinness OP, Hutton JC, Powell DR, O'Brien RM (2009) Deletion of the mouse Slc30a8 gene encoding zinc transporter-8 results in impaired insulin secretion. Biochem J 421(3):371–376. doi:10.1042/BJ20090530

Pound LD, Sarkar SA, Ustione A, Dadi PK, Shadoan MK, Lee CE, Walters JA, Shiota M, McGuinness OP, Jacobson DA, Piston DW, Hutton JC, Powell DR, O'Brien RM (2012) The physiological effects of deleting the mouse SLC30A8 gene encoding zinc transporter-8 are influenced by gender and genetic background. PLoS ONE 7(7), e40972. doi:10.1371/journal.pone.0040972

Robertson RP, Zhou H, Slucca M (2011) A role for zinc in pancreatic islet beta-cell cross-talk with the alpha-cell during hypoglycaemia. Diabetes Obes Metab 13(Suppl 1):106–111. doi:10.1111/j.1463-1326.2011.01448.x

Rutter GA (2010) Think zinc: new roles for zinc in the control of insulin secretion. Islets 2 (1):49–50. doi:10.4161/isl.2.1.10259

Rutter GA, Chimienti F (2015) SLC30A8 mutations in type 2 diabetes. Diabetologia 58(1):31–36. doi:10.1007/s00125-014-3405-7

Saxena R, Voight BF, Lyssenko V, Burtt NP, de Bakker PI, Chen H, Roix JJ, Kathiresan S, Hirschhorn JN, Daly MJ, Hughes TE, Groop L, Altshuler D, Almgren P, Florez JC, Meyer J, Ardlie K, Bengtsson K, Isomaa B, Lettre G, Lindblad U, Lyon HN, Melander O, Newton-Cheh C, Nilsson P, Orho-Melander M, Rastam L, Speliotes EK, Taskinen MR, Tuomi T, Guiducci C, Berglund A, Carlson J, Gianniny L, Hackett R, Hall L, Holmkvist J, Laurila E, Sjogren M, Sterner M, Surti A, Svensson M, Svensson M, Tewhey R, Blumenstiel B, Parkin M, Defelice M, Barry R, Brodeur W, Camarata J, Chia N, Fava M, Gibbons J, Handsaker B, Healy C, Nguyen K, Gates C, Sougnez C, Gage D, Nizzari M, Gabriel SB, Chirn GW, Ma Q, Parikh H, Richardson D, Ricke D, Purcell S (2007) Genome-Wide Association Analysis Identifies Loci for Type 2 Diabetes and Triglyceride Levels. Science 316:1331–1336

Saxena R, Hivert MF, Langenberg C, Tanaka T, Pankow JS, Vollenweider P, Lyssenko V, Bouatia-Naji N, Dupuis J, Jackson AU, Kao WH, Li M, Glazer NL, Manning AK, Luan J, Stringham HM, Prokopenko I, Johnson T, Grarup N, Boesgaard TW, Lecoeur C, Shrader P, O'Connell J, Ingelsson E, Couper DJ, Rice K, Song K, Andreasen CH, Dina C, Kottgen A, Le Bacquer O, Pattou F, Taneera J, Steinthorsdottir V, Rybin D, Ardlie K, Sampson M, Qi L, van Hoek M, Weedon MN, Aulchenko YS, Voight BF, Grallert H, Balkau B, Bergman RN, Bielinski SJ, Bonnefond A, Bonnycastle LL, Borch-Johnsen K, Bottcher Y, Brunner E, Buchanan TA, Bumpstead SJ, Cavalcanti-Proenca C, Charpentier G, Chen YD, Chines PS, Collins FS, Cornelis M, JC G, Delplanque J, Doney A, Egan JM, Erdos MR, Firmann M, Forouhi NG, Fox CS, Goodarzi MO, Graessler J, Hingorani A, Isomaa B, Jorgensen T, Kivimaki M, Kovacs P, Krohn K, Kumari M, Lauritzen T, Levy-Marchal C, Mayor V, McAteer JB, Meyre D, Mitchell BD, Mohlke KL, Morken MA, Narisu N, Palmer CN, Pakyz R, Pascoe L, Payne F, Pearson D, Rathmann W, Sandbaek A, Sayer AA, Scott LJ, Sharp SJ, Sijbrands E, Singleton A, Siscovick DS, Smith NL, Sparso T, Swift AJ, Syddall H, Thorleifsson G, Tonjes A, Tuomi T, Tuomilehto J, Valle TT, Waeber G, Walley A, Waterworth DM, Zeggini E, Zhao JH, Illig T, Wichmann HE, Wilson JF, van Duijn C, Hu FB, Morris AD, Frayling TM, Hattersley AT, Thorsteinsdottir U, Stefansson K, Nilsson P, Syvanen AC, Shuldiner AR, Walker M, Bornstein SR, Schwarz P, Williams GH, Nathan DM, Kuusisto J, Laakso M, Cooper C, Marmot M, Ferrucci L, Mooser V, Stumvoll M, Loos RJ, Altshuler D, Psaty BM, Rotter JI, Boerwinkle E, Hansen T, Pedersen O, Florez JC, McCarthy MI, Boehnke M, Barroso I, Sladek R, Froguel P, Meigs JB, Groop L, Wareham NJ, Watanabe RM (2010) Genetic variation in GIPR influences the glucose and insulin responses to an oral glucose challenge. Nat Genet 42(2):142–148, doi:ng.521 [pii] 10.1038/ng.521

Scott LJ, Mohlke KL, Bonnycastle LL, Willer CJ, Li Y, Duren WL, Erdos MR, Stringham HM, Chines PS, Jackson AU, Prokunina-Olsson L, Ding CJ, Swift AJ, Narisu N, Hu T, Pruim R, Xiao R, Li XY, Conneely KN, Riebow NL, Sprau AG, Tong M, White PP, Hetrick KN, Barnhart MW, Bark CW, Goldstein JL, Watkins L, Xiang F, Saramies J, Buchanan TA, Watanabe RM, Valle TT, Kinnunen L, Abecasis GR, Pugh EW, Doheny KF, Bergman RN, Tuomilehto J, Collins FS, Boehnke M (2007) A Genome-Wide Association Study of Type 2 Diabetes in Finns Detects Multiple Susceptibility Variants. Science 316:1341–1345

Shan Z, Bao W, Zhang Y, Rong Y, Wang X, Jin Y, Song Y, Yao P, Sun C, Hu FB, Liu L (2013) Interactions between Zinc Transporter-8 Gene (SLC30A8) and Plasma Zinc Concentrations for Impaired Glucose Regulation and Type 2 Diabetes. Diabetes. doi:10.2337/db13-0606

Sladek R, Rocheleau G, Rung J, Dina C, Shen L, Serre D, Boutin P, Vincent D, Belisle A, Hadjadj S, Balkau B, Heude B, Charpentier G, Hudson TJ, Montpetit A, Pshezhetsky AV, Prentki M, Posner BI, Balding DJ, Meyre D, Polychronakos C, Froguel P (2007) A genome-wide association study identifies novel risk loci for type 2 diabetes. Nature 445(7130):881–885

Smidt K, Pedersen SB, Brock B, Schmitz O, Fisker S, Bendix J, Wogensen L, Rungby J (2007) Zinc-transporter genes in human visceral and subcutaneous adipocytes: lean versus obese. Mol Cell Endocrinol 264(1–2):68–73. doi:10.1016/j.mce.2006.10.010

Staiger H, Machicao F, Stefan N, Tschritter O, Thamer C, Kantartzis K, Schafer SA, Kirchhoff K, Fritsche A, Haring HU (2007) Polymorphisms within novel risk loci for type 2 diabetes determine beta-cell function. PLoS ONE 2(9), e832. doi:10.1371/journal.pone.0000832

Stancakova A, Kuulasmaa T, Paananen J, Jackson AU, Bonnycastle LL, Collins FS, Boehnke M, Kuusisto J, Laakso M (2009) Association of 18 confirmed susceptibility loci for type 2 diabetes with indices of insulin release, proinsulin conversion, and insulin sensitivity in 5,327 nondiabetic Finnish men. Diabetes 58(9):2129–2136. doi:10.2337/db09-0117

Stein EA, Mellis S, Yancopoulos GD, Stahl N, Logan D, Smith WB, Lisbon E, Gutierrez M, Webb C, Wu R, Du Y, Kranz T, Gasparino E, Swergold GD (2012) Effect of a monoclonal antibody to PCSK9 on LDL cholesterol. N Engl J Med 366(12):1108–1118. doi:10.1056/NEJMoa1105803

Steinthorsdottir V, Thorleifsson G, Reynisdottir I, Benediktsson R, Jonsdottir T, Walters GB, Styrkarsdottir U, Gretarsdottir S, Emilsson V, Ghosh S, Baker A, Snorradottir S, Bjarnason H, Ng MC, Hansen T, Bagger Y, Wilensky RL, Reilly MP, Adeyemo A, Chen Y, Zhou J, Gudnason V, Chen G, Huang H, Lashley K, Doumatey A, So WY, Ma RC, Andersen G, Borch-Johnsen K, Jorgensen T, van Vliet-Ostaptchouk JV, Hofker MH, Wijmenga C, Christiansen C, Rader DJ, Rotimi C, Gurney M, Chan JC, Pedersen O, Sigurdsson G, Gulcher JR, Thorsteinsdottir U, Kong A, Stefansson K (2007) A variant in CDKAL1 influences insulin response and risk of type 2 diabetes. Nat Genet 39:770–775

Strawbridge RJ, Dupuis J, Prokopenko I, Barker A, Ahlqvist E, Rybin D, Petrie JR, Travers ME, Bouatia-Naji N, Dimas AS, Nica A, Wheeler E, Chen H, Voight BF, Taneera J, Kanoni S, Peden JF, Turrini F, Gustafsson S, Zabena C, Almgren P, Barker DJ, Barnes D, Dennison EM, Eriksson JG, Eriksson P, Eury E, Folkersen L, Fox CS, Frayling TM, Goel A, Gu HF, Horikoshi M, Isomaa B, Jackson AU, Jameson KA, Kajantie E, Kerr-Conte J, Kuulasmaa T, Kuusisto J, Loos RJ, Luan J, Makrilakis K, Manning AK, Martinez-Larrad MT, Narisu N, Nastase Mannila M, Ohrvik J, Osmond C, Pascoe L, Payne F, Sayer AA, Sennblad B, Silveira A, Stancakova A, Stirrups K, Swift AJ, Syvanen AC, Tuomi T, van't Hooft FM, Walker M, Weedon MN, Xie W, Zethelius B, Ongen H, Malarstig A, Hopewell JC, Saleheen D, Chambers J, Parish S, Danesh J, Kooner J, Ostenson CG, Lind L, Cooper CC, Serrano-Rios M, Ferrannini E, Forsen TJ, Clarke R, Franzosi MG, Seedorf U, Watkins H, Froguel P, Johnson P, Deloukas P, Collins FS, Laakso M, Dermitzakis ET, Boehnke M, McCarthy MI, Wareham NJ, Groop L, Pattou F, Gloyn AL, Dedoussis GV, Lyssenko V, Meigs JB, Barroso I, Watanabe RM, Ingelsson E, Langenberg C, Hamsten A, Florez JC (2011) Genome-wide association identifies nine common variants associated with fasting proinsulin levels and provides new insights into the pathophysiology of type 2 diabetes. Diabetes 60 (10):2624–2634, doi:db11-0415 [pii] 10.2337/db11-0415

Sun Q, van Dam RM, Willett WC, Hu FB (2009) Prospective study of zinc intake and risk of type 2 diabetes in women. Diabetes Care 32(4):629–634. doi:10.2337/dc08-1913

Tabara Y, Osawa H, Kawamoto R, Onuma H, Shimizu I, Miki T, Kohara K, Makino H (2009) Replication study of candidate genes associated with type 2 diabetes based on genome-wide screening. Diabetes 58(2):493–498, doi:db07-1785 [pii]10.2337/db07-1785

Tamaki M, Fujitani Y, Hara A, Uchida T, Tamura Y, Takeno K, Kawaguchi M, Watanabe T, Ogihara T, Fukunaka A, Shimizu T, Mita T, Kanazawa A, Imaizumi MO, Abe T, Kiyonari H,

Hojyo S, Fukada T, Kawauchi T, Nagamatsu S, Hirano T, Kawamori R, Watada H (2013) The diabetes-susceptible gene SLC30A8/ZnT8 regulates hepatic insulin clearance. J Clin Invest 123(10):4513–4524. doi:10.1172/JCI68807

Walford GA, Green T, Neale B, Isakova T, Rotter JI, Grant SF, Fox CS, Pankow JS, Wilson JG, Meigs JB, Siscovick DS, Bowden DW, Daly MJ, Florez JC (2012) Common genetic variants differentially influence the transition from clinically defined states of fasting glucose metabolism. Diabetologia 55(2):331–339. doi:10.1007/s00125-011-2353-8

Weijers RN (2010) Three-dimensional structure of beta-cell-specific zinc transporter, ZnT-8, predicted from the type 2 diabetes-associated gene variant SLC30A8 R325W. Diabetol Metab Syndr 2(1):33. doi:10.1186/1758-5996-2-33

Wenzlau JM, Juhl K, Yu L, Moua O, Sarkar SA, Gottlieb P, Rewers M, Eisenbarth GS, Jensen J, Davidson HW, Hutton JC (2007) The cation efflux transporter ZnT8 (Slc30A8) is a major autoantigen in human type 1 diabetes. Proc Natl Acad Sci U S A 104(43):17040–17045, doi:0705894104 [pii]10.1073/pnas.0705894104

Wenzlau JM, Liu Y, Yu L, Moua O, Fowler KT, Rangasamy S, Walters J, Eisenbarth GS, Davidson HW, Hutton JC (2008) A common nonsynonymous single nucleotide polymorphism in the SLC30A8 gene determines ZnT8 autoantibody specificity in type 1 diabetes. Diabetes 57 (10):2693–2697. doi:10.2337/db08-0522

Wijesekara N, Chimienti F, Wheeler MB (2009) Zinc, a regulator of islet function and glucose homeostasis. Diabetes Obes Metab 11(Suppl 4):202–214. doi:10.1111/j.1463-1326.2009.01110.x

Wijesekara N, Dai FF, Hardy AB, Giglou PR, Bhattacharjee A, Koshkin V, Chimienti F, Gaisano HY, Rutter GA, Wheeler MB (2010) Beta cell-specific Znt8 deletion in mice causes marked defects in insulin processing, crystallisation and secretion. Diabetologia 53(8):1656–1668. doi:10.1007/s00125-010-1733-9

Xu K, Zha M, Wu X, Yu Z, Yu R, Xu X, Chen H, Yang T (2011) Association between rs13266634 C/T polymorphisms of solute carrier family 30 member 8 (SLC30A8) and type 2 diabetes, impaired glucose tolerance, type 1 diabetes--a meta-analysis. Diabetes Res Clin Pract 91 (2):195–202. doi:10.1016/j.diabres.2010.11.012

Zeggini E, Weedon MN, Lindgren CM, Frayling TM, Elliott KS, Lango H, Timpson NJ, Perry JR, Rayner NW, Freathy RM, Barrett JC, Shields B, Morris AP, Ellard S, Groves CJ, Harries LW, Marchini JL, Owen KR, Knight B, Cardon LR, Walker M, Hitman GA, Morris AD, Doney AS, McCarthy MI, Hattersley AT (2007) Replication of Genome-Wide Association Signals in U.K. Samples Reveals Risk Loci for Type 2 Diabetes. Science 316:1336–1341

Zhou H, Zhang T, Harmon JS, Bryan J, Robertson RP (2007) Zinc, not insulin, regulates the rat alpha-cell response to hypoglycemia in vivo. Diabetes 56(4):1107–1112. doi:10.2337/db06-1454

Chapter 19
From Association to Function: *MTNR1B*

Amélie Bonnefond and Valeriya Lyssenko

Abstract The discovery that variants in the melatonin receptor 2 (*MTNR1B*) gene were associated with glucose levels, insulin secretion, and risk for type 2 diabetes (T2D) in genome-wide association studies (GWAS) reinforced the previously suggested link between glucose homeostasis and circadian rhythmicity. Diurnal secretion of melatonin has reported to be altered in people with diabetes and rodent models of T2D. The proposed underlying mechanisms by which altered melatonin signaling could predispose to progression to T2D and gestational diabetes mellitus (GDM) involve altered expression of *MTNR1B* in pancreatic beta cells, leading to impaired insulin secretion, consequent increased fasting glucose concentrations, and eventually overt T2D. Thus blocking the inhibition of insulin secretion may have potential clinical implications, and these effects could be more pronounced in individuals carrying risk genotypes. Finally, given that melatonin could emerge as an attractive treatment for a variety of conditions including pregnancies associated with GDM, preeclampsia, and intrauterine growth retardation, pharmacogenetic studies are warranted to determine treatment response and side effects according to genotype.

A. Bonnefond (✉)
European Genomic Institute for Diabetes (EGID), Lille, France

CNRS-UMR8199, Lille Pasteur Institute, Lille, France

Lille 2 University, Lille, France
e-mail: amelie.bonnefond@good.ibl.fr

V. Lyssenko
Department of Clinical Sciences, Diabetes and Endocrinology, Clinical Research Centre, Lund University, Malmö, Sweden

Steno Diabetes Center, Gentofte, Denmark
e-mail: valeriya.lyssenko@med.lu.se

J.C. Florez (ed.), *The Genetics of Type 2 Diabetes and Related Traits*,
DOI 10.1007/978-3-319-01574-3_19

19.1 *MTNR1B* as a Primary Genome-Wide Association Signal for Fasting Plasma Glucose Levels

Type 2 diabetes (T2D) is a complex genetic metabolic disorder which has developed into a major health problem responsible for early morbidities (including severe micro- and/or macro-vascular complications) and mortality, with a burden increasing globally, particularly in developing countries (World Health Organization 2013). T2D results from progressive dysfunction of insulin secretion from pancreatic beta cells on the background of resistance to insulin action (American Diabetes Association 2014a). Since 2007, exponential progress has been made in identifying genetic determinants of T2D through the use of cutting-edge DNA microarrays allowing for large-scale genome-wide association studies (GWAS) and their meta-analyses. After analyzing T2D as a binary disorder in case–control GWAS, a logical next step was to investigate new variants contributing to the variation of metabolic quantitative traits linked to the pathophysiological processes leading to T2D, as this may provide new insights into the etiological mechanisms of this highly complex disorder. Among such quantitative traits, fasting plasma glucose (FPG) levels were of major interest as hyperglycemia in the fasting state remains one of the criteria used by the American Diabetes Association (ADA) to define T2D (American Diabetes Association 2014b). Furthermore, elevated FPG levels within the range specified by the ADA was known to be an independent risk factor for T2D (Tirosh et al. 2005) and for cardiovascular and all-cause mortality (Barr et al. 2007). Of note, approximately one-third of the variation of FPG was shown to be genetic (Watanabe et al. 1999; Pilia et al. 2006).

In 2008, via GWAS, two independent groups identified a new genome-wide association signal with FPG levels closed to *G6PC2* and *ABCB11* genes, in nondiabetic European participants (Bouatia-Naji et al. 2008; Chen et al. 2008). Of note, the single nucleotide polymorphism (SNP; rs560887) in *G6PC2* which significantly contributed to variation of FPG levels was not associated with T2D risk (Bouatia-Naji et al. 2008) (see Chap. 17).

One year later, using larger sample sizes in GWAS meta-analyses which markedly increased the statistical power, a new locus associated with FPG levels was identified in *MTNR1B* (encoding melatonin receptor 2 [MT2]) in nondiabetic European individuals (Bouatia-Naji et al. 2009; Prokopenko et al. 2009). The association signal was subsequently refined by genotyping eight SNPs in strong linkage disequilibrium ($r^2 > 0.70$) within the association block (Sparsø et al. 2009). This study showed that rs10830963, located in the middle of the unique *MTNR1B* intron, carried most of the effect on FPG variation (Sparsø et al. 2009). Of note, rs10830963 was the strongest SNP associated with FPG levels in the first and most recent publications by the Meta-Analyses of Glucose and Insulin-related traits Consortium (MAGIC) in 36,610 and 133,010 nondiabetic individuals of European descent, respectively (Fig. 19.1) (Prokopenko et al. 2009; Scott et al. 2012). The *MTNR1B* locus was shown to be robustly associated with FPG levels in nondiabetic individuals from other various ethnicities including

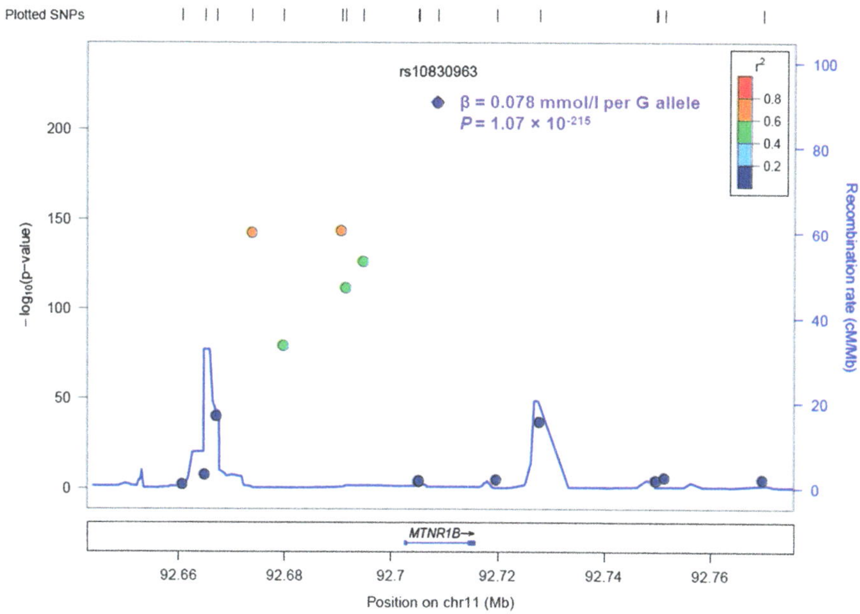

Fig. 19.1 Regional plots showing the association between SNPs and FPG levels at the *MTNR1B* locus in the lastly published MAGIC study performed in 133,010 nondiabetic individuals of European descent (Scott et al. 2012)

African-Americans, Koreans, Japanese, Sri Lankans, Han Chinese, and Indian Asians (Rönn et al. 2009; Chambers et al. 2009; Liu et al. 2010; Takeuchi et al. 2010; Kan et al. 2010; Ramos et al. 2011; Kim et al. 2011; Rasmussen-Torvik et al. 2012), suggesting a putatively causal role of the *MTNR1B* gene in the association with FPG levels. Furthermore, studies in adolescents and children suggested that variants in *MTNR1B* most likely influence FPG levels early from childhood onward (Barker et al. 2011).

19.2 Effect of *MTNR1B* Genetic Variants on T2D Risk

In addition to the variation of FPG levels, the *MTNR1B* locus was found to be significantly associated with T2D risk (Bouatia-Naji et al. 2009; Lyssenko et al. 2009; Prokopenko et al. 2009), demonstrating that studies of continuous glycemic phenotypes in nondiabetic individuals can successfully complement the genetic analyses of T2D as a dichotomous trait. Dupuis et al. demonstrated a genome-wide significant association between *MTNR1B* SNP rs10830963 and T2D risk (Dupuis et al. 2010), which was confirmed by the Diabetes Genetics Replication And Meta-analysis (DIAGRAM) consortium (Voight et al. 2010). The association between *MTNR1B* locus and T2D risk was also reported in individuals from various ethnicities including Han Chinese, Japanese, and Indian Asians (Rönn et al. 2009; Chambers et al. 2009; Kan et al. 2010; Ling et al. 2011; Ohshige et al. 2011; Zhao et al. 2014), even though evidence of heterogeneity due to ethnicity was found in two meta-analyses: the association seemed to be stronger in Europeans, compared with East or South Asians (Xia et al. 2012; Wang et al. 2013).

The causal link between the *MTNR1B* locus and T2D risk has been demonstrated by a large-scale exon resequencing study of *MTNR1B* in combination with systematic functional investigations of each identified non-synonymous variant (Bonnefond et al. 2012). The sequencing of *MTNR1B* coding exons in 7632 Europeans, including 2186 individuals with T2D, identified 40 non-synonymous variants of which 36 variants were very rare (with a minor allele frequency below 0.1 %) and significantly associated with T2D, while the four frequent or rare non-synonymous variants (with a minor allele frequency above 1 %) did not contribute to T2D risk (Bonnefond et al. 2012). After functional investigations (including proper surface expression study, melatonin binding study, and ERK or Gi pathway study of each mutant), Bonnefond et al. demonstrated that 13 partial- or total-loss-of-function very rare variants in combination yield more than fivefold increased risk for T2D, while the 23 neutral very rare variants did not exhibit any effect on T2D (Bonnefond et al. 2012). This study established a firm functional link between the *MTNR1B* gene and T2D risk, highlighting that the discovery of functional coding mutations in a previously associated locus can help establish a specific gene as the molecular cause of the association signal.

19.3 Effect of *MTNR1B* Variants on Early Phase Insulin Secretion

The mechanisms by which variation at *MTNR1B* raises FPG and T2D risk were explored by the Diabetes Genetics Initiative GWAS for insulin secretion (Lyssenko et al. 2009). Lyssenko et al. demonstrated that the glucose-raising allele of the *MTNR1B* rs10830963 was associated with impaired early insulin release to both oral (insulinogenic and disposition index) and intravenous (first phase insulin response) glucose loads (Lyssenko et al. 2009). In addition, the risk allele carriers showed deterioration in insulin secretion over time as compared to non-risk allele carriers (Lyssenko et al. 2009). The same study showed that in the risk allele carriers, elevated FPG levels and reduced phase insulin response to glucose were translated into 11 % increased risk for future T2D in two large prospective studies of more than 18,000 individuals, of whom 2201 developed T2D during a mean follow-up period of 23.5 years (Lyssenko et al. 2009). Very recently, a largest to date GWAS meta-analyses for dynamic measurements of insulin secretion during an oral glucose tolerance test (OGTT) in more than 10,000 nondiabetic individuals confirmed *MTNR1B* rs10830963 as the strongest signal for first phase insulin secretion (Prokopenko et al. 2014). Importantly, in the Diabetes Prevention Program, the association of *MTNR1B* rs10830963 with impaired early insulin release persisted at 1 year despite adjustment for the baseline trait, suggesting a progressive deterioration of the effect at this locus (Florez et al. 2012).

Notably, Lyssenko et al. demonstrated that *MTNR1B* mRNA was expressed in human pancreatic islets and more specifically that nondiabetic individuals carrying the risk allele and patients with T2D display increased expression of the receptor in pancreatic islets (Lyssenko et al. 2009). These observations were further confirmed by a large gene-expression analysis of human pancreatic islets (Taneera et al. 2012). Exogenously administered melatonin has been shown to inhibit insulin secretion in rodents (Bailey et al. 1974). In line with these observations, Lyssenko et al. showed that melatonin inhibited insulin release in response to glucose in INS-1 rat beta cells (Lyssenko et al. 2009).

Of note, *MTNR1B* rs10830963 significantly increased the risk of isolated impaired fasting glucose but not the risk of isolated impaired glucose tolerance (Sparsø et al. 2009), and the same SNP was shown to impact the rate of progression from normal fasting glucose to impaired fasting glucose, but not the rate of progression from impaired fasting glucose to T2D (Walford et al. 2012). In the GLACIER study which is a population-based prospective cohort study from northern Sweden, *MTNR1B* rs10830963 was reported to increase risk of developing impaired fasting glucose during 10-year follow-up but to be protective of worsening glucose tolerance (Renström et al. 2011), which confirmed a strong effect of *MTNR1B* locus on impaired fasting glucose, independently of impaired glucose tolerance measured 2 h after a meal.

In addition to its effects on insulin secretion and isolated impaired fasting glucose, variants in the *MTNR1B* gene were also reported to be associated with hepatic insulin resistance (Sparsø et al. 2009; Vangipurapu et al. 2011). Notably, the risk carriers of *MTNR1B* were a clear outlier when known hyperbolic relationship between insulin secretion and the degree of insulin sensitivity was plotted, showing a strong insulin-resistant phenotype for the given impairment in insulin secretion (Jonsson et al. 2013). An association with insulin sensitivity could involve effects of *MTNR1B* on energy expenditure. In this vein, it has been recently demonstrated that *MTNR1B* variant could modify effects of dietary fat intake on changes in energy expenditure during a 2-year period (Mirzaei et al. 2014). Additionally, this could involve insular-incretin axis as *MTNR1B* variants were shown to be associated with incretin-stimulated insulin secretion (Simonis-Bik et al. 2010).

19.4 Effect of *MTNR1B* Variants on the Risk for Gestational Diabetes Mellitus

In addition to the risk of T2D, there is compelling robust evidence for an association of the *MTNR1B* locus with gestational diabetes mellitus (GDM) in several ethnic populations (Kwak et al. 2012; Vlassi et al. 2012; Huopio et al. 2013). A GWAS in a Korean population consisting of 468 women with GDM and 1242 nondiabetic women reported a variant near *MTNR1B*, SNP rs10830962, to be associated with GDM at a genome-wide significance level (Kwak et al. 2012). In a smaller study from Greece, SNP rs10830963 also conferred association with GDM (Vlassi et al. 2012), while in one study of Chinese pregnant women, several variants in the *MTNR1B* gene were associated with elevated glucose concentrations (Liao et al. 2012). An independent Chinese study demonstrated reduced beta-cell function as measured with HOMA-B index (Wang et al. 2011). Recently, a large Finnish study showed that the risk genotypes of the *MTNR1B* rs10830963 were associated with GDM risk, increased fasting plasma glucose, and reduced insulin secretion (Huopio et al. 2013). Notably, melatonin crosses the human placenta easily and rapidly (Okatani et al. 1998) and has been suggested to play an important role during pregnancy through its antioxidant properties but also as a regulator of normal growth and development of fetal organs (Reiter et al. 2014). The underlying mechanisms were attributed to the melatonin effect on the epigenetic modifications of genes implicated in placental development, fetal growth, and intrauterine programming (Korkmaz et al. 2012). However, there have been no any studies thus far on whether variants in the melatonin receptor(s) could contribute to these effects.

19.5 Melatonin Secretion and Physiological Functions

Melatonin (N-acetyl-5-methoxytryptamine), also known as the hormone of darkness, is an indoleamine synthesized from the amino acid tryptophan via serotonin. Tryptophan is converted by 5-tryptophan hydroxylase to 5-hydroxytryptophan, which then in the pineal gland is converted through hydroxylation and decarboxylation into serotonin. Serotonin is subsequently acetylated by arylalkylamine-N-acetyltransferase (AANAT, also called serotonin N-acetyltransferase), the rate-limiting step in melatonin biosynthesis, and, finally, through a methylation reaction converted into melatonin by acetylserotonin O-methyltransferase (ASMT) (Axelrod and Weissbach 1960). Melatonin is a key mediator of the entrainment of biological rhythms in the body, a “zeitgeber” (German: “time giver”), in which biological rhythmic secretion was initially linked to regulation of seasonal reproduction in photoperiodic species (Hoffman and Reiter 1965).

The hormone was traditionally thought to emanate from rhythmicity pinealocytes located in the pineal gland. This endocrine organ is located in the midline of the brain, just above the posterior commissure at the dorsal edge of the third ventricle, and receives a rich supply of adrenergic innervation from the superior cervical ganglion. However, melatonin remains detectable after pinealectomy in some species (Axelrod and Weissbach 1960), leading to the realization that the hormone is produced in neuroendocrine cells of both central (pineal gland, retina, Harderian glands) and peripheral origins, comprising enterochromaffin cells dispersed in a number of organs, e.g., the gastrointestinal tract, pancreas, and many more (Kvetnoy et al. 1997). Melatonin is also produced by numerous non-endocrine cells, e.g., immune cells. Thus, while substantial local biosynthesis also occurs in retina and in some other organs, e.g., the gastrointestinal tract, the diurnal rhythm of the circulating melatonin in blood exclusively accounts for its secretion from the pineal gland.

Light has a dual effect on the production of melatonin. First, it entrains the circadian clock, making melatonin production occur during the night. The duration of melatonin production varies with time over the year, because the onset and offset of melatonin secretion are controlled by the clock and can move closer (summer) or apart (winter) (Illnerová and Sumová 1997). In fact, in the absence of light, and in blind people, the melatonin rhythm persists following a circadian rhythm and cycle length governed by the suprachiasmatic nucleus (SCN) (Klerman et al. 2002). Light at night can also have an acute suppressive effect on melatonin production. In rodents, light at night decreases melatonin within minutes, while RNA levels remain high for hours. Circulating levels of melatonin peak at 80–100 pg/ml in the middle of the night and drop to 10–20 pg/ml during day time; the half-life of the hormone in the circulation is less than 20 min.

The hormone exerts its effects both through activation of its receptors (melatonin receptor 1 [MT1], MT2, and the orphan GPR50 receptor that can regulate the function of MT1 and MT2 through receptor heterodimerization) (Boutin et al. 2005; Jockers et al. 2008) but also via receptor-independent mechanisms such as its capacity to act as an antioxidant (Hardeland 2005). These effects can either be through the circulating levels of the hormone or in a more autocrine/paracrine fashion near target tissues (Kvetnoy et al. 1997; Peschke 2008).

Melatonin is widely known to affect the CNS, where it alters hormone release and phase shifts neuronal firing both in the 24 h rhythm and seasonal changes (Dubocovich and Markowska 2005). The phase-advancing effects of melatonin are taken advantage of treating insomnia or limiting jet lag when traveling across time zones (Arendt et al. 1997; Zhdanova and Wurtman 1997; Arendt 2006). Recently, the effects of melatonin in synergizing maternal and fetal circadian rhythms during pregnancy, the development of normal placentation, and its potential beneficial properties in the treatment of compromised pregnancies associated with GDM, preeclampsia, and intrauterine growth retardation have been emphasized (Reiter et al. 2014).

In the periphery, melatonin promotes vasoconstriction through MT1 and vasodilation through MT2 (Masana et al. 2002). In the adrenal cortex, it lowers cortisol secretion, an action shared with insulin (Weitzman et al. 1971; Peschke 2008). Interestingly, human adipocytes, a major target tissue for insulin, express MT2 and have been shown to reduce the expression of the insulin-dependent glucose transporter GLUT4 after melatonin stimulation (Brydon et al. 2001). In muscle cells, melatonin stimulates glucose uptake by phosphorylation of the insulin receptor substrate-1 (IRS-1) through suggested MT2 signaling (Ha et al. 2006). MT2 is also expressed in hepatocytes, and melatonin injections elevated glucose release from the liver in mice (Poon et al. 2001). Melatonin receptors are also widely expressed in the gut and could thus have an effect on incretin hormones like glucose insulinotropic polypeptide (GIP) and glucagon-like peptide 1 (GLP-1) (Chen et al. 2011).

19.6 Melatonin Receptors: A Putative Role in Pancreatic Islets?

The genetic studies described above have strongly suggested a putatively direct role of *MTNR1B* variants in the dysfunction of insulin secretion from pancreatic beta cells, leading to T2D. The discovery that genetic variants in the *MTNR1B* gene were associated with FPG levels, insulin secretion, and risk for T2D reinforced the previously suggested link between glucose homeostasis and circadian rhythmicity. It has been previously observed that plasma insulin levels exhibit a circadian rhythm and that disturbance in the oscillation affects plasma glucose and hormone levels (Scheer et al. 2009). Diurnal secretion of melatonin has reported to be altered in patients and rodent models of T2D and that expression of the melatonin receptor is increased (Peschke et al. 2006b). Importantly, it has recently been demonstrated that lower nocturnal melatonin secretion was independently associated with a higher risk of developing T2D in nondiabetic women (McMullan et al. 2013). However, the specific role of the melatonin receptors in these observations remains an open question. How did cell studies and animal models indicate a putative role of melatonin receptors in pancreatic islets and T2D?

19.6.1 Melatonin Receptors Are G-Protein-Coupled Receptors (GPCRs)

Melatonin receptors belong to the GPCR superfamily which represents the largest family of cell membrane receptors in humans with approximately 800 GPCRs (>1 % of the total protein-coding human genome) (Jassal et al. 2010). GPCRs are characterized by seven alpha-helical transmembrane domains which are connected by intra- and extracellular loops of various lengths. They sense extracellular signals (including photons, metabolites, amino acids, lipids, odor molecules, hormones, neurotransmitters, peptides) and activate various intracellular signaling pathways, through the interaction with G proteins (including Gs proteins which activate the adenylyl cyclase pathway, Gi/o proteins which inhibit this pathway, Gq/11 proteins which activate the phospholipase C pathway, and G12/13 proteins which are involved in the activation of small G proteins and cytoskeleton rearrangements) and/or beta-arrestin (Kristiansen 2004; Venkatakrishnan et al. 2013).

Melatonin receptors constitute a subfamily of rhodopsin-like (type A or 1) GPCRs which include three members: MT1, MT2, and the orphan GPR50 receptor (Jockers et al. 2008; Jassal et al. 2010). They share specific short amino acid sequences and approximately 55 % overall sequence homology (Reppert et al. 1994, 1995). Melatonin receptors have been neglected for a long time by academic research due to the lack of specific pharmacological tools, apparently mild phenotypes of knockout mice, and poorly defined functions. Notably, rodent animal models are limited for the study of the link between melatonin and metabolism, as rodents are nocturnal animals (therefore the circadian rhythms of food intake and metabolism are shifted by 12 h with respect to humans, despite the nocturnal secretion of melatonin by humans and rodents). Furthermore, some mouse strains used in laboratory are known to exhibit very low or even undetectable circulating melatonin levels due to impaired melatonin production within the pineal gland (Kennaway et al. 2002). Therefore, in the studies of melatonin and its receptors, the extrapolation of results obtained from rodents to humans must be done cautiously (Karamitri et al. 2013).

19.6.2 Distribution of MT1 and MT2 in Humans

Melatonin receptors are widely expressed throughout the human body. Notably, both *MTNR1A* and *MTNR1B* are expressed in the brain (in particular the SCN which is the master circadian pacemaker, the hippocampus, and the thalamus) and in the retina (Reppert et al. 1995; Jockers et al. 2008). Interestingly, insulin-target tissues express *MTNR1B*: adipocytes, muscle cells, hepatocytes, and the gut. Furthermore, both *MTNR1A* mRNA and *MTNR1B* mRNA have been detected in human islets (Ramracheya et al. 2008). The same study showed that *MTNR1A* mRNA was primarily detected in pancreatic alpha cells and the level of *MTNR1B* mRNA was globally much lower than the level of *MTNR1A* mRNA in human islets (Ramracheya et al. 2008). This last result was not confirmed by Lyssenko et al., who did not detect any differences in the levels of MT1 and MT2 in human islets (Lyssenko et al. 2009). MT2 was predominantly observed in human pancreatic beta cells, while MT1 was mostly detected in a population of peripherally located beta cells (Lyssenko et al. 2009).

19.6.3 Melatonin Receptor Signaling in Pancreatic Beta Cells

Most of the existing literature (based on rodent models) has shown that melatonin inhibits insulin secretion from pancreatic beta cells (Peschke 2008). However, in human pancreatic islets, a study demonstrated that melatonin stimulates insulin secretion, without affecting cAMP levels (Ramracheya et al. 2008).

Melatonin receptors can actually modulate insulin secretion from pancreatic beta cells via different signaling pathways which can lead to opposite effects (Fig. 19.2). First, activated melatonin receptors coupled with $G\alpha_i$ proteins inhibit cAMP production (via adenylyl cyclase [AC]) and activation of cAMP-dependent protein kinase A (PKA) and therefore decrease insulin granules exocytosis (Fig. 19.2) (Kemp et al. 2002; Peschke et al. 2002, 2006a). Furthermore, activated MT2 receptor coupled with $G\alpha_i$ proteins can decrease cGMP levels (via guanylate cyclase), leading to reduced insulin secretion (Fig. 19.2) (Stumpf et al. 2008, 2009). In addition, activated melatonin receptors coupled with $G\alpha_{q/11}$ enhance phospholipase C (PLC), increasing the release of Ca^{2+} from intracellular stores (endoplasmic reticulum) by the stimulation of inositol 1,4,5-triphosphate (IP3) receptor, which induces insulin granules exocytosis (Fig. 19.2) (Bach et al. 2005). All these results were obtained from rodent models which can have some limitations. More investigations based on human pancreatic islets would be of major interest for the field. Of note, melatonin receptors (in particular MT1) may also play a role in pancreatic alpha cells, which could regulate glucose homeostasis via the glucagon. However, the mechanisms are not clear as there are some discrepancies between studies, and investigations based on human islets are also lacking (Karamitri et al. 2013).

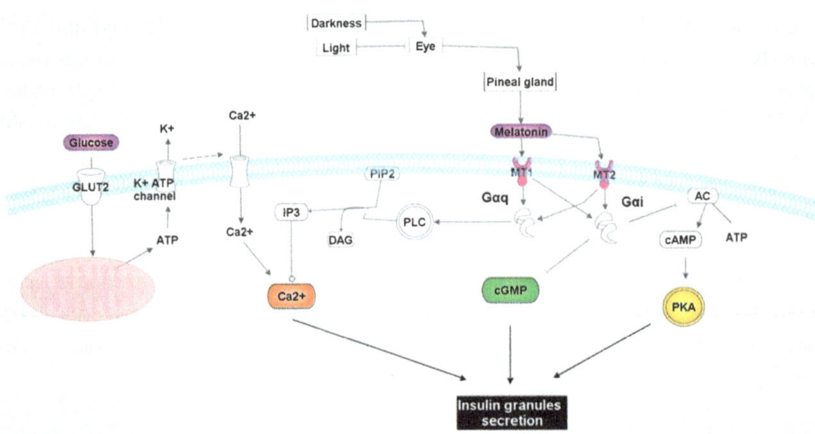

Fig. 19.2 Signal transduction of melatonin in pancreatic beta cells. *AC*, adenylyl cyclase; *ATP*, adenosine triphosphate; *cAMP*, cyclic adenosine monophosphate; *cGMP*, cyclic guanosine monophosphate; *DAG*, diacylglycerol; *IP3*, inositol 1,4,5-triphosphate; *PIP2*, phosphatidylinositol 4,5-bisphosphate; *PKA*, protein kinase A; *PLC*, phospholipase C

19.7 Dysfunction of Circadian Rhythm and Metabolic Disorders

Organisms, ranging from bacteria to mammals, possess accurate internal biological rhythms that time various daily events from photosynthesis in plants to sleep/wakefulness in humans (Takahashi et al. 2008). In mammals, many physiological processes are actually regulated by these inherent clocks, including glucose homeostasis, body temperature, feeding behavior, and hormone release (Takahashi et al. 2008). The circadian system (defined as its approximately 24 h cycles in the absence of environmental clues) represents the hierarchy of these multiple biological clocks, which is controlled by the SCN known as the "master clock" or "master circadian pacemaker," located in the anterior hypothalamus (Mohawk et al. 2012). The SCN is synchronized by the daily light–dark cycle in mammals, and in turn, it regulates output pathways controlling various biological functions and overt rhythms (e.g., hormone release, feeding, body temperature) which can synchronize peripheral biological clocks including circadian rhythms in pancreatic islets (Takahashi et al. 2008). Of note, these peripheral clocks can feed back and interfere with the master clock SCN (Mohawk et al. 2012).

At the molecular level, circadian clocks are mostly controlled by an autoregulatory transcriptional feedback loop (the "core circadian clock") involving the activator transcription factors CLOCK and BMAL1 and their target genes *period* (*PER1*, *PER2*, *PER3*) and *cryptochrome* (*CRY1*, *CRY2*) whose gene products form proteins that complex, translocate into the nucleus, and repress CLOCK and BMAL1-mediated transcription until the degradation of the Per/Cry repressor complex (Takahashi et al. 2008; Mohawk et al. 2012). This loop takes roughly 24 h and operates in most cells of the body to control either ubiquitous or tissue-specific physiological functions. Of note, there are additional (less essential) loops involving other clock genes or proteins (e.g., *Rev-erbα*, *Rorα*) which influence the core circadian clock. All these feedback loops interact with electrical and metabolic oscillations which modulate physiological function (Evans and Davidson 2013).

Compelling evidence has linked a disturbed circadian rhythm to metabolic syndrome, including T2D and obesity (Bass and Takahashi 2010; Shi et al. 2013). First, in the early 2000s, an outstanding study showed that a significant part of the transcriptome follows circadian rhythms in the mouse and that most of the signaling pathways regulated by these circadian clocks are involved in fundamental metabolism (Panda et al. 2002). Furthermore, several mutant animal models for clock genes (*Clock*, *Bmal1*, *Rev-erbα*, *Rev-erbβ*) have been shown to present with metabolic disorders including obesity, diabetic phenotypes (insulin resistance, hyperglycemia, impaired pancreatic function), hypertension, hyperlipidemia, and/or hepatic steatosis (Turek et al. 2005; Takahashi et al. 2008; Lamia and Evans 2010; Marcheva et al. 2010; Cho et al. 2012). Importantly, two studies demonstrated that pancreatic islets (probably pancreatic beta cells) possess self-sustained circadian oscillations (Marcheva et al. 2010; Sadacca et al. 2011). This intrinsic pancreatic circadian clock is required for normal insulin secretion and glucose homeostasis (Sadacca et al. 2011) and, when disrupted, leads to hypoinsulinemia and diabetes (Marcheva et al. 2010). In humans, a considerable number of epidemiological studies have reported that shift work is associated with increased risk for metabolic syndrome (including T2D and obesity) and cardiovascular disease, relative to day work (Kivimäki et al. 2011; Pan et al. 2011; Vyas et al. 2012; Buxton et al. 2012; Monk and Buysse 2013). Three main factors were known to explain these negative effects: circadian misalignment, sleep deprivation, and exposure to light at night (Evans and Davidson 2013). However, lastly, Eve Van Cauter's group has reported that circadian misalignment may increase risk of T2D and obesity-related phenotypes, independently of sleep deprivation (Leproult et al. 2014). Finally, some genetic studies reported nominal or genome-wide significant associations between common SNPs near clock genes (*CRY2*, *PER3*, *ARNTL* [*BMAL1*], *CLOCK*, *NR1D1* [*REV-ERBA*]) and risk of T2D/obesity or variation of related metabolic traits (including FPG, lipid levels) (Woon et al. 2007; Dupuis et al. 2010; Garaulet et al. 2010, 2013; Below et al. 2011; Kelly et al. 2012; Goumidi et al. 2013). Interestingly, a functional study showed that expression of *PER2*, *PER3*, and *CRY2* genes was significantly decreased in pancreatic islets of patients with T2D, and this low expression was positively correlated with increased levels of HbA1c and decreased insulin secretion (Stamenkovic et al. 2012).

Therefore, it is highly probable that dysfunction of central and/or peripheral circadian clocks leads to T2D and other cardiometabolic studies. However, the specific involvement of the melatonin and its receptors into this dysfunction remains an open question.

19.8 Clinical and Pharmacological Implications

Taken together, as most observations available support an inhibitory effect of melatonin on insulin secretion, selective blocking of the melatonin ligand-receptor system in islets would be an attractive potential pharmacological target for the treatment of T2D. An individual carrying a risk variant may thus be more sensitive to the inhibitory melatonin effect than an individual without the risk allele, with a normal level of *MTNR1B* expression in islets. Such a restraining effect of melatonin fits with the impairment of early phase insulin secretion that was observed in risk carriers (Lyssenko et al. 2009). Thus, assessment of plasma levels of melatonin during different stages of glucose intolerance (NGT, IFG, and T2D) but also during pregnancy in GDM women in risk and non-risk genotype carriers of the *MTNR1B* gene deserves future studies to support or reject this notion. Furthermore, it remains to be elucidated whether ascribed protective properties of melatonin during pregnancy could be altered in the individuals carrying the risk allele in the *MTNR1B* gene. Finally, whether the ascribed effects of melatonin on epigenetic intrauterine programming contribute to the early defects in abnormal glucose metabolism and risk for T2D later in life warrants further investigation.

19.9 Conclusions

The proposed mechanisms by which altering melatonin signaling could predispose to progression to T2D and putatively GDM involve altering expression of *MTNR1B* in pancreatic beta cells leading to impaired insulin secretion. As a result, this leads to increased fasting glucose concentrations and eventually overt T2D. Thus, it would be of great clinical value to evaluate the combined risk for T2D but also for GDM of a model comprising *MTNR1B* common and rare functional variants in the gene, together with melatonin metabolites reflective of the activity of key enzymes in the melatonin pathway: in an intriguing finding that supports this notion, Illig et al. have shown that the tryptophan to phenylalanine ratio is affected by genetic variants in the *MTNR1B* gene (Illig et al. 2010). Whether this observation is reproducible and clinically translatable awaits the integration of independent genomic, metabolomic, and prospective data sets.

References

American Diabetes Association (2014a) Standards of medical care in diabetes--2014. Diabetes Care 37(Suppl 1):S14–S80. doi:10.2337/dc14-S014

American Diabetes Association (2014b) Diagnosis and classification of diabetes mellitus. Diabetes Care 37(Suppl 1):S81–S90. doi:10.2337/dc14-S081

Arendt J (2006) Does melatonin improve sleep? Efficacy of melatonin. BMJ 332:550. doi:10.1136/bmj.332.7540.550

Arendt J, Skene DJ, Middleton B et al (1997) Efficacy of melatonin treatment in jet lag, shift work, and blindness. J Biol Rhythms 12:604–617

Axelrod J, Weissbach H (1960) Enzymatic O-methylation of N-acetylserotonin to melatonin. Science 131:1312

Bach AG, Wolgast S, Mühlbauer E, Peschke E (2005) Melatonin stimulates inositol-1,4,5-trisphosphate and Ca2+ release from INS1 insulinoma cells. J Pineal Res 39:316–323. doi:10.1111/j.1600-079X.2005.00253.x

Bailey CJ, Atkins TW, Matty AJ (1974) Melatonin inhibition of insulin secretion in the rat and mouse. Horm Res 5:21–28

Barker A, Sharp SJ, Timpson NJ et al (2011) Association of genetic Loci with glucose levels in childhood and adolescence: a meta-analysis of over 6,000 children. Diabetes 60:1805–1812. doi:10.2337/db10-1575

Barr ELM, Zimmet PZ, Welborn TA et al (2007) Risk of cardiovascular and all-cause mortality in individuals with diabetes mellitus, impaired fasting glucose, and impaired glucose tolerance: the Australian Diabetes, Obesity, and Lifestyle Study (AusDiab). Circulation 116:151–157. doi:10.1161/CIRCULATIONAHA.106.685628

Bass J, Takahashi JS (2010) Circadian integration of metabolism and energetics. Science 330:1349–1354. doi:10.1126/science.1195027

Below JE, Gamazon ER, Morrison JV et al (2011) Genome-wide association and meta-analysis in populations from Starr County, Texas, and Mexico City identify type 2 diabetes susceptibility loci and enrichment for expression quantitative trait loci in top signals. Diabetologia 54:2047–2055. doi:10.1007/s00125-011-2188-3

Bonnefond A, Clément N, Fawcett K et al (2012) Rare MTNR1B variants impairing melatonin receptor 1B function contribute to type 2 diabetes. Nat Genet 44:297–301. doi:10.1038/ng.1053

Bouatia-Naji N, Rocheleau G, Van Lommel L et al (2008) A polymorphism within the G6PC2 gene is associated with fasting plasma glucose levels. Science 320:1085–1088. doi:10.1126/science.1156849

Bouatia-Naji N, Bonnefond A, Cavalcanti-Proença C et al (2009) A variant near MTNR1B is associated with increased fasting plasma glucose levels and type 2 diabetes risk. Nat Genet 41:89–94. doi:10.1038/ng.277

Boutin JA, Audinot V, Ferry G, Delagrange P (2005) Molecular tools to study melatonin pathways and actions. Trends Pharmacol Sci 26:412–419. doi:10.1016/j.tips.2005.06.006

Brydon L, Petit L, Delagrange P et al (2001) Functional expression of MT2 (Mel1b) melatonin receptors in human PAZ6 adipocytes. Endocrinology 142:4264–4271. doi:10.1210/endo.142.10.8423

Buxton OM, Cain SW, O'Connor SP et al (2012) Adverse metabolic consequences in humans of prolonged sleep restriction combined with circadian disruption. Sci Transl Med 4:129ra43. doi:10.1126/scitranslmed.3003200

Chambers JC, Zhang W, Zabaneh D et al (2009) Common genetic variation near melatonin receptor MTNR1B contributes to raised plasma glucose and increased risk of type 2 diabetes among Indian Asians and European Caucasians. Diabetes 58:2703–2708. doi:10.2337/db08-1805

Chen W-M, Erdos MR, Jackson AU et al (2008) Variations in the G6PC2/ABCB11 genomic region are associated with fasting glucose levels. J Clin Invest 118:2620–2628. doi:10.1172/JCI34566

Chen C-Q, Fichna J, Bashashati M et al (2011) Distribution, function and physiological role of melatonin in the lower gut. World J Gastroenterol 17:3888–3898. doi:10.3748/wjg.v17.i34.3888

Cho H, Zhao X, Hatori M et al (2012) Regulation of circadian behaviour and metabolism by REV-ERB-α and REV-ERB-β. Nature 485:123–127. doi:10.1038/nature11048

Dubocovich ML, Markowska M (2005) Functional MT1 and MT2 melatonin receptors in mammals. Endocrine 27:101–110. doi:10.1385/ENDO:27:2:101

Dupuis J, Langenberg C, Prokopenko I et al (2010) New genetic loci implicated in fasting glucose homeostasis and their impact on type 2 diabetes risk. Nat Genet 42:105–116. doi:10.1038/ng.520

Evans JA, Davidson AJ (2013) Health consequences of circadian disruption in humans and animal models. Prog Mol Biol Transl Sci 119:283–323. doi:10.1016/B978-0-12-396971-2.00010-5

Florez JC, Jablonski KA, McAteer JB et al (2012) Effects of genetic variants previously associated with fasting glucose and insulin in the Diabetes Prevention Program. PLoS ONE 7, e44424. doi:10.1371/journal.pone.0044424

Garaulet M, Corbalán MD, Madrid JA et al (2010) CLOCK gene is implicated in weight reduction in obese patients participating in a dietary programme based on the Mediterranean diet. Int J Obes 34:516–523. doi:10.1038/ijo.2009.255

Garaulet M, Smith CE, Gomez-Abellán P et al (2013) REV-ERB-ALPHA circadian gene variant associates with obesity in two independent populations: Mediterranean and North American. Mol Nutr Food Res. doi:10.1002/mnfr.201300361

Goumidi L, Grechez A, Dumont J et al (2013) Impact of REV-ERB alpha gene polymorphisms on obesity phenotypes in adult and adolescent samples. Int J Obes 37:666–672. doi:10.1038/ijo.2012.117

Ha E, Yim S-V, Chung J-H et al (2006) Melatonin stimulates glucose transport via insulin receptor substrate-1/phosphatidylinositol 3-kinase pathway in C2C12 murine skeletal muscle cells. J Pineal Res 41:67–72. doi:10.1111/j.1600-079X.2006.00334.x

Hardeland R (2005) Antioxidative protection by melatonin: multiplicity of mechanisms from radical detoxification to radical avoidance. Endocrine 27:119–130

Hoffman RA, Reiter RJ (1965) Pineal gland: influence on gonads of male hamsters. Science 148:1609–1611
Huopio H, Cederberg H, Vangipurapu J et al (2013) Association of risk variants for type 2 diabetes and hyperglycemia with gestational diabetes. Eur J Endocrinol 169:291–297. doi:10.1530/EJE-13-0286
Illig T, Gieger C, Zhai G et al (2010) A genome-wide perspective of genetic variation in human metabolism. Nat Genet 42:137–141. doi:10.1038/ng.507
Illnerová H, Sumová A (1997) Photic entrainment of the mammalian rhythm in melatonin production. J Biol Rhythms 12:547–555
Jassal B, Jupe S, Caudy M et al (2010) The systematic annotation of the three main GPCR families in Reactome. Database (Oxford) 2010:baq018. doi:10.1093/database/baq018
Jockers R, Maurice P, Boutin JA, Delagrange P (2008) Melatonin receptors, heterodimerization, signal transduction and binding sites: what's new? Br J Pharmacol 154:1182–1195. doi:10.1038/bjp.2008.184
Jonsson A, Ladenvall C, Ahluwalia TS et al (2013) Effects of common genetic variants associated with type 2 diabetes and glycemic traits on α- and β-cell function and insulin action in humans. Diabetes 62:2978–2983. doi:10.2337/db12-1627
Kan MY, Zhou DZ, Zhang D et al (2010) Two susceptible diabetogenic variants near/in MTNR1B are associated with fasting plasma glucose in a Han Chinese cohort. Diabet Med 27:598–602. doi:10.1111/j.1464-5491.2010.02975.x
Karamitri A, Renault N, Clement N et al (2013) Minireview: toward the establishment of a link between melatonin and glucose homeostasis: association of melatonin MT2 receptor variants with type 2 diabetes. Mol Endocrinol 27:1217–1233. doi:10.1210/me.2013-1101
Kelly MA, Rees SD, Hydrie MZI et al (2012) Circadian gene variants and susceptibility to type 2 diabetes: a pilot study. PLoS ONE 7, e32670. doi:10.1371/journal.pone.0032670
Kemp DM, Ubeda M, Habener JF (2002) Identification and functional characterization of melatonin Mel 1a receptors in pancreatic beta cells: potential role in incretin-mediated cell function by sensitization of cAMP signaling. Mol Cell Endocrinol 191:157–166
Kennaway DJ, Voultsios A, Varcoe TJ, Moyer RW (2002) Melatonin in mice: rhythms, response to light, adrenergic stimulation, and metabolism. Am J Physiol Regul Integr Comp Physiol 282:R358–R365. doi:10.1152/ajpregu.00360.2001
Kim YJ, Go MJ, Hu C et al (2011) Large-scale genome-wide association studies in East Asians identify new genetic loci influencing metabolic traits. Nat Genet 43:990–995. doi:10.1038/ng.939
Kivimäki M, Batty GD, Hublin C (2011) Shift work as a risk factor for future type 2 diabetes: evidence, mechanisms, implications, and future research directions. PLoS Med 8, e1001138. doi:10.1371/journal.pmed.1001138
Klerman EB, Shanahan TL, Brotman DJ et al (2002) Photic resetting of the human circadian pacemaker in the absence of conscious vision. J Biol Rhythms 17:548–555
Korkmaz A, Rosales-Corral S, Reiter RJ (2012) Gene regulation by melatonin linked to epigenetic phenomena. Gene 503:1–11. doi:10.1016/j.gene.2012.04.040
Kristiansen K (2004) Molecular mechanisms of ligand binding, signaling, and regulation within the superfamily of G-protein-coupled receptors: molecular modeling and mutagenesis approaches to receptor structure and function. Pharmacol Ther 103:21–80. doi:10.1016/j.pharmthera.2004.05.002
Kvetnoy I, Sandvik AK, Waldum HL (1997) The diffuse neuroendocrine system and extrapineal melatonin. J Mol Endocrinol 18:1–3
Kwak SH, Kim S-H, Cho YM et al (2012) A genome-wide association study of gestational diabetes mellitus in Korean women. Diabetes 61:531–541. doi:10.2337/db11-1034
Lamia KA, Evans RM (2010) Metabolism: tick, tock, a beta-cell clock. Nature 466:571–572. doi:10.1038/466571a

Leproult R, Holmbäck U, Van Cauter E (2014) Circadian misalignment augments markers of insulin resistance and inflammation, independently of sleep loss. Diabetes. doi:10.2337/db13-1546

Liao S, Liu Y, Tan Y et al (2012) Association of genetic variants of melatonin receptor 1B with gestational plasma glucose level and risk of glucose intolerance in pregnant Chinese women. PLoS ONE 7, e40113. doi:10.1371/journal.pone.0040113

Ling Y, Li X, Gu Q et al (2011) A common polymorphism rs3781637 in MTNR1B is associated with type 2 diabetes and lipids levels in Han Chinese individuals. Cardiovasc Diabetol 10:27. doi:10.1186/1475-2840-10-27

Liu C, Wu Y, Li H et al (2010) MTNR1B rs10830963 is associated with fasting plasma glucose, HbA1C and impaired beta-cell function in Chinese Hans from Shanghai. BMC Med Genet 11:59. doi:10.1186/1471-2350-11-59

Lyssenko V, Nagorny CLF, Erdos MR et al (2009) Common variant in MTNR1B associated with increased risk of type 2 diabetes and impaired early insulin secretion. Nat Genet 41:82–88. doi:10.1038/ng.288

Marcheva B, Ramsey KM, Buhr ED et al (2010) Disruption of the clock components CLOCK and BMAL1 leads to hypoinsulinaemia and diabetes. Nature 466:627–631. doi:10.1038/nature09253

Masana MI, Doolen S, Ersahin C et al (2002) MT(2) melatonin receptors are present and functional in rat caudal artery. J Pharmacol Exp Ther 302:1295–1302

McMullan CJ, Schernhammer ES, Rimm EB et al (2013) Melatonin secretion and the incidence of type 2 diabetes. JAMA 309:1388–1396. doi:10.1001/jama.2013.2710

Mirzaei K, Xu M, Qi Q et al (2014) Variants in glucose- and circadian rhythm-related genes affect the response of energy expenditure to weight-loss diets: the POUNDS LOST Trial. Am J Clin Nutr 99:392–399. doi:10.3945/ajcn.113.072066

Mohawk JA, Green CB, Takahashi JS (2012) Central and peripheral circadian clocks in mammals. Annu Rev Neurosci 35:445–462. doi:10.1146/annurev-neuro-060909-153128

Monk TH, Buysse DJ (2013) Exposure to shift work as a risk factor for diabetes. J Biol Rhythms 28:356–359. doi:10.1177/0748730413506557

Ohshige T, Iwata M, Omori S et al (2011) Association of new loci identified in European genome-wide association studies with susceptibility to type 2 diabetes in the Japanese. PLoS ONE 6, e26911. doi:10.1371/journal.pone.0026911

Okatani Y, Okamoto K, Hayashi K et al (1998) Maternal-fetal transfer of melatonin in pregnant women near term. J Pineal Res 25:129–134

Pan A, Schernhammer ES, Sun Q, Hu FB (2011) Rotating night shift work and risk of type 2 diabetes: two prospective cohort studies in women. PLoS Med 8, e1001141. doi:10.1371/journal.pmed.1001141

Panda S, Antoch MP, Miller BH et al (2002) Coordinated transcription of key pathways in the mouse by the circadian clock. Cell 109:307–320

Peschke E (2008) Melatonin, endocrine pancreas and diabetes. J Pineal Res 44:26–40. doi:10.1111/j.1600-079X.2007.00519.x

Peschke E, Mühlbauer E, Musshoff U et al (2002) Receptor (MT(1)) mediated influence of melatonin on cAMP concentration and insulin secretion of rat insulinoma cells INS-1. J Pineal Res 33:63–71

Peschke E, Bach AG, Mühlbauer E (2006a) Parallel signaling pathways of melatonin in the pancreatic beta-cell. J Pineal Res 40:184–191. doi:10.1111/j.1600-079X.2005.00297.x

Peschke E, Frese T, Chankiewitz E et al (2006b) Diabetic Goto Kakizaki rats as well as type 2 diabetic patients show a decreased diurnal serum melatonin level and an increased pancreatic melatonin-receptor status. J Pineal Res 40:135–143. doi:10.1111/j.1600-079X.2005.00287.x

Pilia G, Chen W-M, Scuteri A et al (2006) Heritability of cardiovascular and personality traits in 6,148 Sardinians. PLoS Genet 2, e132. doi:10.1371/journal.pgen.0020132

Poon AM, Choy EH, Pang SF (2001) Modulation of blood glucose by melatonin: a direct action on melatonin receptors in mouse hepatocytes. Biol Signals Recept 10:367–379

Prokopenko I, Langenberg C, Florez JC et al (2009) Variants in MTNR1B influence fasting glucose levels. Nat Genet 41:77–81. doi:10.1038/ng.290

Prokopenko I, Poon W, Mägi R et al (2014) A central role for GRB10 in regulation of islet function in man. PLoS Genet 10, e1004235. doi:10.1371/journal.pgen.1004235

Ramos E, Chen G, Shriner D et al (2011) Replication of genome-wide association studies (GWAS) loci for fasting plasma glucose in African-Americans. Diabetologia 54:783–788. doi:10.1007/s00125-010-2002-7

Ramracheya RD, Muller DS, Squires PE et al (2008) Function and expression of melatonin receptors on human pancreatic islets. J Pineal Res 44:273–279. doi:10.1111/j.1600-079X.2007.00523.x

Rasmussen-Torvik LJ, Guo X, Bowden DW et al (2012) Fasting glucose GWAS candidate region analysis across ethnic groups in the Multiethnic Study of Atherosclerosis (MESA). Genet Epidemiol 36:384–391. doi:10.1002/gepi.21632

Reiter RJ, Tan DX, Korkmaz A, Rosales-Corral SA (2014) Melatonin and stable circadian rhythms optimize maternal, placental and fetal physiology. Hum Reprod Update 20:293–307. doi:10.1093/humupd/dmt054

Renström F, Shungin D, Johansson I et al (2011) Genetic predisposition to long-term nondiabetic deteriorations in glucose homeostasis: ten-year follow-up of the GLACIER study. Diabetes 60:345–354. doi:10.2337/db10-0933

Reppert SM, Weaver DR, Ebisawa T (1994) Cloning and characterization of a mammalian melatonin receptor that mediates reproductive and circadian responses. Neuron 13:1177–1185

Reppert SM, Godson C, Mahle CD et al (1995) Molecular characterization of a second melatonin receptor expressed in human retina and brain: the Mel1b melatonin receptor. Proc Natl Acad Sci USA 92:8734–8738

Rönn T, Wen J, Yang Z et al (2009) A common variant in MTNR1B, encoding melatonin receptor 1B, is associated with type 2 diabetes and fasting plasma glucose in Han Chinese individuals. Diabetologia 52:830–833. doi:10.1007/s00125-009-1297-8

Sadacca LA, Lamia KA, deLemos AS et al (2011) An intrinsic circadian clock of the pancreas is required for normal insulin release and glucose homeostasis in mice. Diabetologia 54:120–124. doi:10.1007/s00125-010-1920-8

Scheer FAJL, Hilton MF, Mantzoros CS, Shea SA (2009) Adverse metabolic and cardiovascular consequences of circadian misalignment. Proc Natl Acad Sci USA 106:4453–4458. doi:10.1073/pnas.0808180106

Scott RA, Lagou V, Welch RP et al (2012) Large-scale association analyses identify new loci influencing glycemic traits and provide insight into the underlying biological pathways. Nat Genet 44:991–1005. doi:10.1038/ng.2385

Shi S, Ansari TS, McGuinness OP et al (2013) Circadian disruption leads to insulin resistance and obesity. Curr Biol 23:372–381. doi:10.1016/j.cub.2013.01.048

Simonis-Bik AM, Nijpels G, van Haeften TW et al (2010) Gene variants in the novel type 2 diabetes loci CDC123/CAMK1D, THADA, ADAMTS9, BCL11A, and MTNR1B affect different aspects of pancreatic beta-cell function. Diabetes 59:293–301. doi:10.2337/db09-1048

Sparsø T, Bonnefond A, Andersson E et al (2009) G-allele of intronic rs10830963 in MTNR1B confers increased risk of impaired fasting glycemia and type 2 diabetes through an impaired glucose-stimulated insulin release: studies involving 19,605 Europeans. Diabetes 58:1450–1456. doi:10.2337/db08-1660

Stamenkovic JA, Olsson AH, Nagorny CL et al (2012) Regulation of core clock genes in human islets. Metabolism 61:978–985. doi:10.1016/j.metabol.2011.11.013

Stumpf I, Mühlbauer E, Peschke E (2008) Involvement of the cGMP pathway in mediating the insulin-inhibitory effect of melatonin in pancreatic beta-cells. J Pineal Res 45:318–327. doi:10.1111/j.1600-079X.2008.00593.x

Stumpf I, Bazwinsky I, Peschke E (2009) Modulation of the cGMP signaling pathway by melatonin in pancreatic beta-cells. J Pineal Res 46:140–147. doi:10.1111/j.1600-079X.2008.00638.x
Takahashi JS, Hong H-K, Ko CH, McDearmon EL (2008) The genetics of mammalian circadian order and disorder: implications for physiology and disease. Nat Rev Genet 9:764–775. doi:10.1038/nrg2430
Takeuchi F, Katsuya T, Chakrewarthy S et al (2010) Common variants at the GCK, GCKR, G6PC2-ABCB11 and MTNR1B loci are associated with fasting glucose in two Asian populations. Diabetologia 53:299–308. doi:10.1007/s00125-009-1595-1
Taneera J, Lang S, Sharma A et al (2012) A systems genetics approach identifies genes and pathways for type 2 diabetes in human islets. Cell Metab 16:122–134. doi:10.1016/j.cmet.2012.06.006
Tirosh A, Shai I, Tekes-Manova D et al (2005) Normal fasting plasma glucose levels and type 2 diabetes in young men. N Engl J Med 353:1454–1462. doi:10.1056/NEJMoa050080
Turek FW, Joshu C, Kohsaka A et al (2005) Obesity and metabolic syndrome in circadian Clock mutant mice. Science 308:1043–1045. doi:10.1126/science.1108750
Vangipurapu J, Stančáková A, Pihlajamäki J et al (2011) Association of indices of liver and adipocyte insulin resistance with 19 confirmed susceptibility loci for type 2 diabetes in 6,733 non-diabetic Finnish men. Diabetologia 54:563–571. doi:10.1007/s00125-010-1977-4
Venkatakrishnan AJ, Deupi X, Lebon G et al (2013) Molecular signatures of G-protein-coupled receptors. Nature 494:185–194. doi:10.1038/nature11896
Vlassi M, Gazouli M, Paltoglou G et al (2012) The rs10830963 variant of melatonin receptor MTNR1B is associated with increased risk for gestational diabetes mellitus in a Greek population. Hormones (Athens) 11:70–76
Voight BF, Scott LJ, Steinthorsdottir V et al (2010) Twelve type 2 diabetes susceptibility loci identified through large-scale association analysis. Nat Genet 42:579–589. doi:10.1038/ng.609
Vyas MV, Garg AX, Iansavichus AV et al (2012) Shift work and vascular events: systematic review and meta-analysis. BMJ 345, e4800
Walford GA, Green T, Neale B et al (2012) Common genetic variants differentially influence the transition from clinically defined states of fasting glucose metabolism. Diabetologia 55:331–339. doi:10.1007/s00125-011-2353-8
Wang Y, Nie M, Li W et al (2011) Association of six single nucleotide polymorphisms with gestational diabetes mellitus in a Chinese population. PLoS ONE 6, e26953. doi:10.1371/journal.pone.0026953
Wang H, Liu L, Zhao J et al (2013) Large scale meta-analyses of fasting plasma glucose raising variants in GCK, GCKR, MTNR1B and G6PC2 and their impacts on type 2 diabetes mellitus risk. PLoS ONE 8, e67665. doi:10.1371/journal.pone.0067665
Watanabe RM, Valle T, Hauser ER et al (1999) Familiality of quantitative metabolic traits in Finnish families with non-insulin-dependent diabetes mellitus. Finland-United States Investigation of NIDDM Genetics (FUSION) Study investigators. Hum Hered 49:159–168, doi: 22865
Weitzman ED, Fukushima D, Nogeire C et al (1971) Twenty-four hour pattern of the episodic secretion of cortisol in normal subjects. J Clin Endocrinol Metab 33:14–22. doi:10.1210/jcem-33-1-14
Woon PY, Kaisaki PJ, Bragança J et al (2007) Aryl hydrocarbon receptor nuclear translocator-like (BMAL1) is associated with susceptibility to hypertension and type 2 diabetes. Proc Natl Acad Sci USA 104:14412–14417. doi:10.1073/pnas.0703247104
World Health Organization (2013) http://www.who.int/mediacentre/factsheets/fs312/. Fact sheet N°312
Xia Q, Chen Z-X, Wang Y-C et al (2012) Association between the melatonin receptor 1B gene polymorphism on the risk of type 2 diabetes, impaired glucose regulation: a meta-analysis. PLoS ONE 7, e50107. doi:10.1371/journal.pone.0050107

Zhao Q, Xiao J, He J et al (2014) Cross-sectional and longitudinal replication analyses of genome-wide association Loci of type 2 diabetes in han chinese. PLoS ONE 9, e91790. doi:10.1371/journal.pone.0091790

Zhdanova IV, Wurtman RJ (1997) Efficacy of melatonin as a sleep-promoting agent. J Biol Rhythms 12:644–650

Part IV
Clinical Translation

Chapter 20
Type 2 Diabetes Prediction

Shuai Wang, Frank B. Hu, and Josée Dupuis

Abstract In this chapter, we review prediction models for type 2 diabetes, starting with models without genetic information and moving on to models with increasing complexity in terms of incorporating genetic information. We discuss the strength of current prediction models and indicate how we might be able to improve diabetes prediction in the future. For a complementary discussion of prediction models in the context of obesity, please see Chap. 23.

20.1 Introduction

One of the promises of the investigation of the genetic causes of type 2 diabetes is the ability to translate the genetic knowledge into better type 2 diabetes treatments. One could envision personalized medicine, where therapy would be tailored based on an individual genetic makeup. However, another aspect that may have a greater impact would be to use the genetics knowledge in the prevention of type 2 diabetes. Such an approach would involve determining individuals who are most at risk of developing type 2 diabetes based on their risk factors, including genetic profile, and developing interventions to prevent or delay the development of the disease. The advantage of genetic information for prediction over other risk factors is that some genetic information, such as alleles at a genetic risk locus, can be measured at birth and does not change throughout a person's lifetime. Therefore, an accurate risk prediction algorithm based solely on genetic factors could determine type 2 diabetes risk at birth and a prevention program could be implemented early in life, when it is most likely to have a lifelong impact.

S. Wang • J. Dupuis (✉)
Department of Biostatistics, Boston University School of Public Health, Crosstown Center, 3rd floor, 801 Massachusetts Avenue, Boston, MA 02118, USA
e-mail: dupuis@bu.edu

F.B. Hu
Departments of Nutrition and Epidemiology, Harvard T.H. Chan School of Public Health, Boston, MA, USA

J.C. Florez (ed.), *The Genetics of Type 2 Diabetes and Related Traits*,
DOI 10.1007/978-3-319-01574-3_20

Before the advent of the Human Genome Project and the ensuing technology development allowing the measurements of thousands of genetic variants in large samples in a cost-efficient manner, epidemiologists and clinicians have attempted to develop models for predicting patients most at risk to develop type 2 diabetes (Morsiani et al. 1985). The best predictive factors increasing risk of type 2 diabetes include advanced age, high body mass index (BMI), smoking, physical inactivity, diet in rich *trans*-fat, and glycemic load but lower in polyunsaturated fats and fiber, while moderate alcohol consumption has been found to be protective (Hu et al. 2001). Even before one could measure genotypes efficiently, family history was found to be a strong predictor of type 2 diabetes incidence (Pierce et al. 1995), with approximately a twofold increase in risk of developing diabetes in individuals with diabetic parents (Pierce et al. 1995; Wilson et al. 2007). In the next section, we first review the statistical approaches used to predict type 2 diabetes and to evaluate prediction models. We then present earlier prediction models and compare them with more recent models that incorporate genetic information. We conclude with a few thoughts on improving type 2 diabetes prediction.

20.2 Prediction Models

Most prediction models have been developed in prospective cohort studies, where characteristics at baseline are evaluated for association and predictive ability of type 2 diabetes incidence some years later, with most models in published reports to date looking at predictive ability 5–10 years in the future (Noble et al. 2011). Logistic regression is often used to assess predictive ability, although survival models such as Cox proportional hazard regression have been used either to supplement the logistic models (Wilson et al. 2007) or as the primary analysis (Kahn et al. 2009).

In a logistic model, if we assume that the affection status y_i is coded as either 0 (unaffected) or 1 (affected) and that x_{ik} is the kth of K risk factors, such as sex, age, hypertension or genotypes, measured on ith individual, the probability that individual i will be affected with type 2 diabetes is modeled as

$$p_i = \Pr(y_i = 1) = \frac{e^{\beta_0 + \beta_1 X_{i1} + \ldots + \beta_K X_{iK}}}{1 + e^{\beta_0 + \beta_1 X_{i1} + \ldots + \beta_K X_{iK}}}.$$

Evaluation of the importance of each risk factor in predicting type 2 diabetes can be performed by formally testing the hypothesis that $\beta_k = 0$. Once risk factors have been established, one can use the above equation to obtain a "risk score" for each individual by taking a person's covariate values (x_{ki}'s) and the predicted effect sizes $\hat{\beta}_k$. A risk score can be computed for each person $r_i = \hat{\beta}_0 + \sum_{k=1}^{K} \hat{\beta}_k x_{ik}$, and a prediction probabilities obtained as

$$\hat{p}_i = \frac{e^{r_i}}{1 + e^{r_i}}.$$

Note that a model cannot be evaluated in the sample used to estimate the β- coefficients because it would lead to an overly optimistic evaluation of performance. Ideally, an external cohort is used to assess the prediction equation (external validation). If an independent cohort is not available, another approach is to derive the prediction score in a subset of the cohort, for example, 75 % of the sample, and evaluate in the remaining individuals who are not part of the sample used to estimate the coefficients (internal validation) (Kahn et al. 2009).

20.3 Prediction Model Evaluation

Prediction models are often evaluated based on three characteristics. The first characteristic, *calibration*, is the ability of a model to accurately estimate someone's risk. For example, in a subset of individuals assigned 80 % risk of developing diabetes, the observed proportion of individuals who would eventually develop diabetes should be close to 80 %. The second characteristic, *discrimination*, is the ability of a prediction score to accurately distinguish high-risk individuals who are likely to develop type 2 diabetes from low-risk patients who are less likely to eventually develop the disease. The third characteristic, *generalization*, is the ability of a prediction model to have high prediction accuracy in a population other than the one in which the score was derived. This last property is especially useful for the score to have any clinical relevance.

There are many measures that have been used to assess predictive models, including the area under a receiver operating characteristic (ROC) curve, often referred to as AUC_{ROC} or c-statistic (Zweig and Campbell 1993), the net reclassification (Pencina et al. 2008), the Hosmer-Lemeshow test, and the pseudo-R^2(Windmeijer 1995; Veall and Zimmermann 1996).

AUC_{ROC}, the area under the ROC curve, measures the performance of a classifier and is a measure of discrimination (see Chap. 23). A ROC curve is constructed by varying the threshold used to discriminate between cases and controls, computing the sensitivity (proportion of cases accurately classified as cases) and specificity (proportion of controls accurately classified as controls) for each threshold used. For example, in Table 20.1, when all participants with a probability of developing diabetes greater than 15.4 % were assigned to be cases, we accurately classified 99.2 % of the cases but also misclassified 93.4 % ((100–6.6) %) of the controls. The higher the threshold, the fewer cases are correctly classified but the larger the proportion of controls who will be correctly assigned. A perfect classifier would have a threshold where sensitivity = specificity = 1. To measure the quality of a classifier, the sensitivity is plotted against 1-specificity as the threshold for assigning individuals a case status is varied, as in Fig. 20.1. The resulting plot is called a ROC curve, and the area under the ROC curve provides a measure of the

Table 20.1 Sensitivity and specificity for various probability thresholds

Probability threshold (%)	Sensitivity (%)	Specificity (%)
15.4	99.2	6.6
19.9	97.0	17.0
28.0	90.6	32.5
31.7	87.8	41.9
35.7	83.6	50.9
48.2	70.9	69.1
51.2	67.1	72.3
63.8	50.1	86.2
70.2	39.9	90.2
75.9	26.1	93.2
81.6	15.6	96.0
85.8	5.0	99.0

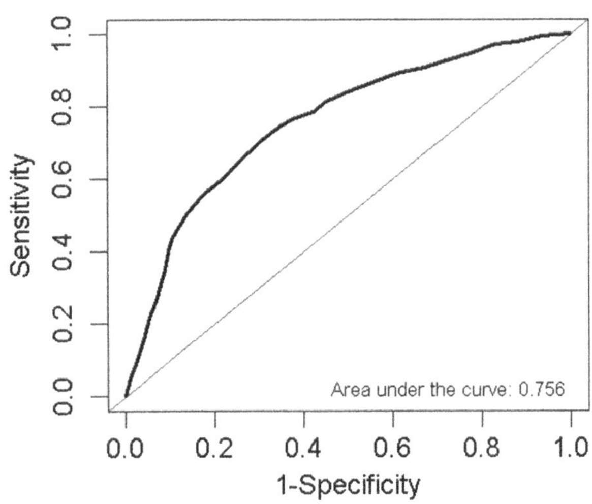

Fig. 20.1 Receiver operating characteristic (ROC) *curve*

accuracy in prediction of the model, with an area of 1 representing perfect discrimination and an area of 0.5 being no better than a random assignment. Mathematically, AUC_{ROC} measures the probability that a randomly sampled affected individual will have a higher risk score than a randomly sampled unaffected individual.

In practice, the AUC_{ROC} will fall somewhere between the perfect classification ($AUC_{ROC} = 1$) and the random classification ($AUC_{ROC} = 0.5$). In Fig. 20.1, $AUC_{ROC} = 0.756$ meaning that there is a 75.6 % chance that the predictive score of a randomly selected individual with type 2 diabetes is larger than the predictive score of a randomly selected unaffected individual.

In order to determine if a particular set of variables, such as genotypes, improves the predictive ability of a prediction model, one can compare the AUC_{ROC} of models with and without the added variables. The AUC_{ROC} is widely used to evaluate the incremental improvement resulting from addition of new biomarkers, such as genotypes, to a prediction model. A great advantage of the AUC_{ROC} is that it does not rely on arbitrary risk categorization because it is computed based on the continuous risk score, such as predicted probabilities from a logistic model. However, when there are strong predictors in the baseline model, the AUC_{ROC} is insensitive to predictive gains from adding a new biomarker, especially when some of the new biomarkers are correlated with the original predictors (Herder et al. 2013).

Another approach for comparing two prediction models is the net reclassification improvement (NRI) (Pencina et al. 2008) proposed by Pencina and colleagues in the context of evaluating performance improvement with newly added information (e.g., a new biomarker). Because the AUC_{ROC} is not sensitive to the addition of new predictors, especially when there are strong predictors already in the model, NRI considers the change in the risk assigned by the prediction models to affected and unaffected individuals separately. For an affected individual, an increase in the probability of being affected is regarded as an improvement, whereas for an unaffected subject, a decrease in risk probability is regarded as an improvement. NRI quantifies the change in risk for affected and unaffected individuals by comparing the increase in risk probabilities in cases and the decrease in risk probabilities in controls as follows:

$$\frac{\sum_{\{i \text{ in events}\}}[p_{\text{new}}(i) - p_{\text{old}}(i)]}{\#\text{ of events}} + \frac{\sum_{\{j \text{ in nonevents}\}}[p_{\text{old}}(j) - p_{\text{new}}(j)]}{\#\text{ of nonevents}}.$$

The null hypothesis that NRI = 0 is evaluated using a simple chi-squared test for paired data (McNemar's statistic) to determine if the added variables significantly improve the prediction model. Improvement can be tested in the affected and unaffected separately.

The NRI is more sensitive to detect added value from new biomarkers compared to the AUC_{ROC}. Another advantage of the NRI is that the improvement in predictive ability of the models with added variables can be assessed in affected and unaffected individuals separately. The general recommendation is to use both AUC_{ROC} and NRI in the evaluation of added predictive ability from a new biomarker (Pencina et al. 2008).

The Hosmer-Lemeshow test is a widely used measure on how close the fitted risk probabilities from a model match the observed risk (calibration) (Hosmer et al. 1988). It is a goodness-of-fit test for logistic regression.

For quantitative traits, the most commonly used measure of goodness-of-fit and predictive ability of a prediction model is the R^2 value, which measures how much of the variance in the trait can be explained by the variables in the model. A value of R^2 of 100 % represents perfect prediction, and a value of R^2 close to 0 would indicate that the variables included in the prediction model have no predictive value.

The concept has been extended to binary traits with the understanding that these proposed pseudo-R^2 measure the variance in the underlying disease risk that is explained by the prediction model. The most popular pseudo-R^2 measures include Efron's (Windmeijer 1995), McFadden's (Windmeijer 1995), McFadden's adjusted (Agrawal et al. 2010), Cox and Snell's (Lowensohn et al. 2007), and Nagelkerke's (Nagelkerke 1991), among others.

Efron's pseudo-R^2 shares the same form mathematically with the R^2 for continuous traits (Windmeijer 1995). For binary traits, Efron's measures are $R^2{}_{\text{Efron}} = 1 - \frac{\sum (y_i - \hat{p}_i)^2}{\sum (y_i - \overline{y})^2}$ with $\hat{p}_i$ standing for the fitted probability that the ith observation is a case, and $y_i = 0$ or 1.

Other pseudo-R^2 measures compare the likelihood of models with and without the covariates of interest. It should be noted that the upper limit of 1 is often not achievable for most pseudo-R^2 measures. However, Nagelkerke's pseudo-R^2 has been adapted from other R^2 measures (Lowensohn et al. 2007) so that the upper limit of 1 is achievable. In addition, Nagelkerke's pseudo-R^2 has a nice interpretation: $1 - R^2{}_{\text{Nagelkerke}}$ accounts for the proportion of variance unexplained by the variables included in the prediction model. Although Nagelkerke's pseudo-R^2 has been reported to assess predictive model for psychiatric disorders (International Schizophrenia Consortium et al. 2009; Ligthart et al. 2014), its use in type 2 diabetes prediction has been limited.

There is a direct relationship between the R^2 and expected AUC_{ROC} (Wray et al. 2010) that depends solely on the population disease prevalence. It is easy to see that for a model with $R^2 = 100$ %, all of the population variance in disease risk is explained by the predictors included in the model, and we would expect $AUC_{ROC} = 1$. On the other hand, if the predictors explain none of the variance in disease risk ($R^2 = 0$), then the prediction model has no predictive value ($AUC_{ROC} = 0.5$). Expected AUC_{ROC} values for a range of R^2 are presented in Table 20.2 for a disease prevalence of 8 %. The R^2 measure has a relatively simple interpretation as the proportion of population variance in disease risk explained by

Table 20.2 Expected AUC_{ROC} for a range of R^2 for a disease prevalence of 8 %.

R^2 (%)	Expected AUC_{ROC}
1	0.55
5	0.62
10	0.67
20	0.74
20	0.79
40	0.83
50	0.86
70	0.91
90	0.95

the predictors in the model. One can also compare R^2 between two models to quantify the added predictive values of new biomarkers.

20.4 Prediction Models Without Genotype Information

Long before the availability of low-cost high-throughput genotyping and sequencing technologies, there had been great interest in trying to predict individuals who were more likely to develop type 2 diabetes. As a proxy for genetic information, such models often included family history of diabetes in the form of either (1) variables for both maternal and paternal diabetes status, (2) number of parents with type 2 diabetes, or (3) a simple indicator variable to indicate if a person had one or more parents affected with type 2 diabetes. In addition to family history of diabetes, the simplest models include risk factors such as age; one or more measures of adiposity, such as BMI; waist circumference (WC) or waist/hip ratio; and some measure of hypertension, either in the form of systolic blood pressure, hypertension diagnosis, or use of antihypertensive drugs and may or may not include smoking history. Such basic models have been shown to have reasonable 5- to 10-year predictive ability, with AUC_{ROC} ranging from 0.71 to 0.75. For example, Kahn and colleagues (Kahn et al. 2009) reported a AUC_{ROC} of 0.71 in their validation sample that consisted of 25 % of the Atherosclerosis Risk in Communities Study (ARIC) for their risk scores that included only the following clinical variables: WC, diabetic mother, diabetic father, hypertension (140/90 mmHg), short stature, black race, age >55 years, weight, rapid pulse, and ever smoking. Wilson and colleagues (Wilson et al. 2007) obtained a similar value of the AUC_{ROC} (0.72) in the Framingham Heart Offspring Study with a basic model that only included age, sex, BMI, and parental history of diabetes. Similar values have also been obtained for other ethnic groups, including Thai samples ($AUC_{ROC} = 0.74$) (Aekplakorn et al. 2006), an Iranian cohort ($AUC_{ROC} = 0.75$) (Bozorgmanesh et al. 2013), and a Chinese population from Taiwan ($AUC_{ROC} = 0.71$) (Chuang et al. 2011). Addition of laboratory results such as fasting blood glucose and lipid levels (high-density lipoprotein (HDL) or triglycerides (TG)) increases the AUC_{ROC} to around 0.79 to 0.85, giving very good predictive ability, which is not surprising given that high blood glucose levels is part of the definition of type 2 diabetes (World Health Organization 2011). In the ARIC cohort, addition of fasting glucose, triglyceride, HDL, and uric acid levels improved the AUC_{ROC} from 0.71 to 0.79 (Kahn et al. 2009); in the Framingham Heart Offspring Study, a model with the basic risk factors and fasting glucose, triglyceride, and HDL levels yielded a prediction model with $AUC_{ROC} = 0.85$ (Wilson et al. 2007). Addition of additional covariates yielded similar improvement in the AUC_{ROC} in other ethnic groups ($AUC_{ROC} = 0.79$ for the Thai study (Aekplakorn et al. 2006), $AUC_{ROC} = 0.85$ for Iranians (Bozorgmanesh et al. 2013), and $AUC_{ROC} = 0.84$ for the Taiwanese study (Chuang et al. 2011)). In the Nurses' Health Study and Health Professionals' Follow-Up Study, a prediction model including age, sex, BMI, family history of diabetes, smoking, alcohol

Table 20.3 AUC_{ROC} from prior reports for various models and study populations of diverse origins

Study	Population	Basic model covariates	Basic model AUC_{ROC}	Additional covariates	AUC_{ROC} for basic + other covariates
Wilson et al. (2007)	US white	Age, sex, BMI, T2D family history	0.72	FG, HDL, TG, hypertension	0.85
Kahn et al. (2009)	US black and white	WC, T2D family history, hypertension, race, weight, rapid pulse, smoking, short stature	0.71	FG, TG, HDL, uric acid, alcohol	0.79
Aekplakorn et al. (2006)	Thailand	Age, sex, BMI, WC, hypertension, T2D family history	0.75	GTT, TG, HDL	0.79
Bozorgmanesh et al. (2013)	Iran	Age, SBP, WHR, waist-height ratio, T2D family history	0.75	FG, TG/HDL	0.85
Chuang et al. (2011)	Taiwan	Age, sex, education, alcohol, BMI, WC, BP, T2D family history	0.73	FG, TG	0.84

FG fasting glucose, *HDL* high-density lipoprotein, *TG* triglycerides, *GTT* 2 h glucose tolerance test, *BP* blood pressure, *SBP* systolic blood pressure, *WHR* waist-to-hip ratio, *T2D* type 2 diabetes

intake, a diet quality score, and physical activity yielded an AUC_{ROC} of 0.78, without adding biochemical markers (Cornelis et al. 2009). See Table 20.3 for a summary of published results in various populations, restricted to reports with type 2 diabetes family history included in the prediction model.

Many of the studies discussed above predate the discovery of tens of loci associated with type 2 diabetes via genome-wide association studies (GWAS; see Chap. 2). In the previously described studies, family history was included in the prediction models as a surrogate for genotypes at the predisposing loci. However, with better knowledge of the genetic architecture of the disease, many investigators have attempted to improve on earlier prediction models by including the genotypic information at risk loci, and we describe some of these attempts in the next section.

20.5 Adding Genotype Information in the Prediction of Type 2 Diabetes

When genotypes at known risk loci are available, one can incorporate such genotypes in the prediction of type 2 diabetes. The hope is that genotypes alone could predict risk and hence be used at birth to identify individuals at risk of developing type 2 diabetes. However, because non-genetic risk factors play a crucial role in the development of type 2 diabetes, the best prediction models are likely to include both genetic and non-genetic risk factors.

When a small number of risk variants are available, each variant can be added to the prediction model as a separate covariate. However, because of the complexity of a model that would include dozens of risk loci, it is more common to compute a risk score and to include the value of the risk score in the prediction model. A genetic risk score comprised of m risk variants is computed for each individual i as $s_i = \sum_{k=1}^{m} w_k g_{ik}$ where g_{ik} is the (expected) number of risk alleles carried by individual i for the kth genetic variant and w_k $(k = 1, \ldots, m)$ are the pre-specified weights. The risk score s_i can then be added as an additional covariate in the prediction equation. How to specify the weights and which variants to include in such a risk score are discussed below.

For the weight specification (w_k), the most common options include regression coefficients from prior reports (e.g., weights proportional to the effect size in the original GWAS (Meigs et al. 2008)) or equal weights ($w_k = 1$ for all k). If good estimates of effect sizes are available, and they can reasonably be assumed to be similar in the population under study, a weighted score will offer potentially better predictive power by appropriately giving more emphasis to loci with higher effect sizes than simply counting the number of risk alleles. However, if the estimates of effect sizes are expected to be very different in the population under study, an unweighted score is a reasonable alternative.

In some instances, for example, prediction based on sequencing data, weights may depend on the allele frequency. Because rare variants may have a larger effect than common variants that have survived natural selection, one may want to weigh rarer variants with potentially larger effect on the disease more heavily. For example, the Madsen-Browning weights (Madsen and Browning 2009) and the Wu weights (Wu et al. 2011) have been widely used in the association test of rare variants, and these weighting schemes assign weights that are roughly proportional to the inverse of the allele frequency, with rarer alleles being upweighted while more common alleles are downweighted. Whether this assumption applies to type 2 diabetes, where the deleterious effects of natural selection may take place once the individual is past his/her reproductive age, remains a matter of speculation.

Most risk scores have been constructed from known type 2 diabetes risk variants. However, it is often of interest to also evaluate variants known to influence other type 2 diabetes risk factors, such as fasting glucose or BMI, and determine their impact of type 2 diabetes prediction.

When evaluating risk scores, there are three main questions of interest: (1) What is the predictive ability of a risk score that includes solely genetic factors without any other risk factors? (2) what is the added value of a genetic risk score above and beyond the typical risk factors? and (3) what is the predictive ability of a genetic risk score above and beyond the information provided in type 2 diabetes family history?

So far, 81 GWAS variants have been found to be associated with type 2 diabetes susceptiblity (DIAbetes Genetics Replication And Meta-analysis (DIAGRAM) Consortium 2014; Hara et al. 2014; Steinthorsdottir et al. 2014; SIGMA Type 2 Diabetes Consortium et al. 2014), and together these variants account for about 10 % of the population variance in disease (Pal and McCarthy 2013) (see Chap. 2). Many risk scores based on a partial list of these GWAS variants have been created and evaluated. Typically, AUC_{ROC} serves as a metric to evaluate the performance of genetic risk models in predicting type 2 diabetes, although one study estimated the pseudo-R^2 of 63 loci to be around 5.7 %(Morris et al. 2012) . In a recent systematic review (Bao et al. 2013), genetic risk models performed poorly in type 2 diabetes prediction, with the median AUC_{ROC} being about 0.6, which was just a little higher than the AUC_{ROC} of a random prediction. The median AUC_{ROC} was the highest in case–control studies and the lowest in cross-sectional studies. The median AUC_{ROC} tended to be positively correlated with the number of cases in a study, and the median AUC_{ROC} tended to be higher in Asian studies, compared to studies of European-ancestry participants. Moreover, the increment in median AUC_{ROC} by including additional GWAS markers was small in magnitude. In the context of determining the added value of predictors, the AUC_{ROC} may not be a good measure of the added value of a genetic risk score (Cook 2007). Instead, the NRI may be a more appropriate measure of the added value of additional risk factors to a prediction model. In the Framingham Heart Study (de Miguel-Yanes et al. 2011), a genetic score composed of 40 genetic variants achieved a NRI of around 10 % in addition to clinical risk factors among participants younger than 50 years old, although the increment to the AUC_{ROC} from adding the risk score to the model was negligible. Moreover, in the Framingham Heart Study the NRI was much smaller in the full Offspring cohort (NRI = 1.8 %) or when restricting the analysis to participants older than 50 (NRI = 0.4 %). The NRI for other reports range from −2.19 % to 10.2 % (Echouffo-Tcheugui et al. 2013).

Overall, the increment evaluated by AUC_{ROC} in the predictive performance by adding GWAS loci to models with clinical risk factors and family history of type 2 diabetes is minimal, regardless of the study design, ethnicity of the participants, and the number of GWAS markers included in the model (Bao et al. 2013). It should be noted that some of the strong clinical risk factors are also influenced by genetic factors, such as BMI and fasting glucose, and the clinical risk factors are often studied as traits in GWAS. By including these traits along with family history in the risk prediction model, investigators are already capturing some of the genetic susceptibility, albeit in an inaccurate manner. For example, in the EPIC-InterAct study, the genetic score alone explained only 2 % of the family history-associated excess risk of type 2 diabetes (InterAct Consortium 2013). In the Nurses' Health Study and Health Professionals' Follow-Up Study, family history of diabetes and the genetic risk score independently and jointly predict risk of type 2 diabetes (Cornelis et al. 2009). The fact that GWAS markers account for only a small amount of the variance in disease susceptibility may explain why they add very minimal predictive value beyond conventional type 2 diabetes risk factors.

Theoretically, there is room for improvement in the predictive performance of models based on genetic factors alone. Janssens (Bao et al. 2013; Janssens et al. 2006) and colleagues suggested that in order to increase AUC_{ROC} of genetic profiling to 0.8 or higher, at least 400 genetic variants with minor allele frequency of 0.1 and odds ratio being 1.25 or greater are needed. Hence, it is not surprising that the AUC_{ROC} of a model including ~40 genetic variants is low. Moreover, because the current identified risk loci only explain about 10 % of variance in disease susceptibility (Pal and McCarthy 2013), it has been estimated that there are hundreds of risk loci yet undiscovered that could be added to predictive models to improve the ability to predict type 2 diabetes from genetic variants alone or to improve prediction based on clinical and genetic risk factors. For example, Stahl and colleagues estimated that there are approximately 3000 type 2 diabetes risk loci (Stahl et al. 2012). However, any newly identified common variants through GWAS are likely to have even smaller effect sizes than known type 2 diabetes risk loci given the statistical power already accrued in extant GWAS meta-analyses. In addition, the incremental value of adding rare genetic variants to the prediction model developed for the general population is unlikely to be substantial since rare variants by definition are present in only a small proportion of affected individuals (Mihaescu et al. 2013).

The maximum achievable AUC_{ROC} based on genetic risk factors alone can be computed from estimates of disease prevalence and the recurrence risk to offspring (Wray et al. 2010). If we assume a population prevalence 8 %, a recurrence risk to siblings of 1.75 (Wilson et al. 2007), the maximum achievable AUC_{ROC} is 0.86 based on genetic factors alone (Wray et al. 2010), a value that is slightly lower than the value of maximum $AUC_{ROC} = 0.94$ obtained by Wray and colleagues (Wray et al. 2010) using a higher sibling recurrence ratio. It should be noted that this is the maximum achievable AUC_{ROC} if all genetic risk factors were known and included in the model. A model including genetic variants explaining 10 % of the variance in disease susceptibility has a maximum AUC_{ROC} of 0.67 (see Table 20.2). Hence, it is not surprising that genetic risk scores evaluated to date that include a subset of the 81 known risk loci reach AUC_{ROC} below 0.67, the maximum attainable with the inclusion of all 81 known loci. The addition of gene-environment interaction effects (see Chap. 13) may also improve prediction models and may be worth exploring in the future (Bao et al. 2013).

20.6 Predictive Model with Genetic Information: Beyond Genotypes

One future research avenue consists of the assessment of the added predictive ability from other biomarkers, such as RNA (e.g., mRNA and miRNAs), proteins, and metabolites (Herder et al. 2013). Until recently, only a very small proportion of biomarkers have been investigated for their usefulness in predicting development of type 2 diabetes.

The quantification of proteins and peptides (Herder et al. 2013) in blood samples on a large scale is still technologically challenging, and this challenge has limited the exploration of these biomarkers as predictor of type 2 diabetes. Despite the comparative ease of measures of mRNA expression, their collection has been mainly restricted to large cross-sectional cohorts, limiting their evaluation as predictors of type 2 diabetes development. Some miRNAs have been associated with insulin resistance and beta-cell function (Fernandez-Valverde et al. 2011; Williams and Mitchell 2012), although their utility in predicting type 2 diabetes above and beyond clinical measurements has yet to be evaluated. Metabolites have also been explored as a way to identify individuals at risk of developing type 2 diabetes. A recent analysis from the Framingham Offspring Study suggested that incorporating metabolomics data to clinical and genetic factors moderately but significantly improved accuracy of type 2 diabetes prediction (Walford et al. 2014). However, addition of metabolites to the prediction models has only yielded modest improvement in prediction, as was the case for the addition of genotypes from known type 2 diabetes loci (Wang et al. 2011). Whether the addition of novel biomarkers such as metabolite levels can further improve diabetes prediction warrants more investigations.

One aspect that has not been fully exploited is the use of repeated biomarker measurements to improve prediction (Herder et al. 2013). Trajectories of fasting glucose and the hemoglobin A1c (HbA1c) taken over time have been shown to be different in individuals who go on to develop type 2 diabetes, but this information has not been incorporated in models for diabetes prediction (Heianza et al. 2012). Moreover, a model based on multiple measures taken over a long period may be harder to implement in the clinic and might delay intervention that could prevent the development of the disease.

Further research on the potential of novel biomarkers to improve our ability to predict type 2 diabetes is needed. Evaluation of existing and novel biomarkers in prospective cohort studies, as opposed to cross-sectional studies, with multiple measurements over time, would enable a better evaluation of the added value of these biomarkers to the prediction of type 2 diabetes. Studies from multiple ethnic groups, such as African Americans, Hispanics, and Asians, would be most helpful given that type 2 diabetes prevalence varies by ancestry and the most useful biomarkers may also vary across ethnicities. So far, with limited progress in exploring the added value of novel biomarkers in type 2 diabetes prediction, the conventional risk models still serve as a good and reliable benchmark (Bao et al. 2013; Vassy and Meigs 2012).

20.7 The Future of Type 2 Diabetes Prediction

The promise of using genotypes to predict type 2 diabetes development rests on the fact that such an approach would permit targeting individuals at a young age, before they develop some of other risk factors such as obesity and hypertension. A lifestyle intervention has been shown to be quite successful for type 2 diabetes (Knowler et al. 2002), and individuals most at risk are more likely to benefit from such intervention programs. While we can reasonably predict individuals at high risk of developing type 2 diabetes in their next 5 to 10 years of life using commonly measured risk factors such as familial history of type 2 diabetes, BMI, blood pressure, smoking, and fasting glucose, the addition of genotypes at known risk loci does not currently improve our short-term predictive ability substantially. This is not surprising given that identified loci to date only explain about 10 % of the variance in type 2 diabetes risk (Pal and McCarthy 2013). Moreover, inclusion of family history in the risk model captures some of the genetic risk factors.

Another consideration is that the majority of identified diabetes variants are in noncoding intronic or intergenic regions of the genome and thus these variants may not be true causal variants, but only proxies in linkage disequilibrium with the causal variants. Theoretically, the true causal variants may have larger effect sizes and the use of these variants may improve prediction of the cumulative genetic information. It is conceivable that a genotype risk score that would include all true disease loci would have the potential to improve current short-term prediction models but also to make long-term prediction model based on genotypes more accurate and useful. Expanding the time horizon of diabetes risk prediction beyond 10 years may improve the predictive power of genetic information because such information represents a lifetime risk, long before conventional diabetes risk factors such as obesity and unhealthy lifestyle are developed (Vassy and Meigs 2012). Lyssenko and colleagues observed that the predictive ability of 16 known risk loci improves as the duration of follow-up increases, with the highest AUC_{ROC} of 0.623 achieved for individuals with a mean follow-up time of 28 years (Lyssenko et al. 2008). In the Framingham Offspring Study, the genetic risk score significantly improved the NRI in participants younger than 50 years old at baseline, but not in older individuals (de Miguel-Yanes et al. 2011). Future prediction models may need to use the life course approach that considers the role of genetic prediction at different stages of life.

Current evidence indicates that we have not reached a time when we can use an individual's genotypes to accurately predict their type 2 diabetes risk beyond traditional risk factors. Therefore, we conclude that at present, genetic screening or testing cannot be recommended for type 2 diabetes risk prediction in clinical or public health settings. It is possible that we may be able to use genotypes in the near future to identify subgroups of the population who are most at risk for developing type 2 diabetes for targeted interventions, but more research in this area is warranted as well. Whether providing genetic risk counseling based on currently available variants can motivate individuals to improve their diet and lifestyle remains uncertain (see Chaps. 23 and 27). Limited evidence suggests that genetic risk counseling does not significantly improve adherence to diabetes prevention programs among overweight individuals at high risk for diabetes (Grant et al. 2013). However, this may change with more education and greater awareness about genetic risk of diabetes in the health professionals and the general public. Despite the impressive progress that has been made in the discovery of genetic variants of type 2 diabetes in the past decade, we still have a long way to go before genomics information can be translated into prediction and personalized prevention of type 2 diabetes in clinical practice and public health settings.

References

Aekplakorn W, Bunnag P, Woodward M et al (2006) A risk score for predicting incident diabetes in the thai population. Diabetes Care 29(8):1872–1877. doi:10.2337/dc05-2141

Agrawal Y, Niparko JK, Dobie RA (2010) Estimating the effect of occupational noise exposure on hearing thresholds: the importance of adjusting for confounding variables. Ear Hear 31 (2):234–237. doi:10.1097/AUD.0b013e3181c6b9fd

Bao W, Hu FB, Rong S et al (2013) Predicting risk of type 2 diabetes mellitus with genetic risk models on the basis of established genome-wide association markers: a systematic review. Am J Epidemiol 178(8):1197–1207. doi:10.1093/aje/kwt123

Bozorgmanesh M, Hadaegh F, Azizi F (2013) Transportability of the updated diabetes prediction model from atherosclerosis risk in communities study to a middle eastern adult population: community-based cohort study. Acta Diabetol 50(2):175–181. doi:10.1007/s00592-010-0241-1

Chuang SY, Yeh WT, Wu YL, Chang HY, Pan WH, Tsao CK (2011) Prediction equations and point system derived from large-scale health check-up data for estimating diabetic risk in the chinese population of taiwan. Diabetes Res Clin Pract 92(1):128–136. doi:10.1016/j.diabres. 2010.12.022

Cook NR (2007) Use and misuse of the receiver operating characteristic curve in risk prediction. Circulation 115(7):928–935. doi:10.1161/CIRCULATIONAHA.106.672402

Cornelis MC, Qi L, Zhang C et al (2009) Joint effects of common genetic variants on the risk for type 2 diabetes in U.S. men and women of european ancestry. Ann Intern Med 150(8):541–550

de Miguel-Yanes JM, Shrader P, Pencina MJ et al (2011) Genetic risk reclassification for type 2 diabetes by age below or above 50 years using 40 type 2 diabetes risk single nucleotide polymorphisms. Diabetes Care 34(1):121–125. doi:10.2337/dc10-1265

DIAbetes Genetics Replication And Meta-analysis (DIAGRAM) Consortium (2014) Genome-wide trans-ancestry meta-analysis provides insight into the genetic architecture of type 2 diabetes susceptibility. Nat Genet 46(3):234–244. doi:10.1038/ng.2897; 10.1038/ng.2897

Echouffo-Tcheugui JB, Dieffenbach SD, Kengne AP (2013) Added value of novel circulating and genetic biomarkers in type 2 diabetes prediction: a systematic review. Diabetes Res Clin Pract 101(3):255–269. doi:10.1016/j.diabres.2013.03.023

Fernandez-Valverde SL, Taft RJ, Mattick JS (2011) MicroRNAs in β-cell biology, insulin resistance, diabetes and its complications. Diabetes 60(7):1825–1831. doi:10.2337/db11-0171

Grant RW, O'Brien KE, Waxler JL et al (2013) Personalized genetic risk counseling to motivate diabetes prevention: a randomized trial. Diabetes Care 36(1):13–19. doi:10.2337/dc12-0884

Hara K, Fujita H, Johnson TA et al (2014) Genome-wide association study identifies three novel loci for type 2 diabetes. Hum Mol Genet 23(1):239–246. doi:10.1093/hmg/ddt399

Heianza Y, Arase Y, Fujihara K et al (2012) Longitudinal trajectories of HbA1c and fasting plasma glucose levels during the development of type 2 diabetes: the toranomon hospital health management center study 7 (TOPICS 7). Diabetes Care 35(5):1050–1052. doi:10.2337/dc11-1793

Herder C, Kowall B, Tabak AG, Rathmann W (2013) The potential of novel biomarkers to improve risk prediction of type 2 diabetes. Diabetologia. doi:10.1007/s00125-013-3061-3

Hosmer DW, Lemeshow S, Klar J (1988) Goodness-of-fit testing for the logistic regression model when the estimated probabilities are small. Biom J 30(8):911–924. doi:10.1002/bimj.4710300805

Hu FB, Manson JE, Stampfer MJ et al (2001) Diet, lifestyle, and the risk of type 2 diabetes mellitus in women. N Engl J Med 345(11):790–797. doi:10.1056/NEJMoa010492

InterAct Consortium (2013) The link between family history and risk of type 2 diabetes is not explained by anthropometric, lifestyle or genetic risk factors: the EPIC-InterAct study. Diabetologia 56(1):60–69. doi:10.1007/s00125-012-2715-x

International Schizophrenia Consortium, Purcell SM, Wray NR et al (2009) Common polygenic variation contributes to risk of schizophrenia and bipolar disorder. Nature 460(7256):748–752. doi:10.1038/nature08185

Janssens AC, Aulchenko YS, Elefante S, Borsboom GJ, Steyerberg EW, van Duijn CM (2006) Predictive testing for complex diseases using multiple genes: fact or fiction? Genet Med 8 (7):395–400, doi:10.109701.gim.0000229689.18263.f4

Kahn HS, Cheng YJ, Thompson TJ, Imperatore G, Gregg EW (2009) Two risk-scoring systems for predicting incident diabetes mellitus in U.S. adults age 45 to 64 years. Ann Intern Med 150 (11):741–751

Knowler WC, Barrett-Connor E, Fowler SE et al (2002) Reduction in the incidence of type 2 diabetes with lifestyle intervention or metformin. N Engl J Med 346(6):393–403. doi:10.1056/NEJMoa012512

Ligthart L, Hottenga JJ, Lewis CM et al (2014) Genetic risk score analysis indicates migraine with and without comorbid depression are genetically different disorders. Hum Genet 133 (2):173–186. doi:10.1007/s00439-013-1370-8

Lowensohn S, Johnson LE, Elder RJ, Davies SP (2007) Auditor specialization, perceived audit quality, and audit fees in the local government audit market. J Account Public Policy 26 (6):705–732. doi:10.1016/j.jaccpubpol.2007.10.004

Lyssenko V, Jonsson A, Almgren P et al (2008) Clinical risk factors, DNA variants, and the development of type 2 diabetes. N Engl J Med 359(21):2220–2232. doi:10.1056/NEJMoa0801869

Madsen BE, Browning SR (2009) A groupwise association test for rare mutations using a weighted sum statistic. PLoS Genet 5(2), e1000384. doi:10.1371/journal.pgen.1000384

Meigs JB, Shrader P, Sullivan LM et al (2008) Genotype score in addition to common risk factors for prediction of type 2 diabetes. N Engl J Med 359(21):2208–2219. doi:10.1056/NEJMoa0804742

Mihaescu R, Pencina MJ, Alonso A et al (2013) Incremental value of rare genetic variants for the prediction of multifactorial diseases. Genome Med 5(8):76. doi:10.1186/gm480

Morris AP, Voight BF, Teslovich TM et al (2012) Large-scale association analysis provides insights into the genetic architecture and pathophysiology of type 2 diabetes. Nat Genet 44 (9):981–990. doi:10.1038/ng.2383

Morsiani M, Beretta P, Pareschi PL, Manservigi D, Bottoni L (1985) Long-term results in preventive medicine for type II diabetes. Acta Diabetol Lat 22(3):191–202

Nagelkerke NJD (1991) A note on a general definition of the coefficient of determination. Biometrika 78(3):691–692. doi:10.1093/biomet/78.3.691

Noble D, Mathur R, Dent T, Meads C, Greenhalgh T (2011) Risk models and scores for type 2 diabetes: systematic review. BMJ 343:d7163. doi:10.1136/bmj.d7163

Pal A, McCarthy MI (2013) The genetics of type 2 diabetes and its clinical relevance. Clin Genet 83(4):297–306. doi:10.1111/cge.12055

Pencina MJ, D'Agostino RBS, D'Agostino RB Jr, Vasan RS (2008) Evaluating the added predictive ability of a new marker: from area under the ROC curve to reclassification and beyond. Stat Med 27(2):157–172. doi:10.1002/sim.2929, discussion 207–12

Pierce M, Keen H, Bradley C (1995) Risk of diabetes in offspring of parents with non-insulin-dependent diabetes. Diabet Med 12(1):6–13

SIGMA Type 2 Diabetes Consortium, Williams AL, Jacobs SB et al (2014) Sequence variants in SLC16A11 are a common risk factor for type 2 diabetes in mexico. Nature 506(7486):97–101. doi:10.1038/nature12828; 10.1038/nature12828

Stahl EA, Wegmann D, Trynka G et al (2012) Bayesian inference analyses of the polygenic architecture of rheumatoid arthritis. Nat Genet 44(5):483–489. doi:10.1038/ng.2232

Steinthorsdottir V, Thorleifsson G, Sulem P et al (2014) Identification of low-frequency and rare sequence variants associated with elevated or reduced risk of type 2 diabetes. Nat Genet 46 (3):294–298. doi:10.1038/ng.2882

Vassy JL, Meigs JB (2012) Is genetic testing useful to predict type 2 diabetes? Best Pract Res Clin Endocrinol Metab 26(2):189–201. doi:10.1016/j.beem.2011.09.002

Veall MR, Zimmermann KF (1996) Pseudo-R2 measures for some common limited dependent variable models. J Econ Surv 10(3):241–259. doi:10.1111/j.1467-6419.1996.tb00013.x

Walford GA, Porneala BC, Dauriz M, Vassy JL, Cheng S, Rhee EP, Wang TJ, Meigs JB, Gerszten RE, Florez JC (2014) Metabolite traits and genetic risk provide complementary information for the prediction of future type 2 diabetes. Diabetes Care 37(9):2508–2514

Wang TJ, Larson MG, Vasan RS et al (2011) Metabolite profiles and the risk of developing diabetes. Nat Med 17(4):448–453. doi:10.1038/nm.2307

WHO (2011) Use of glycated haemoglobin (HbA1c) in the diagnosis of diabetes mellitus. Abbreviated report of a WHO consultation;WHO/NMH/CHP/CPM/11.1

Williams MD, Mitchell GM (2012) MicroRNAs in insulin resistance and obesity. Exp Diabetes Res 2012:484696. doi:10.1155/2012/484696

Wilson PW, Meigs JB, Sullivan L, Fox CS, Nathan DM, D'Agostino RBS (2007) Prediction of incident diabetes mellitus in middle-aged adults: the framingham offspring study. Arch Intern Med 167(10):1068–1074. doi:10.1001/archinte.167.10.1068

Windmeijer FAG (1995) Goodness-of-fit measures in binary choice models. Econ Rev 14 (1):101–116

Wray NR, Yang J, Goddard ME, Visscher PM (2010) The genetic interpretation of area under the ROC curve in genomic profiling. PLoS Genet 6(2), e1000864. doi:10.1371/journal.pgen.1000864

Wu MC, Lee S, Cai T, Li Y, Boehnke M, Lin X (2011) Rare-variant association testing for sequencing data with the sequence kernel association test. Am J Hum Genet 89(1):82–93. doi:10.1016/j.ajhg.2011.05.029

Zweig MH, Campbell G (1993) Receiveroperating characteristic (ROC) plots: a fundamental evaluation tool in clinical medicine. Clin Chem 39(4):561–577

Chapter 21
The Value of Genetic Variation in the Prediction of Obesity

Ruth J.F. Loos and A. Cecile J.W. Janssens

Abstract Obesity is predominantly caused by an unhealthy lifestyle, but genetic factors also contribute to people's susceptibility to gain weight. Rare high-risk mutations have been identified that cause extreme and early-onset obesity in a fraction of the population, while numerous common low-risk loci have been identified through genome-wide association studies that contribute to obesity in the general population. As insights into the contribution of genetic variation to obesity increase, the interest in using genetic variants to predict who is at risk to gain weight has also increased. Before constructing risk models, however, one needs to have a clear view on what form of obesity needs to be predicted, why, in whom, and for what purpose. Obesity is multifactorial, in that it results from an interplay between genes and environmental risk factors, such that models solely based on genetic variants will be unlikely to reach high predictive ability, as we illustrate using the literature on currently identified BMI-associated loci. Furthermore, it seems that irrespective of the poor predictive ability, communicating genetic information does not seem effective in making people adopt a healthy lifestyle. While using genetic information in the prediction of obesity is a legitimate aim, we believe that the most valuable contribution of gene discovery studies lies in their contribution to elucidate new physiological pathways that underlie obesity susceptibility, which in turn could lead to the identification of therapeutic targets and make its way into mainstream health care.

R.J.F. Loos (✉)
The Genetics of Obesity and Related Metabolic Traits Program, The Charles Bronfman Institute for Personalized Medicine, The Mindich Child Health and Development Institute, Department of Preventive Medicine, The Icahn School of Medicine at Mount Sinai, One Gustave L. Levy Place, Box 1003, New York, NY 10029, USA
e-mail: ruth.loos@mssm.edu

A.C.J.W. Janssens
Department of Epidemiology, Rollins School of Public Health, Emory University, Atlanta, GA 30322, USA

J.C. Florez (ed.), *The Genetics of Type 2 Diabetes and Related Traits*,
DOI 10.1007/978-3-319-01574-3_21

21.1 Introduction

The prevalence of obesity in the USA has more than doubled since the 1970s in adults and children alike (Flegal et al. 1998; Ogden et al. 2014), and similar sharp increases have been reported worldwide (Finucane et al. 2011; Popkin et al. 2012; IASO 2014). While a westernized lifestyle that promotes excessive intake of energy-dense foods and discourages physical activity has been a major driver of the epidemic (Swinburn et al. 2011), heritability studies have provided strong evidence that genetic variation contributes to interindividual differences in obesity susceptibility as well (Maes et al. 1997; Elks et al. 2012).

Several types of genetic variation have been linked to obesity. These include high-risk rare mutations, chromosomal abnormalities, and structural variants that contribute to extreme and early-onset forms of obesity that affect only a fraction of the population (Chung 2012) and low-risk common variants that contribute to the multifactorial forms of obesity that is widely present in the general population (see Chap. 4) (Lu and Loos 2013).

As gene discovery efforts continue to identify more disease-associated loci and as the cost of high-throughput genotyping decreases, there is growing interest in applying genetic testing to the general population to identify who will be at risk of developing disease, allowing opportunities for personalized preventive strategies. Several companies have already seized this as a business opportunity and have developed genetic tests that are offered directly to consumers. The assumption is that knowledge of one's lifetime disease risk will motivate people to adopt risk-reducing behavior early.

Here, we review whether genetic variation can be used to effectively identify individuals at risk of obesity. We have organized our chapter along the key questions in prediction research: what needs to be predicted, in whom, how, and for what purpose. The purpose of testing determines what needs to be predicted and in whom, and this in turn determines the predictors or risk factors that are available to be considered. We address these key questions with a main focus on multifactorial obesity that affects a large proportion of the population. We close the chapter with reflections on the role of genetic variation in obesity prediction.

21.2 Why Predict Obesity?

Obesity is a primary target for prevention as it not only imposes a substantial economic burden on societies (Cawley and Meyerhoefer 2012) but also causes serious personal health problems (Must et al. 1999). Specifically, obesity increases risk of type 2 diabetes (Vazquez et al. 2007), cardiovascular disease (Yusuf et al. 2005), some cancers (Renehan et al. 2008), and depression (de Wit et al. 2010), among others, and increases risk of premature mortality (Adams et al. 2006). Even obesity in childhood has been shown to have longtime effects on cardiovascular and metabolic health in adult life (Baker et al. 2007; Park et al. 2012).

Early prediction of obesity allows identifying individuals at high risk who can be offered timely intervention to prevent excess weight gain, or at least contain it, through health recommendations or regular surveillance of diet and physical activity (Khan et al. 2009). These health-promoting behaviors are generally effective short term, but the main challenge is to sustain them over years, adopting them as a lifestyle. This is particularly difficult in western societies that are notoriously obesogenic. Awareness of one's risk of obesity is assumed to enhance motivation to sustain healthy behaviors.

21.3 What to Predict?

Obesity is defined as excess accumulation of body fat, which is assessed by the body mass index (BMI). In general, adults with a BMI of 30 kg/m^2 or higher are considered obese (WHO 2000), whereas a lower cutoff point is used in Asian adults (World Health Organisation 2000). In children and adolescents, a BMI cutoff at or above the 95th percentile of sex-specific BMI-for-age growth charts is used (Ogden and Flegal 2010; Cole and Lobstein 2012).

Despite a clear-cut definition, obesity is a heterogeneous condition. For example, the degree of obesity can vary widely, and subclasses I (BMI 30–34.99 kg/m^2), II (BMI 35–39.99 kg/m^2), and III (BMI $\geq$ 40 kg/m^2) distinguish the less from the more extreme obese cases (WHO 2000). The extremely obese individuals are believed to be more genetically susceptible than the less extreme, more common, obese individuals (Katzmarzyk et al. 1999). Also the age at which one becomes obese ranges with some individuals becoming obese from early on in childhood, whereas others gain weight throughout their life course. While it has been speculated that individuals with early-onset and extreme obesity might be enriched for common variants that also contribute to less extreme forms of obesity and thus share the same genetic susceptibility to some extent (Meyre et al. 2009), there is evidence that at least a fraction of early-onset and extreme obesity is due to rare single gene defects or chromosomal abnormalities (Chung 2012; Ramachandrappa and Farooqi 2011).

In addition, the biological (genetic) mechanisms and environmental causes that underlie common obesity susceptibility will vary from one person to another, as well as the obesity-related comorbidities. A respectable proportion of obese individuals even seems to be protected from cardio-metabolic abnormalities longer than others (Wildman et al. 2008).

Different forms of obesity may require different risk models. From a genetic perspective, the distinction between extreme (monogenic) and common (multifactorial) obesity is the most important one. Other distinctions that could be relevant from a prevention perspective include early- vs. late-onset obesity, metabolically healthy vs. metabolically unhealthy obesity, or obesity classes (I, II, III). These distinctions are not mutually exclusive; for example, while extreme and monogenic obesity almost always has an early onset and common multifactorial forms typically are seen later in life, due to unhealthier lifestyles, common multifactorial obesity is nowadays seen at younger ages as well.

Different forms of obesity may result from different risk factors, which should be accounted for in risk modeling; i.e., these risk factors become either variables in the risk models or reasons to make separate models. The latter approach is recommended when many of the other risk factors have different effects between the types. For example, as the genetic architecture and environmental risk factors for monogenic obesity differ from those of multifactorial obesity, separate risk models will be needed that include different genetic variants and environmental variables. If the effects of the risk factors are similar between subgroups (e.g., class I vs. class II obesity), but one type has higher risks than the others, the subgroup classifier (risk factor (e.g., obesity class)) is best added as a variable in the risk model.

21.4 In Whom Should We Predict Obesity?

Predicting obesity is particularly valuable for those who are not (yet) obese in whom prevention is most effective, or in overweight individuals in whom the early consequences are still reversible. For example, screening the population for mutations that cause monogenic obesity might successfully identify individuals at high risk, but for most of those cases, no effective preventive strategies or treatments exist. In contrast, multifactorial obesity, which is common in the populations and caused by an interplay of multiple genetic and environmental factors, might be harder to predict, but will identify high-risk individuals who could benefit from early behavioral intervention.

These considerations imply that children and adolescents, overweight nonobese adults, and/or individuals with a family history might benefit most from risk prediction and targeted prevention and may be more responsive to risk information about obesity. Predicting obesity in adults without a family history who have been of normal weight for many decades seems superfluous.

The target population for prediction needs to be defined before (genetic) risk models are being constructed as the predictive ability of the risk models might differ between populations. The risk factors of obesity may be the same across different subpopulations, but not necessarily. Also, even when risk factors are the same, their effects on obesity risk might differ. For example, a test may be predictive for childhood obesity, but does not predict obesity in adults, or vice versa. Or, one mutation may be fairly predictive of monogenic obesity, but it does not predict (multifactorial) obesity in the general population.

Thus, the predictive ability of a model needs to be investigated in populations that are representative for the population in which the implementation of the test is ultimately intended.

21.5 How Should We Predict Obesity?

How obesity should be predicted will depend on the answers to the previous questions, i.e., what to predict, in whom, and for what purpose. Unlike for type 2 diabetes (Chap. 22), few studies have examined the effectiveness of using genetic variation in risk prediction models for obesity. So far, all have used common genetic variants identified through genome-wide association studies (GWAS) in individuals of mainly European ancestry to predict multifactorial obesity ($BMI \geq 30\ kg/m^2$) (i.e., what?) in adults, which has a high prevalence in the general population of high-income countries (in whom?) (Chap. 4). An important feature of this type of common obesity is that variation in susceptibility is due to genetic (explained variance: 40–70 %) and environmental factors, such that risk prediction models will need to include predictors that capture both genes and environment to effectively distinguish those at risk from those who are not at risk.

The effectiveness of genetic prediction varies across diseases and depends on a number of factors (Table 21.1). The maximum predictive ability that can be obtained, assuming complete understanding of the genetic origins, is determined by the heritability of the disease and the complexity of its causes. When the disease is very heritable and has a simple cause, such as the case for monogenic forms of obesity, the predictive ability of genetic determinants can be very high. Whether that predictive ability is achieved in practice depends on whether the mutations are known and what their penetrance is. When the disease is less heritable and the cause is complex, such as for multifactorial forms of obesity where genetic and lifestyle factors interact, the predictive ability of a *genetic* test will unlikely be very high: when lifestyle factors play a role, these have to be considered to increase the predictive ability of the risk model. The predictive ability of genetic tests for multifactorial forms of disease is determined by the number of variants that are known, their frequency in the population, and their impact on disease risk (effect size). The incidence determines the absolute risks of getting the disease.

Here, we review the available literature that has examined the predictive value of the established obesity-susceptibility loci and their ability to discriminate between individuals who are at high risk of obesity in adult life versus those who are at low risk. We focus on BMI-associated loci only (Chap. 4), as currently no data is available on the predictive ability of any of the other obesity-susceptibility loci, such as those identified for waist-to-hip ratio, childhood and adult obesity, or body fat percentage.

To properly assess the predictive ability of genetic information, we first consider the predictive ability of traditional, nongenetic, risk factors of obesity, which can serve as a reference or a basis for genetic prediction.

Table 21.1 Factors that determine the effectiveness of genetic prediction

Maximum predictive ability
1. Heritability of the disease: The extent to which genetic predisposition contributes to the variation in disease risk. The higher the heritability, the better the predictive ability based on genetic variants can be
2. Complexity of its cause:
– Monogenic: One mutation solely or largely causes the disease. There may be thousands of different mutations for a disease, but each of them is sufficient to substantially increase the risk of disease
– Multifactorial: Multiple susceptibility variants interact with nongenetic risk factors to cause the disease; each single genetic variant only has a minor impact on disease risk
Predictive ability in practice
1. Monogenic:
– Penetrance of mutations: The risk of disease associated with carrying the mutation. A mutation that has a low penetrance will have a lower predictive ability than one with high penetrance
– Coverage: How many mutations are known to cause disease? If the disease can be cause by multiple different mutations, but when only a small number of the mutations are known, their predictive ability will be low
2. Multifactorial:
– Number of susceptibility variants: How many genetic variants are known to be associated with disease risk? If only few genetic variants have been identified, genetic risk models will be inaccurate
– Prevalence of the variants: frequency of the risk allele
– Effect size: strength of association between variant and disease risk
– Incidence of the disease: risk of disease in population, percentage of new cases during a period of time

21.5.1 Traditional Prediction of Obesity Risk

Several longitudinal studies have established that parental obesity and personal childhood obesity are the strongest *traditional* risk factors of obesity in adulthood (Whitaker et al. 1997; Lake et al. 1997; Burke et al. 2001; Magarey et al. 2003; Cheung et al. 2004; Toschke et al. 2007; Li et al. 2009). The advantage of these predictors is that the collection of data on family history and childhood obesity is typically easy, inexpensive, and noninvasive. Importantly, both of these predictors capture genetic as well as familial environmental susceptibility.

While effects reported across studies are typically consistent, Whitaker et al. (1997) are one of the few groups who assessed the influence of both family history and childhood obesity on obesity risk in (young) adulthood. They also reported the required statistics to allow us to calculate the discriminative accuracy of these traditional risk factors (Table 21.2), and therefore, we used the data in this study as a representative example to estimate discrimination based on family history and childhood obesity.

A total of 854 parent-offspring trios participated in this longitudinal study to assess the offspring's adult obesity risk, based on their own and their parents'

Table 21.2 Indicators of predictive ability, explained for the prediction of obesity

Discriminative accuracy
For binary predictors, such as gender or a genetic mutation that can be present or not:
Sensitivity
Sensitivity is the proportion of individuals who will become obese that are correctly classified by the test as being at higher risk
Specificity
Specificity is the proportion of nonobese individuals that are correctly classified by the test as being at lower risk
For continuous predictors, such as age or risk models:
Area under the receiver operating characteristic curve (AUC_{ROC}) (or C-statistic)
When a predictor can have many values, a threshold is needed to identify all individuals above the threshold as at higher risk of becoming obese and below the threshold as at lower risk. Because the choice of threshold is arbitrary, sensitivity and specificity are calculated for all possible cutoff thresholds. When these combinations of sensitivity and specificity are plotted in a so-called ROC plot, the area under the curve (AUC_{ROC}) is indicative for the average sensitivity across all values of the specificity. It is the probability that a random individual who will develop obesity has a higher obesity risk than a random individual who will not develop obesity. The AUC_{ROC} ranges from 0.50 (equal to tossing a coin) to 1.0 (perfect prediction)
Predictive value
For binary predictors:
Positive predictive value (PPV)
PPV is the proportion of individuals who will become obese among all individuals who are at higher risk. It is the risk of obesity in individuals who are at higher risk
Negative predictive value (NPV)
NPV is the proportion of individuals who will not become obese among all individuals who are at lower risk. It is 1-risk of obesity in individuals who are at lower risk
For continuous predictors, such as age or risk models:
A summary statistic of the predictive ability across all possible thresholds of a continuous predictor does not exist. The risks associated with specific values of the continuous predictor can be graphically presented in a risk distribution

obesity status during childhood. Height and weight of the offspring were measured five times throughout childhood and adolescence and once more during adult life (21–29 years). Height and weight of the parents were measured at the time their offspring's data was collected. Parental obesity during one's childhood was consistently associated with a greater risk of obesity as a young adult. Having had one obese parent throughout childhood and adolescence increased the risk of adult obesity by 2.2- to 3.2-fold compared to someone whose parents were not obese (Fig. 21.1). When both parents were obese during one's childhood and adolescence (with the exception of parental obesity at the age of 10–14 years), the risk of being obese during adulthood was increased by 5- to 15.3-fold (Fig. 21.1). As expected, the effect of one's own obesity status during childhood and adolescence on their obesity risk in adulthood increased steadily with the age at which they were obese. For example, having been obese during infancy (1–2 years) did not increase risk of adult obesity, compared to someone who was not obese as an infant. However,

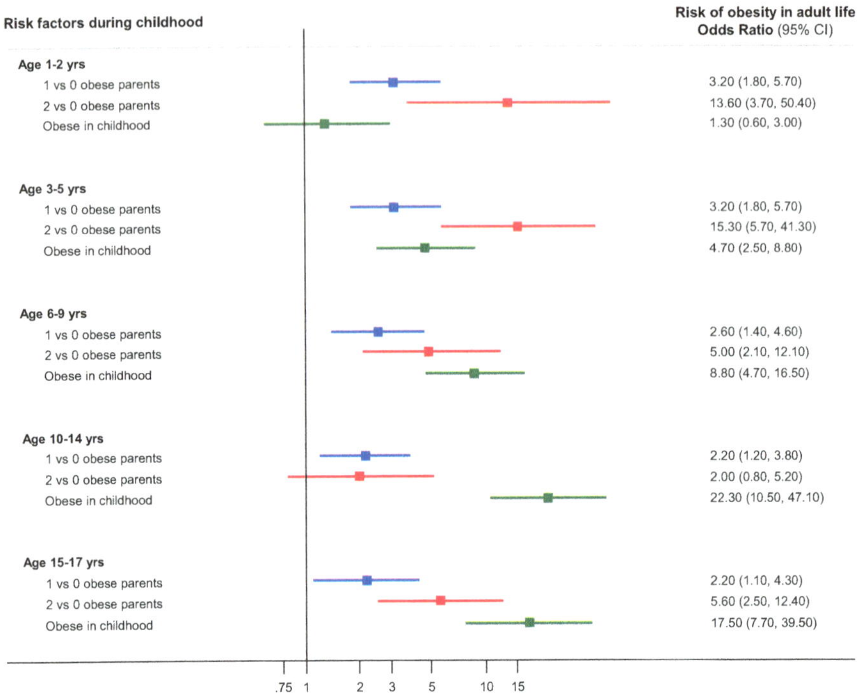

Fig. 21.1 Risk of obesity in adult life (21–29 years) given one parent (*blue*) or both parents (*red*) were obese during the individuals' childhood or adolescence, or given the individual was themselves obese during childhood or adolescence (*green*). For example, if both parents were obese when the individual was 3–5 years old, their risk of obesity was 15.30-fold increased compared to someone whose parents were not obese. Note that adult obesity was defined when BMI was $\geq$27.8 kg/m^{-2} [Data were derived from Whitaker et al. (1997). Figure adapted from Loos (2012)]

having been obese in adolescence (15–17 years) increased the risk of being an obese adult by 17.5-fold compared to someone who was not obese at that age (Fig. 21.1). While these *associations* support the fact that family history and one's own childhood obesity increase the risk of being obese as an adult, they do not tell us whether these risk factors can effectively *predict* who will be obese and who will not be obese.

While association results on the relationship between childhood obesity and/or family history and future adult obesity are widely available, data on their discriminative accuracy are sparse (Table 21.2). Discriminative accuracy is indicated by the sensitivity and specificity of a predictor. In this analysis, sensitivity is the percentage of individuals with adult obesity that had a parent with obesity at younger age, and specificity is the percentage of nonobese adults that had no parents with obesity at younger age. Discriminative accuracy should be distinguished from predictive value, which is the risk of adult obesity in children with or without parents who are obese. The positive predictive value (PPV) is the probability that children with at least one obese parent will become obese in adulthood, and the negative predictive value (NPV) is the probability that children whose parents are not obese will not be obese in adulthood. Predictive value is assessed prospectively, while discriminative accuracy is assessed retrospectively. Taken together they indicate the predictive ability of a risk factor.

We calculated sensitivity, specificity, PPV, and NPV using the Whitaker et al. (1997) data to illustrate the ability of parental obesity to predict their offspring's adult obesity. For example, parental obesity status has a sensitivity of 0.73 and specificity of 0.59 to predict adult obesity in adolescents (15–17 years) (Fig. 21.2, gray diamond). In other words, using parental obesity (i.e., having had at least one parent with obesity at age 15–17 years) to predict whether adolescents will become obese in adulthood (at age 21–29 years) will correctly identify 73 % of the obese adults and 59 % of the nonobese adults. Parental obesity status has a PPV of 0.24 and a NPV of 0.93 during adolescence. This means that 24 % of adolescents with at least one obese parent will become obese adults themselves, whereas 93 % of adolescents whose parents are not obese, will not become obese adults. The sensitivity, specificity, PPV, and NPV of parental obesity to predict one's adult obesity for the five age groups are shown in Fig. 21.2. The sensitivity and specificity of parental obesity to predict adult obesity risk vary with one's age at the time the obesity status of the parents was assessed. Overall, it seems that the sensitivity of parental obesity as a risk factor for one's own adult obesity increases with the age at which parental obesity was assessed, whereas the specificity decreases with age. The PPV and NPV of parental obesity seem more stable, with a generally low PPV (0.24–0.29) and a high NPV (0.90–0.93) (Fig. 21.2).

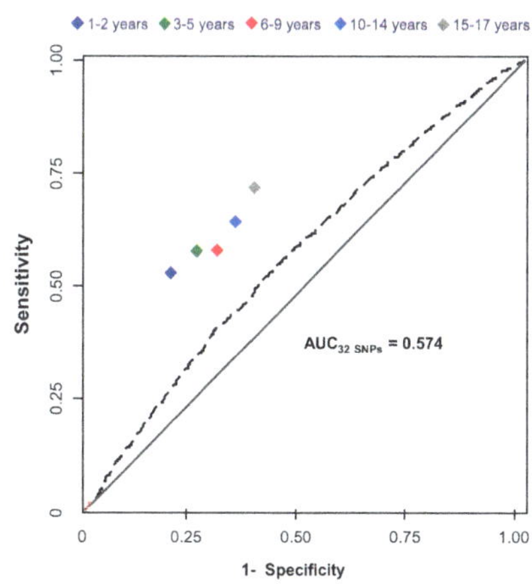

***Sensitivity, Specificity, Positive Predictive Value (PPV) and Negative Predictive Value (NPV)* of having at least one obese parent.**

Age	Specificity	Sensitivity	PPV	NPV
1-2yrs	0.78	0.52	0.29	0.90
3-5yrs	0.74	0.58	0.29	0.91
6-9yrs	0.70	0.58	0.27	0.90
10-14yrs	0.64	0.66	0.26	0.91
15-17yrs	0.59	0.73	0.24	0.93

Fig. 21.2 The AUC_{ROC} for the 32 BMI-associated loci to predict obesity in 8120 individuals from the ARIC study (Speliotes et al. 2010). For comparison, sensitivity, 1-specificity, and corresponding AUC_{ROC} are shown for "parental obesity" as a test at various ages during childhood and adolescence (1–2 years in *dark blue*, 3–5 years in *green*, 6–9 years in *red*, 10–14 years in *light blue*, 15–17 years in *gray*) to predict obesity in adult life, with data derived from Whitaker et al. (1997) [Figure adapted and updated from Loos (2012)]

While Whitaker et al. (1997) did not report required values to calculate the predictive ability of one's own obesity status during childhood, the British 1958 Birth Cohort ($n = 12,327$) found that BMI at age 11 years was a good predictor of obesity at age 33 years. In this study, the area under the receiver operator characteristics curve (AUC_{ROC}) was used. This metric results from plotting the sensitivity of a test against 1-specificity and indicates the probability that a random individual who will develop obesity has a higher obesity risk than a random individual who will not develop obesity (Table 21.2). Here, the AUC_{ROC} was 0.78 for men and 0.80 for women (Cheung et al. 2004).

These two traditional predictors, childhood obesity and parental obesity, appear to be reasonably good predictors of obesity risk in young adulthood. Data on how well they predict obesity risk together, as well as on how well they predict obesity at older ages (e.g., 40–60 years), is not available. Also, prospective analysis in a representative population is needed to quantify the predictive value, the risk of future obesity in children based on their own and their parents' obesity status.

21.5.2 Genetic Prediction of Obesity Risk

Collecting genetic variation is more expensive, invasive, and cumbersome than obtaining data on parental and childhood obesity. Thus, genetic prediction of obesity should be better or at least substantially improve the predictive ability of traditional noninvasive risk factors. This poses a major challenge on genetic prediction. Given that 40–70 % of the variation in obesity susceptibility is due to genetic variation and that the remainder is explained by lifestyle factors, we know a priori that prediction of obesity solely based on genetic information will never be perfect (Janssens et al. 2006; Wray et al. 2011). Therefore, genetic prediction should ideally be considered in the context of supplementing traditional predictors of obesity, rather than replacing them. Commercially available genome profiling tests that predict a range of diseases are exclusively based on customers' genotypes and typically use only one locus (*FTO*, see Chap. 20) or occasionally more BMI-associated loci to predict obesity.

Unfortunately, so far few longitudinal studies that have data on both genetic and nongenetic risk factors have examined which risk model is most effective in discriminating obese and nonobese individuals. Four large-scale cross-sectional studies have examined the ability of the BMI-associated loci to predict adult obesity (Li et al. 2010; Speliotes et al. 2010; Locke et al. 2015; Belsky et al. 2013) (Table 21.3). As these studies had no data on parental or childhood obesity, their results reflect the predictive ability of the commercially available tests that are also based on genotypes only. We describe the studies according to the number of BMI-associated loci that were included in the risk prediction models (from one locus up to 97 loci) and compare the predictive ability of the models using the AUC_{ROC} (Table 21.2).

Table 21.3 Overview of studies that have assessed risk prediction models using BMI-associated loci to predict adult obesity

Study[a]	Sample size	Ancestry	Outcome	Cumulative number of BMI-associated loci included in the model	Explained variance in BMI	AUC_{ROC} for obesity	Other predictors in the risk model	Reference
EPIC-Norfolk	6452	European	Normal-weight vs. obese	One locus (FTO)	0.34 %	0.55	None	Li et al. (2010)
EPIC-Norfolk	6452	European	Normal-weight vs. obese	Two loci (FTO, near MC4R)	0.59 %	0.55	None	Li et al. (2010)
ARIC	8286	European	Non-obese vs. obese	Two loci (FTO, near MC4R)	0.59 %	0.54	None	Belsky et al. (2013)
						0.55	Age, age^2, sex, and study center	
ARIC	2442	African-American	Non-obese vs. obese	Two loci (FTO, near MC4R)	0.00 %	0.52	None	Belsky et al. (2013)
						0.61	age, age^2, sex, and study center	
EPIC-Norfolk	6452	European	Normal-weight vs. obese	12 BMI-associated loci	0.98 %	0.57	None	Li et al. (2010)
						0.60	Age and sex	Li et al. (2010)
ARIC	8120	European	Non-obese vs. obese	32 BMI-associated loci	1.45 %	0.57	None	Speliotes et al. (2010)
						0.58	Age, age^2 and sex	Speliotes et al. (2010)

(continued)

Table 21.3 (continued)

Study[a]	Sample size	Ancestry	Outcome	Cumulative number of BMI-associated loci included in the model	Explained variance in BMI	AUC_{ROC} for obesity	Other predictors in the risk model	Reference
ARIC	8286	European	Non-obese vs. obese	32 BMI-associated loci	–	0.57	Age, age^2, sex, and study center	Belsky et al. (2013)
ARIC	2442	African-American	Non-obese vs. obese	32 BMI-associated loci	–	0.61	Age, age^2, sex, and study center	Belsky et al. (2013)
HRS	8164	European	Non-obese vs. obese	97 BMI-associated loci	2.70 %	0.60	age, age^2, sex and 4 principal components	Locke et al. (2015)

Table adapted and updated from Loos (2012)
[a]Note that there is overlap between the Speliotes et al. (2010) and Belsky et al. (2013) paper that both reported on the ARIC study; the European-ancestry subpopulation largely overlap

Li et al. (2010) examined the predictive ability of three risk prediction models with each consecutive model including a cumulative number of BMI-associated loci, which allowed assessing whether more loci in the model would improve prediction. Models compared normal-weight ($18.5\ \text{kg/m}^2 \leq \text{BMI} < 25\ \text{kg/m}^2$) vs. obese ($\text{BMI} \geq 30\ \text{kg/m}^2$) individuals ($n = 6452$, age 39–79 years) from the population-based EPIC-Norfolk study (Li et al. 2010). The *FTO* locus was the first BMI-associated locus to be identified by GWAS in 2007 (Frayling et al. 2007; Scuteri et al. 2007) and has since been widely replicated across populations of diverse ancestry and in children and adults alike (Loos and Yeo 2014) (see Chaps. 4 and 20). Of all GWAS-identified obesity-susceptibility loci, the *FTO* locus has the largest effect on BMI, is common in populations of European ancestry (risk allele frequency ~42 %), and yet explains only 0.34 % of the variation in BMI in the EPIC-Norfolk study, consistent with other reports. For a long time, the *FTO* locus was the only variant on commercially available direct-to-consumer (DTC) tests to predict obesity. However, Li et al. show that *FTO* genotypes are not able to predict who will be obese and who will not be obese, with an AUC_{ROC} of 0.55 (Table 21.3) (Li et al. 2010).

The locus near *MC4R* was the second BMI-associated locus identified (Loos et al. 2008; Chambers et al. 2008). Together the *FTO* and near-*MC4R* loci explain 0.59 % of the BMI variation, and a risk prediction model based on both loci does not improve the ability to discriminate between obese and nonobese individuals (AUC_{ROC} of 0.55), as compared to using only *FTO* in the model (Table 21.3). The results from the ARIC study (age 45–64 years) are remarkably similar to those of the EPIC-Norfolk study; a model based on both loci has an equally poor predictive ability (AUC_{ROC} of 0.54) in the European ancestry subpopulation and an even worse predictive ability (AUC_{ROC} of 0.52) in the African American subpopulation (Belsky et al. 2013). The latter might be due to the fact that both loci were first identified in European ancestry populations; their risk allele frequency, effect size, and explained variance are less pronounced in African than in European ancestry populations (Monda et al. 2013). Of note is that adding age, sex, and study center in the models improves the predictive ability in the African American (AUC_{ROC} of 0.61), but not in the European American (AUC_{ROC} of 0.55), subpopulation of ARIC (Table 21.3). These observations support the notion that risk prediction models will require customization by ancestry, not only because of differences in genetic background but also in environments (lifestyle, culture) as risk factors of obesity.

In subsequent risk prediction models, an increasing number of BMI-associated loci were added according to the number of loci that had been identified by the time the model was established. The loci are typically combined into a genetic risk score that represents the number of risk alleles an individual carries; the higher the score, the more genetically susceptible one is assumed to be. So far, models with 12, 32, and 97 BMI-associated loci have been tested. Despite the increasing number of loci included, the models' ability to discriminate between obese and nonobese individuals has not improved substantially, and with AUC_{ROC} of 0.61 or less, they have no clinical utility (Table 21.3). For example, the AUC_{ROC} of a model that included the

12 most strongly BMI-associated loci by 2009 (including *FTO* and near-*MC4R*) was exactly the same (AUC_{ROC} of 0.57) (Li et al. 2010) as that of a model that included an additional 20 BMI-associated loci (i.e., 32 loci in total) that were identified in 2010 (Speliotes et al. 2010) (Table 21.3). It should be noted that while the predictive ability of the two genetic predisposition scores (12 loci vs. 32 loci) is the same, the outcome at the individual level might be different. For example, an individual who carries many of the first 12 BMI-associated risk alleles but much fewer of the subsequent 20 BMI-associated risk alleles may be classified as at high risk of obese when the score based on 12 loci is used, but as at low risk if the score of 32 loci is used. Thus, tests based on a different configuration of genetic variants might provide discrepant risk predictions for the same person (Kalf et al. 2014).

As shown in Fig. 21.2, the specificity and sensitivity for parental obesity on its own, as derived from Whitaker et al. (1997), are better than that of the genetic prediction score based on 32 BMI loci. Furthermore, also the AUC_{ROC} for childhood BMI at age 11 (AUC_{ROC} of ~0.80) is substantially higher than that of the genetic predisposition scores (Cheung et al. 2004). Unfortunately, none of the studies that assessed the risk models with genetic loci had data available on parental obesity and/or childhood obesity to allow comparing the predictive ability of traditional risk factors versus genetic risk factors or to assess the added accuracy of including genetic risk factors in traditional risk models.

Risk models that include other predictors, such as age, sex, study center, or principal components, perform marginally better than those based on only genetic loci (Table 21.3). However, even a model that includes age, sex, and principal components and a genetic risk score of 97 BMI-associated loci will, with an AUC_{ROC} of 0.60, often wrongly classify individuals as obese or nonobese.

Taken together, the predictive ability of the currently available genetic information is poor and does not allow accurate discrimination between those at high risk of obesity and those at low risk. To design the best predictive models for the targeted population, large-scale studies with a longitudinal design, data on genotypes, childhood BMI, parental obesity, and other relevant risk factors, as well as clear answers on what to predict, in whom, and why, will be needed.

21.6 Does Knowledge of One's Genetic Obesity Risk Change Lifestyle?

Irrespective of the generally poor performance of risk prediction models based on genetic information, DTC genomic profiling that assesses the customer's disease risk might change behavior. Proponents of genetic testing believe that knowledge of one's genetic susceptibility to disease will motivate them to adopt health-promoting behaviors. Skeptics fear that the same information might cause anxiety and depression in others.

The behavioral and psychological responses to personal genetic risk assessments will likely be disease specific, e.g., a test that suggests a high genetic risk for lung cancer will affect a person in a different way compared to a high genetic risk for restless leg syndrome. Unlike for type 2 diabetes (Chap. 22), we speculate that a genetic test that assesses one's risk of obesity will likely have little effect on a person's lifestyle or anxiety levels. After all, unlike many other common conditions, obesity and overweight develop early in adulthood or even during childhood or adolescence, such that by the time a genetic test is undertaken, typically in adulthood, a person has already a good appreciation of their susceptibility and has adopted a lifestyle that keeps them relatively weight stable.

While data is still limited, the first insights from early reports suggest that genetic testing evokes limited or no beneficial or harmful behavioral responses. A Cochrane review identified 13 studies that examined the effects of communicating genotype-based disease risk estimates on risk-reducing behaviors and on the motivation to undertake such behaviors (Marteau et al. 2010). Overall, the communication of genotype-based disease risk estimates had no effect on smoking cessation or physical activity, whereas there may be a small effect on improved dietary habits and on intentions to change behavior. Two of the 13 studies focused specifically on obesity and found only suggestive evidence that higher risk estimates increase people's motivation to adopt a healthier lifestyle (Frosch et al. 2005; Sanderson et al. 2010; Marteau et al. 2010).

A large-scale longitudinal cohort study by Bloss et al. (2011, 2013) examined whether DTC genome-wide profiling, used to estimate individuals' lifetime risk for a variety of health conditions, results in psychological and behavioral changes. Participants (mean age [SD]: 46.7 [12]) purchased the genetic test at a reduced rate and were informed about their estimated lifetime risk for 23 conditions, including obesity. At baseline ($n = 3639$), 3 months ($n = 2037$, 56 % of baseline) and 1 year ($n = 1325$, 36 %) later at follow-up, their anxiety levels, dietary fat intake, and exercise behavior was assessed by self-report using a web-based survey. At 3 months of follow-up, the participants' overall anxiety level, dietary fat intake, or exercise behavior had not changed (Bloss et al. 2011). Furthermore, the composite lifetime risk score, which combines the risk estimates of the 23 conditions, was not associated with changes in anxiety level, diet, and exercise. Nevertheless, for obesity in particular, a higher estimated lifetime risk score was associated with a significant increase in dietary fat intake. Thus, rather than motivating to adopt a healthy lifestyle, the genetic information seems to have given participants a sense of lack of control over their obesity susceptibility. The fact that this is only seen for obesity may not be a surprise as, more than for any of the other 22 conditions, many of the participants may have already been "affected" (i.e., be obese or overweight) at the time of the genetic testing, which may have exacerbated a fatalistic perception. However, after 1 year of follow-up, the association between lifetime risk of obesity and increased fat intake was substantially attenuated and not significant anymore (Bloss et al. 2013). This suggests that the effects of communicating risk estimates might only be short lived, at least for obesity. It should be noted that there was a high dropout at 3 months and again at 1 year and that those who persisted throughout the whole study tended to be a health-minded bias subsample, who may be more conscious of their true risk of disease.

The study by Bloss et al. (2011, 2013) is the first large-scale prospective study to provide valuable insights in the behavioral and psychological responses to genotyped-based lifetime risk estimates. More large-scale prospective studies, including randomized control trials, will be needed to examine the effects of communicating personal genetic information in greater detail.

21.7 Conclusions and Future Directions

The predictive ability of models that use genetic variants to determine who will become obese and who will remain of normal weight depends on a number of factors. Before constructing the predictive models, one will need to have a clear view on what form of obesity needs to be predicted, why, in whom, and for what purpose. Multifactorial obesity is highly prevalent in the general population and results from an interplay between genes and environmental risk factors, such that models solely based on genetic variants will likely never reach high predictive ability, as we have illustrated using the literature on currently identified BMI-associated loci.

Whether genetic profiling should be used to change people's behavior, even when we know that for common obesity such profiles are often incorrect, opens up an ethical debate that we should engage in. In the case of common obesity, the first studies suggest that communicating genetic information may not be effective in making people adopt a healthy lifestyle.

While using genetic information in the prediction of obesity is an honorable aim, we believe that the most valuable contribution of gene discovery studies lies in their contribution to elucidate new physiological pathways that underlie obesity susceptibility, which in turn could lead to the identification of therapeutic targets and make its way into mainstream health care.

References

Adams KF, Schatzkin A, Harris TB, Kipnis V, Mouw T, Ballard-Barbash R, Hollenbeck A, Leitzmann MF (2006) Overweight, obesity, and mortality in a large prospective cohort of persons 50 to 71 years old. N Engl J Med 355(8):763–778. doi:10.1056/NEJMoa055643

Baker JL, Olsen LW, Sorensen TI (2007) Childhood body-mass index and the risk of coronary heart disease in adulthood. N Engl J Med 357(23):2329–2337. doi:10.1056/NEJMoa072515

Belsky DW, Moffitt TE, Sugden K, Williams B, Houts R, McCarthy J, Caspi A (2013) Development and evaluation of a genetic risk score for obesity. Biodemogr Soc Biol 59(1):85–100. doi:10.1080/19485565.2013.774628

Bloss CS, Schork NJ, Topol EJ (2011) Effect of direct-to-consumer genomewide profiling to assess disease risk. N Engl J Med 364(6):524–534

Bloss CS, Wineinger NE, Darst BF, Schork NJ, Topol EJ (2013) Impact of direct-to-consumer genomic testing at long term follow-up. J Med Genet 50(6):393–400. doi:10.1136/jmedgenet-2012-101207

Burke V, Beilin LJ, Dunbar D (2001) Family lifestyle and parental body mass index as predictors of body mass index in Australian children: a longitudinal study. Int J Obes 25(2):147–157

Cawley J, Meyerhoefer C (2012) The medical care costs of obesity: an instrumental variables approach. J Health Econ 31(1):219–230. doi:10.1016/j.jhealeco.2011.10.003

Chambers JC, Elliott P, Zabaneh D, Zhang W, Li Y, Froguel P, Balding D, Scott J, Kooner JS (2008) Common genetic variation near MC4R is associated with waist circumference and insulin resistance. Nat Genet 40(6):716–718. doi:10.1038/ng.156

Cheung YB, Machin D, Karlberg J, Khoo KS (2004) A longitudinal study of pediatric body mass index values predicted health in middle age. J Clin Epidemiol 57(12):1316–1322

Chung WK (2012) An overview of mongenic and syndromic obesities in humans. Pediatr Blood Cancer 58(1):122–128. doi:10.1002/pbc.23372

Cole TJ, Lobstein T (2012) Extended international (IOTF) body mass index cut-offs for thinness, overweight and obesity. Pediatr Obes 7(4):284–294. doi:10.1111/j.2047-6310.2012.00064.x

de Wit L, Luppino F, van Straten A, Penninx B, Zitman F, Cuijpers P (2010) Depression and obesity: a meta-analysis of community-based studies. Psychiatry Res 178(2):230–235. doi:10.1016/j.psychres.2009.04.015

Elks CE, den Hoed M, Zhao JH, Sharp SJ, Wareham NJ, Loos RJ, Ong KK (2012) Variability in the heritability of body mass index: a systematic review and meta-regression. Front Endocrinol 3:29. doi:10.3389/fendo.2012.00029

Finucane MM, Stevens GA, Cowan MJ, Danaei G, Lin JK, Paciorek CJ, Singh GM, Gutierrez HR, Lu Y, Bahalim AN, Farzadfar F, Riley LM, Ezzati M (2011) National, regional, and global trends in body-mass index since 1980: systematic analysis of health examination surveys and epidemiological studies with 960 country-years and 9.1 million participants. Lancet 377 (9765):557–567. doi:10.1016/S0140-6736(10)62037-5

Flegal KM, Caroll MD, Kuczmarski RJ, Johnson CL (1998) Overweight and obesity in the United States: prevalence and trends, 1960–1994. Int J Obes Relat Metab Disord 22(1):39–47

Frayling TM, Timpson NJ, Weedon MN, Zeggini E, Freathy RM, Lindgren CM, Perry JRB, Elliott KS, Lango H, Rayner NW, Shields B, Harries LW, Barrett JC, Ellard S, Groves CJ, Knight B, Patch AM, Ness AR, Ebrahim S, Lawlor DA, Ring SM, Ben-Shlomo Y, Jarvelin MR, Sovio U, Bennett AJ, Melzer D, Ferrucci L, Loos RJF, Barroso I, Wareham NJ, Karpe F, Owen KR, Cardon LR, Walker M, Hitman GA, Palmer CNA, Doney ASF, Morris AD, Smith GD, Hattersley AT, McCarthy MI (2007) A common variant in the FTO gene is associated with body mass index and predisposes to childhood and adult obesity. Science 316(5826):889–894

Frosch DL, Mello P, Lerman C (2005) Behavioral consequences of testing for obesity risk. Cancer Epidemiol Biomark Prev 14(6):1485–1489

IASO (2014) International Association for the Study of Obesity http://www.iaso.org/resources/obesity-data-portal/resources/trends/. (2013) http://www.iotf.org/

Janssens AC, Aulchenko YS, Elefante S, Borsboom GJ, Steyerberg EW, van Duijn CM (2006) Predictive testing for complex diseases using multiple genes: fact or fiction? Genet Med 8 (7):395–400, doi:10.109701.gim.0000229689.18263.f4

Kalf RR, Mihaescu R, Kundu S, de Knijff P, Green RC, Janssens AC (2014) Variations in predicted risks in personal genome testing for common complex diseases. Genet Med 16 (1):85–91. doi:10.1038/gim.2013.80

Katzmarzyk PT, Perusse L, Rao DC, Bouchard C (1999) Familial risk of obesity and central adipose tissue distribution in the general Canadian population. Am J Epidemiol 149:933–942

Khan LK, Sobush K, Keener D, Goodman K, Lowry A, Kakietek J, Zaro S (2009) Recommended community strategies and measurements to prevent obesity in the United States. MMWR recommendations and reports: morbidity and mortality weekly report recommendations and reports. Centers Dis Control 58(RR-7):1–26

Lake JK, Power C, Cole TJ (1997) Child to adult body mass index in the 1958 British birth cohort: associations with parental obesity. Arch Dis Child 77(5):376–380

Li L, Law C, Lo Conte R, Power C (2009) Intergenerational influences on childhood body mass index: the effect of parental body mass index trajectories. Am J Clin Nutr 89(2):551–557

Li S, Zhao JH, Ja L, Luben RN, Rodwell SA, Khaw KT, Ong KK, Wareham NJ, Loos RJF (2010) Cumulative effects and predictive value of common obesity-susceptibility variants identified by genome-wide association studies. Am J Clin Nutr 91(1):184–190

Locke AE, Kahali B, Berndt SI, Justice AE, Pers TH, Day FR, Powell C, Vedantam S, Buchkovich ML, Yang J, Croteau-Chonka DC, Esko T, Fall T, Ferreira T, Gustafsson S, Kutalik Z, Luan J, Mägi R, Randall JC, Winkler TW, Wood AR, Workalemahu T, Faul JD, Smith JA, Hua Zhao J, Zhao W, Chen J, Fehrmann R, Hedman ÅK, Karjalainen J, Schmidt EM, Absher D, Amin N, Anderson D, Beekman M, Bolton JL, Bragg-Gresham JL, Buyske S, Demirkan A, Deng G, Ehret GB, Feenstra B, Feitosa MF, Fischer K, Goel A, Gong J, Jackson AU, Kanoni S, Kleber ME, Kristiansson K, Lim U, Lotay V, Mangino M, Mateo Leach I, Medina-Gomez C, Medland SE, Nalls MA, Palmer CD, Pasko D, Pechlivanis S, Peters MJ, Prokopenko I, Shungin D, Stančáková A, Strawbridge RJ, Ju Sung Y, Tanaka T, Teumer A, Trompet S, van der Laan SW, van Setten J, Van Vliet-Ostaptchouk JV, Wang Z, Yengo L, Zhang W, Isaacs A, Albrecht E, Ärnlöv J, Arscott GM, Attwood AP, Bandinelli S, Barrett A, Bas IN, Bellis C, Bennett AJ, Berne C, Blagieva R, Blüher M, Böhringer S, Bonnycastle LL, Böttcher Y, Boyd HA, Bruinenberg M, Caspersen IH, Ida Chen YD, Clarke R, Daw EW, de Craen AJ, Delgado G, Dimitriou M, Doney AS, Eklund N, Estrada K, Eury E, Folkersen L, Fraser RM, Garcia ME, Geller F, Giedraitis V, Gigante B, Go AS, Golay A, Goodall AH, Gordon SD, Gorski M, Grabe HJ, Grallert H, Grammer TB, Gräßler J, Grönberg H, Groves CJ, Gusto G, Haessler J, Hall P, Haller T, Hallmans G, Hartman CA, Hassinen M, Hayward C, Heard-Costa NL, Helmer Q, Hengstenberg C, Holmen O, Hottenga JJ, James AL, Jeff JM, Johansson Å, Jolley J, Juliusdottir T, Kinnunen L, Koenig W, Koskenvuo M, Kratzer W, Laitinen J, Lamina C, Leander K, Lee NR, Lichtner P, Lind L, Lindström J, Sin Lo K, Lobbens S, Lorbeer R, Lu Y, Mach F, Magnusson PK, Mahajan A, McArdle WL, McLachlan S, Menni C, Merger S, Mihailov E, Milani L, Moayyeri A, Monda KL, Morken MA, Mulas A, Müller G, Müller-Nurasyid M, Musk AW, Nagaraja R, Nöthen MM, Nolte IM, Pilz S, Rayner NW, Renstrom F, Rettig R, Ried JS, Ripke S, Robertson NR, Rose LM, Sanna S, Scharnagl H, Scholtens S, Schumacher FR, Scott WR, Seufferlein T, Shi J, Vernon Smith A, Smolonska J, Stanton AV, Steinthorsdottir V, Stirrups K, Stringham HM, Sundström J, Swertz MA, Swift AJ, Syvänen AC, Tan ST, Tayo BO, Thorand B, Thorleifsson G, Tyrer JP, Uh HW, Vandenput L, Verhulst FC, Vermeulen SH, Verweij N, Vonk JM, Waite LL, Warren HR, Waterworth D, Weedon MN, Wilkens LR, Willenborg C, Wilsgaard T, Wojczynski MK, Wong A, Wright AF, Zhang Q, LifeLines Cohort Study, Brennan EP, Choi M, Dastani Z, Drong AW, Eriksson P, Franco-Cereceda A, Gådin JR, Gharavi AG, Goddard ME, Handsaker RE, Huang J, Karpe F, Kathiresan S, Keildson S, Kiryluk K, Kubo M, Lee JY, Liang L, Lifton RP, Ma B, McCarroll SA, McKnight AJ, Min JL, Moffatt MF, Montgomery GW, Murabito JM, Nicholson G, Nyholt DR, Okada Y, Perry JR, Dorajoo R, Reinmaa E, Salem RM, Sandholm N, Scott RA, Stolk L, Takahashi A, Tanaka T, Van't Hooft FM, Vinkhuyzen AA, Westra HJ, Zheng W, Zondervan KT, ADIPOGen Consortium, AGEN-BMI Working Group, CARDIOGRAMplusC4D Consortium, CKDGen Consortium, GLGC, ICBP, MAGIC Investigators, MuTHER Consortium, MIGen Consortium, PAGE Consortium, ReproGen Consortium, GENIE Consortium, International Endogene Consortium, Heath AC, Arveiler D, Bakker SJ, Beilby J, Bergman RN, Blangero J, Bovet P, Campbell H, Caulfield MJ, Cesana G, Chakravarti A, Chasman DI, Chines PS, Collins FS, Crawford DC, Cupples LA, Cusi D, Danesh J, de Faire U, den Ruijter HM, Dominiczak AF, Erbel R, Erdmann J, Eriksson JG, Farrall M, Felix SB, Ferrannini E, Ferrières J, Ford I, Forouhi NG, Forrester T, Franco OH, Gansevoort RT, Gejman PV, Gieger C, Gottesman O, Gudnason V, Gyllensten U, Hall AS, Harris TB, Hattersley AT, Hicks AA, Hindorff LA, Hingorani AD, Hofman A, Homuth G, Hovingh GK, Humphries SE, Hunt SC, Hyppönen E, Illig T, Jacobs KB, Jarvelin MR, Jöckel KH, Johansen B, Jousilahti P, Jukema

JW, Jula AM, Kaprio J, Kastelein JJ, Keinanen-Kiukaanniemi SM, Kiemeney LA, Knekt P, Kooner JS, Kooperberg C, Kovacs P, Kraja AT, Kumari M, Kuusisto J, Lakka TA, Langenberg C, Le Marchand L, Lehtimäki T, Lyssenko V, Männistö S, Marette A, Matise TC, McKenzie CA, McKnight B, Moll FL, Morris AD, Morris AP, Murray JC, Nelis M, Ohlsson C, Oldehinkel AJ, Ong KK, Madden PA, Pasterkamp G, Peden JF, Peters A, Postma DS, Pramstaller PP, Price JF, Qi L, Raitakari OT, Rankinen T, Rao DC, Rice TK, Ridker PM, Rioux JD, Ritchie MD, Rudan I, Salomaa V, Samani NJ, Saramies J, Sarzynski MA, Schunkert H, Schwarz PE, Sever P, Shuldiner AR, Sinisalo J, Stolk RP, Strauch K, Tönjes A, Trégouët DA, Tremblay A, Tremoli E, Virtamo J, Vohl MC, Völker U, Waeber G, Willemsen G, Witteman JC, Zillikens MC, Adair LS, Amouyel P, Asselbergs FW, Assimes TL, Bochud M, Boehm BO, Boerwinkle E, Bornstein SR, Bottinger EP, Bouchard C, Cauchi S, Chambers JC, Chanock SJ, Cooper RS, de Bakker PI, Dedoussis G, Ferrucci L, Franks PW, Froguel P, Groop LC, Haiman CA, Hamsten A, Hui J, Hunter DJ, Hveem K, Kaplan RC, Kivimaki M, Kuh D, Laakso M, Liu Y, Martin NG, März W, Melbye M, Metspalu A, Moebus S, Munroe PB, Njølstad I, Oostra BA, Palmer CN, Pedersen NL, Perola M, Pérusse L, Peters U, Power C, Quertermous T, Rauramaa R, Rivadeneira F, Saaristo TE, Saleheen D, Sattar N, Schadt EE, Schlessinger D, Slagboom PE, Snieder H, Spector TD, Thorsteinsdottir U, Stumvoll M, Tuomilehto J, Uitterlinden AG, Uusitupa M, van der Harst P, Walker M, Wallaschofski H, Wareham NJ, Watkins H, Weir DR, Wichmann HE, Wilson JF, Zanen P, Borecki IB, Deloukas P, Fox CS, Heid IM, O'Connell JR, Strachan DP, Stefansson K, van Duijn CM, Abecasis GR, Franke L, Frayling TM, McCarthy MI, Visscher PM, Scherag A, Willer CJ, Boehnke M, Mohlke KL, Lindgren CM, Beckmann JS, Barroso I, North KE, Ingelsson E, Hirschhorn JN, Loos RJ, Speliotes EK (2015) Genetic studies of body mass index yield new insights for obesity biology. Nature 518(7538):197–206. doi:10.1038/nature14177

Loos RJ (2012) Genetic determinants of common obesity and their value in prediction. Best Pract Res Clin Endocrinol Metab 26(2):211–226. doi:10.1016/j.beem.2011.11.003

Loos RJ, Lindgren CM, Li S, Wheeler E, Zhao JH, Prokopenko I, Inouye M, Freathy RM, Abecasis GR, Albai G, Attwood AP, Beckmann JS, Berndt SI, Boehnke M, Chanock SJ, Hayes RB, Bergmann S, Bennett AJ, Bingham SA, Bochud M, Brown M, Cauchi S, Chanock SJ, Collins FS, Connell JM, Cooper C, Smith GD, Day I, Dina C, De S, Dermitzakis ET, Doney AS, Elliott KS, Elliott P, Evans DM, Sadaf FI, Froguel P, Ghori J, Gieger C, Groop L, Groves CJ, Gwilliam R, Hadley D, Hall AS, Hankinson SE, Hattersley AT, Hayes RB, Hebebrand J, Heid IM, Illig T, Lamina C, Gieger C, Illig T, Kraft P, Meitinger T, Wichmann HE, Herrera B, Hinney A, Hu FB, Hunt SE, Hunter DJ, Jackson AU, Jacobs KB, Jarvelin MR, Johnson T, Jolley JD, Karpe F, Keniry A, Khaw KT, Lamina C, Luben RN, Lyon HN, Mangino M, Marchini J, McArdle WL, McGinnis R, Meitinger T, Meyre D, Munroe PB, Mohlke KL, Morris AD, Nagaraja R, Ness AR, Neville MJ, Nica AC, Ong KK, O'Rahilly S, Owen KR, Palmer CN, Papadakis K, Potter S, Pouta A, Qi L, Randall JC, Rayner NW, Ridderstrale M, Ring SM, Sandhu MS, Sanna S, Scherag A, Scheet P, Sims MA, Schlessinger D, Song K, Soranzo N, Speliotes EK, Syddall HE, Teichmann SA, Timpson NJ, Tobias JH, Tuomilehto J, Uda M, Vogel CI, Voight BF, Wallace C, Waterworth DM, Weedon MN, Wichmann HE, Willer CJ, Wraight, Yuan X, Zeggini E, Hirschhorn JN, Strachan DP, Ouwehand WH, Caulfield MJ, Samani NJ, Frayling TM, Vollenweider P, Waeber G, Mooser V, Deloukas P, McCarthy MI, Wareham NJ, Barroso I (2008) Common variants near MC4R are associated with fat mass, weight and risk of obesity. Nat Genet 40(6):768–775

Loos RJ, Yeo GS (2014) The bigger picture of FTO--the first GWAS-identified obesity gene. Nat Rev Endocrinol 10(1):51–61. doi:10.1038/nrendo.2013.227

Lu Y, Loos RJ (2013) Obesity genomics: assessing the transferability of susceptibility loci across diverse populations. Genome Med 5(6):55. doi:10.1186/gm459

Maes HH, Neale MC, Eaves LJ (1997) Genetic and environmental factors in relative body weight and human adiposity. Behav Genet 27(4):325–351

Magarey AM, Daniels LA, Boulton TJC, Cockington RA (2003) Predicting obesity in early adulthood from childhood and parental obesity. Int J Obes 27(4):505–513

Marteau TM, French DP, Griffin SJ, Prevost AT, Sutton S, Watkinson C, Attwood S, Hollands GJ (2010) Effects of communicating DNA-based disease risk estimates on risk-reducing behaviours. Cochrane Database Syst Rev 10, CD007275

Meyre D, Delplanque J, Chevre JC, Lecoeur C, Lobbens S, Gallina S, Durand E, Vatin V, Degraeve F, Proenca C, Gaget S, Korner A, Kovacs P, Kiess W, Tichet J, Marre M, Hartikainen AL, Horber F, Potoczna N, Hercberg S, Levy-Marchal C, Pattou F, Heude B, Tauber M, McCarthy MI, Blakemore AIF, Montpetit A, Polychronakos C, Weill J, Coin LJM, Asher J, Elliott P, Jarvelin MR, Visvikis-Siest S, Balkau B, Sladek R, Balding D, Walley A, Dina C, Froguel P (2009) Genome-wide association study for early-onset and morbid adult obesity identifies three new risk loci in European populations. Nat Genet 41(2):157–159

Monda KL, Chen GK, Taylor KC, Palmer C, Edwards TL, Lange LA, Ng MC, Adeyemo AA, Allison MA, Bielak LF, Chen G, Graff M, Irvin MR, Rhie SK, Li G, Liu Y, Liu Y, Lu Y, Nalls MA, Sun YV, Wojczynski MK, Yanek LR, Aldrich MC, Ademola A, Amos CI, Bandera EV, Bock CH, Britton A, Broeckel U, Cai Q, Caporaso NE, Carlson CS, Carpten J, Casey G, Chen WM, Chen F, Chen YD, Chiang CW, Coetzee GA, Demerath E, Deming-Halverson SL, Driver RW, Dubbert P, Feitosa MF, Feng Y, Freedman BI, Gillanders EM, Gottesman O, Guo X, Haritunians T, Harris T, Harris CC, Hennis AJ, Hernandez DG, McNeill LH, Howard TD, Howard BV, Howard VJ, Johnson KC, Kang SJ, Keating BJ, Kolb S, Kuller LH, Kutlar A, Langefeld CD, Lettre G, Lohman K, Lotay V, Lyon H, Manson JE, Maixner W, Meng YA, Monroe KR, Morhason-Bello I, Murphy AB, Mychaleckyj JC, Nadukuru R, Nathanson KL, Nayak U, N'Diaye A, Nemesure B, Wu SY, Leske MC, Neslund-Dudas C, Neuhouser M, Nyante S, Ochs-Balcom H, Ogunniyi A, Ogundiran TO, Ojengbede O, Olopade OI, Palmer JR, Ruiz-Narvaez EA, Palmer ND, Press MF, Rampersaud E, Rasmussen-Torvik LJ, Rodriguez-Gil JL, Salako B, Schadt EE, Schwartz AG, Shriner DA, Siscovick D, Smith SB, Wassertheil-Smoller S, Speliotes EK, Spitz MR, Sucheston L, Taylor H, Tayo BO, Tucker MA, Van Den Berg DJ, Edwards DR, Wang Z, Wiencke JK, Winkler TW, Witte JS, Wrensch M, Wu X, Yang JJ, Levin AM, Young TR, Zakai NA, Cushman M, Zanetti KA, Zhao JH, Zhao W, Zheng Y, Zhou J, Ziegler RG, Zmuda JM, Fernandes JK, Gilkeson GS, Kamen DL, Hunt KJ, Spruill IJ, Ambrosone CB, Ambs S, Arnett DK, Atwood L, Becker DM, Berndt SI, Bernstein L, Blot WJ, Borecki IB, Bottinger EP, Bowden DW, Burke G, Chanock SJ, Cooper RS, Ding J, Duggan D, Evans MK, Fox C, Garvey WT, Bradfield JP, Hakonarson H, Grant SF, Hsing A, Chu L, Hu JJ, Huo D, Ingles SA, John EM, Jordan JM, Kabagambe EK, Kardia SL, Kittles RA, Goodman PJ, Klein EA, Kolonel LN, Le Marchand L, Liu S, McKnight B, Millikan RC, Mosley TH, Padhukasahasram B, Williams LK, Patel SR, Peters U, Pettaway CA, Peyser PA, Psaty BM, Redline S, Rotimi CN, Rybicki BA, Sale MM, Schreiner PJ, Signorello LB, Singleton AB, Stanford JL, Strom SS, Thun MJ, Vitolins M, Zheng W, Moore JH, Williams SM, Ketkar S, Zhu X, Zonderman AB, Kooperberg C, Papanicolaou GJ, Henderson BE, Reiner AP, Hirschhorn JN, Loos RJ, North KE, Haiman CA (2013) A meta-analysis identifies new loci associated with body mass index in individuals of African ancestry. Nat Genet. doi:10.1038/ng.2608

Must A, Spadano J, Coakley EH, Field AE, Colditz G, Dietz WH (1999) The disease burden associated with overweight and obesity. JAMA 282(16):1523–1529

Ogden CL, Flegal KM (2010) Changes in terminology for childhood overweight and obesity. Natl Health Stat Rep 25:1–5

Ogden CL, Carroll MD, Kit BK, Flegal KM (2014) Prevalence of childhood and adult obesity in the United States, 2011–2012. JAMA 311(8):806–814. doi:10.1001/jama.2014.732

Park MH, Falconer C, Viner RM, Kinra S (2012) The impact of childhood obesity on morbidity and mortality in adulthood: a systematic review. Obes Rev 13(11):985–1000. doi:10.1111/j.1467-789X.2012.01015.x

Popkin BM, Adair LS, Ng SW (2012) Global nutrition transition and the pandemic of obesity in developing countries. Nutr Rev 70(1):3–21. doi:10.1111/j.1753-4887.2011.00456.x

Ramachandrappa S, Farooqi IS (2011) Genetic approaches to understanding human obesity. J Clin Invest 121(6):2080–2086. doi:10.1172/JCI46044

Renehan AG, Tyson M, Egger M, Heller RF, Zwahlen M (2008) Body-mass index and incidence of cancer: a systematic review and meta-analysis of prospective observational studies. Lancet 371(9612):569–578. doi:10.1016/S0140-6736(08)60269-X

Sanderson SK, Persky S, Michie S (2010) Psychological and behavioral responses to genetic test results indicating increased risk of obesity: does the causal pathway from gene to obesity matter? Public Health Genomics 13(34):47

Scuteri A, Sanna S, Chen W-M, Uda M, Albai G, Strait J, Najjar S, Nagaraja R, Orru M, Usala G, Dei M, Lai S, Maschio A, Busonero F, Mulas A, Ehret GB, Fink AA, Weder AB, Cooper RS, Galan P, Chakravarti A, Schlessinger D, Cao A, Lakatta E, Abecasis GR (2007) Genome-wide association scan shows genetic variants in the FTO gene are associated with obesity-related traits. PLoS Genet 3(7), e115

Speliotes EK, Willer CJ, Berndt SI, Monda KL, Thorleifsson G, Jackson AU, Allen HL, Lindgren CM, Luan J, Magi R, Randall JC, Vedantam S, Winkler TW, Qi L, Workalemahu T, Heid IM, Steinthorsdottir V, Stringham HM, Weedon MN, Wheeler E, Wood AR, Ferreira T, Weyant RJ, Segre AV, Estrada K, Liang L, Nemesh J, Park JH, Gustafsson S, Kilpelainen TO, Yang J, Bouatia-Naji N, Esko T, Feitosa MF, Kutalik Z, Mangino M, Raychaudhuri S, Scherag A, Smith AV, Welch R, Zhao JH, Aben KK, Absher DM, Amin N, Dixon AL, Fisher E, Glazer NL, Goddard ME, Heard-Costa NL, Hoesel V, Hottenga JJ, Johansson A, Johnson T, Ketkar S, Lamina C, Li S, Moffatt MF, Myers RH, Narisu N, Perry JRB, Peters MJ, Preuss M, Ripatti S, Rivadeneira F, Sandholt C, Scott LJ, Timpson NJ, Tyrer JP, van Wingerden S, Watanabe RM, White CC, Wiklund F, Barlassina C, Chasman DI, Cooper MN, Jansson JO, Lawrence RW, Pellikka N, Prokopenko I, Shi J, Thiering E, Alavere H, Alibrandi MTS, Almgren P, Arnold AM, Aspelund T, Atwood LD, Balkau B, Balmforth AJ, Bennett AJ, Ben-Shlomo Y, Bergman RN, Bergmann S, Biebermann H, Blakemore AIF, Boes T, Bonnycastle LL, Bornstein SR, Brown MJ, Buchanan TA, Busonero F, Campbell H, Cappuccio FP, Cavalcanti-Proenca C, Chen YDI, Chen CM, Chines PS, Clarke R, Coin L, Connell J, Day INM, Heijer M, Duan J, Ebrahim S, Elliott P, Elosua R, Eiriksdottir G, Erdos MR, Eriksson JG, Facheris MF, Felix SB, Fischer-Posovszky P, Folsom AR, Friedrich N, Freimer NB, Fu M, Gaget S, Gejman PV, Geus EJC, Gieger C, Gjesing AP, Goel A, Goyette P, Grallert H, Graszler J, Greenawalt DM, Groves CJ, Gudnason V, Guiducci C, Hartikainen AL, Hassanali N, Hall AS, Havulinna AS, Hayward C, Heath AC, Hengstenberg C, Hicks AA, Hinney A, Hofman A, Homuth G, Hui J, Igl W, Iribarren C, Isomaa B, Jacobs KB, Jarick I, Jewell E, John U, Jorgensen T, Jousilahti P, Jula A, Kaakinen M, Kajantie E, Kaplan LM, Kathiresan S, Kettunen J, Kinnunen L, Knowles JW, Kolcic I, Konig IR, Koskinen S, Kovacs P, Kuusisto J, Kraft P, Kvaloy K, Laitinen J, Lantieri O, Lanzani C, Launer LJ, Lecoeur C, Lehtimaki T, Lettre G, Liu J, Lokki ML, Lorentzon M, Luben RN, Ludwig B, Manunta P, Marek D, Marre M, Martin NG, McArdle WL, McCarthy A, McKnight B, Meitinger T, Melander O, Meyre D, Midthjell K, Montgomery GW, Morken MA, Morris AP, Mulic R, Ngwa JS, Nelis M, Neville MJ, Nyholt DR, O'Donnell CJ, O'Rahilly S, Ong KK, Oostra B, Pare G, Parker AN, Perola M, Pichler I, Pietilainen KH, Platou CGP, Polasek O, Pouta A, Rafelt S, Raitakari O, Rayner NW, Ridderstrale M, Rief W, Ruokonen A, Robertson NR, Rzehak P, Salomaa V, Sanders AR, Sandhu MS, Sanna S, Saramies J, Savolainen MJ, Scherag S, Schipf S, Schreiber S, Schunkert H, Silander K, Sinisalo J, Siscovick DS, Smit JH, Soranzo N, Sovio U, Stephens J, Surakka I, Swift AJ, Tammesoo ML, Tardif JC, Teder-Laving M, Teslovich TM, Thompson JR, Thomson B, Tonjes A, Tuomi T, van Meurs JBJ, van Ommen GJ (2010) Association analyses of 249,796 individuals reveal 18 new loci associated with body mass index. Nat Genet 42(11):937–948

Swinburn BA, Sacks G, Hall KD, McPherson K, Finegood DT, Moodie ML, Gortmaker SL (2011) The global obesity pandemic: shaped by global drivers and local environments. Lancet 378 (9793):804–814

Toschke AM, Ruckinger S, Reinehr T, von Kries R (2007) Growth around puberty as predictor of adult obesity. Eur J Clin Nutr 62(12):1405–1411

Vazquez G, Duval S, Jacobs DR Jr, Silventoinen K (2007) Comparison of body mass index, waist circumference, and waist/hip ratio in predicting incident diabetes: a meta-analysis. Epidemiol Rev 29:115–128. doi:10.1093/epirev/mxm008

Whitaker RC, Wright JA, Pepe MS, Seidel KD, Dietz WH (1997) Predicting obesity in young adulthood from childhood and parental obesity. N Engl J Med 337(13):869–873

WHO (2000) Obesity: preventing and managing the global epidemic. Report of a WHO consultation. World Health Organ Tech Rep Ser 894:i–xii, 1–253

Wildman RP, Muntner P, Reynolds K, McGinn AP, Rajpathak S, Wylie-Rosett J, Sowers MR (2008) The obese without cardiometabolic risk factor clustering and the normal weight with cardiometabolic risk factor clustering: prevalence and correlates of 2 phenotypes among the US population (NHANES 1999–2004). Arch Intern Med 168(15):1617–1624. doi:10.1001/archinte.168.15.1617

World Health Organisation (2000) The Asia-Pacific perspective: redefining obesity and its treatment. In: International Obesity Taskforce (ed) International associates for the study of obesity. Health Communications Australia, Melbourne

Wray NR, Purcell SM, Visscher PM (2011) Synthetic associations created by rare variants do not explain most GWAS results. PLoS Biol 9(1), e1000579

Yusuf S, Hawken S, Ounpuu S, Bautista L, Franzosi MG, Commerford P, Lang CC, Rumboldt Z, Onen CL, Lisheng L, Tanomsup S, Wangai P Jr, Razak F, Sharma AM, Anand SS (2005) Obesity and the risk of myocardial infarction in 27,000 participants from 52 countries: a case–control study. Lancet 366(9497):1640–1649. doi:10.1016/S0140-6736(05)67663-5

Chapter 22
Pharmacogenetics of Metformin

Sook Wah Yee, Kaixin Zhou, and Kathleen M. Giacomini

Abstract Metformin is the most widely prescribed antidiabetic drug in the world. Despite its beneficial effects in reducing the risk for developing vascular complications associated with diabetes, the glycemic response to metformin is highly variable. Genetic factors, along with factors such as various comorbidities and body weight, contribute to this variability. In this chapter, we focus on genetic polymorphisms that associate with metformin pharmacokinetics as well as poor glycemic response to the drug. In particular, genetic polymorphisms in membrane transporters that play a role in metformin absorption, disposition, and response are highlighted. Studies in healthy volunteers, prediabetic and diabetic patients, and patients with polycystic ovary disease are described. Using genome-wide data, it is estimated that the heritability of glycemic response to metformin is around 30 %. The first genome-wide association study of metformin glycemic response in patients with type 2 diabetes reveals a locus in chromosome 11. Finally, we provide an overview of future directions for metformin pharmacogenomic studies to further elucidate genetic loci and targets for metformin action.

22.1 Metformin History

The biguanides, metformin (1,1-dimethylbiguanide), and its more hydrophobic analog, phenformin (2-(*N*-phenethylcarbamimidoyl)guanidine), are synthetic derivatives of naturally occurring guanide compounds found in the French lilac. For centuries, the French lilac was used as an herbal remedy for the treatment of type 2 diabetes (Watanabe 1918). Synthesized in the 1920s, metformin was

S.W. Yee • K.M. Giacomini
Department of Bioengineering and Therapeutic Sciences, Schools of Pharmacy and Medicine, University of California San Francisco, San Francisco, CA, USA

K. Zhou (✉)
School of Medicine, University of Dundee, Dundee, UK
e-mail: K.Zhou@dundee.ac.uk

J.C. Florez (ed.), *The Genetics of Type 2 Diabetes and Related Traits*,
DOI 10.1007/978-3-319-01574-3_22

approved for use in Europe in 1947 and in the United States in 1995. Unlike phenformin, which was associated with profound life-threatening lactic acidosis and withdrawn from the market in the 1970s, metformin is extremely safe with an extraordinarily low incidence of lactic acidosis (3 in 100,000) (see review; Bailey and Turner 1996). Currently the first-line therapy for type 2 diabetes, metformin is used in over 120 million people worldwide. In addition to its use in adults diagnosed with type 2 diabetes, metformin is also used to prevent diabetes in individuals at high risk and has been suggested for use in treating women who develop gestational diabetes or polycystic ovary syndrome (PCOS). Metformin is used off-label in other diseases associated with metabolic syndrome such as nonalcoholic fatty liver disease and steatohepatitis. Moreover, recent studies suggest that metformin may be effective in other diseases beyond those associated with metabolic syndrome. In particular, the drug is in clinical trials for prevention of tumor recurrence in breast, prostate, colorectal, and pancreatic cancer. Metformin is also being tested for use in mild cognitive impairment and for improving biomarkers associated with Alzheimer's disease. Whether metformin will be the panacea that it is purported to be is difficult to predict and will be revealed as the pertinent clinical trials are published. Nevertheless, studies of metformin use in animal models of various diseases have been highly promising, and a recent study in mice fed with metformin since middle age shows that the drug prolongs life span and health span (see reviews Viollet et al. 2012; Viollet and Foretz 2013).

22.2 Metformin Pharmacokinetics

Metformin is a polar (log $P = -0.5$), basic drug (pKa = 12.4), which at physiologic pHs is largely (>99 %) present in the ionized form. As such, the drug does not readily diffuse across biological membranes and requires membrane transporters to cross. The primary plasma membrane transporters involved in metformin tissue distribution are in the solute carrier superfamily (SLC) and specifically in two families: SLC22 and SLC47. SLC22 transporters involved in metformin biodistribution are OCT1 (encoded by *SLC22A1*), OCT2 (*SLC22A2*), and OCT3 (*SCL22A3*). Metformin is an excellent substrate of all three OCTs. Recently, metformin was reported to be a substrate of OCTN1 (*SLC22A4*), although weaker than the above OCTs (Nakamichi et al. 2013). OCT1 is responsible for the uptake of metformin into hepatocytes (Shu et al. 2007), whereas OCT2 is responsible for the uptake of metformin into proximal tubule cells of the kidney, the primary organ involved in metformin elimination (Chen et al. 2009). The role of OCT3 in metformin tissue distribution is not established, but the transporter is the primary organic cation transporter in adipose tissue and skeletal muscle, two tissues thought to play major roles in metformin pharmacologic action (Chen et al. 2010). MATE1 (*SLC47A1*), multidrug and toxin extrusion protein 1, is expressed on the apical membrane of the renal proximal tubule and is involved in the transport of metformin from the proximal tubule cell to the tubule lumen. The transporter is also expressed on the canalicular membrane of the hepatocyte and may play a role in metformin biliary excretion. MATE2 (*SLC47A2*) and its splice variant (Komatsu et al. 2011), MATE2K, are expressed in the proximal tubule on the apical membrane and appear to mediate metformin renal elimination (see Figs. 22.1 and 22.2).

Metformin pharmacokinetic properties have been well-established in many clinical studies (Tucker et al. 1981; Sirtori et al. 1978; Graham et al. 2011). In particular, the drug is not metabolized and is eliminated by renal excretion with an overall renal clearance in individuals with good renal function of about 500 ml/min. Active tubular secretion involving transporters represents about 80 % of its overall renal clearance. Some studies estimate higher renal clearances in diabetic patients, perhaps reflecting a disease-dependent regulation in renal transporters (Tucker et al. 1981; Graham et al. 2011). Recovery studies suggest that metformin may undergo some intestinal or hepatic elimination since 20 % of an oral dose is not accounted for in the urine (Tucker et al. 1981). Metformin tissue distribution is broad, consistent with its interaction with organic cation transporters, OCT1, OCT2, and OCT3, which collectively are expressed in many tissues in the body. In particular, the drug is highly distributed to the liver, lung, intestine, and kidney with an apparent volume of distribution ranging from 63 to 276 L after intravenous doses of the drug (Graham et al. 2011). In line with its hydrophilicity, the drug exhibits minimal binding to plasma proteins. The half-life of metformin is around 5 h, and the absolute bioavailability of metformin is 55 ± 16 % (Graham et al. 2011).

22.3 Metformin Pharmacodynamics

Metformin pharmacodynamics have been recently summarized in several excellent review articles (Viollet et al. 2012; Viollet and Foretz 2013; Pernicova and Korbonits 2014). The drug exhibits well-documented clinical effects with respect to lowering blood glucose levels in patients with type 2 diabetes. In particular, it reduces hepatic glucose production and enhances insulin sensitivity (Shu et al. 2007). In addition, studies have suggested that it antagonizes glucagon-mediated glycogenolysis through cAMP-mediated effects. It also produces beneficial effects on lipids and enhances oxidation of fatty acids while reducing their synthesis. The drug's primary action is through inhibition of complex I in the mitochondria though the exact target remains elusive. The drug appears to accumulate in mitochondria and reduces the proton gradient across the mitochondrial membrane, which is essential for synthesis of ATP. As a result, ATP synthesis is reduced, and AMP accumulates. Changes in the levels of these two nucleotides ultimately appear to account for most of the beneficial effects of metformin on glucose metabolism and disposition. In particular, because of its accumulation, AMP effectively binds to AMP kinase (AMPK), the cell's energy sensor, resulting in its phosphorylation or activation. P-AMPK results in a host of regulatory effects on cellular kinases and transcription factors involved in cellular glucose

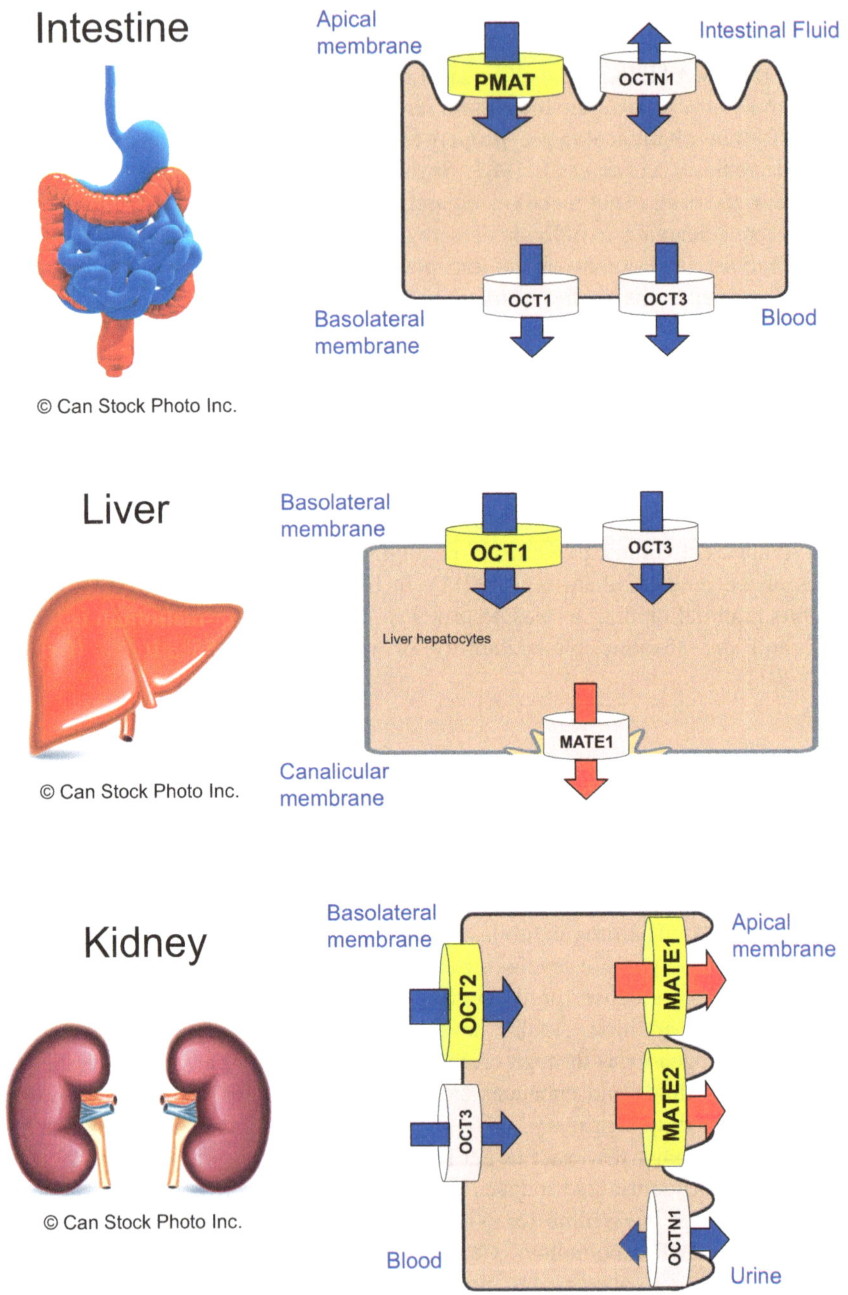

Fig. 22.1 Organic cation transporters play major roles in metformin disposition and response. Metformin is a substrate of several organic cation transporters: OCT1 (*SLC22A1*), OCT2 (*SLC22A2*), OCT3 (*SLC22A3*), OCTN1 (*SLC22A4*), PMAT (*SLC29A4*), MATE1 (*SLC47A1*), and MATE2 (*SLC47A2*). The localization of SLC transporters in three major tissues: the intestine, liver and kidney (Motohashi et al. 2013; Masuda et al. 2006; Nies et al. 2009; Giacomini et al. 2010). The major organic cation transporters that are expressed at high levels are highlighted

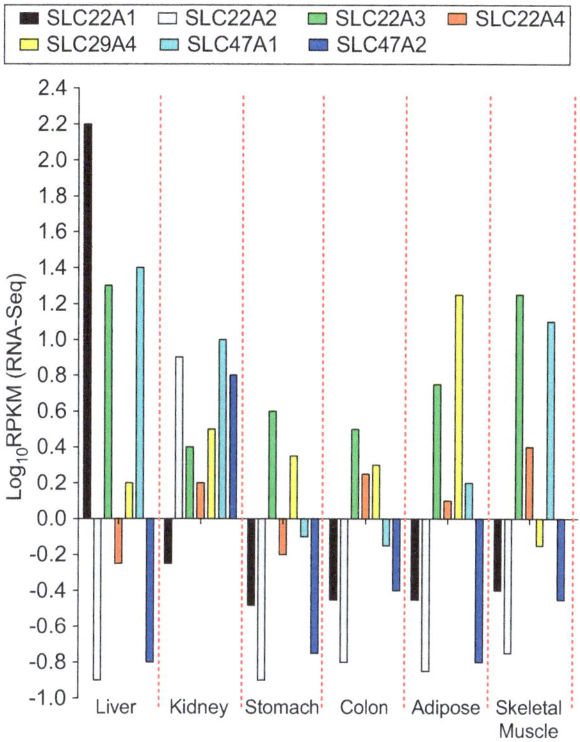

Fig. 22.2 The transcript levels of the seven organic cation transporters known to transport metformin by in vitro, in vivo, and/or genetic studies (see Table 22.1). The expression levels of these transporters are obtained from the GTEx Portal (http://www.broadinstitute.org/gtex/searchGenes) (GTEx Consortium 2013). The median transcript levels were plotted in this panel, and the sample sizes for each tissue are as follows: the liver ($n = 106$); kidney cortex ($N = 94$); stomach ($N = 181$); colon, transverse ($N = 181$); adipose, subcutaneous ($N = 189$); and skeletal muscle ($N = 188$). The individual transcript levels were not publicly available, and hence the median levels were obtained from the GTEx Portal and plotted in Fig. 22.2

disposition, most notably disruption of transcription factor complexes that are responsible for gluconeogenesis. Independent of AMPK, the reduced ATP levels directly result in reduced glucose production as ATP is required in synthesis of glucose molecules.

Fig. 22.1 (continued) in *yellow*. The arrow shows the major directions for the transport of metformin into or away from the cells. Please note that in the intestine, the *arrows* for OCT3 and OCT1 could be in the reverse directions depending on the route of administration. After oral administration of metformin, metformin enters the intestine and is secreted into the blood via OCT1 and OCT3 (as shown in figure). However, after intravenous injection, metformin enters into the intestine from the blood via OCT3 and OCT1. On the other hand, OCTN1 could play important role in bidirectional transport of solutes including metformin across cells (Nakamichi et al. 2013; Kato et al. 2010)

Metformin in vivo pharmacologic effects are dependent on dosage, with increasing doses through 2000 mg associated with improved response with respect to reduction in fasting glucose levels and glycated hemoglobin (HbA1c) (Garber et al. 1997). In addition, the drug shows route-dependent effects in rats. That is, intraduodenal and intra-portal doses are associated with greater effects than intravenous doses (at similar systemic plasma concentrations) suggesting that metformin may produce its pharmacologic effects on the first pass through the intestine and liver (Stepensky et al. 2002). Variation in metformin response is high with about 35 % of patients with diabetes failing to achieve clinically acceptable HbA1c levels on metformin monotherapy (Cook et al. 2007; Kahn et al. 2006). Nongenetic factors that relate to variation in response to metformin are baseline HbA1c levels, with higher baseline levels associating with better response (more HbA1c reduction). Increasing dose and higher serum creatinine levels are also associated with significantly greater response presumably reflecting greater exposure to metformin (or time-averaged systemic plasma levels). The remainder of the chapter will focus on genetic determinants of response to metformin.

22.4 Metformin Pharmacogenomics

22.4.1 Heritability of Metformin Glycemic Response

Before performing pharmacogenomics studies, it is essential to address the question of how much response variation in patients on metformin can be explained by genetic factors, which is often termed as heritability. Historically, the heritability of drug response has been rarely established as traditional twin and family-based study designs are impractical for investigating drug response phenotypes. Recently, alternative methods using population-based GWAS data for heritability estimation have been developed. One of these methods, the Genome-wide Complex Trait Analysis (GCTA) (Yang et al. 2011), can estimate the distant genetic relationship between conventionally unrelated individuals using GWAS SNP data and correlate the genetic similarity to the phenotypic similarity in order to partition the total phenotypic variance into genetic and environmental causes.

The GCTA method was applied to GWAS data from 2058 metformin-treated patients, and the heritability of glycemic response to metformin was estimated (Zhou et al. 2014). Univariate analyses showed that for commonly used definitions of HbA1c-based treatment efficacy measurements, heritability estimates ranged from 20 to 34 %. These heritability estimates are comparable to the heritability of 21 % for BMI and 30 % for Alzheimer's disease as estimated by the GCTA method, suggesting that glycemic response to metformin in patients with type 2 diabetes is reasonably heritable. Chromosome-wise heritability analysis, which was performed to understand the genetic architecture of glycemic response, revealed a significant linear correlation between heritability estimates and chromosome length, suggesting that many variants with small to moderate effects on metformin response are scattered throughout the genome. Response to metformin is associated with baseline glucose level (pretreatment) as well as to the direct effects of the drug (on-treatment glucose level); both components may be determined by genetic factors. To this end, a novel bivariate analysis was applied to jointly estimate the heritability of pretreatment and on-treatment HbA1c measures. The results showed that variants contributing to almost half of the heritability had differential impact on pretreatment and on-treatment HbA1c. Though all of the variants will be informative in predicting treatment outcome, only those having differential impact on pretreatment and on-treatment HbA1c will be informative to the drug action mechanism. It is worth noting that interpretation of all these heritability estimates from the GCTA analyses and those secondary results can only be made in the context of the common SNPs captured by the GWAS arrays. Contributions from the rare variants that are poorly covered by the GWAS panels will not form part of the heritability estimated by GCTA, but will remain in the environmental component. Thus, the heritability of glycemic response to metformin may have been underestimated. Further, sequencing-based genomic studies with an emphasis on the rare metformin response variants may be worth conducting irrespective of heritability estimates from GWAS SNPs.

Collectively, the heritability analyses of metformin glycemic response in type 2 diabetes patients suggests that a moderate proportion of the variance is genetic and reflects underlying biological difference between individuals. As summarized below, the known metformin response variants identified through candidate gene studies and the first GWAS study can only explain a small proportion of the heritability. In future metformin pharmacogenomics, performing GWAS analysis with larger samples could find more genetic variants that enable us make better predictions for personalized or stratified medicine, as well as unravel novel mechanisms of metformin action at reducing hyperglycemia.

22.4.2 Candidate Gene Studies

Over the last decade, a number of candidate gene metformin pharmacogenetic studies have been published. Most of these studies focused on the known variants in transporter genes, which play important roles in metformin accumulations in tissues such as the liver, which are responsible for the pharmacologic action of the drug, and in the kidney, which plays a major role in metformin pharmacokinetics. In Table 22.1, key results from these candidate genes studies related to transporters are summarized. Included in the table are studies focused on healthy volunteers, prediabetic patients, and patients with type 2 diabetes. Findings from knockout mice of these transporters are also included to provide further support of the effects of the associated genes. These candidate gene studies have been recently reviewed (Zolk 2012; Becker et al. 2013). Many of the studies investigated variants in *SLC22A1* (OCT1), *SLC22A2* (OCT2), *SLC47A1* (MATE1), and *SLC47A2*

Table 22.1 Effects of genetic polymorphisms of organic cation transporters on metformin disposition and response

Transporters (gene name)	Study populations (ethnicity)	Phenotypes	Knockout mouse
OCT1 (*SLC22A1*)	Healthy volunteers (Caucasians)	OCT1-reduced function variants ((R61C (rs12208357), G401S (rs34130495), M420del (rs202220802), G465R (rs34059508)) were associated with higher metformin AUC and reduced response to metformin (Watanabe 1918; Bailey and Turner 1996)	Intravenous injection of metformin showed Oct1 knockout mice having higher plasma concentrations and significantly lower metformin levels in the liver, duodenum, jejunum, and ileum. However, there was no difference in the kidney and urinary excretion (Viollet et al. 2012) Oral dose of metformin to mice showed *Oct1* −/− mice have higher AUC in initial phase (0–1 h) (Viollet et al. 2012). There was a trend that Oct1 −/− mice have greater metformin AUC from 0–24 h and smaller oral volume of distribution; however, this is not significant (Bailey and Turner 1996) After 8 weeks of high-fat diet, *Oct1* +/+ mice showed significant reduction in fasting plasma glucose after 5 days of metformin treatment; however, *Oct1* −/− mice showed no effect (Watanabe 1918)
OCT1 (*SLC22A1*)	Healthy volunteers (Caucasians)	OCT1-reduced function variants ((R61C (rs12208357), G401S (rs34130495), M420del (rs202220802), G465R (rs34059508)) were associated with reduce renal clearance (Viollet and Foretz 2013)	
OCT1 (*SLC22A1*)	Healthy volunteers (Korean)	OCT1-reduced function variant (P134L (rs2282143)) was associated with higher metformin AUC (Nakamichi et al. 2013)	
OCT1 (*SLC22A1*)	People with type 2 diabetes (Caucasians)	OCT1-reduced function variants (R61C (rs12208357) and M420del (rs202220802)) were not significantly associated with metformin response (Shu et al. 2007)	
OCT1 (*SLC22A1*)	People with type 2 diabetes (Caucasians)	OCT1-reduced function variants were not significantly associated with metformin response (Chen et al. 2009)	
OCT1 (*SLC22A1*)	People with type 2 diabetes (Caucasians)	The number of OCT1-reduced function alleles (R61C (rs12208357), S189L (rs34104736), G401S (rs34130495), M420del (rs72552763), and G465R (rs34059508)) was correlated with metformin trough levels and HbA1c levels (Chen et al. 2010)	

(continued)

Table 22.1 (continued)

Transporters (gene name)	Study populations (ethnicity)	Phenotypes	Knockout mouse
OCT1 (*SLC22A1*)	People with type 2 diabetes (Caucasians)	An intronic variant (rs622342) associated with HbA1c levels, and also there is an interaction of this variant with MATE1 intronic variant (rs2289669) (Komatsu et al. 2011)	
OCT1 (*SLC22A1*)	People with type 2 diabetes (Caucasians)	OCT1 non-synonymous variant Met408Val (rs628031) was significantly associated with metformin-induced gastrointestinal side effects (Tucker et al. 1981)	
OCT1 (*SLC22A1*)	Polycystic ovarian syndrome (PCOS)	Reduced function OCT1 variants were significantly associated with reduced total cholesterol and triglyceride levels in patients on metformin for PCOS (Sirtori et al. 1978)	
OCT1 (*SLC22A1*)	Postmenopausal diabetic patients with or without cancer (Caucasians)	Investigation of OCT1 variants (rs12208357 and rs622342) and other candidate variants in diabetic patients with and without cancer (Graham et al. 2011)	
OCT2 (*SLC22A2*)	Healthy volunteers (multiple ethnic groups)	OCT2 A270S (rs316019) was associated with increased metformin renal clearance and secretory clearance (Pernicova and Korbonits 2014)	Oct1/Oct2 double knockout mice showed significantly lower liver- and kidney-to-plasma concentration ratios. However, the double knockout mice were not significantly different from wild type for their response to metformin's glucose-lowering effect (Garber et al. 1997)
OCT2 (*SLC22A2*)	People with type 2 diabetes (Asians)	OCT2 A270S was associated with increase plasma lactate in patients treated with metformin (Stepensky et al. 2002)	
OCT2 (*SLC22A2*)	Healthy volunteers (Korean)	OCT2 A270S was associated with higher AUC and Cmax (Nakamichi et al. 2013)	
OCT2 (*SLC22A2*)	People with type 2 diabetes (multiple ethnic groups)	OCT2 A270S (rs316019) was not significantly associated with metformin response (Chen et al. 2009)	

(continued)

Table 22.1 (continued)

Transporters (gene name)	Study populations (ethnicity)	Phenotypes	Knockout mouse
OCT2 (*SLC22A2*)	Healthy volunteers (multiple ethnic groups)	OCT2 A270S (rs316019) was associated with metformin response after adjusted for the MATE1 5′UTR variant g.-66 T > C (Cook et al. 2007)	
OCT2 (*SLC22A2*)	Healthy volunteers (Korean)	OCT2 A270S was associated with decreased renal excretion and increased plasma concentration of metformin (Kahn et al. 2006)	
OCT2 (*SLC22A2*)	Healthy volunteers (Korean)	No significant difference in metformin renal clearance and secretory clearance (Yang et al. 2011)	
MATE1 (*SLC47A1*)	People with type 2 diabetes (Japanese)	Blood lactate levels were not significantly different after metformin treatment in patients with or without MATE1 non-synonymous variants (Zhou et al. 2014)	Mate1 (−/−) is associated with metformin-induced lactic acidosis (Zhou et al. 2014) Mate1 (−/−) has higher metformin levels in the liver (Zhou et al. 2014) Mate1 (−/−) has higher metformin AUC levels, lower urinary excretion of metformin (Zolk 2012)
MATE1 (*SLC47A1*)	Healthy volunteers (multiple ethnic groups)	The reduced function MATE1 5′UTR variant g.-66 T > C (rs2252281) associated with enhanced response to metformin effects on oral glucose tolerance test (Becker et al. 2013)	
MATE1 (*SLC47A1*)	People with type 2 diabetes (Caucasians)	MATE1 intron variant (rs2289669) was associated with metformin response (Zhou et al. 2007; Yoon et al. 2013)	
MATE1 (*SLC47A1*)	Prediabetic patients (multiple ethnic groups)	MATE1 intron variant (rs8065082, in high linkage disequilibrium to rs2289669) was associated with metformin response (Zhou et al. 2009)	
MATE1 (*SLC47A1*)	People with type 2 diabetes (multiple ethnic groups)	MATE1 intron variant (rs2289669) was not significantly associated with metformin response (Chen et al. 2009)	

(continued)

Table 22.1 (continued)

Transporters (gene name)	Study populations (ethnicity)	Phenotypes	Knockout mouse
MATE2 (*SLC47A2*)	Healthy volunteers (Caucasians and Korean)	MATE2K 5′UTR variant g.-130G > A (rs12943590) significantly associated with reduced response to metformin effects on oral glucose tolerance test and change in HbA1c (Chen et al. 2009; Yang et al. 2011)	No mouse studies
OCTN1 (*SLC22A4*)	Healthy volunteers (Korean)	OCTN1 T306I (rs272893) showed significant effect on metformin pharmacokinetics. Individuals with variant allele showed higher peak concentration and greater AUC levels (Nakamichi et al. 2013)	The maximum plasma concentration (Cmax) after oral administration of metformin (50 mg/kg) in Octn1−/− was higher than that in wild-type mice, However, Cmax in Octn1−/− mice given at higher dose (175 mg/kg) was lower than that in wild-type mice (Stocker et al. 2013)
OCT2 (*SLC22A2*) and MATE1 (*SLC47A1*)	Healthy volunteers (Caucasians)	OCT2 A270S was not significantly associated with renal or secretory clearance of metformin. However, renal and secretory clearance increased with minor allele A270S in subjects with homozygous reference allele of MATE1 promoter variant, g.-66T > C (rs2252281) (Cook et al. 2007)	No mouse studies
OCT1 (*SLC22A1*) and MATE1 (*SLC47A1*)	People with type 2 diabetes (Caucasians)	A genetic interaction between OCT1 intron variant (rs622342) and MATE1 intron variant (rs2289669) was reported (Komatsu et al. 2011)	No mouse studies

This table summarizes the results from studies of (i) genetic polymorphisms and (ii) knockout mice in organic cation transporters (*SLC22A1*, *SLC22A2*, *SLC22A4*, *SLC47A1*, and *SLC47A2*) that have been investigated in metformin disposition and response. References for these studies are available in the footnote

Footnote: References from Table 22.1

1. Shu, Y. et al. Effect of genetic variation in the organic cation transporter 1 (OCT1) on metformin action. *J Clin Invest* **117**, 1422–31 (2007).
2. Shu, Y. et al. Effect of genetic variation in the organic cation transporter 1, OCT1, on metformin pharmacokinetics. *Clin Pharmacol Ther* **83**, 273–80 (2008).
3. Wang, D.S. et al. Involvement of organic cation transporter 1 in hepatic and intestinal distribution of metformin. *J Pharmacol Exp Ther* **302**, 510–5 (2002).

(continued)

Table 22.1 (continued)

Transporters (gene name)	Study populations (ethnicity)	Phenotypes	Knockout mouse

4. Tzvetkov, M.V. et al. The effects of genetic polymorphisms in the organic cation transporters OCT1, OCT2, and OCT3 on the renal clearance of metformin. *Clin Pharmacol Ther* **86**, 299–306 (2009).
5. Yoon, H., Cho, H.Y., Yoo, H.D., Kim, S.M. & Lee, Y.B. Influences of organic cation transporter polymorphisms on the population pharmacokinetics of metformin in healthy subjects. *AAPS J* **15**, 571–80 (2013).
6. Zhou, K. et al. Reduced-function SLC22A1 polymorphisms encoding organic cation transporter 1 and glycemic response to metformin: a GoDARTS study. *Diabetes* **58**, 1434–9 (2009).
7. Choi, J.H. et al. A common 5′-UTR variant in MATE2-K is associated with poor response to metformin. *Clin Pharmacol Ther* **90**, 674–84 (2011).
8. Christensen, M.M. et al. The pharmacogenetics of metformin and its impact on plasma metformin steady-state levels and glycosylated hemoglobin A1c. *Pharmacogenet Genomics* **21**, 837–50 (2011).
9. Becker, M.L. et al. Interaction between polymorphisms in the OCT1 and MATE1 transporter and metformin response. *Pharmacogenet Genomics* **20**, 38–44 (2010).
10. Tarasova, L. et al. Association of genetic variation in the organic cation transporters OCT1, OCT2 and multidrug and toxin extrusion 1 transporter protein genes with the gastrointestinal side effects and lower BMI in metformin-treated type 2 diabetes patients. *Pharmacogenet Genomics* **22**, 659–66 (2012).
11. Gambineri, A. et al. Organic cation transporter 1 polymorphisms predict the metabolic response to metformin in women with the polycystic ovary syndrome. *J Clin Endocrinol Metab* **95**, E204-8 (2010).
12. Berstein, L.M., Iyevleva, A.G., Vasilyev, D., Poroshina, T.E. & Imyanitov, E.N. Genetic polymorphisms potentially associated with response to metformin in postmenopausal diabetics suffering and not suffering with cancer. *Cell Cycle* **12**, 3681–8 (2013).
13. Chen, Y. et al. Effect of genetic variation in the organic cation transporter 2 on the renal elimination of metformin. *Pharmacogenet Genomics* **19**, 497–504 (2009).
14. Higgins, J.W., Bedwell, D.W. & Zamek-Gliszczynski, M.J. Ablation of both organic cation transporter (OCT)1 and OCT2 alters metformin pharmacokinetics but has no effect on tissue drug exposure and pharmacodynamics. *Drug Metab Dispos* **40**, 1170–7 (2012).
15. Li, Q. et al. SLC22A2 gene 808 G/T variant is related to plasma lactate concentration in Chinese type 2 diabetics treated with metformin. *Acta Pharmacol Sin* **31**, 184–90 (2010).
16. Christensen, M.M. et al. A gene-gene interaction between polymorphisms in the OCT2 and MATE1 genes influences the renal clearance of metformin. *Pharmacogenet Genomics* **23**, 526–34 (2013).
17. Song, I.S. et al. Genetic variants of the organic cation transporter 2 influence the disposition of metformin. *Clin Pharmacol Ther* **84**, 559–62 (2008).
18. Chung, J.Y. et al. Functional characterization of MATE2-K genetic variants and their effects on metformin pharmacokinetics. *Pharmacogenet Genomics* **23**, 365–73 (2013).
19. Toyama, K. et al. Loss of multidrug and toxin extrusion 1 (MATE1) is associated with metformin-induced lactic acidosis. *Br J Pharmacol* **166**, 1183–91 (2012).
20. Tsuda, M. et al. Targeted disruption of the multidrug and toxin extrusion 1 (mate1) gene in mice reduces renal secretion of metformin. *Mol Pharmacol* **75**, 1280–6 (2009).
21. Stocker, S.L. et al. The effect of novel promoter variants in MATE1 and MATE2 on the pharmacokinetics and pharmacodynamics of metformin. *Clin Pharmacol Ther* **93**, 186–94 (2013).
22. Becker, M.L. et al. Genetic variation in the multidrug and toxin extrusion 1 transporter protein influences the glucose-lowering effect of metformin in patients with diabetes: a preliminary study. *Diabetes* **58**, 745–9 (2009).
23. Tkac, I. et al. Pharmacogenomic association between a variant in SLC47A1 gene and therapeutic response to metformin in type 2 diabetes. *Diabetes Obes Metab* **15**, 189–91 (2013).

(continued)

Table 22.1 (continued)

Transporters (gene name)	Study populations (ethnicity)	Phenotypes	Knockout mouse

24. Jablonski, K.A. et al. Common variants in 40 genes assessed for diabetes incidence and response to metformin and lifestyle intervention in the diabetes prevention program. *Diabetes* **59**, 2672–81 (2010).
25. Nakamichi, N. et al. Involvement of carnitine/organic cation transporter OCTN1/SLC22A4 in gastrointestinal absorption of metformin. *J Pharm Sci* **102**, 3407–17 (2013).

(MATE2) in pharmacodynamic and pharmacokinetic studies in healthy volunteers and in participants with type 2 diabetes. More recently, *SLC22A3* (OCT3), *SLC29A4* (ENT4), and *SLC22A4* (OCTN1) have been shown to transport metformin (Nakamichi et al. 2013; Chen et al. 2010; Zhou et al. 2007); however, only variants in OCTN1 have been associated with metformin disposition in healthy volunteers and knockout mice (Nakamichi et al. 2013; Yoon et al. 2013). Collectively, the studies shed important light on the clinical pharmacokinetic mechanisms of metformin.

However, genetic variants in transporters generally have small effects on metformin pharmacokinetics and pharmacodynamics, and as a result, their effects may not be reproducible among studies (Table 22.1). For example non-synonymous reduced function variants in OCT1 have been shown to significantly reduce metformin uptake in cellular assays and in healthy volunteers associated with reduced metformin response. However, *SLC22A1* variants had no impact on treatment efficacy in 1531 patients with type 2 diabetes in Europe (Zhou et al. 2009). Clinical confounders such as polypharmacy and gene-by-gene interactions between various transporters involved in metformin response may mask the genetic impact of *SLC22A1* variants. Recent studies in Caucasian and African American patients with type 2 diabetes demonstrated that the impact of a genetic variant in *SLC47A1* could only be observed in individuals who harbor the *SLC22A1* reference allele (Stocker et al. 2013). The study highlights the need to investigate the variants in these transporters jointly in pharmacogenomic studies of people with type 2 diabetes, on the basis of known pharmacokinetic information. However, larger samples will be required as the statistical power diminishes quickly with the increasing number of gene-by-gene tests.

As noted in Table 22.1, most of the current candidate gene studies were performed in healthy volunteers or diabetes patients with ancestry in Europe. Clearly there is an enormous need to study other populations, as type 2 diabetes is a global epidemic and metformin is used in millions of people worldwide. Though metformin is very safe, it does exhibit gastrointestinal side effects in some patients. Of the studies performed to date, there is only one study that focused on the identification of genetic variants that associate with metformin gastrointestinal side effects (Table 22.1) (Tarasova et al. 2012).

In addition to identification of genetic variants that associate with glycemic response to metformin in diabetic patients, several studies have focused on identifying genetic variants that associate with metformin efficacy in different diseases. These are briefly summarized below.

Individuals with Prediabetes The Diabetes Prevention Program (DPP) includes patients at increased risk for diabetes, who are randomized to metformin, lifestyle changes, or placebo for prevention. Though intense lifestyle changes have associated with the best outcome, metformin has been clearly shown to prevent diabetes in the DPP. A large candidate gene study carried out in 2994 individuals in the DPP involved genotyping 1590 SNPs in 40 genes encoding transporters, proteins involved in metformin action, and proteins associated with type 2 diabetes (Jablonski et al. 2010). A SNP in *SLC47A1* (MATE1) intron (rs8065082) was associated with metformin response and replicated the finding from a previous study in patients with type 2 diabetes, which reported an association of the *SLC47A1* intronic SNP rs2289669 in high linkage disequilibrium (LD) with rs8065082, with response to metformin (Table 22.1) (Becker et al. 2009). Consistent evidence has also been obtained by Tkac et al. (2013).

Women with PCOS Metformin is also used to treat women with PCOS due to its effect on improving insulin resistance (Pau et al. 2014). Recent study suggests that metformin results in symptomatic improvement of patients with PCOS by exerting effects on lowering glucose, testosterone, and androstenedione (Pau et al. 2014). A group in Italy reported that OCT1-reduced function variants predict metformin response in patients with PCOS (Gambineri et al. 2010).

Cancer Patients Metformin use has been shown to improve survival in patients with cancer (Zhang and Li 2014). Although its mechanisms are not fully understood, there are several ongoing clinical trials to evaluate the effect of metformin on improving treatment and survival in cancer patients. The effect of OCT1 variants with response to metformin in diabetic patients with or without cancer is under investigation (Berstein et al. 2013).

Transporters have come under scrutiny in drug development as being associated with clinical drug-drug interactions. As such, inhibitors of transporters may phenocopy genetic variants. In particular, prescription drugs that inhibit metformin transporters in the liver or kidney could potentially modulate metformin effects. Although it is not the focus of this chapter, it is worth mentioning that several clinical studies have been performed in healthy subjects and have observed effects on metformin disposition with administration of prescription drugs that are inhibitors of organic cation transporters in the SLC22 and 47 families (Ding et al. 2014; Grun et al. 2013; Kusuhara et al. 2011).

22.4.3 Genome-Wide Association Study

Only one genome-wide association study (GWAS) of metformin glycemic response has been published to date (Zhou et al. 2011). The authors screened 700,000 polymorphisms in 1024 metformin users from the GoDARTS cohort and replicated the top hit in a further 1783 GoDARTS participants as well as 1113 UKPDS participants. They identified variant rs11212617 near the *ATM* gene that associated with achieving a treatment target of 7 % HbA1c ($P = 2 \times 10^{-9}$) in the combined analysis of the three data sets. The same variant was also associated with a per-allele change difference in HbA1c of 0.11 % ($P = 6.7 \times 10^{-7}$). The variant could explain up to 2.5 % of the variance in HbA1c reduction among the UKPDS participants. Subsequently, a meta-analysis of three other independent cohorts of type 2 diabetes also replicated the association (van Leeuwen et al. 2012).

Despite the concrete evidence of association in people with type 2 diabetes, in an attempt to replicate the finding in prediabetes patients, no association was observed between rs11212617 and diabetes incidence in 988 metformin-treated DPP participants (Florez et al. 2012). Secondary analyses also found no association with HbA1c change in the DPP. Further stratified analysis of Caucasian sample in the DPP showed that ethnic differences were an unlikely explanation to the discrepancy in genetic impacts of rs11212617 on type 2 diabetes and people with prediabetes. Given the sample size in the DPP, this study had enough statistical power (>99 %) to detect an allelic effect of 0.61 % HbA1c change as reported in the replication sample of UKPDS in the initial GWAS report. The differential genetic effect of rs11212617 on metformin efficacy in these two large pharmacogenetic studies suggests that some genetic variants might be expected to have an impact on glycemic response in certain physiological conditions or a narrow range of glucose levels. Even without more confirmed genetic variants showing differential impact on response to metformin between people with type 2 diabetes and prediabetes, such expectation is not unrealistic given the latest results in GWAS analyses of type 2 diabetes by the DIAGRAM consortium and fasting glucose in MAGIC. Clearly, there is a significant overlap between the genetic determinants of glucose levels in the normal physiological state as reported by the MAGIC and those in the pathophysiological state of type 2 diabetes, suggesting common avenues of glycemic control. However, there are also a significant number of variants specific to each state suggesting certain levels of difference in glycemic control between the two physiological states. Although prediabetes is an intermediate state between normal physiological conditions and type 2 diabetes state, differences may exist in the glycemic control between type 2 diabetes and prediabetes states. Therefore, the genetic impact on metformin treatment efficacy may also vary regardless of the measurements of treatment outcome. On the other hand, considering the fact that variants in *SLC47A1* have been shown to affect metformin treatment efficacy in similar manner in both people with prediabetes and those with type 2 diabetes, joint analyses of data from cohorts of both types are still warranted as long as the treatment outcomes are defined in the same framework.

Compared to the candidate gene studies based on specific prior biological or functional knowledge, GWAS analysis is hypothesis-free and bears the potential to reveal novel mechanism of drug action. Indeed the first metformin response GWAS identified metformin efficacy associated with a group of variants in strong LD spanning seven genes, none of which has been previously implicated in the pathway of metformin action, and identification of the causal gene at this locus is likely to shed new light on the action mechanism of this old drug. Previous reports showed that individuals carrying the recessive mutations of *ATM*, one of the seven genes in the LD block, developed the ataxia telangiectasia syndrome and were insulin resistant. The authors of the GWAS showed that activation of AMPK by metformin was reduced by treatment with an ATM inhibitor, KU-55933, and suggested that *ATM* was the causal gene for metformin response with a direct role in the activation of AMPK by metformin. However, as shown by two other groups independently, KU-55933, the supposedly specific inhibitor of ATM, has an off-target effect of altering metformin uptake by the organic cation transporter OCT1 (Woods et al. 2012; Yee et al. 2012). Thus the question of which of the seven genes is causal in metformin glycemic response remains unanswered, and several genes in the loci may represent viable candidates.

22.5 Conclusions and Future Directions

In conclusion, many candidate gene studies of metformin disposition and glycemic response have been performed and have added to the body of knowledge of mechanisms of metformin action and disposition in people with type 2 diabetes. A single GWAS has revealed new genes that may potentially be involved in metformin pharmacologic action in people with type 2 diabetes. However, the study was not replicated in people with prediabetes perhaps because of differences in pathological states of the populations. Nevertheless, the GWAS identified a locus with novel genes that need to be studied for their effects on metformin response. Future human genetic studies should be in four major directions.

- First and foremost, there is an enormous need to identify genetic variants that predict glycemic response to metformin in other ethnic groups in addition to those of European ancestries. For example, metformin is commonly prescribed in African Americans, a population with the highest prevalence of type 2 diabetes among the four major ethnic groups in the United States (African Americans, Asian Americans, European Americans, and Hispanics). Moreover, metformin is increasingly being used on a global basis as the epidemic of metabolic syndrome continues to grow. In particular, East Asian countries are experiencing alarming increases in the rates of type 2 diabetes. Moreover, South Asians are at high risk for type 2 diabetes, and other populations globally have not been studied. Clearly pharmacogenomics studies identifying genetic variants that associate with glycemic response to metformin are needed in all ethnic groups.
- Second, large sample sets are needed. To date, only common variants have been identified, and only a limited number of patients have been studied. Heritability estimates suggest that many more common variants that associate with glycemic response to metformin have not been identified because of issues related to small sample sizes. Consortia and other mechanisms to pool samples are clearly

needed. Further, gene-by-gene interactions can only be identified in larger sample sets.

- Third, rare variants have not been ascertained because of limited sample sizes. Rare variants with large effect sizes may be identified in larger sample sets or in studies of extreme responders and nonresponders.
- Finally, studies need to extend beyond initial glycemic response to metformin. Diabetes ultimately leads to a host of life-threatening morbidities, which provide the rationale for glycemic control with metformin and other diabetic agents. Studies focused on susceptibility to diabetic complications on metformin such as cardiovascular events, microvessel disease, and other pathologies are clearly needed. These studies should extend to complications on metformin combination therapies, which are commonly used as diabetes progresses. Genetic determinants of resistance to metformin have not been identified. It is clear that some individuals are initially resistant to metformin, whereas others develop resistance over time. What are the genetic factors that underlie longer-term resistance to the drug?

Though not a part of this chapter, genetic determinants that underlie response to metformin for other diseases beyond diabetes such as cancer and PCOS need to be identified. We envision a future in which investigators worldwide come together and share samples and data involving metformin disposition and therapeutic response across a spectrum of human disease. Only in this setting can we begin to understand the mechanisms of action and predictors of response to this widely prescribed and highly beneficial therapeutic agent.

References

Bailey CJ, Turner RC (1996) Metformin. N Engl J Med 334:574–579

Becker ML et al (2009) Genetic variation in the multidrug and toxin extrusion 1 transporter protein influences the glucose-lowering effect of metformin in patients with diabetes: a preliminary study. Diabetes 58:745–749

Becker ML, Pearson ER, Tkac I (2013) Pharmacogenetics of oral antidiabetic drugs. Int J Endocrinol 2013:686315

Berstein LM, Iyevleva AG, Vasilyev D, Poroshina TE, Imyanitov EN (2013) Genetic polymorphisms potentially associated with response to metformin in postmenopausal diabetics suffering and not suffering with cancer. Cell Cycle 12:3681–3688

Chen Y et al (2009) Effect of genetic variation in the organic cation transporter 2 on the renal elimination of metformin. Pharmacogenet Genomics 19:497–504

Chen L et al (2010) Role of organic cation transporter 3 (SLC22A3) and its missense variants in the pharmacologic action of metformin. Pharmacogenet Genomics 20:687–699

Cook MN, Girman CJ, Stein PP, Alexander CM (2007) Initial monotherapy with either metformin or sulphonylureas often fails to achieve or maintain current glycaemic goals in patients with Type 2 diabetes in UK primary care. Diabet Med 24:350–358

Ding Y et al (2014) The effect of lansoprazole, an OCT inhibitor, on metformin pharmacokinetics in healthy subjects. Eur J Clin Pharmacol 70:141–146

Florez JC et al (2012) The C allele of ATM rs11212617 does not associate with metformin response in the Diabetes Prevention Program. Diabetes Care 35:1864–1867

Gambineri A et al (2010) Organic cation transporter 1 polymorphisms predict the metabolic response to metformin in women with the polycystic ovary syndrome. J Clin Endocrinol Metab 95:E204–E208

Garber AJ, Duncan TG, Goodman AM, Mills DJ, Rohlf JL (1997) Efficacy of metformin in type II diabetes: results of a double-blind, placebo-controlled, dose–response trial. Am J Med 103:491–497

Giacomini KM et al (2010) Membrane transporters in drug development. Nat Rev Drug Discov 9:215–236

Graham GG et al (2011) Clinical pharmacokinetics of metformin. Clin Pharmacokinet 50:81–98

Grun B et al (2013) Trimethoprim-metformin interaction and its genetic modulation by OCT2 and MATE1 transporters. Br J Clin Pharmacol 76:787–796

GTEx Consortium (2013) The Genotype-Tissue Expression (GTEx) project. Nat Genet 45:580–585

Jablonski KA et al (2010) Common variants in 40 genes assessed for diabetes incidence and response to metformin and lifestyle intervention in the diabetes prevention program. Diabetes 59:2672–2681

Kahn SE et al (2006) Glycemic durability of rosiglitazone, metformin, or glyburide monotherapy. N Engl J Med 355:2427–2443

Kato Y et al (2010) Gene knockout and metabolome analysis of carnitine/organic cation transporter OCTN1. Pharm Res 27:832–840

Komatsu T et al (2011) Characterization of the human MATE2 proton-coupled polyspecific organic cation exporter. Int J Biochem Cell Biol 43:913–918

Kusuhara H et al (2011) Effects of a MATE protein inhibitor, pyrimethamine, on the renal elimination of metformin at oral microdose and at therapeutic dose in healthy subjects. Clin Pharmacol Ther 89:837–844

Masuda S et al (2006) Identification and functional characterization of a new human kidney-specific H+/organic cation antiporter, kidney-specific multidrug and toxin extrusion 2. J Am Soc Nephrol 17:2127–2135

Motohashi H et al (2013) Precise comparison of protein localization among OCT, OAT, and MATE in human kidney. J Pharm Sci 102:3302–3308

Nakamichi N et al (2013) Involvement of carnitine/organic cation transporter OCTN1/SLC22A4 in gastrointestinal absorption of metformin. J Pharm Sci 102:3407–3417

Nies AT et al (2009) Expression of organic cation transporters OCT1 (SLC22A1) and OCT3 (SLC22A3) is affected by genetic factors and cholestasis in human liver. Hepatology 50:1227–1240

Pau CT, Keefe C, Duran J, Welt C (2014) Metformin improves glucose effectiveness, not insulin sensitivity: predicting treatment response in women with polycystic ovary syndrome in an open-label, interventional study. J Clin Endocrinol Metab 99(5):1870–1878

Pernicova I, Korbonits M (2014) Metformin-mode of action and clinical implications for diabetes and cancer. Nat Rev Endocrinol 10:143–156

Shu Y et al (2007) Effect of genetic variation in the organic cation transporter 1 (OCT1) on metformin action. J Clin Invest 117:1422–1431

Sirtori CR et al (1978) Disposition of metformin (N, N-dimethylbiguanide) in man. Clin Pharmacol Ther 24:683–693

Stepensky D, Friedman M, Raz I, Hoffman A (2002) Pharmacokinetic-pharmacodynamic analysis of the glucose-lowering effect of metformin in diabetic rats reveals first-pass pharmacodynamic effect. Drug Metab Dispos 30:861–868

Stocker SL et al (2013) The effect of novel promoter variants in MATE1 and MATE2 on the pharmacokinetics and pharmacodynamics of metformin. Clin Pharmacol Ther 93:186–194

Tarasova L et al (2012) Association of genetic variation in the organic cation transporters OCT1, OCT2 and multidrug and toxin extrusion 1 transporter protein genes with the gastrointestinal side effects and lower BMI in metformin-treated type 2 diabetes patients. Pharmacogenet Genomics 22:659–666

Tkac I et al (2013) Pharmacogenomic association between a variant in SLC47A1 gene and therapeutic response to metformin in type 2 diabetes. Diabetes Obes Metab 15:189–191

Tucker GT et al (1981) Metformin kinetics in healthy subjects and in patients with diabetes mellitus. Br J Clin Pharmacol 12:235–246

van Leeuwen N et al (2012) A gene variant near ATM is significantly associated with metformin treatment response in type 2 diabetes: a replication and meta-analysis of five cohorts. Diabetologia 55:1971–1977

Viollet B, Foretz M (2013) Revisiting the mechanisms of metformin action in the liver. Ann Endocrinol (Paris) 74:123–129

Viollet B et al (2012) Cellular and molecular mechanisms of metformin: an overview. Clin Sci (Lond) 122:253–270

Watanabe CK (1918) Studies in the metabolic changes induced by administration of guanidine bases. J Biol Chem 33:253

Woods A, Leiper JM, Carling D (2012) The role of *ATM* in response to metformin treatment and activation of AMPK. Nat Genet 44:360–361

Yang J, Lee SH, Goddard ME, Visscher PM (2011) GCTA: a tool for genome-wide complex trait analysis. Am J Hum Genet 88:76–82

Yee SW, Chen L, Giacomini KM (2012) The role of *ATM* in response to metformin treatment and activation of AMPK. Nat Genet 44:359–360

Yoon H, Cho HY, Yoo HD, Kim SM, Lee YB (2013) Influences of organic cation transporter polymorphisms on the population pharmacokinetics of metformin in healthy subjects. AAPS J 15:571–580

Zhang ZJ, Li S (2014) The prognostic value of metformin for cancer patients with concurrent diabetes: a systematic review and meta-analysis. Diabetes Obes Metab 16:707–710

Zhou M, Xia L, Wang J (2007) Metformin transport by a newly cloned proton-stimulated organic cation transporter (plasma membrane monoamine transporter) expressed in human intestine. Drug Metab Dispos 35:1956–1962

Zhou K et al (2009) Reduced-function SLC22A1 polymorphisms encoding organic cation transporter 1 and glycemic response to metformin: a GoDARTS study. Diabetes 58:1434–1439

Zhou K et al (2011) Common variants near ATM are associated with glycemic response to metformin in type 2 diabetes. Nat Genet 43:117–120

Zhou K et al (2014) Heritability of variation in glycaemic response to metformin: a genome-wide complex trait analysis. Lancet Diabetes Endocrinol 2(6):481–487

Zolk O (2012) Disposition of metformin: variability due to polymorphisms of organic cation transporters. Ann Med 44:119–129

Chapter 23
Pharmacogenetics of Sulfonylureas

Peter Kovacs and Ewan Pearson

Abstract Despite tremendous efforts and recent advances in the understanding of the molecular biology and genetics of type 2 diabetes (T2D), its translation into clinical practice remains limited primarily to monogenic forms of diabetes. Pharmacogenetics, which focuses on the relationship between individual gene variants and their influence on drug action or side effect, may substantially improve patients' health by enabling application of therapies targeted to patient subgroups. The present review focuses on the pharmacogenetic aspects of treatment with the widely used oral antidiabetic drugs—sulfonylureas (SUs). Initially, the review addresses two dramatic clinical applications of pharmacogenetics in diabetes—extreme sensitivity to SUs in patients with *HNF1A* MODY and SU treatment in patients with neonatal diabetes due to mutations in the potassium channel genes *KCNJ11* (encoding Kir6.2) and *ABCC8* (encoding SUR1). In addition to monogenic forms, pharmacogenetic aspects of treatment with SUs in polygenic T2D will be elaborated as well. Whereas pharmacodynamic variation in polygenic T2D is attributed to genetic variants in *ABCC8*/*KCNJ11* and *TCF7L2*, known to be strongly associated with T2D, variants in *CYP2C9* encoding sulfonylurea metabolising cytochrome P450 isoenzyme 2C9 play a major role in pharmacokinetic variation in polygenic T2D. Despite currently lacking large-scale population studies addressing the pharmacogenomics of SUs, it is our hope that given the recently established international collaborative efforts in this field, the next few years will see pharmacogenomics of SUs in T2D mirror the success seen in monogenic diabetes.

P. Kovacs, PhD (✉)
IFB Adiposity Diseases, University of Leipzig, Liebigstrasse 21, 04103 Leipzig, Germany
e-mail: peter.kovacs@medizin.uni-leipzig.de

E. Pearson, PhD (✉)
Medical Research Institute, Ninewells Hospital and Medical School, University of Dundee, Mail Box 12, Level 5, Dundee DD1 9SY, UK
e-mail: E.Z.Pearson@dundee.ac.uk

J.C. Florez (ed.), *The Genetics of Type 2 Diabetes and Related Traits*,
DOI 10.1007/978-3-319-01574-3_23

23.1 Introduction

Personalised medicine, as one of the major challenges of modern medicine, may greatly benefit from better knowledge of genetic mechanisms underlying human diseases, which is essential to improve diagnostics, prognostics and treatment strategies. The role of human as well as animal genetics in illuminating molecular processes in the pathophysiology of human diseases has been well documented. For instance, it has been reported that variants in the disintegrin and metalloprotease 33 gene (*ADAM33*) confer a 1.4-fold increased risk for asthma (Blakey et al. 2005). This may mean little to the individual patient when considering the complexity of disease aetiology and numerous additional factors contributing to the manifestation of the clinical outcome. However, it was exactly these studies which ultimately helped to clarify the exact role of metalloproteases in the function of fibroblasts and smooth muscle cells. Likewise, genome-wide association studies (GWAS) for diabetes and related traits revealed multiple variants with modest associations which pointed to the beta cell as central to the pathophysiology of type 2 diabetes (T2D) (Pal and McCarthy 2013). Yet, the contribution of genetics should not be limited to better understanding of biological functions and disease mechanisms. From a clinical point of view, it is potentially more relevant to derive clinical application for the individual patient, i.e., personalising their management. In this context, the link between breast cancer and mutations in the tumour suppressor genes *BRCA1* and *BRCA2* is a good example. Women with specific mutations in these genes are at high risk of developing breast cancer and thus are advised to undergo a prophylactic, bilateral mastectomy (Meijers-Heijboer et al. 2001). In diabetes, it is particularly the monogenic forms such as maturity onset diabetes of the young (MODY) whose treatment may strongly depend on the underlying molecular mechanisms. There are clear differences between the diabetes caused by the different MODY genes, which result in discrete clinical entities. For instance, patients with glucokinase (GCK) mutations have mild fasting hyperglycaemia from birth which deteriorates very little with age. Consequently, pharmacological treatment is rarely required, and it is only occasionally associated with microvascular complications (Hatterslay 2005) (see also Chap. 11).

Despite tremendous efforts and recent advances in the understanding of the molecular biology and genetics of T2D, its translation into clinical practice remains limited, primarily to monogenic forms of diabetes. This may be attributed to the physiological and genetic complexity of T2D, but also to genetic heterogeneity of the studied populations, which makes judging the effects of genetic variants in a single patient with diabetes highly challenging. Nevertheless, it is reasonable to assume that there is likely to be greater clinical application resulting from genetic studies of T2D as our knowledge of its genetic basis continues to expand. In particular, pharmacogenetics, which focuses on the relationship between individual gene variants and their influence on drug action or side effect, may substantially improve patients' health by enabling a stratified application of current therapies or development of novel therapies targeted to patient subgroups (Roden et al. 2006).

The present review focuses on the pharmacogenetic aspects of treatment with widely used oral antidiabetic drugs—sulfonylureas (SUs).

23.2 Mechanism of Action of Sulfonylureas

Sulfonylureas have been in clinical use for more than 60 years and are a potent anti-hyperglycaemic drug established to reduce microvascular complications of diabetes [UK Prospective Diabetes Study (UKPDS) Group 1998]. They have two major side effects: insulin secretion is stimulated even in the presence of normal or low glucose and thus hypoglycaemia is a significant risk; hyperinsulinemia results in an increase in body weight. SUs act on the pancreatic beta cell to promote insulin secretion by bringing about closure of ATP-sensitive potassium channels at the pancreatic beta-cell membrane (Fig. 23.1). The ATP-sensitive potassium channels are heteromultimers including an inward rectifying potassium channel ($K_{ir}6.1$ or $K_{ir}6.2$) in close association with SU binding sites. Various tissue-specific isoforms of SU-receptors (SUR) are known: SUR-1 in pancreatic beta cells, SUR-2A in cardiac cells and SUR-2B in smooth muscle cells (Seino 1999). The different SUs bind with differing affinity to these SUR subtypes. Tolbutamide, chlorpropamide, gliclazide and glipizide bind specifically to SUR-1, whereas glibenclamide and glipizide bind to both SUR-1 and SUR-2 subunits (Evans et al. 2008). See Chap. 18 for further details.

23.3 Variability of Glycaemic Response to Sulfonylureas and Potential Mechanisms

There is considerable variation in efficacy of action of SUs. Whilst some of this variation in response could reflect variation in adherence or lifestyle, it is likely that some of the variation is intrinsic to the individual. These intrinsic differences may reflect differences in diabetes aetiology or differences in SU pharmacokinetics or variation in the pharmacodynamics of SUs. These will be discussed in turn, initially focusing on the two dramatic clinical applications of pharmacogenetics in diabetes—extreme sensitivity to SUs in patients with MODY 3 and SU treatment in patients with neonatal diabetes due to mutations in the potassium channel genes *KCNJ11* (encoding $K_{ir}6.2$; potassium inwardly rectifying channel, subfamily J, member 11) and *ABCC8* [encoding SUR1; ATP-binding cassette, subfamily C (CFTR/MRP), member 8].

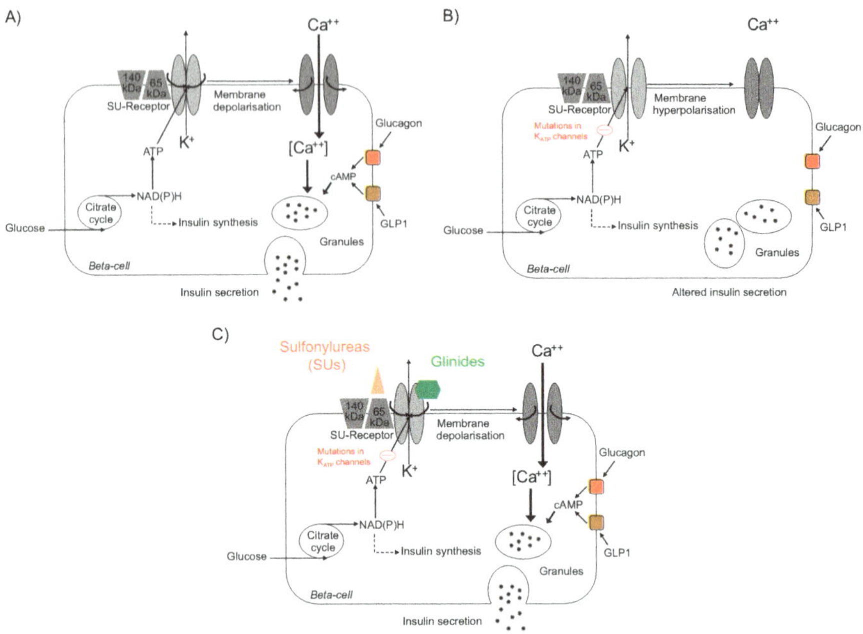

Fig. 23.1 Proposed model of the glycaemic response to sulfonylureas in beta cells with mutations in the Kir6.2 subunit of the K_{ATP} channels.
(**a**) Beta-cell regulation of nutrient stimulated insulin secretion under physiologic conditions. Glucose entering the beta-cell is metabolised and the increased ATP leads to closure of the K_{ATP} channels, which results in depolarisation of the membrane. This keeps the voltage-gated calcium channels open, allowing calcium influx which stimulates exocytosis of insulin-containing secretory granules. (**b**) Since mutations in the Kir6.2 subunit of the K_{ATP} channel are less sensitive to ATP inhibition, K_{ATP} channels remain open in the presence of glucose, which keeps the plasma membrane hyperpolarized. The calcium channels remain closed and insulin cannot be released from the granules. Also stimuli such as GLP1 are ineffective. (**c**) Sulfonylureas can bind to SUR1 subunit of the K_{ATP} channels, close them and so, depolarize the membrane. Consequently, the voltage-gated calcium channels open, allowing calcium influx and insulin release [Adapted from Pearson et al. (2006)]

23.3.1 Diabetes Aetiology and Sulfonylurea Response: HNF1A *and* HNF4A *Maturity Onset Diabetes of the Young (MODY)*

As discussed in Chap. 11, MODY type 1 and MODY type 3 are caused by mutations in genes encoding the hepatic nuclear factor 4 alpha (HNF4α) and the hepatic nuclear factor 1 alpha (HNF1α), respectively. Both MODY types are characterised by progressive beta-cell dysfunction (Yamagata et al. 1996a, b), and unlike MODY 2, they are frequently associated with diabetic complications (Gloyn and Ellard 2006).

Subjects with MODY 3 are extremely sensitive to the glucose-lowering effects of SUs. Their response to gliclazide is markedly greater than their response to metformin, and this is even more pronounced than in the patients with T2D (Pearson et al. 2003). The mechanism for this SU sensitivity is likely to reflect the underlying beta-cell defect seen in these patients. Patients with *HNF1A* MODY have poor insulin secretory response to intravenous glucose but maintain a robust insulin secretory response to intravenous tolbutamide (a SU) (Pearson et al. 2003). This is consistent with the *HNF1A* models where loss of *HNF1A* results in impaired glucose metabolism but does not alter the K_{ATP} channel expression or the post-K_{ATP} channel mechanism mediating insulin release (Wang et al. 2000).

Many patients with *HNF1A* mutations are often assumed to have type 1 diabetes (T1D) due to the age at which they present and are treated with insulin from diagnosis. Once a diagnosis of *HNF1A* MODY is made, most of these patients can transition off insulin onto SUs to control their blood glucose even after insulin treatment lasting several decades (Shepherd et al. 2009). However, patients with a long duration of diabetes often fail quickly after transition to SUs. Here, the combination of long-acting insulin with SUs can be an effective therapy (Shepherd et al. 2009). SU sensitivity is also seen with other 'SU-like' drugs such as nateglinide, with low-dose nateglinide being reported to be more effective than glibenclamide at limiting acute postprandial glucose rises (Tuomi et al. 2006).

Although randomised clinical studies have not been carried out in patients with *HNF4A* mutations, they too appear to be SU sensitive (Pearson et al. 2005).

In summary, identification and characterisation of the various MODY genes have improved our understanding of the associated clinical outcome and allowed clinicians to offer specific treatments based on genetic aetiology. Replacing insulin treatment by oral antidiabetic drugs in patients with *HNF1A* or *HNF4A* mutations leads to an improved quality of life with a stable or even better glycaemic control. These insights strongly demonstrate the relevance of genetic testing in all potential patients with monogenic forms of diabetes.

23.3.2 Neonatal Diabetes

Neonatal diabetes is characterised by diabetes onset before 6 months of age, and these infants usually present with marked hyperglycaemia, insulin deficiency and in many cases diabetic ketoacidosis. As a result, they are treated with insulin injections. Multiple studies have reported that activating mutations in the *KCNJ11* and *ABCC8* cause neonatal diabetes. As the K_{ATP} channel is the site of action of SUs, this raised the possibility that patients with K_{ATP} channel mutations could be treated with SUs (Fig. 23.1). Early work showed a robust insulin secretory response to intravenous tolbutamide in 3 patients (who had minimal insulin response to glucose) (Gloyn et al. 2004; Hattersley and Ashcroft 2005; Flanagan et al. 2007). Subsequently, it has been robustly established that 90 % of patients with neonatal diabetes due to K_{ATP} channel mutations can successfully change from insulin to high-dose oral SU therapy, with improved glycaemic control; the blood glucose can often be normalised (to that of a nondiabetic) without risk of hypoglycaemia (Pearson et al. 2006). Physiological studies carried out in these patients before and after transition off insulin onto SUs suggest that the SU treatment is not having a marked 'direct effect' on insulin secretion; rather, it is enabling the beta cell to respond to the potentiating effect of other stimuli such as the incretin peptides GLP-1 and GIP, which may explain why these patients can achieve normoglycaemia without hypoglycaemia (Fig. 23.1). The specific response of these patients to SUs clearly illustrates the value of genetics in assigning the best therapeutic strategy based on identification of subgroups of patients that might benefit from targeted treatment. Considering these therapeutic consequences, it is strongly recommended to sequence the *KCNJ11* and *ABCC8* for mutations in all patients who are diagnosed with diabetes within the first 6 months of life (see www.diabetesgenes.org). Admittedly, this form of diabetes is extremely rare, and it cannot be anticipated that this study would result in a dramatic increase of patients offered genetic screening in T2D. Nevertheless, these key examples from monogenic diabetes provide hope that further genetic studies could lead to classification of subgroups for a specific treatment even in patients with polygenic T2D.

23.3.3 Pharmacodynamic Variation in Polygenic Type 2 Diabetes

As SUs are fairly specific in their site of action—the K_{ATP} channel of the pancreatic beta cell—it seems likely that variation in the beta-cell regulation of nutrient-stimulated insulin secretion will alter glycaemic response to SUs. Genetic studies of diabetes risk have implicated many variants in beta-cell genes predicted to alter mass or function (Ahlqvist et al. 2011). Each of these variants is a potential candidate for SU response. Unfortunately, there have yet to be any systematic assessment of these diabetes variants, or indeed a composite genetic risk score, on SU response. Nevertheless, it is possible that depending on the point at which a given genetic variant interferes with this pathway, the carriers may respond better or worse to the action of SUs. If the genetic defect is proximal to the action of SUs on SUR/Kir6.2, the SUs could make the system work, such that carriers would show an enhanced response (as in neonatal diabetes); however, if the genetic defect is distal to the point of entry of SUs or affect beta-cell mass more generally, then carriers may respond worse (e.g., *TCF7L2*). Thus, SU perturbations can also help clarify the site of action of novel genetic variants.

23.3.3.1 ABCC8/KCNJ11

Functional polymorphisms in genes such as *KCNJ11* which control beta-cell function might result in alterations in response to SU therapy in patients with T2D. Sesti et al. showed an impairment of glibenclamide-induced insulin secretion after 24 h exposure to high glucose concentration in islets isolated from carriers of the E23K variant of *KCNJ11* which might at least partially explain the observed increased risk of secondary failure to SUs in patients with T2D carrying this variant (Sesti et al. 2006). Although it has to be noted that the study of Sesti et al. describes rather the failure of the combination of SU and metformin treatment than of SU itself, Holstein et al. supported its data by showing that patients with T2D who carry the K variant of the E23K polymorphism in *KCNJ11* had reduced response to SU therapy, which resulted in increased glycated haemoglobin (HbA1c) and consequently in lower risk for severe hypoglycaemia (Holstein et al. 2009). In contrast, a recent study by Javorsky et al. demonstrated that carriers of the *KCNJ11* K-allele had a better therapeutic response to gliclazide (Javorsky et al. 2012). A significant greater reduction in HbA1c following gliclazide treatment has also been shown in Chinese carriers of A-allele of *ABCC8* S1369A polymorphism (Zhang et al. 2007; Feng et al. 2008). This is noteworthy as *ABCC8* S1369A is in high linkage disequilibrium with *KCNJ11* E23K, and so, it does support the findings by Javorsky et al. in Caucasian subjects (Javorsky et al. 2012). It appears that the discrepancies may at least in part be attributed to different structural and pharmacokinetic properties of the investigated drugs. Lang et al. investigated the pharmacogenomics of SUs to determine the structure–activity relationships of SUs in K_{ATP} channels containing either the E23/S1369 nonrisk or K23/A1369 risk haplotypes (Lang et al. 2012). They demonstrated that K_{ATP} channels carrying the K23/A1369 haplotype were more sensitive to gliclazide and mitiglinide inhibition, whereas channels containing the E23/S1369 haplotype were more sensitive to tolbutamide, chlorpropamide and glimepiride. This is most likely to be explained by the distinct drug structure motif (ring-fused pyrrole moiety) in several A-site drugs which may underlie the observed inhibitory potency of these drugs on K_{ATP} channels containing the K23/A1369 risk haplotype. Data by Lang et al. have potential clinical application which includes either the selection of drug on the basis of patient genotype/haplotype or enhancing drug efficacy by a dose escalation for patients who are less sensitive to specific drugs, although the magnitude of effect reported to date is too small to be of clinical utility.

23.3.3.2 Transcription Factor 7-Like 2 (*TCF7L2*)

Although numerous susceptibility loci for T2D have been discovered recently (see Chap. 2), from a clinical point of view, their effects are rather moderate. Besides genes such as *KCNJ11* (Gloyn et al. 2003) or *PPARG* (peroxisome proliferator-activated receptor gamma) (Altshuler et al. 2000), *TCF7L2* is one of the few genes whose variants seem to have notable effects on the genetic predisposition to T2D (Grant et al. 2006) (see Chap. 15). The *TCF7L2* gene encodes a transcription factor (Tcf-4) that is involved in the regulation of cellular proliferation and differentiation (Clatworthy and Subramanian 2001). Variants in *TCF7L2* have initially been shown to be associated with T2D in the population isolate of Iceland (Grant et al. 2006) and have subsequently been replicated in independent studies in multiple ethnic populations (Cauchi et al. 2007). The diabetes risk associated with *TCF7L2* variants reaches an allelic odds ratio (OR) near 1.40 which is impressive for a polygenic disease, but still with limited relevance in a clinical context. Clinically, it would be desirable to have a possibility to find those yet nondiabetic individuals who are at higher risk to manifest the disease since they carry the disease risk genotype.

Although relevant and noteworthy, an odds ratio of 1.40 is relatively negligible for such a variant, when considering that a body mass index change from 20 to 30 kg/m^2 results in a tenfold increase of risk to develop diabetes. Therefore, when it comes to predicting a disease, clinical information may still be of more value for a clinician than various genetic models (Lyssenko et al. 2008; Meigs et al. 2008). Moreover, we have to bear in mind that conclusions of such studies need to be seen with caution as they are just pieces of a complicated puzzle in the context of environmental factors, lifestyle habits and other, not yet known risk genotypes. This complex interaction is particularly evident in the Diabetes Prevention Program (Florez et al. 2006). Here, it was demonstrated that within an observation period of 3 years, an intensive lifestyle improvement can significantly reduce the diabetes risk independent of the *TCF7L2* risk genotype. Although this does not really help in the decision to propagate intensive lifestyle change, as it helps in all genotypic groups, another study by Pearson et al. clearly demonstrated the pharmacogenetic potential of *TCF7L2* (Pearson et al. 2007). In 911 sulfonylurea users of 4469 patients with T2D from the DARTS/MEMO (Diabetes Audit and Research Tayside/Medicines Monitoring Unit) collaboration database, who were recruited to GoDARTS between 1997 and 2006, the authors observed that homozygous carriers of the *TCF7L2* risk alleles for rs1225372 and rs7903146 were twice as likely not to respond to SUs as patients homozygous for the nonrisk alleles. Forty-two percent of SU users with the risk genotype failed to respond to the therapy which suggested that variation in *TCF7L2* influences therapeutic response to SUs. In contrast to SUs, no association was observed between metformin response and either single nucleotide polymorphism (SNP), after adjusting for baseline HbA1c. This result has subsequently been replicated in small studies (Holstein et al. 2011; Schroner et al. 2011), but widespread replication is limited. From a mechanistic point of view, failure to respond to SUs in patients with *TCF7L2* diabetes risk alleles might be partially explained by the role of the TCF7L2 protein in insulin secretion. It has been shown that a small fraction of the insulin granules is situated in close proximity of the voltage-gated calcium channels and is immediately available for release in the presence of TCF7L2 (da Silva Xavier et al. 2009; Gloyn et al. 2009). In the absence of TCF7L2 however, the calcium channels may detach from the secretory granules resulting in reduced glucose-stimulated

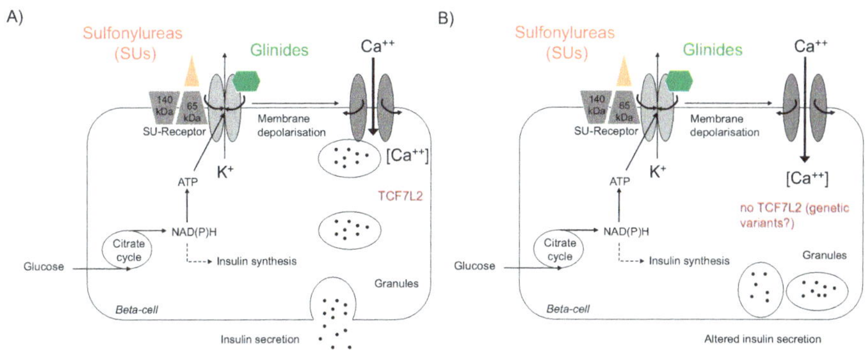

Fig. 23.2 Proposed model of changes in glycaemic response to sulfonylureas mediated by TCF7L2.
(**a**) A small fraction of the insulin granules is situated in close proximity of the voltage-gated calcium channels and are immediately available for release in the presence of TCF7L2 [Adapted from Gloyn et al. (2009)]. (**b**) In the absence of TCF7L2, the calcium channels may detach from the secretory granules resulting in reduced glucose-stimulated insulin secretion [Adapted from Gloyn et al. (2009)]. Consequently, it may lead to altered glycaemic response to SUs

insulin secretion (da Silva Xavier et al. 2009; Gloyn et al. 2009) (Fig. 23.2). This results in reduced glucose-stimulated insulin secretion under physiologic conditions and might consequently lead to altered glycaemic response to SUs. Again, this illustrates that whether SUs lead to better or worse glycaemic response in risk genotype carriers depends on the specific site along the insulin secretion pathway at which the relevant gene product exerts its molecular action, relative to the point of entry of SUs into this pathway.

23.3.4 Pharmacokinetic Variation in Polygenic Type 2 Diabetes

SUs are mainly metabolised by the cytochrome enzyme CYP2C9, a major factor for drug clearance and drug response. More than 30 known variant alleles are known for the *CYP2C9* gene. Each named *CYP2C9* star (*) allele denotes a genotype at one or more SNPs. Individuals homozygous for the reference *CYP2C9* allele (*CYP2C9*1*) have the 'normal metaboliser' phenotype, and *CYP2C9*2* (rs1799853) and *CYP2C9*3* (rs1057910) are the two most common variants with reduced enzyme activity among individuals of European ancestry (Lee et al. 2002). Whereas the CYP2C9*2 polymorphism seems to have moderate effects, the CYP2C9*3 polymorphism has been shown to be associated with strong reduction of the enzyme activity in healthy volunteers. For instance, compared with the CYP2C9*1/*1 genotype, the tolbutamide clearance in subjects with the CYP2C9*2/*2 genotype was reduced by 25 % and in those with the CYP2C9*3/*3 genotype by 84 %

(Kirchheiner et al. 2002a). Furthermore, pharmacokinetic studies in healthy subjects demonstrated that the clearance of glibenclamide and glimepiride in CYP2C9*3/*3 carriers was only 20 % of that in the wild-type carriers (Kirchheiner et al. 2002b). Consistently, even the CYP2C9*2/*3 genotype has been shown to be associated with significantly lower clearance of glibenclamide and glimepiride in nondiabetic subjects (Niemi et al. 2002). Wang et al. reported 1.3–2.8-fold increase in the SUs exposure in volunteers with a CYP2C9*3 allele compared with carriers of the CYP2C9*1/*1 genotype (Wang et al. 2005; Suzuki et al. 2006). However, a population-based study by Becker et al. revealed variation in SU response among patients with different CYP2C9 polymorphisms only for tolbutamide but not for glibenclamide, gliclazide and glimepiride. Albeit not statistically significant, the mean decrease in fasting serum glucose levels after initiation of tolbutamide was larger in CYP2C9*3 allele carriers when compared with patients with the wild-type genotype (Becker et al. 2008). Also recent data from the GoDARTS Study including 1073 SU-treated patients, 80 % of whom received gliclazide, showed that patients with two copies of the CYP2C9*2 or CYP2C9*3 loss-of-function alleles were 3.4 times more likely to reach HbA1c levels of <7 % as compared with wild-type carriers (Zhou et al. 2010). In summary, there is convincing evidence that the CYP2C9*2/*2, CYP2C9*2/*3 and CYP2C9*3/*3 genotypes impair SU metabolism and so influence the glycaemic response to SUs.

The proposed effects on glycaemic response to SUs would imply that the CYP2C9 genotypes might affect the risk of hypoglycaemia as well. Although limited by a small sample size, Holstein et al. reported a significant overrepresentation of the CYP2C9 genotypes *3/*3 and *2/*3 in subjects treated with glimepiride and glibenclamide who had experienced severe hypoglycaemia (Holstein et al. 2005). Another retrospective study indicated that the CYP2C9*3 allele might be a determinant for mild hypoglycaemia in patients treated with glimepiride or gliclazide. However, it is of note that in this study, the definition of hypoglycaemia did not comply with accepted standards as it included self-reported unspecific symptoms with partially missing confirmation of a blood glucose concentration <65 mg/dl (Ragia et al. 2009).

23.4 Pharmacogenetics of Sulfonylurea-Like Drugs: Meglitinide Derivatives

Repaglinide, a benzoic acid derivative, and nateglinide, a d-phenylalanine derivative, are the major meglitinide derivatives. Similarly to SUs, they stimulate insulin release via ATP-sensitive K+ channels and voltage-sensitive Ca2+ −channels but through different binding sites on the beta-cell receptor. In contrast to SUs, the meglitinides are characterised by a rapid onset and shorter duration of action.

Repaglinide and nateglinide are metabolised in the liver (repaglinide to 100 % and nateglinide to >85 %), whereas they are eliminated via bile (repaglinide to

~90 %) and via urine (nateglinide to ~80 %). Strong interindividual and interethnic differences in pharmacokinetics suggest that genetic factors might underlie meglitinides disposition in humans. In addition to CYP3A4, nateglinide is metabolised mainly via CYP2C9, whilst repaglinide metabolism involves predominantly CYP2C8 (Scheen 2007). Functional polymorphisms in CYP2C8 and CYP2C9 drug-metabolising genes have been described to be associated with meglitinides pharmacokinetics. For instance, repaglinide AUC was 45 % lower in healthy subjects with the CYP2C8*1/*3 genotype when compared with the wild-type CYP2C8*1 homozygotes (Niemi et al. 2003). Also nateglinide AUC was twofold higher in CYP2C9*3 homozygotes compared to wild-type CYP2C9*1 homozygotes (Kirchheiner et al. 2004). In addition to the metabolising enzymes, genes encoding transporters of meglitinides have to be considered. Indeed, numerous functional variants in *SLCO1B1* have been described. The gene encodes the organic anion-transporting polypeptide 1B1 (OATP1B1) which is the principal transporter of meglitinides. However, preliminary data indicate that the definitive role of these *SLCO1B1* polymorphisms still warrants further investigations to clearly validate their pharmacogenetic potential (Pacanowski et al. 2008).

23.5 Conclusions

SUs are a commonly used diabetes drugs. In monogenic diabetes, the SU drug class provides some of the clearest clinical examples of an individual's genotype altering treatment response seen anywhere in the pharmacogenomic literature, with genetic identification of *HNF1A/HNF4A* MODY and neonatal diabetes due to K_{ATP} channel mutation having a profound impact on patient care. In polygenic diabetes, like all other polygenic disease, the clinical translation is proving challenging. Yet SUs offer considerable potential for pharmacogenomics as the biology of action is relatively well understood; the beta cell is the predominant site of defect in T2D, and SUs expose risk of patients to a severe side effect—hypoglycaemia. Given this, it is disappointing that to date no clinical translation has occurred for SU therapy in T2D. This mostly reflects the limited sample sizes available for study. For example, in the population-based resources, metformin is used first line, and the sample sizes available (e.g., in GoDARTS) are nearly threefold greater than for SUs. However, unlike metformin, SUs are often used as comparator drugs in clinical trials, where clear efficacy endpoints and adverse reactions are properly documented. To make progress in the pharmacogenomics of SUs in T2D, it is clear that there needs to be greater access to and coordination of clinical trials. A number of initiatives may enable this: firstly, the greater access to clinical trial data, led by GSK (www.clinicalstudyrequest.com), although access to this is limited to data rather than samples; secondly, the increasing collaboration between pharma and academia, as exemplified in the EU Innovative Medicines Initiative, which in diabetes has a focus on stratified medicines' approaches in the consortium IMI-DIRECT (www.direct-diabetes.org/). These collaborative efforts would not only allow testing the effects of candidate genes (preferentially those known to be involved in the regulation of insulin secretion) on glycaemic response to SUs but also running GWAS for discovery of novel players in pharmacogenomics of SUs. Even though GWAS are unlikely to yield large-effect variants, they might suggest novel mechanisms of how SUs work in humans. Hopefully, the next few years will see pharmacogenomics of SUs in T2D mirror the success seen in monogenic diabetes.

References

Ahlqvist E, Ahluwalia TS, Groop L (2011) Genetics of type 2 diabetes. Clin Chem 57:241–254

Altshuler D, Hirschhorn JN, Klannemark M, Lindgren CM, Vohl MC, Nemesh J, Lane CR, Schaffner SF, Bolk S, Brewer C, Tuomi T, Gaudet D, Hudson TJ, Daly M, Groop L, Lander ES (2000) The common PPAR gamma Pro12Ala polymorphism is associated with decreased risk of type 2 diabetes. Nat Genet 26:76–80

Becker ML, Visser LE, Trienekens PH, Hofman A, van Schaik RH, Stricker BH (2008) Cytochrome P450 2C9 *2 and *3 polymorphisms and the dose and effect of sulfonylurea in type II diabetes mellitus. Clin Pharmacol Ther 83:288–292

Blakey J, Halapi E, Bjornsdottir US, Wheatley A, Kristinsson S, Upmanyu R, Stefansson K, Hakonarson H, Hall IP (2005) Contribution of ADAM33 polymorphisms to the population risk of asthma. Thorax 60:274–276

Cauchi S, El Achhab Y, Choquet H, Christian D, Krempler F, Weitgasser R, Nejjari C, Patsch W, Chikri M, Meyre D, Froguel P (2007) TCF7L2 is reproducibly associated with type 2 diabetes in various ethnic groups: a global meta-analysis. J Mol Med 85:777–782

Clatworthy JP, Subramanian V (2001) Stem cells and the regulation of proliferation, differentiation and patterning in the intestinal epithelium: emerging insights from gene expression patterns, transgenic and gene ablation studies. Mech Dev 101:3–9

da Silva Xavier G, Loder MK, McDonald A, Tarasov AI, Carzaniga R, Kronenberger K, Barg S, Rutter GA (2009) TCF7L2 regulates late events in insulin secretion from pancreatic islet beta-cells. Diabetes 58:894–905

Evans JM, Ogston SA, Reimann F, Gribble FM, Morris AD, Pearson ER (2008) No differences in mortality between users of pancreatic-specific and non-pancreatic-specific sulphonylureas: a cohort analysis. Diabetes Obes Metab 10:350–352

Feng Y, Mao G, Ren X, Xing H, Tang G, Li Q, Li X, Sun L, Yang J, Ma W, Wang X, Xu X (2008) Ser1369Ala variant in sulfonylurea receptor gene ABCC8 is associated with antidiabetic efficacy of gliclazide in Chinese type 2 diabetic patients. Diabetes Care 31:1939–1944

Flanagan SE, Patch AM, Mackay DJG, Edghill EL, Gloyn AL, Robinson D, Shield JPH, Temple K, Ellard S, Hattersley AT (2007) Mutations in ATP-sensitive K+ channel genes cause transient neonatal diabetes and permanent diabetes in childhood or adulthood. Diabetes 56:1930–1937

Florez JC, Jablonski KA, Bayley N, Pollin TI, de Bakker PIW, Shuldiner AR, Knowler WC, Nathan DM, Altshuler D (2006) TCF7L2 polymorphisms and progression to diabetes in the diabetes prevention program. N Engl J Med 355:241–250

Gloyn AL, Ellard S (2006) Defining the genetic aetiology of monogenic diabetes can improve treatment. Expert Opin Pharmacother 7:1759–1767

Gloyn AL, Weedon MN, Owen KR, Turner MJ, Knight BA, Hitman G, Walker M, Levy JC, Sampson M, Halford S, McCarthy MI, Hattersley AT, Frayling TM (2003) Large-scale association studies of variants in genes encoding the pancreatic beta-cell KATP channel subunits Kir6.2 (KCNJ11) and SUR1 (ABCC8) confirm that the KCNJ11 E23K variant is associated with type 2 diabetes. Diabetes 52:568–572

Gloyn AL, Pearson ER, Antcliff JF, Proks P, Bruining GJ, Slingerland AS, Howard N, Srinivasan S, Silva JMCL, Molnes J, Edghill EL, Frayling TM, Temple IK, Mackay D, Shield JPH, Sumnik Z, van Rhijn A, Wales JKH, Clark P, Gorman S, Aisenberg J, Ellard S, Njolstad PR, Ashcroft FM, Hattersley AT (2004) Activating mutations in the gene encoding the ATP-sensitive potassium-channel subunit Kir6.2 and permanent neonatal diabetes. N Engl J Med 350:1838–1849

Gloyn AL, Braun M, Rorsman P (2009) Type 2 diabetes susceptibility gene TCF7L2 and its role in beta-cell function. Diabetes 58:800–802

Grant SF, Thorleifsson G, Reynisdottir I, Benediktsson R, Manolescu A, Sainz J, Helgason A, Stefansson H, Emilsson V, Helgadottir A, Styrkarsdottir U, Magnusson KP, Walters GB, Palsdottir E, Jonsdottir T, Gudmundsdottir T, Gylfason A, Saemundsdottir J, Wilensky RL, Reilly MP, Rader DJ, Bagger Y, Christiansen C, Gudnason V, Sigurdsson G, Thorsteinsdottir U, Gulcher JR, Kong A, Stefansson K (2006) Variant of transcription factor 7-like 2 (TCF7L2) gene confers risk of type 2 diabetes. Nat Genet 38:320–323

Hatterslay AT (2005) Molecular genetics goes to the diabetes clinic. Clin Med 5:476–481

Hattersley AT, Ashcroft FM (2005) Activating mutations in Kir6.2 and neonatal diabetes—new clinical syndromes, new scientific insights, and new therapy. Diabetes 54:2503–2513

Holstein A, Plaschke A, Ptak M, Egberts EH, El Din J, Brockmoller J, Kirchheiner J (2005) Association between CYP2C9 slow metabolizer genotypes and severe hypoglycaemia on medication with sulphonylurea hypoglycaemic agents. Br J Clin Pharmacol 60:103–106

Holstein A, Hahn M, Stumvoll M, Kovacs P (2009) The E23K variant of KCNJ11 and the risk for severe sulfonylurea-induced hypoglycemia in patients with type 2 diabetes. Horm Metab Res 41:387–390

Holstein A, Hahn M, Korner A, Stumvoll M, Kovacs P (2011) TCF7L2 and therapeutic response to sulfonylureas in patients with type 2 diabetes. BMC Med Genet 12:30

Javorsky M, Klimcakova L, Schroner Z, Zidzik J, Babjakova E, Fabianova M, Kozarova M, Tkacova R, Salagovic J, Tkac I (2012) KCNJ11 gene E23K variant and therapeutic response to sulfonylureas. Eur J Intern Med 23:245–249

Kirchheiner J, Bauer S, Meineke I, Rohde W, Prang V, Meisel C, Roots I, Brockmoller J (2002a) Impact of CYP2C9 and CYP2C19 polymorphisms on tolbutamide kinetics and the insulin and glucose response in healthy volunteers. Pharmacogenetics 12:101–109

Kirchheiner J, Brockmoller J, Meineke I, Bauer S, Rohde W, Meisel C, Roots I (2002b) Impact of CYP2C9 amino acid polymorphisms on glyburide kinetics and on the insulin and glucose response in healthy volunteers. Clin Pharmacol Ther 71:286–296

Kirchheiner J, Meineke I, Muller G, Bauer S, Rohde W, Meisel C, Roots I, Brockmoller J (2004) Influence of CYP2C9 and CYP2D6 polymorphisms on the pharmacokinetics of nateglinide in genotyped healthy volunteers. Clin Pharmacokinet 43:267–278

Lang VY, Fatehi M, Light PE (2012) Pharmacogenomic analysis of ATP-sensitive potassium channels coexpressing the common type 2 diabetes risk variants E23K and S1369A. Pharmacogenet Genomics 22:206–214

Lee CR, Goldstein JA, Pieper JA (2002) Cytochrome P450 2C9 polymorphisms: a comprehensive review of the in-vitro and human data. Pharmacogenetics 12:251–263

Lyssenko V, Jonsson A, Almgren P, Pulizzi N, Isomaa B, Tuomi T, Berglund G, Altshuler D, Nilsson P, Groop L (2008) Clinical risk factors, DNA variants, and the development of type 2 diabetes. N Engl J Med 359:2220–2232

Meigs JB, Shrader P, Sullivan LM, McAteer JB, Fox CS, Dupuis J, Manning AK, Florez JC, Wilson PW, D'Agostino RB, Cupples LA (2008) Genotype score in addition to common risk factors for prediction of type 2 diabetes. N Engl J Med 359:2208–2219

Meijers-Heijboer H, van Geel B, van Putten WL, Henzen-Logmans SC, Seynaeve C, Menke-Pluymers MB, Bartels CC, Verhoog LC, van den Ouweland AM, Niermeijer MF, Brekelmans CT, Klijn JG (2001) Breast cancer after prophylactic bilateral mastectomy in women with a BRCA1 or BRCA2 mutation. N Engl J Med 345:159–164

Niemi M, Cascorbi I, Timm R, Kroemer HK, Neuvonen PJ, Kivisto KT (2002) Glyburide and glimepiride pharmacokinetics in subjects with different CYP2C9 genotypes. Clin Pharmacol Ther 72:326–332

Niemi M, Leathart JB, Neuvonen M, Backman JT, Daly AK, Neuvonen PJ (2003) Polymorphism in CYP2C8 is associated with reduced plasma concentrations of repaglinide. Clin Pharmacol Ther 74:380–387

Pacanowski MA, Hopley CW, Aquilante CL (2008) Interindividual variability in oral antidiabetic drug disposition and response: the role of drug transporter polymorphism. Expert Opin Drug Metab Toxicol 4:529–544

Pal A, McCarthy MI (2013) The genetics of type 2 diabetes and its clinical relevance. Clin Genet 83:297–306

Pearson ER, Starkey BJ, Powell RJ, Gribble FM, Clark PM, Hattersley AT (2003) Genetic cause of hyperglycaemia and response to treatment in diabetes. Lancet 362:1275–1281

Pearson ER, Pruhova S, Tack CJ, Johansen A, Castleden HA, Lumb PJ, Wierzbicki AS, Clark PM, Lebl J, Pedersen O, Ellard S, Hansen T, Hattersley AT (2005) Molecular genetics and phenotypic characteristics of MODY caused by hepatocyte nuclear factor 4alpha mutations in a large European collection. Diabetologia 48:878–885

Pearson ER, Flechtner I, Njolstad PR, Malecki MT, Flanagan SE, Larkin B, Ashcroft FM, Klimes I, Codner E, Iotova V, Slingerland AS, Shield J, Robert J, Holst JJ, Clark PM, Ellard S, Sovik O, Polak M, Hattersley AT (2006) Switching from insulin to oral sulfonylureas in patients with diabetes due to Kir6.2 mutations. N Engl J Med 355:467–477

Pearson ER, Donnelly LA, Kimber C, Whitley A, Doney ASF, McCarthy MI, Hattersley AT, Morris AD, Palmer CNA (2007) Variation in TCF7L2 influences therapeutic response to sulfonylureas—a GoDARTs study. Diabetes 56:2178–2182

Ragia G, Petridis I, Tavridou A, Christakidis D, Manolopoulos VG (2009) Presence of CYP2C9*3 allele increases risk for hypoglycemia in Type 2 diabetic patients treated with sulfonylureas. Pharmacogenomics 10:1781–1787

Roden DM, Altman RB, Benowitz NL, Flockhart DA, Giacomini KM, Johnson JA, Krauss RM, McLeod HL, Ratain MJ, Relling MV, Ring HZ, Shuldiner AR, Weinshilboum RM, Weiss ST (2006) Pharmacogenomics: challenges and opportunities. Ann Intern Med 145:749–757

Scheen AJ (2007) Drug-drug and food-drug pharmacokinetic interactions with new insulinotropic agents repaglinide and nateglinide. Clin Pharmacokinet 46:93–108

Schroner Z, Javorsky M, Tkacova R, Klimcakova L, Dobrikova M, Habalova V, Kozarova M, Zidzik J, Rudikova M, Tkac I (2011) Effect of sulphonylurea treatment on glycaemic control is related to TCF7L2 genotype in patients with type 2 diabetes. Diabetes Obes Metab 13:89–91

Seino S (1999) ATP-sensitive potassium channels: a model of heteromultimeric potassium channel/receptor assemblies. Annu Rev Physiol 61:337–362

Sesti G, Laratta E, Cardellini M, Andreozzi F, Del Guerra S, Irace C, Gnasso A, Grupillo M, Lauro R, Hribal ML, Perticone F, Marchetti P (2006) The E23K variant of KCNJ11 encoding the pancreatic beta-cell adenosine 5′-triphosphate-sensitive potassium channel subunit Kir6.2 is associated with an increased risk of secondary failure to sulfonylurea in patients with type 2 diabetes. J Clin Endocrinol Metab 91:2334–2339

Shepherd M, Shields B, Ellard S, Rubio-Cabezas O, Hattersley AT (2009) A genetic diagnosis of HNF1A diabetes alters treatment and improves glycaemic control in the majority of insulin-treated patients. Diabet Med 26:437–441

Suzuki K, Yanagawa T, Shibasaki T, Kaniwa N, Hasegawa R, Tohkin M (2006) Effect of CYP2C9 genetic polymorphisms on the efficacy and pharmacokinetics of glimepiride in subjects with type 2 diabetes. Diabetes Res Clin Pract 72:148–154

Tuomi T, Honkanen EH, Isomaa B, Sarelin L, Groop LC (2006) Improved prandial glucose control with lower risk of hypoglycemia with nateglinide than with glibenclamide in patients with maturity-onset diabetes of the young type 3. Diabetes Care 29:189–194

UK Prospective Diabetes Study (UKPDS) Group (1998) Intensive blood-glucose control with sulphonylureas or insulin compared with conventional treatment and risk of complications in patients with type 2 diabetes (UKPDS 33). Lancet 352:837–853

Wang H, Antinozzi PA, Hagenfeldt KA, Maechler P, Wollheim CB (2000) Molecular targets of a human HNF1 alpha mutation responsible for pancreatic beta-cell dysfunction. EMBO J 19:4257–4264

Wang R, Chen K, Wen SY, Li J, Wang SQ (2005) Pharmacokinetics of glimepiride and cytochrome P450 2C9 genetic polymorphisms. Clin Pharmacol Ther 78:90–92

Yamagata K, Furuta H, Oda N, Kaisaki PJ, Menzel S, Cox NJ, Fajans SS, Signorini S, Stoffel M, Bell GI (1996a) Mutations in the hepatocyte nuclear factor-4alpha gene in maturity-onset diabetes of the young (MODY1). Nature 384:458–460

Yamagata K, Oda N, Kaisaki PJ, Menzel S, Furuta H, Vaxillaire M, Southam L, Cox RD, Lathrop GM, Boriraj VV, Chen X, Cox NJ, Oda Y, Yano H, Le Beau MM, Yamada S, Nishigori H, Takeda J, Fajans SS, Hattersley AT, Iwasaki N, Hansen T, Pedersen O, Polonsky KS, Bell GI (1996b) Mutations in the hepatocyte nuclear factor-1alpha gene in maturity-onset diabetes of the young (MODY3). Nature 384:455–458

Zhang H, Liu X, Kuang H, Yi R, Xing H (2007) Association of sulfonylurea receptor 1 genotype with therapeutic response to gliclazide in type 2 diabetes. Diabetes Res Clin Pract 77:58–61

Zhou K, Donnelly L, Burch L, Tavendale R, Doney AS, Leese G, Hattersley AT, McCarthy MI, Morris AD, Lang CC, Palmer CN, Pearson ER (2010) Loss-of-function CYP2C9 variants improve therapeutic response to sulfonylureas in type 2 diabetes: a Go-DARTS study. Clin Pharmacol Ther 87:52–56

Chapter 24
Causal Inference in Medicine via Mendelian Randomization

Timothy M. Frayling and Benjamin F. Voight

Abstract The central aim of preventative care is to manage or avoid entirely life-threatening and costly disease endpoints. Success in this broad goal requires researchers and clinicians to correctly distinguish between biomarkers that *cause* disease from those that are simply correlated with outcome. The randomized controlled trial is a scientifically valid approach to assess causal relationships, but is time-consuming and expensive, and success is not a guaranteed endpoint. Recently, a statistical approach has been translated from the econometrics literature, a strategy which utilizes genetic information identified from human subjects as "instruments" to generate an assessment of causality between biomarker and disease. This methodology, dubbed Mendelian Randomization, is directly analogous to that of the controlled trial, circumventing the issues of confounding and reverse causation that precludes conventional epidemiological studies from making causal assessments. Owing to the growing dissection of genetically heritable traits in the literature, Mendelian Randomization has emerged as a high-value tool for efficient translation of genetics research to the bedside. In the following chapter, we present the framework of Mendelian Randomization and motivation for causal assessment, the analogy of Mendelian Randomization to the randomized controlled trial, discuss general considerations for study design and assumptions of the approach, and exemplify case studies from the literature of applications of MR to type 2 diabetes and other clinical endpoints.

T.M. Frayling
Genetics of Complex Traits, Royal Devon and Exeter NHS Foundation Trust, University of Exeter, RILD Building, Barrack Road, Exeter EX2 5DW, UK
e-mail: t.m.frayling@exeter.ac.uk

B.F. Voight (✉)
Department of Systems Pharmacology and Translational Therapeutics, University of Pennsylvania—Perelman School of Medicine, Philadelphia, PA, USA

Department of Genetics, University of Pennsylvania—Perelman School of Medicine, Philadelphia, PA, USA
e-mail: bvoight@upenn.edu

J.C. Florez (ed.), *The Genetics of Type 2 Diabetes and Related Traits*,
DOI 10.1007/978-3-319-01574-3_24

24.1 Overview and Scope

The central aim of preventative care is to manage at an early stage—or avoid entirely—life-threatening and costly disease endpoints, like type 2 diabetes (T2D), coronary heart disease (CHD), or stroke. Precise and early diagnosis of pre-disease states with subsequent monitoring is a critical first step in this process, achieved through predictive strategies to identify high-risk patient populations. Once high-risk groups have been identified, a second step applies an intervention to a modifiable exposure (e.g., diet, exercise, lifestyle, etc.) or by other means (e.g., pharmacologic) at early stages, to ameliorate or avoid significant morbidity in later stages.

For success to be achieved, knowledge of the biological, physiological, and molecular factors that *cause* disease is required, as well as possession of a suitable, therapeutically beneficial intervention for the established cause. Thus, two key questions must be addressed: First, which exposures are causal for—as distinct from those merely correlated with—a disease state? Second, does an intervention for that exposure modify disease risk in a beneficial way? Addressing these two questions are active areas of research, and in human subjects this is achieved through randomized controlled trials, an expensive and time-consuming experiment to undertake.

Recently, an approach analogous to the study design of the controlled trial but that utilizes genetic information from human subjects to make causal inference has received much deserved attention (Katan 1986; Smith and Ebrahim 2003; Sheehan et al. 2008). Termed *Mendelian Randomization* (MR), the approach is intrinsically translational, owing to the increasing abundance of genetic information in large numbers of individuals; the increasing study of genetically heritable, but modifiable, biomarkers prognostic for disease endpoints; and the relative efficiency of the approach to evaluate causality, compared to a controlled trial. As such, we posit that MR studies will be an important tactic deployed in an overall strategy to comprehensively understand the biology of human disease, one that maximizes the pace in which new beneficial interventions successfully reach the bedside.

In this chapter, we present the framework of MR first with a discussion of the limits of epidemiology and motivation for causal assessment, the model underlying causal inference, and an analogy of MR to the randomized controlled trial (24.2). Then, we discuss general considerations for study design and assumptions of the approach (24.3). Finally, we exemplify case studies from the literature of applications of MR to T2D and other clinical endpoints (24.4). Awareness and understanding of the MR design, approach, assumptions, and methodology will be a useful keystone for translational scientists keen on pursuing interventional studies in humans in the near future.

24.2 Causal Inference: Rationale, Models, and Mendelian Randomization

To achieve the promise of preventive care, one must begin with knowledge of exposures that are *predictive* for the disease endpoint in question. These factors are generally obtained from epidemiological studies, whose primary purpose is to measure the correlation between an exposure and hazard to disease over time. These observations are typically made from longitudinal studies of population cohorts such as the Framingham Heart Study (Splansky et al. 2007) and numerous studies beyond this are ongoing. Classic *prognostic* biomarkers obtained from such studies include (not limited to) blood pressure and cholesterol levels and risk to coronary heart disease (Kannel et al. 1964), C-reactive protein for CHD (Danesh et al. 1998), as well as many others. For research primarily interested in predicting the hazard of disease, any biomarkers, variables, and environmental exposures are relevant.

24.2.1 Correlation Does Not Imply Causation: Limits of Epidemiology

The underlying hypothesis tested in prospective cohort studies is that an identified marker is *correlated,* or associated, with hazard to a disease endpoint. An inappropriate conclusion from a significant correlation—particularly those with prior biological significance—is that the exposure also causes disease. Unfortunately, the causation assumption is a common one. Most (if not all) epidemiological monitoring studies do not provide a hypothesis test of causality, because most studies lack a formal intervention required for causality to be tested.

But why does one require an intervention to provide determination of causality? First off, determination of the *direction* of the associated variables—which is the cause and which is the effect—may be challenging (Fig. 24.1). In a prospective cohort study where a risk factor is measured years before disease onset, one might argue that reverse causation from the disease or disease treatment might be controlled. However, for many diseases, and particularly metabolic diseases, the disease process may have started many years before the disease is diagnosed, even as early as age 10 in the case of raised lipid levels (Whincup et al. 2002). In another example, increasing serum levels of an inflammatory biomarker may appear to precede a coronary event and, thus, correctly predict such events. But instead, unfavorable conditions in arterial walls (e.g., coronary plaques or calcification) could instead produce such an inflammatory marker, as damage to the coronary artery progresses over time. Specific inflammatory markers may not actually be the underlying cause of such events, even though such markers are substantially prognostic. A classic example of such confusion with an inflammatory marker is the much-studied association between raised high-sensitivity C-reactive

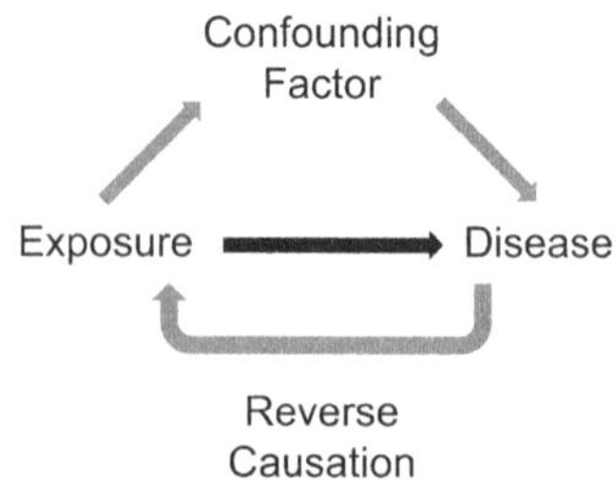

Fig. 24.1 The challenges in performing causal inference. Imagine one wants to determine if a specific exposure is causally related to a disease state (*black arrow*). Epidemiologic studies often cannot address the issue of *reverse causation*, where the disease leads to a change in the exposure, or *confounding*, where a second factor which is correlated to the exposure is related to the disease

protein (hsCRP) and CHD. MR using genetic variants that alter circulating hsCRP provided robust evidence, in several large studies, that hsCRP is not causal for CHD (Casas et al. 2006, CCGC et al. 2011).

Second, epidemiological studies may not fully capture—or control for—all exposures related to the endpoint of interest (Fig. 24.1). This is problematic only if such exposures are also associated with the marker of study. In this situation, our inference is *confounded* by that unmeasured factor. For metabolic diseases like type 2 diabetes, many variables are often correlated with the disease, e.g., body mass index (BMI). Changes in BMI result in many perturbations in serum biomarkers and anthropometric traits; but correcting for BMI may not fully account for all of these additional effects. For example, correcting for BMI does not fully correct for adiposity and, in particular, visceral adiposity including fatty liver that is associated with and likely the cause of many adverse metabolic features. Without precise control of the direction of the prognostic marker, as well as addressing potential unmeasured factors that may confound, we are limited in our abilities to correctly interpret the manner in which an exposure impacts an important clinical outcome.

While epidemiology yields a large, baseline collection of biomarkers or exposures that are worth testing, correlational observation does not constitute *etiologic* evidence. In some cases, intervention based on epidemiologic evidence alone was successful. That said, we should temper these examples with the larger number of cases where causality was strongly refuted after controlled trial (Tatsioni et al. 2007).

24.2.2 Approach to Causal Inference: The Randomized Clinical Trial

In order to directly test for causality, any approach needs to address the issues of reverse causality and confounding as far as possible within feasible design parameters. Once established, the general framework in which these factors are

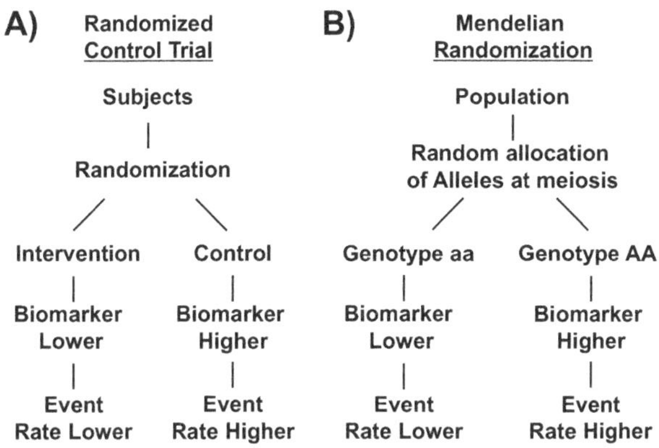

Fig. 24.2 Mendelian randomization is "nature's clinical trial." (**a**) A summarized depiction of a placebo-based clinical trial to evaluate efficacy of a therapeutic agent. A population is divided at random into two groups that are disease-free at the start of the study, which are indistinguishable from each other, except that one group is administered a drug which reduces a biomarker and the other is given a placebo. The prediction is that dosage of drug lowers the biomarker, which in turn lowers hazard to the endpoint if the biomarker is causally related to the endpoint. (**b**) A summarized version of the Mendelian Randomization design. A genetic variant is selected with a strong impact on the biomarker of interest. As a result of meiosis, a sampled population is "naturally" divided into two groups: carriers and noncarriers of the gene. These two groups are equivalent to having experienced reduced or elevated levels of the biomarker—analogous to having received a "weak drug" since birth. Association between variant and endpoint rejects the null hypothesis that the biomarker is not causally related to the disease endpoint

controlled—and causality evaluated—is the randomized controlled trial (RCT). For the purpose of exposition, imagine a pharmacologic therapeutic which lowers a particular biomarker, in which previous epidemiological study has correlated reproducibly with hazard to a disease endpoint (Fig. 24.2a). Here, we are interested in testing the hypothesis that lowering the biomarker through drug intervention reduces hazard to the disease endpoint of study. Patients are collected and randomized for the intervention who are disease-free at baseline, with biomarker and outcomes over time measured and compared between groups—this is the general design for trials used to determine clinical benefit (Gray et al. 2006). This design controls for reverse causality, as the intervention occurs before measurement of the outcome. This design also controls for confounding, with the assumption that the groups (intervention and placebo) are indistinguishable from one another for measured (as well as potentially unmeasured) factors.

The barriers to activate this line of experimental inquiry are daunting. First, the approach relies upon possession a suitable, proven intervention for the biomarker of interest, and such an intervention might not actually be readily available. For drug intervention trials, a great deal of time and expense are required for any drug to enter this stage of testing (i.e., Phase I and II trials). However, even with an intervention in hand, this trial takes years to deploy and a great deal of cost to complete, in order to obtain a final result. The low-throughput and increasingly abundant prognostic biomarkers that await formal causal evaluation now represent a significant barrier to progress in translational science.

24.2.3 The Case for Mendelian Randomization

MR studies will not replace RCTs, but rather serve as a complement to them. Beyond the technical feasibility of the MR study design in addressing reverse causality and nongenetic confounding (Fig. 24.3), application of the MR design offers other advantages. Perhaps the main advantage is cost and pace of discovery: multiple phases of RCT take years and millions of dollars to complete, and success is not guaranteed at the end of the trial (Arrowsmith and Miller 2013). Approaches that can help prioritize trials most likely to succeed would be of tremendous benefit in terms of cost saving and development of safe and effective therapeutics. Furthermore, high-throughput technologies developed in the last 10 years have increased the rate of assay and characterization of human phenotypes and molecular markers in genetic and epidemiologic studies. Such growth makes RCT on all such discoveries infeasible; thus, screening tools like MR can help to separate prognostic from high-priority biomarkers with etiologic support from human subjects, where the latter are moved into RCT.

MR studies are also uniquely powerful beyond the designs of the RCT. First, MR has the advantage that it usually tests the effects of lifelong exposure to a subtle change in a potential risk factor. In contrast, RCTs test the effects of much more acute, short-term, and stronger interventions. Second, MR also provides a strategy for scientific advance, whereby in a standard RCT, it might be unethical to provide such an intervention, for example, evaluating the causal relationship between

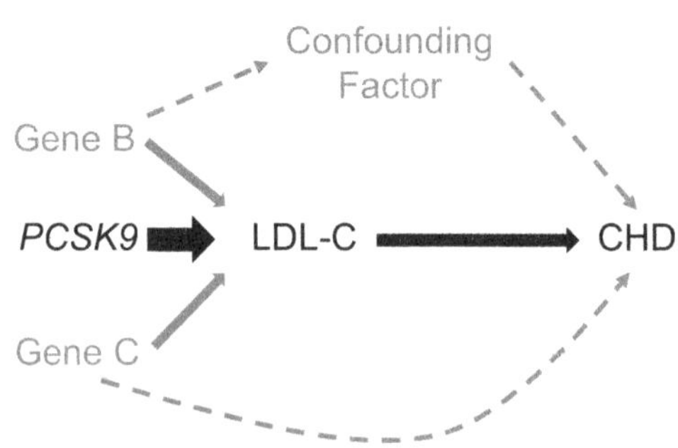

Fig. 24.3 The Mendelian randomization design in graphical terms. Here, we depict that a gene (in this case, *PCSK9*) influences the exposure (LDL-C). As a result, a polymorphism that strongly influences LDL-C which segregates randomly in the population can be used to test the hypothesis that LDL-C causally impacts risk to CHD (*black arrow*). Other variants which may associate with confounding factors, which impact the endpoint beyond LDL-C, violate the assumptions of MR and are not considered for analysis (*gray arrows*). Individually weak variants may be used in a combined score, if assumptions for each variant are met

developmental stunting and infection to strains of *Plasmodium* known to cause life-threatening malaria (one is ethically obligated to treat the infection!).

Owing to the efficiency of design, availability of data, and utility as a screening tool, MR is a powerful experimental approach, one which can help to clarify cases where an RCT has not been successful or to motivate new RCTs where evidence is altogether unclear.

24.2.4 The Analogy of MR to Randomized Clinical Trial

Despite their differences, the RCT is a useful analogy for the Mendelian Randomization design (Fig. 24.2b). Intuitively, MR is "nature's randomized trial" (Thanassoulis and O'Donnell 2009) whereby individuals in the population have been randomly assigned into carrier and noncarrier genotype groups, at meiosis. Ideally, these two groups are indistinguishable from one another, except for genotype at the given locus of interest. The genetic variant that is used to stratify groups is carefully selected, such that carriage of a specific allele modifies the level of the trait exposure one wants to test a causal role for to the endpoint. For example, carriage of loss-of-function alleles at the proprotein convertase subtilisin/kexin type 9 (*PCSK9*) gene results in lower LDL cholesterol (LDL-C) compared to noncarriers, and these two genotype groups are otherwise indistinguishable from one another. Thus, a loss-of-function genotype would facilitate a formal test of the hypothesis that LDL-C is causality related to CHD or stroke (Fig. 24.3) and would make the prediction that carriage of *PCSK9* loss-of-function would protect against hazard to these endpoints, via LDL-C lowering (Cohen et al. 2006). Intuitively, one can conceptualize allele carriage (dosage) as analogous to the intervention in a typical RCT, as a *weak* drug perturbation administered over a lifetime. And because genotypes are almost always going to be independent of nongenetic confounding and are unmodified by disease processes, reverse causality is controlled for appropriately.

24.3 Considerations and Approaches for Mendelian Randomization

Given the increasing abundance of genetic information available today and the power of the approach, it is perhaps unsurprising that the number of MR publications has skyrocketed, with over 550 citations after 2006, compared to <30 before hand. MR studies can be initiated in many ways, varying in their approach to study design and selection of genetic variants, which may come with additional required assumptions. We turn our attention next to these details.

24.3.1 Study Design Considerations for Mendelian Randomization

A primary study design for MR can be initiated in a large prospective cohort, where each variable of interest is measured directly in all participants. These measures include the monitored endpoint, the intermediate trait or biomarker, genetic variables which influence the tested intermediate trait, and any additional confounding variables (Fig. 24.3). This design has the advantage of the availability of individualized genotypes and phenotypes and, thus, allows the most direct control over the analysis and interpretation. The main limitation of this design is low statistical power owing to small sample sizes. The probability of obtaining a meaningfully interpretable result is proportional to the number of participants surveyed, particularly the number who have suffered an endpoint. Prospective cohort studies monitor thousands of individuals, of which only hundreds of whom progress to a disease condition. One important question to determine is if observed correlation between trait and endpoint—estimated from epidemiologic studies—is the same as a causal estimate through a genetic instrument. While it is plausible to combine evidence across multiple studies through meta-analysis, the numbers of disease endpoints may still not be sufficient to generate a conclusive result for even this straightforward comparison.

To overcome this limitation, a class of MR statistical approaches have been developed that take advantage of "summary-level" association data. Summary-level data are distinct from individual-level genotype or phenotype data, in that the estimates of effects between SNP and traits that are used for the MR study are pooled across a large number of individuals (often hundreds of thousands), though no individual relationships are identifiable from such data. Summary level data are available for a large number of traits and disease endpoints, with traits that may potentially confound. This "two-stage" approach generates a substantial boost in statistical power to estimate causal relationships, but at the costs of some control over the testing procedure. Importantly, the statistical test for causality is valid for this data type (Burgess and Thompson 2013), though the standard assumptions still must be met.

24.3.2 Development of Genetic Instruments for Mendelian Randomization

The most easily interpretable MR analyses begin with a single genetic instrument, one with a profound impact on the intermediate trait of interest. Often, these are coding mutations that result in partial or complete loss of function, i.e., see examples described below in *PCSK9* or *LIPG* for LDL and HDL cholesterol, respectively (Cohen et al. 2006; Voight et al. 2012). While it is advantageous to select instruments where the mechanism between genotype and phenotype has been established, this is not strictly required for MR to provide a meaningful result. Individual genetic instruments that stratify populations into two groups (carriers vs. noncarriers) allow for a straightforward evaluation of effects, particularly confounding associations, and fit readily into the controlled trial analogy.

Unfortunately, individual genetic instruments that have a strong impact on the intermediate trait are the exception, rather than the rule. Often, genetic variants with a strong trait effect tend to be rare in the population. Catalogs of low-frequency human genetic variation, also tending toward rare, have only been made available recently (1000 Genomes Project Consortium et al. 2012), and much work remains to relate this spectrum of variation to human phenotypes. That said, even if rare variants and their association to disease were known, owing to their frequency, the statistical power to make causal inference would still be challenging; one might still need to combine evidence across multiple variants into a single statistical test to make appropriate inference.

It should also be noted that an individual, specific genetic instrument tests only a single mechanism and trait/disease pathway. In some cases, this might be precisely what one requires. In other cases, a stronger argument can be made through examination of multiple gene targets and mechanisms of action (Voight et al. 2012). Consequently, to address both of these concerns, statistical approaches that aggregate multiple genetic instruments into a single assessment of causality have been developed (Johnson 2012; Burgess and Thompson 2013), resulting in a valid assessment if certain assumptions are satisfied. While this approach is typically based on summary-level data, it can also be applied in the individual-level data types, via a genetic risk score, a variable that linearly combines allele dosage with effect over a collection of genetic variants (Voight et al. 2012; Johnson 2012).

24.3.3 Assumptions Underlying Mendelian Randomization

For all MR study designs, there are three primary assumptions that must be held in order for the test to be valid (Smith and Ebrahim 2003; Hernán and Robins 2006; Sheehan et al. 2008):

1. *Confounding*. The genetic instrument is not related to another, causal variable which has a measurable effect on the endpoint (confounder). A confounding variable could be another trait or other factor that associated with the trait, SNP, and/or endpoint measurements. This is often referred to as pleiotropy in the genetics literature—meaning a genetic variant influences two or more independent traits.
2. *Potency*. The genetic instrument selected must have a *strong* and reliable impact on the intermediate trait tested.
3. *Exclusivity*. The effect on the endpoint must occur through the intermediate trait of interest, with no additional impact outside the two variables measured. Put

another way, the genetic instrument selected should not have a *direct* effect on the endpoint above and beyond the effect through the intermediate trait.

In typical designs, genetic instruments are selected to have a strong reproducible effect on the trait of interest. Confounding is typically addressed by measuring additional associations with variables that may potentially confound. For example, if using a genetic instrument to test the causal relationship between raised triglycerides and coronary artery disease, we would want to know that the genetic variant is not also associated with raised blood pressure or increased BMI. In addition, instruments must also not be subject to *genetic* confounding, from unaccounted for differences in allele frequency due to population ancestry. Those instruments that have no obvious or reproducible association with confounders are subsequently advanced into further MR analysis (Fig. 24.3).

Variants identified by genetic studies of common variation are individually weak, which may raise a concern if the potency assumption is upheld. To address this concern, multiple genetic instruments are often combined together to perform a specific test. If multiple instruments are utilized, there are additional required assumptions (Johnson 2012):

1. *Uncorrelated*. Each variant included in the genetic score should be mutually uncorrelated, i.e., should not be in strong linkage disequilibrium with one another (inherited on the same ancestral chromosomal segment).
2. *Equanimity*. Individual variants included should not contribute *excessively* relative to other included SNPs.
3. *Additivity*. For some tests, the dosage of alleles at each variant should influence the intermediate trait in an additive manner.

While this approach can address the potency assumption, using multiple variants still requires care to construct valid tests. The use of multiple variants increases the likelihood of pleiotropy confounding the results—that the accumulated genetic instrument will be related to additional traits. Furthermore, situations could arise where individual variants analyzed separately are more powerful or useful than collecting multiple instruments and evaluating collectively. Overall, the challenges and benefits of using multiple instruments have not been fully described, though research along these lines continues actively.

24.4 Highlighted Applications of Mendelian Randomization

With a broad overview of the approach and its underlying assumptions, we now turn to specific examples of MR studies from the literature, describing the question, approach, inferences made, and residual challenges. While here, our focus is on cardiovascular and T2D phenotypes, we point curious readers to additional examples of MR studies outside of these traits which exemplify the approach (Chen et al. 2008; Pfister et al. 2011).

24.4.1 Serum Plasma Cholesterol Levels and Heart Disease

Epidemiological studies have demonstrated reproducible association with serum levels of cholesterol and risk of CHD (Kannel et al. 1964). Specifically, elevation of LDL-C is associated with increased risk to CHD, while elevation of serum high-density lipoprotein cholesterol (HDL-C) is associated with protection against atherosclerosis. With the accumulated body of knowledge of genetic factors related to LDL-C and HDL-C (Teslovich et al. 2010), the stage was set to apply the framework of MR to both of these traits. For LDL-C, in one MR study of nine polymorphisms strongly associated with LDL-C levels (Ference et al. 2012), they reported a causal 54.5 % estimated risk reduction to CHD for each mmol/l lower LDL-C, consistent with another MR study (Voight et al. 2012). Collectively, these results are consistent with randomized controlled trials for LDL-C lowering therapies, which have been successful at reducing cardiovascular events (Baigent et al. 2005).

In contrast, evidence for a causal role for HDL-C is quite uncertain. Controlled trials for drug interventions (e.g., torcetrapib, dalcetrapib, niacin, and others) that raise plasma HDL-C over a range of mechanisms have not met with success in lowering rates of myocardial infarction (MI). To address the hypothesis that HDL-C is causally related to MI, one study applied a two-pronged MR approach (Voight et al. 2012). First, the authors took advantage of a low-frequency coding mutation in endothelial lipase, *LIPG* N396S, strongly associated with higher HDL-C (0.14 mmol/L), but not associated with other MI-related confounding factors. Association analysis in prospective cohorts, combined with data from case/control studies of MI, found no association ($P = 0.85$). Second, the authors collected a set of 14 genetic variants associated exclusively with HDL-C and performed a combined variant MR analysis. In contrast to LDL-C, the HDL-C instrument was not causally associated with risk to MI ($P = 0.63$). A conservative conclusion from this data is that some genetic mechanisms that elevate plasma HDL-C do not seem to confer protection against MI. These observations challenge the simple hypothesis that HDL-C elevation—by any means—will systematically lower risk to MI.

24.4.2 The Controversy of hsCRP

Subclinical inflammation is associated with many metabolic diseases and higher levels of adverse metabolic traits including raised cholesterol, fatty liver markers, and insulin resistance (Brunner et al. 2008). Furthermore, animal studies provide evidence that some inflammatory markers may causally influence obesity

(Bachmaier et al. 1999). Studies in humans have identified a number of genetic variants associated with several markers for inflammation, including interleukin 6 signaling, C-reactive protein, soluble intercellular adhesion molecule-1, and P-selectin (Raman et al. 2013). One recent MR study has provided evidence that the interleukin 6 signaling pathway is causally linked to heart disease (IL6R MR Consortium et al. 2012). These observations strongly motivate translation work in humans, that is, do markers of inflammation, particularly hsCRP levels, *cause* metabolic disease or adverse metabolic conditions?

Mendelian Randomization studies have now provided a very strong case that increased levels of subclinical (non-acute phase) C-reactive protein do not causally increase the risk of heart disease (Casas et al. 2006, CCGC et al. 2011). Known to be a marker of heart disease and other traits, considerable research efforts have been made to understand its role as both a biomarker and causal factor. Fortunately, the presence of common genetic variants that alter circulating CRP levels, in the gene that encodes the C-reactive protein, provided an exquisite, early example of the power of the MR approach (Timpson et al. 2005). Additional MR studies have examined the role of hsCRP in a range of metabolic diseases. These variants are not associated with any metabolic traits, and the instrumental variable analyses using them provide no evidence of a causal relationship between higher hsCRP and metabolic disease and traits (Brunner et al. 2008).

24.4.3 Serum Urate, Heart Disease, and Type 2 Diabetes

An end product of purine metabolism is uric acid, which circulates in blood serum as the anion urate. Patients with gout have shown elevated levels of serum urate, and consistent with a causal role in gout, randomized clinical trials with urate lowering therapies have demonstrated these drugs as effective therapeutics (Tayar et al. 2012). Epidemiologic studies have also correlated serum urate levels with a number of cardiometabolic traits, including blood pressure, lipids, as well as endpoint disease (Hozawa et al. 2006; Holme et al. 2009; Kim et al. 2010). A number of genetic variants that result in population level variation in serum urate levels have been identified (Köttgen et al. 2013). Predictably, these variants are also risk factors for gout. With such genetic factors in hand, several studies have looked for association at markers related to serum urate levels with heart disease (Stark et al. 2009; Yang et al. 2010). Altogether, these studies indicate no causal relationship. A particularly strong variant (rs12398742) at the *SLC2A9* locus is associated with a profound change in serum urate levels (0.37 mg/dl). However, this variant is not associated with T2D (odds ratio (OR) = 0.99, $P = 0.52$). This contrasts to the effect expected from epidemiology (expected OR = 1.06, given a 0.37 mg/dl change in serum urate), which have demonstrated a positive correlation between urate levels and T2D (Kodama et al. 2009). These data strongly de-emphasize the need for a randomized clinical trial for urate versus CHD and T2D endpoints. But additional work to evaluate other endpoints—including kidney disease, stroke, or heart failure, with MR analyses that utilize multiple genetic variants for urate levels—is certainly possible and source of active investigation.

24.4.4 Adiponectin and Insulin Resistance

MR studies have recently provided much needed insight into the relationship between insulin resistance and adiponectin. Adiponectin is a large, multimeric, circulating protein produced and secreted exclusively from adipose tissue (Stefan and Stumvoll 2002). Adiponectin is strongly and inversely correlated with insulin resistance (Stefan and Stumvoll 2002). Studies in mice, including the adipose tissue-specific knockout of the adiponectin receptors, has led many to suggest adiponectin is an important mediator of insulin resistance (Okada-Iwabu et al. 2013). In human subjects, however, the evidence is less clear cut. For example, individuals with severe mutations that cause a primary disorder of insulin resistance have reduced adiponectin levels, providing evidence that altered adiponectin levels are a consequence, not cause of insulin resistance (Semple et al. 2008). Several studies have identified common genetic variants within the gene that encodes adiponectin (*ADIPOQ*) as associated with circulating levels. A combination of four variants at this locus explain approximately 4 % of the population variance in adiponectin levels, more than sufficient to meet the *potency* assumption required for MR studies (Yaghootkar et al. 2013). Furthermore, because these variants are situated in and around the gene that encodes the protein of interest, the chances that they have pleiotropic effects on other phenotypes—which are not a consequence of altered adiponectin levels—are greatly reduced compared to variants elsewhere in the genome. Consequently, these genetic variants provide a great tool to evaluate the causal consequences of altered adiponectin levels on important clinical endpoints and are excellent proof of principle examples of MR. A recent MR study (Yaghootkar et al. 2013) of more than 29,000 individuals with both adiponectin and fasting insulin measurements found no causal evidence between lower adiponectin and higher insulin resistance ($P = 0.60$), or with T2D ($P = 0.77$). While studies of proportionally fewer samples have suggestive evidence to the contrary (Gao et al. 2013), the overall current weight of evidence appears to be that in humans, altered adiponectin levels are more likely to be a consequence rather than cause of insulin resistance in the general population.

24.4.5 Circulating Sex Hormone-Binding Globulin and Type 2 Diabetes

Common genetic variants in the gene that encodes an important carrier protein—sex hormone-binding globulin (*SHBG*)—provide another example of an excellent instrument for measuring the causal effects of the *SHBG* protein. *SHBG* binds to and transports testosterone and, to a lesser extent, estrogen, around the circulation to its target tissues (Petra 1991). Produced by the liver, it is inversely correlated with insulin resistance (Ding et al. 2006), and as with adiponectin, higher levels are associated with a healthy metabolic profile and with lower risk of T2D. In contrast to adiponectin, most people had assumed that lower *SHBG* levels were merely a consequence of insulin resistance and T2D, not a cause. However, two independent studies have recently shown that common variants in the *SHBG* gene are associated with T2D (Ding et al. 2009; Perry et al. 2010). MR effect estimates (OR per SHBG raising allele: 0.94, 95 % confidence interval (CI): 0.91–0.97) are consistent to the extent expected given the association between the gene variants and circulating SHBG and the phenotypic association between *SHBG* and T2D as observed from epidemiologic studies (expected OR 0.92, 95 % CI: 0.88–0.96) (Perry et al. 2010). While further studies are needed to establish the mechanism behind this association, MR studies have changed conventional thinking about the role of *SHBG* in diabetes.

24.4.6 FTO: A Good Instrument for Evaluating Adiposity to Related Traits and Endpoints?

In 2007, several groups described the first reproducible association between a common genetic variant and normal variation in BMI. An allele in an intron of the "fatso" gene, *FTO*, was associated with a 0.5 kg/m^2 increase in BMI. Carriers of two copies of the BMI-increasing allele (approximately 16 % of Europeans) were 1 kg/m^2 larger than individuals who carried zero copies of this allele. This finding potentially offered a superb opportunity for Mendelian Randomization studies. Epidemiologic data has shown that variation in BMI is correlated with multiple metabolic traits and disease endpoints, including susceptibility to T2D, CHD, some cancers, hypertension, raised circulating inflammatory markers, lower adiponectin, and *SHBG*. The difficulty in dissecting cause and effect in obesity is further illustrated by the fact that more than half of all human genes are differentially expressed in the adipose tissue of larger individuals compared to smaller individuals, presumably, mostly as a consequence rather than cause of obesity (Emilsson et al. 2008). Furthermore, many of these measures are correlated with each other independently of BMI. Multiple, phenotypic correlations between many (and potentially confounding) traits make understanding the causal factors that link obesity to increased risk of disease difficult, at best. A genetic variant that effectively randomizes people to higher or lower BMI could be extremely useful for disentangling some of these relationships.

To this end, several studies have performed MR experiments using *FTO* as an instrument for altered BMI (Fall et al. 2013; Holmes et al. 2014). These studies have suggested that higher BMI is causally related to susceptibility to T2D, heart failure, and several circulating biomarkers of public health importance including lower vitamin D levels (Vimaleswaran et al. 2013). However, some scientists have raised concerns that we must be cautious when using *FTO* as an instrument for BMI. First of all BMI is used as a surrogate of adiposity, but more importantly, little is known about how *FTO* alters BMI—we are still uncertain as to whether or not the genetic variant targets *FTO* itself or a nearby gene, and even if *FTO* is the target, we do not know exactly how it alters BMI. These are legitimate concerns, but several lines of evidence suggest that the variant in *FTO* can help improve our knowledge of causal relationships between adiposity and other traits. First, DEXA scans in children show that it is specifically associated with an alteration in fat mass, not increased skeletal or muscle mass (Frayling et al. 2007). Second, most studies show that the associations between the *FTO* variant and related traits are entirely consistent with that expected given the association between the *FTO* variant and BMI and the association between BMI and the "outcome" related trait. For example, *FTO* variants are associated with raised triglycerides and insulin levels (a marker of insulin resistance) to a greater extent than blood pressure and cholesterol levels—in keeping with the stronger phenotypic associations between BMI and triglycerides and insulin compared to BMI and blood pressure and cholesterol (Freathy et al. 2008).

24.4.7 Triglycerides and Fasting Insulin, Glucose, or Type 2 Diabetes

Elevated circulating triglyceride levels are strongly correlated with a poor metabolic state, including insulin resistance, higher levels of fasting glucose, and overt type 2 diabetes. Observational data suggest that elevated circulating triglycerides could be causal for this state, for example, through accumulation in adipose or liver resulting in lipotoxicity or impairment of hepatic insulin signaling resulting in insulin resistance (Trauner et al. 2010). Because genome-wide studies had accumulated associations with plasma triglycerides along with lipid levels, a Mendelian Randomization study was once again possible. De Silva et al. (2011) identified ten genetic variants strongly associated with triglyceride levels ($P < 2 \times 10^{-72}$) and, when aggregated in a score, resulted in a 0.59 SD change in triglyceride levels (upper vs. lower quintile). But in contrast to observational epidemiology (the Go-DARTS study), where a 1-SD increase in $\log_{10}$-triglyceride level was associated with elevated fasting glucose levels, HOMA-B, fasting insulin, HOMA-IR, and type 2 diabetes, the genetic risk score was not associated with any of these outcomes in diabetic or in control populations. Furthermore, the score was also not associated with frank type 2 diabetes (De Silva et al. 2011). While additional clarifying work is required, particularly to consider confounding in the context of this analysis, this raises significant questions about the expected efficacy of triglyceride control in prediabetic populations as a strategy to control or prevent type 2 diabetes onset.

24.5 Closing Summary

The simple idea underlying Mendelian Randomization is that a genetic variant, strongly related to a biomarker of interest, by the transitive property facilitates an appropriate framework to test the hypothesis that the given biomarker is *causal* for the endpoint of interest. Instrumental variable analysis—a procedure utilized in great detail in the field of econometrics—provides the precise framework by which this test can be formally implemented. Today, the wide abundance of human genetics data across a great range of phenotypes provides a unique opportunity to test a range of hypotheses that, until today, could not have been evaluated rigorously by the clinical research community, owing to the prohibitive and time-consuming costs of randomized controlled trials. For example, a potential application of great relevance to the T2D and CHD field, as yet untested, is whether hyperglycemia is pathogenic for T2D, given the controversy of conflicting RCTs in T1D and T2D.

Excitingly, genetic information across human subjects—hundreds of thousands of individuals—is accruing at a rapid pace. This growth is following in step with separate advances in sequencing technologies, which is enabling the characterization of *less-common* genetic variation and association studies with human disease. This next-generation of data sets will facilitate the application of a great deal of useful and important MR studies to understand—and develop targeted therapies for—disease endpoints. While the initial model and framework for MR is a solid advance for the field, a great deal of methodological work still remains to improve the approach to widen applicability, to take advantage of these upcoming data sets. That said, MR studies have been, and will continue to be, an essential tool in the arsenal of translational studies in human disease.

References

1000 Genomes Project Consortium, Abecasis GR, Auton A, Brooks LD, DePristo MA, Durbin RM, Handsaker RE, Kang HM, Marth GT, McVean GA (2012) An integrated map of genetic variation from 1,092 human genomes. Nature 491(7422):56–65

Arrowsmith J, Miller P (2013) Trial watch: phase II and phase III attrition rates 2011–2012. Nat Rev Drug Discov 12(8):569

Bachmaier K, Neu N, Yeung RS, Mak TW, Liu P, Penninger JM (1999) Generation of humanized mice susceptible to peptide-induced inflammatory heart disease. Circulation 99 (14):1885–1891

Baigent C, Keech A, Kearney PM, Blackwell L, Buck G, Pollicino C, Kirby A, Sourjina T, Peto R, Collins R, Simes R, Cholesterol Treatment Trialists' (CTT) Collaborators (2005) Efficacy and safety of cholesterol-lowering treatment: prospective meta-analysis of data from 90,056 participants in 14 randomised trials of statins. Lancet 366(9493):1267–1278

Brunner EJ, Kivimäki M, Witte DR, Lawlor DA, Davey Smith G, Cooper JA, Miller M, Lowe GD, Rumley A, Casas JP, Shah T, Humphries SE, Hingorani AD, Marmot MG, Timpson NJ, Kumari M (2008) Inflammation, insulin resistance, and diabetes—Mendelian randomization using CRP haplotypes points upstream. PLoS Med 5(8), e155. doi:10.1371/journal.pmed.0050155

Burgess S, Thompson SG (2013) Use of allele scores as instrumental variables for Mendelian randomization. Int J Epidemiol 42(4):1134–1144

Casas JP, Shah T, Cooper J, Hawe E, McMahon AD, Gaffney D, Packard CJ, O'Reilly DS, Juhan-Vague I, Yudkin JS, Tremoli E, Margaglione M, Di Minno G, Hamsten A, Kooistra T, Stephens JW, Hurel SJ, Livingstone S, Colhoun HM, Miller GJ, Bautista LE, Meade T, Sattar N, Humphries SE, Hingorani AD (2006) Insight into the nature of the CRP-coronary event association using Mendelian randomization. Int J Epidemiol 35(4):922–931

Chen L, Davey Smith G, Harbord RM, Lewis SJ (2008) Alcohol intake and blood pressure: a systematic review implementing a Mendelian randomization approach. PLoS Med 5(3), e52

Cohen JC, Boerwinkle E, Mosley TH Jr, Hobbs HH (2006) Sequence variations in PCSK9, low LDL, and protection against coronary heart disease. N Engl J Med 354(12):1264–1272

C-Reactive Protein Coronary Heart Disease Genetics Collaboration (CCGC), Wensley F, Gao P, Burgess S, Kaptoge S, Di Angelantonio E, Shah T, Engert JC, Clarke R, Davey-Smith G, Nordestgaard BG, Saleheen D, Samani NJ, Sandhu M, Anand S, Pepys MB, Smeeth L, Whittaker J, Casas JP, Thompson SG, Hingorani AD, Danesh J (2011) Association between C reactive protein and coronary heart disease: Mendelian randomisation analysis based on individual participant data. BMJ 15(342):d548

Danesh J, Collins R, Appleby P, Peto R (1998) Association of fibrinogen, C-reactive protein, albumin, or leukocyte count with coronary heart disease: meta-analyses of prospective studies. JAMA 279(18):1477–1482

De Silva NM, Freathy RM, Palmer TM, Donnelly LA, Luan J, Gaunt T, Langenberg C, Weedon MN, Shields B, Knight BA, Ward KJ, Sandhu MS, Harbord RM, McCarthy MI, Smith GD, Ebrahim S, Hattersley AT, Wareham N, Lawlor DA, Morris AD, Palmer CN, Frayling TM (2011) Mendelian randomization studies do not support a role for raised circulating triglyceride levels influencing type 2 diabetes, glucose levels, or insulin resistance. Diabetes 60 (3):1008–1018

Ding EL, Song Y, Malik VS, Liu S (2006) Sex differences of endogenous sex hormones and risk of type 2 diabetes: a systematic review and meta-analysis. JAMA 295(11):1288–1299

Ding EL, Song Y, Manson JE, Hunter DJ, Lee CC, Rifai N, Buring JE, Gaziano JM, Liu S (2009) Sex hormone-binding globulin and risk of type 2 diabetes in women and men. N Engl J Med 361(12):1152–1163

Emilsson V, Thorleifsson G, Zhang B, Leonardson AS, Zink F, Zhu J, Carlson S, Helgason A, Walters GB, Gunnarsdottir S, Mouy M, Steinthorsdottir V, Eiriksdottir GH, Bjornsdottir G, Reynisdottir I, Gudbjartsson D, Helgadottir A, Jonasdottir A, Jonasdottir A, Styrkarsdottir U, Gretarsdottir S, Magnusson KP, Stefansson H, Fossdal R, Kristjansson K, Gislason HG, Stefansson T, Leifsson BG, Thorsteinsdottir U, Lamb JR, Gulcher JR, Reitman ML, Kong A, Schadt EE, Stefansson K (2008) Genetics of gene expression and its effect on disease. Nature 452(7186):423–428

Fall T, Hägg S, Mägi R, Ploner A, Fischer K, Horikoshi M, Sarin AP, Thorleifsson G, Ladenvall C, Kals M, Kuningas M, Draisma HH, Ried JS, van Zuydam NR, Huikari V, Mangino M, Sonestedt E, Benyamin B, Nelson CP, Rivera NV, Kristiansson K, Shen HY, Havulinna AS,

Dehghan A, Donnelly LA, Kaakinen M, Nuotio ML, Robertson N, de Bruijn RF, Ikram MA, Amin N, Balmforth AJ, Braund PS, Doney AS, Döring A, Elliott P, Esko T, Franco OH, Gretarsdottir S, Hartikainen AL, Heikkilä K, Herzig KH, Holm H, Hottenga JJ, Hyppönen E, Illig T, Isaacs A, Isomaa B, Karssen LC, Kettunen J, Koenig W, Kuulasmaa K, Laatikainen T, Laitinen J, Lindgren C, Lyssenko V, Läärä E, Rayner NW, Männistö S, Pouta A, Rivadeneira F, Ruokonen A, Savolainen MJ, Sijbrands EJ, Small KS, Smit JH, Steinthorsdottir V, Syvänen AC, Taanila A, Tobin MD, Uitterlinden AG, Willems SM, Willemsen G, Witteman J, Perola M, Evans A, Ferrières J, Virtamo J, Kee F, Tregouet DA, Arveiler D, Amouyel P, Ferrario MM, Brambilla P, Hall AS, Heath AC, Madden PA, Martin NG, Montgomery GW, Whitfield JB, Jula A, Knekt P, Oostra B, van Duijn CM, Penninx BW, Davey Smith G, Kaprio J, Samani NJ, Gieger C, Peters A, Wichmann HE, Boomsma DI, de Geus EJ, Tuomi T, Power C, Hammond CJ, Spector TD, Lind L, Orho-Melander M, Palmer CN, Morris AD, Groop L, Järvelin MR, Salomaa V, Vartiainen E, Hofman A, Ripatti S, Metspalu A, Thorsteinsdottir U, Stefansson K, Pedersen NL, McCarthy MI, Ingelsson E, Prokopenko I (2013) European network for genetic and genomic epidemiology (ENGAGE) consortium. The role of adiposity in cardiometabolic traits: a Mendelian randomization analysis. PLoS Med 10(6), e1001474

Ference BA, Yoo W, Alesh I, Mahajan N, Mirowska KK, Mewada A, Kahn J, Afonso L, Williams KA Sr, Flack JM (2012) Effect of long-term exposure to lower low-density lipoprotein cholesterol beginning early in life on the risk of coronary heart disease: a Mendelian randomization analysis. J Am Coll Cardiol 60(25):2631–2639

Frayling TM, Timpson NJ, Weedon MN, Zeggini E, Freathy RM, Lindgren CM, Perry JR, Elliott KS, Lango H, Rayner NW, Shields B, Harries LW, Barrett JC, Ellard S, Groves CJ, Knight B, Patch AM, Ness AR, Ebrahim S, Lawlor DA, Ring SM, Ben-Shlomo Y, Jarvelin MR, Sovio U, Bennett AJ, Melzer D, Ferrucci L, Loos RJ, Barroso I, Wareham NJ, Karpe F, Owen KR, Cardon LR, Walker M, Hitman GA, Palmer CN, Doney AS, Morris AD, Smith GD, Hattersley AT, McCarthy MI (2007) A common variant in the FTO gene is associated with body mass index and predisposes to childhood and adult obesity. Science 316(5826):889–894

Freathy RM, Timpson NJ, Lawlor DA, Pouta A, Ben-Shlomo Y, Ruokonen A, Ebrahim S, Shields B, Zeggini E, Weedon MN, Lindgren CM, Lango H, Melzer D, Ferrucci L, Paolisso G, Neville MJ, Karpe F, Palmer CN, Morris AD, Elliott P, Jarvelin MR, Smith GD, McCarthy MI, Hattersley AT, Frayling TM (2008) Common variation in the FTO gene alters diabetes-related metabolic traits to the extent expected given its effect on BMI. Diabetes 57 (5):1419–1426

Gao H, Kim YM, Chen P, Igase M, Kawamoto R, Kim MK, Kohara K, Lee J, Miki T, Ong RT, Onuma H, Osawa H, Sim X, Teo YY, Tabara Y, Tai ES, van Dam RM (2013) Genetic variation in CDH13 is associated with lower plasma adiponectin levels but greater adiponectin sensitivity in East Asian populations. Diabetes 62(12):4277–4283

Gray R, Manola J, Saxman S, Wright J, Dutcher J, Atkins M, Carducci M, See W, Sweeney C, Liu G, Stein M, Dreicer R, Wilding G, DiPaola RS (2006) Phase II clinical trial design: methods in translational research from the Genitourinary Committee at the Eastern Cooperative Oncology Group. Clin Cancer Res 12(7 Pt 1):1966–1969

Hernán MA, Robins JM (2006) Instruments for causal inference. An epidemiologist's dream? Epidemiology 17(4):360–372

Holme I, Aastveit AH, Hammar N, Jungner I, Walldius G (2009) Uric acid and risk of myocardial infarction, stroke and congestive heart failure in 417,734 men and women in the apolipoprotein MOrtality RISk study (AMORIS). J Intern Med 266(6):558–570

Holmes MV, Lange LA, Palmer T, Lanktree MB, North KE, Almoguera B, Buxbaum S, Chandrupatla HR, Elbers CC, Guo Y, Hoogeveen RC, Li J, Li YR, Swerdlow DI, Cushman M, Price TS, Curtis SP, Fornage M, Hakonarson H, Patel SR, Redline S, Siscovick DS, Tsai MY, Wilson JG, van der Schouw YT, Fitzgerald GA, Hingorani AD, Casas JP, de Bakker PI, Rich SS, Schadt EE, Asselbergs FW, Reiner AP, Keating BJ (2014) Causal effects of body mass index on cardiometabolic traits and events: a Mendelian randomization analysis. Am J Hum Genet 94(2):198–208, pii: S0002-9297(13)00584-3

Hozawa A, Folsom AR, Ibrahim H, Nieto FJ, Rosamond WD, Shahar E (2006) Serum uric acid and risk of ischemic stroke: the ARIC study. Atherosclerosis 187(2):401–407

Interleukin-6 Receptor Mendelian Randomisation Analysis (IL6R MR) Consortium, Hingorani AD, Casas JP (2012) The interleukin-6 receptor as a target for prevention of coronary heart disease: a mendelian randomisation analysis. Lancet 379(9822):1214–1224

Johnson T (2012) Efficient calculation for Multi-SNP genetic risk scores. Poster presented at the American society of human genetics annual meeting, San Francisco, 6–10 November 2012

Kannel WB, Dawber TR, Friedman GD, Glennon WE, McNamara PM (1964) Risk factors in coronary heart disease. An evaluation of several serum lipids as predictors of coronary heart disease; the Framingham Study. Ann Intern Med 61:888–899

Katan MB (1986) Apolipoprotein E isoforms, serum cholesterol, and cancer. Lancet 1 (8479):507–508

Kim SY, Guevara JP, Kim KM, Choi HK, Heitjan DF, Albert DA (2010) Hyperuricemia and coronary heart disease: a systematic review and meta-analysis. Arthritis Care Res (Hoboken) 62(2):170–180

Kodama S, Saito K, Yachi Y, Asumi M, Sugawara A, Totsuka K, Saito A, Sone H (2009) Association between serum uric acid and development of type 2 diabetes. Diabetes Care 32 (9):1737–1742

Köttgen A, Albrecht E, Teumer A, Vitart V, Krumsiek J, Hundertmark C, Pistis G, Ruggiero D, O'Seaghdha CM, Haller T, Yang Q, Tanaka T, Johnson AD, Kutalik Z, Smith AV, Shi J, Struchalin M, Middelberg RP, Brown MJ, Gaffo AL, Pirastu N, Li G, Hayward C, Zemunik T, Huffman J, Yengo L, Zhao JH, Demirkan A, Feitosa MF, Liu X, Malerba G, Lopez LM, van der Harst P, Li X, Kleber ME, Hicks AA, Nolte IM, Johansson A, Murgia F, Wild SH, Bakker SJ, Peden JF, Dehghan A, Steri M, Tenesa A, Lagou V, Salo P, Mangino M, Rose LM, Lehtimäki T, Woodward OM, Okada Y, Tin A, Müller C, Oldmeadow C, Putku M, Czamara D, Kraft P, Frogheri L, Thun GA, Grotevendt A, Gislason GK, Harris TB, Launer LJ, McArdle P, Shuldiner AR, Boerwinkle E, Coresh J, Schmidt H, Schallert M, Martin NG, Montgomery GW, Kubo M, Nakamura Y, Tanaka T, Munroe PB, Samani NJ, Jacobs DR Jr, Liu K, D'Adamo P, Ulivi S, Rotter JI, Psaty BM, Vollenweider P, Waeber G, Campbell S, Devuyst O, Navarro P, Kolcic I, Hastie N, Balkau B, Froguel P, Esko T, Salumets A, Khaw KT, Langenberg C, Wareham NJ, Isaacs A, Kraja A, Zhang Q, Wild PS, Scott RJ, Holliday EG, Org E, Viigimaa M, Bandinelli S, Metter JE, Lupo A, Trabetti E, Sorice R, Döring A, Lattka E, Strauch K, Theis F, Waldenberger M, Wichmann HE, Davies G, Gow AJ, Bruinenberg M, LifeLines Cohort Study, Stolk RP, Kooner JS, Zhang W, Winkelmann BR, Boehm BO, Lucae S, Penninx BW, Smit JH, Curhan G, Mudgal P, Plenge RM, Portas L, Persico I, Kirin M, Wilson JF, Mateo Leach I, van Gilst WH, Goel A, Ongen H, Hofman A, Rivadeneira F, Uitterlinden AG, Imboden M, von Eckardstein A, Cucca F, Nagaraja R, Piras MG, Nauck M, Schurmann C, Budde K, Ernst F, Farrington SM, Theodoratou E, Prokopenko I, Stumvoll M, Jula A, Perola M, Salomaa V, Shin SY, Spector TD, Sala C, Ridker PM, Kähönen M, Viikari J, Hengstenberg C, Nelson CP, CARDIoGRAM Consortium, DIAGRAM Consortium, ICBP Consortium, MAGIC Consortium, Meschia JF, Nalls MA, Sharma P, Singleton AB, Kamatani N, Zeller T, Burnier M, Attia J, Laan M, Klopp N, Hillege HL, Kloiber S, Choi H, Pirastu M, Tore S, Probst-Hensch NM, Völzke H, Gudnason V, Parsa A, Schmidt R, Whitfield JB, Fornage M, Gasparini P, Siscovick DS, Polašek O, Campbell H, Rudan I, Bouatia-Naji N, Metspalu A, Loos RJ, van Duijn CM, Borecki IB, Ferrucci L, Gambaro G, Deary IJ, Wolffenbuttel BH, Chambers JC, März W, Pramstaller PP, Snieder H, Gyllensten U, Wright AF, Navis G, Watkins H, Witteman JC, Sanna S, Schipf S, Dunlop MG, Tönjes A, Ripatti S, Soranzo N, Toniolo D, Chasman DI, Raitakari O, Kao WH, Ciullo M, Fox CS, Caulfield M, Bochud M, Gieger C (2013) Genome-wide association analyses identify 18 new loci associated with serum urate concentrations. Nat Genet 45(2):145–154

Okada-Iwabu M, Yamauchi T, Iwabu M, Honma T, Hamagami K, Matsuda K, Yamaguchi M, Tanabe H, Kimura-Someya T, Shirouzu M, Ogata H, Tokuyama K, Ueki K, Nagano T, Tanaka A, Yokoyama S, Kadowaki T (2013) A small-molecule AdipoR agonist for type 2 diabetes and short life in obesity. Nature 503(7477):493–499

Perry JR, Weedon MN, Langenberg C, Jackson AU, Lyssenko V, Sparsø T, Thorleifsson G, Grallert H, Ferrucci L, Maggio M, Paolisso G, Walker M, Palmer CN, Payne F, Young E, Herder C, Narisu N, Morken MA, Bonnycastle LL, Owen KR, Shields B, Knight B, Bennett A, Groves CJ, Ruokonen A, Jarvelin MR, Pearson E, Pascoe L, Ferrannini E, Bornstein SR, Stringham HM, Scott LJ, Kuusisto J, Nilsson P, Neptin M, Gjesing AP, Pisinger C, Lauritzen T, Sandbaek A, Sampson M, MAGIC, Zeggini E, Lindgren CM, Steinthorsdottir V, Thorsteinsdottir U, Hansen T, Schwarz P, Illig T, Laakso M, Stefansson K, Morris AD, Groop L, Pedersen O, Boehnke M, Barroso I, Wareham NJ, Hattersley AT, McCarthy MI, Frayling TM (2010) Genetic evidence that raised sex hormone binding globulin (SHBG) levels reduce the risk of type 2 diabetes. Hum Mol Genet 19 (3):535–544

Petra PH (1991) The plasma sex steroid binding protein (SBP or SHBG). A critical review of recent developments on the structure, molecular biology and function. J Steroid Biochem Mol Biol 40(4–6):735–753

Pfister R, Sharp S, Luben R, Welsh P, Barroso I, Salomaa V, Meirhaeghe A, Khaw KT, Sattar N, Langenberg C, Wareham NJ (2011) Mendelian randomization study of B-type natriuretic peptide and type 2 diabetes: evidence of causal association from population studies. PLoS Med 8(10), e1001112

Raman K, Chong M, Akhtar-Danesh GG, D'Mello M, Hasso R, Ross S, Xu F, Paré G (2013) Genetic markers of inflammation and their role in cardiovascular disease. Can J Cardiol 29 (1):67–74

Semple RK, Cochran EK, Soos MA, Burling KA, Savage DB, Gorden P, O'Rahilly S (2008) Plasma adiponectin as a marker of insulin receptor dysfunction: clinical utility in severe insulin resistance. Diabetes Care 31(5):977–979

Sheehan NA, Didelez V, Burton PR, Tobin MD (2008) Mendelian randomisation and causal inference in observational epidemiology. PLoS Med 5(8), e177

Smith GD, Ebrahim S (2003) 'Mendelian randomization': can genetic epidemiology contribute to understanding environmental determinants of disease? Int J Epidemiol 32(1):1–22

Splansky GL, Corey D, Yang Q, Atwood LD, Cupples LA, Benjamin EJ, D'Agostino RB Sr, Fox CS, Larson MG, Murabito JM, O'Donnell CJ, Vasan RS, Wolf PA, Levy D (2007) The third generation cohort of the national heart, lung, and blood institute's Framingham Heart Study: design, recruitment, and initial examination. Am J Epidemiol 165(11):1328–1335

Stark K, Reinhard W, Grassl M, Erdmann J, Schunkert H, Illig T, Hengstenberg C (2009) Common polymorphisms influencing serum uric acid levels contribute to susceptibility to gout, but not to coronary artery disease. PLoS One 4(11), e7729

Stefan N, Stumvoll M (2002) Adiponectin—its role in metabolism and beyond. Horm Metab Res 34(9):469–474

Tatsioni A, Bonitis NG, Ioannidis JPA (2007) Persistence of contradicted claims in the literature. JAMA 298:2517–2526

Tayar JH, Lopez-Olivo MA, Suarez-Almazor ME (2012) Febuxostat for treating chronic gout. Cochrane Database Syst Rev 11, CD008653

Teslovich TM, Musunuru K, Smith AV, Edmondson AC, Stylianou IM, Koseki M, Pirruccello JP, Ripatti S, Chasman DI, Willer CJ, Johansen CT, Fouchier SW, Isaacs A, Peloso GM, Barbalic M, Ricketts SL, Bis JC, Aulchenko YS, Thorleifsson G, Feitosa MF, Chambers J, Orho-Melander M, Melander O, Johnson T, Li X, Guo X, Li M, Shin Cho Y, Jin Go M, Jin Kim Y, Lee JY, Park T, Kim K, Sim X, Twee-Hee Ong R, Croteau-Chonka DC, Lange LA, Smith JD, Song K, Hua Zhao J, Yuan X, Luan J, Lamina C, Ziegler A, Zhang W, Zee RY, Wright AF, Witteman JC, Wilson JF, Willemsen G, Wichmann HE, Whitfield JB, Waterworth DM, Wareham NJ, Waeber G, Vollenweider P, Voight BF, Vitart V, Uitterlinden AG, Uda M,

Tuomilehto J, Thompson JR, Tanaka T, Surakka I, Stringham HM, Spector TD, Soranzo N, Smit JH, Sinisalo J, Silander K, Sijbrands EJ, Scuteri A, Scott J, Schlessinger D, Sanna S, Salomaa V, Saharinen J, Sabatti C, Ruokonen A, Rudan I, Rose LM, Roberts R, Rieder M, Psaty BM, Pramstaller PP, Pichler I, Perola M, Penninx BW, Pedersen NL, Pattaro C, Parker AN, Pare G, Oostra BA, O'Donnell CJ, Nieminen MS, Nickerson DA, Montgomery GW, Meitinger T, McPherson R, McCarthy MI, McArdle W, Masson D, Martin NG, Marroni F, Mangino M, Magnusson PK, Lucas G, Luben R, Loos RJ, Lokki ML, Lettre G, Langenberg C, Launer LJ, Lakatta EG, Laaksonen R, Kyvik KO, Kronenberg F, König IR, Khaw KT, Kaprio J, Kaplan LM, Johansson A, Jarvelin MR, Janssens AC, Ingelsson E, Igl W, Kees Hovingh G, Hottenga JJ, Hofman A, Hicks AA, Hengstenberg C, Heid IM, Hayward C, Havulinna AS, Hastie ND, Harris TB, Haritunians T, Hall AS, Gyllensten U, Guiducci C, Groop LC, Gonzalez E, Gieger C, Freimer NB, Ferrucci L, Erdmann J, Elliott P, Ejebe KG, Döring A, Dominiczak AF, Demissie S, Deloukas P, de Geus EJ, de Faire U, Crawford G, Collins FS, Chen YD, Caulfield MJ, Campbell H, Burtt NP, Bonnycastle LL, Boomsma DI, Boekholdt SM, Bergman RN, Barroso I, Bandinelli S, Ballantyne CM, Assimes TL, Quertermous T, Altshuler D, Seielstad M, Wong TY, Tai ES, Feranil AB, Kuzawa CW, Adair LS, Taylor HA Jr, Borecki IB, Gabriel SB, Wilson JG, Holm H, Thorsteinsdottir U, Gudnason V, Krauss RM, Mohlke KL, Ordovas JM, Munroe PB, Kooner JS, Tall AR, Hegele RA, Kastelein JJ, Schadt EE, Rotter JI, Boerwinkle E, Strachan DP, Mooser V, Stefansson K, Reilly MP, Samani NJ, Schunkert H, Cupples LA, Sandhu MS, Ridker PM, Rader DJ, van Duijn CM, Peltonen L, Abecasis GR, Boehnke M, Kathiresan S (2010) Biological, clinical and population relevance of 95 loci for blood lipids. Nature 466(7307):707–713

Thanassoulis G, O'Donnell CJ (2009) Mendelian randomization: nature's randomized trial in the post-genome era. JAMA 301(22):2386–2388

Timpson NJ, Lawlor DA, Harbord RM, Gaunt TR, Day IN, Palmer LJ, Hattersley AT, Ebrahim S, Lowe GD, Rumley A, Davey Smith G (2005) C-reactive protein and its role in metabolic syndrome: Mendelian randomisation study. Lancet 366(9501):1954–1959

Trauner M, Arrese M, Wagner M (2010) Fatty liver and lipotoxicity. Biochim Biophys Acta 1801:299–310

Vimaleswaran KS, Berry DJ, Lu C, Tikkanen E, Pilz S, Hiraki LT, Cooper JD, Dastani Z, Li R, Houston DK, Wood AR, Michaëlsson K, Vandenput L, Zgaga L, Yerges-Armstrong LM, McCarthy MI, Dupuis J, Kaakinen M, Kleber ME, Jameson K, Arden N, Raitakari O, Viikari J, Lohman KK, Ferrucci L, Melhus H, Ingelsson E, Byberg L, Lind L, Lorentzon M, Salomaa V, Campbell H, Dunlop M, Mitchell BD, Herzig KH, Pouta A, Hartikainen AL, Genetic Investigation of Anthropometric Traits-GIANT Consortium, Streeten EA, Theodoratou E, Jula A, Wareham NJ, Ohlsson C, Frayling TM, Kritchevsky SB, Spector TD, Richards JB, Lehtimäki T, Ouwehand WH, Kraft P, Cooper C, März W, Power C, Loos RJ, Wang TJ, Järvelin MR, Whittaker JC, Hingorani AD, Hyppönen E (2013) Causal relationship between obesity and vitamin D status: bi-directional Mendelian randomization analysis of multiple cohorts. PLoS Med 10(2):e1001383

Voight BF, Peloso GM, Orho-Melander M, Frikke-Schmidt R, Barbalic M, Jensen MK, Hindy G, Hólm H, Ding EL, Johnson T, Schunkert H, Samani NJ, Clarke R, Hopewell JC, Thompson JF, Li M, Thorleifsson G, Newton-Cheh C, Musunuru K, Pirruccello JP, Saleheen D, Chen L, Stewart A, Schillert A, Thorsteinsdottir U, Thorgeirsson G, Anand S, Engert JC, Morgan T, Spertus J, Stoll M, Berger K, Martinelli N, Girelli D, McKeown PP, Patterson CC, Epstein SE, Devaney J, Burnett MS, Mooser V, Ripatti S, Surakka I, Nieminen MS, Sinisalo J, Lokki ML, Perola M, Havulinna A, de Faire U, Gigante B, Ingelsson E, Zeller T, Wild P, de Bakker PI, Klungel OH, der Zee AH M-v, Peters BJ, de Boer A, Grobbee DE, Kamphuisen PW, Deneer VH, Elbers CC, Onland-Moret NC, Hofker MH, Wijmenga C, Verschuren WM, Boer JM, van der Schouw YT, Rasheed A, Frossard P, Demissie S, Willer C, Do R, Ordovas JM, Abecasis GR, Boehnke M, Mohlke KL, Daly MJ, Guiducci C, Burtt NP, Surti A, Gonzalez E, Purcell S, Gabriel S, Marrugat J, Peden J, Erdmann J, Diemert P, Willenborg C, König IR, Fischer M, Hengstenberg C, Ziegler A, Buysschaert I, Lambrechts D, Van de Werf F, Fox KA, El

Mokhtari NE, Rubin D, Schrezenmeir J, Schreiber S, Schäfer A, Danesh J, Blankenberg S, Roberts R, McPherson R, Watkins H, Hall AS, Overvad K, Rimm E, Boerwinkle E, Tybjaerg-Hansen A, Cupples LA, Reilly MP, Melander O, Mannucci PM, Ardissino D, Siscovick D, Elosua R, Stefansson K, O'Donnell CJ, Salomaa V, Rader DJ, Peltonen L, Schwartz SM, Altshuler D, Kathiresan S (2012) Plasma HDL cholesterol and risk of myocardial infarction: a Mendelian randomisation study. Lancet 572(9841):572–580

Whincup PH, Gilg JA, Papacosta O, Seymour C, Miller GJ, Alberti KG, Cook DG (2002) Early evidence of ethnic differences in cardiovascular risk: cross sectional comparison of British South Asian and white children. BMJ 324(7338):635

Yaghootkar H, Lamina C, Scott RA, Dastani Z, Hivert MF, Warren LL, Stancáková A, Buxbaum SG, Lyytikäinen LP, Henneman P, Wu Y, Cheung CY, Pankow JS, Jackson AU, Gustafsson S, Zhao JH, Ballantyne CM, Xie W, Bergman RN, Boehnke M, el Bouazzaoui F, Collins FS, Dunn SH, Dupuis J, Forouhi NG, Gillson C, Hattersley AT, Hong J, Kähönen M, Kuusisto J, Kedenko L, Kronenberg F, Doria A, Assimes TL, Ferrannini E, Hansen T, Hao K, Häring H, Knowles JW, Lindgren CM, Nolan JJ, Paananen J, Pedersen O, Quertermous T, Smith U, GENESIS Consortium, RISC Consortium, Lehtimäki T, Liu CT, Loos RJ, McCarthy MI, Morris AD, Vasan RS, Spector TD, Teslovich TM, Tuomilehto J, van Dijk KW, Viikari JS, Zhu N, Langenberg C, Ingelsson E, Semple RK, Sinaiko AR, Palmer CN, Walker M, Lam KS, Paulweber B, Mohlke KL, van Duijn C, Raitakari OT, Bidulescu A, Wareham NJ, Laakso M, Waterworth DM, Lawlor DA, Meigs JB, Richards JB, Frayling TM (2013) Mendelian randomization studies do not support a causal role for reduced circulating adiponectin levels in insulin resistance and type 2 diabetes. Diabetes 62(10):3589–3598

Yang Q, Köttgen A, Dehghan A, Smith AV, Glazer NL, Chen MH, Chasman DI, Aspelund T, Eiriksdottir G, Harris TB, Launer L, Nalls M, Hernandez D, Arking DE, Boerwinkle E, Grove ML, Li M, Linda Kao WH, Chonchol M, Haritunians T, Li G, Lumley T, Psaty BM, Shlipak M, Hwang SJ, Larson MG, O'Donnell CJ, Upadhyay A, van Duijn CM, Hofman A, Rivadeneira F, Stricker B, Uitterlinden AG, Paré G, Parker AN, Ridker PM, Siscovick DS, Gudnason V, Witteman JC, Fox CS, Coresh J (2010) Multiple genetic loci influence serum urate levels and their relationship with gout and cardiovascular disease risk factors. Circ Cardiovasc Genet 3 (6):523–530

Chapter 25
Diabetes Prevention

Marie-France Hivert and William C. Knowler

Abstract Type 2 diabetes (T2D) can be prevented or delayed in individuals at high risk by adoption of healthful lifestyle and moderate weight loss or via medication—the most commonly used drug being metformin. This chapter focuses primarily on genetics in the context of lifestyle interventions to prevent T2D. We cover the main findings about the impact of risk variants at candidate genes and known T2D loci (derived from genome-wide association studies) on response to the lifestyle interventions and the few studies that have addressed genetic counseling for risk of T2D on motivation and behavior changes. We also summarize the limited data available on pharmacogenetics of metformin and acarbose in the diabetes prevention trials that included genetics investigations. For a fuller treatment of the pharmacogenetics of metformin, the reader is referred to Chap. 24.

25.1 Lifestyle Interventions for Diabetes Prevention: Evidence from Trials

Behavioral interventions to support healthful lifestyle and moderate weight loss have shown great success in preventing T2D in high-risk populations. These interventions have worked consistently in different racial/ethnic groups in several randomized controlled trials (Knowler et al. 2002; Lindstrom et al. 2003; Pan et al. 1997; Ramachandran et al. 2006; Saito et al. 2011). Good reviews on the subject have been published (Crandall et al. 2008; Yoon et al. 2013), and the relationship of genetic susceptibility to diabetes prevention has been reviewed

M.-F. Hivert
Department of Population Medicine, Harvard Medical School, Harvard Pilgrim Health Care Institute, 133 Brookline Avenue, Boston, MA 02215

Diabetes Unit, Massachusetts General Hospital, 50 Staniford Street, Boston, MA 02114
e-mail: mhivert@partners.org

W.C. Knowler, MD, DrPH (✉)
Diabetes Epidemiology and Clinical Research Section, National Institute of Diabetes and Digestive and Kidney Diseases, 1550 E. Indian School Rd., Phoenix, AZ 85014, USA
e-mail: knowler@nih.gov; wknowler@phx.niddk.nih.gov

J.C. Florez (ed.), *The Genetics of Type 2 Diabetes and Related Traits*,
DOI 10.1007/978-3-319-01574-3_25

(Hivert et al. 2014). We summarize here the main results of the original trials while emphasizing genetics findings with implications for prevention.

The Diabetes Prevention Program (DPP) recruited 3234 overweight or obese adults with elevated fasting glucose and glucose intolerance representing five common self-identified racial/ethnic groups in the US (White, African American, Hispanic, American Indian, Asian) and randomly assigned them to participate in an intensive lifestyle intervention (ILI) or to receive metformin or placebo (Knowler et al. 2002). The ILI aimed to achieve and maintain a weight loss of at least 7 % by dietary modification and increasing physical activity. Case managers provided behavioral support on a one-to-one basis during 16 sessions over the first 24 weeks, followed by monthly encounters to reinforce maintenance of healthful lifestyle. After a mean follow-up of 2.8 years, individuals in the ILI group had lost 5.6 kg compared with 0.1 kg in the placebo group and 2.1 kg in the metformin group ($P < 0.001$). The weight loss and adoption of healthful lifestyle in the ILI group led to a reduction of 58 % [95 % confidence interval = 48–66 %] in diabetes incidence compared with the placebo group. The effect was consistent across subgroups stratified either by age, sex, race/ethnicity, or baseline BMI or fasting glucose. During long-term follow-up of DPP participants, the risk reduction related to the initial ILI was still present, but diminished, 10 years after randomization, despite the fact that the lifestyle intervention was offered to all participants after the end of the active phase of the trial (Diabetes Prevention Program Research Group et al. 2009).

The Finnish Diabetes Prevention Study (DPS) reported strikingly similar findings (Lindstrom et al. 2003). The lifestyle intervention aimed to achieve and maintain 5 % weight loss through dietary modification and physical activity. After 2 years, individuals assigned to the lifestyle intervention lost 3.5 ± 5.5 kg compared with 0.8 ± 4.4 kg in the control group ($P < 0.001$) leading to a 58 % reduction in diabetes incidence at 4-year follow-up. In other populations, similar lifestyle interventions convincingly reduced diabetes incidence, albeit with slightly lower effect sizes. In the Da Qing, China, study, persons with IGT were assigned by clinic to one of four interventions: diet alone, exercise alone, or diet + exercise intervention, compared with only follow-up for diabetes (Pan et al. 1997). The diabetes hazard rates were reduced by 31–46 % by the three interventions. In India, a lifestyle intervention based on the DPP experience led to a 38 % risk reduction over 3 years in glucose-intolerant individuals (Ramachandran et al. 2006). In Japan, a frequent-contact intervention compared with low-frequency contact intervention to support healthful lifestyle reduced diabetes incidence rate by 44 % over 3 years in overweight individuals with impaired fasting glucose at baseline (Saito et al. 2011). Following publications of these controlled trials, many programs have implemented similar approaches in communities that seem to be effective (Ali et al. 2012). Lacking randomized comparison groups, however, the effects of these programs are hard to evaluate (Knowler and Ackermann 2013).

Overall, today there is no doubt that intensive intervention supporting lifestyle modifications such as reduction of caloric intake and increase in physical activity leading to modest weight loss is the most effective way to prevent diabetes in individuals at high risk, and this seems to be consistent across many factors including ethnic origin and country of residence. Nevertheless, there is a wide variation in response to the intervention, either defined as the amount of weight loss or the risk of developing diabetes according to weight loss. For example, in the DPP, participants who responded "highly" (top quartile of weight loss) to the intervention lost a mean of 10 kg at 1 year, while the participants who were "low" responders (bottom quartile) showed a weight loss of about 2.6 kg over the first year and about 10 % participants even gained weight. The different responses to the intervention might be partly explained by genetic differences.

Moreover, despite the large benefit and cost-effectiveness of diabetes prevention, intensive lifestyle interventions are not offered in most health-care settings because of lack of resources (dietician, kinesiologist, behavioral counselor). One hope of the "personalized medicine era" in the field of T2D is to identify patients that are more likely to response to or to benefit from intensive lifestyle intervention.

25.2 Lifestyle Intervention and Genetic Risk

The effect of lifestyle intervention according to genetic risk has been investigated first looking at candidate T2D genes and known T2D-associated loci (based on GWAS results) and then using many known loci to capture an aggregated genetic risk for T2D.

25.2.1 TCF7L2 (See Table 25.1)

Variants at the *TCF7L2* gene have been most often replicated and have one the largest effect sizes for increasing the risk of T2D in European populations (OR $\approx$ 1.4 per risk allele, most often reported being the T allele at rs7903146; see Chap. 15). Many lifestyle intervention studies have, therefore, investigated the impact of risk variants at *TCF7L2* on the effect of interventions in individuals at high risk of T2D (summarized in Table 25.1).

In the DPP, individuals carrying two copies of the T risk allele at rs7903146 had a higher risk of developing T2D in the placebo group (HR = 1.81; $P = 0.004$), as expected from the observational studies (Florez et al. 2006). In contrast, the risk allele was not associated with progression to T2D in the ILI arm (HR = 1.15; $P = 0.60$) suggesting that the lifestyle intervention overcame the excess risk conferred by the genetic variant (T risk allele). Similar results were found in the Finnish DPS (Wang et al. 2007). Together, these two studies strongly suggest that the intensive lifestyle intervention trumped the genetic progression toward T2D attributable to the risk allele at *TCF7L2*.

Another diabetes prevention randomized clinical trial tested a 1-year lifestyle intervention in individuals with metabolic syndrome at baseline (Bo et al. 2009). The 1-year intervention had a significant effect on weight and glycemic indices at 1 year, but the impact was not maintained at the 4-year follow-up (3 years after the

Table 25.1 Effect of T2D increasing risk allele at rs7903146 (T) or rs12255372 (T) at *TCF7L2* in participants undergoing lifestyle interventions aiming at T2D prevention (randomized control trials or one-arm intervention)

Study (author, ref)	Design	Population	Intervention	Main effect of the lifestyle intervention	Main findings on genetics analyses of *TCF7L2* risk allele
DPP Florez et al. (2006)	RCT 3 arms; comparing lifestyle (ILI), metformin, and placebo (control)	$n = 2994$ IGT; US multiethnic; 67 % women; age = 51 ± 11 years; BMI = 34 ± 7 kg/m^2	DPP: ILI = lifestyle counseling, diet and PA, aiming at >7 % weight loss; 16 sessions in 24 weeks and then monthly follow-up Mean follow-up = 3 years	Reduction of T2D incidence by 58 %	In the placebo group, homozygotes TT at rs7903146 were at higher risk of T2D (HR = 1.81, 95 % CI = 1.21–2.70; $P = 0.004$) In the ILI group, homozygotes TT at rs7903146 were not at significantly higher risk T2D (HR = 1.15, 95 % CI = 0.68–1.94; $P = 0.60$) Similar finding for rs12255372 (T)
Finnish DPS Wang et al. (2007)	RCT: lifestyle intervention vs. control	$n = 507$ IGT; Finland 67 % women; age = 55 ± 7 years; BMI = 31 ± 5 kg/m^2	Finnish Diabetes Prevention Study: individualized support for healthier diet and PA Mean follow-up = 3.9 years	Reduction of T2D incidence by 58 %	In the control group, homozygotes TT at rs12255372 were at increase risk of T2D (HR = 2.85, 95 % CI = 11.17–6.95; $P = 0.02$) In the intervention group, homozygotes TT at rs12255372 were not at increase risk of T2D (HR = 0.61, 95 % CI = 0.08–4.52; $P = 0.62$)

					Risk allele (T) at rs7903146 was not associated with T2D risk in either groups
Bo et al. (2009)	RCT: lifestyle intervention vs. control	$n = 335$ Met Syn 59 % women; age = mid-1950s	1-year intervention, lifestyle intervention program supported by trained health professionals (Italy) Follow-up = 4 years	At 4 years = effect on weight (+0.6 kg in intervention vs. +3.1 kg in control; $P < 0.001$) but did not significantly reduce T2D risk (5.4 % vs. 10.2 % in control; $P = 0.09$)	In the overall study, T allele at rs7903146 was borderline associated with T2D incidence at 4 years in multi-adjusted model (OR = 2.6, 95 % CI = 1.0–6.9; $P = 0.05$) with no interaction with intervention
TULIP 1 Haupt et al. (2010) 2 Heni et al. (2010)	One-arm intervention (pre-post)	$n = 309$ at high risk of T2D Germany Age = 46 ± 11 years BMI = 30 ± 6 kg/m^2	TULIP: intervention adapted from DPS, diet, and PA Follow-up 9 months	Weight loss = −0.9 unit of BMI at 9 months ($P < 0.001$). Insulin secretion = no significant improvement	1. Risk carriers of T allele at rs7903146 displayed a lower reduction in BMI ($P < 0.01$). Similar results were found with rs12255372 2. Participants with IGT at baseline who carried the T risk allele at rs7903146 showed a significant increase in insulin secretion during lifestyle intervention compared with NGT participants

intervention ended). The T risk allele at rs7903146 was nominally associated with T2D incidence at 4 years ($P = 0.05$) but there was no overall effect of the intervention and no interaction between the intervention and presence of the T risk allele. The lack of main effect of the intervention at 4 years was suggested to be due to the short duration of the intervention (1 year), making it difficult to compare these findings with DPP and Finnish DPS.

In the Tübingen Lifestyle Intervention Program (TULIP), *TCF7L2* risk allele (T) carriers had less reduction in BMI over the course of the 9-month lifestyle intervention compared with the non-carriers (Haupt et al. 2010). These results are difficult to compare to the findings from DPP and DPS since the TULIP study had a single-arm design (pre/post-intervention), included "at-risk" individuals with and without IGT (60 % with NGT), and investigated intervention over a shorter period than DPS or DPP trials. Interestingly, in a subsequent investigation of the TULIP participants those with IGT at baseline who carried the T risk allele at rs7903146 had a significant increase in insulin secretion during lifestyle intervention compared with NGT participants (Heni et al. 2010). This observation is consistent with the findings from DPP and DPS suggesting that lifestyle intervention may mitigate the progression to T2D in *TCF7L2* risk carriers with IGT.

In summary, intensive lifestyle interventions seem to overcome the increased risk of T2D conferred by the now well-established T2D risk variants at *TCF7L2*. Whether this effect is restricted to individuals with IGT at very high risk of progression to T2D remains to be investigated. Mechanisms by which the lifestyle intervention can limit the progression toward T2D in risk allele carriers are also unknown.

25.2.2 *PPARG* (See Table 25.2)

Variant Pro12Ala at rs1801282 within the *PPARG* gene was one of the first T2D risk loci identified and among the few genes derived from candidate genes that were subsequently confirmed in GWAS. The Pro allele is associated with increased risk of T2D with an odds ratio around 1.2 per risk allele (Morris et al. 2012). Consistent with the observational data, the Pro allele was associated (but not significantly) with the risk of developing T2D (HR = 1.24; $P = 0.07$) in DPP participants, and there was no significant genotype by intervention interaction (Florez et al. 2007a). The protective Ala allele was associated with more favorable weight changes at 1 year in all 3 main arms (placebo, metformin, and ILI) of the DPP (Florez et al. 2007a). By contrast, in the one-arm TULIP study, the Pro12Ala variant was not associated with difference in weight change over 9 months of intervention (Rittig et al. 2007).

In the Finnish DPS, the protective Ala allele was also associated with greater weight loss in the intervention arm, consistent with DPP findings (Franks et al. 2007), while the weight change in the control arm was not associated with the Pro12Ala variant (Lindi et al. 2002). Nevertheless, the report concerning this *PPARG* variant in the DPS participants was puzzling by the observation that the

Table 25.2 Effect of genetic variants at rs1801282 at *PPARG* in participants undergoing lifestyle interventions aiming at T2D prevention (randomized control trials or one-arm intervention)

Study (author, ref)	Design	Population	Intervention	Main effect of the lifestyle intervention	Main findings on genetics analyses of *PPARG* risk variant (Pro = higher T2D risk)
DPP Florez et al. (2007a)	RCT 4 arms; comparing lifestyle (ILI), metformin, troglitazone, and placebo (control)	$n = 3548$ IGT including 554 in troglitazone arm US multiethnic 67 % women Age = 51 years BMI = 34 kg/m^2	DPP: ILI = lifestyle counseling, diet and PA, aiming at >7 % weight loss; 16 sessions in 24 weeks then monthly follow-up Mean follow-up = 3 years	Reduction of T2D incidence by 58 %	Carriers of the Pro allele at rs1801282 seem to progress more rapidly from IGT to T2D (HR = 1.24, 95 % CI = 0.99–1.57; $P = 0.07$) in the overall DPP study; no difference between arms
Finnish DPS Lindi et al. (2002)	RCT: lifestyle intervention vs. control	$n = 522$ IGT; Finland 67 % women; age = 40–68 years; BMI = 31 ± 5 kg/m^2	Finnish Diabetes Prevention Study: individualized support for healthier diet and PA Follow-up = 3 years	Reduction of T2D incidence by 58 %	In the control group, the protective Ala allele at rs1801282 was associated with **higher** risk of T2D (OR = 2.36, 95 % CI = 1.21–4.60) In the intervention group, the Pro12Ala variant was not significantly associated with T2D risk (Ala risk OR = 1.90, 95 % CI = 0.70–5.18)
TULIP Rittig et al. (2007)	One-arm intervention (pre-post)	$n = 166$ at high risk of T2D Germany 65 % women Age = 46 ± 11 years BMI > 27 kg/m^2	TULIP: intervention adapted from DPS, diet, and PA Follow-up 9 months	Decrease in BMI 29.3 to 28.6 kg/m^2 over 9 months ($P < 0.001$)	Pro12Ala was not associated with difference in BMI change over time

known "protective" Ala allele was associated with higher risk of developing T2D over the 3 years of the study in the control group (OR = 2.36; CI = 1.21–4.60), and this remained significant even after adjustment for weight at baseline and weight change over time. In the intervention arm, risk of developing T2D was not significantly associated with Pro12Ala, but the direction of effect still suggested that the Ala allele was positively associated with risk of T2D (OR = 1.90 CI = 0.70–5.18) opposite to the direction of effect expected from observational studies. In a subsequent report (Kilpelainen et al. 2008), the Ala allele was again associated with increased risk of T2D in the overall DPS participant analyses, but the association was attenuated and not significant when adjusted for multiple factors including baseline fasting glucose.

In light of all these reports, it is difficult to draw conclusions about the impact of Pro12Ala variant in response to lifestyle intervention in individuals at risk of T2D. The current body of evidence suggests that Ala carriers (protective allele) might experience a greater weight loss in the context of lifestyle intervention and that the levels of initial BMI might modulate the risk attributed to the Pro12Ala variant (Florez et al. 2007a; Lindi et al. 2002). Yet, the relatively smaller effect size (compared with *TCF7L2*) and low minor allele frequency at rs1801282 limited the power in each study reported so far (power being even lower for detecting interaction). Other variants within or near *PPARG* have also been investigated in DPP and DPS (Florez et al. 2007a; Kilpelainen et al. 2008), but results were null or inconsistent.

25.2.3 *ENPP1*

Ectoenzyme nucleotide pyrophosphatase phosphodiesterase 1 (*ENPP1*) reduces insulin signaling by direct inhibition of the insulin receptor's tyrosine kinase activity. The genetic variant K121Q (rs1044498) in *ENPP1* may result in a gain-of-function mutation leading to greater inhibition of the insulin receptor and clinical insulin resistance, prompting many investigations of the association between K121Q and T2D or glycemic traits.

In the DPP, participants carrying at least one copy of the Q risk variant at K121Q in *ENPP1* were more likely to develop T2D in the placebo group (HR = 1.38; $P = 0.03$), but this risk was abolished in the intensive lifestyle arm (HR = 0.89; $P = 0.51$) (Moore et al. 2009). In contrast to the lack of genetic effect in the intervention arm of the DPP, improvement in insulin sensitivity was greater in KK homozygotes in the TULIP single-arm intervention while it decreased in QQ individuals (Mussig et al. 2010). The differences between studies are difficult to explain and underline the importance of replication in genetics, especially with respect to gene-lifestyle interactions. Overall the role of *ENPP1* in T2D is still obscure since the current largest GWAS meta-analyses of T2D cases have not revealed *ENPP1* as one of the T2D loci (Morris et al. 2012), but previous reports suggested possible interaction with BMI in testing the association with T2D (McAteer et al. 2008).

25.2.4 Investigations of Biological Candidate Genes

Many other candidate genes for T2D or weight/glycemic regulation have been tested for interaction with lifestyle intervention before the wave of GWAS discovery, with mostly mixed and contradictory results, again illustrating the challenges of testing gene-lifestyle interaction. For example, 1536 single-nucleotide polymorphisms (SNPs) located in 40 candidate genes selected for potential biological relevance were investigated in the DPP participants to reveal genotype by treatment interactions. Twenty-three loci had nominally significant interactions with the lifestyle intervention, but none of the loci maintained significance after adjustment for multiple testing (Jablonski et al. 2010).

25.2.5 Known Loci Derived from GWAS: T2D and Obesity

25.2.5.1 T2D Loci

Since the early 2000s, technological and scientific advances have led to identification of more and more genetic variants associated with T2D. As of early 2014, more than 60 loci are accepted and "established" T2D risk loci based on associations that reached the genome-wide significance level ($P < 5\text{x}10^{-8}$; see Chap. 2). In DPP, many of the known variants were tested for their effect on T2D incidence, overall and within each treatment arm (Jablonski et al. 2010; Hivert et al. 2011; Moore et al. 2008). Among the 34 loci known at the time, association with T2D was detected at *TCF7L2, HNF1B*, and *PROX1* in the overall DPP study participants, independent of treatment allocation (Hivert et al. 2011). Only 3 loci were suggestive of gene-treatment interaction: *KCNJ11, HNF1A*, and *PLEHF2*—and only at *PLEHF2* was the risk variant associated with higher risk in the placebo group (HR = 1.28; CI = 1.06–1.54; $P = 0.009$) and that risk was overcome by the lifestyle intervention (HR = 0.91; CI = 0.71–1.16; $P = 0.45$). Again, this needs to be interpreted with caution given multiple testing. The lack of replication of loci detected in T2D case–control studies is likely due to the difference in study design, the limited power, and the fact that DPP participants were already glucose intolerant at baseline, decreasing the effect size that genetic risk variants might have on progression to T2D.

25.2.5.2 Obesity/Adiposity Loci

Genome-wide association studies have also revealed genetic variants associated with obesity and adiposity traits, and some of these variants might also play a role in diabetes and glycemic regulation. The most commonly known genetic variant for risk of obesity is located near *FTO* and was initially identified as a T2D locus until its effect on BMI was unmasked. The risk allele A at rs9939609 near *FTO* was not associated with weight changes over >3 years in DPP (Franks et al. 2008) or in the Finnish DPS (Lappalainen et al. 2009) and no significant interaction with lifestyle intervention was detected on change in BMI. In a subsample of DPP participants with CT scan measurements of body composition, a possible gene-treatment interaction ($P = 0.05$) was detected for change in 1-year subcutaneous tissue, possibly due to the A risk allele that tended to be associated with greater increase in subcutaneous tissue at 1 year in the placebo group (Franks et al. 2008).

Other known obesity loci were investigated in DPP for effects on weight change (Delahanty et al. 2012). The Ala allele at rs1801282 near *PPARG* was associated with greater weight loss at 6-month and 2-year follow-up, irrespective of treatment allocation. For the 2-year weight change, there was a suggestive interaction between treatment arms and genetic variants at *NEGR1* in addition to *FTO*, in line with the suggestive interaction on the 1-year change in subcutaneous fat previously reported (Franks et al. 2008). Genetic variants at *PPARG* and *BDNF* were associated with weight regain after 6 months independent of treatment arm, and possible interactions were suggested at *TMEM18* and *KTCD15*. Yet, as the authors underlined, these results must be interpreted with caution and are mainly hypothesis generating.

25.2.6 Genetic Risk Scores for Risk of T2D

Another question that can be addressed from the T2D prevention trials is whether individuals at high risk based on their global genotype profile for risk of T2D are likely to benefit from an intensive lifestyle intervention. The impact of an aggregated T2D genetic risk score on the risk of developing T2D over the course of the lifestyle intervention to prevent T2D was assessed in both the DPP and DPS studies.

Based on 34 T2D loci known at the time, DPP participants were classified according to their genetic risk profile: the 34 GRS predicted the risk of developing T2D in the overall cohort (HR = 1.02; CI = 1.00–1.05 per risk allele; $P = 0.03$ in fully adjusted model) and a lower chance to regression to normoglycemia (HR = 0.95; CI = 0.93–0.98; $P < 0.001$) (Hivert et al. 2011). Individuals at highest genetic risk (top quartile of the GRS) clearly benefited from the intensive lifestyle intervention; in this subgroup, diabetes incidence was 12 cases/100 person-years in the placebo vs. 5 cases/100 person-years in lifestyle arm ($P < 0.001$). The results also suggested that the benefit of the lifestyle intervention might be greater in the individuals at higher genetic risk, but the interaction between GRS and treatment arm was not significant. Moreover, the lifestyle intervention significantly increased the chance of regression to normoglycemia ($P < 0.001$) in participants with the greater T2D genetic burden (top quartile of GRS) (Hivert et al. 2011).

In the DPS, the lifestyle intervention prevented T2D independently of a GRS composed of 19 known T2D loci (Uusitupa et al. 2011). There was no significant interaction between the GRS and treatment arm on the risk of T2D ($P = 0.65$). None of the 19 individual loci was significantly associated with T2D incidence in the overall DPS participants, which was not surprising given the relatively small sample size ($N = 522$).

Overall, those studies are reassuring and can help clinicians to have a positive message to communicate with their patients: T2D can be prevented by lifestyle modification and modest weight loss, no matter how "genetically" at risk an individual may be. On the other hand, whether knowing ones "T2D genetic risk profile" will increase the likelihood of sustained behavior changes leading to health benefits is a separate question addressed in the next section.

25.3 T2D Genetics for Motivating Health Behavior Change

Aside from its ability to improve the risk stratification of individuals, it has been suggested that personal knowledge of being at higher genetic susceptibility to T2D would increase motivation for behavior changes. In surveys and questionnaires, patients have reported that they would be more motivated to adopt healthier eating habits and be more physically active if they learned they were at greater genetic risk of T2D (Grant et al. 2009; Vassy et al. 2012, 2013). In a primary care practice population, 53 % of patients reported that they would be "very likely" to order a genetic test to know their risk of T2D and 71 % of them reported they would be "highly motivated" to make behavioral modification to their lifestyle if they were found to be at higher risk (Grant et al. 2009). However, translating intentions into actual behaviors is another leap, especially when complex behaviors such as healthy eating and physical activity have to be maintained over a long period of time.

So far, observational and experimental studies do not support that knowledge of personal genetic risk of T2D leads to sustained behavior change. In an assessment of direct-to-consumer genome-wide genetic profiling for 22 conditions (including T2D), the Scripps Genomic Health Initiative followed 2037 people after they received the results of their genetic profile. In an examination of T2D risk in particular, the magnitude of the genetic risk estimate for T2D (expressed as risk over lifetime or compared with the general population) provided to each individual was not associated with a change in fat intake or exercise at 3 months after receiving the results (Bloss et al. 2011).

The Genetic Counseling and Lifestyle Change for Diabetes Prevention (GC/LC) Study randomized patients at risk of T2D to genetic testing and counseling or not (control group) (Grant et al. 2013). For participants randomized to the genetic testing/counseling arm, genetic susceptibility to T2D was estimated from a genetic risk score composed of 36 established T2D genetic variants. Persons either at "higher" or "lower" risk of T2D based on the score were invited to participate to the lifestyle intervention program. A certified genetic counselor provided the genetic counseling session and used visual cues to explain "higher" and "lower" risk of developing T2D compared with the general population. Participants randomized to control group did not undergo genetic testing and were not seen by the genetic counselor before entering the lifestyle intervention program. All participants were then enrolled in a 12-week lifestyle modification program based on the DPP curriculum. The program succeeded in achieving a mean weight loss of 3.9 kg, but this did not differ between the group that received genetic testing and counseling or not. There was also no difference between participants who learned they were at higher or lower genetic risk.

As of early 2014, a few other trials are currently testing whether genetic counseling for risk of T2D enhances adoption of a healthier lifestyle to prevent T2D, mainly in the primary care setting (Voils et al. 2012; Cho et al. 2012; Vorderstrasse et al. 2013). The results of these trials are yet to be published and should inform clinical practice in the future.

25.4 Pharmacogenetics in Diabetes Prevention

25.4.1 Metformin

As of early 2014, investigations concerning metformin-gene interactions in prevention of T2D are exclusively derived from reports in the DPP participants (Florez et al. 2006, 2007b, 2008, 2012; Jablonski et al. 2010; Moore et al. 2008; Chiasson et al. 2002). Because of the current lack of replication available, readers should be cautious in interpretation of the findings, despite many being biologically possible. For a full overview of metformin pharmacogenetics, please refer to Chap. 24.

Potential interactions were discovered at genes encoding metformin transporters (*SLC22A1*, *SLC22A2*, and *SLC47A1*). Effects of genotypes on T2D incidence were detected only in the metformin group, but not in the placebo or lifestyle groups (Jablonski et al. 2010). Analyses also suggested interactions in genes encoding proteins in gluconeogenesis pathways (*PRKAA1*, *PRKAB2*, *PRKAA2*, *STK11*, and *PCK1*) (Jablonski et al. 2010). In pathways related to peroxisome proliferator-activated receptors, genetic variants at *PPARA* (G at rs4253652) and *PPARGC1A* (C at rs10213440) associated with increased risk of T2D in the metformin group appeared protective in the placebo group (Jablonski et al. 2010).

Among the loci known for risk of T2D in the general population, risk variants at *TCF7L2*, *WFS1*, *CDKAL1*, *IGF2BP2*, *HHEX*, and *SCL30A8* did not interact with metformin treatment (Florez et al. 2006; Moore et al. 2008; Chiasson et al. 2002).

At *KCNJ11*, no significant interaction was detected at the T2D-associated SNP rs5219 (Florez et al. 2007b), but an interaction was suggested at rs7124355 (Jablonski et al. 2010). Interactions were also suggested in a few other known T2D genes: *CDKN2A/B*, *HNF4A*, *HNF1B*, and *GCK* (Jablonski et al. 2010; Moore et al. 2008). Other loci associated with fasting glucose or fasting insulin (*G6PC2*, *FADS1*, *ADRA2*, *DGKB*, *IGF1*, *PROX1*, *GCK*, *GCKR*, *GLIS3*, *IRS1*, *ADCY5*, *MTNR1B*) had no significant interactions with treatment on change in fasting glucose at 1 year of treatment, except for *MADD* and *C2CD4B* loci (each $P = 0.04$ for interaction with metformin treatment) (Florez et al. 2012). Interactions were suggested at candidate genes selected for their known or suspected role in glycemic regulation such as *ABCC8*, *ENPP1*, and *ADIPOR2* (Jablonski et al. 2010).

25.4.2 Acarbose

In the STOP-NIDDM randomized controlled trial, acarbose reduced the risk of developing T2D by 25 % compared with placebo in persons with glucose intolerance (Andrulionyte et al. 2007). Genetic investigations in candidate genes suggested interactions in *PPARA* (2 out of 11 SNPs investigated) (Andrulionyte et al. 2006), *HNF4A* (2 out 6 SNPs) (Zacharova et al. 2005), *LIPC* (Andrulionyte et al. 2004), and *PPARGC1A* [49], but not at *PPARG* locus [49]. Those results were derived from a single clinical trial, with no validation or replication available at this time, warranting caution in interpretation.

25.5 Future Perspectives

As described in Sect. 25.2 of this chapter, genetic investigations to determine who is more likely to response physiologically to an intensive lifestyle intervention such as DPP or DPS has not been highly successful so far. The genetics of "high responders" to a lifestyle intervention are unlikely to lie within genes identified in GWAS since they detected main genetic effects based on case–control studies. Indeed, the "known loci" approaches have mainly told us that adoption of healthful lifestyle and modest weight loss can successfully prevent T2D largely independently of genetic burden. With health-care resource allocation in mind, we could argue that it would be worth identifying individuals at the higher end of the spectrum for T2D genetic burden since they seem to benefit at least as much from the intensive lifestyle intervention. This will require cost-effectiveness analyses that take into account the cost of genotyping, counseling, and the number of cases of T2D avoided or delayed in that subgroup, in addition to the cost of the lifestyle intervention.

One of the next challenges will lie in identifying "responders" genes to target individuals that would have higher response to our known successful interventions or who would need less intensive versions of behavioral support for lifestyle modifications. This will require larger sample sizes to investigate gene-lifestyle (or gene-intervention) interactions or acceptance of less stringent thresholds to define interaction. Other "omics" scientific advances are likely to contribute to our understanding of "response" to dietary and physical activity modifications. Large-scale epigenetic (e.g., DNA methylation), transcriptomic (expression), and metabolomic studies may help identify pathways that "respond" to nutritional and exercise challenges over short (meal test, exercise testing) and long periods of time (weeks or years of intervention). If such "responder" genes are identified based on samples and data collected in previous trials, their interaction with interventions should be tested in independent T2D prevention trials before we can hope to influence clinical practice and witness clinical applications of genetics in prevention of T2D. Further, it is hoped that better understanding of the pathways leading to T2D will lead to development of new behavioral or pharmacological interventions for prevention.

References

Ali MK, Echouffo-Tcheugui J, Williamson DF (2012) How effective were lifestyle interventions in real-world settings that were modeled on the Diabetes Prevention Program? Health Aff (Millwood) 31:67–75

Andrulionyte L, Zacharova J, Chiasson JL, Laakso M, STOP-NIDDM Study Group (2004) Common polymorphisms of the PPAR-gamma2 (Pro12Ala) and PGC-1alpha (Gly482Ser) genes are associated with the conversion from impaired glucose tolerance to type 2 diabetes in the STOP-NIDDM trial. Diabetologia 47:2176–2184

Andrulionyte L, Laukkanen O, Chiasson JL, Laakso M, STOP-NIDDM Study Group (2006) Single nucleotide polymorphisms of the HNF4alpha gene are associated with the conversion to type 2 diabetes mellitus: the STOP-NIDDM trial. J Mol Med (Berl) 84:701–708

Andrulionyte L, Kuulasmaa T, Chiasson JL, Laakso M, STOP-NIDDM Study Group (2007) Single nucleotide polymorphisms of the peroxisome proliferator-activated receptor-alpha gene (PPARA) influence the conversion from impaired glucose tolerance to type 2 diabetes: the STOP-NIDDM trial. Diabetes 56:1181–1186

Bloss CS, Schork NJ, Topol EJ (2011) Effect of direct-to-consumer genomewide profiling to assess disease risk. N Engl J Med 364:524–534

Bo S, Gambino R, Ciccone G et al (2009) Effects of TCF7L2 polymorphisms on glucose values after a lifestyle intervention. Am J Clin Nutr 90:1502–1508

Chiasson JL, Josse RG, Gomis R et al (2002) Acarbose for prevention of type 2 diabetes mellitus: the STOP-NIDDM randomised trial. Lancet 359:2072–2077

Cho AH, Killeya-Jones LA, O'Daniel JM et al (2012) Effect of genetic testing for risk of type 2 diabetes mellitus on health behaviors and outcomes: study rationale, development and design. BMC Health Serv Res 12:16

Crandall JP, Knowler WC, Kahn SE et al (2008) The prevention of type 2 diabetes. Nat Clin Pract Endocrinol Metab 4:382–393

Delahanty LM, Pan Q, Jablonski KA et al (2012) Genetic predictors of weight loss and weight regain after intensive lifestyle modification, metformin treatment, or standard care in the Diabetes Prevention Program. Diabetes Care 35:363–366

Diabetes Prevention Program Research Group, Knowler WC, Fowler SE et al (2009) 10-year follow-up of diabetes incidence and weight loss in the Diabetes Prevention Program Outcomes Study. Lancet 374:1677–1686

Florez JC, Jablonski KA, Bayley N et al (2006) TCF7L2 polymorphisms and progression to diabetes in the Diabetes Prevention Program. N Engl J Med 355:241–250

Florez JC, Jablonski KA, Sun MW et al (2007a) Effects of the type 2 diabetes-associated PPARG P12A polymorphism on progression to diabetes and response to troglitazone. J Clin Endocrinol Metab 92:1502–1509

Florez JC, Jablonski KA, Kahn SE et al (2007b) Type 2 diabetes-associated missense polymorphisms KCNJ11 E23K and ABCC8 A1369S influence progression to diabetes and response to interventions in the Diabetes Prevention Program. Diabetes 56:531–536

Florez JC, Jablonski KA, McAteer J et al (2008) Testing of diabetes-associated WFS1 polymorphisms in the Diabetes Prevention Program. Diabetologia 51:451–457

Florez JC, Jablonski KA, McAteer JB et al (2012) Effects of genetic variants previously associated with fasting glucose and insulin in the Diabetes Prevention Program. PLoS ONE 7, e44424

Franks PW, Jablonski KA, Delahanty L et al (2007) The Pro12Ala variant at the peroxisome proliferator-activated receptor gamma gene and change in obesity-related traits in the Diabetes Prevention Program. Diabetologia 50:2451–2460

Franks PW, Jablonski KA, Delahanty LM et al (2008) Assessing gene-treatment interactions at the FTO and INSIG2 loci on obesity-related traits in the Diabetes Prevention Program. Diabetologia 51:2214–2223

Grant RW, Hivert M, Pandiscio JC, Florez JC, Nathan DM, Meigs JB (2009) The clinical application of genetic testing in type 2 diabetes: a patient and physician survey. Diabetologia 52:2299–2305

Grant RW, O'Brien KE, Waxler JL et al (2013) Personalized genetic risk counseling to motivate diabetes prevention: a randomized trial. Diabetes Care 36:13–19

Haupt A, Thamer C, Heni M et al (2010) Gene variants of TCF7L2 influence weight loss and body composition during lifestyle intervention in a population at risk for type 2 diabetes. Diabetes 59:747–750

Heni M, Ketterer C, Thamer C et al (2010) Glycemia determines the effect of type 2 diabetes risk genes on insulin secretion. Diabetes 59:3247–3252

Hivert MF, Jablonski KA, Perreault L et al (2011) Updated genetic score based on 34 confirmed type 2 diabetes Loci is associated with diabetes incidence and regression to normoglycemia in the diabetes prevention program. Diabetes 60:1340–1348

Hivert MF, Vassy JL, Meigs JB (2014) Susceptibility to type 2 diabetes mellitus-from genes to prevention. Nat Rev Endocrinol 10:198–205

Jablonski KA, McAteer JB, de Bakker PI et al (2010) Common variants in 40 genes assessed for diabetes incidence and response to metformin and lifestyle intervention in the diabetes prevention program. Diabetes 59:2672–2681

Kilpelainen TO, Lakka TA, Laaksonen DE et al (2008) SNPs in PPARG associate with type 2 diabetes and interact with physical activity. Med Sci Sports Exerc 40:25–33

Knowler WC, Ackermann RT (2013) Preventing diabetes in American Indian communities. Diabetes Care 36:1820–1822

Knowler WC, Barrett-Connor E, Fowler SE et al (2002) Reduction in the incidence of type 2 diabetes with lifestyle intervention or metformin. N Engl J Med 346:393–403

Lappalainen TJ, Tolppanen AM, Kolehmainen M et al (2009) The common variant in the FTO gene did not modify the effect of lifestyle changes on body weight: the Finnish Diabetes Prevention Study. Obesity (Silver Spring) 17:832–836

Lindi VI, Uusitupa MI, Lindstrom J et al (2002) Association of the Pro12Ala polymorphism in the PPAR-gamma2 gene with 3-year incidence of type 2 diabetes and body weight change in the Finnish Diabetes Prevention Study. Diabetes 51:2581–2586

Lindstrom J, Louheranta A, Mannelin M et al (2003) The Finnish Diabetes Prevention Study (DPS): lifestyle intervention and 3-year results on diet and physical activity. Diabetes Care 26:3230–3236

McAteer JB, Prudente S, Bacci S et al (2008) The ENPP1 K121Q polymorphism is associated with type 2 diabetes in European populations: evidence from an updated meta-analysis in 42,042 subjects. Diabetes 57:1125–1130

Moore AF, Jablonski KA, McAteer JB et al (2008) Extension of type 2 diabetes genome-wide association scan results in the diabetes prevention program. Diabetes 57:2503–2510

Moore AF, Jablonski KA, Mason CC et al (2009) The association of ENPP1 K121Q with diabetes incidence is abolished by lifestyle modification in the diabetes prevention program. J Clin Endocrinol Metabol 94:449–455

Morris AP, Voight BF, Teslovich TM, Ferreira T, Segre AV, Steinthorsdottir V, Strawbridge RJ, Khan H, Grallert H, Mahajan A, Prokopenko I, Kang HM, Dina C, Esko T, Fraser RM, Kanoni S, Kumar A, Lagou V, Langenberg C, Luan J, Lindgren CM, Muller-Nurasyid M, Pechlivanis S, Rayner NW, Scott LJ, Wiltshire S, Yengo L, Kinnunen L, Rossin EJ, Raychaudhuri S, Johnson AD, Dimas AS, Loos RJ, Vedantam S, Chen H, Florez JC, Fox C, Liu CT, Rybin D, Couper DJ, Kao WH, Li M, Cornelis MC, Kraft P, Sun Q, van Dam RM, Stringham HM, Chines PS, Fischer K, Fontanillas P, Holmen OL, Hunt SE, Jackson AU, Kong A, Lawrence R, Meyer J, Perry JR, Platou CG, Potter S, Rehnberg E, Robertson N, Sivapalaratnam S, Stancakova A, Stirrups K, Thorleifsson G, Tikkanen E, Wood AR, Almgren P, Atalay M, Benediktsson R, Bonnycastle LL, Burtt N, Carey J, Charpentier G, Crenshaw AT, Doney AS, Dorkhan M, Edkins S, Emilsson V, Eury E, Forsen T, Gertow K, Gigante B, Grant GB, Groves CJ, Guiducci C, Herder C, Hreidarsson AB, Hui J, James A, Jonsson A, Rathmann W, Klopp N, Kravic J, Krjutskov K, Langford C, Leander K, Lindholm E, Lobbens S, Mannisto S, Mirza G, Muhleisen TW, Musk B, Parkin M, Rallidis L, Saramies J, Sennblad B, Shah S, Sigurethsson G, Silveira A, Steinbach G, Thorand B, Trakalo J, Veglia F, Wennauer R, Winckler W, Zabaneh D, Campbell H, van Duijn C, Uitterlinden AG, Hofman A, Sijbrands E, Abecasis GR, Owen KR, Zeggini E, Trip MD, Forouhi NG, Syvanen AC, Eriksson JG, Peltonen L, Nothen MM, Balkau B, Palmer CN, Lyssenko V, Tuomi T, Isomaa B, Hunter DJ, Qi L, Wellcome Trust Case Control Consortium, Meta-Analyses of Glucose and Insulin-related traits Consortium (MAGIC) Investigators, Genetic Investigation of ANthropometric Traits (GIANT) Consortium, Asian Genetic Epidemiology Network-Type 2 Diabetes (AGEN-T2D) Consortium, South Asian Type 2 Diabetes (SAT2D) Consortium, Shuldiner AR, Roden M, Barroso I, Wilsgaard T, Beilby J, Hovingh K, Price JF, Wilson JF, Rauramaa R, Lakka TA, Lind L, Dedussis G, Njolstad I, Pedersen NL, Khaw KT, Wareham NJ, Keinanen-Kiukaanniemi SM, Saaristo TE, Korpi-Hyovalti E, Saltevo J, Laakso M, Kuusisto J, Metspalu A, Collins FS, Mohlke KL, Bergman RN, Tuomilehto J, Boehm BO, Gieger C, Hveem K, Cauchi S, Froguel P, Baldassarre D, Tremoli E, Humphries SE, Saleheen D, Danesh J, Ingelsson E, Ripatti S, Salomaa V, Erbel R, Jockel KH, Moebus S, Peters A, Illig T, de Faire U, Hamsten A, Morris AD, Donnelly PJ, Frayling TM, Hattersley AT, Boerwinkle E, Melander O, Kathiresan S, Nilsson PM, Deloukas P, Thorsteinsdottir U, Groop LC, Stefansson K, Hu F, Pankow JS, Dupuis J, Meigs JB, Altshuler D, Boehnke M, McCarthy MI, DIAbetes Genetics Replication And Meta-analysis (DIAGRAM) Consortium (2012) Large-scale association analysis provides insights into the genetic architecture and pathophysiology of type 2 diabetes. Nat Genet 44:981–990

Mussig K, Heni M, Thamer C et al (2010) The ENPP1 K121Q polymorphism determines individual susceptibility to the insulin-sensitising effect of lifestyle intervention. Diabetologia 53:504–509

Pan XR, Li GW, Hu YH et al (1997) Effects of diet and exercise in preventing NIDDM in people with impaired glucose tolerance. The Da Qing IGT and Diabetes Study.see comment. Diabetes Care 20:537–544

Ramachandran A, Snehalatha C, Mary S et al (2006) The Indian Diabetes Prevention Programme shows that lifestyle modification and metformin prevent type 2 diabetes in Asian Indian subjects with impaired glucose tolerance (IDPP-1). Diabetologia 49:289–297

Rittig K, Thamer C, Machicao F et al (2007) The Pro12Ala polymorphism in PPARG2 increases the effectiveness of primary prevention of cardiovascular disease by a lifestyle intervention. Diabetologia 50:1345–1347

Saito T, Watanabe M, Nishida J et al (2011) Lifestyle modification and prevention of type 2 diabetes in overweight Japanese with impaired fasting glucose levels: a randomized controlled trial. Arch Intern Med 171:1352–1360

Uusitupa MI, Stancakova A, Peltonen M et al (2011) Impact of positive family history and genetic risk variants on the incidence of diabetes: the Finnish Diabetes Prevention Study. Diabetes Care 34:418–423

Vassy JL, O'Brien KE, Waxler JL et al (2012) Impact of literacy and numeracy on motivation for behavior change after diabetes genetic risk testing. Med Decis Making 32:606–615

Vassy JL, Hivert MF, Porneala B, Florez JC, Dupuis J, Meigs JB (2013) Type 2 diabetes prediction with 17-, 40-, and 62-variant genotype risk scores: the Framnigham Offspring Study. American Heart Association meetings; Nutrition, Physical Activity, and Metabolism; Epidemiology Councils Abstract (Abstract)

Voils CI, Coffman CJ, Edelman D et al (2012) Examining the impact of genetic testing for type 2 diabetes on health behaviors: study protocol for a randomized controlled trial. Trials 13:121

Vorderstrasse AA, Ginsburg GS, Kraus WE, Maldonado MC, Wolever RQ (2013) Health coaching and genomics-potential avenues to elicit behavior change in those at risk for chronic disease: protocol for personalized medicine effectiveness study in air force primary care. Glob Adv Health Med 2:26–38

Wang J, Kuusisto J, Vanttinen M et al (2007) Variants of transcription factor 7-like 2 (TCF7L2) gene predict conversion to type 2 diabetes in the Finnish Diabetes Prevention Study and are associated with impaired glucose regulation and impaired insulin secretion. Diabetologia 50:1192–1200

Yoon U, Kwok LL, Magkidis A (2013) Efficacy of lifestyle interventions in reducing diabetes incidence in patients with impaired glucose tolerance: a systematic review of randomized controlled trials. Metabolism 62:303–314

Zacharova J, Todorova BR, Chiasson JL, Laakso M, STOP-NIDDM Study Group (2005) The G-250A substitution in the promoter region of the hepatic lipase gene is associated with the conversion from impaired glucose tolerance to type 2 diabetes: the STOP-NIDDM trial. J Intern Med 257:185–193

Chapter 26
Nutrigenetics of Type 2 Diabetes

Lu Qi and Paul W. Franks

Abstract Type 2 diabetes has become a leading health problem throughout the world. The escalating epidemic of type 2 diabetes is believed to result from a collision between inherent biological susceptibilities (genotypes) and a shift toward dietary habits and lifestyle that promote obesity over the past several decades; the transition from "traditional" to modern "obesogenic" lifestyles is characterized by widespread access to highly palatable, nutrient-deficient, and calorie-dense foods and beverages, as well as circumstances that promote sedentary behaviors and inhibit physical activity. In the past decades, a large body of epidemiological studies has associated various dietary factors with type 2 diabetes risk. In the meantime, genetic studies have made great strides in unraveling the genetic basis of type 2 diabetes by identifying more about 100 common genetic loci related to the disease. Nutrigenetics, a relatively new branch of nutrition science, focuses on determining the interplay between dietary exposures and genetic factors in the etiology of many diseases including type 2 diabetes. Even with many hundreds of gene-diet/lifestyle interaction studies on diabetes-related traits published over the past two decades, few examples have been adequately replicated or validated. By contrast, a number of replicated examples of interactions between lifestyle factors (e.g., consumption of sugar sweetened beverages and fried foods and low physical activity and sedentary lifestyles) and genetic factors in obesity (a major risk factor for diabetes) have recently emerged. Further advances are likely to come from the optimization of methods and study designs for nutrigenetic analyses. The develop-

L. Qi
Department of Epidemiology, Tulane University School of Public Health and Tropical Medicine, New Orleans, LA, USA

Harvard School of Public Health, Boston, MA, USA

P.W. Franks (✉)
Harvard School of Public Health, Boston, MA, USA

Genetic and Molecular Epidemiology Unit, Department of Clinical Sciences, Clinical Research Center, Skåne University Hospital Malmö, Lund University, Building 60, Level 12, Jan Waldenströms gata 35, 205 02 Malmö, Sweden

Department of Public Health and Clinical Medicine, Umeå University, SE-90188 Umeå, Sweden
e-mail: Paul.Franks@med.lu.se

J.C. Florez (ed.), *The Genetics of Type 2 Diabetes and Related Traits*,
DOI 10.1007/978-3-319-01574-3_26

ment of methods to integrate genetic, transcriptomic, epigenomic, proteomic, and metabolomic data to help define the functional mechanisms that might underlie observations of gene-lifestyle interactions is an especially exciting, yet challenging, area. Nutrigenetic studies hold great promise to inform personalized diet and lifestyle interventions to reduce type 2 diabetes risk and improve human health; however, deriving replicated examples of such interactions and determining how best to translate these findings into public health practice and medical intervention remain major challenges.

Abbreviations

AAA	Aromatic amino acid
BCAA	Branched chained amino acid
BMI	Body mass index
GWAS	Genome-wide association studies
SNP	Single-nucleotide polymorphism
SSB	Sugar-sweetened beverage

26.1 Introduction

Susceptibility to type 2 diabetes varies within and between populations, with American Indians and people descending from the Middle East, South Asia, and the Pacific Islands being at high risk and people of northern European ancestry at considerably lower risk (Kooperberg and Leblanc 2008). At diagnosis, patients are often overweight or obese (American Diabetes Association 2010); in these individuals, intensive lifestyle modification centered on dietary changes that lead to substantial long-term weight loss can halve the risk of developing diabetes, regardless of ethnic background (Gillies et al. 2007). Although in many people overweight and obesity are prerequisites for diabetes to develop, the extent to which dietary modification influences diabetes risk in lean persons is unclear, as the large clinical trials exploring the effects of lifestyle in diabetes incidence have been undertaken in overweight and obesity populations. Although dietary modification and weight loss are now frontline therapies for diabetes prevention, the effects of these interventions vary greatly from one person to the next, as do patients' responses to antidiabetic drugs (Knowler et al. 2002). The variable physiological characteristics, susceptibilities, treatment requirements, and treatment responses of people at risk of or with diabetes strongly suggest the presence of different pathophysiological subclasses. These subclasses may be determined to some extent by genetic differences between individuals. Understanding the genetic basis to diabetes and the extent to which genotypes modify the response to risk factors and preventive interventions might help tackle the rising prevalence of diabetes, improve treatment, reduce unnecessary side effects, and lower costs directly and indirectly related to the disease. Although gene-environment interactions have been clearly defined in animal (Baron 1935) and plant models (Meyer and Purugganan 2013; Des Marais et al. 2013), and to some extent also in human Mendelian diseases caused by single gene mutations (Horner and Streamer 1956), as we explain below, evidence of gene-diet interactions in type 2 diabetes in human populations is fairly limited, but stronger evidence of such interactions exists for diabetes-related traits such as obesity.

The past decades have witnessed a systematic transition from "traditional" lifestyles, where fibrous, nutrient-rich foods and occupational physically actives were commonplace, to an "obesogenic pattern" characterized by increased consumption of energy-dense foods and beverages and sedentary lifestyles (Zimmet et al. 2001; Polonsky 2012). In parallel, the prevalence of type 2 diabetes has increased at an alarming rate in both developed and developing countries (Whiting et al. 2011). Compelling evidence from epidemiological studies and clinical trials clearly indicates these obesogenic habits and obesity are among the most important modifiable risk factors for type 2 diabetes (Gillies et al. 2007). However, susceptibility to environmental risk factors and response to clinical interventions varies considerably between individuals. Such difference in susceptibility to obesogenic environments and responsiveness to lifestyle interventions appears to have a heritable component (Bouchard 2012; Bouchard et al. 1990). A widely discussed explanation for this phenomenon is that selective pressures throughout human evolution have, to some degree, determined genetic diversity as it relates to a person's predisposition to metabolic diseases. Studies of common variation in the human genome have so far resulted in the discovery of hundreds of variants associated with type 2 diabetes, glycemic traits, and obesity (Lindgren et al. 2009; Randall et al. 2013; Speliotes et al. 2010; Berndt et al. 2013; Morris et al. 2012; Scott et al. 2012). Nonetheless, even in aggregate, these variants explain a fairly small proportion of the heritability of these metabolic disorders, and there is little evidence to suggest these variants are present in the genome because of selective pressures (Ayub et al. 2014), which forms the basis to some hypotheses about gene-environment interactions (the thrifty genotype hypothesis) (Neel 1962); alternative explanations propose that allelic variation is a consequence of genomic drift (Ohta and Kimura 1969) and hence unaffected by strong selective forces throughout most of human evolution or migration to environments contrasting those within which populations primarily evolved (Sellayah et al. 2014).

"Nutrigenetics" is an emerging branch of nutrition science that seeks to address the interplay between genetic factors and diet or lifestyle factors. Many hope that by identifying genetic variants that interact with dietary or lifestyle factors, it may be possible to use this information to stratify nutritional advice for the prevention or treatment of type 2 diabetes by targeting specific population subgroups classified by genotypes.

This chapter will review the rationale for testing gene-diet/lifestyle interactions and summarize the recent advances in this rapidly growing area. This chapter will also briefly discuss the implication of nutrigenetics in public health and medical practice and the potential challenges and possibilities that lie ahead.

26.2 Nutrigenetics: Interaction Between Nutrition/Diet and Human Genome

Throughout the twentieth century, nutrition science concentrated on the discovery of nutrients such as vitamins and minerals, as well as defining their roles in preventing diseases caused by deficiency of these nutrients. As nutrition-related health problems of the developed world shifted from undernutrition to overnutrition and technologies that allow detailed characterization of biologic systems (e.g., at a genomic, transcriptomic, proteomic, metabolomic level) have improved manifold, the focus of nutrition research has changed accordingly (Isaak and Siow 2013) .

Type 2 diabetes is one of the major overnutrition-related metabolic disorders of our time, and its rapid rise in the United States and many other countries has followed the widespread adoption of dietary patterns that favor consumption of energy-rich processed foods and beverages, accompanied by reduced physical activity and more sedentary lifestyles (Polonsky 2012). For example, high intakes of red meat and sugar-sweetened beverage (SSB) are related to an increased risk (Aune et al. 2009; Malik et al. 2010a), while high consumption of coffee and whole grains is associated with a reduced risk of type 2 diabetes (Jiang et al. 2014; Cho et al. 2013). Data from randomized clinical trials also show that reducing caloric density of the diet and increasing the consumption of fruits, vegetables, and fibrous foods reduce type 2 diabetes risk (Gillies et al. 2007).

An implicit assumption of most nutrition research is that the risk of disease conferred by dietary exposures is uniform within populations. However, the high degree of variability in the associations between dietary exposures and disease outcomes and the heterogeneous responses to dietary interventions observed in clinical trials undermines this view. Such heterogeneity may reflect complex interactions between dietary factors and a person's biological characteristics, such as their genotypes. These observations have motivated efforts to investigate the interplay between genetic and dietary factors and to understand how nutrients affect the transcription and translation of genes. This area of research, often called "nutrigenomics" or "nutrigenetics" (Junien 2001), has become popular since the turn of the twenty-first century but was alluded to almost a century earlier by Garrod when he observed that "even those idiosyncrasies with regard to drugs and articles of food......presumably have a chemical [genetic] basis" (Garrod 1909).

26.3 Statistical Considerations

Multiplicative interactions are defined by the departure from an additive effect of two or more independent variables, usually in a regression model, on a selected outcome. Early efforts to detect gene-diet interactions largely focused on candidate genes that were selected on the basis of prior biological information that supported their role in a given interaction effect. More recently, the field has been dominated by interaction studies that follow up on established disease-associated variants, usually discovered through genome-wide association studies (GWAS). For type 2 diabetes, about 100 such variants had been identified by 2015 (see Chap. 2). Although only one study to date has undertaken a comprehensive assessment of these loci in relation to gene-lifestyle interactions in diabetes incidence (Langberg et al. 2014), a handful of other interaction studies have explored subsets of these variants in relation to diabetes (several selected studies are shown in Table 26.1).

Hundreds of publications have reported on gene-diet/lifestyle interactions for diabetes-related traits since the mid-1990s (reviewed in: Ahmad et al. 2013a; Franks et al. 2013); the majority of these studies were performed in relatively small epidemiological cohorts, case-control studies, and clinical trials. With few exceptions, these reports have not been accompanied by compelling replication data, suggesting that many of these are likely to be false positive, the implications of which we discuss elsewhere (Franks and Nettleton 2010). The examples of gene-diet/lifestyle interactions that have been convincingly replicated suggest that for common gene variants, diet/lifestyle exposures, and outcomes, such interaction effects are likely to relatively small in magnitude. Hence, statistical power is a major barrier to the detection of interactions (Luan et al. 2001a). For this reason, great emphasis is being placed on developing methods and approaches that help overcome this problem (reviewed in Cornelis et al. (2012)). Because most analyses on gene-diet/lifestyle interactions are undertaken in existing materials (rather than by conducting new studies specifically designed to detect interactions), maximizing sample size by pooling cohorts and developing more statistically powerful analytical tools are the two main strategies for enhancing statistical power.

Almost all studies of gene-diet/lifestyle interactions in diabetes-related traits published prior to 2005 involved sample sizes of <2000 participants, with many totaling <500 participants (Franks et al. 2007). However, the availability of affordable, massively parallel genotyping technologies, and the emergence of consortia that leverage large cohort collections for genetic association studies have raised expectations across all areas of population genetics research. Thus, small studies of gene-diet/lifestyle interactions are now difficult to publish in reputable journals without robust replication data, and scrutiny of statistical methods and data reporting practices has increased considerably.

Only a handful of large individual cohorts ($n > 10{,}000$ participants) exist within which studies of gene-diet/lifestyle interactions have been performed; hence, epidemiologists often pool results from many cohorts using an approach called "meta-analysis." This approach is appealing because very large sample collections can be constructed, and sharing of individual level data between cohorts is not necessary, as the meta-analysis is performed on summary statistics generated locally by cohort analysts. Indeed, several very large meta-analyses have been reported recently, with a recent study of gene-physical activity interactions in obesity including ~220,000 adults (Kilpelainen et al. 2011).

However, the gains in statistical power that one might anticipate by increasing the total sample size by pooling cohorts are often partially offset by residual error

Table 26.1 Selected studies of gene-diet/lifestyle interaction on type 2 diabetes

Candidate gene	Study design	Sample size	Environment	Main findings
PPARG-Pro12Ala	Cross-sectional (Nelson et al. 2007)	1482	Physical activity	*PPARG* Pro12Ala interacted with physical activity on risk of type 2 diabetes (P for interaction = 0.022). The Pro12 allele was significantly associated with T2D in those with low physical activity
PPARG-Pro12Ala and 1431C > T	Cohort (Lamri et al. 2012)	4676	Dietary fat	Dietary fat intake modulated the association of polymorphisms at the *PPARG* locus with type 2 diabetes risk (P for interaction = 0.05). A high fat consumption was associated with an increased type 2 diabetes risk among ProPro and CC homozygotes, but not in Ala and T carriers
GWAS identified loci	Study design	Sample size	Environment	Main findings
TCF7L2-rs7903146	Case-control (Fisher et al. 2009)	3042	Dietary whole grain	The *TCF7L2* rs7903146 T allele modified the inverse association between whole-grain intake and type 2 diabetes risk (P for interaction =0.016). Whole-grain intake was inversely associated with type 2 diabetes risk among rs7903146 CC homozygote carriers, the T allele negated the protective effect of whole-grain intake
SLC30A8-rs13266634	Case-control (Shan et al, 2014)	1796	Plasma zinc concentration	Each 10 μg/dl increment of plasma zinc was associated with 22 % (OR, 0.78; 95 % CI, 0.72–0.85) lower odds of type 2 diabetes in TT genotype carriers, 17 % (0.83; 0.80–0.87) lower odds in CT genotype carriers, and 7 % (0.93; 0.90–0.97) lower

(continued)

Table 26.1 (continued)

GWAS identified loci	Study design	Sample size	Environment	Main findings
				odds in CC genotype carriers (P for interaction = 0.01)
GIPR-rs10423928	Cohort (Sonestedt et al. 2012)	24,840	Dietary carbohydrate and fat	There is significant interaction between *GIPR* genotype and dietary carbohydrate (P for interaction = 0.0005) and fat intake (P for interaction = 0.0006) on incident type 2 diabetes. The TT genotype carriers within the highest compared with the lowest carbohydrate quintile were at 23 % decreased T2D risk. In contrast, AA genotype carriers in the highest compared with the lowest fat quintile were at 69 % decreased type 2 diabetes risk
HNF1B-rs4430796	Cohort (Brito et al. 2009)	16,003	Physical activity	Physical activity interacted with the *HNF1B* rs4430796 variant (P for interaction = 0.0004) in determining type 2 diabetes risk. Subjects with GG genotype have lower risk of T2D than those with GA or AA genotypes
FTO-rs9939609 and *MC4R*-rs17782313	Case-control (Ortega-Azorin et al. 2012)	3430 T2D patients, 3622 control subjects	Adherence to the Mediterranean diet	The associations of the *FTO*-rs9939609 and the *MC4R* rs17782313 with type 2 diabetes depended on a high adherence to the Mediterranean diet. When adherence to the Mediterranean diet was low, carriers of the variant alleles had higher type 2 diabetes risk (P for interaction = 0.019 for *FTO*-rs9939609 and P for interaction = 0.035 for *MC4R*-rs17782313) than wild-type subjects

that is introduced by combining data from cohorts that vary in the methods used, data structures, and participant characteristics (Ahmad et al. 2013b). Thus, while meta-analysis of data on gene-diet/lifestyle interactions from multiple cohorts may seem intuitive on the surface, the careful selection of cohorts will maximize the benefits of this approach, whereas meta-analyzing all available materials, which is often done, may prove counterproductive when it comes to statistical power.

Statistical methods for modeling gene-environment interaction effects are evolving rapidly, driven by the recognition that conventional statistical approaches are underpowered for the detection of interactions when, as is often so, multiple hypotheses are tested in relatively small cohorts that include crudely characterized nutritional and lifestyle exposures. Those conventional approaches include pairwise tests of interaction, when the product of the genetic and environmental exposures is modeled (e.g., SNP x dietary fiber) against a quantitative (e.g., blood glucose concentrations) or categorical (with or without diabetes) outcome. The objective of these tests is usually to numerically characterize complex biologic phenomena; however, these simple statistical approaches may be poorly suited to this task. Hence, a range of new methods are now coming online that aim to integrate several layers of biological information (systems genetics) (Civelek and Lusis 2014) and to do this on a genome-wide scale; there are several major barriers to doing this though, which include the impact on multiple testing on type 1 error rates and the challenges of aligning diverse data structures. To address this problem, several innovative genome-wide interaction tests have been proposed recently; these include approaches that collapse the tests of main effect and interaction effect into a single test (Aschard et al. 2010; Manning et al. 2011), as well as inferential tests that model genetic effects on the signatures of an interaction, as such as phenotypic variance (Pare et al. 2010; Visscher and Posthuma 2010).

To date, no studies have reported on genome-wide gene-diet interaction analysis in type 2 diabetes incidence, although a large European study is under way to address this issue (InterAct et al. 2011). The primary challenges with conventional approaches include the need to test a large number of hypotheses (upward of one million), the requisite for a sufficiently powered sample size, and prospective design. Several methods have been developed to address the statistical challenges, for example, multistage procedures and Bayesian model selection, among others (Kooperberg and Leblanc 2008; Zhang and Liu 2007). However, application of these methods may not match the theoretical expectations on which they are based, and to our knowledge, no replicated interaction has been published to date using these methods. Recently, a "variance prioritization" method was proposed to identify genetic variants that are sensitive to the environment and thus prioritize variants for interaction studies (Pare et al. 2010). The prioritization procedure is based on the analysis of phenotypic variance per genotype. By using the variance prioritization procedure, Pare et al. identified gene-environment interactions on several CVD risk factors. However, whether the approach could lead to the identification of gene-diet/lifestyle interactions on type 2 diabetes has yet to be validated.

26.4 Gene-Diet Interaction in Type 2 Diabetes

26.4.1 Observational Studies

The *PPARG* Pro12Ala variant is an established diabetes-associated locus. Of all the published studies focused on interactions between dietary factors and non-GWAS candidate genes, the *PPARG* Pro12Ala variant is probably the most intensively explored locus (Table 26.1). Several small studies have examined interactions between this locus and dietary factors such as the total fat, ratio of polyunsaturated fat to saturated fat (P:S ratio) (Luan et al. 2001b), oleic acid (Soriguer et al. 2006), fish intake and marine n-3 fatty acids (Ylonen et al. 2008), and lifestyle factors such as physical activity (Nelson et al. 2007; Brito et al. 2009) with type 2 diabetes risk or blood glucose concentrations. In studies from the United Kingdom and the United States, the association of dietary fats with plasma insulin levels and BMI differed by Pro12Ala genotype, with an inverse association between these variables observed in carriers of the minor Ala12 allele and no detectable association in Pro12Pro homozygotes. Numerous follow-up studies have been reported (Franks et al. 2007); however, attempts to summarize this evidence failed (Palla et al. 2010), in part because publications on gene-environment interactions generally lack key details needed for pooled analyses and dietary exposures are often measured with approaches that cannot be adequately standardized across studies.

Several studies have assessed interactions between SNPs in other GWAS-identified loci, such as *TCF7L2*, *SLC30A8*, *GIPR*, *HNF1B*, and *FTO*, and dietary factors in relation to type 2 diabetes (Table 26.1). Among these diabetes-related genes, some are involved in nutrient metabolism and therefore interesting candidates for testing gene-diet interactions. For example, zinc is an essential trace element found in most foods that facilitates catalytic, structural, and transcriptional actions (Prasad 2007). In epidemiology studies, plasma zinc concentrations have been associated with impaired glucose regulation (IGR) and type 2 diabetes risk. The type 2 diabetes-associated zinc transporter-8 (*SLC30A8*) gene is exclusively expressed in pancreatic beta cells and encodes a protein that transports zinc from the cytoplasm into insulin secretory vesicles, an important step in insulin synthesis and secretion. Shan et al. recently assessed the interaction of the variants in this gene with plasma zinc levels in relation to type 2 diabetes in 1796 Chinese: 218 participants with newly diagnosed IGR, 785 participants with newly diagnosed type 2 diabetes, and 793 individuals with normal glucose tolerance (NGT) (Shan et al. 2014). In all the study samples, the multivariable OR of type 2 diabetes associated with a 10 μg/dl higher plasma zinc level was 0.87 (0.85–0.90). It was found that the association of plasma zinc concentrations with type 2 diabetes was modified by *SLC30A8* rs13266634. Each 10 μg/dl increment of plasma zinc was associated with 22 % (OR, 0.78; 95 % CI, 0.72–0.85) lower odds of type 2 diabetes in TT genotype carriers, 17 % (0.83; 0.80–0.87) lower odds in CT genotype carriers, and 7 % (0.93; 0.90–0.97) lower odds in CC genotype carriers (*P* for interaction = 0.01).

The Cohorts for Heart and Aging Research in Genomic Epidemiology (CHARGE) consortium has published several reports on gene-diet interactions for diabetes-related quantitative traits. Analyses have focused on exploring whether established glucose- and insulin-associated loci modify the relationships of whole grains, zinc, and magnesium and overall dietary patterns with fasting blood glucose and insulin concentrations. These studies yielded tentative evidence of interactions for variants in *GCKR* and dietary whole grains in insulin concentrations (Nettleton et al. 2010) and in *SLC30A8* and total zinc intake on fasting glucose concentrations (Kanoni et al. 2011); however, no evidence of interactions between the selected genetic variants and dietary magnesium or dietary patterns was observed. In a large prospective cohort study from southern Sweden, Ericson et al. (2013) and Sonestedt et al. (2012) reported interactions between *IRS1* and *GIPR* variants, respectively, and dietary fat and carbohydrate intakes in relation to incident type 2 diabetes.

The *TCF7L2* locus harbors the strongest common variant association signals for type 2 diabetes; hence, the possibility that *TCF7L2* variants interact with dietary factors, such as whole grains, fiber, fat, protein, glycemic index, and Mediterranean diet, has also attracted extensive attention. Hindy et al. assessed the interactions between *TCF7L2* rs7903146 and intakes of carbohydrate, fat, protein, or fiber in a case-control study of 1649 diabetes cases and 5216 nondiabetic controls from the Malmö Diet and Cancer Study (MDCS) (Hindy et al. 2012). It was found that the genetic associations with type 2 diabetes risk increased with higher intake of dietary fiber, with ORs ranging from 1.24 (95 % CI, 1.04, 1.47) to 1.56 (95 % CI, 1.31, 1.86) from the lowest to highest quintile (P for interaction $= 0.049$). Interestingly, it was also found that high intake of dietary fiber was inversely associated with diabetes incidence only among CC genotype carriers (OR 0.74, 95 % CI, 0.58, 0.94 per quintile, $P = 0.03$). Whole grain consumption has been inversely associated with diabetes risk, and the protective effect is thought to be partly attributable to the fiber component. In the European Prospective Investigation into Cancer and Nutrition (EPIC)-Potsdam cohort, Fisher et al. found that *TCF7L2* rs7903146 genotype modified the inverse association between whole-grain intake and diabetes risk (P for interaction $= 0.02$). While whole-grain intake was significantly associated with a reduced diabetes risk among the CC homozygotes (hazard ratio [HR] for 50 g portion/day $= 0.86$; 95 % CI, 0.75, 0.99), the T allele attenuated the associations (HR $= 1.08$; 95 % CI, 0.96, 1.23) (Fisher et al. 2009). In the Diabetes Prevention Program (DPP), (Florez et al. 2006) a stronger association between the TT genotype of *TCF7L2* rs12255372 and the diabetes risk was found in the placebo group (hazard ratio $= 1.81$; 95 % CI, 1.19–2.75) than in the lifestyle-intervention groups (reduced intakes of total fat and saturated fat and increased intake of fiber; moderate exercise for at least 30 min per day; HR $= 1.24$; 0.73–2.12). Although the test of interaction was not statistically significant ($P > 0.10$), these data suggest that healthy diet/lifestyle intervention may attenuate the diabetogenic effects of this locus. In a prospective study of 7018 individuals with a median follow-up of 4.8 years, Corella et al. found that adherence to the Mediterranean diet reduced the increase in blood levels of fasting glucose and lipids associated with *TCF7L2* rs7903146 (Corella et al. 2013). Of note, replication data are largely lacking for these reported gene-diet interactions, most have not accounted for the effects of multiple testing on type 1 error, and they do not all provide complementary findings in terms of the direction of interaction effects.

Most studies of gene-diet interactions have focused on single nutrients or foods. However, there are several conceptual and methodological drawbacks to this approach. First, people do not consume single nutrients or foods. Rather, meals consist of a variety of foods with complex combinations of nutrients that are likely to be interactive or synergistic. Thus, the "single nutrient/food" approach may not adequately account for complicated nutrient interactions in free-living populations. Second, while the effect of a single nutrient/food may be too small to detect, cumulative effects of multiple nutrients/foods, such as those comprising a dietary pattern or quality index, may be sufficiently large to identify. Third, individually analyzing a large number of single nutrients or foods may produce statistically significant associations simply by chance, and few studies adequately account for this. Therefore, overall dietary patterns may be more appropriate dietary variables in studying gene-diet interaction. Qi et al. examined the interactions between the genetic predisposition and dietary patterns in a prospective cohort of US men from the Health Professionals Follow-Up Study (HPFS) (Qi et al. 2009). A group of type 2 diabetes-associated loci *HHEX*, *CDKAL1*, *IGF2BP2*, *SLC30A8*, *WFS1*, *CDKN2A/B*, *TCF7L2*, *PPARG*, and *KCNJ11* were selected from the GWAS, and a genetic risk score (GRS) was calculated using a simple count method to feature the overall genetic susceptibility to type 2 diabetes (Cornelis et al. 2009). The participants were grouped into low (GRS $<$ 10 alleles), median (10–11 alleles), and high ($>$12 alleles) genetic risk. Based on the baseline dietary information, the factor analysis identified two major dietary patterns (van Dam et al. 2002): a *prudent* dietary pattern was loaded heavily with vegetables, legumes, and whole grains and a *Western* dietary pattern. The GRS interacted with Western dietary pattern in relation to diabetes risk (P for interaction = 0.02; Fig. 26.1). The multivariable ORs of type 2 diabetes across increasing quartiles of the Western dietary pattern were 1.00, 1.23 (95 % CI, 0.88, 1.73), 1.49 (1.06, 2.09), and 2.06 (1.48, 2.88) among men with a high GRS only. Further analyses indicate that intakes of red meat, processed meat, and heme iron might be the major foods/nutrients driving the interactions. These findings suggest that the adoption of a Westernized diet may increase diabetes risk especially among the genetically high-risk population.

26.4.2 Randomized Controlled Trials

Although observational studies often prove extremely valuable for generating hypotheses around gene-diet interactions, such data are prone to bias, confounding, and reverse causation. On the other hand, randomized controlled trials are fairly robust to many of these limitations and provide a powerful adjunct through which observations of interactions can be causally evaluated.

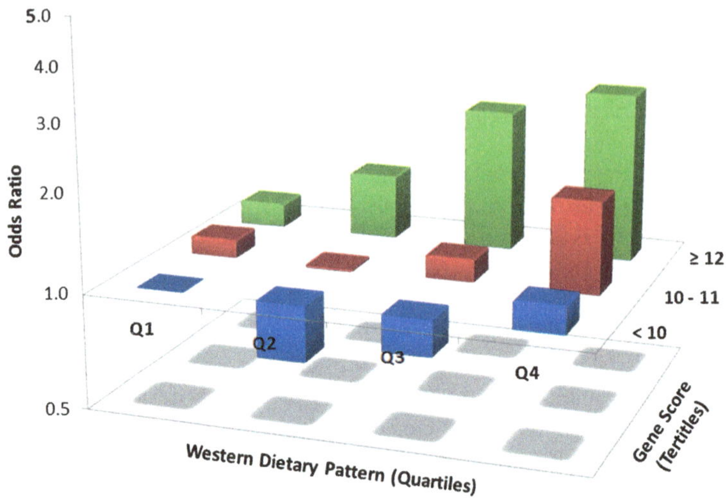

Fig. 26.1 Interaction between genetic variants and dietary patterns on risk of type 2 diabetes. The analyses were performed in the Health Professionals Follow-Up Study. The figure shows the odds ratios of type 2 diabetes risk according to joint classifications of the Western dietary pattern score (in quartiles) and the genetic risk score for type 2 diabetes (GRS, <10, 10–11, and ≥12). The analyses were adjusted for age, body mass index, smoking, alcohol consumption, physical activity, family history of diabetes, and total energy intakes

Only one randomized controlled trial (the DPP) has reported on gene-lifestyle interactions in type 2 diabetes incidence. The DPP is the largest randomized controlled trial of lifestyle modification for diabetes prevention. The DPP investigators have examined interactions between variants at *TCF7L2* (Florez et al. 2006), *PPARG* (Moore et al. 2008), *WFS1* (Florez et al. 2008), other established type 2 diabetes-associated loci (Moore et al. 2008; Hivert et al. 2011), established glucose- and insulin-associated loci (Florez et al. 2012), common and rare variants at the *MC4R* (Pan et al. 2013) and *SLC80A8* loci (Billings et al. 2014; Moore et al. 2009), and other candidate loci (Moore et al. 2009; Mather et al. 2012; Jablonski et al. 2010). Gene-centric tests have also highlighted possible gene-lifestyle interactions at the *PARGC1A* locus for several cardiometabolic traits in the DPP (Franks et al. 2014).

26.5 Gene-Diet Interaction in Obesity and Weight Change

Obesity is the predominant modifiable risk factor for type 2 diabetes. The diabetogenic role of obesity has also been demonstrated in genetic studies. In fact, the identification of the strongest obesity locus *FTO* is a by-product of GWAS of type 2 diabetes (Frayling et al. 2007). In addition, it has been demonstrated that many other obesity-related genetic variants also affect risk of type 2 diabetes or its risk factors (Speliotes et al. 2010). There have been major efforts to develop diet interventions to improve weight loss and maintenance and ameliorate type 2 diabetes risk (Sacks et al. 2009; Schellenberg et al. 2013). Compelling evidence has shown that modification of diet is effective in reducing body weight. However, considerable interindividual heterogeneity in participants' response has been noted, possibly because of genetic differences.

During the past decades, the temporal patterns in the increasing consumption of sugar-sweetened beverages (SSBs) have paralleled the rise in the prevalence of obesity in the United States; both have more than doubled since the late 1970s. Compelling evidence supports a positive link between the consumption of SSBs and the risk of obesity (Mozaffarian et al. 2011; Malik et al. 2010b; Schulze et al. 2004). However, there is scarce evidence concerning whether the risk of obesity is modified by the genetic background. Qi et al. (2012) analyzed the interaction between a genetic risk score (GRS)—an index for genetic predisposition to obesity calculated by adding up the number of risk alleles of 32 BMI-associated SNPs—and the intake of SSBs in relation to BMI and obesity risk in 6934 women from the Nurses' Health Study (NHS) and in 4423 men from the Health Professionals Follow-Up Study (HPFS) and in a replication cohort of 21,740 women from the Women's Genome Health Study (WGHS). The genetic association with adiposity, a long-term BMI increase, and the risk of incident obesity appeared to be more pronounced with greater intake of SSBs. When combining the three cohorts, the pooled relative risks (RR) of incident obesity for each increment of ten risk alleles were 1.35 (95 % CI, 1.18 to 1.54), 1.59 (95 % CI, 1.33 to 1.91), 1.56 (95 % CI, 1.26 to 1.92), and 3.35 (95 % CI, 2.22 to 5.05) across the four categories of intake ($P < 0.001$ for interaction). In a similar analysis among the three cohorts NHS, HPFS, and WGHS, consistent interactions were found between the obesity GRS and fried food consumption in relation to BMI (Qi et al. 2014). In the NHS and HPFS (P for interaction ≤ 0.001), among participants in the highest tertile of the GRS, the differences in BMI between individuals who consumed fried foods ≥ 4 times/week and those consumed <1 time/week amounted to 1.0 (SE 0.2) in women and 0.7 (0.2) kg/m^2 in men, whereas the corresponding differences were 0.5 (0.2) and 0.4 (0.2) kg/m^2 in the lowest tertile of the GRS. The gene-diet interaction was replicated in the WGHS (P for interaction < 0.001). In addition to SSBs and fried food intakes, numerous studies have examined interactions of various dietary factors such as total fat, P:S ratio, and carbohydrate (Qi and Cho 2008) with preselected genetic variants (in candidate genes or GWAS-identified obesity genes) in relation to body weight, BMI, and obesity risk. However, few findings are reproducible. By the end of year 2013, no study has reported significant gene-diet interactions on obesity-related traits on the whole-genome scale.

The Preventing Overweight Using Novel Dietary Strategies (Pounds Lost) is a clinical trial including in total of 811 overweight (BMI ≥ 25 kg/m^2) and obese (BMI ≥ 30 kg/m^2) adults who were randomly assigned to one of four weight-loss diets varying in macronutrient contents (fat, protein, and carbohydrates) for 2 years (Sacks et al. 2009). At 6 months, participants assigned to each diet had lost an average of 6 kg, which represents 7 % of their initial weight. The participants began to regain weight after 12 months. By 2 years, weight loss remained similar in those who were assigned to various diets. In the Pounds Lost trial, a series of analyses have been performed to test gene-diet interactions in relation to weight loss. In the Pounds Lost trial, the type 2 diabetes-associated *TCF7L2* SNP rs12255372 (strongly correlated with the causal SNP rs7903146 in Europeans) significantly interacted with fat intake on changes in BMI, total fat mass, and trunk fat mass (all $P < 0.05$) at 6 months, with nonsignificant larger decreases for TT (risk genotype) carriers on a low-fat diet (Mattei et al. 2012). Elsewhere (Haupt et al. 2010), T allele (vs. C) at rs7903146 in *TCF7L2* was associated with greater weight loss following lifestyle intervention in the Tübingen Lifestyle Intervention Program (TULIP), which consisted of exercise and diet intervention with decreased intake of fat and increased intake of fibers (>15 g fiber per 1000 kcal), but not in the control arm (Haupt et al. 2010). However, the findings were not replicated in the DPP, possibly because fiber intake was not explicitly part of the DPP intervention, although consumption of whole grains was. In a follow-up study of 304 participants from TULIP, the rs7903146 SNP CC genotype of the *TCF7L2* SNP was associated with significantly greater weight loss in participants with high fiber intake, but not in those with low fiber intake (Heni et al. 2012).

Other type 2 diabetes- and obesity-associated genetic variants have been also found to modify diet interventions on weight change. For example, the *FTO* SNP rs1558902 was found to interact with dietary protein on 2-year changes in fat-free mass, total percentage of fat mass, and total-, visceral-, and superficial adipose tissue mass in the Pounds Lost trial (Zhang et al. 2012). Also in this trial, Qi et al. found significant interactions between the *IRS1* SNP rs2943641 and carbohydrate intake in relation to changes in weight loss and insulin resistance (Qi et al. 2011). At 6 months, participants with the risk-conferring CC genotype had greater decreases in weight loss ($P = 0.02$) than those without this genotype in the highest-carbohydrate diet group, whereas the genetic effect was not significant in participants assigned to the lowest-carbohydrate diet group (P for interaction $= 0.03$). The gene-diet interaction was attenuated at 2 years due to weight regain. These data highlight that nutrigenetic analyses in the randomized clinical trials may generate data reflecting direct response to dietary intervention. However, how to validate the findings of gene-diet interactions remains a major challenge, considering the relatively small sample size in most of the existing trials and that studies are generally different in designs.

26.6 Interactions Between Genetic Variants and Prenatal Nutrition

Reduced birth weight is associated with adulthood risk of T2D; however, the mechanisms underlying the association remain unclear. The thrifty phenotype hypothesis (sometimes known as Barker's hypothesis) states that malnutrition during fetal development leads to poor fetal and infant growth and predisposes to adverse health outcomes (Hales and Barker 2001). According to the hypothesis, fetal adaptation to a deprived intrauterine environment may produce permanent changes that affect insulin secretion and resistance, and such metabolic changes lead to elevated risk of late-onset diseases such as T2D. Of note, an alternative explanation, known as the fetal insulin hypothesis, is that genetic variants that reduce insulin secretion or action may predispose to T2D and also reduce birth weight, since insulin is a key fetal growth factor. Such hypothesis has been evidenced by observations that diabetes risk alleles are associated with low birth weight in several studies including recent GWAS in the EGG Consortium (Freathy et al. 2009; Freathy et al. 2007; Horikoshi et al. 2013).

Small birth size (birth weight), which is widely used as an indicator for fetal malnutrition and growth retardation, has been associated with glucose intolerance, impaired beta-cell secretory function, and an increased type 2 diabetes risk (Barker et al. 1993; Whincup et al. 2008). These associations may be modulated by type 2 diabetes variants, which are involved in glucose metabolism and insulin secretion (Grarup et al. 2007). An earlier study reported significant interactions between early malnutrition during midgestation and the *PPARG* Pro12Ala polymorphism in relation to impaired glucose tolerance and type 2 diabetes (de Rooij et al. 2006). Several recent studies analyzed the interactions between the GWAS-identified variants and birth weight—a marker of prenatal nutrition status—in relation to adulthood diabetes risk and provided further evidence to the hypothesis.

In one study, variants of nine type 2 diabetes loci, *TCF7L2*, *HHEX*, *PPARG*, *KCNJ11*, *SLC30A8*, *IGF2BP2*, *CDKAL1*, *CDKN2A/2B*, and *JAZF1*, were tested (Pulizzi et al. 2009). The study includes 928 men and 1075 women. 15.6 % of the participants had developed type 2 diabetes. Risk variants at the *HHEX*, *CDKN2A/2B*, and *JAZF1* loci significantly interacted with birth weight to predict future type 2 diabetes ($P = 0.04$, 0.03 and 0.02, respectively). A lower birth weight amplified the risk conferred by the pooled variants in nine type 2 diabetes genes, and a higher birth weight compensated the genetic effects. In another study, Van Hoek et al. (van Hoek et al. 2009) assessed the interactions between type 2 diabetes susceptibility genes and fetal exposure to famine. Seven SNPs at genes *CDKAL1*, *CDKN2AB*, *HHEX*, *IGF2BP2*, *KCNJ11*, *SLC30A8*, and *TCF7L2* were determined in 772 participants of the Dutch Famine Birth Cohort, which is composed of individuals born around the time of the Dutch famine during World War II. Prenatal exposure to famine was defined as a daily food ration of the mother $<$1000 calories during any 13-week period of gestation. The *IGF2BP2* polymorphism showed an interaction with prenatal exposure to famine on glucose level ($P = 0.009$). However, none of the polymorphisms interacted with birth weight. Taken together, these data may provide a clue that an individual's genetic background may modulate the response to prenatal nutrition and subsequently affect type 2 diabetes risk caused by hypercaloric environment in later life.

In two nested case-control studies of 2591 type 2 diabetes cases and 3052 healthy controls from the NHS and HPFS (Li et al. 2012), Li et al. tested the interactions between genetic susceptibility to obesity or type 2 diabetes, evaluated on the basis of 32 BMI-associated variants or 35 type 2 diabetes-associated variants and birth weight in relation to adulthood risk of type 2 diabetes. A significant interaction was observed between birth weight and obesity GRS, which showed a stronger association with type 2 diabetes risk in individuals with low birth weight. In low-birth-weight individuals, the multivariable-adjusted odds ratio (OR) was 2.55 (95 % CI, 1.34–4.84) by comparing the extreme quartiles of the obesity GRS, while the OR was 1.27 (1.04–1.55) among individuals with birth weight >2.5 kg (P for interaction = 0.02). The type 2 diabetes-related GRS, however, did not interact with birth weight. The data suggest that low birth weight and genetic susceptibility to obesity may synergistically affect adulthood risk of type 2 diabetes. In a Chinese cohort (Hong et al. 2013), Hong et al. calculated an obesity GRS based on six SNPs associated with obesity risk in Chinese. The obesity GRS showed significant interaction with birth weight in relation to obesity (P for interaction < 0.001). The genetic effect appeared to be more pronounced in individuals with normal range of birth weight (25–75 %) than those with either low (<25 %) or high (>75 %) birth weight. The data suggest birth weight may significantly modify genetic susceptibility to obesity risk.

26.7 Summary and Future Directions

A collection of dietary factors has been related to type 2 diabetes risk; in the meanwhile, genetic research has identified numerous variants in human genome that also affect type 2 diabetes and its risk factors. These findings have paved new avenue for extensive investigations on gene-diet interactions. However, data from this fast-moving area, namely, nutrigenetics, are still preliminary, and major challenges exist in study design and analytical strategies. Although GWAS have demonstrated the validity of cross-sectional design in the identification of disease-predisposing variants, inherent bias of cross-sectional analysis such as confounding and reverse causation becomes paramount in testing gene-diet interaction (Qi 2012). In addition, the lack of replication is also a serious limitation. An epidemiologic framework for evaluating gene-diet/lifestyle interaction has yet to be well established, which should include a method to account for the unbounded universe of potential hypotheses. In contrast to the genomic space, where the number of independent tests among common variants in the human genome is finite and can be quantified, the number of nutritional components being evaluated by nutrigenetics investigators is far more diffuse. Nevertheless, several recent large-scale studies with prospective design and replication have emerged to shed light on the potential ways to move forward. Previous studies largely focused on preselected candidate genes or GWAS-identified genes; a usually very productive next step would be to perform genome-wide analysis. In addition, to integrate genomic studies with investigations on other global features of the human body, such as epigenomics, proteomics, and metabolomics, as well as functional studies is essential to provide insights into the potential mechanisms underlying the interactions between genetic and dietary factors. These should include the putative effects of genetic variation on human dietary behaviors.

One of the ultimate goals of nutrigenetics is to develop personalized diet and lifestyle interventions based on genetic profile that are better tailored to meet the individuals' needs. Understanding gene-diet interactions in relation to type 2 diabetes risk holds great promise for achieving such a goal. However, how statistically significant interaction may reflect biological interaction has been a controversial topic. In addition, the interactions determined for populations cannot be applied directly to the individuals. Developing individual risk or benefit factors in light of the individual's genetic makeup and the complexity of foods remain significant challenges for personalizing dietary advice for the prevention of the disease. In addition, curbing the epidemics of obesity and related disorders calls not only for changes in diet habits but also changes in policy, physical and social environment, and lifestyles. In the following years, public health practice targeting type 2 diabetes will not be able to ignore the impact of nutrigenetics, although it is still a long journey to better appreciate its relevance to the practice of preventive approaches for delaying onset of the disease, diminishing its severity, and optimizing human health.

Acknowledgment Dr. Qi is supported by grants from the National Heart, Lung, and Blood Institute (HL071981, HL034594), the National Institute of Diabetes and Digestive and Kidney Diseases (DK091718), the Boston Obesity Nutrition Research Center (DK46200), and United States-Israel Binational Science Foundation Grant 2011036. Dr. Qi was a recipient of the American Heart Association Scientist Development Award (0730094N). Dr. Franks is supported by grants from the National Institute of Child Health and Human Development, the Swedish Research Council, Novo Nordisk, the Swedish Heart-Lung Foundation, and the European Union. We than Mr. Yan Chen (Genetic & Molecular Epidemiology Unit, Department of Clinical Sciences, Lund University, Malmö, Sweden) for his help with reference formatting.

References

Ahmad S, Varga TV, Franks PW (2013a) Gene x environment interactions in obesity: the state of the evidence. Hum Hered 75(2-4):106–115

Ahmad S et al (2013b) Gene x physical activity interactions in obesity: combined analysis of 111,421 individuals of European ancestry. PLoS Genet 9(7), e1003607

American Diabetes Association (2010) Diagnosis and classification of diabetes mellitus. Diabetes Care 33(Suppl 1):S62–S69

Aschard H et al (2010) Genome-wide meta-analysis of joint tests for genetic and gene-environment interaction effects. Hum Hered 70(4):292–300

Aune D, Ursin G, Veierod MB (2009) Meat consumption and the risk of type 2 diabetes: a systematic review and meta-analysis of cohort studies. Diabetologia 52(11):2277–2287

Ayub Q et al (2014) Revisiting the thrifty gene hypothesis via 65 loci associated with susceptibility to type 2 diabetes. Am J Hum Genet 94(2):176–185

Barker DJ et al (1993) Type 2 (non-insulin-dependent) diabetes mellitus, hypertension and hyperlipidaemia (syndrome X): relation to reduced fetal growth. Diabetologia 36(1):62–67

Baron AL (1935) Facet number in Drosophila melanogaster as influenced by certain genetic and environmental factors. J Exp Zool 70:461–490

Berndt SI et al (2013) Genome-wide meta-analysis identifies 11 new loci for anthropometric traits and provides insights into genetic architecture. Nat Genet 45(5):501–512

Billings LK et al (2014) The INFLUENCE of rare genetic variation in SLC30A8 on diabetes incidence and beta-cell function. J Clin Endocrinol Metab 99(5):E926–E930. doi:10.1210/jc.2013-2378

Bouchard C (2012) Genomic predictors of trainability. Exp Physiol 97(3):347–352

Bouchard C et al (1990) The response to long-term overfeeding in identical twins. N Engl J Med 322(21):1477–1482

Brito EC et al (2009) Previously associated type 2 diabetes variants may interact with physical activity to modify the risk of impaired glucose regulation and type 2 diabetes: a study of 16,003 Swedish adults. Diabetes 58(6):1411–1418

Cho SS et al (2013) Consumption of cereal fiber, mixtures of whole grains and bran, and whole grains and risk reduction in type 2 diabetes, obesity, and cardiovascular disease. Am J Clin Nutr 98(2):594–619

Civelek M, Lusis AJ (2014) Systems genetics approaches to understand complex traits. Nat Rev Genet 15(1):34–48

Corella D et al (2013) Mediterranean diet reduces the adverse effect of the TCF7L2-rs7903146 polymorphism on cardiovascular risk factors and stroke incidence: a randomized controlled trial in a high-cardiovascular-risk population. Diabetes Care 36(11):3803–3811

Cornelis MC et al (2009) Joint effects of common genetic variants on the risk for type 2 diabetes in U.S. men and women of European ancestry. Ann Intern Med 150(8):541–550

Cornelis MC et al (2012) Gene-environment interactions in genome-wide association studies: a comparative study of tests applied to empirical studies of type 2 diabetes. Am J Epidemiol 175 (3):191–202

de Rooij SR et al (2006) The effects of the Pro12Ala polymorphism of the peroxisome proliferator-activated receptor-gamma2 gene on glucose/insulin metabolism interact with prenatal exposure to famine. Diabetes Care 29(5):1052–1057

Des Marais DL, Hernandez KM, Juenger TE (2013) Genotype-by-environment interaction and plasticity: exploring genomic responses of plants to the abiotic environment. Annu Rev Ecol Evol Syst 44(1):5–29

Ericson U et al (2013) Sex-specific interactions between the IRS1 polymorphism and intakes of carbohydrates and fat on incident type 2 diabetes. Am J Clin Nutr 97(1):208–216

Fisher E et al (2009) Whole-grain consumption and transcription factor-7-like 2 (TCF7L2) rs7903146: gene-diet interaction in modulating type 2 diabetes risk. Br J Nutr 101(4):478–481

Florez JC et al (2006) TCF7L2 polymorphisms and progression to diabetes in the Diabetes Prevention Program. N Engl J Med 355(3):241–250

Florez JC et al (2008) Testing of diabetes-associated WFS1 polymorphisms in the Diabetes Prevention Program. Diabetologia 51(3):451–457

Florez JC et al (2012) Effects of genetic variants previously associated with fasting glucose and insulin in the Diabetes Prevention Program. PLoS ONE 7(9), e44424

Franks PW, Nettleton JA (2010) Invited commentary: gene X lifestyle interactions and complex disease traits--inferring cause and effect from observational data, sine qua non. Am J Epidemiol 172(9):992–997, discussion 998-9

Franks PW et al (2007) Gene-lifestyle interaction on risk of type 2 diabetes. Nutr Metab Cardiovasc Dis 17(2):104–124

Franks PW, Pearson E, Florez JC (2013) Gene-environment and gene-treatment interactions in type 2 diabetes: progress, pitfalls, and prospects. Diabetes Care 36(5):1413–1421

Franks PW et al (2014) Common variation at PPARGC1A/B and change in body composition and metabolic traits following preventive interventions: the Diabetes Prevention Program. Diabetologia 57(3):485–490

Frayling TM et al (2007) A common variant in the FTO gene is associated with body mass index and predisposes to childhood and adult obesity. Science 316(5826):889–894

Freathy RM et al (2007) Type 2 diabetes TCF7L2 risk genotypes alter birth weight: a study of 24,053 individuals. Am J Hum Genet 80(6):1150–1161

Freathy RM et al (2009) Type 2 diabetes risk alleles are associated with reduced size at birth. Diabetes 58(6):1428–1433

Garrod AE (1909) Inborn errors of metabolism: the Croonian lectures delivered before the Royal College of Physicians of London in June 1908. Frowde; Hodder and Stoughton, London

Gillies CL et al (2007) Pharmacological and lifestyle interventions to prevent or delay type 2 diabetes in people with impaired glucose tolerance: systematic review and meta-analysis. BMJ 334(7588):299

Grarup N et al (2007) Studies of association of variants near the HHEX, CDKN2A/B, and IGF2BP2 genes with type 2 diabetes and impaired insulin release in 10,705 Danish subjects: validation and extension of genome-wide association studies. Diabetes 56(12):3105–3111

Hales CN, Barker DJ (2001) The thrifty phenotype hypothesis. Br Med Bull 60:5–20

Haupt A et al (2010) Gene variants of TCF7L2 influence weight loss and body composition during lifestyle intervention in a population at risk for type 2 diabetes. Diabetes 59(3):747–750

Heni M et al (2012) Dietary fiber intake modulates the association between variants in TCF7L2 and weight loss during a lifestyle intervention. Diabetes Care 35(3):e24

Hindy G et al (2012) Role of TCF7L2 risk variant and dietary fibre intake on incident type 2 diabetes. Diabetologia 55(10):2646–2654

Hivert MF et al (2011) Updated genetic score based on 34 confirmed type 2 diabetes Loci is associated with diabetes incidence and regression to normoglycemia in the diabetes prevention program. Diabetes 60(4):1340–1348

Hong J et al (2013) Genetic susceptibility, birth weight and obesity risk in young Chinese. Int J Obes (Lond) 37(5):673–677

Horikoshi M et al (2013) New loci associated with birth weight identify genetic links between intrauterine growth and adult height and metabolism. Nat Genet 45(1):76–82

Horner FA, Streamer CW (1956) Effect of a phenylalanine-restricted diet on patients with phenylketonuria; clinical observations in three cases. J Am Med Assoc 161(17):1628–1630

InterAct C et al (2011) Design and cohort description of the InterAct Project: an examination of the interaction of genetic and lifestyle factors on the incidence of type 2 diabetes in the EPIC Study. Diabetologia 54(9):2272–2282

Isaak CK, Siow YL (2013) The evolution of nutrition research. Can J Physiol Pharmacol 91 (4):257–267

Jablonski KA et al (2010) Common variants in 40 genes assessed for diabetes incidence and response to metformin and lifestyle intervention in the diabetes prevention program. Diabetes 59(10):2672–2681

Jiang X, Zhang D, Jiang W (2014) Coffee and caffeine intake and incidence of type 2 diabetes mellitus: a meta-analysis of prospective studies. Eur J Nutr 53(1):25–38

Junien C (2001) [Colon cancer and nutritional genetics: modifier genes]. Ann Med Interne (Paris) 152(5):337–351

Kanoni S et al (2011) Total zinc intake may modify the glucose-raising effect of a zinc transporter (SLC30A8) variant: a 14-cohort meta-analysis. Diabetes 60(9):2407–2416

Kilpelainen TO et al (2011) Physical activity attenuates the influence of FTO variants on obesity risk: a meta-analysis of 218,166 adults and 19,268 children. PLoS Med 8(11), e1001116

Knowler WC et al (2002) Reduction in the incidence of type 2 diabetes with lifestyle intervention or metformin. N Engl J Med 346(6):393–403

Kooperberg C, Leblanc M (2008) Increasing the power of identifying gene x gene interactions in genome-wide association studies. Genet Epidemiol 32(3):255–263

Lamri A et al (2012) Dietary fat intake and polymorphisms at the PPARG locus modulate BMI and type 2 diabetes risk in the D.E.S.I.R. prospective study. Int J Obes (Lond) 36(2):218–224

Langberg C et al (2014) Gene-lifestyle interaction and type 2 diabetes. PLoS Med. doi:10.1371/journal.pmed.1001647

Li Y et al (2012) Birth weight, genetic susceptibility, and adulthood risk of type 2 diabetes. Diabetes Care 35(12):2479–2484

Lindgren CM et al (2009) Genome-wide association scan meta-analysis identifies three Loci influencing adiposity and fat distribution. PLoS Genet 5(6), e1000508

Luan JA et al (2001a) Sample size determination for studies of gene-environment interaction. Int J Epidemiol 30(5):1035–1040

Luan J et al (2001b) Evidence for gene-nutrient interaction at the PPARgamma locus. Diabetes 50 (3):686–689

Malik VS et al (2010a) Sugar-sweetened beverages and risk of metabolic syndrome and type 2 diabetes: a meta-analysis. Diabetes Care 33(11):2477–2483

Malik VS, Popkin BM, Bray GA, Després JP, Hu FB (2010b) Sugar-sweetened beverages, obesity, type 2 diabetes mellitus, and cardiovascular disease risk. Circulation 121(11):1356–1364

Manning AK et al (2011) Meta-analysis of gene-environment interaction: joint estimation of SNP and SNP x environment regression coefficients. Genet Epidemiol 35(1):11–18

Mather KJ et al (2012) Common variants in genes encoding adiponectin (ADIPOQ) and its receptors (ADIPOR1/2), adiponectin concentrations, and diabetes incidence in the Diabetes Prevention Program. Diabet Med 29(12):1579–1588

Mattei J et al (2012) TCF7L2 genetic variants modulate the effect of dietary fat intake on changes in body composition during a weight-loss intervention. Am J Clin Nutr 96(5):1129–1136

Meyer RS, Purugganan MD (2013) Evolution of crop species: genetics of domestication and diversification. Nat Rev Genet 14(12):840–852

Moore AF et al (2008) Extension of type 2 diabetes genome-wide association scan results in the diabetes prevention program. Diabetes 57(9):2503–2510

Moore AF et al (2009) The association of ENPP1 K121Q with diabetes incidence is abolished by lifestyle modification in the diabetes prevention program. J Clin Endocrinol Metab 94(2):449–455

Morris AP et al (2012) Large-scale association analysis provides insights into the genetic architecture and pathophysiology of type 2 diabetes. Nat Genet 44(9):981–990

Mozaffarian D, Hao T, Rimm EB, Willett WC, Hu FB (2011) Changes in diet and lifestyle and long-term weight gain in women and men. N Engl J Med 364(25):2392–2404

Neel JV (1962) Diabetes mellitus: a "thrifty" genotype rendered detrimental by "progress"? Am J Hum Genet 14:353–362

Nelson TL et al (2007) Association of the peroxisome proliferator-activated receptor gamma gene with type 2 diabetes mellitus varies by physical activity among non-Hispanic whites from Colorado. Metabolism 56(3):388–393

Nettleton JA et al (2010) Interactions of dietary whole-grain intake with fasting glucose- and insulin-related genetic loci in individuals of European descent: a meta-analysis of 14 cohort studies. Diabetes Care 33(12):2684–2691

Ohta T, Kimura M (1969) Linkage disequilibrium at steady state determined by random genetic drift and recurrent mutation. Genetics 63(1):229–238

Ortega-Azorin C et al (2012) Associations of the FTO rs9939609 and the MC4R rs17782313 polymorphisms with type 2 diabetes are modulated by diet, being higher when adherence to the Mediterranean diet pattern is low. Cardiovasc Diabetol 11:137

Palla L et al (2010) Challenges in the use of literature-based meta-analysis to examine gene-environment interactions. Am J Epidemiol 171(11):1225–1232

Pan Q et al (2013) Variation at the melanocortin 4 receptor gene and response to weight-loss interventions in the diabetes prevention program. Obesity (Silver Spring) 21(9):E520–E526

Pare G et al (2010) On the use of variance per genotype as a tool to identify quantitative trait interaction effects: a report from the Women's Genome Health Study. PLoS Genet 6(6), e1000981

Polonsky KS (2012) The past 200 years in diabetes. N Engl J Med 367(14):1332–1340

Prasad AS (2007) Zinc: mechanisms of host defense. J Nutr 137(5):1345–1349

Pulizzi N et al (2009) Interaction between prenatal growth and high-risk genotypes in the development of type 2 diabetes. Diabetologia 52(5):825–829

Qi L (2012) Gene-diet interactions in complex disease: current findings and relevance for public health. Curr Nutr Rep 1(4):222–227

Qi L, Cho YA (2008) Gene-environment interaction and obesity. Nutr Rev 66(12):684–694

Qi L et al (2009) Genetic predisposition, Western dietary pattern, and the risk of type 2 diabetes in men. Am J Clin Nutr 89(5):1453–1458

Qi Q et al (2011) Insulin receptor substrate 1 gene variation modifies insulin resistance response to weight-loss diets in a 2-year randomized trial: the Preventing Overweight Using Novel Dietary Strategies (POUNDS LOST) trial. Circulation 124(5):563–571

Qi Q et al (2012) Sugar-sweetened beverages and genetic risk of obesity. N Engl J Med 367 (15):1387–1396

Qi Q et al (2014) Fried food consumption, genetic risk, and body mass index: gene-diet interaction analysis in three US cohort studies. BMJ 348

Randall JC et al (2013) Sex-stratified genome-wide association studies including 270,000 individuals show sexual dimorphism in genetic loci for anthropometric traits. PLoS Genet 9(6), e1003500

Sacks FM et al (2009) Comparison of weight-loss diets with different compositions of fat, protein, and carbohydrates. N Engl J Med 360(9):859–873

Schellenberg ES et al (2013) Lifestyle interventions for patients with and at risk for type 2 diabetes: a systematic review and meta-analysis. Ann Intern Med 159(8):543–551

Schulze MB, Manson JE, Ludwig DS, Colditz GA, Stampfer MJ, Willett WC, Hu FB (2004) Sugar-sweetened beverages, weight gain, and incidence of type 2 diabetes in young and middle-aged women. JAMA 292(8):927–934

Scott RA et al (2012) Large-scale association analyses identify new loci influencing glycemic traits and provide insight into the underlying biological pathways. Nat Genet 44(9):991–1005

Sellayah D, Cagampang FR, Cox RD (2014) On the evolutionary origins of obesity: a new hypothesis. Endocrinology 155(5):1573–1588. doi:10.1210/en.2013-2103

Shan Z et al (2014) Interactions between zinc transporter-8 gene (SLC30A8) and plasma zinc concentrations for impaired glucose regulation and type 2 diabetes. Diabetes 63(5):1796–1803

Sonestedt E et al (2012) Genetic variation in the glucose-dependent insulinotropic polypeptide receptor modifies the association between carbohydrate and fat intake and risk of type 2 diabetes in the Malmo Diet and Cancer cohort. J Clin Endocrinol Metab 97(5):E810–E818

Soriguer F et al (2006) Pro12Ala polymorphism of the PPARG2 gene is associated with type 2 diabetes mellitus and peripheral insulin sensitivity in a population with a high intake of oleic acid. J Nutr 136(9):2325–2330

Speliotes EK et al (2010) Association analyses of 249,796 individuals reveal 18 new loci associated with body mass index. Nat Genet 42(11):937–948

van Dam RM et al (2002) Dietary patterns and risk for type 2 diabetes mellitus in U.S. men. Ann Intern Med 136(3):201–209

van Hoek M et al (2009) Genetic variant in the IGF2BP2 gene may interact with fetal malnutrition to affect glucose metabolism. Diabetes 58(6):1440–1444

Visscher PM, Posthuma D (2010) Statistical power to detect genetic Loci affecting environmental sensitivity. Behav Genet 40(5):728–733

Whincup PH et al (2008) Birth weight and risk of type 2 diabetes: a systematic review. JAMA 300 (24):2886–2897

Whiting DR et al (2011) IDF diabetes atlas: global estimates of the prevalence of diabetes for 2011 and 2030. Diabetes Res Clin Pract 94(3):311–321

Ylonen SK et al (2008) The Pro12Ala polymorphism of the PPAR-gamma2 gene affects associations of fish intake and marine n-3 fatty acids with glucose metabolism. Eur J Clin Nutr 62 (12):1432–1439

Zhang Y, Liu JS (2007) Bayesian inference of epistatic interactions in case-control studies. Nat Genet 39(9):1167–1173

Zhang X et al (2012) FTO genotype and 2-year change in body composition and fat distribution in response to weight-loss diets: the POUNDS LOST trial. Diabetes 61(11):3005–3011
Zimmet P, Alberti KG, Shaw J (2001) Global and societal implications of the diabetes epidemic. Nature 414(6865):782–787

Chapter 27
Epilogue: What the Future Holds: Genomic Medicine at the Heart of Diabetes Management

Mark I. McCarthy, David Altshuler, and Michael Boehnke

Abstract Individual predisposition to type 2 diabetes is influenced by the combination of genetic variants, environmental exposures, behaviour and chance. Human genetics offers a method to identify specific genetic variants that influence disease risk and thereby the pathways and mechanisms through which they operate. These pathways provide a powerful lens through which to develop biological insights into metabolism and disease and have the potential to inform diagnosis and treatment. Indeed, this potential is already being realised in precision medical management of monogenic and syndromic forms of diabetes. While substantial progress has been made identifying genetic variants for the common, multifactorial forms of type 2 diabetes, major challenges remain before we gain insight and translational benefit. The difficulty derives from the genetic architecture of type 2 diabetes and other common diseases, which involves a large number of variants of modest effect, many non-coding and presumably regulatory in nature. The ability to define more complete individual inventories of genetic risk and environmental exposure is only

---

This is a revised and expanded version of McCarthy, MI: 'Genomic medicine at the heart of diabetes management', published in *Diabetologia* 2015 Aug;58(8):1725–9.

M.I. McCarthy (✉)
Robert Turner Professor of Diabetic Medicine, Oxford Centre for Diabetes, Endocrinology and Metabolism, University of Oxford, Churchill Hospital, Old Road, Oxford OX3 7LJ, UK

Wellcome Trust Centre for Human Genetics, University of Oxford, Roosevelt Drive, Oxford OX3 7BN, UK

Oxford NIHR Biomedical Research Centre, Churchill Hospital, Old Road, Oxford OX3 7LJ, UK

Department of Genetic Medicine and Development, University of Geneva Medical School, 1 Rue Michel-Servet, Geneva 1211, Switzerland
e-mail: mark.mccarthy@drl.ox.ac.uk

D. Altshuler
Vertex Pharmaceuticals, Boston, MA 02210, USA

M. Boehnke
Department of Biostatistics and Center for Statistical Genetics, University of Michigan, Ann Arbor, MI 48109-2029, USA

J.C. Florez (ed.), *The Genetics of Type 2 Diabetes and Related Traits*,
DOI 10.1007/978-3-319-01574-3_27

a start towards understanding the complex molecular pathophysiology that underlies disease, understanding that will support the development of integrative readouts that track causal pathogenetic mechanisms and the invention of new therapies that restore homeostasis through these pathways.

Abbreviations

GLP-1	Glucagon-like peptide-1
GWAS	Genome-wide association studies

An individual's risk of developing type 2 diabetes reflects a mix of genetic predisposition, environmental influence, behaviours and their interactions. The rising global prevalence of type 2 diabetes highlights the limitations of current preventative strategies, and high complication rates attest to the deficiencies of available treatments. The increase in prevalence and complications speaks to the importance of understanding the heterogeneous aetiology and complex pathophysiology of type 2 diabetes.

For many diseases, human genetics has provided transformative insights into the biology of disease and in some cases, defined entirely novel translational opportunities. However, despite the robust identification of scores of genetic loci influencing risk of type 2 diabetes over the last decade (Morris et al. 2012; Mahajan et al. 2014), the impact of these discoveries on clinical management has so far been negligible. This is not specific to type 2 diabetes: the same is true for most other common chronic diseases in which the last decade has seen the mapping of many new loci for each disease. Most large-scale efforts to accelerate the implementation of genomic medicine into clinical care (such as the UK's plan to sequence 100,000 genomes and the USA's efforts towards a Precision Medicine Cohort) are, at least initially, focussed on rare diseases and cancer and barely recognise the potential value for common chronic diseases. Some observe this long timeline and have concluded that the role of genetics as an engine for clinical advances for common diseases has been massively oversold (Joyner & Paneth 2015).

Such sentiments combine natural impatience with a failure to appreciate the true timescales for translation: in the few cases of Mendelian diseases where new treatments emerged, the timeline from gene discovery to new therapy has been long. For example, it took over 20 years to develop new treatments in the case of cystic fibrosis, and most Mendelian diseases lack any targeted therapy. Similarly, despite the great success identifying genetic underpinnings of cancer, only a handful of targeted therapies are widely used, and most cancers lack mechanistic treatment. After all, most genetic discovery in type 2 diabetes is less than a decade old. In addition to the timeline, there is the greater complexity of type 2 diabetes and common diseases that makes the path harder still. This begs the question: What will it take to make genomic medicine a reality for common complex diseases, what would it look like, and what impact will it have in the coming decades?

27.1 An Assessment of the Genetics of Type 2 Diabetes

As described earlier in this volume, prior to the 2000s, there was little genetic insight into the common forms of type 2 diabetes. Linkage studies and candidate gene studies had failed to provide durable findings, and understanding was limited to Mendelian forms. The recent advances in unravelling the genetics of type 2 diabetes and other common complex diseases date to the implementation of genome-wide association studies (GWAS) and consequent development of the scientific and collaborative framework that allowed large-scale data aggregation and sharing. These efforts have rapidly generated a catalogue of more than 100 loci with robust and reproducible associations to type 2 diabetes (Morris et al. 2012; Mahajan et al. 2014), each surpassing stringent thresholds for significance (typically $P < 10^{-8}$ or better) and reproducibility. Due to the technologies available at the time, almost all the risk alleles discovered to date have been common (minor allele frequency >5 %), and most are widely represented across human populations. Both these observations are consistent with the model that these type 2 diabetes risk alleles arose long ago in human prehistory and suggest a degree of tolerance to their evolutionary impact that is consistent with their modest effects on glucose homeostasis. Cumulatively, these 100 loci explain 5–10 % of the variation in type 2 diabetes predisposition (Morris et al. 2012; Mahajan et al. 2014) and thus 10–20 % of the heritability of the disease.

The modest fraction of heritability explained by these statistically highly significant loci has prompted considerable speculation regarding the nature and extent of the remaining component of genetic variance. The two main models are based around (a) additional common alleles (the so-called infinite alleles model) or (b) rarer variants of large effect (or a mixture of the two). Analyses that incorporate all common variants (including those with statistical significance that fails the reach the thresholds that are taken to constitute unequivocal association) consistently show much higher fractions of heritability explained, indicating a long tail of common variants of more modest effects (Morris et al. 2012). To date, sequencing studies have failed to document more than a handful of lower-frequency variants (see Chap. 5), and those found have had modest rather than Mendelian effects. Larger-scale sequencing studies are in process and will more definitively assess the contribution of rare variants, but there is little evidence that they will explain much of the variance, even if discovering such variation may prove to be useful in its own right. Based on analyses of available GWAS data, little of the undiscovered variance resides in the interaction between genetic variants themselves or between genes and the environment (Langenberg et al. 2014). No doubt, larger studies with extensive and consistent data on lifestyle factors will increase power to address this possibility.

The picture emerging is one of great complexity in the genetic basis of type 2 diabetes—as was expected for a late-onset disease with a high population prevalence involving core processes of energy homeostasis that are redundant and highly conserved. Before the advent of GWAS, some argued that the genetics of type 2 diabetes was likely so complex that no significant findings would be found.

Having observed that a large but finite number of genetic risk factors can be found, the key question is whether new biology can be learned and new interventions developed based on this insight. Where it has been possible to transmute those signals into biological inference, the data point to the involvement of multiple etiological pathways, reflecting the diverse mechanisms through which the complex machinery of glucose homeostasis can be perturbed.

27.2 From Genetics to Biology and Interventions

It is our hope that in a generation or two, the management of type 2 diabetes will be significantly advanced as compared today. The focus will be on preserving normal glucose homeostasis in those demonstrated at substantial risk, rather than dealing retrospectively with the consequences of prolonged hyperglycaemia. It is tempting to imagine a greater emphasis on public health-based prevention and education focussed on an individual's identified risk profile and observed lifestyle and behaviour, although that will require interventions much more effective than those currently available. This will include new drugs capable of effective prevention based on reversing the underlying causes of disease. Where intervention is required in those who have escaped prevention and developed disease, effective approaches may include surgical or cell-based therapies in addition to pharmacological agents: these should aim to reverse the disease process (to restore beta-cell mass and/or function and/or to correct the primary defects in insulin sensitivity) rather than, as now, attempting merely to control the hyperglycaemia that is a manifestation of disease.

For human genetics to play a major role in these developments, risk variants revealed by population-level sequencing must be used to identify targets and pathways causally implicated in the development of human diabetes (Plenge et al. 2013). The overlap between genetics and treatment is already evident: the genes encoding the targets of sulfonylureas and thiazolidinediones harbour variants associated with type 2 diabetes (Morris et al. 2012; Mahajan et al. 2014), and variants in and around the genes encoding the receptors for gastric inhibitory polypeptide and glucagon-like peptide-1 (GLP-1) are associated with diabetes risk or related glycaemic traits, paralleling the clinical value of therapeutic manipulation of this pathway using GLP-1 receptor agonists and dipeptidyl peptidase-4 (DPP-4) inhibitors (Morris et al. 2012; Wessel et al. 2015; Mahajan et al. 2015). A similar story holds for the targets of statins and ezetimibe, both of which appear amongst the GWAS signals for lipid traits (Teslovich et al. 2010; Global Lipids Genetics Consortium et al. 2013). These examples provide evidence of the overlap between genetic causation and therapeutic benefit.

If we are to translate findings from GWAS into therapeutic hypotheses, the effort will require identification of the genes responsible for mediating the effects of each variant, assembling these genes into pathways and understanding the directional relationship between pathway activity and disease risk. Connecting non-coding variants to the genes they regulate is a major challenge, but the ability to generate hypotheses has been made more tractable by emerging maps of epigenetic regulatory factors and of tissue-specific gene regulatory networks. The ability to test these hypotheses has been powerfully enabled by the advent of gene editing to rigorously test the role of individual gene variants in their native genomic context. Rapid advances in stem cell differentiation may offer human cell models in which the cellular context as well as the genomic context can be tailored to the human disease state.

Given the many technical hurdles required to understand non-coding variants, at present the most direct route to novel targets lies in the detection, through sequencing, of rare, large-effect alleles. Those with protective phenotypic effects can point to targets to inhibit, with a primary example being *PCSK9* for LDL-cholesterol and coronary artery disease (CAD). It is little more than a decade since it was shown that loss-of-function variants in this gene have profound effects on lipid levels and CAD risk, and two different potent proprotein convertase subtilisin/kexin type 9 (PCSK9) inhibitors have been shown to lower LDL-cholesterol and reduce risk of CAD (Cohen et al. 2006). Evidence that individuals with only one functional copy of the *SLC30A8* gene have a 70 % reduction in diabetes risk offers similar potential (Flannick et al. 2014). The latter example illustrates the overlap of genes with common and rare variants, as the *SLC30A8* gene was selected for deep targeted sequencing because it harbours a common variant signal of modest effect identified initially by GWAS.

While it is unclear how many such examples remain to be discovered, the numerical relationship between the per-base mutation rate ($\sim 1 \times 10^{-8}$ per base per generation) and the size of the human population ($\sim 1.5 \times 10^{10}$ haploid genomes) means that most mutations compatible with survival are present in hundreds of living individuals. The phenotypic consequences of losing one or both copies of a given gene are typically being played out in the medical histories of thousands of our fellow humans. Harnessing that information should be practical with the combination of electronic health records, genome sequencing and combined analyses—although overcoming the social and ethical barriers to large-scale data collection and sharing is one of the great challenges of our age. Whether sooner or later, it is likely that many millions of people will have genome sequencing data combined with information on diabetes, with success bringing exquisite biological insights and additional interventional targets.

27.3 From Genetics to Prediction

One can view the delivery of mechanistic insights leading to novel interventional targets as knowledge gained through the joint analysis of many human sequences. However, there is a second, parallel imperative, which is focused on understanding how analysis of a single human sequence (yours, mine, your child's) can provide clinically useful information, be that through stratification of future disease risk, definition of disease subtype or prediction of response to a diversity of potential interventions.

Neonatal diabetes provides an example of what is possible. The combination of an extreme phenotype (diabetes diagnosed in the first 6 months of life) and penetrant mutations in a well-understood gene (*KCNJ11*, encoding the beta-cell's K_{ATP} channel) has led directly to genetically driven individualised therapy in the form of high-dose sulfonylureas (Gloyn et al. 2004; Pearson et al. 2006).

The common, complex form of type 2 diabetes presents a much greater challenge. As we have seen, known risk variants account for only 5–10 % of variation in risk, and, in tests of diagnostic or predictive accuracy, genetic tests are comfortably outperformed by classical risk factors (age, BMI, ethnicity) (Meigs et al. 2008; Walford et al. 2014). There is some evidence that the relative performance of genetic prediction improves when performed early in life or in lower risk individuals (de Miguel-Yanes et al. 2011), but predictive accuracy still falls a long way short of that required for clinical utility.

It should not be assumed that the limited predictive value simply reflects the incomplete coverage of existing discovery efforts and that further expansion of the inventory of predisposing variants will solve the problem. Rather, there are two major reasons it can be assumed that the accuracy of risk estimation is likely to remain constrained, even in the face of a more complete inventory of predisposing variants.

First, genome sequence variation represents only one of several contributors to an individual's risk of developing a late-onset disease such as type 2 diabetes. Full specification of that risk requires that information on genetic predisposition be integrated with all the other factors that impinge on metabolic performance over a lifetime. These include the impact of early life events (such as intrauterine malnutrition), mediated through epigenetic modifications, as well as the constellation of environmental and lifestyle experiences (diet, physical activity, illness, drugs, social conflict, etc.) from infancy through to senescence, not to mention the play of chance, mediated, for example, through somatic mutation, for which there is growing evidence of a relationship to diabetes (Jaiswal et al. 2014) (see Fig. 27.1a). The certainty of limited prediction is evidenced best by studies of identical twins, who despite showing high concordance in type 2 diabetes over a lifetime, can develop the disease decades apart or in some cases remain discordant altogether.

Second, in a polygenic disease, the independent assortment of many risk variants leads to a continuous rather than dichotomous distribution of risk. Thus, even if it proves possible to enumerate the extended list of variants that explain most of the genetic variance, this information will provide a graded distribution of genetic risk, with most individuals at near-average levels of risk. Moreover, variants in the tail of the distribution (which collectively may explain a substantial proportion of the variance) will not be particularly useful in determining individual predisposition, because their wide confidence intervals dilute the precision of their large effect size. Additional complexity derives from possible epistatic relationships that may be too numerous to be identified and too context-dependent to be informative (Visscher et al. 2010).

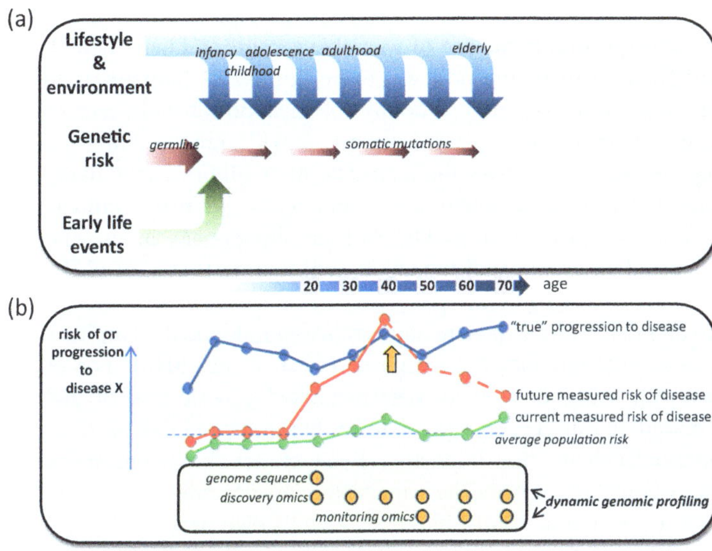

Fig. 27.1 (**a**) Progress of an individual from health to disease is dependent on the dynamic interplay of genetic, epigenetic and environmental factors during life. Whole genome sequence gathered in early life captures only a small component of risk. (**b**) Consider the progress of an individual from health to a given diseased state. Based on perfect information, this theoretical individual's 'true' trajectory follows the blue line. With current knowledge (e.g. GWAS data, limited clinical and biochemical data), the 'visible' trajectory for that same individual tracks this only poorly (*green*). With more complete information (whole genome sequence plus dynamic genomic profiling for a robust molecular signature every 10 years), tracking of disease progression is much improved (*red*), prompting (in this example) intervention at age ~40 years (*orange arrow*) that is successful at reducing that individual's risk of subsequent disease

Nonetheless, the cost of genome sequencing is falling fast, and the value of these data needs not be transformative to become a regular part of the medical record. It seems likely that in a decade or two, when capturing and interpreting a whole genome sequence may costs less than a course of antibiotics or an outpatient appointment, the value of depositing this information within an individual's medical record will no longer be a matter of debate. Once universal genome-wide medical sequencing is in place—justified primarily by the obvious benefits for the management of rare diseases, and of cancer—the proposition for common, chronic disease changes. The question becomes not whether to collect the data, but rather what additional information is needed to interpret, refine and augment the clinical value of these data for a disease like type 2 diabetes?

In a generation from now, we will have implemented, for many diseases, strategies for measuring individual risk that integrate genetic information from a multiplicity of risk variants with information from the diversity of environmental and behavioural impacts on risk. These strategies will be partly based around better metrics of the risk factors themselves, achieved through advances in human genetics and by an explosion in the use of wearable sensors and hand-held devices.

An additional critical step is likely to be the use of biomarkers that directly capture disease progression and pathophysiological profile and which can, through periodic measurement (e.g. every 5 years), provide readouts of an individual's fluctuating risk of disease over the course of their lifetime. It is likely that these biomarkers will reflect the integrated output of the pathways influenced by the genetic risk factors, just as serum LDL-cholesterol integrates the summary action of many genes and environmental factors in contributing to risk of CAD. Such biomarkers would reflect genetic potential (as imperfectly revealed by genome sequence) and risk factor exposure and capture an individual's empirical trajectory as they travel through molecular 'space' (Chen et al. 2012). For example, an individual with a higher than average estimated genetic risk of diabetes, who takes up marathon running, should see his or her measured risk of disease progression biomarkers ebb and flow in parallel. Because these biomarker measures would be repeated at regular intervals, the trajectories they describe could be judged not so much against population norms, but against one's own personal history (i.e. using the 'historical self' as control), in much the same way that the authorities monitor athletes and cyclists for evidence of doping. To be informative of risk throughout life, both before and after intervention, the best biomarkers will represent direct readouts of causal pathways (see Fig. 27.1b).

If the goal is to find an 'LDL-cholesterol' for type 2 diabetes, where should we look? One path is to work 'up' from the genetics—identifying the molecular pathways regulated by scores of disease-associated genetic variants. The second is to work 'down' from the phenotype: capturing the burgeoning range of global endophenotypes within a pathological state allows us to search beyond classical biomarkers and admits the possibility of complex molecular signatures of risk, based around networks of genetic, transcriptomic, proteomic and metabolomic features. Access to large, well-characterised cohorts, with longitudinal sampling and linkage to electronic health records, will generate rich datasets that can be mined to identify them.

There is no guarantee that such a biomarker exists for type 2 diabetes or for any other particular chronic disease for that matter. The potential for pathophysiological heterogeneity of type 2 diabetes may represent an obstacle that proves impossible to dissect. A more optimistic view is that the many genetic risk factors reflect many components of a more limited number of pathways. It is possible that the longitudinal integration of diverse data types may actually empower efforts to shatter the type 2 diabetes monolith and expose the distinct processes that contribute to individual risk of disease, each of which may have its own molecular signature. This hypothesised reclassification of type 2 diabetes is often simplistically described in terms of defining discrete subtypes (type 2a, 2b, 2c, etc.). It seems more likely, given the genetic architecture and pervasiveness of relevant exposures, that each individual has his or her own pathophysiological multicoloured palette of type 2 diabetes risk. If so, the aim should be to describe what contribution each of these processes is, at any given time, making to the evolution of their disease and to tailor their management accordingly.

27.4 Diabetes Care in the Future

If all this comes to pass, the focus of diabetes care will have shifted substantially. The disease may not have been eliminated, even in wealthy countries, but the emphasis, and the funds, will have migrated from treatment to prevention. Reliable, actionable, real-time information on type 2 diabetes risk and subtype profile will be available through the integration of baseline genome sequence information and enhanced measures of exposures, with complex molecular biomarkers gathered as part of a universal programme of genomic disease prediction. Such information will guide individualised preventative strategies that draw upon a far wider range of effective interventions. This will include more precise behavioural modifications, as well as smarter drugs, optimised for efficacy and safety, which can be targeted to individuals on the basis of their component pathophysiological profile. If so, genomic medicine, already a reality for monogenic forms of diabetes, will have proven equally transformative for more complex forms of the disease.

Acknowledgements The authors wish to acknowledge helpful discussions of the issues described in this article with Timothy Spector and Kerrin Small (London, UK), Sylvia Richardson (Cambridge, UK), Chris Holmes (Oxford, UK), Emmanoulis Dermitzakis (Geneva, Switzerland) and Mike Snyder (Stanford, USA) and with members of the DIRECT consortium (www.direct-diabetes.org/) including Ewan Pearson (Dundee, UK), Paul Franks and Leif Groop (both at the University of Lund, Sweden).

Duality of Interest The authors declare that there is no duality of interest associated with this manuscript. D.A. is currently employed by Vertex Pharmaceuticals.

References

Chen R et al (2012) Personal omics profiling reveals dynamic molecular and medical phenotypes. Cell 148:1293–1307

Cohen JC, Boerwinkle E, Mosley TH Jr, Hobbs HH (2006) Sequence variations in *PCSK9*, low LDL, and protection against coronary heart disease. N Engl J Med 354:1264–1272

de Miguel-Yanes JM et al (2011) Genetic risk reclassification for type 2 diabetes by age below or above 50 years using 40 type 2 diabetes risk single nucleotide polymorphisms. Diabetes Care 34:121–125

Flannick J et al (2014) Loss-of-function mutations in SLC30A8 protect against type 2 diabetes. Nat Genet 46:357–363

Global Lipids Genetics Consortium et al (2013) Discovery and refinement of loci associated with lipid levels. Nat Genet 45:1274–1283

Gloyn AL et al (2004) Activating mutations in the gene encoding the ATP-sensitive potassium-channel subunit Kir6.2 and permanent neonatal diabetes. N Engl J Med 350:1838–1849

Jaiswal S et al (2014) Age-related clonal hematopoiesis associated with adverse outcomes. N Engl J Med 371:2488–2498
Joyner MJ, Paneth N (2015) Seven questions for personalized medicine. JAMA. doi:10.1001/jama.2015.7725
Langenberg C et al (2014) Gene-lifestyle interaction and type 2 diabetes: the EPIC interact case-cohort study. PLoS Med 11, e1001647
Mahajan A et al (2014) Genome-wide trans-ancestry meta-analysis provides insight into the genetic architecture of type 2 diabetes susceptibility. Nat Genet 46:234–244
Mahajan A et al (2015) Identification and functional characterization of *G6PC2* coding variants influencing glycemic traits define an effector transcript at the *G6PC2-ABCB11* locus. PLoS Genet 11, e1004876
Meigs JB et al (2008) Genotype score in addition to common risk factors for prediction of type 2 diabetes. N Engl J Med 359:2208–2219
Morris AP et al (2012) Large-scale association analysis provides insights into the genetic architecture and pathophysiology of type 2 diabetes. Nat Genet 44:981–990
Pearson ER et al (2006) Switching from insulin to oral sulfonylureas in patients with diabetes due to Kir6.2 mutations. N Engl J Med 355:467–477
Plenge RM, Scolnick EM, Altshuler D (2013) Validating therapeutic targets through human genetics. Nat Rev Drug Discov 12:581–594
Teslovich TM et al (2010) Biological, clinical and population relevance of 95 loci for blood lipids. Nature 466:707–713
Visscher PM, Yang J, Goddard ME (2010) A commentary on 'common SNPs explain a large proportion of the heritability for human height' by Yang et al. (2010). Twin Res Hum Genet 13:517–524
Walford GA et al (2014) Metabolite traits and genetic risk provide complementary information for the prediction of future type 2 diabetes. Diabetes Care 37:2508–2514
Wessel J et al (2015) Low-frequency and rare exome chip variants associate with fasting glucose and type 2 diabetes susceptibility. Nat Commun 6:5897

Index

J.C. Florez (ed.), *The Genetics of Type 2 Diabetes and Related Traits*,
DOI 10.1007/978-3-319-01574-3

W

Lightning Source UK Ltd.
Milton Keynes UK
UKOW06n2141070616

275794UK00003B/16/P